Complexity of Seismic Time Series

Complexity of Seismic Time Series
Measurement and Application

Edited by

Tamaz Chelidze
M. Nodia Institute of Geophysics, Tbilisi, Georgia

Filippos Vallianatos
Technological Educational Institute of Crete, Laboratory of
Geophysics and Seismology, UNESCO Chair on Solid Earth Physics and
Geohazards Risk Reduction, Crete, Greece

Luciano Telesca
National Research Council, Tito, Italy

ELSEVIER

Elsevier
Radarweg 29, PO Box 211, 1000 AE Amsterdam, Netherlands
The Boulevard, Langford Lane, Kidlington, Oxford OX5 1GB, United Kingdom
50 Hampshire Street, 5th Floor, Cambridge, MA 02139, United States

Notices
Knowledge and best practice in this field are constantly changing. As new research and experience broaden our
understanding, changes in research methods, professional practices, or medical treatment may become necessary.

Practitioners and researchers must always rely on their own experience and knowledge in evaluating and using any
information, methods, compounds, or experiments described herein. In using such information or methods they
should be mindful of their own safety and the safety of others, including parties for whom they have a professional
responsibility.

To the fullest extent of the law, neither the Publisher nor the authors, contributors, or editors, assume any liability
for any injury and/or damage to persons or property as a matter of products liability, negligence or otherwise, or
from any use or operation of any methods, products, instructions, or ideas contained in the material herein.

British Library Cataloguing-in-Publication Data
A catalogue record for this book is available from the British Library

Library of Congress Cataloging-in-Publication Data
A catalog record for this book is available from the Library of Congress

ISBN: 978-0-12-813138-1

For Information on all Elsevier publications
visit our website at https://www.elsevier.com/books-and-journals

Working together
to grow libraries in
developing countries

www.elsevier.com • www.bookaid.org

Publisher: Candice Janco
Acquisition Editor: Marisa LaFleur
Editorial Project Manager: Katerina Zaliva
Production Project Manager: Nilesh Kumar Shah
Cover Designer: Victoria Pearson

Typeset by MPS Limited, Chennai, India

Contents

Part III Complexity in Earthquake Generation and Seismic Hazard Assessment 321

10. Complexity and Time-Dependent Seismic Hazard Assessment: Should We Use Fuzzy, Approximate and Prone-to-Errors Prediction Models to Overcome the Limitations of Time-Independent Models? 323

COSTAS B. PAPAZACHOS, DOMENIKOS A. VAMVAKARIS,
GEORGE F. KARAKAISIS, CHRISTOS A. PAPAIOANNOU,
EMMANUEL M. SCORDILIS AND BASIL C. PAPAZACHOS

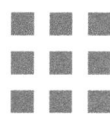

List of Contributors

Tamaz Chelidze   M. Nodia Institute of Geophysics, Tbilisi, Georgia; Ivane Javakhishvili Tbilisi State University, Tbilisi, Georgia

Zurab Chelidze   Ivane Javakhishvili Tbilisi State University, Tbilisi, Georgia

Yiannis Contoyiannis   University of West Attica, Athens, Greece

Zbigniew Czechowski   Polish Academy of Sciences, Warsaw, Poland

Angeliki Efstathiou   National and Kapodistrian University of Athens, Athens, Greece

Konstantinos Eftaxias   National and Kapodistrian University of Athens, Athens, Greece

Zurab Javakhishvili   Ilia State University, Tbilisi, Georgia

Nato Jorjiashvili   Ilia State University, Tbilisi, Georgia

George F. Karakaisis   Aristotle University of Thessaloniki, Thessaloniki, Greece

Eugenio Lippiello   University of Campania "L: Vanvitelli", Caserta, Italy

Alexey Lyubushin   Institute of Physics of the Earth, Russian Academy of Sciences, Moscow, Russia

Teimuraz Matcharashvili   M. Nodia Institute of Geophysics, Tbilisi, Georgia; Ilia State University, Tbilisi, Georgia

Temur Matcharashvili   Ivane Javakhishvili Tbilisi State University, Tbilisi, Georgia

Ekaterine Mepharidze   M. Nodia Institute of Geophysics, Tbilisi, Georgia; Ivane Javakhishvili Tbilisi State University, Tbilisi, Georgia

Georgios Michas Technological Educational Institute of Crete and UNESCO Chair on Solid Earth Physics and Geohazards Risk Reduction, Chania, Crete, Greece

Giorgos Papadakis Technological Educational Institute of Crete and UNESCO Chair on Solid Earth Physics and Geohazards Risk Reduction, Chania, Crete, Greece

Christos A. Papaioannou Institute of Engineering Seismology and Earthquake Engineering (ITSAK), Thessaloniki, Greece

Basil C. Papazachos Aristotle University of Thessaloniki, Thessaloniki, Greece

Costas B. Papazachos Aristotle University of Thessaloniki, Thessaloniki, Greece

Stelios M. Potirakis University of West Attica, Athens, Greece

Vassilis Saltas Technological Educational Institute of Crete, UNESCO Chair on Solid Earth Physics and Geohazards Risk Reduction, Chania, Crete, Greece

Nicholas V. Sarlis National and Kapodistrian University of Athens, Athens, Greece

Aleksandre Sborshchikovi M. Nodia Institute of Geophysics, Tbilisi, Georgia

Emmanuel M. Scordilis Aristotle University of Thessaloniki, Thessaloniki, Greece

Ia Shengelia Ilia State University, Tbilisi, Georgia

Efthimios S. Skordas National and Kapodistrian University of Athens, Athens, Greece

Ilias Stavrakas University of West Attica, Athens, Greece

Luciano Telesca National Research Council, Institute of Methodologies for Environmental Analysis, Tito (PZ), italy

Dimitri Tephnadze Ivane Javakhishvili Tbilisi State University, Tbilisi, Georgia

Dimos Triantis   University of West Attica, Athens, Greece

Andreas Tzanis   National and Kapodistrian University of Athens, Athens, Greece

Filippos Vallianatos   Technological Educational Institute of Crete and UNESCO Chair on Solid Earth Physics and Geohazards Risk Reduction, Chania, Crete, Greece

Domenikos A. Vamvakaris   Aristotle University of Thessaloniki, Thessaloniki, Greece

Nodar Varamashvili   Ivane Javakhishvili Tbilisi State University, Tbilisi, Georgia

Panayiotis A. Varotsos   National and Kapodistrian University of Athens, Athens, Greece

Natalya Zhukova   M. Nodia Institute of Geophysics, Tbilisi, Georgia; Ivane Javakhishvili Tbilisi State University, Tbilisi, Georgia

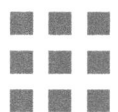

Foreword

Complex systems are those exhibiting complexity. But what is complexity? Like beauty, complexity is hard to define but relatively easy to identify. Complexity requires some sort of essential nonlinearity, caused by long-range correlations at nearly all space–time scales. Typically, complex systems include living or living-like matter and tend towards nontrivial stationary or quasistationary states, whereas simple systems include inanimate matter and tend towards standard thermal equilibrium. From the quantitative viewpoint, most of the properties of simple systems asymptotically depend exponentially on time, space, energy, momentum, and other basic variables, whereas complex systems tend to exhibit asymptotic subexponential behaviours, typically power-laws.

This collection of contributions, *Complexity of Seismic Time Series: Measurement and Application*, certainly satisfies the above requirements, with seismicity being a prototypical and very important example of complexity. These contributions have been selected by distinguished authorities in the subject: Filippos Vallianatos, Luciano Telesca and Tamaz Chelidze. They address both the causes and consequences of earthquakes, as well as various approaches for analysing the associated information. Natural and artificial, as well as discrete and continuous, time series are focused on. Earthquakes in several regions of the globe are examined, as illustrations of the concepts and procedures that are currently being used. Electromagnetic and acoustic signals, aftershock sequences, stick–slip along fractures, fractal structures and other related phenomena are addressed using various methods, including mesoscopic equations.

We should, in particular, mention that a special type of recent unified approach, based on nonadditive entropies and their associated nonextensive statistical mechanics, is also applied. One of the aspects of this approach consists of numerically establishing quantities, such as the q-triplet, which characterize a variety of crucial properties including the entropy time evolution, relaxation phenomena, description of the basic distributions emerging from stationary or quasistationary states, and others.

Summarizing, we have in our hands an excellent series of chapters which should be useful to both the recently initiated and senior researchers in the field of seismology: just enjoy it!

Constantino Tsallis, 15 March 2018
Centro Brasileiro de Pesquisas Fisicas and National Institute
of Science and Technology for Complex Systems, Rio de Janeiro, Brazil
Santa Fe Institute, Santa Fe, NM, United States
Complexity Science Hub Vienna, Vienna, Austria

Introduction

Over recent years, a fast-growing number of studies have concerned an approach to seismicity and other natural hazards based on the science of complex systems (e.g., Bak et al., 1988; Vallianatos and Telesca, 2012). Seismicity exhibits complexity that is strongly related to the deformation and sudden rupture of the Earth's brittle crust. Moreover, the lithosphere of the Earth is considered to be a nonlinear dynamic system. Consequently, earthquakes interact over a wide range of spatial and temporal scales exhibiting scale-invariance and fractality (Turcotte, 1997; Turcotte et al., 2009).

Complex systems are featured by several characteristics such as nonlinearity, criticality, long-range coherence, scaling, self-similarity, fractality/multifractality in the space and time domains, high sensitivity to small impacts, synchronization by weak forcing, etc., which are ubiquitous in nature from the subnuclear scale to cosmology.

Seismicity is a clear example of complex systems (Sornette, 2000; Chelidze and Matcharashvili, 2015). The first evidence of complexity in seismicity was revealed by Omori (1894), who obtained the first empirical power law for aftershock rate decay in time. Then, the empirical law of Gutenberg and Richter (1954) for earthquake magnitude distribution reinforced such a view of seismic processes as complex; a concept that found its mathematical basis in the fractal geometry of nature (Mandelbrot, 1967) leading to many investigations focused on the fractal and multifractal analysis of earthquake spatial and temporal distributions.

Nonextensive Statistical Physics (NESP) (Tsallis, 2009) seems a suitable framework for studying complex systems exhibiting phenomena such as fractality, long-range interactions, and memory effects (Vallianatos and Telesca, 2012). Therefore, in order to analyse the behaviour of earthquake sequences and faulting systems with fractal or multifractal distribution of their elements, the concept of NESP, originally introduced by Tsallis (1988), is suggested as an appropriate tool. NESP is based on a generalization of Boltzmann–Gibbs (BG) entropy and has the main advantage of offering a consistent theoretical framework based on the thermodynamic principle of entropy. Moreover, the concept of NESP has been applied to various fields of Earth sciences including seismicity, plate tectonics, fault length distributions, natural hazards, geomagnetic reversals and rock physics.

The dynamic friction instability of active tectonic faults (stick–slip) is considered to be the main mechanism, explaining the seismic process and recurrence of earthquakes since the basic works of Brace and Byerlee (1966) and Burridge and Knopoff (1967). Taking into account the complexity of stick–slip leads to the formulation of nonlinear rate-and-state friction law (Dieterich, 1979). Note that stick–slip is a typical example of nonlinear integrate-and-fire complex physical systems with two time scales that are ubiquitous in nature, where the slow nucleation (integrate, accumulation, stick) phase terminates, after approaching a threshold value, by a short fire (slip, stress drop) phase (Pikovsky et al., 2001). The stick—slip process, as all integrate-and-fire phenomena, is highly sensitive to a weak external forcing, which results in triggering and synchronization phenomena (Pikovsky et al., 2001) (see also Chapter 9: Complexity and Synchronization Analysis in Natural and Dynamically Forced Stick—Slip: A Review).

Modern tools of complexity analysis reveal in seismic data sets hidden nonlinear temporospatial structures, such as the recurrence and clustering of earthquakes, which are essential features that could vary with time. The development of effective tools of complexity theory, such as Shannon or Tsallis entropy, recurrence quantification analysis, detrended fluctuation analysis (DFA), singular spectral analysis (SSA), phase space plot analysis, algorithmic complexity measures, etc., makes it possible to measure accurately the complexity of the seismic process and its variability with time.

The aim of this book is to provide the reader with the most advanced and recent theoretical as well as observational and methodological developments in the analysis of seismic processes in the context of complex system theory.

In Chapter 1, Analysis of the Complexity of Seismic Data Sets: Case Study for Caucasus, Matcharashvili et al. present an overview of the complexity analysis of the seismic process using earthquake catalogues and seismic noise recordings, obtained in Caucasus and adjacent territories. The authors investigate the scaling properties of the series of waiting times and earthquake interdistances between consecutive seismic events in the earthquake catalogue of Caucasus by several data analysis methods such as power spectrum regression, DFA and multifractal detrended fluctuation analysis (MF-DFA). Finally, long seismic noise data sets were analysed by the Langevin equation method. The application of complexity theory methods allows for the discovery of new dynamic features of the earthquake generation process.

In Chapter 2, Nonextensive Statistical Seismology: An Overview, Vallianatos et al. present an overview of nonextensive Statistical Seismology, introducing the concept of nonextensive statistical mechanics and its usefulness in the investigation of phenomena exhibiting fractality and long-range interactions such as earthquake activity. Through a review of the empirical scaling relations widely used in seismology and the statistical mechanics approaches to seismicity they arrived at the unique properties of nonextensive statistical mechanics for the description of natural processes exhibiting extreme behaviour.

In Chapter 3, Spatiotemporal Clustering of Seismic Occurrence and Its Implementation in Forecasting Models, Lipiello presents the Epidemic Type Aftershock Sequence (ETAS) model, addressing the main critique towards the model and its limits in the description of seismic occurrence. The author discusses how to incorporate in the ETAS model the short-term aftershock incompleteness which strongly affects seismic statistical features the first few days after large shocks. Moreover, the author shows that seismicity before mainshock occurrence exhibits spatiotemporal patterns not fully captured by the ETAS model.

In Chapter 4, Fractal, Informational and Topological Methods for the Analysis of Discrete and Continuous Seismic Time Series: An Overview, Telesca presents an overview of the most advanced and robust statistical methodologies used for describing the properties of seismic time series: fractal, informational and topological, used to emphasize different features of the same seismic phenomenon and for a deep understanding of seismic time fluctuations. The fractal methods are based on the concept of self-similarity of the seismic phenomenon and are used to identify and quantify correlation properties; the informational methods permit highlighting of the organization or order of a seismic time series; the visibility graph method allows for the conversion of a seismic sequence in a graph with several topological properties that furnish a nonstandard description of the seismic phenomenon. The integration of all these methods would be useful to get a deeper and multiperspective picture of the seismic process.

In Chapter 5, Modelling of Persistent Time Series by the Nonlinear Langevin Equation, Czechowski introduces the modified Langevin equation, which, as distinct from the standard approach, can describe some classes of non-Markov processes. The novel generalized Langevin equation and the associated Fokker−Planck equation can be treated as a nonlinear model of persistent/antipersistent processes. Two efficient procedures of reconstruction of the model from the observed time series, the purely numerical one and the semianalytical one, are proposed and tested on synthetic data. The modified Langevin equation can present a macroscopic stochastic nonlinear model of many geophysical processes. Accounting for the drift and diffusion functions enables derivation of short-time transition probability, which can be useful in forecasting.

In Chapter 6, Synchronization of Geophysical Field Fluctuations, Lyubushin presents methods for investigating the synchronization of multiple geophysical monitoring time series, which are based on using wavelet-based and spectral measures of coherence, estimated within a moving time window. Synchronization of noise measurements, obtained on an extended monitoring network, is an indicator of the approach of complex systems to a drastic change in its properties by virtue of their own dynamics. The author investigates precursory properties of coherence and multifractal structure of seismic noise at global and regional networks. The analysis of seismic noise in Japan gave the possibility for prediction of the Tohoku earthquake on 11 March 2011 and for forecasting of the next mega-earthquake in the region of Nankai Trough.

In Chapter 7, Natural Time Analysis of Seismic Time Series, the ideas of natural time analysis of seismic time series are presented by Sarlis et al. Natural time is a new time domain introduced by the authors almost 15 years ago. Here, they review the analysis of seismic time series in this new time domain. This analysis unveils novel dynamic features that are hidden in seismic time series and enable the introduction of an order parameter for seismicity, which finds useful applications in various occasions. Examples are given for seismic-prone areas including California, Japan and Greece.

In Chapter 8, Complexity in Laboratory Seismology: From Electrical and Acoustic Emissions to Fracture, Saltas et al. review, within the frame of the complexity approach, the generation and behaviour of electrical and acoustic signal emissions, mainly when geomaterials are subjected to mechanical stress. The experimental results are examined within the framework of their capacity to provide information regarding the initial stages of microcrack generation, propagation and coalescence, aiming at their use as fracture precursors. The similarities with the observations associated with fracture are viewed in relation to the electrical and acoustic signal laboratory results.

In Chapter 9, Complexity and Synchronization Analysis in Natural and Dynamically Forced Stick—Slip: A Review, Chelidze et al. analyse stick—slip motion, regarded as a model of earthquake generation. The authors investigate acoustic emission time series, accompanying the stick—slip movement of basalt samples in a laboratory slider-spring device under different experimental conditions, including weak mechanical or electromagnetic forcing of various intensities and frequencies. Different methods of complexity analysis are used to assess changes that occur under external forcing in time series of slips. The phase space plots of periodically forced stick—slip at different intensities and frequencies of forcing reveal the synchronization area. Two models of synchronization area plots (Arnold's tongues and 'nucleation' phase space plot) are considered. The results of laboratory modelling point to some new geophysical applications.

In Chapter 10, Complexity and Time-Dependent Seismic Hazard Assessment: Should We Use Fuzzy, Approximate and Prone-to-Errors Prediction Models to Overcome the Limitations of Time-Independent Models?, Papazachos et al. present the most recent ideas along with an historical overview on 'Complexity and time-dependent SHA: Should we use fuzzy, approximate and prone-to-errors prediction models to overcome the limitations of time-independent models?' based exclusively on spatiotemporal changes of seismicity (rates, patterns, etc.). They discuss the main model assumptions and performance, as well as their historical evolution. Finally, in order to examine the possible impact of such models in SHA, a parametric analysis of uncertainties related to a generic prediction model is presented.

In Chapter 11, Are Seismogenetic Systems Random or Organized? A Treatise of Their Statistical Nature Based on the Seismicity of the North-Northeast Pacific Rim, Efstathiou et al. examin the question 'Are seismogenetic systems random or organized? A treatise of their statistical nature based on the seismicity of the north-northeast Pacific Rim', introducing ideas of NESP, that suggest that crustal seismogenetic systems along the Pacific—North American plate boundaries in California, Alaska and the Aleutian Arc are invariably subextensive; they exhibit prominent operative long-range interaction and long-term memory, therefore they are self-organized and possibly critical.

In Chapter 12, Phase Space Portraits of Earthquake Time Series of Caucasus: Signatures of Strong Earthquake Preparation, Chelidze et al. analyse the spatiotemporal parameters of seismic rate using a phase space plot compilation method. They reveal nonlinear structures in the phase space plots constructed for several regions of the Caucasus, including the areas with the two strongest Caucasian earthquakes: 1988 Spitak and 1991 Racha events. The seismic phase space portraits were constructed for different time windows, epicentral distances and magnitude thresholds. The trajectories on phase space plots form a 'noisy attractor' with diffuse source area, corresponding to the background seismicity and anomalous orbit-like deviations from the source area related to clusters and strong earthquake occurrences (foreshock and aftershock activity). The phase portraits reveal some patterns of seismic process dynamics that are possibly related to precursors and after-effects of strong earthquakes.

In Chapter 13, Four-Stage Model of Earthquake Generation in Terms of Fracture-Induced Electromagnetic Emissions: A Review, Eftaxias et al. review a 'Four-stage model of earthquake generation in terms of fracture-induced electromagnetic emissions', where the initially observed MHz electromagnetic (EM) anomaly is due to the fracture of the highly heterogeneous system that surrounds the formation of strong brittle and high-strength entities (asperities) distributed along the rough surfaces of the main fault sustaining the system.

The knowledge and application of modern tools of complexity theory will have a strong impact on understanding the basic rules governing the seismic process and, maybe, in future will help in solving the problems of earthquake forecasting and prediction, which in our times is 'The Holy Grail of Seismology'.

References

Bak, P., Tang, C., Wiesenfeld, K., 1988. Self-organized criticality. Phys. Rev. A 38, 364–374.

Brace, W.E., Byerlee, I.D., 1966. Stick slip as a mechanism for Earthquakes. Science 153, 990–992.

Burridge, R., Knopoff, L., 1967. Model and theoretical seismicity. Bull. Seism. Soc. Am. 57, 341–371.

Chelidze, T., Matcharashvili, T., 2015. Dynamical patterns in seismology. In: Webber, C., Marwan, N. (Eds.), Recurrence Quantification Analysis: Theory and Best Practices. Springer, Heidelberg, pp. 291–335.

Dieterich, J.H., 1979. Modeling of rock friction 1. Experimental results and constitutive equations. J. Geophys. Res. 84B, 2161–2168.

Gutenberg, B., Richter, C.F., 1954. Seismicity of the Earth and Associated Phenomena. Princeton University Press, Princeton, NJ.

Mandelbrot, B., 1967. How long is the Coast of Britain? Statistical self-similarity and fractional dimension. Sci. New Ser. 156, 636–638.

Omori, F., 1894. On the aftershocks of earthquakes. J. College Sci. Imperial Univ. Tokyo 7, 111–200.

Pikovsky, A., Rosenblum, M.G., Kurths, J., 2001. Synchronization: Universal Concept in Nonlinear Science. Cambridge University Press, Cambridge.

Sornette, D., 2000. Critical Phenomena in Natural Sciences. Springer, Berlin.

Tsallis, C., 1988. Possible generalization of Boltzmann-Gibbs statistics. J. Stat. Phys. 52, 479–487.

Tsallis, C., 2009. Introduction to Nonextensive Statistical Mechanics: Approaching a Complex World.. Springer Verlag, Berlin.

Turcotte, D., 1997. Fractals and Chaos in Geology and Geophysics. Cambridge University Press, Cambridge.

Turcotte, D., Scherbakov, R., Rundle, J., 2009. Complexity and earthquakes. In: Kanamori, H. (Ed.), Earthquake Seismology. Elsevier, Amsterdam, pp. 676–696.

Vallianatos, F., Telesca, L., 2012. Application of statistical physics in earth sciences and natural hazards. Acta Geophys. 60, 2012.

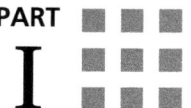

Complexity Measurement in Seismograms and Natural and Artificial Time Series of EQs (Catalogs)

Analysis of the Complexity of Seismic Data Sets: Case Study for Caucasus

Teimuraz Matcharashvili[1,2], Tamaz Chelidze[1], Zurab Javakhishvili[2], Natalya Zhukova[1], Nato Jorjiashvili[2], Ia Shengelia[2], Ekaterine Mepharidze[1], Aleksandre Sborshchikovi[1]

[1]*M. NODIA INSTITUTE OF GEOPHYSICS, TBILISI, GEORGIA* [2]*ILIA STATE UNIVERSITY, TBILISI, GEORGIA*

CHAPTER OUTLINE

1.1 Introduction

Several decades ago, geophysical objects and events in Earth sciences were mainly considered as either random or deterministic. Complexity analysis reveals the enormous domain of structures and processes ranging from complete randomness to behaviour that can be considered as more or less deterministic. Modern tools of data analysis help to assess the extent of complexity in natural processes (Abarbanel and Tsimring, 1993; Kantz and Schreiber, 1997; Strogatz, 2000; Sornette, 2000; Sprott, 2003; Webber and Marwan, 2015). There are many definitions of complexity. Almost all stress the following main properties of a complex system (CS): (1) a CS consists of many components, which interact nonlinearly; (2) a CS emerges due to the nonlinear (nonadditive) interaction between the components; and (3) these collective interactions lead to the phenomenon of 'emergence', i.e., the appearance of a new state of the system, which cannot develop from the simple addition of components. This new state reveals universal properties of a CS, such as nonlinearity, criticality, long-range correlations, scaling (power law behaviour), self-similarity, fractality/multifractality in the space and time domains, recurrence (ordering), high sensitivity to small impacts, synchronization by weak forcing, etc., which are ubiquitous in nature from subnuclear scales to cosmology.

According to current views, earthquakes are regarded as one of the most dramatic phenomena occurring in nature, causing enormous human and economic losses. Earthquakes are complicated processes and serious debate about earthquake precursors and moreover about earthquake predictability is still ongoing. We will not go deeper into this discussion, however we underline that, at present, there are many contrasting arguments concerning this issue. Meanwhile, researches aimed at the investigation of different features of the complex process of earthquake generation are still underway. This is quite logical because it is absolutely clear that without understanding the dynamic features of earthquake generation, further progress in this field cannot be achieved. Therefore, in recent decades interest in the qualitative and quantitative investigation of the complexity of earthquake generation phenomena has become one of the main targets in the analysis of these complex processes (Scholz, 1990; Keilis-Borok and Soloviev, 2002; Tabar et al., 2006).

Several tools of complexity theory have already been used in the analysis of seismic data sets of the Caucasus, such as fractal dimensions (Matcharashvili et al., 2000, 2002), recurrence quantitative analysis (RQA) (Chelidze and Matcharashvili, 2014), extensive (Shannon) and nonextensive (Tsallis) statistical analysis (Matcharashvili et al., 2011), information statistics methods (Chelidze and Matcharashvili, 2007), etc.

In this chapter, we present an overview of the analysis of seismic processes in the Caucasus using earthquake catalogues and seismic noise recordings as sources of considered data sets. Such researches are important because the Caucasus is a seismically active zone and in recent decades it has been stricken by strong earthquakes, such as Spitak 07 December 1988 (M6.9), Racha 21 April 1991 (M6.9), Barisakho 23 November 1992 (M6.5), and Racha 07 September 2009 (M6.1).

We start with an assessment of the scaling properties of the series of waiting times and earthquake interdistances (EIDs) between consecutive seismic events in the earthquake catalogue of the Caucasus and adjacent territories by using several data analysis methods such as power spectrum regression, detrended fluctuation analysis (DFA), and multifractal detrended fluctuation analysis (MF-DFA). Furthermore, the Langevin equation method has been used for long seismic noise data sets.

The presented results indicate that modern methods of complex time series analysis can be successfully used for seismic data sets.

1.2 Data

1.2.1 Waiting Times and Earthquake Interdistances

We used data sets from two sources of seismic databases. In particular, waiting times and earthquake interdistance data sets have been compiled from earthquake catalogues and data sets of seismic noises obtained from seismograms. We have used earthquake catalogues and seismograms from the M. Nodia Institute of Geophysics, Tbilisi State University and the Institute of Earth Sciences of Ilia State University, Tbilisi, Georgia.

The area of analysis (Fig. 1−1) included the segment of the Mediterranean Alpine Belt, located between the still converging Eurasian and Africa-Arabian lithosphere plates and represents a typical continent−continent collision zone. From the beginning of 1960, the former USSR observation network created for the Caucasus region was equipped with highly sensitive analogue seismographs of different types (see also Matcharashvili et al., 2013b, 2016). The short period seismograph (SKM) type, medium period seismograph (Kyrnos type) (SK) and long period seismograph (SKD) types were the most common. At that time data from the adjacent territories of Turkey and Iran were available for the same type of network. Later the number of seismic stations was decreased. For example, in Georgia, instead of 40 stations in 1991, there are now only 27 digital seismic stations operating, of which nine are the broadband type.

Waiting times and earthquake interdistance data sets were extracted from the above-mentioned earthquake catalogue for the period from 01 January 1960 to 31 December 2014 (Fig. 1−1).

According to the Guttenberg−Richter relationship analysis this catalogue can be considered complete for $M \geq 2.2$. At the same time, taking into account the results of time

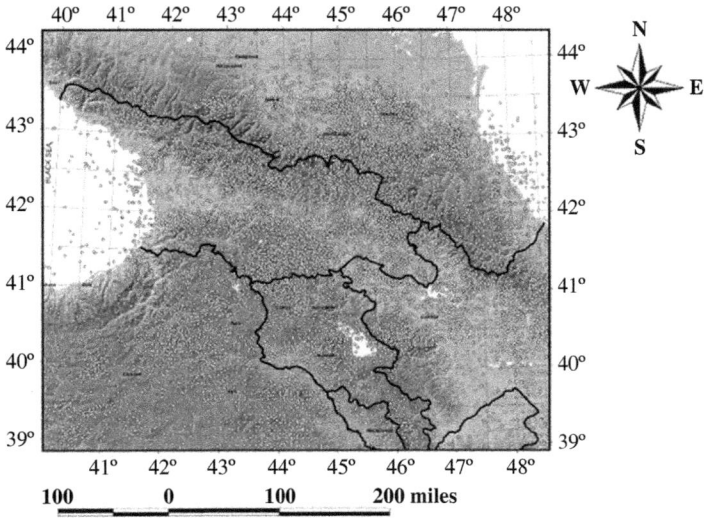

FIGURE 1−1 Map of the seismicity of the Caucasus and adjacent areas considered in this chapter.

completeness analysis (not shown here), for the whole catalogue and the entire time of observation, the $M \geq 3.0$ magnitude threshold has been used. This was done to avoid possible problems related to conditions of the seismic network (for details see Telesca et al., 2012; Matcharashvili et al., 2013b). Then, in order to remove any bias due to the presence of aftershocks, we declustered the catalogue using Reasenberg's algorithm (1985). From the declustered catalogue, we calculated sequences of waiting times or interevent times (IET) in minutes, as well as EIDs in kilometres.

1.2.2 Ambient Seismic Noise

Noise data sets have been compiled from digital seismograms recorded by a broadband permanent station located in the Greater Caucasus mountains near the town of Oni (42.5905N, 43.4525E), Georgia (Fig. 1−2). We mainly focused on the time fluctuations of the Earth's vertical velocity, V_z. The data were recorded at a sampling frequency of 100 Hz with a dynamic range over 140 dB. The station has a flat velocity response from 0.01 to 100 Hz frequency band. The seismograms were corrected for instrument response before analysis to get the ground velocity. Seismic station Oni, where analysed waveforms were recorded, is part of the seismic network operated by the Ilia State University, Seismic Monitoring Centre of Georgia (Matcharashvili et al., 2012).

In order to compare scaling characteristics of ambient noise data sets at different levels of local seismic activity, we selected data sets for different periods. Firstly the 4-day recordings, preceding Racha M6.0 earthquake (22:41:35 (UTC) on 07 September 2009, Lat. 42.5727, Long. 43.4825) were investigated. In these recordings, waveforms arriving from two remote earthquakes were visible. Namely, M4.9 occurred in Afghanistan (09:01:53 (UTC) on 07 September 2009, Lat. 36.45, Long. 70.73) and M6.2 occurred in Indonesia (16:12:22 (UTC) on 07 September 2009, Lat. 10.20, Long. 110.63). In addition, two M1.6 (14:06:35 (UTC) on 03 September 2009, Lat. 42.5414, Long. 43.5282) and M2.1 (14:17:31 (UTC) on 03 September 2009, Lat. 42.5508, Long. 43.528) foreshocks of Racha earthquake occurred during this 4-day

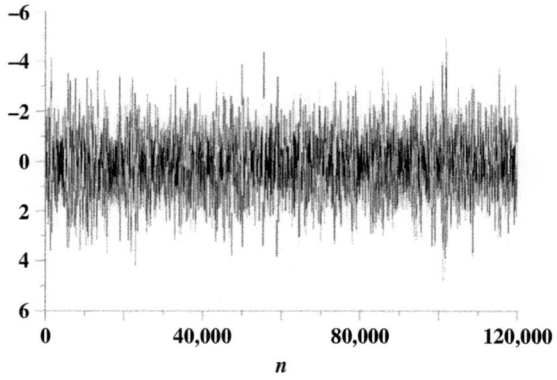

FIGURE 1–2 Typical record of Earth's vertical velocity at the Oni seismic station.

time period. We analysed the seismic waveforms from all these events. Therefore, ambient fluctuations at Oni station in this case were influenced by strong and weaker local and remote seismic activities. The next series were seismic record data sets for the period without local seismic activity in March 2011. At the end of this period the arrival of waveforms from Japan M9.0 (05:46:24 (UTC) on 11 March 2011, Lat. 38.322°N, Long. 142.369°E) earthquake was recorded by Oni station. Additionally, seismic records were considered from 23.59 (UTC) on 21 January 2009 to 19.00 (UTC) on 22 January 2009, when no local or remote seismic activity was detected by broadband Oni station. We also considered the period from 00.00 (UTC) to 18.59 (UTC) on 30 October 2010, when slight local seismic activity (series of M1.6, M1.7 events) was detected and wavetrains arrived from M5.2 earthquake, that occurred in Japan (19:06:19 (UTC) on 30 October 2010, Lat. 34.38N, Long. 141.33E) (for further details on the used seismic noise data sets, see Matcharashvili et al., 2013a).

1.3 Methods of Analysis

The measurement of complexity in experimental time series is currently possible using a modern data analysis toolbox, involving modern linear and nonlinear dynamics methods (Abarbanel and Tsimring, 1993; Kantz and Schreiber, 1997; Strogatz, 2000; Sornette, 2000; Sprott, 2003; Webber and Marwan, 2015). These methods help to reveal important hidden dynamic features of complex processes. In this research, we have used several of these methods, such as power spectrum regression, DFA, MF-DFA, and RQA. In addition to scaling features testing for long seismic noise recordings, we have also used the Langevin equation method.

Calculation of the power spectrum regression exponent enables elucidation of the scaling features of ambient noise in the frequency domain. By this method a fractal property of time series is reflected as a power law dependence between the spectral power $S(f)$ and the frequency (f) with a spectral exponent β:

$$S(f) \sim \frac{1}{f^\beta}$$

The spectral exponent β is a measure of the strength of the persistence or antipersistence, which is related to the type of correlation present in the time series (Malamud and Turcotte, 1999; Munoz-Diosdado et al., 2004). For example, $\beta \approx 0$ corresponds to the uncorrelated white noise, whereas short-range correlated noise or Brownian motion has $\beta \approx 2$ and processes.

In order to quantify long-range time correlations in the investigated data sets we have used the DFA method (Peng et al., 1993a,b, 1995). This analysis technique provides a simple quantitative parameter (DFA scaling exponent) representing the correlation properties of the time series. DFA helps to avoid the spurious detection of apparent long-range correlations that are artifacts of nonstationarity. The method consists of three steps (Peng et al., 1993a,b). First, the initial time series $x(k)$ (of length N) is integrated and the 'profile' $Y(i)$ is determined. After this the resulting series $Y(i)$ is divided into boxes of size n. In each box of length n, the local trend, $Y_{n(i)}$ is calculated and subtracted from the integrated series $Y(i)$ in each box. The root mean square fluctuation of the integrated and detrended series is then calculated:

$$F(n) = \sqrt{\frac{1}{N} \sum_{i=1}^{N} [Y(i) - Y_n(i)]^2}$$

This process is repeated for different scales (box sizes) to reveal a power law behaviour between $F(n)$ and n. If the signal follows scaling law, a power law behaviour for the function $F(n)$ is observed:

$$F(n) \sim n^{\alpha}$$

The scaling exponent α gives the information about the long-range power law correlations of the signal. The scaling exponent $\alpha = 0.5$ corresponds to a white noise (noncorrelated signal), whereas at $\alpha < 0.5$ the correlation in the signal is antipersistent, and if $\alpha > 0.5$ the correlation in the signal is persistent. $\alpha = 1$ points to a uniform power law behaviour of $1/f$ noise and $\alpha = 1.5$ corresponds to a Brownian motion (Peng et al., 1993a,b, 1995). The value $\alpha = 1.5$ corresponds to long-range correlations that may be related to both stochastic and deterministic correlations (Peng et al., 1995; Rodriguez et al., 2007). It may often happen that the correlations of recorded data do not follow the same scaling law for all considered n timescales. In such cases, in double logarithmic plots of the DFA fluctuation function, one or more crossovers can be observed separating regimes with different scaling exponents (Peng et al., 1995; Kantelhardt et al., 2002a,b). The relationship between spectral exponents β and α is given by the formula $\alpha = (1 + \beta)/2$ (Peng et al., 1993a,b; Iyengar et al., 1996; Penzel et al., 2003).

Additionally, in order to test intrinsic scaling properties of earthquake time and space distributions we used the DFA method for subseries obtained by decomposition of the original IET and EID data sets into magnitude and sign series according to Ashkenazy et al. (2001). In the frame of this approach, any observable (of analysed process) in given data series is considered as the product of a magnitude and a sign. First we generate increment series $\Delta x(i) = x(i) - x(i - 1)$ and then decompose it into increment magnitudes $|\Delta x(i)|$ and sign series. To avoid trends, we subtract respective average values from the magnitude and sign series. Then, the second-order detrended fluctuation analysis, DFA(2) is performed and the scaling exponent of subseries is calculated as $\alpha = a^{\text{int}} - 1$, where a^{int} is the scaling exponent calculated for integrated subseries (Ashkenazy et al., 2001, 2003; Ivanov et al., 2003). The essence of this method is that the long-range correlations of magnitude series indicate nonlinear behaviour and the sign time series mainly relate to linear properties of the original series. Long-range correlation testing, based on the second-order DFA of magnitude and sign subseries, has already been successfully applied to diverse fields (see, e.g., Zheng et al., 2012; Telesca et al., 2004; Matcharashvili et al., 2015; Ashkenazy et al., 2001, 2003, etc.). Our analysis was done for sliding windows of different length.

The above methods are often used to describe features of CSs behaviour. At the same time, when the dynamics is not characterized by a sole scaling exponent, but by a multitude of scaling exponents, we deal with a multifractal process and special methods should be used. For example, one can use a multifractal MF-DFA algorithm (Kantelhardt et al., 2002a).

MF-DFA presumes two additional steps to the standard DFA (Kantelhardt et al., 2002a). At first we average over all n segments to obtain the q-th order fluctuation function,

$$F_q(n) = \left[\frac{1}{N} \sum_{i=1}^{N} [Y(i) - Y_n(i)]^q \right]^{1/q}$$

where, in general, the index variable q can take any real nonzero value. For $q = 2$, the standard DFA procedure is retrieved. As far as we are interested, how the generalized q-dependent fluctuation functions $F_q(n)$ depend on the timescale n, calculation should be repeated for different values of q and for different timescales. Multifractal fluctuation analysis (MDFA) has been proposed to study multifractality in nonstationary signals when they are long-range power-law correlated:

$$F_q(n) \sim n^{\alpha(q)},$$

where $\alpha(q)$ is the generalized scaling exponent. For monofractal time series, $\alpha(q)$ is independent of q, and only if small and large fluctuations scale differently, will there be a significant dependence of α on q for multifractal data sets.

We also used the RQA approach (Zbilut and Webber, 1992; Webber and Zbilut, 1994; Marwan et al., 2007), which helps in studying the temporal dynamics of a different time series obtained from nonstationary processes. RQA is a quantitative extension of the recurrent plot (RP) construction method, which is based on the fact that return (recurrence) to the certain condition of the system or to the corresponding state space location, is a fundamental property of any deterministic dynamic system (Eckmann et al., 1987). Recurrence property holds also for systems that are not exactly deterministic but have nonrandom dynamic structures. RQA calculations, to be successfully fulfilled, at first necessitate reconstruction of the phase space trajectory from given scalar data sets — the proximity of the phase trajectory points should be tested and marked by the condition that the distance between them is less than a specified threshold ε (Eckmann et al., 1987). In this way, we obtain a two-dimensional representation of the recurrence features of dynamics, embedded in a high-dimensional phase space. Then the small-scale structure of recurrence plots can be quantified by a recurrence quantification method (Zbilut and Webber, 1992; Webber and Zbilut, 1994, 2005; Marwan et al., 2007; Webber et al., 2009). This technique quantifies visual features in an $N \times N$ distance matrix recurrence plot and defines several measures of complexity. RQA provides exact measures of complexity based on the quantification of diagonally and vertically oriented lines in the recurrence plot. In this research we have calculated several measures mentioned above. However, these measures as a rule are not contradictory. Thus, we present only determinism-DET, the ration of reccurence points forming diagonal structures to all recurrence points.

As was mentioned earlier, in addition to these methods, in order to tackle long digital seismograms recorded by the broadband permanent station located in the Great Caucasus mountains near the town of Oni (42.5905N, 43.4525E), Georgia (Fig. 1–1), we used the Langevin equation method. This is a method for retrieving features of a stochastic dynamic system from measured data (Friedrich et al., 2000; Renner et al., 2001). A basic assumption of this approach to the analysis of fluctuating data is the presence of a Markovian property, which for real systems can be valid above a certain time or length scale (Gottschall and Peinke, 2008; Langner et al., 2010). For such systems, prediction of future evolution requires only knowledge of the actual situation and this requirement is given formally by a probability:

$$p(x(t+\tau)|x(t), x(t-\tau)) = p(x(t+\tau)|x(t)), \tag{1.1}$$

where t is the actual time and τ is a time increment.

For analysis of features of Earth surface seismic fluctuations we applied a stochastic differential equation (Langevin equation) reconstruction method. This method has already been successfully used for the analysis of complex processes in different fields (see, e.g., Renner et al., 2001; Langner et al., 2010; Czechowski and Telesca, 2011; Czechowski and Rozmarynowska, 2008). For a finite small time step τ, the Langevin equation is given as

$$x(t) = x(t-\tau) + \tau D_1(x) + \sqrt{\tau D_2(x)}\Gamma(t), \tag{1.2}$$

Here $D_1(x)$, the so-called drift term, gives the deterministic contribution to the process evolution, $D_2(x)$, which is a diffusion term that is related to the amplitude of the noise, and $\Gamma(t)$ denotes delta-correlated white noise (Friedrich et al., 2000; Langner et al., 2010). Methods developed in the last decade (Friedrich et al., 2000; Gottschall and Peinke, 2008; Langner et al., 2010) enable the estimation of drift and diffusion terms directly from the data:

$$D_n(x) = \frac{1}{n!}\lim_{\tau \to 0}\frac{1}{\tau}M_n(x, \tau), \tag{1.3}$$

In Eq. (1.3), conditional moments $M_n(x, \tau)$ are given as:

$$M_n(x, \tau) = \langle [x(t+\tau) - x(t)]^n \rangle|_{x(t)=x}, \tag{1.4}$$

In practice, the first and second conditional moments ($n = 1$ and $n = 2$) can be calculated by binning the data $x(t)$ and using different time steps, τ (Siegert et al., 1998; Siefert et al., 2003; Gottschall and Peinke, 2008; Langner et al., 2010).

To investigate dynamic features of fluctuations of the Earth's vertical velocity, we have analysed windows of 120,000 data, which is long enough to ensure the appropriate population of each bin during the moments'calculation procedure. We moved 120,000 data length windows by the 60,000 data step, throughout considered Earth vertical velocity data. Analysis was accomplished for data sets recorded in each seismically quiet day, as well as for long data sets consisting of pooled recordings of all 12 seismically quiet days. As far as variation of $D_1(x)$ provides important information about dynamic changes in seismic noise data, we will be restricted to these results.

1.4 Results of Analysis

1.4.1 Waiting Times and Earthquake Interdistance Analysis

As was described in Section 1.3, we analysed IET and EID data sets obtained from the Caucasian earthquake catalogue from 1960 to 2014. In Fig. 1−3, we present DFA fluctuation curves calculated for the entire data sets of IET and EID sequences for a different order of polynomial fitting.

DFA scaling exponents of IET and EID series, calculated for the entire time span of the Caucasian catalogue, are larger than 0.5 at any used order of polynomial fitting. Exact DFA scaling exponents at fifth-order polynomial fitting for interearthquake times are 0.65 ± 0.01 and for interearthquakes distances 0.62 ± 0.01. This means that earthquake spatial and temporal distributions are correlated nonrandom processes. Significantly smaller was the calculated scaling exponent of the Caucasian magnitude sequences (0.52 ± 0.01).

These results, in agreement with our earlier findings, show that the extent of order in earthquake time and space distribution is higher compared to earthquake energy (magnitude) distribution (Matcharashvili et al., 2000, 2002; Chelidze and Matcharashvili, 2007; Telesca et al., 2012).

Similar dynamic features of earthquake temporal, spatial and energy distributions were found also for different seismically active regions of the globe (Goltz, 1998; German, 2006; Li and Xu, 2010, 2013; Iliopoulos et al., 2012).

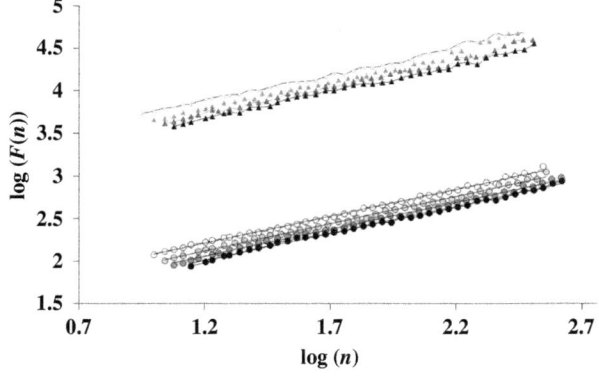

FIGURE 1–3 DFA fluctuation curves of IET (top) and EID (bottom) sequences obtained from the declustered, 1960−2014, Caucasian earthquake catalogue at M3.0 threshold. The order of the polynomial fitting is given by hollow to black circles from $p = 2$ to $p = 5$, respectively. n indicates the timescale (given here as the sequential number of data in analysed IET and EID series).

Further, second-order DFAs of magnitude subseries of IET and EID data sequences show positive correlations ($\alpha = 0.87 \pm 0.02$ and $\alpha = 0.66 \pm 0.01$, respectively); fluctuation curves in both cases do not indicate crossovers. This points to the nonlinear character of correlations, underlying the dynamics of earthquake temporal and spatial distributions. As for linear features, signs of subseries of IET and EID data sets, reveal anticorrelated behaviour ($\alpha = 0.22 \pm 0.01$ and $\alpha = 0.26 \pm 0.01$, respectively) at small scales. At larger scales we observed persistent behaviour ($\alpha = 0.63 \pm 0.02$) for IET data sets and randomlike behaviour ($\alpha = 0.49 \pm 0.01$) for EID data series. Thus, according to sign-magnitude decomposition and DFA(2) analysis of data sets from the catalogue of the entire time span, the internal dynamic structure of earthquake time distribution reveals persistent behaviour, in which nonlinear correlations clearly prevail, but at larger scales (larger DFA box sizes), linear correlations may contribute too. At the same time, the persistent behaviour of earthquake spatial distribution detects only nonlinear correlations.

The sliding window procedure helps to reveal time-varying long-range dependence in the considered data sets. For the used unevenly sampled data sets it is not possible to carry out calculations for the predefined time step, here 500 data windows were shifted by one data step. Five hundred data length windows were regarded as the most appropriate (Matcharashvili et al., 2016).

The results of calculations at different orders of polynomial fitting were quite similar, so for illustration we present the results of DFA scaling exponent calculations only for $p = 5$. As follows from Fig. 1−4, scaling exponents varied over a wide range during the observation period. Variation is especially noticeable for waiting time data sets. In most of the windows, waiting time sequences revealed clearly persistent behaviour, with scaling exponents larger

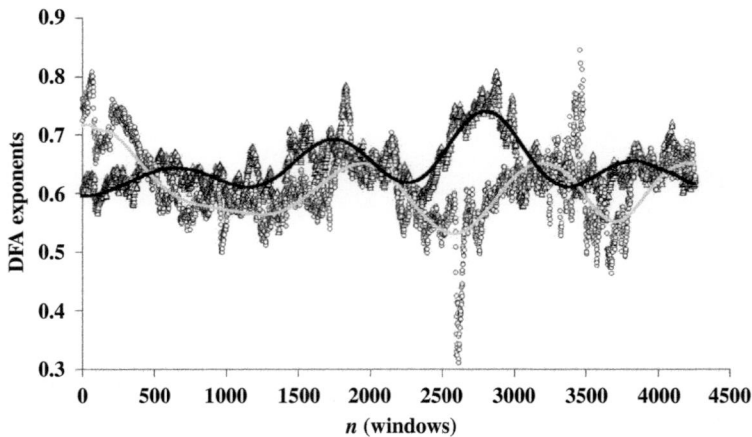

FIGURE 1–4 DFA scaling exponents' variation of IET (circles) and EID (triangles) data sets calculated for 500 consecutive data windows with one data step. The order of the polynomial fitting is $p = 5$. DFA scaling exponents reconstructed from the first four main singular spectrum analysis (SSA) components of EID data sets are given by the bold black curve and for IET data sets are given by the grey curves.

than 0.5 and in many less windows, antipersistent behaviour. In a few windows, approximately coinciding with periods when stronger earthquakes occurred, DFA scaling exponents for IET were close to 0.5. In contrast, for the EIDs in all the windows we observed, scaling exponent values were larger than 0.5 and smaller than 1. These results for IET and EID data sets signify a clear prevalence of persistent long-range correlations in the Caucasian earthquake temporal and spatial distributions.

The results presented in Fig. 1−4 indicate variability in the long-range correlation features of IET and EID data sets through the entire observation period. It is noticeable that the character of variations of scaling exponents calculated for waiting times and EID data sets was different. To assess this difference further, we analysed these data sequences in the presence of hidden cyclic components using a SSA approach (Broomhead and King, 1986). The aim of SSA is to decompose the observed series into the sum of independent components, trends, and different cyclical and noise components via an eigen-decomposition procedure. After the decomposition of the original signal into the sum of components, a new time series can be reconstructed depending on which feature of the original signal one wants to highlight (for details see, e.g., Vautard et al., 1992; Elsner and Tsonis, 1996; Golyandina et al., 2001).

We used SSA decomposition of sequences of scaling exponents calculated for 500 consecutive data windows of waiting times and EIDs. Actually, for many geophysical records, only a few leading SSA components correspond to the record's dominant oscillatory and/or trend modes, while the rest are just a noise (Ghil et al., 2002). In our case, reconstructed data sets of DFA scaling exponents of IET and EID sequences consisted of the first four main components, explaining at least 99% variance of the original data sets. As we were mainly interested in the longest possible cycles in the analysed process, we performed calculations for SSA window length close to a half-length of the analysed data sets, according to the method of Golyandina (2010). Reconstructed sequences, shown in Fig. 1−4 as bold black curves, represent slowly varying trends in the scaling exponent series, calculated for the 500 consecutive data length windows of original IET and EID data sets.

The results, presented in Fig. 1−4, show that long-range correlations in earthquake space and time distributions are characterized by slow, almost periodically recurring, variations over the analysed time period. These cycles in scaling exponent values (calculated for 500 consecutive data of IET and EID sequences) are of approximately 1000 data (scaling exponent) length, and are clearly visible starting from about 1200th window, which approximately corresponds to the sliding window in the mid-1970s. It is interesting that cycles of variation of long-range correlations of IET and EID do not coincide and sometimes even vary in antiphase (see Fig. 1−4). This may be related to different features of earthquakes' spatial and temporal clustering in the Caucasus.

Further analysis showed that antipersistent scaling exponents were found for periods when the strongest regional earthquakes (Spitak, 1988 and Racha, 1991) occurred (Matcharashvili et al., 2016).

The results of RQA are in general agreement with the results of the above analysis. As can be seen in Fig. 1−5, the extent of determinism in both IET and EID data sets clearly changed

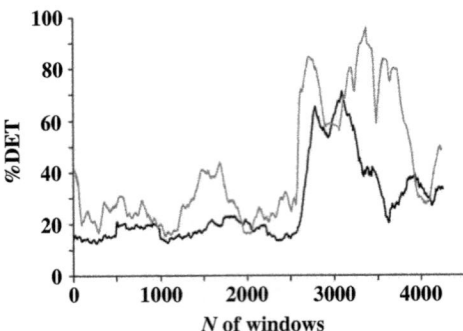

FIGURE 1–5 RQA %DET of IET (grey) and EID (black) data sets calculated for 500 consecutive data windows by one data step.

in different parts of the analysed period. The antiphase character of changes was also mostly preserved. At the same time differences between the results of DFA and RQA can be seen. Namely, on one side the extent of regularity in spatial distribution increases in the period of increased regional seismic activity in the 1990s, which is in accordance with long-range correlation analysis. On the other hand, we do not observe a decrease in the regularity of time distribution in the same period, which contradicts the DFA analysis results. It cannot be excluded that the observed discrepancies may be related to the relatively short windows that were used and, obviously, this analysis should be continued in the future.

1.4.2 Ambient Seismic Noise Data Analysis

The total length of the considered ambient noise time series was in the range of 10–35 million readings. First, we analysed the scaling properties of these time series in the frequency domain. Thus, the spectral scaling properties of the consecutive nonoverlapping 10-min segments of ambient noise time series were calculated (see Fig. 1–6).

Since the values of the calculated spectral exponents were varied greatly, we presented their distribution (Fig. 1–7): the ambient noise fluctuations mainly look like a combination of nonrandom, short- and long-range correlated noise (the slope varies from −1 to −2). It is important to mention that the scaling exponent of all considered time series after shuffling came close to zero. We found that there were no differences in the calculated power spectral scaling characteristics for ambient noise data sets, recorded when the local seismic activity increased prior to the Racha M6.0 earthquake and during quiet periods, preceding the arrival of the seismic wavetrains from the Japan M9.0 earthquake that occurred on 11 March 2011.

We then investigated the long-range correlation characteristics of the ambient noise data sets by using the DFA (Fig. 1–8).

In order to better visualize general shapes of the $F(n)$ vs. n relationship for these groups, the curves of averaged fluctuation functions are presented in Fig. 1–8. Here we see the crossover timescale at about 10 s in ambient noise fluctuations, both in the quiet time

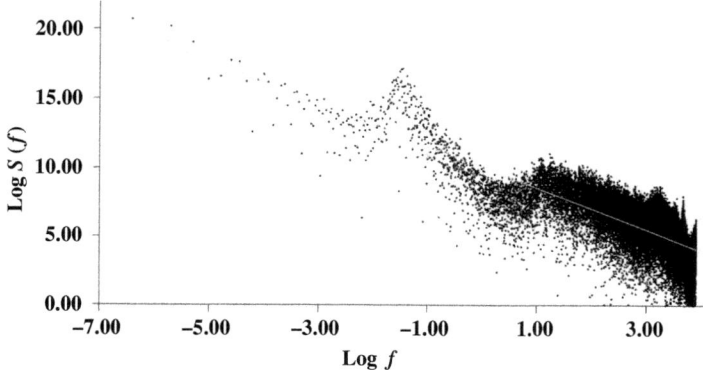

FIGURE 1–6 Typical plot of the log—log *S(f)* versus *f* relation of ambient noise time series, calculated for one of the 60,000 data windows prior to the Racha earthquake.

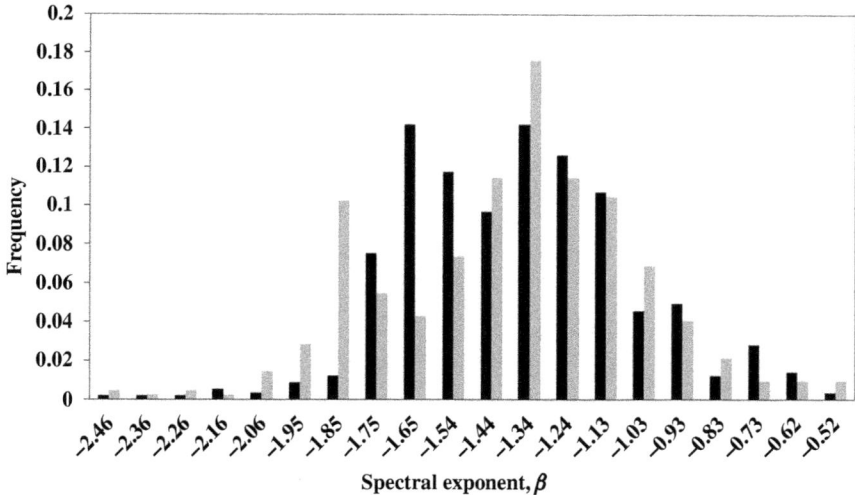

FIGURE 1–7 Histograms of the power spectral exponents calculated for consecutive windows. The sequence of 60,000 data of EW(001) seismic noise components, recorded at the Oni seismic station, was used. Dark columns correspond to windows of seismic noise prior to Racha M6.1 EQ 2009, grey columns correspond to windows, prior to and during the arrival of Japan M7.9 EQ, 2011 wavetrains.

windows and in the time windows when wavetrains from remote earthquakes arrived. Above this crossover, the scaling exponent drastically decreases, indicating the strong antipersistent character of seismic noise data sets, which is close to the lack of power law scaling behaviour (in this case scaling exponent values are 0.07 and 0.1, respectively).

At smaller timescales, we observe different behaviour of ambient noise fluctuations in the 'quiet' windows and in the time windows when remote seismic wavetrains arrived (grey and black curves). In the last case, there are no clear crossovers and the scaling exponent (> 1.5)

indicates long-range correlations, which may be related to a stochastic process. In contrast to this, in the quiet windows (lower curve in Fig. 1–8), crossover occurs at about a 0.5 s timescale, below which the process looks like Brownian motion (see also Matcharashvili et al., 2012).

Next, we compare the scaling properties of ambient noise data sets for different time periods. In Fig. 1–9, histograms of the scaling exponents are calculated for consecutive windows of 60,000. There data are presented for about a 90-h time period containing the Racha M6.1 earthquake with aftershocks, as well as the 80-h time period involving arrival of seismic wavetrains from the Japan M9.0 earthquake. According to our results, the original time series, by their DFA scaling exponents, is always markedly different from a random walk or

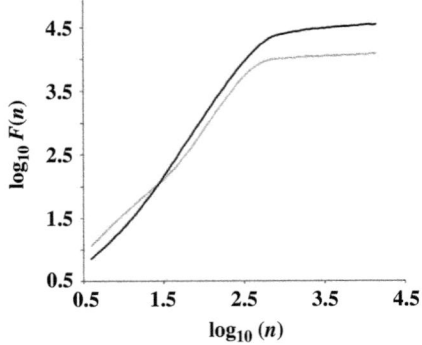

FIGURE 1–8 Averaged DFA fluctuation curves obtained for the z-component of seismic noise records at Oni seismic station. The grey curve corresponds to time windows of quiet periods prior to the M6.1 Racha earthquake (2009), the black curve corresponds to windows, when the arrival of wavetrains from a remote earthquake were being registered.

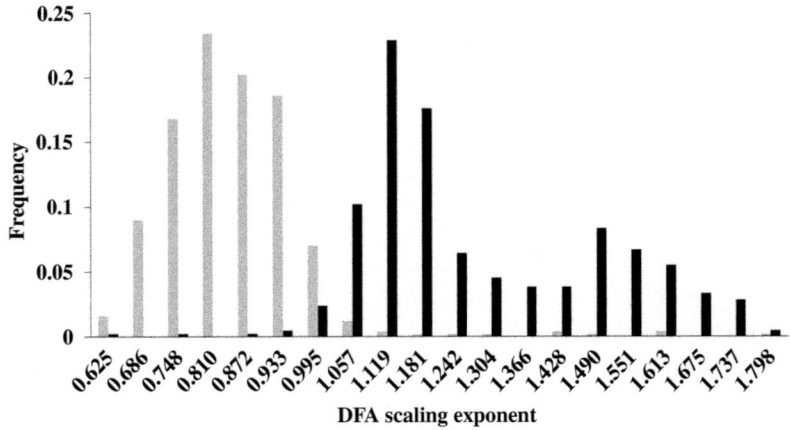

FIGURE 1–9 The histograms of the integral scaling exponents, calculated for consecutive windows of 60,000 data seismic noise z-components record at Oni seismic station. Grey columns correspond to 500 time windows prior to the Racha M6.1 earthquake; dark columns correspond to 400 windows in March 2011, involving the period before arrival of seismic waves from the Japan M7.9 earthquake.

antipersistent behaviour with scaling exponent close to 0.5. Indeed, time histograms indicate important differences in the Earth surface fluctuation properties for the two considered periods. We can see that not less than 90% of values of calculated DFA scaling exponents, for the period of the strong local Racha earthquake, correspond to a persistent long-range correlated process (grey columns). On the other hand, we observe an essential shift to the larger DFA exponents for seismic noise data, recorded at a locally quiet period, when teleseismic waves from the remote Japan M9.0 earthquakes arrived. In this case more than 97% of calculated DFA scaling exponents exceed a value of 1, spanning a wide range up to about 1.8. These different kinds of fluctuation characteristics, with respect to scaling behaviour, demonstrate different stochastic structures in ambient noise data sets for different patterns of local/remote seismic activity. This result indicates that during quiet periods ambient seismic noise comprises processes from uniform power law behaviour of $1/f$ type ($\alpha = 1$) to long-range correlations, which may be of stochastic fractional Brownian motion or even of a deterministic nature ($\alpha > 1.5$), also involving Brownian motion type processes ($\alpha = 1.5$). Differences in the mean values of integral scaling exponents for these two groups ($\alpha_{\text{avg } R} = 0.92 \pm 0.10$, $\alpha_{\text{avg } I} = 1.28 \pm 0.19$, $P < 0.001$) also provide an additional argument in favour of a viewpoint that the dynamics of seismic noises in the two considered periods are different. The difference is not caused by the influence of a seismic component of ambient noise (i.e., by the waves from local or remote earthquakes), because the results were practically the same, when we analysed only quiet time windows, when arrivals of seismic waves from local or remote earthquakes had not been detected at Oni station (see also Matcharashvili et al., 2012). We speculate that the correlated behaviour at increased local seismic activity is related to a decrease in the complexity in the Earth surface vibration dynamics during earthquake preparation processes.

Our finding that ambient noise has a higher fractal dimension than the seismic signals and that the probability density function (PDF) of ambient noise may undergo the transition from a Gaussian to a long tailed non-Gaussian prior to moderate and large earthquakes, is in agreement with results presented elsewhere (e.g., Padhy, 2004; Tabar et al., 2006; Manshour et al., 2009, 2010). It seems that the shift to a more correlated behaviour prior to increased seismic activity can be also detected in other nonseismic processes, related to earthquake preparation. Indeed, recent analysis of the temporal evolution of the fractal characteristics of preseismic electromagnetic emission indicates that approaching the earthquake nucleation phase is accompanied by a significant reduction in the complexity and transition to the persistent behaviour (Karamanos et al., 2006).

The behaviour of ambient noise fluctuations described above is typical for the multifractal process that we investigated in ambient noise time series by using an MF-DFA analysis technique (Kantelhardt et al., 2002a).

This approach is based on the identification of scaling of a q-th order moment depending on signal segment length and it is generalization of the standard DFA method in which $q = 2$. In Fig. 1−10, where the results of MF-DFA are presented, we see dependence of the generalized Hurst exponent $H(q)$ on q for data sets from both seismically active and relatively quiet periods, which is typical for multifractal sets. At the same time, the multifractal pattern is more pronounced in the case of a locally quiet period prior to the arrival of seismic

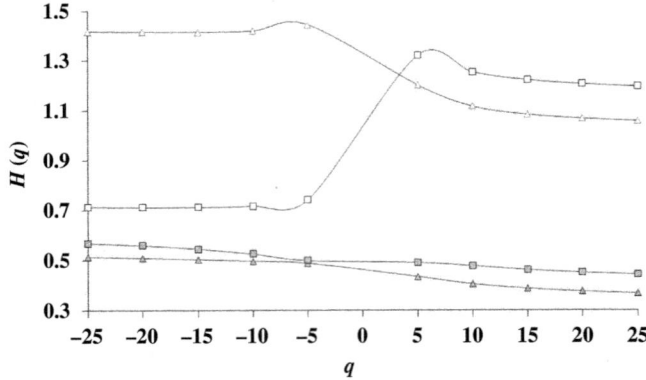

FIGURE 1–10 The generalized Hurst exponent $H(q)$ versus q calculated for the seismic noise Z-component time series. Open triangles correspond to the period prior to the M6.0 Racha earthquake (2009) and the squares to the period of arrival of wavetrains of the Japan M9.0 earthquake (2011). Grey triangles and squares correspond to shuffled data sets.

waves from the Japan earthquake in March 2011 (open squares in Fig. 1–10). According to the MF-DFA method, positive values of q, $H(q)$ correspond to the scaling behaviour of time segments with large fluctuations, i.e., with large variance of $F(n)$ or at large deviation from the corresponding fit. For negative values of q, $H(q)$ describes the scaling behaviour of time segments with small fluctuations. We can conclude from Fig. 1–10 that at increased local seismic activity small fluctuations in ambient noise dominate, while at decreased local seismic activity larger fluctuations prevail.

One of the subjects of special interest in scaling behaviour analysis is the determination of the source of multifractality. The easiest way to do this is to analyse the corresponding shuffled and surrogate time series. In general, two types of multifractality in time series can be distinguished: (1) multifractality due to a fatness of PDF of the time series and (2) multifractality due to different correlation levels in small- and large-scale fluctuations. When multifractality is related to fatness of PDF, it cannot be removed by a shuffling procedure, while in the second case corresponding shuffled time series will exhibit monofractal scaling, since all long-range correlations are destroyed by shuffling procedure (for details, see Matcharashvili et al., 2012). It may also happen that both types of multifractality are present. In such cases, a shuffled series will show weaker multifractality than the original series (Kantelhardt et al., 2002a).

According to Fig. 1–10, the shuffling procedure destroys the correlation structure of ambient noise data sets. $H(q)$ values calculated for shuffled series are concentrated in the vicinity of 0.5, exhibiting nonmultifractal scaling (Kantelhardt et al., 2002a). At the same time, observed scaling cannot be regarded as clearly monofractal, rather there is a weaker type of multifractality in shuffled ambient noise data. This indicates that the observed multifractality is mainly related to correlations and to a lesser degree to the distributional features of analysed data sets.

From the analysis mentioned above, we conclude that in both considered cases of ambient noises we deal with process, characterized by multifractal scaling. At the same time, in spite of this qualitative similarity, ambient noises for periods of increased and decreased local seismic activities reveal quantitative differences in their long-range correlation properties.

For further analysis we used the digital seismograms recorded by the broadband permanent station located in the Great Caucasus mountains near the town of Oni (42.5905N, 43.4525E), Georgia. In particular, we investigated data sets of the Earth's vertical velocity V_z. In order to trace the temporal behaviour of stochastic features of ambient noises, we analysed 8-h-long seismograms recorded at night (from 0000 to 0800, local time) in the area around Oni seismic station for seismically quiet days. It should be emphasized that in this area seismicity is currently active and, during the observation period from 2005 to 2012, two strong earthquakes occurred in the vicinity of Oni seismic station; namely an M5.2 earthquake in 2006 (04:08:1.3 (UTC), 06 February 2006, Lat. 42.520, Long. 43.545) and an M6.0 earthquake in 2009 (22:41:35 (UTC) on 07 September 2009, Lat. 42.5727, Long. 43.4825). The epicentres of these earthquakes were located 10 and 4 km from Oni seismic station, respectively. The selection conditions of days that we regarded as seismically quiet, were rather strong. We started from the supposition that the day would be regarded as seismically quiet if no seismic wavetrains from local or remote earthquakes were registered by the seismograph at station Oni. However, in practice, finding such quiet days at the selected location in the seismically active Caucasian mountains appeared impossible. Moreover, in recent years, when several strong earthquakes occurred in Racha region, seismic activity in the Oni seismic station area increased significantly and small earthquakes occur very regularly. Therefore, we regarded a day as a quiet one, in the sense of local seismic activity, if in the 20 km area around Oni station no earthquakes greater than M0.9 occurred, even if no epicentre location was defined. We succeeded in finding 12 such days from 2005 to 2012.

In order to investigate stochastic features of local ambient noises, time series of the described Earth's vertical velocity V_z (recorded on these quiet days) were analysed. It is important to mention that Oni is located in a rural area with low urban and industrial noise. At the same time, in order to further decrease possible unwanted influences we analysed seismograms, recorded at night (from 00.00 to 08.00). Thus, we used data sets of Earth vertical velocity recorded at 8 night hours of selected seismically quiet days from 2005 to 2012. We analysed the daily data sets (recordings of 8 night hours of each selected day) as well as a long time series (pooled 8 night hours recordings of 12 selected days). In both cases calculations were performed for consecutive windows of 120,000 data (20 min long segments of seismogram) shifted by 60,000 data steps.

For each selected quiet day, 48 windows and, consequently, 576 windows for the 12-day time series were analysed. For each window the drift coefficients were calculated according to the method described in Section 1.3 (see also Matcharashvili et al., 2013a). Here we present the results of drift term (D_1) calculations.

As can be seen from Fig. 1−11, averaged for 48 consecutive windows, the D_1 vs. amplitude relation generally contains more or less linear segments. We observe clear differences in the slopes of D_1 for different windows.

These differences are more obvious in Fig. 1−12, where the slopes of D_1 are shown. As described earlier, averaging was fulfilled for 48 consecutive windows of 20-min long night seismic records.

Changes, found in the Earth vertical velocity time series both for seismically quiet as well as for selected days, show that the dynamics of ambient noise undergo noticeable changes

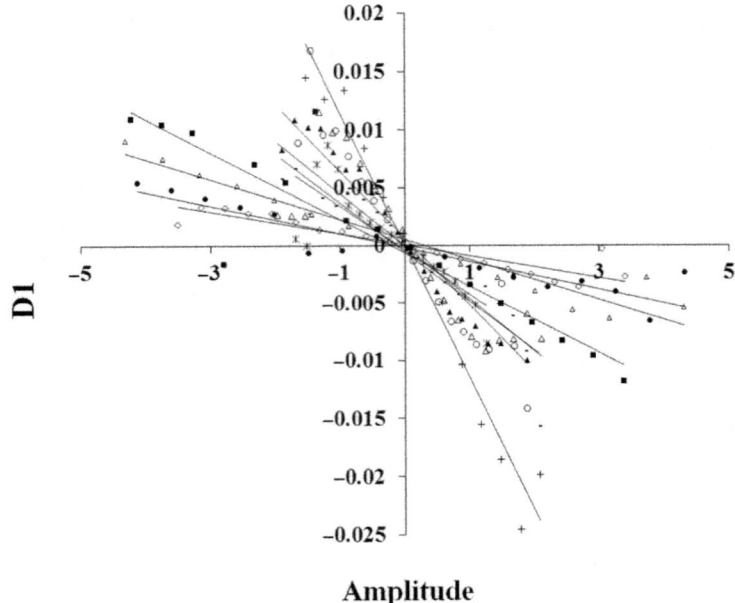

FIGURE 1–11 Typical view of drift coefficients (D_1) plot calculated for Earth vertical velocity time series recorded at night.

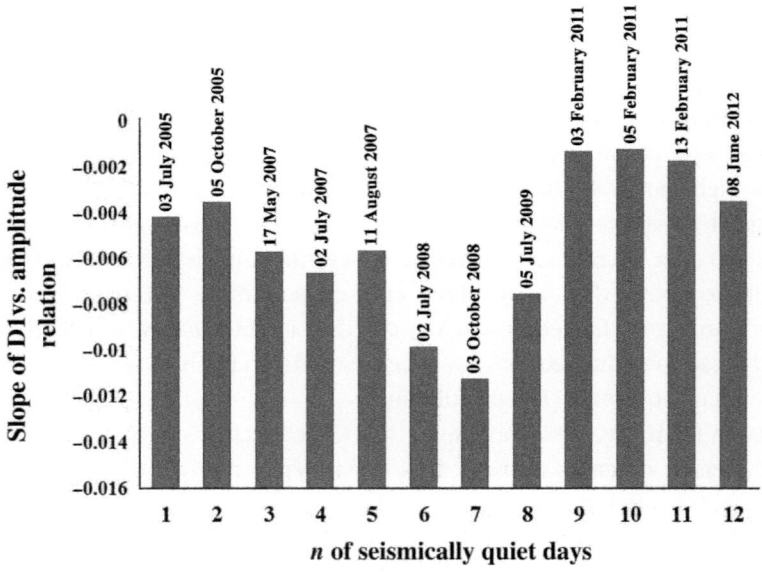

FIGURE 1–12 Slopes of averaged D_1 vs. amplitude relation calculated for night data of 12 nonequidistant seismically quiet days around Oni seismic station from 2005 to 2012.

for the analysed period. These changes can be regarded as being related to the local seismic activity process, because they coincide in time with the period prior to and after the last strongest Caucasian earthquake (M6.0, 07 September 2009), the epicentre of which was located 4 km from Oni seismic station.

The changes before the strong M6.0 earthquake may be related to the increased extent of correlations in the tectonic system during strong earthquake preparation. In our previous research, it was shown that at Oni seismic station in seismically calm periods, in the periods preceding essentially increased local seismic activity, ambient noises revealed more long-range correlated behaviour compared to periods of strong earthquake and aftershock occurrence (Matcharashvili et al., 2012, 2015). This explains the increase in the D_1 slopes for seismically quiet periods, when collective behaviour increases and the system approaches a critical point of highly likely earthquake occurrence. When seismic processes leading to more correlated behaviour in the Earth surface vibration become weaker, distributional and dynamic features of ambient noises should return to the initial conditions, which was observed in our analysis.

1.5 Conclusions

In this chapter we have shown the usefulness of modern data analysis methods for investigating dynamic features of the earthquake generation process. In particular, we have used earthquake catalogues of the Caucasus as well as seismic noise recordings as the sources of different seismic data sets.

Derived from the Caucasian seismic catalogue, IETs as well as earthquake interdistance data sets have been analysed to assess long-range correlations of features in earthquake temporal and spatial distributions. The presented results show changes in the persistence as well as in the recurrence of the data including in different observation periods.

In the case of seismic noise data sets it was shown that at shorter timescales the seismic process looks like a combination of different stochastic structures, which is typical for multifractal behaviour. The scaling features of seismic noise data sets are noticeably different for seismically active and relatively quiet periods. At increased seismic activity Earth surface vibrations are closer to persistent long-range correlations, while at quiet periods we observed a multitude of stochastic behaviours.

Using the method of stochastic model reconstruction based on real data we discovered changes in fluctuation features of Earth vertical velocity time series at Oni seismic station. We conclude that this method may help to discern subtle dynamic changes in the local seismic noise patterns.

The presented results indicate that modern methods of complex time series analysis can be successfully used in the analysis of seismic data sets.

Acknowledgement

This work was supported by Shota Rustaveli National Science Foundation (SRNSF), grant 217838 'Investigation of dynamics of earthquake's temporal distribution'.

References

Abarbanel, H., Tsimring, L.S., 1993. The analysis of observed chaotic data in physical systems. Rev. Mod. Phys. 65, 1331−1392.

Ashkenazy, Y., Ivanov, P.C., Havlin, S., Peng, C.K., Goldberger, A.L., Stanley, H.E., 2001. Magnitude and sign correlations in heartbeat fluctuations. Phys. Rev. Lett. 86, 1900−1903.

Ashkenazy, Y., Havlin, S., Ivanov, P.C., Peng, C.K., Schulte-Frohlinde, F., Stanley, H.E., 2003. Magnitude and sign scaling in power-law correlated time series. Physica A 323, 19−41.

Broomhead, D.S., King, G.P., 1986. Extracting qualitative dynamics from experimental data. Physica D 20, 217−236.

Chelidze, T., Matcharashvili, T., 2007. Complexity of seismic process; measuring and applications—a review. Tectonophysics 431, 49−60.

Chelidze, T., Matcharashvili, T., 2014. Dynamical patterns in seismology. In: Webber Jr., C.L., Marwan, N. (Eds.), Recurrence Quantification Analysis: Theory and Best Practices (Understanding Complex Systems). Springer, Berlin, pp. 291−334.

Czechowski, Z., Rozmarynowska, A., 2008. The importance of the privilege for appearance of inverse-power solutions in Ito equations. Physica A 387, 5403−5416.

Czechowski, Z., Telesca, L., 2011. The construction of an Ito model for geoelectrical signals. Physica A 390, 2511−2519.

Eckmann, J.P., Kamphorst, S., Ruelle, D., 1987. Recurrence plots of dynamical systems. Europhys. Lett. 4, 973−977.

Elsner, J.B., Tsonis, A.A., 1996. Singular Spectrum Analysis: A New Tool in Time Series Analysis. Plenum Press, New York.

Friedrich, R., Siegert, S., Peinke, J., Lück, S., Siefert, M., Lindemann, M., et al., 2000. Extracting model equations from experimental data. Phys. Lett. A 271, 217−222.

German, V.I., 2006. Unified scaling theory for distributions of temporal and spatial characteristics in seismology. Tectonophysics 424 (3−4), 167−175.

Ghil, M., Allen, M.R., Dettinger, M.D., Ide, K., Kondrashov, D., Mann, M.E., et al., 2002. Advanced spectral methods for climatic time series. Rev. Geophys. 40 (1), 3-1−3-41.

Goltz, C., 1998. Fractal and chaotic properties of earthquakes. Lect. Notes Earth Sci. Springer, Berlin.

Golyandina, N., 2010. On the choice of parameters in singular spectrum analysis and related subspace-based methods. Stat. Interface 3 (3), 259−279.

Golyandina, N., Nekrutkin, V.V., Zhigljavski, A.A., 2001. Analysis of Time Series Structure: SSA and Related Techniques. CRC Press, Boca Raton.

Gottschall, J., Peinke, J., 2008. On the definition and handling of different drift and diffusion estimates. New J. Phys. 10, 083034. Available from: https://doi.org/10.1088/1367-2630/10/8/083034.

Iliopoulos, A.C., Pavlos, G.P., Papadimitriou, P.P., Sfiris, D.S., Athanasiou, M.A., Tsoutsouras, V.G., 2012. Chaos, selforganized criticality, intermittent turbulence and nonextensivity revealed from seismogenesis in north Aegean area. Int. J. Bifur. Chaos 22 (9), 1250224.

Ivanov, P.C., Ashkenazy, Y., Kantelhardt, J.W., Stanley, H.E., 2003. Quantifying heartbeat dynamics by magnitude and sign correlations. In: Bezrukov, S. (Ed.), Unsolved Problems of Noise and Fluctuations, UPoN 2002: Third International Conference. AIP, pp. 383−391.

Iyengar, N., Peng, C.-K., Morin, R., Goldberger, A.L., Lipsitz, L.A., 1996. Age-related alterations in the fractal scaling of cardiac interbeat interval dynamics. Am. J. Physiol. 271, R1078−R1084.

Kantelhardt, J.W., Zschiegner, S.A., Bunde, A., Havlin, S., Koscielny-Bunde, E., Stanley, H.E., 2002a. Multifractal detrended fluctuation analysis of nonstationary time series. Physica A 316, 87−114.

Kantelhardt, J.W., Ashkenazy, Y., Ivanov, P.C., Bunde, A., Havlin, S., Penzel, T., et al., 2002b. Characterization of sleep stages by correlations in the magnitude and sign of heartbeat increments. Phys. Rev. E 65, 051908.

Kantz, H., Schreiber, T., 1997. Nonlinear Time Series Analysis. Cambridge University Press, Cambridge.

Karamanos, K., Dakopoulos, D., Aloupis, K., Peratzakis, A., Athanasopoulou, L., Nikolopoulos, S., et al., 2006. Preseismic electromagnetic signals in terms of complexity. Phys. Rev. E 74, 016104.

Keilis-Borok, V.I., Soloviev, A.A., 2002. Nonlinear Dynamics of the Lithosphere and Earthquake Prediction. Springer, Heidelberg.

Langner, M., Peinke, J., Flemisch, F., Baumann, M., Beckmann, D., 2010. Drift and diffusion based models of driver behavior. Eur. Phys. J. B 76, 99−107.

Li, Q., Xu, G.M., 2010. Relationship between the characteristic variations of local scaling property and the progress of seismogeny: the revelation of a new physical mechanism or seismicity. Fractals 18, 197.

Li, Q., Xu, G.M., 2013. Scale invariance in complex seismic system and its uses in gaining precursory information before large earthquakes: importance of methodology. Physica A 392 (4), 929−940.

Malamud, B.D., Turcotte, D.L., 1999. Self-affine time series I: generation and analyses. Adv. Geophys. 40, 1−90.

Manshour, P., Saberi, S., Sahimi, M., Peinke, J., Pacheco, A.F., Reza Rahimi Tabar, M., 2009. Turbulent-like behavior of seismic time series. Phys. Rev. Lett. 102, 014101.

Manshour, P., Ghasemi, F., Matsumoto, T., Gómez, J., Sahimi, M., Peinke, J., et al., 2010. Anomalous fluctuations of vertical velocity of Earth and their possible implications for earthquakes. Phys. Rev. E 82, 036105.

Marwan, N., Romano, M.C., Thiel, M., Kurths, J., 2007. Recurrence plots for the analysis of complex system. Phys. Rep. 438, 237−329.

Matcharashvili, T., Chelidze, T., Javakhishvili, Z., 2000. Nonlinear analysis of magnitude and interevent time interval sequences for earthquakes of Caucasian region. Nonlinear Process. Geophys. 7, 9−19.

Matcharashvili, T., Chelidze, T., Javakhishvili, Z., Ghlonti, E., 2002. Detecting differences in temporal distribution of small earthquakes before and after large events. Comput. Geosci. 28, 693−700.

Matcharashvili, T., Chelidze, T., Javakhishvili, Z., Jorjiashvili, N., FraPaleo, U., 2011. Non-extensive statistical analysis of seismicity in the area of Javakhety, Georgia. Comput. Geosci. 37 (10), 1627−1632.

Matcharashvili, T., Chelidze, T., Javakhishvili, Z., Jorjiashvili, N., Zhukova, N., 2012. Scaling features of ambient noise at different levels of local seismic activity: a case study for the Oni seismic station. Acta Geophys. 60 (3), 809−832.

Matcharashvili, T., Chelidze, T., Javakhishvili, Z., Zhukova, N., Jorjiashvili, N., Shengelia, I., 2013a. Discrimination between stochastic dynamics patterns of ambient noises (case study for Oni seismic station). Acta Geophys. 61, 1659.

Matcharashvili, T., Telesca, L., Chelidze, T., Javakhishvili, Z., Zhukova, N., 2013b. Analysis of temporal variation of earthquake occurrences in Caucasus from 1960 to 2011. Tectonophysics 608, 857−865.

Matcharashvili, T., Chelidze, T., Zhukova, N., 2015. Assessment of a ratio of the correlated and uncorrelated waiting times in the Southern California earthquake catalogue. Physica A 433, 291−303.

Matcharashvili, T., Chelidze, T., Javakhishvili, Z., Zhukova, N., 2016. Variation of the scaling characteristics of temporal and spatial distribution of earthquakes in Caucasus. Physica A 449, 136−144.

Munoz-Diosdado, A., Guzman-Vargas, L., Rairez-Rojas, A., Del Rio-Correa, J.L., Angulo-Padhy, F.S., 2004. Rescaled range fractal analysis of a seismogram for identification of signals from an earthquake. Curr. Sci. 87 (5), 637−641.

Padhy, S., 2004. Rescaled range fractal analysis of a seismogram for identification of signals from an earthquake. Curr. Sci. 87 (5), 637−641.

Peng, C.-K., Buldyrev, S.V., Goldberger, A.L., Havlin, S., Simons, M., Stanley, H.E., 1993a. Finite size effects on long-range correlations: implications for analyzing DNA sequences. Phys. Rev. E 47, 3730−3733.

Peng, C.K., Mietus, J., Hausdorff, J., Havlin, S., Stanley, H.E., Goldberger, A.L., 1993b. Long-range anticorrelations and non-Gaussian behavior of the heartbeat. Phys. Rev. Lett. 70, 1343–1346.

Peng, C.K., Havlin, S., Stanley, H.E., Goldberger, A.L., 1995. Quantification of scaling exponents and crossover phenomena in nonstationary heartbeat time series. Chaos 5, 82–87.

Penzel, T., Kantelhardt, J.W., Grote, L., Peter, J.H., Bunde, A., 2003. Comparison of detrended fluctuation analysis and spectral analysis for heart rate variability in sleep and sleep apnea. IEEE Trans. Biomed. Eng. 50 (10), 1143–1151.

Reasenberg, P., 1985. Second-order moment of central California seismicity, 1969–82. J. Geophys. Res. 90, 5479–5495.

Renner, C., Peinke, J., Friedrich, R., 2001. Evidence of Markov properties of high frequency exchange rate data. Physica A 298, 499–520. Available from: https://doi.org/10.1016/S0378-4371(01)00269-2.

Rodriguez, E., Echeverria, J.C., Alvarez-Ramirez, J., 2007. Detrended fluctuation analysis of heart intrabeat dynamics. Phys. A: Stat. Mech. Appl. 384 (2), 429–438.

Scholz, C.H., 1990. The Mechanics of Eathquakes and Faulting. Cambridge University Press, Cambridge.

Siefert, M., Kittel, A., Friedrich, R., Peinke, J., 2003. On a quantitative method to analyze dynamical and measurement noise. Europhys. Lett. 61, 466–472.

Siegert, S., Friedrich, R., Peinke, J., 1998. Analysis of data sets of stochastic systems. Phys. Lett. A 243, 275–280. Available from: https://doi.org/10.1016/S0375-9601(98)00283-7.

Sornette, D., 2000. Critical Phenomena in Natural Sciences. Springer, Berlin.

Sprott, J., 2003. Chaos and Time-Series Analysis. Oxford University Press.

Strogatz, S.H., 2000. Nonlinear Dynamics and Chaos with Applications to Physics, Biology, Chemistry and Engineering. Perseus Books Publishing, New York City, NY.

Tabar, M.R.R., Sahimi, M., Ghasemi, F., Kaviani, K., Allamehzadeh, M., Peinke, J., et al., 2006. Short-term prediction of medium and large-size earthquakes based on Markov and extended self-similarity analysis of seismic data, modelling critical and catastrophic phenomena in geoscience. Lect. Notes Phys. 705, 281–301.

Telesca, L., Cuomo, V., Lapenna, V., Macchiato, M., 2004. Detrended fluctuation analysis of the spatial variability of the temporal distribution of Southern California seismicity. Chaos Solit. Fract. 21, 335–342.

Telesca, L., Matcharashvili, T., Chelidze, T., 2012. Investigation of the temporal fluctuations of the 1960–2010 seismicity of Caucasus. Nat. Hazards Earth Syst. Sci. 12, 1905–1909.

Vautard, R., Yiou, P., Ghil, M., 1992. Singular spectrum analysis: A toolkit for short, noisy chaotic signals. Physica D 58, 95–126.

Webber, C., Marwan, N., 2015. Recurrence Quantification Analysis. Springer, Cham Heidelberg.

Webber, C.L., Zbilut, J.P., 1994. Dynamical assessment of physiological systems and states using recurrence plot strategies. J. Appl. Physiol. 76, 965–973.

Webber, C.L., Zbilut, J.P., 2005. Recurrence quantification analysis of nonlinear dynamical systems. In: Riley, M.A., Van Orden, G.C. (Eds.), Tutorials in Contemporary Nonlinear Methods for the Behavioral Sciences. National Science Foundation, Virginia, pp. 26–94.

Webber, C.L., Marwan, N., Facchini, A., Giuliani, A., 2009. Simpler methods do it better: success of recurrence quantification analysis as a general purpose data analysis tool. Phys. Lett. A 373, 3753–3756.

Zbilut, J.P., Webber, C.L., 1992. Embeddings and delays as derived from quantification of recurrence plots. Phys. Lett. A 171, 199–203.

Zheng, Z., Yamasaki, K., Tenenbaum, J., Podobnik, B., Tamura, Y., Stanley, H.E., 2012. Scaling of seismic memory with earthquake size. Phys. Rev. E 86, 011107.

Nonextensive Statistical Seismology: An Overview

Filippos Vallianatos, Georgios Michas, Giorgos Papadakis

TECHNOLOGICAL EDUCATIONAL INSTITUTE OF CRETE AND UNESCO CHAIR ON SOLID EARTH PHYSICS AND GEOHAZARDS RISK REDUCTION, CRETE, GREECE

CHAPTER OUTLINE

2.1 Introduction

Earthquake physics is one of the most intriguing fields in science. It is not only the abruptness of the phenomenon that attracts our interest and wonder, but also the devastating consequences that earthquakes can have for the anthropogenic environment that require our better understanding of the fundamental physics of this phenomenon in order to mitigate the risks. The earthquake generation process is a complex phenomenon, manifested in the nonlinear dynamics and in the wide range of spatial and temporal scales that are incorporated in the process (Keilis-Borok, 1990; Kagan, 1994; Scholz, 2002; Rundle et al., 2003). Despite the significant progress that has been achieved since the beginning of instrumental seismology in the late 19th century in understanding the mechanisms that lead to the nucleation of individual earthquakes, understanding the exact physics that govern the earthquake generation process and the subsequent prediction of future earthquakes still represent an outstanding challenge for scientists (e.g., Scholz, 2002).

Despite the complexity of the earthquake generation process and our limited knowledge on the physical processes that lead to the initiation and propagation of a seismic rupture giving rise to earthquakes, the collective properties of many earthquakes present patterns that seem universally valid. The most prominent is scale-invariance, which is manifested in the size of faults, the frequency of earthquake sizes and the spatial and temporal scales of seismicity. A variety of fault attributes, such as the distribution of fault trace-lengths or fault displacements, exhibit power-law scaling and (multi)fractal geometries (Bonnett et al., 2001). The frequency−size distribution of earthquakes generally follows the Gutenberg−Richter (GR) law (Gutenberg and Richter, 1944) that resembles power-law scaling in the distribution of dissipated seismic energies and fault rupture areas, limited in each case by the size of the seismogenic system. The aftershock production rate following a main event generally decays as a power-law with time according to the modified Omori formula (Utsu et al., 1995). Scale-invariance and (multi)fractality are also manifested in the temporal evolution of seismicity and the distribution of earthquake epicentres (e.g., Turcotte, 1997).

The organization patterns that earthquakes and faults exhibit have motivated the statistical physics approach to earthquake occurrence (Main, 1996; Rundle et al., 2003; Sornette, 2006; Kawamura et al., 2012; de Arcangelis et al., 2016; Vallianatos et al., 2016). Statistical physics bridges the gap between the underlying physics that characterize seismic rupture at the microscopic scale and the laws that govern friction, chemical reactions, fluid−rock interactions and so on, to the macroscopic scale of large earthquakes and faults (Sornette and Werner, 2009). Within the context of statistical physics, the earthquake generation process has been considered as a critical point phenomenon undergoing a phase transition, characterized by scale-invariance and fractality (Allegre et al., 1982; Chelidze, 1982; Sornette, 2006).

Self-organized criticality (SOC) has been proposed as a possible driving mechanism that produces some of the scale-invariant properties of seismicity (Bak, 1996). Within the SOC context, the Earth's crust self-organizes into a stationary critical or near-critical state, with intermittent fluctuations of individual earthquakes of power-law size distributions that correspond to the GR scaling relation.

Based on statistical physics and the entropy principle, a unified framework that produces the collective properties of earthquakes and faults from the specification of their microscopic elements and their interactions, has recently been introduced. This framework, called nonextensive statistical mechanics (NESM) was introduced by Tsallis (1988), as a generalization of classic statistical mechanics due to Boltzmann and Gibbs (BG), to describe the macroscopic behaviour of complex systems that present strong correlations among their elements, violating some of the essential properties of BG statistical mechanics (Tsallis, 2009). Such complex systems typically present power-law distributions, enhanced by (multi)fractal geometries, long-range interactions and/or large fluctuations between the various possible states, properties that correspond well to the collective behaviour of earthquakes and faults. Many applications during the last decade have highlighted that NESM is a powerful framework for describing the macroscopic behaviour of earthquakes and faults in a wide range of scales (Vallianatos et al., 2016 and references therein), introducing the field of nonextensive statistical seismology (NESS).

In this chapter, we provide an overview on the fundamental properties and applications of NESS. Initially, we provide an overview of the collective properties of earthquake populations and the main empirical statistical models that have been introduced to describe them. In Section 2.3 we briefly describe the main statistical physics models that have been introduced to describe earthquake occurrence and we summarize the classic (BG) statistical mechanics approach to the phenomenology of earthquakes. In Section 2.4 we provide an analytic description of the fundamental theory and the models that have been derived within the NESM framework to describe the collective properties of earthquakes. Section 2.5 provides various applications of NESS to earthquake populations and Section 2.6 summarizes and discusses the present overview.

2.2 The Phenomenology of Earthquake Populations

As already mentioned in the introduction to this chapter, the collective properties of earthquake populations exhibit some universal characteristics, in the sense that these are appearing in a wide range of spatial and temporal scales and tectonic environments. To describe the apparent phenomenology of earthquake populations, simple empirical scaling relations have been developed to describe the fundamental quantities of the earthquake occurrence, i.e., the size of the earthquake, its temporal occurrence and its hypocentral location. In the following subsection, the main empirical scaling relations that describe the earthquake phenomenology are briefly presented.

2.2.1 Frequency–Size Distribution of Earthquakes

Since the beginning of the 20th century and the early years of modern seismology, it became apparent that small earthquakes are considerably more frequent than larger ones (e.g., Utsu, 1999). The number of small to larger size earthquakes, in terms of the earthquake magnitude M, is commonly expressed with the cumulative distribution $N(M)$ of earthquake magnitudes, which indicates the number of earthquakes with magnitude equal to or greater than M. The cumulative distribution $N(M)$ exhibits an exponential decay that is commonly expressed with the well-known GR law (Gutenberg and Richter, 1944):

$$\log N(M) = a - bM, \tag{2.1}$$

where a and b are positive fitting parameters. The parameter a describes the regional level of seismicity and b, known as the seismic b-value, is the slope of the cumulative distribution that estimates the proportion of small to large events. The earthquake magnitude M is estimated from the amplitude of the seismic waves recorded at seismographic stations and includes a variety of measures such as moment (M_w), local (M_L), surface-wave (M_s) and body-wave (m_b) magnitude, depending on the part of the seismic wave that is used to determine its value (Lay and Wallace, 1995).

The frequency–magnitude distribution of earthquakes is an intrinsic part of any earthquake hazard assessment. The parameter *a* varies for different geographical regions, as those exhibit different levels of seismicity, number of earthquakes recorded and ranges of magnitudes. In contrast, the *b*-value seems quite independent of the regional and temporal scales of seismicity and typically falls in the range of 0.7–1.3 (Frohlich and Davis, 1993). In Fig. 2−1 we plot the cumulative distribution of earthquakes $N(M)$ for various geographic regions and earthquake catalogues and the GR law for $b = 1$. The exponential decay of the frequency–magnitude distribution is apparent in all datasets. The lowest magnitude for which the exponential decay of the GR law holds is usually considered as the magnitude of completeness M_c of the catalogue, i.e., the lowest magnitude for which all earthquakes in a given region and time interval are recorded.

An alternative physical measure for the earthquake size is the seismic moment M_0 that expresses the dissipated seismic energy from the source. The seismic moment can be estimated from the spectrum of the seismic signal and can be related to the instrumental earthquake magnitude M with the expression:

$$\log M_o = cM + d, \tag{2.2}$$

with a global average of $c = 1.5$ (Kanamori, 1978). The GR law [Eq. (2.1)] can be alternatively written as:

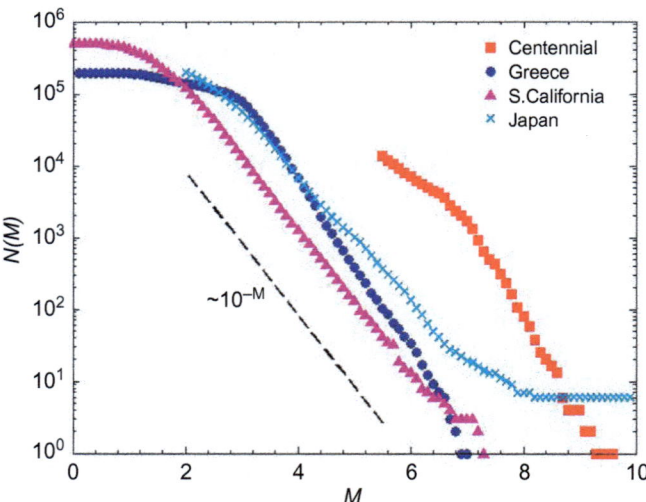

FIGURE 2−1 Discrete cumulative distribution $N(M)$ of earthquake magnitudes M for various geographical regions: Centennial (global) catalogue (1900−2007), Greece (1964−2014), South California (1981−2011) and Japan (1985−98). The dashed line represents the GR law [Eq. (2.3)] for $b = 1$.

$$N(M) = 10^{a-bM}.$$ (2.3)

From Eq. (2.2), Eq. (2.3) can be rewritten in terms of the seismic moment M_o as:

$$N(M_o) \sim M_o^{-\beta-1},$$ (2.4)

with $\beta = (2/3)b$. The latter equation expresses a power-law dependence between the number of earthquakes N and the seismic moment M_o and is considered as the main indication of scale-invariance in the distribution of the dissipated seismic energies (e.g., Main, 1996; Turcotte, 1997). In addition, the seismic moment M_o is related to the surface area of a fault with the expression:

$$M_o = \mu \Delta A,$$ (2.5)

where μ is the shear modulus (rigidity) of the material, A is the surface area of the fault break and Δ is the average slip during the fault break (Kanamori, 1978). Furthermore, experimental data show that the seismic moment M_o of an earthquake and the fault rupture area A scale consistently as (Kanamori and Anderson, 1975):

$$M_o \sim A^{3/2}.$$ (2.6)

Now if we combine the latter equation [Eq. (2.6)] with Eq. (2.4), we can easily show that the number of earthquakes N with rupture areas greater than A has a power-law dependence on A (e.g., Turcotte, 1997). The previous equations and discussion indicate that the GR law is equivalent to a fractal distribution (Aki, 1981).

Regarding the maximum earthquake energy that a seismic fault can release, the power-law form of Eq. (2.4) implies that this is infinite. However, the maximum energy release is not infinite and is associated with the size of the fault. The maximum earthquake energy can be incorporated into Eq. (2.4) by multiplying the right side with an exponential function of the form $\exp(-E/E_c)$, where E is the energy release in terms of the seismic moment and E_c is a characteristic or "corner" earthquake energy that characterizes the finite size-effects (Kagan, 2002a). The new form of Eq. (2.4) will then exhibit a power-law for small and intermediate-size earthquakes and an exponential taper for larger events.

2.2.2 Spatiotemporal Properties of Seismicity

Earthquakes are considered as a complex spatiotemporal phenomenon due to high variability in the time and space of their occurrence. The most prominent feature in the spatiotemporal patterns of seismicity is the presence of clustering. Earthquakes are spatially clustered along the tectonic plate boundaries and regional active fault networks (e.g., Scholz, 2002). The temporal occurrences of earthquakes exhibit variations between periods of relative quiescence, where only few regional earthquakes occur, and sudden seismic bursts, where a significant increase of the regional seismicity rate occurs. This variability can be observed by plotting the regional seismicity rate as a function of time, as shown in Fig. 2–2A for the area of Greece. The periods of increased seismicity are frequently associated with the occurrence of large earthquakes that cause the redistribution of stress in the lithosphere triggering subsequent earthquakes that are named 'aftershocks'. In Fig. 2–2A, the large spikes in the seismicity rate are associated with aftershock sequences triggered by the occurrence of large events (Fig. 2–2B). Intermittency and clustering in the temporal occurrence of seismicity contradicts a Poissonian presumption on the constant seismicity rates and random temporal occurrence.

The aftershock production rate following a large event decays as a power-law with time according to the modified Omori formula (e.g., Utsu et al., 1995):

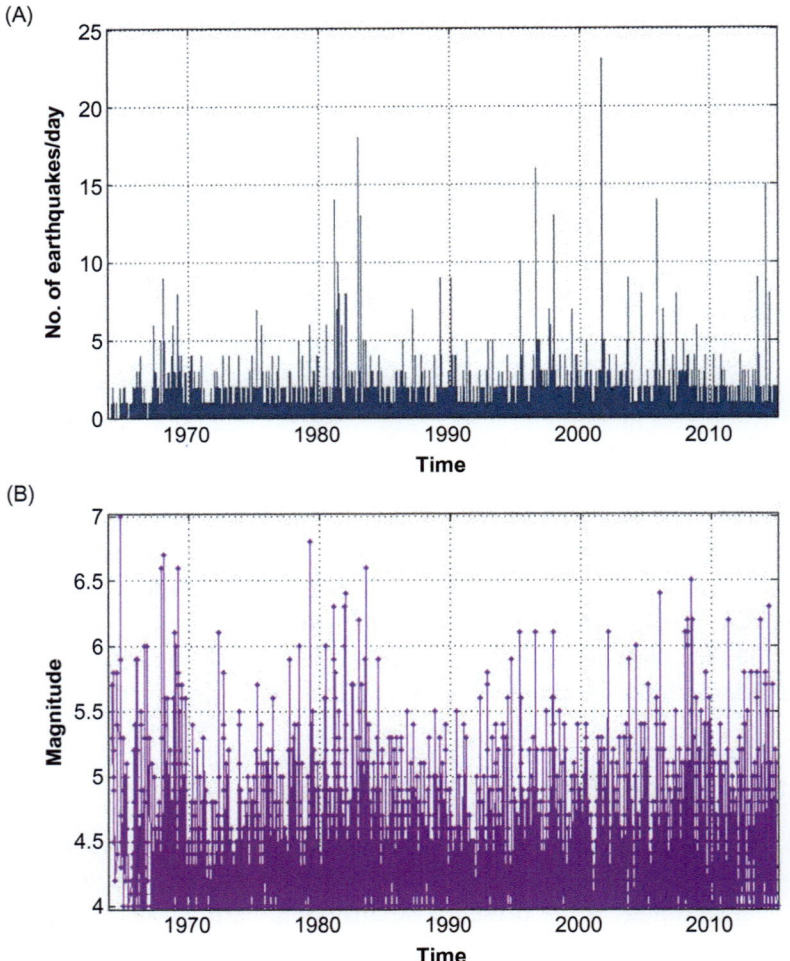

FIGURE 2–2 (A) Number of earthquakes per day and (B) magnitude rate per day for earthquakes with magnitude $M \geq 4$ as function of time for the Greek catalogue (1964–2014).

$$n(t) = K(t+c)^{-p}, \tag{2.7}$$

where $n(t)$ is the production rate, t is the time after the main shock and K, c and p are empirical constants. The power-law exponent p usually takes values close to 1 (e.g., Utsu et al., 1995). The constant K depicts the total number of aftershocks, while the constant c depicts the onset of the power-law decay. The truncation of the power-law regime at short time-scales following the main shock can be attributed to missing early events that are hidden in the coda waves of the main shock or to a real non power-law regime (Narteau et al., 2005). The modified Omori formula expresses a short-term clustering effect associated with the occurrence of large events and their triggered aftershock sequences. However, other functions, such as the exponential function, have been proposed to describe the aftershock production rate (e.g., Narteau et al., 2002).

To extract more detailed information regarding the temporal structure of seismicity in a seismic catalogue we define the distribution of interevent times (or waiting times) between successive earthquakes. If t_i is the time of occurrence of the ith event, then interevent times T are defined as $T_i = t_{i+1} - t_i$, with $i = 1, 2,\ldots, N - 1$, with N being the total number of events. Many functions have been proposed to describe the interevent time distribution including the exponential, lognormal, Weibull and gamma functions. Following the work of Bak et al. (2002) much discussion was stimulated on whether interevent times exhibit universality in their scaling properties. Bak et al. (2002) proposed a unified scaling law for earthquakes that considers the GR frequency–size distribution, the fractal distribution of epicenters and the Omori scaling of aftershocks. By dividing California into spatial cells of various sizes L^2 and by calculating interevent times T in each cell for various cutoff magnitudes, Bak et al. (2002) showed that the resulting interevent time distributions fall onto the same curve, if interevent times T are rescaled according to the expression $S^{-b}L_f^d$ and the probabilities $P_{S,L}(\tau)$ according to T^a, where the parameter a is associated with the Omori exponent [Eq. (2.7)], S is the earthquake size, b the GR b-value [Eq. (2.1)] and d_f the spatial fractal dimension.

The scaling hypothesis of Bak et al. (2002) was further studied by Corral (2003). By using a similar dataset with Bak et al. (2002), Corral (2003) proposed that the interevent time distribution $p(T)$ depends only on the seismicity rate R and showed that $p(T)$ collapses onto the same curve if interevent times are multiplied by the seismicity rate R. When mixing different seismicity rates, Corral (2003) derived a distribution that exhibits two power-law regions, while for stationary periods, where the seismicity rate does not fluctuate, $p(T)$ can be approximated by a generalized gamma distribution of the form:

$$f(T) = CT^{\gamma-1}\exp\left(-\frac{T^\delta}{\beta}\right), \tag{2.8}$$

with C, γ, δ and T_0 being fitting parameters. In a later work, Corral (2004) used several earthquake catalogues to show that interevent time distributions, after rescaling with a time-varying seismic rate, fall onto the same curve and are well described by the generalized gamma distribution [Eq. (2.8)] for the values of $\gamma = 0.67$, $\delta \approx 1$, $\beta = 1.58$ and $C = 0.50$. These results motivated Corral (2004) to claim universality in the temporal occurrence of earthquakes.

Universality in the interevent time distribution of earthquakes was further tested and questioned in several studies (Molchan, 2005; Hainzl et al., 2006; Saichev and Sornette, 2006, 2007; Touati et al., 2009). Molchan (2005) proved mathematically that if such a universal law exists it should be exponential, in strong contradiction to observations at short timescales. Saichev and Sornette (2006, 2007) used the epidemic-type aftershock sequence (ETAS) model to derive the shape of interevent time distribution and found that this is only approximately universal and of a gamma form. Hainzl et al. (2006) analysed both real and synthetic catalogues produced with the ETAS model and concluded that the shape of the distribution is strongly dependent on the rate of background and aftershock activity and that a Poissonian background activity combined with triggered aftershocks that scale according to the modified Omori relation are sufficient to reproduce the observed scaling. Touati et al. (2009) used a similar approach to show that the distribution exhibits a bimodal character of gamma-distributed correlated aftershocks and exponentially distributed uncorrelated activity at short and long interevent times, respectively. Touati et al. (2009) showed that this distribution is not universal and is strongly affected by the size of the area, where larger areas exhibit a higher initiation rate of uncorrelated events and smaller areas exhibit highly nonrandom interevent times.

In accordance with Eq. (2.8), the interevent time distribution exhibits bimodality, where short interevent times, associated with short-term clustering effects due to aftershock sequences, scale as a power-law with exponent $1 - \gamma$ and long interevent times, associated with the background activity of main shocks, exhibit Poissonian behaviour and exponential scaling. This type of scaling is approximately the one predicted from the ETAS model (Hainzl et al., 2006; Saichev and Sornette, 2006; Touati et al., 2009) and has been found to characterize the interevent time distribution for stationary earthquake time series (Corral, 2004; Hainzl et al., 2006). However, power-law scaling and long-term clustering effects have been found to characterize long interevent times of nonstationary earthquake time series (Kagan and Jackson, 1991; Mega et al., 2003, 2013; Michas et al., 2013). The latter indicates clustering effects at both short and long interevent times and implies the existence of memory in the earthquake generation process (Livina et al., 2005; Lennartz et al., 2008).

In Fig. 2–3 we plot the interevent time distribution $p(T)$ for the Southern California earthquake catalogue, where the bimodal character of the interevent time distribution becomes apparent. By rescaling T with the seismicity rate R, we can see that the observed $p(T')$, for various threshold magnitudes M_{th}, approximately collapse onto the same curve. In Fig. 2–3 we can also see that the probability of a subsequent earthquake is high immediately after the occurrence of the previous one and decreases slowly thereafter up to a characteristic T', where a faster decay of $p(T')$ is observed.

In addition, heterogeneous clustering effects in the temporal properties of seismicity are quantitively consistent with multifractal structures (e.g., Geilikman et al., 1990; Telesca and Lapenna, 2006). The multifractal structure of the interevent time series is evident in various

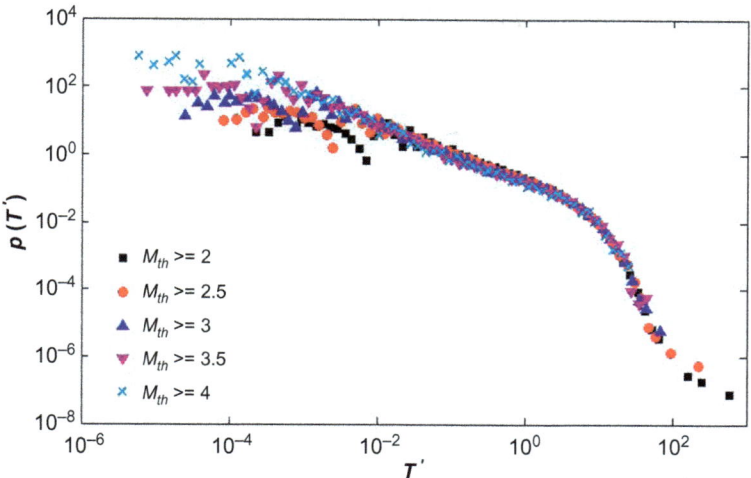

FIGURE 2–3 Normalized probability density function $p(T')$ for the rescaled interevent times $T' = R \cdot T$ for the Southern California earthquake catalogue (1981–2011) and for various threshold magnitudes M_{th}, represented by different colours and symbols.

earthquake sequences, where multifractal analysis is used to quantitively describe the local clustering effects and the temporal correlations of seismicity (Godano and Caruso, 1995; Telesca and Lapenna, 2006; Michas et al., 2015).

The spatial clustering of earthquakes is exemplified by the concentration of hypocentres along the tectonic plate boundaries and the regional fault networks. Another feature is the spatial clustering of aftershocks, which tend to occur close to the epicentre of the main shock. The distribution of aftershock epicentral distances from the main shock has been found to decay as a power-law with distance (e.g., de Arcangelis et al., 2016 and references therein). Scaling has also been found in the interevent distances between successive earthquakes. Davidsen and Paczuski (2005) found power-law scaling at ranges of 20–500 km in the interevent distance distribution of California seismicity, regardless of the threshold magnitude, which might imply a dynamic triggering effect at distances far beyond the aftershock zone scaling (e.g., Kagan, 2002b). Corral (2006) studied the interevent distance distribution for stationary periods in California and global seismicity and found an additional power-law regime with a smaller exponent at large interevent distances indicating bimodality, irrespective of the threshold magnitude. The crossover point between the two power-law regimes is different in the two catalogues (200 km for global and 15 km for California seismicity), implying a scale-dependent maximum triggering distance of correlated aftershocks at shorter distances and uncorrelated events at longer distances (Corral, 2006). Fractal and multifractal geometries have also been found in the spatial distribution of seismicity, originating from local clustering effects in main shock–aftershock zones and regional clustering effects stemming from the (multi)fractal geometry of regional faults (e.g., Kagan and Knopoff, 1980; Turcotte, 1997).

2.3 Statistical Physics of Earthquakes

In Section 2.2, we briefly reviewed some of the fundamental properties that characterize the phenomenology of earthquake populations, with power-laws, scale-invariance and (multi) fractality being integral parts. It is exactly these properties that have motivated the statistical physics approach to earthquake occurrence. In this context, earthquakes are considered as a critical phenomenon undergoing a phase transition and various physical models have been developed to describe the essential properties. In the following subsections, we briefly review the main statistical physics models and the classic statistical mechanics approach to the macroscopic properties of earthquake populations.

2.3.1 Critical Phenomena and Self-Organized Criticality

The universal validity of power-law scaling and (multi)fractality in earthquake properties has led to the development of a variety of physical models of seismogenesis as a critical point phenomenon undergoing a phase transition (e.g., Turcotte, 1997; Rundle et al., 2003; Sornette, 2006). In the context of statistical physics, the critical point is associated with power-law scaling and strong correlations acting at all scales in the system (long-range correlations); properties that are produced as the system approaches the critical point in the order–disorder phase transition. A characteristic example is magnets (e.g., Ising models) where, below the critical temperature (Curie temperature), a more organized state, with power-law size distribution and fractal scaling (ferromagnetic), emerges from random and chaotic (paramagnetic) behaviour (Bruce and Wallace, 1989). Such critical behaviour emerges as repeated long-range interactions between the microscopic elements of the system lead to a macroscopic self-similar state (Sornette, 2006).

In the previous example, the system is driven to the critical point by the precise tuning of some external variables, such as temperature. However, in other complex physical systems there is no external tuning and the system organizes itself to the critical point. Such a mechanism is called SOC, introduced by Bak et al. (1987). Bak et al. (1988) used a cellular automaton model of a sand pile that is supplied with new grains at a constant rate, to illustrate that as the sand pile sufficiently builds up and the pile reaches some critical angle, new added grains cause avalanches of all sizes that obey a power-law size distribution. The dynamic evolution of the sand pile cannot reach equilibrium, but instead remains in a statistically stationary but metastable state, which is characterized by power-law size distributions and fractal geometries (Bak et al., 1988; Bak, 1996).

The strikingly similar properties that the sand pile model and earthquake populations exhibit, such as power-law size distributions and fractality, led various workers to consider the Earth's crust as a SOC system (Bak and Tang, 1989; Sornette and Sornette, 1989). However, earthquake dynamics exhibit far richer responses than the limited behaviour of SOC (e.g., Ben-Zion, 2008) and novel models that present properties that are in better agreement with the organization of seismicity in time, space and magnitude have been introduced (see de Arcangelis et al., 2016 and references therein).

A popular model that has been introduced to simulate the behaviour of seismogenic faults is the slider-block model (Burridge and Knopoff, 1967; Rundle et al., 2003). This type of model typically consists of a set of blocks that are connected to each other via springs and a driver plate connected to the blocks via springs. When the spring force reaches some critical level due to the motion of the driving plate, the blocks slip, representing the slip on faults due to the motion of tectonic plates. The blocks interact with their neighbours, so that slipping on one block can trigger the propagation of slipping across several blocks. Although simplistic, the slider-block model can reproduce some of the empirical properties of seismicity. Its numerical implementation, using the cellular automaton approach, exhibits critical behaviour, similar to SOC, with a frequency-size distribution of slips similar to the GR law (Bak and Tang, 1989; Carlson and Langer, 1989; Olami et al., 1992). Although there is general agreement that this type of model can be used to study the dynamic behaviour of fault systems, in reality slider-blocks represent crude approximations of natural fault systems and the complex interactions that occur on a wide range of scales. Beyond this, various mechanisms can be used to produce power-laws and SOC (e.g., Sornette, 2006), but this property alone does not make them an appropriate model for earthquakes (Kagan, 1994).

2.3.2 Statistical Mechanics and Information Theory

Statistical mechanics and the associated concept of entropy, based on the second law of thermodynamics, constitute one of the cornerstones of contemporary physics that establish the remarkable relationship between the various microscopic states of a system and its macroscopic description. Based on probability theory and statistics, statistical mechanics provide suitable averages for the different microscopic constituents of the system in order to determine its large-scale behaviour. If p_i is the probability of occurrence of the ith microstate with $\sum_{i=1}^{W} p_i = 1$, then the entropy S, which is associated with the macroscopic properties of a system such as the temperature or pressure, is expressed as:

$$S = -k \sum_{i=1}^{W} p_i \ln p_i, \tag{2.9}$$

where W is the total number of all available microstates and k any positive constant, which in thermodynamics takes the value of Boltzmann's constant k_B (1.381×10^{-23} J/K). Eq. (2.9) is the classic definition of entropy, which is also referred to as the BG entropy S_{BG}. In the case of all microstates having equal probabilities of occurrence ($p_i = 1/W, \forall i$), then the entropy takes the celebrated expression:

$$S = k \ln W. \tag{2.10}$$

Eq. (2.10) indicates that the entropy increases logarithmically with the number of all microstates, such that a larger number of microstates corresponds to higher entropy. Entropy can then be regarded as a measure of disorder in the system.

Eqs. (2.9) and (2.10) refer to the discrete case. In the case of a continuous variable x that takes values in the space $[0, \infty]$, then the entropy takes the integral form:

$$S = - k \int_0^\infty p(x) \ln p(x) dx. \tag{2.11}$$

According to statistical mechanics, the most probable macroscopic state of the system is the one that maximizes entropy and satisfies the constraints of normalization of $p(x)$

$$\int_0^\infty p(x) dx = 1 \tag{2.12}$$

and the expectation (average) value of x

$$\langle x \rangle = \int_0^\infty x p(x) dx. \tag{2.13}$$

By using the Lagrange multipliers method and the previous constraints, we can derive the most probable probability distribution of $p(x)$, which takes the form of the well-known Boltzmann distribution:

$$p(x) = \frac{e^{-\beta x}}{\int_0^\infty e^{-\beta x} dx}. \tag{2.14}$$

The term $e^{-\beta x}$ is often referred to as the Boltzmann factor and the denominator of Eq. (2.14) as the partition function. Expressions (2.9) and (2.14) are the milestones of classic statistical mechanics (e.g., Pathria and Beale, 2011). The concept of entropy was later incorporated into information theory by Shannon (1948) as a measure for uncertainty and since the mid-1950s, it has been extended to other fields beyond classic thermodynamics in order to provide a general principle for inferring the least biased probability distribution that describes the macroscopic states of a system, derived from the limited information available for the system (Jaynes, 1957). The latter is commonly known as the maximum entropy principle (MaxEnt). Over the last few decades, the MaxEnt principle has widely been applied in many out-of-equilibrium systems in physics, chemistry, biology and elsewhere, providing novel insights into their macroscopic states, regardless of the complexity that may characterize their microscopic constituents (e.g., Pressé et al., 2013).

In earthquake physics, classic statistical mechanics has been applied as a variational principle for deriving the macroscopic properties of earthquake populations. One of the very first applications of statistical mechanics to earthquakes was that of Berrill and Davis (1980), who applied the maximum entropy principle to derive a truncated exponential distribution for earthquake magnitudes and a Poissonian distribution for the temporal occurrence of earthquakes. The seismic energies distribution was further studied by Main and Burton (1984), who maximized entropy to derive a gamma distribution of seismic energies:

$$p(E) \sim E^{-B-1} e^{-E/\theta}, \tag{2.15}$$

with B a scaling exponent and θ a characteristic seismic energy that corresponds to the probability of occupancy of the different energy states E. Eq. (2.15) consists of a power-law region at small seismic energies E that corresponds to the GR law [Eq. (2.4)] and a Boltzmann (exponential) tail at larger energies. By using earthquake catalogues from the Southern California and the Mediterranean regions, Main and Burton (1984) indicated that Eq. (2.15) is more consistent with real observations. Furthermore, Main and Al-Kindy (2002) used Eq. (2.15) to investigate the proximity of global seismicity to the critical point, characterized by the dissipated seismic energy E and the entropy S and concluded that global seismicity is in a near-critical state. Main and Naylor (2008, 2010) used the Olami–Feder–Christensen model and real earthquake data to study the hypothesis that the Earth's lithosphere is in a state of thermodynamically driven maximum energy production and SOC and concluded that observations were consistent with the hypothesis of entropy production as the driving mechanism for self-organized subcriticality in natural and numerically simulated seismicity.

Other studies considered the Shannon entropy (H) as a measure of disorder in earthquake sequences (Telesca et al., 2004; De Santis et al., 2011). These studies used earthquake data from Italy to show that significant variations of H occur as a response to the occurrence of large events (Telesca et al., 2004; De Santis et al., 2011).

2.4 Nonextensive Statistical Seismology

2.4.1 The Nonadditive Entropy S_q

BG statistical physics is based on the concept of entropy and constitutes a fundamental part of contemporary physics. For physical systems exhibiting short-range correlations in space and time and short-memory (Markovian processes), BG statistical mechanics seem to correctly describe nature. In contrast, for typical complex systems exhibiting long-range correlations in space and time, long memory (non-Markovian processes) and multifractal geometry, the concept of nonextensive statistical physics (NESP) seems appropriate to describe nature (Tsallis, 1988, 2001). Tsallis (2009) provides a detailed review of literature explaining the limitations of the BG entropic functional (S_{BG}). NESP refers to the nonadditive entropy (S_q) (Tsallis, 2009), which is a generalization of BG entropy. The nonadditive entropy S_q, for the case where the states of the system are discrete, reads as:

$$S_q = k \frac{1 - \sum_{i=1}^{W} p_i^q}{q - 1}, (q \in \mathbb{R}) \tag{2.16}$$

with

$$\sum_{i=1}^{W} p_i = 1, \tag{2.17}$$

where k is a positive constant taken to be Boltzmann's constant in thermodynamics, p_i is a set a probabilities, W is the total number of microscopic configurations and q is the entropic index which characterizes the degree of nonextensivity. Expression (2.16) recovers the BG entropy (S_{BG}) in the limit $q \to 1$. Although Tsallis entropy shares many properties with S_{BG} (see Tsallis, 2009), S_{BG} is additive, whereas S_q ($q \neq 1$) is nonadditive. This means that if A and B are two probabilistically independent systems (i.e., $p_{ij}^{A+B} = p_i^A p_j^B (\forall (i,j))$, Tsallis entropy ($S_q$) satisfies:

$$\frac{S_q(A+B)}{k} = \frac{S_q(A)}{k} + \frac{S_q(B)}{k} + (1-q)\frac{S_q(A)}{k}\frac{S_q(B)}{k}. \tag{2.18}$$

Expression (2.18) is the fundamental principle of NESM. Given that $S_q \geq 0$ (nonnegativity), it becomes clear that the cases $q < 1$, $q = 1$ and $q > 1$ correspond, respectively, to superadditivity (superextensivity), additivity (extensivity) and subadditivity (subextensivity). For the case of equal probabilities (for a set of W discrete states), Eq. (2.16) is expressed as:

$$S_q = k \, ln_q W. \tag{2.19}$$

The original idea of introducing the nonadditive entropy (S_q) (Tsallis, 1988) was inspired by multifractals and was based on simple physical principles. Tsallis (2009) describes the procedure for generalizing BG statistical physics and points out that the entropic index q introduces bias in probabilities. Given the fact that $0 < p_i < 1$, then $p_i^q > p_i$ if $q < 1$, and $p_i^q < p_i$ if $q > 1$. Consequently, $q < 1$ enhances the rare events having probabilities close to zero, whereas $q > 1$ enhances the frequent events having probabilities close to unity. Therefore, the entropic form introduced by Tsallis (1988) is based on p_i^q (see also Tsallis, 2009 for a complete description).

2.4.2 Optimizing S_q

The probability distribution $p(X)$ of a continuous variable X is estimated by maximizing the nonadditive entropy (S_q) subject to given constrains. This mathematical optimization is performed using the method of Lagrange multipliers. In seismology, variable X may be the interevent times (T) or distances (D) between successive earthquakes or the surface of a fragment (σ) filling the space between the fault blocks. In an integrated form S_q is defined as:

$$S_q = k \frac{1 - \int_0^{\infty} p^q(X)dX}{q-1}. \tag{2.20}$$

The first constraint refers to the normalization of $p(X)$ and is expressed as:

$$\int_0^\infty p(X)dX = 1. \tag{2.21}$$

The second constraint refers to the generalized expected value X_q, which reads as:

$$X_q = \langle X \rangle_q = \int_0^\infty X P_q(X)dX, \tag{2.22}$$

where $P_q(X)$ is the escort distribution (Abe, 2003; Beck and Schlogl, 1993; Tsallis, 2009) given as:

$$P_q(X) = \frac{p^q(X)}{\int_0^\infty p^q(X)dX}, \text{with} \int_0^\infty P_q(X)dX = 1. \tag{2.23}$$

In the limit $q \to 1$, X_q tends to the ordinary expected value. The concept of escort distributions is used due to the multifractal nature of complex phenomena (Abe, 2003). The next step concerns the use of the Lagrange multipliers and the maximization of the following function:

$$\delta\left(S_q - a^* \int_0^\infty p(X)dX - \beta^* X_q\right) = 0, \tag{2.24}$$

where α^* and β^* are the Lagrange multipliers. Optimization of Eq. (2.24) leads to the expression of the physical probability:

$$p(X) = \frac{\left[1-(1-q)\beta_q X\right]^{1/1-q}}{Z_q} = \frac{exp_q(-\beta_q X)}{Z_q}. \tag{2.25}$$

Furthermore, Tsallis q-exponential function is the numerator of Eq. (2.25) and reads as (see Tsallis, 2009):

$$exp_q(X) = \begin{cases} \left[1+(1-q)X\right]^{1/1-q} & 1+(1-q)X \geq 0 \\ 0 & 1+(1-q)X < 0 \end{cases}. \tag{2.26}$$

Its inverse is the q-logarithmic function given as:

$$ln_q(X) = \frac{1}{1-q}\left(X^{1-q} - 1\right). \tag{2.27}$$

At this point, it is worth mentioning that in the limit $q \to 1$ the exponential and logarithmic functions are recovered respectively. Moreover, the denominator of Eq. (2.25) is called the q-partition function and is expressed as:

$$Z_q = \int_0^\infty exp_q\left(-\beta_q X\right)dX, \tag{2.28}$$

where, $\beta_q = \frac{\beta^*}{C_q+(1-q)\beta^* X_q}$ and $C_q = \int_0^\infty p^q(X)dX$.

The q-exponential distribution is a generalization of the Zipf–Mandelbrot distribution (Mandelbrot, 1982) and exhibits an asymptotic power-law for $q > 1$. On the other hand, for $0 < q < 1$ a cutoff appears at $X_c = 1/(1-q)\beta_q$ (Abe and Suzuki, 2003, 2005).

In order to obtain the cumulative distribution function (CDF) $P(>X)$ integration is performed on the escort distribution $P_q(X)$ and not on the physical probability $p(X)$ (Abe and Suzuki, 2003, 2005; Tsallis, 2009; Vallianatos, 2009). By using Eq. (2.23) the CDF is given as:

$$P(>X) = \int_X^\infty P_q(X)dX. \qquad (2.29)$$

Combining the CDF with the physical probability function $p(X)$ [Eq. (2.25)], the following expression is obtained:

$$P(>X) = exp_q\left(-\beta_q X\right). \qquad (2.30)$$

Furthermore, by setting $\beta_q = 1/X_0$, Eq. (2.30) becomes:

$$P(>X) = exp_q\left(-\frac{X}{X_0}\right), \qquad (2.31)$$

which leads to:

$$\frac{[P(>X)]^{1-q} - 1}{1-q} = -\frac{X}{X_0}. \qquad (2.32)$$

The latter equation corresponds to the q-logarithmic function [Eq. (2.27)] and implies that the latter is linear and its slope equals $-\left(1/X_0\right)$ (Vallianatos, 2011; Vallianatos and Sammonds, 2011).

In classical BG statistical physics, the distribution of X^2 corresponds to the Gaussian distribution. Therefore, optimization of S_q with respect to X^2 results in the q-Gaussian distribution (Picoli et al., 2009), which has the form (Tsallis, 2009):

$$p(X) = \frac{1}{Z_q}\left[1 - (1-q)\left(\frac{X}{X_0}\right)^2\right]^{1/(1-q)}. \qquad (2.33)$$

In the limit $q \to 1$, Eq. (2.33) recovers the normal Gaussian distribution.

2.4.3 The Fragment–Asperity Model for Earthquake Magnitudes

The NESP approach related to the earthquake frequency–magnitude distribution is expressed through a novel cumulative magnitude distribution derived from the fragment–asperity

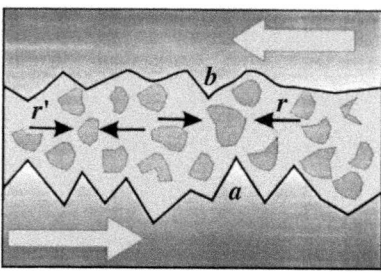

FIGURE 2–4 An illustration of the relative motion of fault blocks in the presence of material filling the space between them. This material hinders the relative motion of the blocks (e.g., between points a and b). *r* (or *r'*) is the size of the hindering fragment. From Sotolongo-Costa and Posadas (2004) (Fig. 1).

model (Sotolongo-Costa and Posadas, 2004). This model, which is consistent with the idea of stick—slip frictional instability in faults, considers the interaction of fault blocks and the fragments filling the gaps between them. It is due to this interaction that an earthquake is triggered. Stress accumulates until a fragment is displaced or an asperity is broken resulting in energy release. Consequently, the displacement of the fault blocks and the released energy are proportional to the size *r* of the hindering fragment (Fig. 2–4). Moreover, Sotolongo-Costa and Posadas (2004) introduced an energy distribution function (EDF) that shows the influence of the size distribution of fragments on the energy distribution of earthquakes, including the GR law as a particular case. Silva et al. (2006) revised the fragment-asperity model using a more realistic relationship between earthquake energy (ε) and fragment size, in accordance with the standard theory of seismic moment scaling with rupture length (Lay and Wallace, 1995). Moreover, Darooneh and Mehri (2010) and Telesca (2011, 2012) refined the fragment—asperity model using approaches that are presented here.

The probability distribution $p(\sigma)$ of finding a fragment with surface (σ) in relation to the nonadditive entropy S_q is given as:

$$S_q = k \frac{1 - \int p^q(\sigma)d\sigma}{q - 1}. \tag{2.34}$$

The probability distribution $p(\sigma)$ of finding a fragment with surface (σ) is obtained by maximizing the nonextensive entropy subject to appropriate constraints, as shown in the previous section (Section 2.4.2). Furthermore, a scale law between the released relative energy (ε) and the volume of the fragments (r^3) has been proposed ($\varepsilon \sim r^3$) by Silva et al. (2006), in agreement with the scaling relationship between seismic moment and rupture length (Lay and Wallace, 1995). The proportionality between the released relative energy (ε) and the three-dimensional size of the fragments (r^3) becomes:

$$\sigma - \sigma_q = \left(\frac{\varepsilon}{A}\right)^{2/3}, \tag{2.35}$$

where σ scales with r^2 and A is proportional to the volumetric energy density (the proportionality constant between ε and r^3). The EDF of the earthquakes can be written as follows (Silva et al., 2006; Telesca, 2011, 2012):

$$p(\varepsilon) = \frac{C_1 \varepsilon^{-1/3}}{\left[1 + C_2 \varepsilon^{2/3}\right]^{1/(q-1)}}, \tag{2.36}$$

with $C_1 = \frac{2}{3A^{2/3}}$ and $C_2 = -(1-q)/(2-q)A^{\frac{2}{3}}$. The probability of the energy is $p(\varepsilon) = n(\varepsilon)/N$, where $n(\varepsilon)$ corresponds to the number of earthquakes with energy ε and N is the total number of earthquakes. The normalized cumulative number of earthquakes is given as:

$$\frac{N(\varepsilon > \varepsilon_{th})}{N} = \int_{\varepsilon_{th}}^{\infty} p(\varepsilon)d\varepsilon, \tag{2.37}$$

where $N(\varepsilon > \varepsilon_{th})$ is the number of earthquakes with energy ε greater than the threshold energy ε_{th}. Combining Eq. (2.36) with Eq. (2.37) the following expression is derived:

$$\frac{N(\varepsilon > \varepsilon_{th})}{N} = \left[1 - \frac{(1-q)}{(2-q)}\left(\frac{\varepsilon_{th}}{A}\right)^{2/3}\right]^{(2-q)/(1-q)}, \tag{2.38}$$

where A is proportional to the volumetric energy density. Furthermore, Telesca (2011) proposed a function between the earthquake magnitude m and the released relative energy ε as (Kanamori, 1978):

$$m \sim \frac{2}{3}\log(\varepsilon). \tag{2.39}$$

Telesca (2012) also proposed a function that relates the cumulative number of earthquakes with earthquake magnitude as

$$\log\left(\frac{N(m > M)}{N}\right) = \frac{2 - q_M}{1 - q_M} \log\left[\frac{1 - ((1 - q_M)/(2 - q_M))\left(10^M/A^{\frac{2}{3}}\right)}{1 - ((1 - q_M)/(2 - q_M))\left(10^{m_0}/A^{\frac{2}{3}}\right)}\right], \tag{2.40}$$

where M is the earthquake magnitude, q_M refers to q related to the earthquake magnitude and m_0 is the threshold magnitude.

Many articles have indicated that the q magnitude parameter (q_M) can be used as a measure of the stability of an active tectonic area (Matcharashvili et al., 2011; Papadakis, 2016; Papadakis et al., 2013, 2015, 2016; Telesca, 2010a,b,c; Valverde-Esparza et al., 2012). A significant increase in q_M indicates strong interactions between the fault blocks and the fragments and implies a transition away from equilibrium. Moreover, in comparison with the GR scaling relation, the fragment–asperity model describes appropriately the earthquake frequency–magnitude distribution over a wider range of magnitudes (Vallianatos et al., 2015) and exhibits an excellent fit to earthquake data sets (Michas, 2016; Vallianatos et al., 2016).

Moreover, the estimation of q_M is stable irrespective of the selection of the threshold magnitude m_0, whereas estimation of the GR b-value is sensitive to the selection of m_0 (Michas, 2016).

2.5 Applications of Nonextensive Statistical Seismology

2.5.1 The Frequency−Magnitude Distribution of Seismicity

The fragment−asperity model has been applied to various earthquake data sets that correspond to a variety of tectonic environments (Michas et al., 2013; Papadakis et al., 2013, 2015, 2016; Silva et al., 2006; Telesca, 2010a,b,c, 2011; Vallianatos et al., 2012, 2014; Vilar et al., 2007) including seismicity in volcanic areas (Telesca, 2010b; Vallianatos et al., 2013).

Sotolongo-Costa and Posadas (2004) calculated the q_M parameter using three different earthquake catalogues for seismicity in California ($q_M = 1.65$), the Iberian Peninsula ($q_M = 1.64$) and the region of Andalusia ($q_M = 1.60$). Similar values have been estimated in Italy ($q_M = 1.66$) (Telesca, 2010b) and Taiwan ($q_M = 1.68$) (Telesca and Chen, 2010), as well as in the USA, Turkey and Brazil ($q_M = 1.60-1.71$) (Silva et al., 2006; Vilar et al., 2007). Matcharashvili et al. (2011) estimated $q_M = 1.80$ for seismicity in the area of the Javakheti highlands (Georgia) for the period 1960−2008.

Michas et al. (2013) applied the fragment−asperity model to examine the earthquake-magnitude distribution [Eq. (2.40)] in West Corinth Gulf, Greece. Moreover, these authors studied the EDF of earthquakes [Eq. (2.38)] and proposed that the model used describes the observed distribution for the lower earthquake energies better than the GR law (Fig. 2−5). In addition, they observed that after some threshold energy the distribution decays as a power-law.

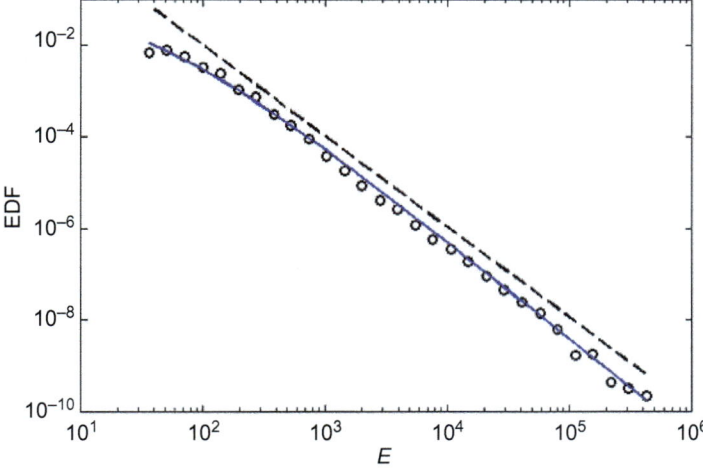

FIGURE 2−5 The EDF (circles) and the fitted curve (solid line). The dashed line represents the GR relation for $b = 1.51$. *EDF,* energy distribution function. *From Michas, G., Vallianatos, F., Sammonds, P., 2013. Non-extensivity and long-range correlations in the earthquake activity at the West Corinth rift (Greece). Nonlinear Processes Geophys. 20, 713−724. (Fig. 3).*

Vallianatos et al. (2013) calculated the nonextensive parameter ($q_M = 1.39$) with regards to the seismicity observed during the 2011−12 unrest at the Santorini volcanic complex (Greece). Telesca (2010b) calculated smaller q_M for the volcanic areas of Vesuvius ($q_M = 1.47$) and Etna ($q_M = 1.56$) compared with the whole Italian territory ($q_M = 1.66$). This author attributes the estimated low q_M values in volcanic areas to the different intensity of seismic activity and the different mechanism of earthquake generation. Moreover, Telesca (2010a, 2011) studied the spatial variation of the nonextensive parameters q_M and A in Italy and the Southern California seismicity ($q_M = 1.50$), respectively.

Antonopoulos et al. (2014) studied seismicity in Greece, whereas Papadakis et al. (2013) calculated the nonextensive parameter q_M along the seismic zones of the Hellenic subduction zone (HSZ) relating its variation to the energy release rate in each seismic zone. Valverde-Esparza et al. (2012) performed an NESP analysis on the earthquake frequency−magnitude distribution in four regions of the Mexican subduction zone during the period 1988−2010.

Papadakis et al. (2016) examined the q_M parameter and the heat flow data in Greece and proposed that high q_M values are consistent with strong earthquakes for focal depths

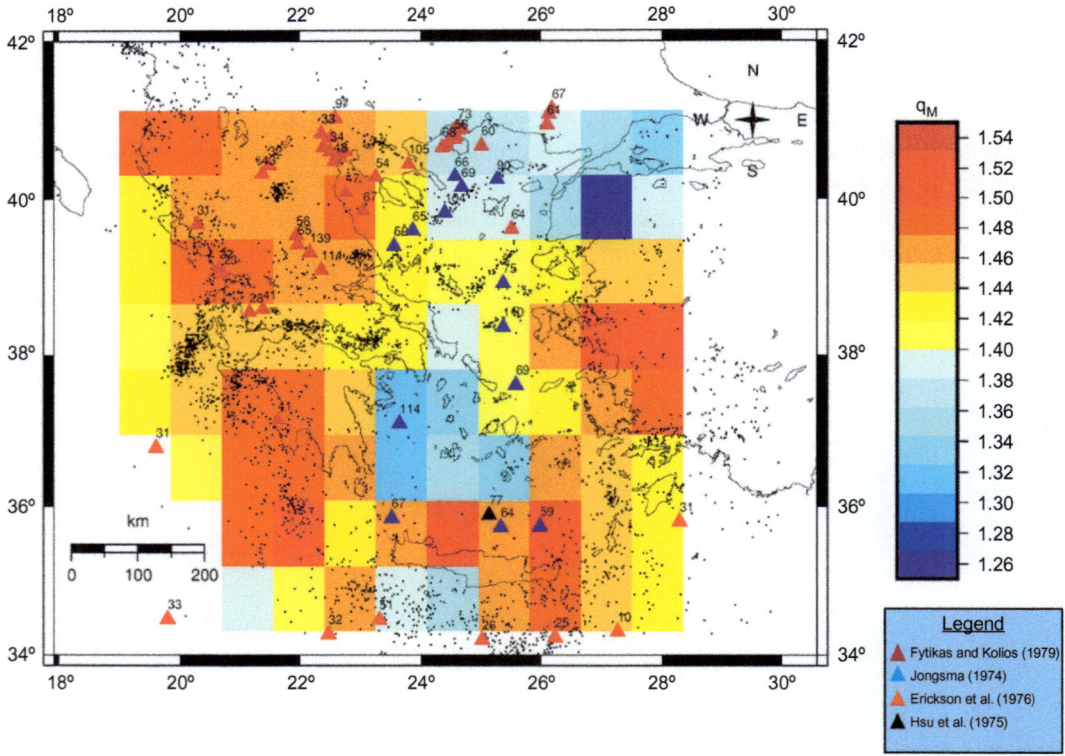

FIGURE 2−6 The spatial distribution of the calculated nonextensive parameter q_M along with the heat flow (triangles with numbers) and seismicity (black dots) for focal depth 0−40 km. Heat flow data provided by Fytikas and Kolios (1979), Jongsma (1974), Erickson et al. (1976) and Hsu et al. (1975). *From Papadakis, G., Vallianatos, F., Sammonds, P., 2016. Non-extensive statistical physics applied to heat flow and the earthquake frequency-magnitude distribution in Greece. Physica A 456, 135−144. (Fig. 5b).*

≤ 40 km. Although they estimated low q_M values in the regions of the central Aegean Sea and the volcanic arc which present high heat flow and the absence of strong earthquakes they noted that the areas southwest of Crete and along the North Aegean Trough (NAT) also present low q_M values, despite the occurrence of strong earthquakes (Fig. 2−6). This finding led the authors to examine further the area of the NAT (Papadakis and Vallianatos, 2017; see also Section 2.5.2).

2.5.2 Temporal Variations and the Evolution of Seismicity

The q magnitude parameter (q_M) derives from the fragment−asperity model (Sotolongo-Costa and Posadas, 2004) and reflects the scale of interactions between fault blocks and the fragments filling the space between them (see Section 2.4.3). An increase in q_M implies that a strong earthquake can be expected as the fault blocks tend away from equilibrium. This section shows that the temporal variation of q_M and its consistency with the evolution of seismicity can be used to distinguish the main characteristics of earthquake dynamics. We present recent published studies regarding the temporal behaviour of parameters q_M and A.

Matcharashvili et al. (2011) examined the temporal variations of the nonextensive parameters q_M and A regarding seismicity in the area of Javakheti, Georgia, during the period 1960−2008. Significant increases of the q_M parameter are observed for those 10-year periods where strong earthquakes occur, whereas the behaviour of the energy density A exhibits high values during seismically quiet periods where accumulated stress energy seems to be released mostly through the relative movement of smaller fragments.

Telesca (2011) studied the temporal variations of q_M and A regarding seismicity in Southern California during the period 1990−2010. This author reported a decrease of q_M with the increase of the minimum magnitude and commented on the importance of small events for the redistribution of stress in the seismogenic system. Telesca (2011) suggested that small and moderate seismic events govern interactions within the seismogenic system.

Vallianatos and Sammonds (2013) used NESP to characterize the global earthquake frequency−magnitude distribution and to interpret the earthquake processes leading up to the 2004 Sumatra-Andaman and 2011 Honshu megaearthquakes. Using the global Centroid moment tensor (CMT) catalogue and the crossover formulation of NESP (see Tsallis, 2009; Vallianatos and Sammonds, 2013 for a complete theoretical description), these authors analysed the seismic moment distribution. Temporal estimation of the nonextensive parameter q, which describes the seismic moment distribution of moderate earthquakes, resulted in a constant value: $q = 1.6$. In addition, temporal examination of the r parameter, which describes the seismic moment distribution of strong earthquakes, revealed that this parameter varies from 1 to 1.5, signifying a shift from the BG exponential distribution $(r = 1)$ to power-law distribution $(r$ varies from 1.4 to 1.5) supporting the concept of the global organization of seismicity (Fig. 2−7).

Telesca (2010c) analysed the frequency−magnitude distribution regarding the L'Aquila area in central Italy during the period 2005−09. The analysis of seismicity reveals that the nonextensive parameter q_M increased days before the occurrence of the strong earthquake

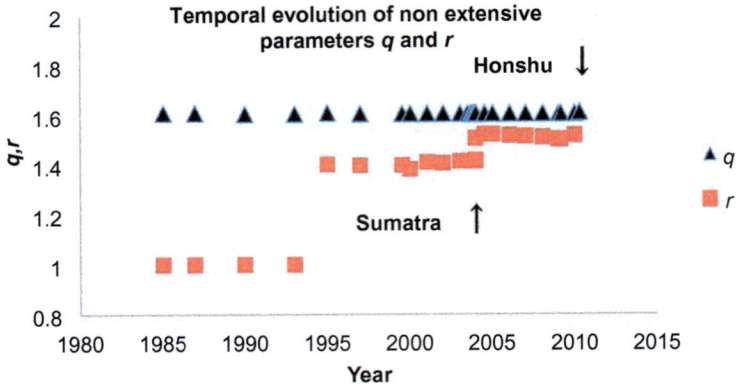

FIGURE 2–7 Temporal evolution of the nonextensive parameters q and r regarding the Sumatra and Honshu megaearthquakes. *From Vallianatos, F., Sammonds, P., 2013. Evidence of non-extensive statistical physics of the lithospheric instability approaching the 2004 Sumatran-Andaman and 2011 Honsu mega-earthquakes. Tectonophysics 590, 52–58 (Fig. 2).*

on 6 April 2009 (M_L = 5.8). According to Telesca (2010c) the increase of the q_M value suggests that the degree of long-range spatial correlations also increases and that the seismogenic system enters a critical state characterized by instability which can cause large-scale reactions and energy release (Kalimeri et al., 2008). Moreover, in accord with critical point behaviour (Zoeller et al., 2001; Zoeller and Hainzl, 2002), long-range correlations are established by redistribution of stress to larger scales signifying that in a highly correlated stress field a small rupture can result in a strong earthquake (Telesca, 2010c).

Vallianatos et al. (2014) applied the ideas of NESP and the method of natural time analysis (Varotsos et al., 2011) to reveal the hidden dynamic features of seismicity before the main shock (M_L = 6.2) occurred on 12 October 2013 in the southwest part of the Hellenic arc (Crete). The NESP analysis of the frequency–magnitude distribution demonstrates that the nonextensive parameter q_M increases significantly two days before the occurrence of the main shock. Furthermore, the natural time analysis shows that foreshock seismicity reaches the critical value k_1 = 0.070 a few days before the occurrence of the strong event (Vallianatos et al., 2014). It is worth mentioning that the nonextensive parameter does not decrease after the strong event (Fig. 2–8). This finding indicates that despite the occurrence of strong events, energy is not fully released and the degree of long-range interactions continues to increase.

Papadakis et al. (2015) examined the frequency–magnitude distribution during the period 1990–98 for a broad area surrounding the epicentre of the 1995 Kobe earthquake. The variations of the q_M parameter show that it significantly increased some months before the strong earthquake on 9 April 1994, manifesting the start of a preparation phase prior to the Kobe earthquake (Fig. 2–9). This increase coincided with the occurrence of six seismic events of magnitude M = 4.1, which change the magnitude pattern. Overall this study showed that the evolution of seismicity is found in accord with the observed variations indicating the system's transition away from equilibrium and its preparation for energy release.

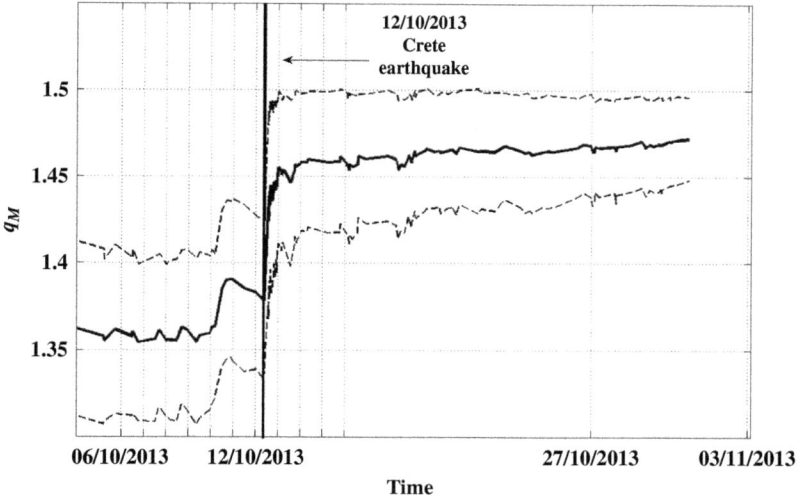

FIGURE 2–8 Temporal evolution of q_M (black continuous line) over increasing (cumulative) event-based windows and the associated standard deviation (black dashed lines). On 10 October 2013, the nonextensive parameter increases significantly leading up to the main shock. *From Papadakis, G., 2016. A Non-extensive Statistical Physics Analysis of Seismic Sequences: Application to the Geodynamic System of the Hellenic Subduction Zone. PhD Thesis. University College, London (Fig. 7.10).*

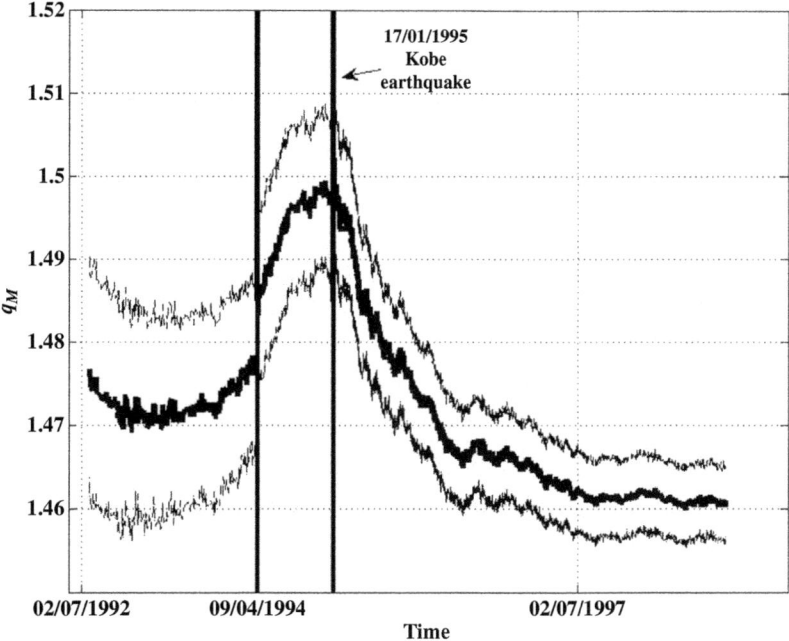

FIGURE 2–9 Temporal variations of the q_M values (black continuous line) over increasing (cumulative) event-based windows and the associated standard deviation (black dashed lines). *From Papadakis, G., Vallianatos, F., Sammonds, P., 2015. A nonextensive statistical physics analysis of the 1995 Kobe, Japan earthquake. Pure Appl. Geophys. 172, 1923–1931 (Fig. 3).*

Through temporal examination of the nonextensive parameters q_M and A, Papadakis and Vallianatos (2017) revealed a continuous high degree of interactions after strong earthquakes for the area of the NAT during 1976–2015 (Fig. 2–10). These authors proposed that if the q_M value does not significantly decrease after a mainshock then the seismogenic system has not returned to a state of equilibrium. Moreover, this study implies that the estimated low q_M value (Papadakis et al., 2016) for the area of NAT and for the whole period 1976–2009 is actually an index of the frequency of appearance of long-range interactions and an indication that the seismic energy has not been fully released.

Examination of the temporal variations of the nonextensive parameter q_M regarding the aforementioned case studies reveals that a significant increase of q_M occurs before strong events. Moreover, this section reveals that after a main shock this parameter can decrease significantly (i.e., the Kobe earthquake) or continue to increase (i.e., Crete earthquake, NAT). The Kobe case indicates that the degree of interactions in the seismogenic system decreases. In the other two cases (i.e., Crete earthquake, NAT) it is observed that despite the occurrence of strong events, energy is not fully released and the degree of long-range correlations continues to increase. In the aforementioned case studies it is also shown that the occurrence of moderate-size events prior to the main shock signifies the start of a transition phase leading to the mainshock. Moreover, examination of parameter A is still at a preliminary stage and more examination is needed to fully understand its behaviour in time.

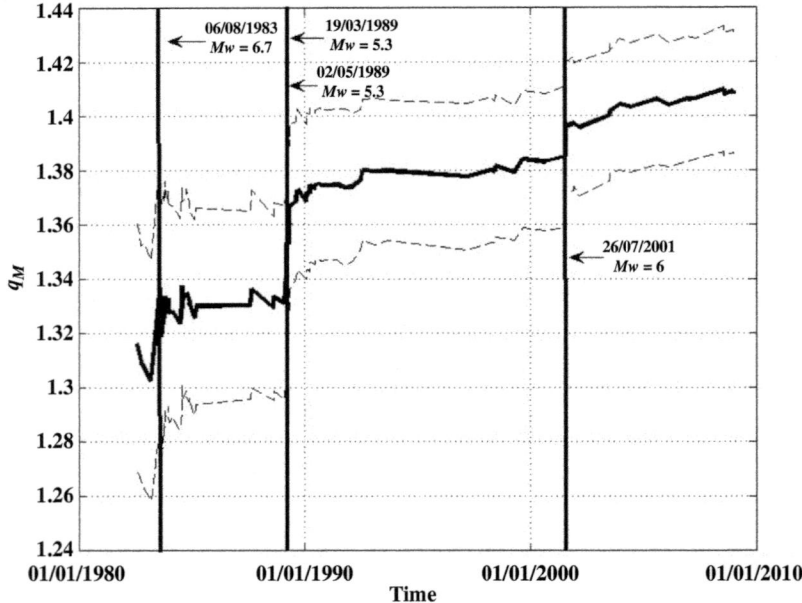

FIGURE 2–10 Temporal variation of q_M (black line) and the standard deviation (black dashed lines). Significant increases of q_M coincide with strong earthquakes (*Mw* > 5) for the area of NAT during 1976–2009. *NAT*, North Aegean Trough. *From Papadakis, G., Vallianatos, F., 2017. Non-extensive statistical physics analysis of earthquake magnitude sequences in North Aegean Trough, Greece. Acta Geophys. DOI 10.1007/s11600-017-0047-4 (Fig. 4).*

Through this section we aimed to show that the variations of nonextensive parameters can be used to improve the modelling of seismicity. Temporal examination of these parameters should be taken into consideration in any earthquake precursory study.

2.5.3 Spatiotemporal Properties of Seismicity and Nonextensive Statistical Mechanics

The concept of NESP was first introduced in Earth sciences by Abe and Suzuki (2003, 2004a, b, 2005). These authors examined the spatiotemporal distribution of seismicity in California and Japan and showed that the interevent distances (Abe and Suzuki, 2003) and the interevent times (Abe and Suzuki, 2005) between successive earthquakes follow a q-exponential distribution [Eq. (2.31)], with $q_T > 1$ and $q_D < 1$ for the interevent times and distances, respectively. Darooneh and Dadashinia (2008), using the concept of NESP, studied the spatial and temporal properties of seismicity in Iran. These authors estimated q_T and q_D to be approximately constant when changing the covering period and the extent of the studied region. Moreover, they found that the calculated nonextensive parameters change significantly with the threshold magnitude. Consequently, q_T increases and q_D decreases with an increasing threshold magnitude, signifying that larger earthquakes trigger aftershocks, leading to a state where long-range correlations appear. The following subsections present further applications to the spatiotemporal distribution of seismicity.

2.5.3.1 Spatial Properties of Seismicity

Abe and Suzuki (2003) investigated the spatial properties of seismicity in California and Japan. These authors found that the three-dimensional distances of successive earthquakes follow the q-exponential distribution and that the q_D values are estimated less than the unity. They also proposed the duality relation: $q_D + q_T \approx 2$. It is worth mentioning that Hasumi (2007, 2009), using the two-dimensional (2D) Burridge–Knopoff spring-block model, estimated similar results ($q_D + q_T \approx 1.5$). However, the physical interpretation of this relation remains open.

Furthermore, Vallianatos et al. (2012) investigated the spatial properties of the 1995 Aigion (Greece) earthquake aftershock sequence. Fig. 2–11 shows the semi-q-log plot of the cumulative inter-event distances distribution for the aftershock sequence of the 1995 Aigion earthquake.

Vallianatos and Sammonds (2013) examined the interevent distance distributions surrounding the 2004 Sumatra-Andaman and 2011 Honsu megaearthquakes. These authors showed that the interevent distances cumulative probability distribution $P(>D)$ does not present changes in respect to the megaearthquakes. Furthermore, Papadakis et al. (2013) examined the interevent distances distribution along the seismic zones of the HSZ and concluded that the q_D variations indicate a different degree of spatial earthquake clustering along the seismic zones (Fig. 2–12).

2.5.3.2 Temporal Properties of Seismicity and the Risk Function

Abe and Suzuki (2005) investigated the temporal properties of seismicity in California and Japan. The cumulative distribution of the interevent times for both regions are described by

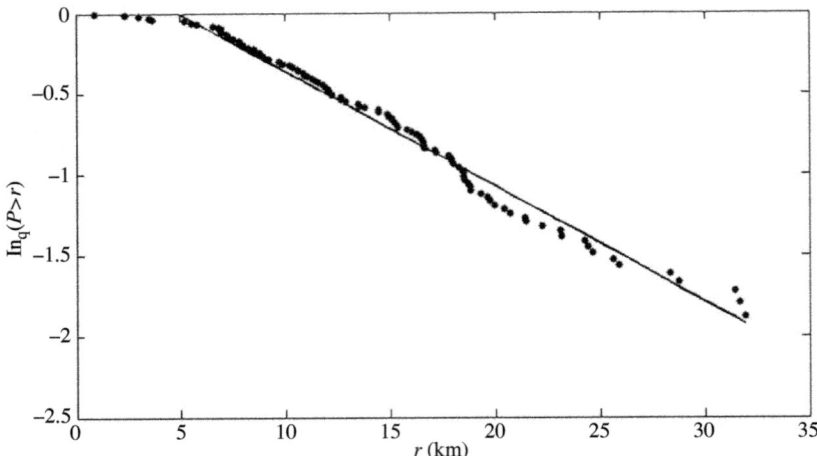

FIGURE 2–11 The semi-q-log plot of the cumulative interevent distances distribution for the 1995 Aigion earthquake aftershock sequence (D is indicated as r in the plot). The straight line represents the q-logarithmic function. The best-fit regression is obtained for $q_D = 0.53$ and the correlation coefficient is equal to $R = -0.9896$. *From Vallianatos, F., Michas, G., Papadakis, G., Sammonds, P., 2012. A non-extensive statistical physics view to the spatiotemporal properties of the June 1995, Aigion earthquake (M6.2) aftershock sequence (West Corinth rift, Greece). Acta Geophys. 60, 758–768 (Fig. 2).*

the q-exponential function with $q_T > 1$ (Fig. 2–13). In addition, these authors found that the q_T value increases monotonically with the value of the threshold magnitude, indicating that the tail of the distribution tends to become heavier for a higher value of threshold magnitude (Abe and Suzuki, 2005).

Moreover, Vallianatos et al. (2012) investigated the temporal properties of the 1995 Aigion (Greece) earthquake aftershock sequence. In this study the q-exponential function fits rather well to the observed distributions, implying the complexity of the temporal properties of seismicity. In addition, Vallianatos et al. (2013) analysed the interevent time distribution during the 2011−12 unrest at the Santorini volcanic complex (Greece), and estimated $q_T =$ 1.52, implying that the swarm-like character of seismicity exhibits complex correlations.

Vallianatos and Sammonds (2013) examined the interevent time in respect to the 2004 Sumatra-Andaman and 2011 Honsu megaearthquakes. These authors, by using global shallow seismicity (CMT catalogue) with $M_W > 5.5$, showed that there is no change in the temporal distribution of seismicity due to the occurrence of the megaevents. On the contrary, significant changes are observed in the frequency−magnitude distribution of global seismicity (see Section 2.5.2).

Furthermore, Antonopoulos et al. (2014) studied the interevent times distribution in Greece, whereas Papadakis et al. (2013) examined the interevent times distribution along the seismic zones of the HSZ showing that the estimated q_T values are related to the time evolution of seismicity, which is a long-term process in the Hellenic arc.

Michas et al. (2013) examined the cumulative interevent time distribution and the probability distribution function for the interevent times of earthquakes in the West Corinth rift,

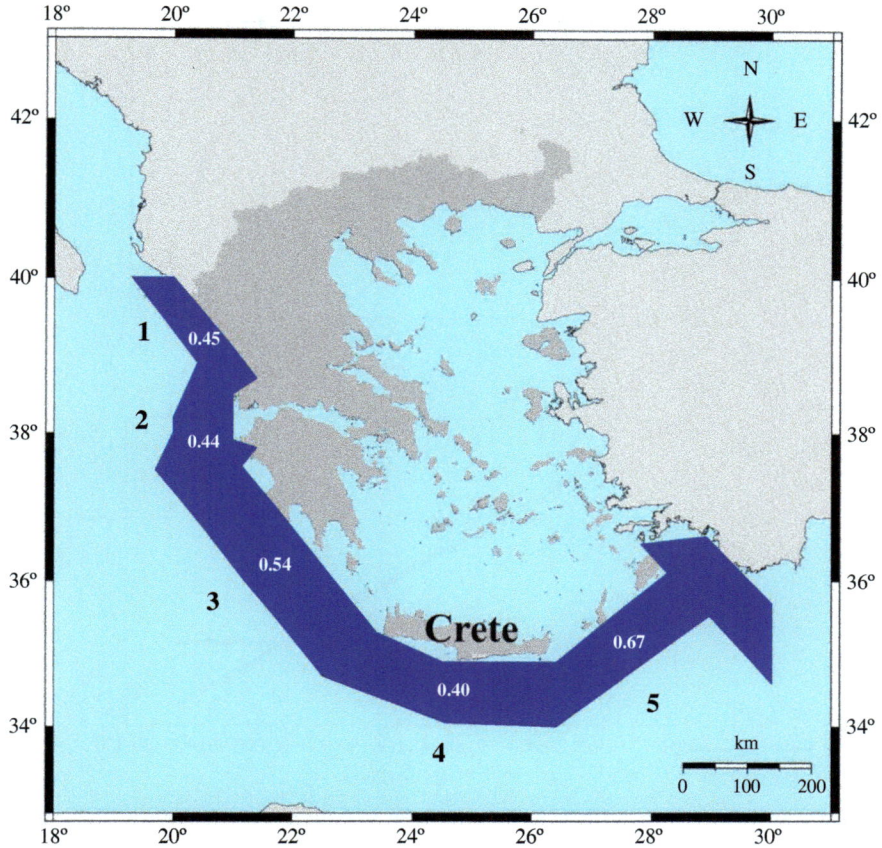

FIGURE 2–12 The q_D variation along the HSZ. *HSZ, Hellenic subduction zone. From Papadakis, G., Vallianatos, F., Sammonds, P., 2013. Evidence of nonextensive statistical physics behavior of the Hellenic subduction zone seismicity. Tectonophysics 608, 1037–1048 (Fig. 8).*

Greece, using a data set that covers the period 2001−08. These authors normalized the interevent times T as $T' = T/\bar{T}$, where T scales with the mean interevent time $\bar{T} = (T_N - T_1)/(N - 1)$ and leads to $P(>T')$. It should be noted that N is the number of events and T_N, T_1 are the occurrence times of the last and the first events in the earthquake catalogue, respectively. In other words, the mean interevent time $\bar{T}$ corresponds to the total time interval divided by the total number of events (see also Section 2.2.2).

Moreover, they propose a generalization of the gamma distribution $(p(T') = CT'^{(\gamma-1)}\exp(-T'/\beta))$ as a possible scaling relation for the probability density. This distribution, which can be called the q-generalized gamma distribution (Queirós, 2005), does not decay exponentially but rather as a power-law: $p(T') = CT'^{(\gamma-1)}\exp_q(-T'/\theta)$. The later expression, known in statistics as the F-distribution (Marchand, 2003), indicates that short and intermediate interevent times, scale with the exponent $\gamma - 1$ and long interevent times scale with the exponent $1/(1 - q)$. It is also worth noting that the q-exponential function appears as the last term of the distribution. In the limit $q \rightarrow 1$ the q-generalized gamma distribution

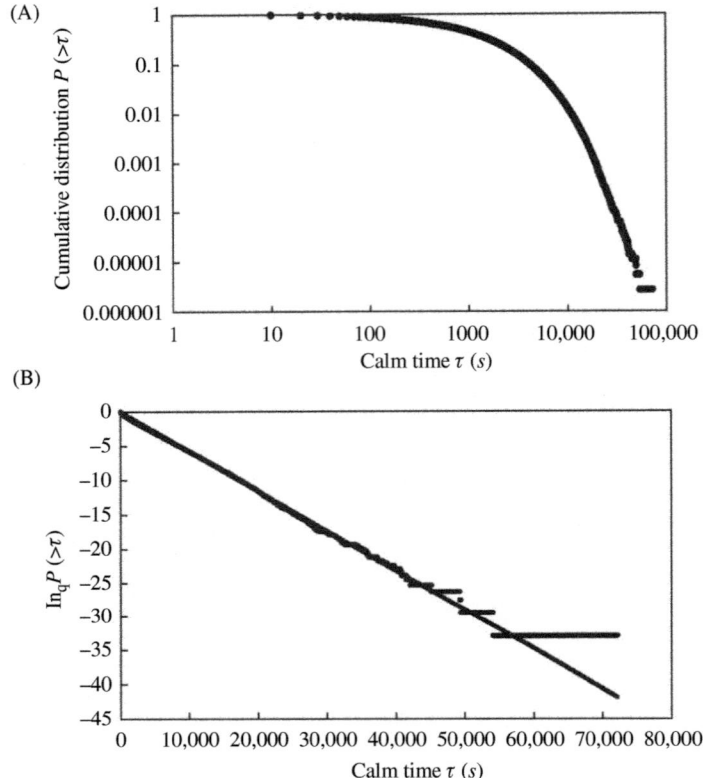

FIGURE 2–13 (A) The log–log plot of the cumulative interevent time distribution in California. (B) The corresponding semi-q-log plot of the cumulative waiting times distribution in California. T is indicated as τ in the plot. The best-fit regression is obtained for $q_T = 1.13$ and $t_0 = 1.724 \times 10^3$ seconds. The associated value of the correlation coefficient is equal to $R = -0.98828$. *From Abe, S., Suzuki, N., 2005. Scale-free statistics of time interval between successive earthquakes. Physica A 350, 588–596 (Fig. 1).*

recovers the ordinary gamma distribution. In Fig. 2–14 we plot the rescaled interevent time distribution for the Southern California seismicity and the corresponding fit according to the q-generalized gamma distribution. There is good agreement between the model and the data, indicating both short- and long-term clustering effects in the temporal evolution of seismicity.

Antonopoulos et al. (2014) proposed a hazard function $W_M (T, \Delta T)$ that represents the probability of an earthquake (with magnitude greater than M) occurring in a time interval (ΔT), considering the time (T) passed since the occurrence of the last event (Fig. 2–15).

The hazard function reads as:

$$W_M(T;\Delta T) = 1 - \left(1 + \frac{\beta(q-1)\Delta T}{1+\beta(q-1)T}\right)^{(q-2)/(q-1)}. \tag{2.41}$$

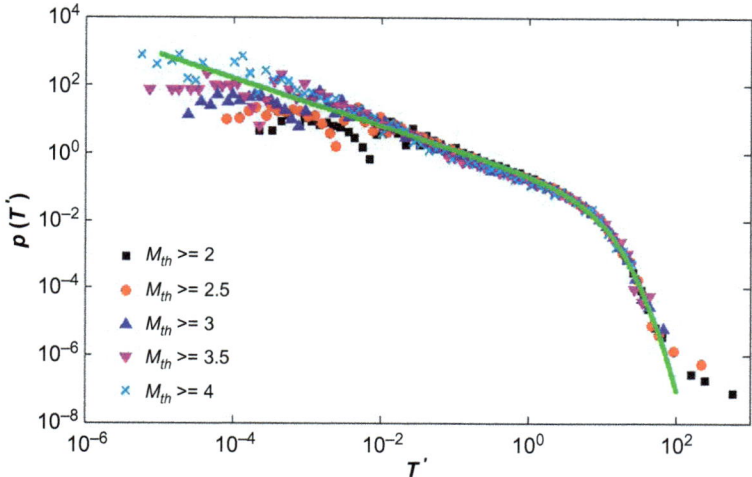

FIGURE 2–14 Normalized probability density $P(T')$ for the rescaled interevent times T' for the Southern California earthquake catalogue (1981–2011) and for various threshold magnitudes. The solid line represents the q-generalized gamma distribution for the values of $C = 0.25$, $\gamma = 0.3$, $\theta = 4.8$ and $q = 1.09$.

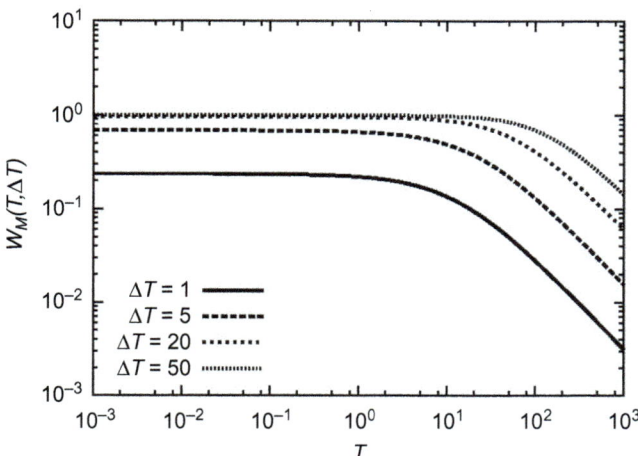

FIGURE 2–15 Plot of the hazard function $W_M (T, \Delta T)$ with the interevent time T (in days) for four different ΔT values. *From Antonopoulos, C.G., Michas, G., Vallianatos, F., Bountis, T., 2014. Evidence of q-exponential statistics in Greek seismicity. Physica A 409, 71–79 (Fig. 3a).*

Using the hazard function, these authors found that as the time interval ΔT decreases the probability of occurrence of at least one earthquake decreases as T increases. However, if the time interval ΔT increases, the probability of earthquake occurrence increases.

2.6 Discussion — Quo Vademus?

Although significant research has been conducted on the complex properties of seismicity including scaling relations, temporal and spatial correlations, critical phenomena and nucleation, many questions regarding earthquake physics remain to be answered in the future. Most of the current research approaches rely on empirical laws rather than on a solid underlying theoretical concept. However, statistical seismology should evolve into a genuine physically based statistical physics of earthquakes. This chapter presents the concept of NESM, which offers a consistent theoretical framework for the macroscopic description of seismicity. This approach, based on the first principles of statistical mechanics, provides a unified framework that produces a range of asymptotic power law distributions for the description of complex phenomena in nature.

As a first step, we review the scaling properties of seismicity, and the classical statistical mechanics approaches that have been developed to describe them. We introduce the theoretical framework of NESM and its application to earthquakes. More particularly, we present an approach that is based on the fundamental principles that lead to the basic law that controls the earthquake frequency—magnitude distribution and the spatiotemporal properties of seismicity.

The various published studies and cases presented here illustrate that in terms of probabilities of the different microstates and their interactions, the large-scale properties of earthquake populations can be deduced by following the principles of NESM. The models derived from Tsallis entropy can be used to describe the evolution of seismicity and to further contribute to seismic hazard assessment. Moreover, it seems that examination of the nonextensive parameters, which derive from the earthquake frequency—magnitude distribution, can be used to describe the organization of seismicity and to reveal its hidden dynamics. In addition, examination of the spatiotemporal distributions reveals the degree of correlation among the elements of the studied system as well as the probabilistic content of earthquake sequences.

Although such results are an important step towards the understanding of earthquake-related phenomena, many questions regarding the earthquake generation process remain wide open.

Using the principles of NESM in a unified approach with the other known laws in fracture mechanics may lead to significant discoveries and may enhance our understanding regarding the physical mechanisms that drive the evolution of seismicity in local, regional and global scales. In addition, a significant number of important questions arise in regard to the applicability of NESM in a broader set of problems in geophysics, where the presence of fat-tail distributions indicates long-range interactions and the existence of memory. NESM should also be used to modify our views in volcanology, theoretical geomorphology, rock and ice physics and applied geophysics, as well in other scientific fields where the system or the medium behaves as a generator of long-range interactions or memory.

Acknowledgements

This work has been partly supported by the project 'HELPOS Hellenic System for Lithosphere Monitoring' (MIS 5002697) of the Operational Programme NSRF 2014-20, co-financed by Greece and the European Union (European Regional Development Fund).

References

Abe, S., 2003. Geometry of escort distributions. Phys. Rev. E 68, 031101.

Abe, S., Suzuki, N., 2003. Law for the distance between successive earthquakes. J. Geophys. Res. 108, 2113.

Abe, S., Suzuki, N., 2004a. Scale-free network of earthquakes. Europhys. Lett. 65 (4), 581−586.

Abe, S., Suzuki, N., 2004b. Statistical similarities between internetquakes and earthquakes. Physica D 193, 310−314.

Abe, S., Suzuki, N., 2005. Scale-free statistics of time interval between successive earthquakes. Phys. A 350, 588−596.

Aki, K., 1981. A probabilistic synthesis of precursory phenomena. In: Simpson, D.W., Richatds, P.G. (Eds.), Earthquake Prediction: An International Review, 4. American Geophysical Union, Washington DC, pp. 566−574.

Allegre, C., Le Mouell, J., Provost, A., 1982. Scaling rules in rock fracture and possible implications for earthquake prediction. Nature 297, 47−49.

Antonopoulos, C.G., Michas, G., Vallianatos, F., Bountis, T., 2014. Evidence of q-exponential statistics in Greek seismicity. Phys. A 409, 71−79.

de Arcangelis, L., Godano, C., Grasso, J.R., Lippiello, E., 2016. Statistical physics approach to earthquake occurrence and forecasting. Phys. Rep. 628, 1−91.

Bak, P., 1996. How Nature Works: The Science of Self-organized Criticality. Springer-Verlag, New York.

Bak, P., Tang, C., 1989. Earthquakes as a self-organized critical phenomenon. J. Geophys. Res. 94, 635−637.

Bak, P., Tang, C., Wiesenfeld, K., 1987. Self-organized criticality: an explanation of 1/f noise. Phys. Rev. Lett. 59, 381−384.

Bak, P., Tang, C., Wiesenfeld, K., 1988. Self-organized criticality. Phys. Rev. A 38, 364−374.

Bak, P., Christensen, K., Danon, L., Scanlon, T., 2002. Unified scaling law for earthquakes. Phys. Rev. Lett. 88, 178501.

Beck, C., Schlogl, F., 1993. Thermodynamics of Chaotic Systems: An Introduction. Cambridge University Press, New York.

Ben-Zion, Y., 2008. Collective behavior of earthquakes and faults: continuum-discrete transitions, progressive evolutionary changes, and different dynamic regimes. Rev. Geophys. 46, RG4006.

Berrill, J.B., Davis, R.O., 1980. Maximum entropy and the magnitude distribution. Bull. Seismol. Soc. Am. 70, 1823−1831.

Bonnet, E., Bour, O., Odling, N.E., Davy, P., Main, I., Cowie, P., et al., 2001. Scaling of fracture systems in geological media. Rev. Geophys. 39, 347−383.

Bruce, A., Wallace, D., 1989. Critical point phenomena: universal physics at large length scales. In: Davies, P. (Ed.), The New Physics. Cambridge University Press, Cambridge, New York, Melbourne, pp. 236−267.

Burridge, L., Knopoff, L., 1967. Model and theoretical seismicity. Bull. Seismol. Soc. Am. 57, 341−371.

Carlson, J.M., Langer, J.S., 1989. Properties of earthquakes generated by fault dynamics. Phys. Rev. Lett. 22, 2632−2635.

Chelidze, T.L., 1982. Percolation and fracture. Phys. Earth Planet. Inter. 28, 93−101.

Corral, A., 2003. Local distributions and rate fluctuations in a unified scaling law for earthquakes. Phys. Rev. E 68, 2003.

Corral, A., 2004. Long-term clustering, scaling, and universality in the temporal occurrence of earthquakes. Phys. Rev. Lett. 92, 108501.

Corral, A., 2006. Universal earthquake-occurrence jumps, correlations with time, and anomalous diffusion. Phys. Rev. Lett. 97, 178501.

Darooneh, A., Mehri, A., 2010. A nonextensive modification of the Gutenberg–Richter law: q-stretched exponential form. Phys. A 389, 509–514.

Darooneh, A.H., Dadashinia, C., 2008. Analysis of the spatial and temporal distributions between successive earthquakes: nonextensive statistical mechanics viewpoint. Phys. A 387, 3647–3654.

Davidsen, J., Paczuski, M., 2005. Analysis of the spatial distribution between successive earthquakes. Phys. Rev. Lett. 94, 048501.

De Santis, A., Cianchini, G., Favali, P., Beranzoli, L., Boschi, E., 2011. The Gutenberg–Richter law and entropy of earthquakes: two case studies in central Italy. Bull. Seismol. Soc. Am. 101, 1386–1395.

Erickson, A.J., Simmons, M.G., Ryan, W.B.F., 1976. Review of heat flow data from the Mediterranean and Aegean Seas. In: Biju-Duval, B., Montadert, L. (Eds.), Int. Symp. on the Strut. Hist. of the Medit. Basin. Split, Yugoslavia, pp. 263–280.

Frohlich, C., Davis, S.D., 1993. Teleseismic b values; or, much ado about 1.0. J. Geophys. Res. 98, 631–644.

Fytikas, M.D., Kolios, N.P., 1979. Preliminary heat flow map of Greece. In: Cermak, V., Rybach, L. (Eds.), Terrestrial Heat Flow in Europe. Springer-Verlag, Berlin Heidelberg, pp. 197–205.

Geilikman, M.B., Golubeva, T.V., Pisarenko, V.F., 1990. Multifractal patterns of seismicity. Earth Planet. Sci. Lett. 99, 127–132.

Godano, C., Caruso, V., 1995. Multifractal analysis of earthquake catalogues. Geophys. J. Int. 121, 385–392.

Gutenberg, B., Richter, C.F., 1944. Frequency of earthquakes in California. Bull. Seismol. Soc. Am. 34, 185–188.

Hainzl, S., Scherbaum, F., Beauval, C., 2006. Estimating background activity based on interevent-time distribution. Bull. Seismol. Soc. Am. 96, 313–320.

Hasumi, T., 2007. Interoccurrence time statistics in the two-dimensional Burridge_Knopoff earthquake model. Phys. Rev. E 76, 026117.

Hasumi, T., 2009. Hypocenter interval statistics between successive earthquakes in the two-dimensional Burridge_Knopoff model. Physica A 388, 477–482.

Hsu, K., Montadert, L., Garrison, R.E., Fabricius, F.H., Bernoulli, D., Melieres, F., et al., 1975. Glomar Challenger returns to the Mediterranean Sea. Geotimes 20, 16–19.

Jaynes, E.T., 1957. Information theory and statistical mechanics. Phys. Rev. 106, 620–630.

Jongsma, D., 1974. Heat flow in the Aegean Sea. Geophys. J. R. Astr. Soc. 37, 337–346.

Kagan, Y.Y., 1994. Observational evidence for earthquakes as a nonlinear dynamic process. Phys. D 77, 160–192.

Kagan, Y.Y., 2002a. Seismic moment distribution revisited: I. Statistical results. Geophys. J. Int. 148, 520–541.

Kagan, Y.Y., 2002b. Aftershock zone scaling. Bull. Seismol. Soc. Am. 92, 641–655.

Kagan, Y.Y., Jackson, D.D., 1991. Long-term earthquake clustering. Geophys. J. Int. 104, 117–133.

Kagan, Y.Y., Knopoff, L., 1980. Spatial distribution of earthquakes: the two-point correlation function. Geophys. J. R. Astron. Soc. 62, 303–320.

Kalimeri, M., Papadimitriou, C., Balasis, G., Eftaxias, K., 2008. Dynamical complexity detection in pre-seismic emissions using nonadditive Tsallis entropy. Phys. A 387, 1161–1172.

Kanamori, H., 1978. Quantification of earthquakes. Nature 271, 411–414.

Kanamori, H., Anderson, D.L., 1975. Theoretical basis of some empirical relations in seismology. Bull. Seismol. Soc. Am. 65, 1073−1095.

Kawamura, H., Hatano, T., Kato, N., Biswas, S., Chakrabarti, B.K., 2012. Statistical physics of fracture, friction and earthquakes. Rev. Mod. Phys. 84, 839−884.

Keilis-Borok, V.I., 1990. The lithosphere of the earth as a nonlinear system with implications for earthquake prediction. Rev. Geophys. 28, 19−34.

Lay, T., Wallace, T., 1995. Modern Global Seismology. Academic Press, New York.

Lennartz, S., Livina, V.N., Bunde, A., Havlin, S., 2008. Long-term memory in earthquakes and the distribution of interoccurrence times. Europhys. Lett. 81, 69001.

Livina, V.N., Havlin, S., Bunde, A., 2005. Memory in the occurrence of earthquakes. Phys. Rev. Lett. 95, 208501.

Main, I., 1996. Statistical physics, seismogenesis, and seismic hazard. Rev. Geophys. 34, 433−462.

Main, I.G., Al-Kindy, F.H., 2002. Entropy, energy, and proximity to criticality in global earthquake populations. Geophys. Res. Lett. 29, 1121.

Main, I.G., Burton, P.W., 1984. Information theory and the earthquake frequency-magnitude distribution. Bull. Seismol. Soc. Am. 74, 1409−1426.

Main, I.G., Naylor, M., 2008. Maximum entropy production and earthquake dynamics. Geophys. Res. Lett. 35, L19311.

Main, I.G., Naylor, M., 2010. Entropy production and self-organized (sub)criticality in earthquake dynamics. Philos. Trans. R. Soc. A 368, 131−144.

Mandelbrot, B.B., 1982. The Fractal Geometry of Nature. W.H. Freeman, New York.

Marchand, J.P., 2003. Distributions. North-Holland, Amsterdam.

Matcharashvili, T., Chelidze, T., Javakhishvili, Z., Jorjiashvili, N., Fra Paleo, U., 2011. Non-extensive statistical analysis of seismicity in the area of Javakheti, Georgia. Comput. Geosci. 37, 1627−1632.

Mega, M.S., Allegrini, P., Grigolini, P., Latora, V., Palatella, L., Rapisarda, A., et al., 2003. Power-law time distribution of large earthquakes. Phys. Rev. Lett. 90, 188501/1−188501/4.

Michas, G., 2016. Generalized Statistical Mechanics Description of Fault and Earthquake Populations in Corinth Rift (Greece), PhD Thesis. University College London.

Michas, G., Vallianatos, F., Sammonds, P., 2013. Non-extensivity and long-range correlations in the earthquake activity at the West Corinth rift (Greece). Nonlinear Processes Geophys. 20, 713−724.

Michas, G., Sammonds, P., Vallianatos, F., 2015. Dynamic multifractality in earthquake time series: insights from the Corinth rift, Greece. Pure Appl. Geophys. 172, 1909−1921.

Molchan, G., 2005. Interevent time distribution in seismicity: a theoretical approach. Pure Appl. Geophys. 162, 1135−1150.

Narteau, C., Shebalin, P., Holschneider, M., 2002. Temporal limits of the power law aftershock decay rate. J. Geophys. Res. 107 (B12), 2359.

Narteau, C., Shebalin, P., Holschneider, M., 2005. Onset of power law aftershock decay rates in southern California. Geophys. Res. Lett. 32, 1−5.

Olami, Z., Feder, H.J., Christensen, K., 1992. Self-organised criticality in a continuous, nonconservative cellular automaton modelling earthquakes. Phys. Rev. Lett. 68, 1244−1247.

Papadakis, G., 2016. A Non-Extensive Statistical Physics Analysis of Seismic Sequences: Application to the Geodynamic System of the Hellenic Subduction Zone, PhD Thesis. University College London.

Papadakis, G., Vallianatos, F., 2017. Non-extensive statistical physics analysis of earthquake magnitude sequences in North Aegean Trough, Greece. Acta Geophys. . Available from: https://doi.org/10.1007/s11600-017-0047-4.

Papadakis, G., Vallianatos, F., Sammonds, P., 2013. Evidence of nonextensive statistical physics behavior of the Hellenic subduction zone seismicity. Tectonophysics 608, 1037−1048.

Papadakis, G., Vallianatos, F., Sammonds, P., 2015. A nonextensive statistical physics analysis of the 1995 Kobe, Japan earthquake. Pure Appl. Geophys. 172, 1923−1931.

Papadakis, G., Vallianatos, F., Sammonds, P., 2016. Non-extensive statistical physics applied to heat flow and the earthquake frequency-magnitude distribution in Greece. Phys. A 456, 135−144.

Pathria, R.K., Beale, P.D., 2011. Statistical Mechanics, third ed. Butterworth-Heinemann, Oxford.

Picoli, S., Mendes, R.S., Malacarne, L.C., Santos, R.P.B., 2009. q-Distributions in complex systems: a brief review. Braz. J. Phys. 39, 468−474.

Pressé, S., Ghosh, K., Lee, J., Dill, K.A., 2013. Principles of maximum entropy and maximum caliber in statistical physics. Rev. Mod. Phys. 85, 1115−1141.

Queirós, S.M.D., 2005. On the emergence of a generalised gamma distribution, application to traded volume in financial markets. Europhys. Lett. 71, 339−345.

Rundle, J.B., Turcotte, D.L., Shcherbakov, R., Klein, W., Sammis, C., 2003. Statistical physics approach to understanding the multiscale dynamics of earthquake fault systems. Rev. Geophys. 41, 4.

Saichev, A., Sornette, D., 2006. "Universal" distribution of interearthquake times explained. Phys. Rev. Lett. 97, 078501.

Saichev, A., Sornette, D., 2007. Theory of earthquake recurrence times. J. Geophys. Res. 112, B04313.

Scholz, C.H., 2002. The Mechanics of Earthquakes and Faulting. Cambridge University Press, Cambridge.

Shannon, C.E., 1948. A mathematical theory of communication. Bell Syst. Tech. J. 27, 379−423.

Silva, R., Franca, G.S., Vilar, C.S., Alcaniz, J.S., 2006. Nonextensive models for earthquakes. Phys. Rev. E 73, 026102.

Sornette, D., 2006. Critical Phenomena in Natural Sciences, second ed. Springer, Heidelberg.

Sornette, A., Sornette, D., 1989. Self-organized criticality and earthquakes. Europhys. Lett. 9, 197−202.

Sornette, D., Werner, M.J., 2009. Statistical physics approaches to seismicity. In: Meyers, R.A. (Ed.), Encyclopedia of Complexity and Systems Science. Springer, New York, pp. 7872−7891.

Sotolongo-Costa, O., Posadas, A., 2004. Fragment-asperity interaction model for earthquakes. Phys. Rev. Lett. 92 (4), 048501.

Telesca, L., 2010a. Analysis of Italian seismicity by using a nonextensive approach. Tectonophysics 494, 155−162.

Telesca, L., 2010b. Nonextensive analysis of seismic sequences. Phys. A 389, 1911−1914.

Telesca, L., 2010c. A non-extensive approach in investigating the seismicity of L' Aquila area (central Italy), struck by the 6 April 2009 earthquake (M_L = 5.8). Terra Nova 22 (2), 87−93.

Telesca, L., 2011. Tsallis-based nonextensive analysis of the southern California seismicity. Entropy 13 (7), 1267−1280.

Telesca, L., 2012. Maximum likelihood estimation of the nonextensive parameters of the earthquake cumulative magnitude distribution. Bull. Seismol. Soc. Am. 102 (2), 886−891.

Telesca, L., Chen, C.C., 2010. Nonextensive analysis of crustal seismicity in Taiwan. Nat. Hazards Earth Syst. Sci. 10, 1293−1297.

Telesca, L., Lapenna, V., 2006. Measuring multifractality in seismic sequences. Tectonophysics 423, 115−123.

Telesca, L., Lapenna, V., Lovallo, M., 2004. Information entropy analysis of seismicity of Umbria−Marche region (central Italy). Nat. Hazards Earth Syst. Sci. 4, 691−695.

Touati, S., Naylor, M., Main, I.G., 2009. Origin and nonuniversality of the earthquake interevent time distribution. Phys. Rev. Lett. 102, 168501.

Tsallis, C., 1988. Possible generalization of Boltzmann−Gibbs statistics. J. Stat. Phys. 52, 479−487.

Tsallis, C., 2001. Nonextensive statistical mechanics and thermodynamics: historical background and present status. In: Abe, S., Okamoto, Y. (Eds.), Nonextensive Statistical Mechanics and its Applications. Lecture Notes in Physics 560, Springer, Berlin-Heildelberg, pp. 3–98. Lecture notes in physics.

Tsallis, C., 2009. Introduction to Nonextensive Statistical Mechanics: Approaching a Complex World. Springer, Berlin.

Turcotte, D.L., 1997. Fractals and Chaos in Geology and Geophysics, second ed. Cambridge University Press, Cambridge, UK.

Utsu, T., 1999. Representation and analysis of the earthquake size distribution: a historical review and some new approaches. Pure Appl. Geophys. 155, 509–535.

Utsu, T., Ogata, Y., Matsu'ura, R.S., 1995. The centenary of the Omori formula for a decay law of aftershock activity. J. Phys. Earth 43, 1–33.

Vallianatos, F., 2009. A non-extensive approach to risk assessment. Nat. Hazards Earth Syst. Sci. 9, 211–216.

Vallianatos, F., 2011. A non-extensive statistical physics approach to the polarity reversals of the geomagnetic field. Phys. A 390, 1773–1778.

Vallianatos, F., Sammonds, P., 2011. A non-extensive statistics of the fault-population at the Valles Marineris extensional province, Mars. Tectonophysics 509, 50–54.

Vallianatos, F., Sammonds, P., 2013. Evidence of non-extensive statistical physics of the lithospheric instability approaching the 2004 Sumatran-Andaman and 2011 Honsu mega-earthquakes. Tectonophysics 590, 52–58.

Vallianatos, F., Michas, G., Papadakis, G., Sammonds, P., 2012. A non-extensive statistical physics view to the spatiotemporal properties of the June 1995, Aigion earthquake (M6.2) aftershock sequence (West Corinth rift, Greece). Acta Geophys 60, 758–768.

Vallianatos, F., Michas, G., Papadakis, G., Tzanis, A., 2013. Evidence of non-extensivity in the seismicity observed during the 2011–2012 unrest at the Santorini volcanic complex, Greece. Nat. Hazards Earth Syst. Sci. 13, 177–185.

Vallianatos, F., Michas, G., Papadakis, G., 2014. Non-extensive and natural time analysis of seismicity before the $M_W6.4$, October 12, 2013 earthquake in the south west segment of the Hellenic arc. Phys. A 414, 163–173.

Vallianatos, F., Michas, G., Papadakis, G., 2015. A description of seismicity based on non-extensive statistical physics: a review. In: D'Amico, S. (Ed.), Earthquakes and their Impact on Society. Springer Natural Hazards, Cham, pp. 1–41.

Vallianatos, F., Papadakis, G., Michas, G., 2016. Generalized statistical mechanics approaches to earthquakes and tectonics. Proc. R. Soc. A 472, 20160497.

Valverde-Esparza, S.M., Ramirez-Rojas, A., Flores-Marquez, E.L., Telesca, L., 2012. Non-extensivity analysis of seismicity within four subduction regions in Mexico. Acta Geophys. 60, 833–845.

Varotsos, P.A., Sarlis, N.V., Skordas, E.S., 2011. Natural Time Analysis: The New View of Time, Precursory Seismic Electric Signals, Earthquakes and Other Complex Time Series, first ed Springer-Verlag, Berlin, Heidelberg.

Vilar, C.S., Franca, G.S., Silva, R., Alcaniz, J.S., 2007. Nonextensivity in geological faults? Phys. A 377, 285–290.

Zoeller, G., Hainzl, S., 2002. A systematic spatiotemporal test of the critical point hypothesis for large earthquakes. Geophys. Res. Lett. 29 (11). Available from: https://doi.org/10.1029/2002GL014856.

Zoeller, G., Hainzl, S., Kurths, J., 2001. Observation of growing correlation length as an indicator for critical point behavior prior to large earthquakes. J. Geophys. Res. 106, 2167–2176.

Spatiotemporal Clustering of Seismic Occurrence and Its Implementation in Forecasting Models

Eugenio Lippiello

UNIVERSITY OF CAMPANIA "L: VANVITELLI", CASERTA, ITALY

CHAPTER OUTLINE

3.1 Introduction

Since the pioneering observations of the Japanese seismologist Omori in 1894 (Omori, 1894) it has been evident that the increase in seismic activity after large events is a striking feature of seismic occurrences. More precisely, Omori formulated his famous law stating that the number of events, hereafter defined as aftershocks, decays as a power law of the time since the occurrence of the main event, i.e., the mainshock. The law for the temporal decay of

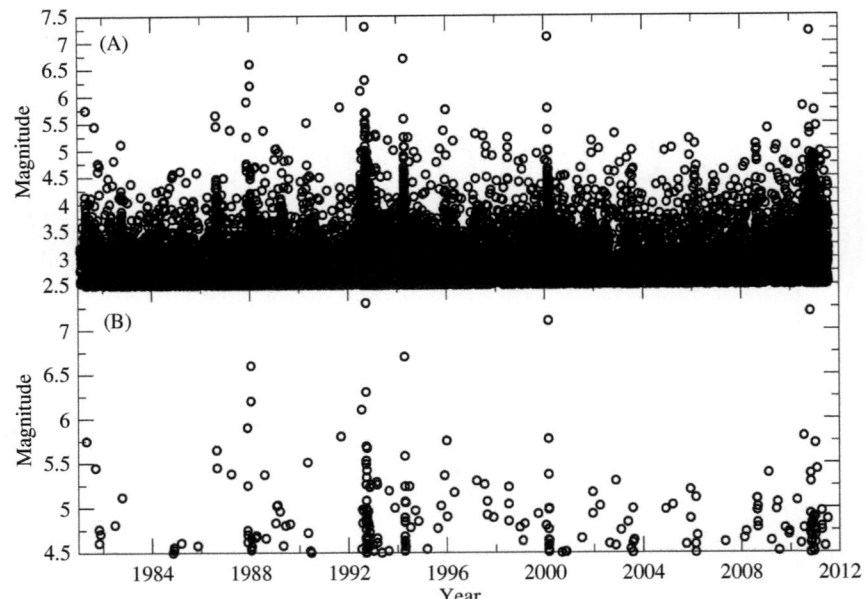

FIGURE 3–1 (A) Seismic temporal series of events recorded in Southern California in the years 1982 and 2012. (B) As for the upper panel, restricted to events with magnitudes larger than 4.5 detectable with the seismographic networks available at the beginning of the 1960s.

aftershocks has been generalized by Utsu et al. (1995) and is now usually called the Omori–Utsu (OU) law. A typical seismic pattern, potentially recorded also with the seismographic networks available at the beginning of the 1960s, is plotted in Fig. 3–1. This figure clearly shows the temporal clustering of seismicity, with many events recorded at small temporal distances after larger ones. Data similar to those in Fig. 3–1 have inspired the possibility of formulating statistical forecasting models in terms of cluster point-processes for seismicity. The central assumption is that the duration of a single earthquake can be neglected with respect to the other temporal scales in the process and therefore each event can be modelled as a single point in time. In the simplest point-process formulation, seismicity can be viewed as the superposition of Omori sequences

$$\lambda(t) = \mu + \sum_{i:t_i < t} \nu(t - t_i), \tag{3.1}$$

where the sum extends over all mainshocks with occurrence time $t_i < t$. In Eq. (3.1), $\lambda(t)$ is the expected seismic rate, μ is a Poisson term corresponding to the contribution of background seismicity and $\nu(t - t_i)$ is the aftershock occurrence rate following the OU law

$$\nu(t) = \frac{K}{(t+c)^p}. \tag{3.2}$$

With the improvement in the efficiency of seismic detection, and the accumulation of experimental observations, it has become clearer that an aftershock can trigger its own (secondary) aftershocks (Felzer et al., 2003). A common example is represented by the magnitude $m = 6.3$ Big-Bear earthquake that occurred on 28 June 1992 in Southern California 3 hours after the $m = 7.3$ Landers earthquake. Because of the short time delay between the two earthquakes and small epicentral distance (the Big-Bear epicentre was within one rupture-length of the Landers mainshock) it appears natural to consider Big-Bear an aftershock of Landers. Interestingly the Big-Bear earthquake was centred on a different fault and it is evident (Fig. 3−2) that a jump of seismic activity, in the area surrounding the Big-Bear epicentre, is observed only immediately after the Big-Bear earthquake and not in correspondence to the Landers mainshock. This suggests that, for a more realistic description of seismic occurrence, Eq. (3.1) must also incorporate second- and higher-order aftershocks. In particular Utsu (1970) discovered that the secondary aftershock sequence also obeys the OU law and therefore the sum in Eq. (3.1) must not be restricted to mainshocks but to all events that occurred at times $t_i < t$. Within this formulation the somewhat ambiguous discrimination between aftershocks and mainshocks is removed and all events can potentially trigger their own aftershocks.

Experimental observations, the so-called productivity law (Reasenberg and Jones, 1989; Helmstetter, 2003; de Arcangelis et al., 2016), indicate that the number of aftershocks

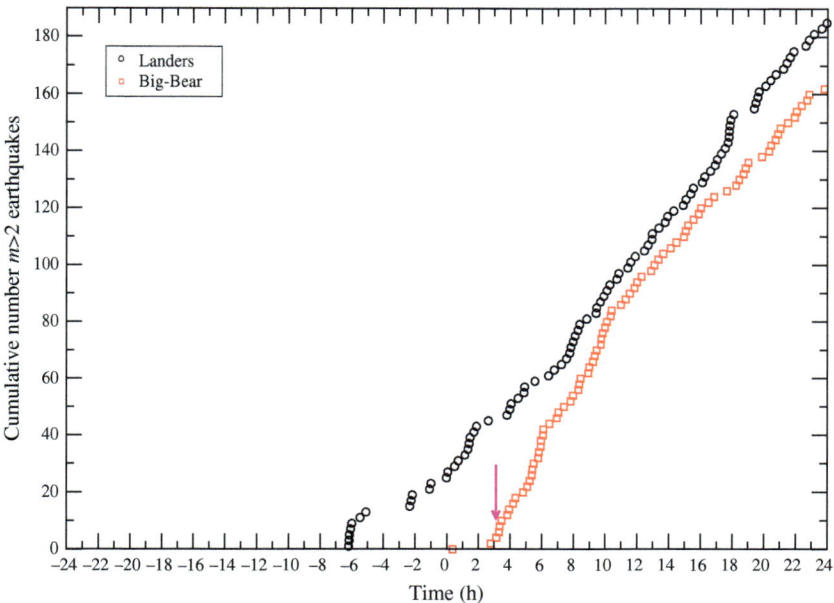

FIGURE 3–2 The cumulative number of $m > 2$ events recorded within a circle of radius 20 km centred in the Landers epicentre (black circles) and in the Big-Bear epicentre [red squares] is plotted as a function of the time t. The time has been shifted so that the Landers earthquake occurred at the time $t = 0$. The vertical magenta arrow indicates the occurrence time of the Big-Bear earthquake.

depends exponentially on the magnitude of the triggering mainshock. This information can be used in Eq. (3.1) writing $\nu(t - t_i)$ in the form

$$\nu(t - t_i) = \frac{A10^{am_i}}{(t - t_i + c)^p} \tag{3.3}$$

where m_i is the magnitude of the triggering mainshock, and A, a, c and p are model parameters. Using this law in Eq. (3.1) one recovers the original point-process formulation by Ogata (1988).

$$\lambda(t) = \mu + \sum_{i:t_i < t} \frac{A10^{am_i}}{(t - t_i + c)^p}. \tag{3.4}$$

It is evident that Eq. (3.4) models seismicity like a branching or epidemic model where independent background events trigger aftershocks, aftershocks trigger second-order generation aftershocks and so on. This is the reason why the model is defined as the epidemic type aftershock sequence (ETAS) model. The final step towards the formulation of a model useful for seismic forecasting was the introduction of spatial dependences in Eq. (3.4). Ogata (1998) proposed two modifications to the temporal ETAS model. The first takes into account that events are mainly located close to active faults which corresponds to a spatially heterogeneous background rate $\mu(\vec{r})$. The second modification originates from the observation of the aftershock spatial clustering: the number of aftershocks decreases as a power law of the epicentral distance from the mainshock epicentre δr

$$n_{aft}(\delta r) \sim \delta r^{-q}. \tag{3.5}$$

This information can be implemented by introducing a spatial kernel

$$G(x) = (x/L_0 + 1)^{-q} \tag{3.6}$$

which depends on the spatial distance $x_{ij} = |\vec{r} - \vec{r}_i|$ between triggering and triggered events. The seismic rate in the position $\vec{r}$ at the time t is then given by

$$\lambda(t, \vec{r}) = \mu(\vec{r}) + \sum_{i:t_i < t} \frac{A10^{am_i}}{(t - t_i + c)^p} G(x_{ij}). \tag{3.7}$$

According to the spatiotemporal ETAS model [Eq. (3.7)] seismicity is also modelled as a point-process in space. This is too crude an approximation since the spatial extent of the aftershock area is often comparable with the rupture-length of the triggering event. Nevertheless, it is still possible to describe earthquakes as single points localized at the epicentral coordinates and to take into account the finite spatial extent by replacing the constant L_0 in $G(x)$ [Eq. (3.6)] with the rupture-length $L(m)$ of the triggering mainshock. This corresponds to replacing $G(x)$ with $G(x, L(m_i))$ in Eq. (3.7) with $L(m_i) \propto Q10^{\gamma m_i}$ (Utsu and Seki, 1954). A more refined description of the spatial clustering is introduced in the ETAS model by means of an asymmetric spatial kernel taking into account that aftershocks are preferentially distributed along the direction parallel to the fault plane (Kagan and Jackson, 1995; Ogata, 1998; Helmstetter et al., 2006; Hainzl et al., 2008; Werner et al., 2011). We will not take this generalization into account and in this chapter we will restrict ourselves to the isotropic ETAS model.

A basic problem of the ETAS formulation is the identification of the best set of model parameters A, a, c, p, Q, γ, q which better reproduces instrumental observations. Several methods based on the log-likelihood maximization have been developed for this purpose (Zhuang et al., 2002, 2004; Bottiglieri et al., 2011; Schoenberg, 2013; Lippiello et al., 2014). Once the appropriate parameter set is identified, the ETAS model usually provides a good description of the spatiotemporal clustering of events in instrumental seismic catalogues. There are, however, some properties of instrumental catalogues not fully captured by the ETAS model. Sections 3.3 and 3.4 are devoted to discussing two of these situations that are very relevant for seismic forecasting. The first concerns deviations from the Gutenberg–Richter (GR) law systematically observed in the first part of instrumental aftershock sequences. Eq. (3.7) for the ETAS model gives the occurrence probability of an event in given spatial position at a given time. The magnitude of the event is assumed to follow the GR law independently of time and space. In the first hours up to some days after large shocks, conversely, instrumental catalogues show significant deviations from the GR law with a deficit of small-magnitude events. In Section 3.3 we show that this deficit, usually attributed to the inefficiency of the seismic network, is an intrinsic property of instrumental catalogues. We will then present two generalizations of the ETAS model, the ETAS incomplete (ETASI) model and the dynamic scaling (DS) ETAS model, which incorporate the observed deviations from the GR law.

The second important difference is observed in the temporal window preceding mainshocks and can be related to the always controversial topic of foreshock occurrence. The ETAS model gives a non-null probability of foreshocks interpreted, within this framework, as small events triggering larger ones. The organization in space and magnitude of foreshocks in instrumental catalogues, however, shows features that cannot be reproduced by the ETAS model. In Section 3.4 we discuss these features and we also suggest how to improve the ETAS model to incorporate foreshock occurrence.

Before discussing these issues, in Section 3.2 we address the main critique towards the ETAS model which is usually considered only a phenomenological model: Empirical observations are simply recovered because they have been put by hand into the model without any physical justification. In Section 3.2 we present a physical derivation of the ETAS model to enlighten the main assumptions used in its formulation. The final section provides information on the algorithm used to perform numerical simulations of the ETAS model presented in this chapter.

3.2 A Physical Interpretation of the ETAS Model

Even if the ETAS model is the most popular statistic model used to describe earthquake occurrence it is often considered only a phenomenological model lacking any physical ingredient. In this section we present a derivation of the ETAS model from a few key physical ingredients and some fundamental assumptions. This section should provide insights into the physical hypotheses below the usual stochastic formulation of the ETAS model.

The main idea is that aftershock occurrence is caused by the stress variation induced in the Earth's crust by previous events. The first very reasonable approximation is that the Earth's crust is an elastic medium. Therefore, stress changes caused by the occurrence of previous earthquakes and those due to the tectonic drive superpose in a linear way. As a consequence the stress $\sigma(t, \vec{r})$, at a given time t and in a given position $\vec{r}$, can be written as

$$\sigma(t, \vec{r}) = \sum_{j:t_j < t} \sigma_j(t, \vec{r}, t_j, \vec{t}_j) + \sigma_D(t, \vec{r}) \tag{3.8}$$

where the sum extends on all previous events. In Eq. (3.8) σ_j represents the stress variation at time t in the reference position $\vec{r}$, produced by the jth earthquake that occurred at time $t_j < t$ in position $\vec{r}_j$. On the other hand, σ_D is the stress accumulated because of the tectonic drive. Rigorously, the sum in Eq. (3.8) is a tensor sum and, indeed, it is well known that stress variations can combine in constructive or destructive ways (Stein, 1999). Our second assumption is that the sum is dominated by close in space events that occurred on the same fault and with very similar focal mechanisms. Therefore, our approximation is that stress interacts only in a constructive way and each event occurrence raises the stress level in the position $\vec{r}$. In other words we assume that σ_j is a positive scalar quantity.

The third hypothesis is time translation invariance. This originates from the observation that the seismic process evolves over geological timescales, whereas the duration of instrumental catalogues is, at most, of some decades. For the sake of simplicity we also neglect spatial heterogeneities and asymmetric behaviour related to the complexity of the fault network. This last approximation can, however, be removed in more refined models. Under the assumption of time translation invariance and space homogeneity, Eq. (3.8) can be written as

$$\sigma(t, \vec{r}) = \sum_{j:t_j < t} \sigma_j(t - t_j, x_{ij}) + \sigma_D(t, \vec{r}), \tag{3.9}$$

with $x_{ij} = |\vec{r} - \vec{r}_i|$.

The final assumption is the proportionality between seismic and stress rate (Schaff et al., 1998; Perfettini and Avouac, 2004). This hypothesis is supported by laboratory friction experiments (Beeler and Lockner, 2003) and theoretical modelization of rate-and-state friction (RSF) (Ruina, 1983; Dieterich, 1994) under the conditions of stress slowly changing in time (Helmstetter and Shaw, 2009). The hypothesis of a stress evolving very slowly in the Earth's crust is very reasonable except for the narrow temporal interval coinciding with the abrupt stress change during an earthquake. Therefore, neglecting the duration of earthquakes, which indeed are considered instantaneous in point-process formulations as in the ETAS model, we then assume

$$\lambda\left(t, \vec{r}\right) \propto \dot{\sigma}(t, \vec{r}), \tag{3.10}$$

where λ is the seismic rate and $\dot{\sigma}$ is the temporal derivative of the local stress. A similar equation $\lambda_j \propto \dot{\sigma}_j$ can be written for the seismic rate induced by the jth earthquake and, using the time derivative of Eq. (3.9) in Eq. (3.10), we obtain

$$\lambda(\vec{r}, t) = \sum_{j:t_j < t} \lambda_j(t - t_j, x_{ij}) + \lambda_D(\vec{r}, t) \tag{3.11}$$

where $\lambda_D \propto \dot{\sigma}_D$ is the seismic rate caused by tectonic drive.

The physical meaning of Eq. (3.11) is straightforward: Because of the linear elasticity assumption, the seismic rate at time t in position $\vec{r}$ is given by the aftershock rate caused by any past earthquake λ_j plus the contribution of events triggered by the tectonic drive λ_D.

It is evident that Eq. (3.11) leads to the ETAS formulation Eq. (3.7) after the identification

$$\lambda_j(t - t_j, x_{ij}) \propto \dot{\sigma}_j(t - t_j, x_{ij}) \propto (t - t_j + c)^{-p} G(x_{ij}). \tag{3.12}$$

The origin of the power law decay of $G(x)$ [Eq. (3.6)] can then be related to the power law decay of stress $\sigma_j(t, x_{ij}) \sim x_{ij}^{-\mu}$ obtained from elastodynamic equations. As a matter of fact, the measure of the exponent q in $G(x)$ has been heavily investigated in the recent years (Felzer and Brodsky, 2006; Lippiello et al., 2009a; Richards-Dinger et al., 2010; Shearer, 2012b; Gu et al., 2013) as a proxy to discriminate between aftershock triggering because of static $\mu = 2$ or dynamic stress $\mu = 1$. Less clear is the origin of the power law decay in time that, according to Eq. (3.12), should reflect the power law temporal decay of the stress rate. Several interpretations have been proposed for the mechanism responsible for the power law decay of λ_j, which is intrinsically assumed in the ETAS model. In the following we briefly comment on two hypotheses: in one case, the power law decay is caused by nontrivial time-dependent features of the friction law. In the second case, conversely, simple friction is assumed, whereas stress relaxation is attributed to coupling with a viscoelastic layer. We discuss the two cases separately.

3.2.1 Rate-and-State Friction

In the Amontons−Coulomb description of friction, the frictional force is proportional to the normal load N by means of a friction coefficient μ assuming two constant values: $\mu = \mu_s$, the static coefficient, when the system is stuck and $\mu = \mu_d < \mu_s$, the dynamic coefficient, when the system moves. Nevertheless, laboratory experiments (Rabinowicz, 1951, 1956, 1958) have shown more complex behaviour. First, the transition from static to dynamic friction does not occur instantaneously and, as a consequence, after the external force overcomes the static friction, stable sliding occurs up to a slip distance D_c. Also, the transition from dynamic to static coefficient does not occur instantaneously, indeed experimental results indicate that if two surfaces are kept in stationary contact under a load for a time t, a healing mechanism is activated producing an increase in the static coefficient μ_s roughly proportional to $\log t$. At the same time, a velocity-weakening effect leads to a decrease in the dynamic coefficient roughly proportional to the logarithm of the slip velocity V. The above three experimental observations can be combined into a friction coefficient $\mu(V, \Theta)$ depending on the slip velocity and on the state variable Θ by means of a RSF formulation (Dieterich, 1972; Ruina, 1983; Chris, 1998):

$$\mu(V, \theta) = \mu_0 + A_0 \log\left(\frac{V}{V_0}\right) + B_0 \log\left(\frac{\Theta V_0}{D_c}\right) \tag{3.13}$$

where V_0 is a reference velocity, A_0 and B_0 are fitting constants and Θ evolves in time according to the equation

$$\frac{d\Theta(t)}{dt} = 1 - \frac{V\Theta}{D_c}. \tag{3.14}$$

The dependence of the static coefficient on the logarithm of the contact time t can be simply obtained in the static limit $V \to 0$. Indeed, from Eq. (3.14) if $V = 0$, $\Theta = t$ and substituting in Eq. (3.13) one immediately obtains $d\mu_s/d(\log t) = B_0$. On the other hand, if we define μ_d as the stationary value of the friction coefficient for a velocity V, in the stationary condition regime $d\Theta(t)/dt = 0$, $\Theta/D_c = 1/V$ and $d\mu_d/d(\log V) = A_0 - B_0$. As a consequence, velocity weakening is observed only if $A_0 < B_0$. It is evident that, for a given V, the transition to this stationary regime occurs if the block slides over a distance D_c. At this point, the friction coefficient drops to μ_d and, if the reduction in the external force is smaller than the reduction in the friction force $(\mu_s - \mu_d)N$, a slip instability is triggered. Describing the fault as frictional blocks and modelling the tectonic drive as an elastic force with coupling K_D, unstable slips occur only if $(\mu_s - \mu_d)N > K_D D_c$. This condition for instability can be reformulated in terms of the coefficients A_0 and B_0, and introducing $K_c = (B_0 - A_0)N/D_c$, instability occurs only if $K_c > K_D$ (Scholz, 2002). Expressing the elastic constant K_D in terms of the Lame' coefficients for the Earth's crust, the stability condition corresponds to a nucleation length L_c inversely proportional to K_C, $L_c \propto 1/D_c$, such that a stable sliding occurs until the slipping region reaches L_c. Generalizing the above description to a seismic fault made of an ensemble of patches under different stress conditions, Dieterich (1972, 1994) has considered the evolution of the seismic rate $\lambda(t)$ after perturbing the system by an increase in the shear stress, or alternatively a reduction in the normal stress. Under the hypothesis that, in the absence of the stress perturbation, the seismic rate is constant λ_0, then after the stress variation the rate $\lambda(t)$ decreases in time consistently with the OU law before relaxing to the stationary value λ_0.

3.2.2 Viscous Coupling With the Asthenosphere

Several studies (Nakanishi, 1990, 1991, 1992; Hainzl et al., 1999; Pelletier, 2000; Lippiello et al., 2015; Landes and Lippiello, 2016) identify the coupling with a viscoelastic medium, representing the asthenosphere, the key ingredient to explaining the power law decay of the aftershock rate. More precisely, the asthenosphere is modelled as a Newton viscous medium of thickness H_a and viscosity η and the lithosphere as an elastic medium of thickness H_L and

Young modulus Y. Under the assumption that viscous flow in the asthenosphere is a linear Couette flow, stress in the lithosphere evolves in time according to a diffusion equation

$$\frac{\partial \sigma(t)}{\partial t} = D \frac{\partial^2 \sigma(t)}{\partial x^2} \tag{3.15}$$

with the diffusion coefficient $D = (Y/\eta)H_l H_a$ (Turcotte and Schubert, 2002). This mechanism for stress relaxation combined with heterogeneous stress distribution in an elastic layer, leads to the OU law for the aftershock decay (Lippiello et al., 2015).

3.3 Short-Term Aftershock Incompleteness and Its Implementation in the ETAS Model

The magnitude of events in the ETAS model is assumed to be independent of the occurrence time and spatial location. As a consequence, the occurrence rate of events with magnitude m, at time t and in position $\vec{r}$, is simply obtained from Eq. (3.7) after multiplying $\lambda(t, \vec{r})$ by the magnitude distribution $P(m)$. Typically one assumes that the magnitude distribution follows the GR law

$$P(m) \propto 10^{-bm}. \tag{3.16}$$

Several experimental studies, for different geographic regions and over long temporal intervals, indicate that the GR law is a stable feature of seismic occurrence with $b \simeq 1$. The common point of view is that the GR law holds for events up to very small magnitudes $m \ll 0$. However, in order to identify an earthquake it is necessary that its signal is significantly above the background noise level. This implies that small earthquakes are typically not reported in experimental catalogues. This corresponds to deviations from the GR law for magnitudes smaller than a completeness magnitude m_c, defined as the magnitude above which all events are identified in the catalogue. A correct estimate of m_c is fundamental in seismic forecasting. Indeed too high a value, discarding usable data, leads to a loss of information by undersampling. Conversely, too low a value leads to an erroneous estimate of parameter values and thus to a biased analysis. A standard way of estimating m_c is to find the minimum magnitude above which the fit with the GR law [Eq. (3.16)] works correctly. The value of m_c clearly depends on the ability to filter noise and on the distance between the earthquake epicentre and the seismic stations necessary to trigger an event declaration in a catalogue. In regions with very dense seismic networks m_c can assume values as small as $m_c = 0$. Nevertheless, even in these regions important deviations from the GR law are observed in the first part of aftershock sequences. In Fig. 3−3 we plot the magnitude distribution evaluated for different temporal intervals after the $m = 7.3$ Landers earthquake in Southern California. The completeness magnitude of the region, before the occurrence of the mainshock, was roughly $m_c = 1.5$. However, restricting to events recorded the first day after the mainshock (black curve in Fig. 3−3), we observe an about flat behaviour of the magnitude distribution which extends up

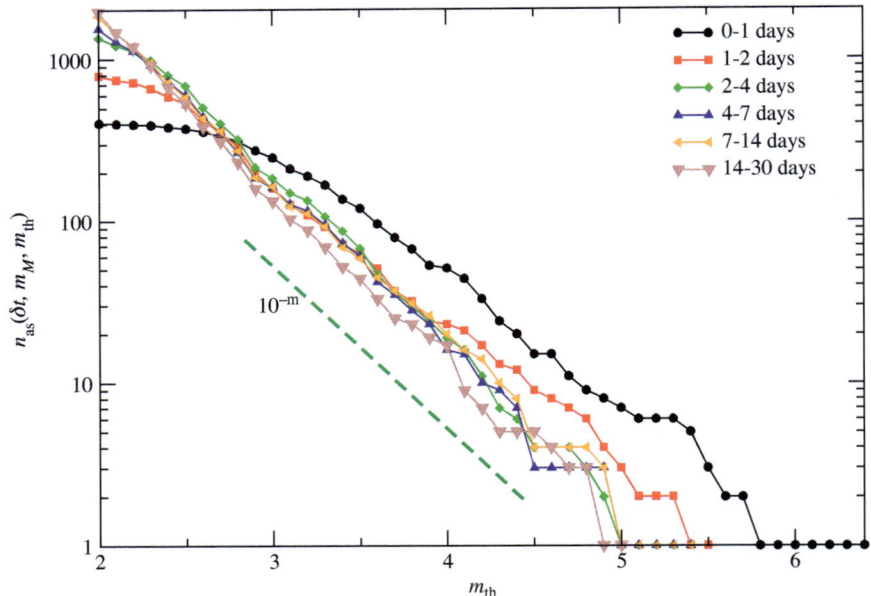

FIGURE 3–3 The number of aftershocks with magnitude larger than m_{th} after the $m_M = 7.3$ Landers earthquake in Southern California, evaluated in different temporal windows δt from the mainshock. The green dashed line is the GR law with $b = 1$.

to magnitudes $m \simeq 3$. A similar behaviour is also observed for data up to $_4$ days after the main-shock with a completeness magnitude m_c, which is a decreasing function of the temporal period since the mainshock. A more quantitative analysis of data in Fig. 3−3 shows that the completeness magnitude $m_c(t)$ depends logarithmically on the time t since the mainshock

$$m_c(t, m_M) = m_M - \frac{1}{d}\log_{10}(t) + \Delta m, \tag{3.17}$$

where m_M is the mainshock magnitude, $d \simeq 1$ and $\Delta m \sim 4$, if time is measured in days, are fitting parameters.

This feature, defined by Kagan (2004) as short-term aftershock incompleteness (STAI), is a distinctive property of seismic occurrences supported by several experimental observations (Kagan, 2004; Shcherbakov et al., 2004; Helmstetter et al., 2005; Enescu et al., 2007). STAI also strongly affects the temporal decay of the number of aftershocks with magnitude larger than a threshold value m_{th}, $N_{as}(t, m_{th})$, producing deviations from the 'true' OU law [Eq. (3.3)]. Indeed, if events smaller than $m_c(t)$ are not reported in instrumental catalogues, a constant behaviour of the aftershock rate $\nu(t)$ extends well above c and the power law decay t^{-p} can be observed only for times larger than a characteristic time c_{meas}. This value can be obtained from Eq. (3.17) after imposing $m_c(c_{\text{meas}}) = m_{th}$ which leads to

$$c_{\text{meas}} = 10^{d(m_M - m_{th} - \Delta m)}. \tag{3.18}$$

The hypothesis of a c_{meas} value which depends on the mainshock magnitude and on the magnitude threshold m_{th} [Eq. (3.18)], can be experimentally verified in instrumental catalogues. More precisely, we measure the rate of aftershocks in the relocated Southern California earthquake catalogue RSCEC (Hauksson et al., 2012) in the years 1981−2013. We apply the Baiesi−Paczusky (BP) declustering criterion (Baiesi and Paczuski, 2005) to identify main−aftershock couples using the same parameters adopted in Moradpour et al. (2014) and Hainzl (2016a). We then define the aftershock daily rate $\rho(t, m_M, m_{th})$ as the number of aftershocks with magnitude larger than m_{th}, occurring at a temporal distance t from their triggering mainshock, with magnitude $m \in [m_M, m_M + 1)$, divided by the number of mainshocks with magnitude $m \in [m_M, m_M + 1)$. The aftershock daily rate is then evaluated for three values of $m_M = (3, 4, 5)$ and for $m_{th} = (1.5, 2.5, 3.5)$. The results (Fig. 3−4) show that the aftershock rate clearly depends on the magnitude difference $m_M - m_{th}$. More precisely, we assume that aftershocks follow the OU law $\rho(t, m_M, m_{th}) = K/(t + c_{meas})^p$ and apply a best-fit procedure to extrapolate the value of c_{meas}, taking $p = 1.1$. Results for c_{meas}, for different values of m_M and m_{th}, are plotted in the lower panel of Fig. 3−4. This shows that the c_{meas} value

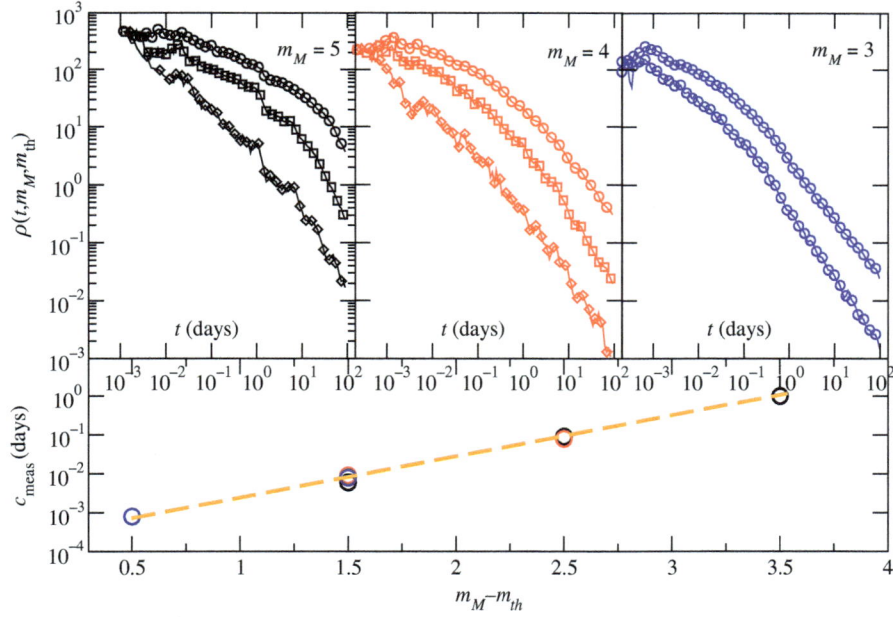

FIGURE 3–4 The number of events identified as aftershocks, by the Baiesi−Paczusky declustering procedure, with magnitude larger than m_{th} and that occurred at a temporal distance t from events identified as mainshocks, with magnitude $m \in [m_M, m_M + 1)$, is divided by the number of identified mainshocks and plotted versus t. Different panels corresponds to different values of the mainshock magnitude class $m \in [m_M, m_M + 1)$. Different symbols indicate different values of the lower threshold: $m_{th} = 1.5$ (circles), $m_{th} = 2.5$ (squares) and $m_{th} = 3.5$ (diamonds). (Lower panel) The time c_{meas} obtained as best fit of the OU law from the upper panels. Different colours correspond to different values of m_M. The dashed orange line is the best fit $y = 10^{x-3.53}$.

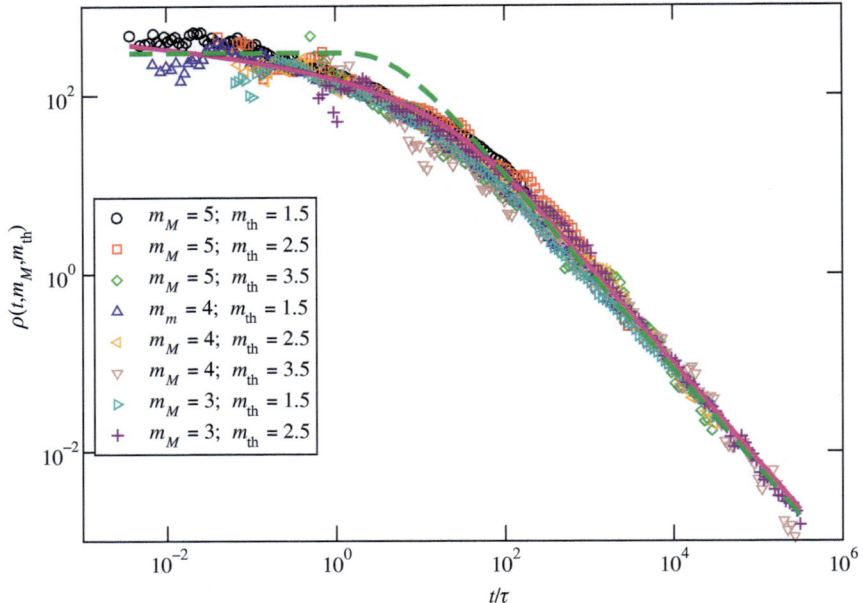

FIGURE 3–5 The same quantity $\rho(t, m_M, m_{th})$ for Fig. 3–4 is plotted as function of t/τ, with $\tau = 10^{d(m_M - m_{th})}$ proportional to c_{meas} [Eq. (3.18)] with $d = 1$, for different values of m_M and m_{th}. The magenta continuous line is the scaling function Eq. (3.28) with $B = 0.8$, $\tau_0 = 0.02$ and $p = 1.1$ whereas the dashed green line is the scaling function Eq. (3.22) with $B = 300$, $\tau_0 = 0.17$ and $p = 1.1$.

follows Eq. (3.18) with $\Delta m = 3.53 \pm 0.05$ and $d = 1 \pm 0.03$. This is confirmed after plotting the same results for Fig. 3–4 as a function of t/τ with $\tau = 10^{d(m_M - m_{th})} \propto c_{meas}$ (Fig. 3–5) obtaining the scaling behaviour

$$\rho(t, m_M, m_{th}) = F\left(\frac{t}{\tau}\right). \tag{3.19}$$

3.3.1 Mechanisms Responsible for STAI

Originally, STAI was interpreted as a technical problem mainly due to the overloading of processing facilities. Indeed it is reasonable to expect that small events are hidden by mainshock coda waves. A typical estimate of coda-wave duration $t_W = 300 \times 10^{0.5(m_M - 4)}$ second (Kagan, 2004) leads to incompleteness temporal windows $c_{meas} \sim t_W$. However this hypothesis implies a typical c_{meas} value significantly smaller than the experimental one obtained from Eq. (3.18), and with a different scaling dependence on m_M. The lack of events over a temporal interval larger than t_W is often attributed to damage in the seismographic network caused by the mainshock as well as to workforce limitations. Since these factors are, at least partially, human activity-related or depend on technical details of the seismic registration, they do not provide a complete explanation for the regular trend [Eq. (3.18)] experimentally observed.

Another possible explanation is that STAI originates from technical problems in the routines for event identification. This interpretation is supported by several studies demonstrating a huge deficit of early-time aftershocks in regional catalogues (Helmstetter et al., 2006; Enescu et al., 2007; Peng et al., 2007; Peng and Zhao, 2009; Omi et al., 2013). For example, Peng and Zhao (2009) applying a match-filtering technique to data signals recorded in the temporal interval of 3 days after the 2004 magnitude 6.0 Parkfield earthquake identified 11,138 events against the 933 listed in the standard catalogue of the Northern California Seismic Network. Similarly, by means of direct inspection of the seismic signals in the first 200 seconds after mainshocks with magnitude $\in [3, 5]$, Peng, Vidale, Ishii and Helmstetter (PVIH) identify about five times as many $m > 0$ aftershocks as are listed in the Japan Meteorological Agency catalogue (Peng et al., 2007). Here we briefly summarize the main steps of the PVIH procedure.

1. The 'best' signal of each mainshock is selected on the basis of the following criteria: (1) it is recorded at a station distant less than 30 km from the mainshock epicentre; (2) there is a low preevent noise level; (3) the signal-to-noise ratio is high; (4) the mainshock coda decays rapidly.
2. Each component of the signal is high-pass filtered by means of a two-pass Butterworth filter with a corner frequency $f_c = 30$ Hz.
3. The envelope of each signal is computed.
4. The signals of the three components are superimposed.
5. The logarithm of the resulting signal is taken.
6. A smoothing procedure by a median operator with a half width of 0.1 second is applied.

We define $\mu(t)$ in the final signal after step (6) and observe that the occurrence of aftershocks must produce double peaks in $\mu(t)$ corresponding to the paired P and S arrivals. Once a double peak is identified, the temporal distance τ_{PS} between the two peaks is measured and compared with the temporal distance between P and S arrivals in the mainshock envelope. In order to restrict to aftershocks occurring at small distances from the mainshock only double peaks with τ_{PS} comparable with the mainshock one are kept in the analysis. Furthermore, the procedure neglects all peaks with amplitude $\mu(t) < \mu_B(t) - 0.3$, where $\mu_B(t)$ is the reference preevent noise level defined as the value of $\mu(t)$ before the peak. Because of the logarithmic nature of $\mu(t)$, the choice $\mu_0 - 0.3$ corresponds a signal-to-noise ratio lower than 2. The selected double peaks are then associated to aftershocks and, in particular, the aftershock origin time is taken from the largest of the two peaks t_i and the peak amplitude $\mu(t_i)$ is used to estimate the event magnitude. More precisely, the magnitude m_i of each aftershock is given by $m_i = \mu_i - \delta m$ where the value of δm is calibrated for each mainshock in order that m_i matches the magnitudes of small events listed in the catalogue.

In Fig. 3−6 we present results of the PVIH procedure applied to the signal of duration two days after the 1999 magnitude 7.1 Hector Mine earthquake recorded at the station CIGSC located at an epicentral distance of 0.83 degrees from the epicentre. In the same

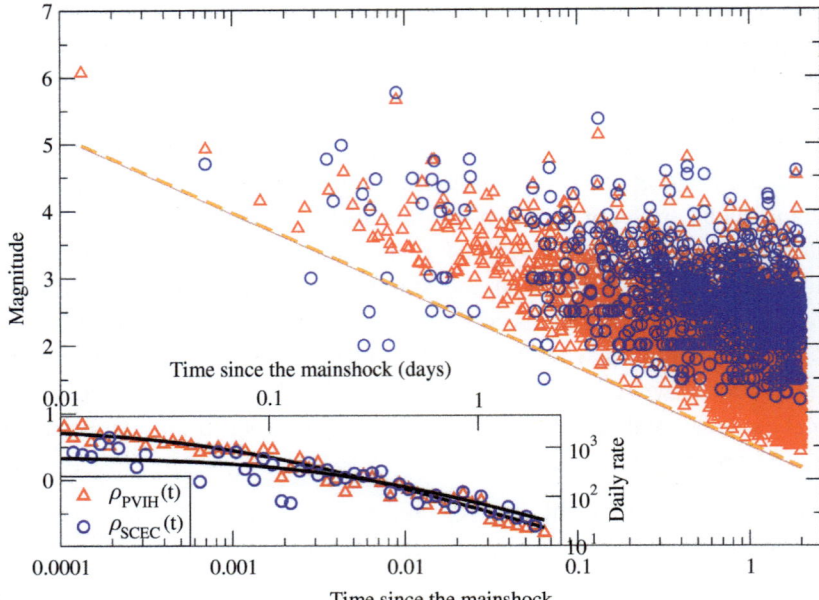

FIGURE 3–6 Comparison between events identified by the PVIH procedure [red triangles] with events listed in the RSCEC [blue circles]. The orange line is the reference minimum magnitude $m_{\min}(t) = m_M - 1.15\log_{10}(t) - 6.6$, with t measured in days. In the inset we plot $\rho_{\text{PVIH}}(t)$ and $\rho_{\text{SCEC}}(t)$ as functions of time. The black lines are the best fit with the OU law $\rho(t) = K(t + c_{\text{meas}})^{-p}$ with $K = 60 \pm 2$, $c_{\text{meas}} = 0.049 \pm 0.012$ days and $p = 1.25 \pm 0.05$ for $\rho_{\text{PVIH}}(t)$, whereas $K = 99 \pm 3$ $c_{\text{meas}} = 0.21 \pm 0.09$ days and $p = 1.25 \pm 0.05$ for $\rho_{\text{SCEC}}(t)$. OU, Omori–Utsu.

figure we also plot events listed in the RSCEC catalogue with epicentral distances less than 100 km from the mainshock epicentre. The PVIH method identifies 3688 $m > 0$ earthquakes compared with the 1029 listed in the RSCEC catalogue. The addition of missing aftershocks to the original data set makes even more clear the existence of a lower magnitude $m_{\min}(t)$, depending on the time since the mainshock, such that there is no event with $m < m_{\min}(t)$. This is, for instance, also evident in Figure S3 of Parsons and Velasco (2009) and in Fig. 6a of Peng et al. (2007). In all cases $m_{\min}(t)$ is consistent with a logarithmic decreasing function of time t since the mainshock (orange line in Fig. 3–6)

$$m_{\min}(t) = m_M - \phi\log_{10}(t) - \Delta\mu. \tag{3.20}$$

In the inset of Fig. 3–6 we plot the daily rate of $m > 3$ aftershocks for both PVIH identified events $\rho_{\text{PVIH}}(t)$ and for events in the RSCEC catalogue $\rho_{\text{SCEC}}(t)$. It is evident that $\rho_{\text{PVIH}}(t) > \rho_{\text{SCEC}}(t)$ at short times, whereas the two distributions roughly coincide for times larger than 0.3 days. This figure suggests that the PVIH significantly reduces the incompleteness of the data set. In particular we obtain, from the decay of $\rho_{\text{PVIH}}(t)$, a c_{meas} value roughly 4.5 times smaller than that obtained in the RSCEC catalogue, but still significantly different than zero.

From the analysis of data like those presented in Fig. 3−6, for different mainshocks in different geographic regions, Lippiello et al. (2016) propose that $m_{min}(t)$ [Eq. (3.20)] is an intrinsic lower bound for aftershock detection which cannot be lowered by improving the quality of the seismic network and/or of the procedure for event identification. This lower bound originates from the existence of a blind time Δt produced by the coda waves of a magnitude m_0 earthquake such that events with magnitude $m < m_0$, occurring at a temporal distance lower than Δt, cannot be detected. Taking into account that the magnitude is proportional to the logarithm of the amplitude $A(t)$ of the seismic signal and that $A(t)$ is a decreasing function of the time, the blind time Δt of events with magnitude smaller than m after an m_0 earthquake depends on the difference $m_0 - m$. In particular, taking into account the power law decay of coda waves, Lippiello et al. (2016) have recovered Eq. (3.20) which leads to a completeness magnitude $m_c(t)$ [Eq. (3.17)] after the identifications $d \simeq \phi$ and $\Delta m \simeq \Delta \mu$. Interestingly Lippiello et al. (2016) have shown that the values of ϕ and $\Delta \mu$ can be related to the values of A and c controlling the 'true' aftershock decay in the OU law [Eq. (3.3)]. In particular, larger A values roughly produce smaller Δm, whereas larger c values roughly correspond to smaller d. These relations have been proposed (Lippiello et al., 2016) as a method to estimate the true values of A and c from the fit of ϕ and $\Delta \mu$, in the first hours after the mainshock. Similar results have been obtained by Hanizl (2016b) under the assumption of a constant blind time Δt which implies the existence of a maximum detectable rate $\lambda_{max} \simeq 1/\Delta t$. The c_{meas} values can therefore be obtained from the OU law [Eq. (3.3)] after imposing $\lambda(c_{meas}) = \lambda_{max}$, which leads to Eq. (3.18) under the assumption $c \ll c_{maes}$, $a \simeq b = dp$ and

$$\Delta m = \frac{1}{p} \log_{10} \left(\frac{\lambda_{max}}{A} \right). \tag{3.21}$$

According to this interpretation, even if c_{meas} is not related to true c value in the OU law it contains important physical information such as the productivity coefficient A.

Hainzl (2016a) has also obtained an analytical expression for the aftershock rate under the assumption of a constant blind time Δt. More precisely, assuming that aftershocks follow the OU law and that each aftershock hides all subsequent smaller events occurring at temporal distances smaller than Δt, Hainzl has obtained the analytical expression for $\rho(t, m_M, m_{th})$ which can be put in the scaling form Eq. (3.19) with

$$F(x) = B \left(1 - e^{-B(x/\tau_0)^{-p}} \right). \tag{3.22}$$

In Fig. 3−5 we compare the scaling function $F(x)$ of Eq. (3.22) with data from the instrumental catalogue. Differences with the theoretical scaling function $F(x)$ [Eq. (3.22)] are observed in particular for $t/\tau \simeq 1$. In the same figure we also plot the scaling form Eq. (3.28) obtained under a DS assumption. This assumption, which roughly corresponds to a blind time $\Delta t \sim c_{meas}$, not constant but depending on the main−aftershock magnitude difference [Eq. (3.18)], provides a better description of experimental data (Fig. 3−5).

3.3.2 The Dynamic Scaling ETAS Model

The DS hypothesis (Lippiello et al., 2007a, 2007b, 2008, 2009a, 2012b; de Arcangelis et al., 2016) assumes that the magnitude difference $m_i - m_j$ fixes a characteristic timescale

$$\tau_{ij} = \tau_0 10^{(b/p)(m_j - m_i)}, \tag{3.23}$$

with τ_0 constant, such that the conditional probability density $p(m_i, t_i | m_j, t_j)$ to have an earthquake of magnitude m_i at time t_i given an earthquake m_j at time t_j, is magnitude-independent when time is rescaled by τ_{ij}

$$p(m_i, t_i | m_j, t_j) = H\left[\frac{t_i - t_j}{\tau_{ij}}\right]. \tag{3.24}$$

Here $H(z)$ is a normalizable function and for a better comparison with the STAI hypothesis we assume

$$H(z) = \frac{B}{z^p + 1} \tag{3.25}$$

where B is fixed by normalization. From Eqs (3.24) and (3.25) one immediately obtains that $p(m, t | m_0, t_0)$ is independent of m for $z^p = \tau_0^{-p}(t - t_0)^p 10^{-b(m_0 - m)} \ll 1$. On the other hand, the GR behaviour $p(m, t | m_0, t_0) \propto 10^{-bm}$ is recovered in the opposite limit $z^p \gg 1$. Hence, the condition $\tau_0^{-p}(t - t_0)^p 10^{-b(m_0 - m^*)} = 1$ fixes a crossover magnitude $m*$ that separates flat from GR behaviour. The quantity m^* can be identified with the completeness magnitude m_c and is in agreement with the empirical Eq. (3.17) with $\delta = p/b$ and $\Delta m = -p/b(\log \tau_0)$.

Next we evaluate the rate of aftershocks with magnitude larger than m_{th} triggered at time t by a magnitude m_M mainshock that occurred at time $t = 0$,

$$\rho(t, m_M, m_{th}) = \int_{m_{th}}^{\infty} p(m, t | m_M, 0) dm, \tag{3.26}$$

and using Eqs (3.24) and (3.25) we obtain

$$\rho(t, m_M, m_{th}) = B \int_{m_{th}}^{\infty} \left(\left(\frac{t}{\tau \tau_0}\right)^p + 1\right)^{-1} dm \tag{3.27}$$

with $\tau = 10^{(b/p)(m_M - m)}$. The above integral can be analytically solved leading to

$$\rho(t, m_M, m_{th}) = \frac{B}{b \log(10)} \log\left(1 + \left(\frac{t}{\tau \tau_0}\right)^{-p}\right). \tag{3.28}$$

It is evident that Eq. (3.28) obeys the scaling form Eq. (3.19) with the scaling function $F(x)$ exhibiting the power law decay $F(x) \propto x^{-p}$ for $t \gg \tau_0 \tau$. Conversely at short times ($t \ll \tau_0 \tau$) Eq. (3.28) predicts a logarithmic decay consistent with experimental data (Fig. 3–5).

3.3.3 The ETAS Incomplete Model

Incompleteness hides the true value of the c value in the OU law, introducing a strong bias in the routines for short-term forecasting (Omi et al., 2013; Lippiello et al., 2016; Omi et al., 2016). This is a problem of huge interest due to the growing interest of public authorities in accurate and real-time aftershock forecasting and this makes it very important to incorporate STAI into the ETAS model. A possibility is via the DS assumption illustrated in the previous section. Another possibility is to explicitly assume the existence of a blind time Δt in the ETAS model. In the simpler ETASI1 model one assumes a constant blind time Δt. Therefore, starting from the original ETAS model, we remove all events occurring at a temporal distance smaller than Δt after a larger event. Synthetic ETASI1 catalogues are then generated according to the numerical protocol described in details in Section 3.5. Here we compare results for the ETASI1 and the RSCEC. In particular we apply to the ETASI1 catalogue the same BP declustering procedure used for the instrumental catalogue in Fig. 3−4. We then identify main−aftershock couples and study $\rho(t, m_M, m_{th})$ for different mainshock magnitude m_M, different thresholds and different values of the productivity coefficient A, implemented in numerical simulations. We always find that the c_{meas} value follows Eq. (3.18) with $d = b/p$. This is shown in Fig. 3−7 where we plot $\rho(t, m_M, m_{th})$ as a function of t/τ with $\tau = 10^{d(m_M - m_{th})}$ and $d = b/p$. Data for different m_M and m_{th} and equal A collapse on the same master curve $F(t/\tau)$. The effect of A roughly corresponds to a translation of $F(x)$ with data for larger A shifted on the right. Fig. 3−7 indicates good agreement between numerical scaling function $F(x)$ and Eq. (3.28) which represents a good fit of instrumental data. This indicates that the hypothesis of a constant blind time implemented in the ETAS model not only reproduces the scaling behaviour [Eq. (3.19)] but also the scaling function $F(x)$. Fig. 3−7, conversely, shows clear deviations from the scaling function Eq. (3.22) obtained under the assumption of a constant blind time but neglecting the cascading aftershock process and restricted to first-order generation aftershocks. It is then reasonable to attributed deviations of $F(x)$ from Eq. (3.22) to the cascading process implemented in the ETAS model. Indeed, aftershocks of higher-order generation are also followed by a blind time, which eventually hides previous-generation aftershocks. This can produce a more gradual decrease in the aftershock number from the initial plateau compared to the situation when higher-order generation aftershocks are not taken into account (Hainzl, 2016a).

In the inset of Fig. 3−7 we also explore the dependence of Δm on the productivity coefficient A implemented in numerical simulations. Data clearly show that Δm becomes more negative for increasing A, confirming the strong correlation between the two quantities. In particular we find that the dependence of Δm on A is consistent with Eq. (3.21) only for small values of A, whereas faster growth is observed for larger A. Also in this case, deviations from Eq. (3.21) can be attributed to the cascading process which causes a total blind time that is longer when compared to the situation when higher-order generation aftershocks are not considered as in Eq. (3.21).

A more accurate description of instrumental data is represented by the ETASI2 model which assumes a blind time $\Delta t \propto c_{\mathrm{meas}}$ given by Eq. (3.18). More precisely, at each

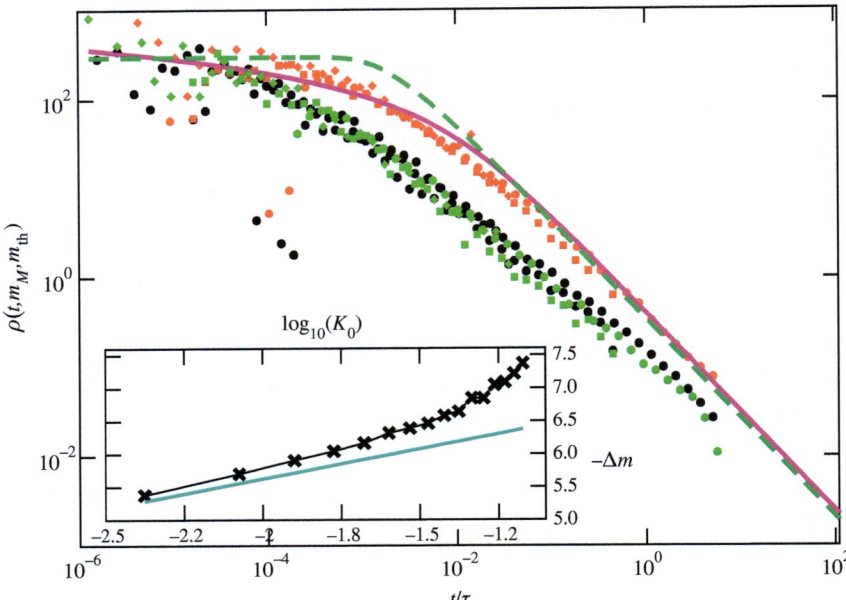

FIGURE 3–7 The aftershock density $\rho(t, m_M, m_{th})$ in the ETASI1 catalogue with a blind time of 1 min is plotted as function of t/τ. Different values of m_M and m_{th} are plotted with different symbols, whereas different colours correspond to different values of A and of the average background rate r_B: $A = 0.035$ and $r_B = 4.38$ days^{-1} (black), $A = 0.035$ and $r_B = 8.3$ days^{-1} [green] and $A = 0.068$ and $r_B = 4.38$ days^{-1} [red]. The continuous magenta line is the scaling function Eq. (3.28) with $B = 0.8$, $\tau_0 = 0.02$ and $p = 1.1$ whereas the dashed green line is the scaling function Eq. (3.22) with $A = 300$, $\tau_0 = 0.17$ and $p = 1.1$, also plotted in Fig. 3–5. (Inset) The value of $-\Delta m$ (Eq. 3.18) as a function of $\log_{10}(A)$ for $r_B = 4.38$. The cyan line is the theoretical prediction [Eq. (3.21)]. *ETASI*, epidemic type aftershock sequence incomplete.

time t we impose the existence of a minimum in the detection magnitude $m_{\min}(t) = \max_{i: t_i < t}(m_i - d\log(t - t_i) - \Delta m)$, where the maximum is evaluated over all events with magnitude m_i that occurred at time $t_i < t$. We then remove from the original ETAS catalogue all events with magnitude m and occurrence time t such that $m < m_{\min}(t)$. The ETASI2 model provides a better description of magnitude correlations in instrumental catalogues than the ETASI1 model (Lippiello et al., 2018). *ETASI*, epidemic type aftershock sequence incomplete.

3.4 Foreshock Occurrence in the ETAS Model

Experimental observations indicate that, often, large shocks are preceded by an increase in the seismic rate in a temporal period extending up to some days before the mainshock (Bouchon et al., 2013). Events responsible for this increment are usually defined foreshocks and the possibility of forecasting some features of the incoming mainshock from foreshock occurrence is a fundamental problem in earthquake forecasting (Brodsky, 2011; Lippiello et al., 2012a; Bouchon et al., 2013; Mignan, 2014; Brodsky and Lay, 2014; Ogata and Katsura, 2014; Felzer et al., 2015; Bouchon and Marsan, 2015). Many features of nontrivial energy-spatiotemporal properties of foreshocks can also be recovered in the ETAS model (Helmstetter and Sornette, 2003; Helmstetter et al., 2003; Felzer et al., 2004; Hardebeck et al., 2008). Indeed, even if the ETAS model implements only two kinds of earthquakes (independent background or triggered events) it also accounts for an increase in seismic activity before large shock occurrence. This is due to a nonzero probability that an event can trigger a larger earthquake. According to this hypothesis, foreshocks are an artefact of the typical erroneous identification of the mainshock as the largest event in a sequence. More precisely, within this interpretation, seismic clustering in time and space, attributed to foreshocks, is a consequence of the fact that events identified as mainshocks, and preceded by close-in-time smaller earthquakes, are aftershocks triggered by smaller events. As a consequence, seismicity before large earthquakes presents no distinct features and does not provide information on the incoming mainshock. In particular, in the first decade of the 21st century, several studies (Helmstetter and Sornette, 2003; Helmstetter et al., 2003; Felzer et al., 2004; Hardebeck et al., 2008) have demonstrated that many foreshock features, first proposed as distinct features of precursory patterns, can be more simply explained by the ETAS model. These results have promoted the interpretation of foreshock occurrence in terms of normal aftershock triggering, i.e., a foreshock is an event which triggers a larger shock with no prognostic role. A striking example is represented (Hough, 2009) by the famous lectures 'The Life and Death of a Prediction Method' at the California State Fullerton University by David Bowman, one of the most active researchers in favour of the accelerated moment release hypothesis (Bufe and Varnes, 1993), who at the end of 2007 certified the 'death of the AMR prediction method'. In the last 5 years, however, several studies have enlightened quantitative differences between foreshock properties observed in instrumental and synthetic ETAS catalogues (Brodsky, 2011; Lippiello et al., 2012a; Bouchon et al., 2013; Shearer, 2012a, b; Hainzl, 2013; Shearer, 2013; Ogata and Katsura, 2014). In particular all the results indicate, in ETAS catalogues, a smaller foreshock-to-aftershock ratio than in instrumental data sets. This should suggest that fore−mainshock clustering may not be primarily caused by the main−aftershock triggering, even if the enhanced ratio can also be attributed to the deficit of aftershocks in instrumental catalogues because of STAI (Brodsky, 2011; Hainzl, 2013). On the other hand, the spatial organization of foreshocks in real catalogues presents not only quantitative but also qualitative differences with respect to the ETAS model (Lippiello et al., 2012a, 2017; Bouchon et al., 2013; Bouchon and Marsan, 2015). Indeed these studies show that, in instrumental catalogues on average, the distance of foreshocks from the mainshock hypocentre is much larger than that expected in the hypothesis of foreshocks triggering the mainshock, for both small $(m < 4)$ (Lippiello et al., 2012a, 2017) and large $(m > 6.5)$ (Bouchon et al., 2013) mainshocks.

In this section we present results indicating a strong similarity in the spatial organization of events between temporal periods preceding and following mainshocks. We also explore the role of the lower-magnitude threshold to support the final conclusion that the magnitude of the incoming mainshock is encoded in the foreshock spatial organization, a feature which cannot be explained by the ETAS model. The first step of this analysis is a clear definition of mainshocks, aftershocks and foreshocks. We define an event as a mainshock if a larger earthquake does not occur in the previous y days and within a distance L. In addition, a larger earthquake must not occur in the selected area in the following y_2 days. We implement typical values, $L = 100$ $L = 100$ km, $y = 3$ and $y_2 = 0.5$ (Felzer and Brodsky, 2006; Lippiello et al., 2009b, 2012a, 2017). Other choices, inside a reasonable range of parameters, provide similar results. Aftershocks and foreshocks are then defined as all events occurring, respectively, in the subsequent or in the preceding time interval $T = 12$ hours and within a circle of radius $R \leq 10$ km centred in the mainshock epicentre. Since we always take $T < y_2 < y$ and $R < L$, by construction aftershocks and foreshocks are smaller than the mainshock. We then group mainshocks in different classes according to their magnitudes $m \in [m_M, m_M + 1)$, and for each class we evaluate the number of mainshocks n_M, the number of aftershocks n_a and the number of foreshocks n_f. We also evaluate the distribution of distances $\rho(\Delta r, m_M)$ between the mainshock epicentre and aftershock or foreshock epicentres. More precisely, $\rho(\Delta r, m_M)$ is defined as the number of aftershocks (foreshocks) with epicentres at a distance in the interval $[\Delta r, 1.2\Delta r)$ from the mainshock, divided by $0.2\Delta r$ and by n_a (n_f). The normalization allows us to directly compare the functional form of the distribution for foreshocks ρ_f and aftershocks ρ_a, even if their number can be very different.

3.4.1 The Foreshock Productivity Law and the Inverse Omori Law

We start by considering the total number of identified aftershocks n_a and foreshocks n_f divided by the total number of mainshocks n_M. These quantities are plotted in Fig. 3–8 for the RSCEC catalogue, using open symbols for aftershocks and filled symbols for foreshocks. In all cases the error bar associated with the measured quantity is smaller than the symbol size. The results (Fig. 3–8) show that aftershocks obey a productivity law $K_a 10^{\alpha_a m_M}$ with $\alpha_a = 0.78 \pm 0.03$. Interestingly also, data for the foreshock number are consistent with a productivity law $K_f 10^{\alpha_f m_M}$, with K_f and $\alpha_f = 0.53 \pm 0.02$ smaller than K_a and α_a, respectively. In the same figure (Fig. 3–8) we also plot the results of numerical simulations of the ETAS model. As discussed in depth in Section 3.5 we set the parameters of the ETAS model in order to reproduce statistical features of events identified as aftershocks in the instrumental RSCEC. The results (Fig. 3–8) indeed show that aftershocks in the ETAS catalogue obey the productivity law with, roughly, the same value of K_a and α_a measured in the RSCEC catalogue. A productivity law is also found for foreshocks in the ETAS catalogue. In this case, however, the best-fit estimates of $\alpha_f \simeq 0.42$ are significantly smaller than that obtained for the RSCEC. This reflects important differences between experimental and numerical data in the ratio n_a/n_f as a function of m_M, with ETAS results always assuming larger values than instrumental results. Since n_a is tuned to reproduce experimental RSCEC data, the above observation implies a deficit of foreshocks in the ETAS catalogue. As already observed (Brodsky, 2011; Lippiello et al., 2012a), the deficit of foreshocks in the numerical catalogue can be related to STAI (Section 3.3). Indeed, since experimental catalogues present a deficit of

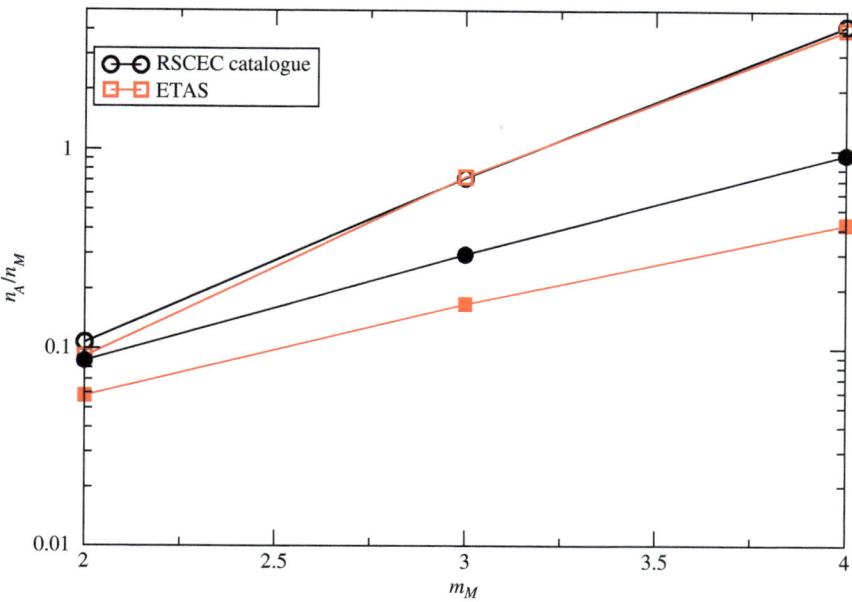

FIGURE 3–8 The number of $m > 2$ aftershocks n_a divided by the number of mainshocks n_M is plotted versus the magnitude class m_M. Black open circles are results obtained for the RSCEC catalogues, and red symbols are used for results for the ETAS catalogue with $a = 0.78$, which provides the best agreement with the RSCEC for the productivity law $n_a/n_M = K_a 10^{\alpha_a m_M}$. Filled symbols are used for the number of $m > 2$ foreshocks, black circles for the RSCEC and red squares for the ETAS catalogue with $a = 0.78$. The best-fit parameter α_f, from the foreshock productivity law $n_f/n_M = K_f 10^{\alpha_f m_M}$, is $\alpha_f = 0.53 \pm 0.02$ for RSCEC and $\alpha_f = 0.42 \pm 0.02$ for the ETAS catalogue. *ETAS*, epidemic type aftershock sequence.

events in the first part of the aftershock sequences (Lippiello et al., 2007b, 2012b; Bottiglieri et al., 2009) the experimental value of K_a is underestimated. This corresponds to underestimating the parameter A in the ETAS model [Eq. (3.7)], producing smaller values of K_f.

We next consider the number of events $n(t)$ as a function of the temporal distance $|t - t_M|$ from the mainshock occurrence time t_m. The results, plotted in Fig. 4.1, show that the number of events increases before t_M consistently with an inverse Omori law $n(t) \sim 10^{\alpha_f m_M}(t_m - t)^{-p_f}$ and decreases afterwards according to the Omori law $n(t) \sim 10^{\alpha_a m_M}(t - t_m)^{-p}$. More precisely we find $p_f \simeq p \simeq 0.7$. The above results are obtained including only events with $m > m_{th} = 2$, but we do not observe significant differences when other choices $m_{th} \in [1, 2.5]$ are considered. Similar results (Fig. 3–9) are recovered if the same analysis is performed on ETAS catalogues, confirming that the inverse Omori law can be an artefact of the erroneous foreshock–mainshock classification (Helmstetter and Sornette, 2003; Helmstetter et al., 2003; Felzer et al., 2004; Hardebeck et al., 2008). The only quantitative difference between instrumental and numerical catalogues is still related to the previously mentioned difference in the value of α_f. *ETAS*, epidemic type aftershock sequence.

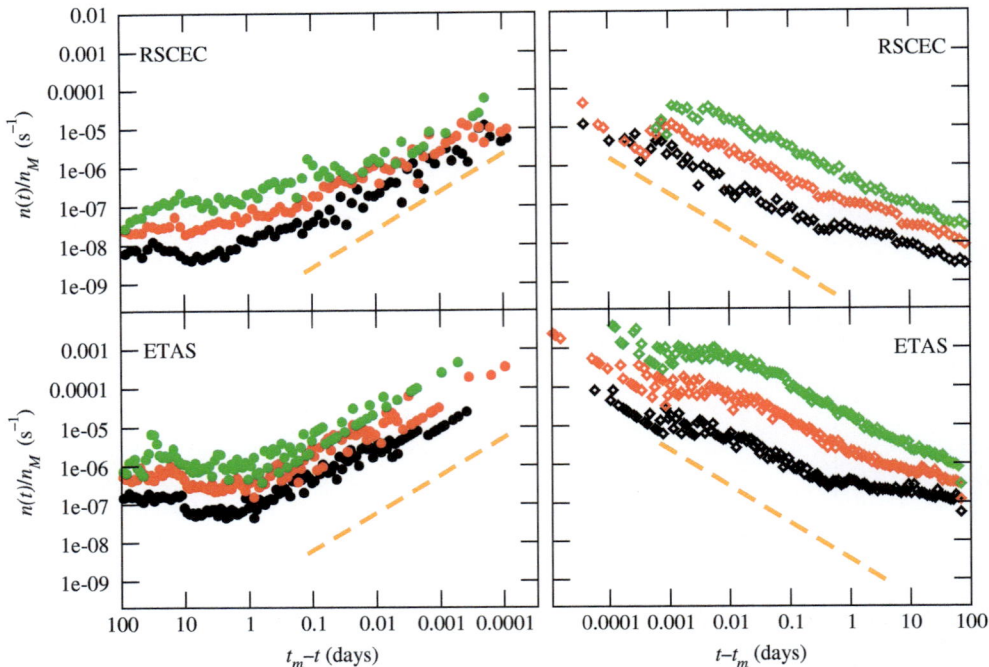

FIGURE 3–9 (Left upper panel) The number of $m > 2$ foreshocks in the RSCEC is divided by the number of mainshocks and plotted as a function of the temporal distance $t_m - t$ from the mainshock. Different colours correspond to different mainshock magnitude classes $m_M = 2$ (black), $m_M = 3$ [red] $m_M = 4$ [green]. (Right upper panel) The number of $m > 2$ aftershocks in the RSCEC is divided by the number of mainshocks and plotted as a function of the temporal distance $t - t_m$ from the mainshock. Different colours correspond to different mainshock magnitude classes $m_M = 2$ (black), $m_M = 3$ [red] $m_M = 4$ [green]. (Left lower panel) The same as the left upper panel for foreshocks in the ETAS catalogue. (Right lower panel) The same as the right upper panel for aftershocks in the ETAS catalogue. In all panels the orange line is the power law $|t - t_m|^{-1}$ drawn as a visual guide. *ETAS*, epidemic type aftershock sequence.

3.4.2 The Foreshock Spatial Distribution

We now turn to consider the aftershock and foreshock epicentral distributions. In Fig. 3–10 we plot $\rho_a(\Delta r, m_M)$ for RSCEC, which clearly depends on the mainshock magnitude m_M. More precisely, it is possible to show that ρ_a follows the scaling relation

$$\rho_a(\Delta r_i, m_i) = L(m_i) M\left(\frac{\Delta r_i}{L(m_i)}\right) \tag{3.29}$$

where $L(m_i) = L(m_c) 10^{\gamma(m_i - m_c)}$ represents the size of the aftershock zone with $\gamma = 0.42$ (Lippiello et al., 2009a). $\rho_a(\Delta r, m_M)$ is obtained considering all events with $m > m_{th} = 1$, but we do not observe significant differences if larger values of m_{th} are considered.

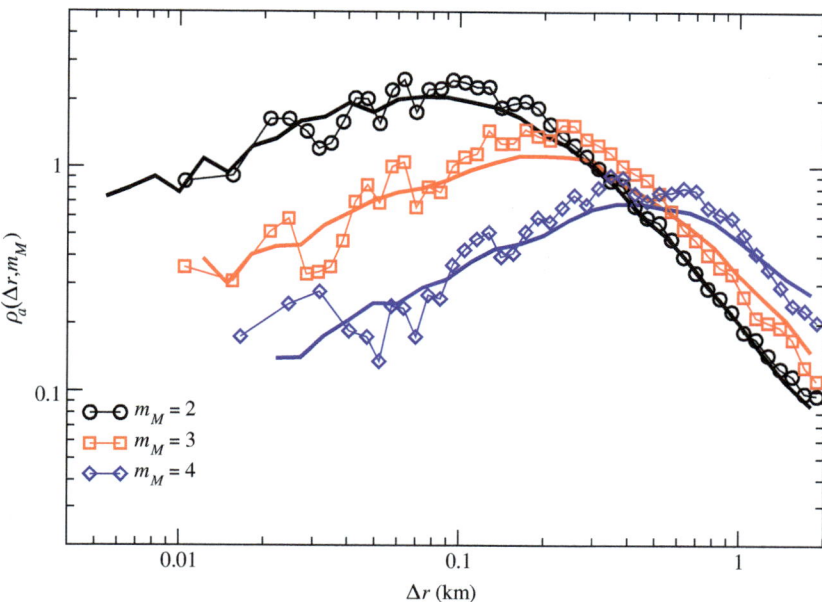

FIGURE 3–10 The aftershock epicentral linear distribution $\rho_a(\Delta r, m_M)$ in the RSCEC is plotted with open symbols for different $m_M = 2, 3, 4$ and $m_{th} = 1$. Continuous lines represent $\rho_a(\Delta r, m_M)$ in the ETAS catalogue for different values of m_M and $m_{th} = 1$. *ETAS*, epidemic type aftershock sequence.

In Fig. 3–11 we plot experimental data for the foreshock distribution $\rho_f(\Delta r, m_M)$. In the upper panel we plot results for both ρ_f and ρ_a for $m_{th} = 1$ and different values of m_M. The results indicate similar distributions of epicentral distances for seismicity before and after mainshocks. Indeed, we find $\rho_f(\Delta r, m_M) \simeq \rho_a(\Delta r, m_M)$ for all values of m_M and the whole considered spatial range. Therefore the epicentral foreshock distribution has a dependence on m_M similar to that [Eq. (3.29)] found for aftershocks. In the lower panel of Fig. 3–11 we plot $\rho_f(\Delta r, m_M)$ for two values of $m_{th} = 1, 2$ and two values of m_M and we observe that $\rho_f(\Delta r, m_M)$ is substantially independent of m_{th}.

We next explore the spatial distribution of aftershocks and foreshocks in the ETAS catalogue. The comparison between the aftershock spatial distribution in the ETAS catalogue and in the RSCEC is given in Fig. 3–10, which shows good agreement for all values of m_M. Both distributions clearly depend on the mainshock magnitude m_M and both follow the scaling relation [Eq. (3.29)]. This is not surprising since the spatial kernel $G(x)$ of the ETAS model [Eq. (3.7)] has been exactly chosen in order to have the same ρ_a in the ETAS and in the instrumental catalogue (see Section 3.5).

In numerical simulations couples of triggered and triggering events are exactly known. More precisely in Fig. 3–12 we plot the quantity $\rho_>(\Delta r, m_M)$ defined as the distribution of epicentral distances of all couples with the magnitude of the triggering earthquake $m \in [m_M, m_M)$ larger than the magnitude of the triggered earthquake. We also consider the

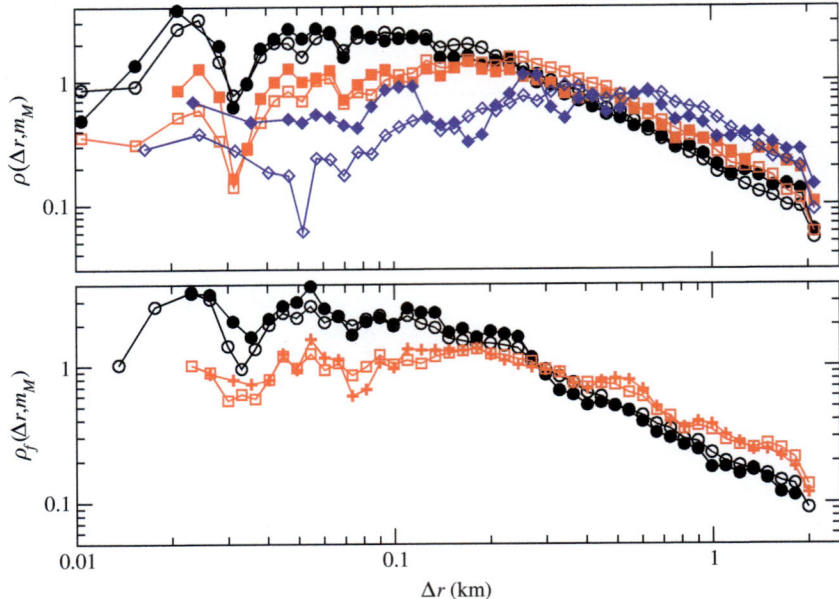

FIGURE 3–11 (Upper panel) The aftershock and foreshock epicentral linear distribution in the RSCEC. We plot with open symbols $\rho_a(\Delta r, m_M)$ and with filled symbols $\rho_f(\Delta r, m_M)$ for different mainshock magnitude classes m_M. The lower magnitude threshold is $m_{th} = 1$. (Lower panel) The foreshock epicentral linear distribution $\rho_f(\Delta r, m_M)$ in the RSCEC for different values of $m_{th} = 1, 2$ and $m_M = 2, 3$.

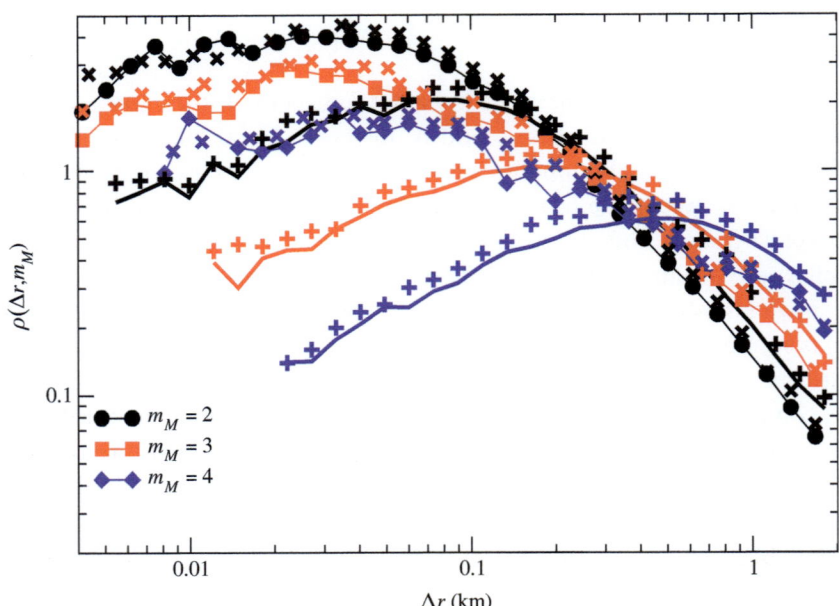

FIGURE 3–12 The linear density distributions $\rho_a(\Delta r, m_M)$ (continuous lines) and $\rho_f(\Delta r, m_M)$ (filled symbols) in the ETAS catalogue for different $m_M = 2, 3, 4$ and $m_{th} = 1$. We plot with pulses (crosses) the distributions $\rho_>(\Delta r, m_M)$ ($\rho_<(\Delta r, m_M)$).

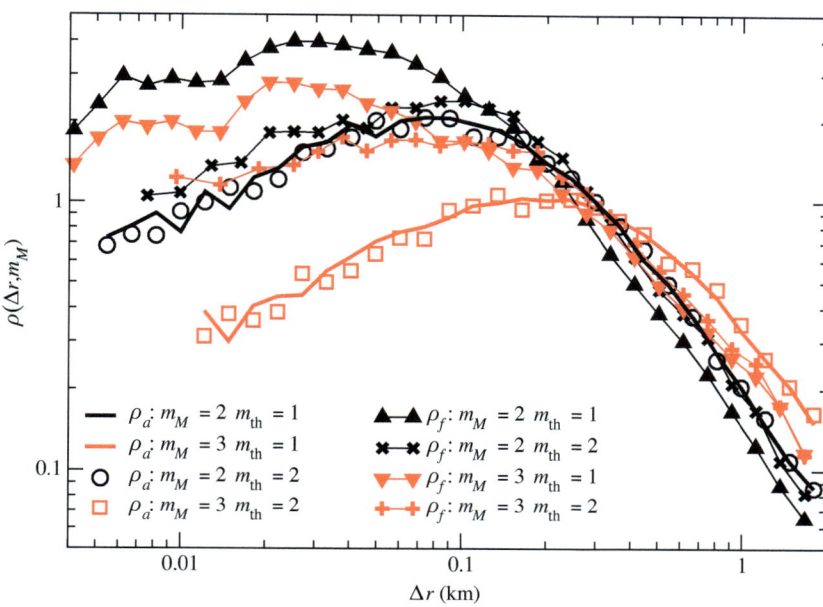

FIGURE 3–13 The aftershock and foreshock epicentral linear distribution in the ETAS catalogue is plotted for different values of m_M and m_{th}. More precisely, filled triangles and crosses are used for the foreshock distribution, whereas open symbols and continuous lines are used for the aftershock distribution. *ETAS*, epidemic type aftershock sequence.

quantity $\rho_<(\Delta r, m_M)$ which includes only couples where the triggered magnitude $m \in [m_M, m_M)$ is larger than the triggering one. The two distributions are compared with $\rho_f(\Delta r, m_M)$ (filled symbols) and $\rho_a(\Delta r, m_M)$ (continuous lines) obtained using the same definition of foreshocks, mainshocks and aftershocks used for the evaluation of ρ_f and ρ_a in the RSCEC. The results of Fig. 3−12 demonstrate that we have used a reasonable definition of aftershocks and indeed $\rho_>(\Delta r, m_M) \simeq \rho_a(\Delta r, m_M)$ for all values of m_M. Furthermore, we find $\rho_f(\Delta r, m_M) \simeq \rho_<(\Delta r, m_M)$, which enlightens that events identified as foreshocks in the ETAS catalogue are really earthquakes triggering a larger event. Fig. 3−12 shows significant differences between the RSCEC and the ETAS catalogue. Indeed, differently from RSCEC, in the ETAS catalogue $\rho_f(\Delta r, m_M) \neq \rho_a(\Delta r, m_M)$ and it only weakly depends on m_M. This is confirmed by results in Fig. 3−13 where we explore the behaviour of the linear distribution in the ETAS synthetic catalogue as a function of m_M and for different values of the lower-magnitude threshold $m_{th} = 1, 2$. The main observation is that $\rho_f(\Delta r, m_M)$ only weakly depends on m_M, whereas it exhibits a clear dependence on the lower-threshold m_{th}. These results have a simple interpretation. In the ETAS model, indeed, events identified as mainshocks and preceded by foreshocks are aftershocks with a magnitude larger than their triggering event, which is one of the earthquakes identified as foreshocks. This is confirmed by Fig. 3−11 indicating that $\rho_<(\Delta r, m_M) \simeq \rho_f(\Delta r, m_M)$. As a consequence, in the ETAS model, the fore−mainshock epicentral distances are not controlled by m_M but by the magnitude of the triggering foreshocks, which is in the interval $[m_{th}, m_M)$ leading to fore−mainshock epicentral distances strongly dependent on m_{th}. Furthermore, since foreshock magnitudes follow the GR law, the majority of foreshocks have small magnitudes $m \simeq m_{th}$ and only for few events $m \simeq m_M$. Therefore the fore−mainshock epicentral distance is only weakly affected by m_M.

3.4.3 The ETAFS Model

Previous results indicate that ETAS catalogues exhibit a systematic deficit of foreshocks with respect to instrumental ones. This discrepancy can be only partially attributed to STAI (Kagan, 2004; Helmstetter et al., 2006; Peng et al., 2006, 2007). However, it is not excluded that a better tuning of model parameters, as well as a more realistic numerical model, could contribute to reduce the discrepancy with experimental findings. A more striking difference between ETAS and instrumental catalogues concerns the foreshock organization in space. Results show that in instrumental catalogues the foreshock epicentral density ρ_f is independent of the lower-magnitude m_{th}, whereas it depends on the mainshock magnitude m_M. Both features are intrinsically impossible to recover in the ETAS model and, therefore, the observed differences in the ρ_f between numerical and instrumental data are expected for any choice of model parameters. Furthermore, the independence of ρ_f on m_{th} suggests that results for the spatial distribution in instrumental catalogues are not affected by incompleteness effects. The above findings therefore support the idea that, in order to provide a more faithful reproduction of instrumental data, synthetic catalogues should also contain, together with the standard main−aftershock triggering, additional earthquakes implementing foreshock occurrence. A more realistic description of instrumental catalogues could be provided by the epidemic type aftershocks and foreshock sequence (ETAFS) model, which implements foreshock occurrence into the usual ETAS model. In the ETAFS model each earthquake can trigger its own aftershocks with a probability λ as in Eq. (3.7) of the ETAS model. The new ingredient is that independent earthquakes, i.e., those occurring with a probability μ in Eq. (3.7), can be anticipated by a number of foreshocks accepted with a probability

$$Q_f(t, \vec{r}) = \frac{A'(p-1)}{c'} \sum_{i:t_i < t} 10^{a'm_i} \left(1 + \frac{t-t_i}{c'}\right)^{-p} G'\left(\left|\vec{r} - \vec{r}_i\right|\right). \tag{3.30}$$

This equation implements an inverse Omori law with the same p as for aftershock occurrence, in order to reduce the number of model parameters. There is no physical justification for it and we expect that very similar results can be recovered with other functional forms of temporal clustering. Analogously, there is no physical constraint on the other parameter A', c', a', which must be chosen in order to have, on average, not only the same number of aftershocks in instrumental and synthetic catalogues, as in ETAS simulations, but also the same number of foreshocks. The central point is that, as in the nucleation scenario (Ohnaka, 1992, 1993; Dodge et al., 1996) with the size of the nucleation zone scaling with the mainshock magnitude, the spatial distribution of foreshocks G' depends on the magnitude of the incoming mainshock. This ingredient can reproduce the statistical features of ρ_f found in the instrumental catalogues.

3.5 Numerical Implementation of the ETAS Model

The fastest procedure to generate synthetic ETAS catalogues is by means of a parallel algorithm (Zhuang et al., 2004). The first step corresponds to the zeroth-order generation, i.e., the generation of independent background events corresponding to the term μ in the right-hand side of Eq. (3.7). For a fixed duration of the catalogue T $N_m = \mu T$ independent main-shocks are randomly located within the temporal interval $[0,T]$. This corresponds to assume a stationary Poisson background rate μ in Eq. (3.4) with $\int d\vec{r} \; \mu(\vec{r}) = N_m/T$. For the spatial position of the mainshocks, one constructs a fine space-covering grid of N_c cells and main-shocks are located within a cell with probability $\mu(\vec{r}_i)$, where $\vec{r}_i$ is the position of the cell centre. The magnitude of each event is randomly extracted from an exponential distribution, i.e., the GR law. The second step corresponds to the generation of first-order aftershocks. Each event i, generated at the previous step, produces a number of aftershocks $n_a(m_i)$ extracted from a Poisson distribution with average $\langle n_a(m_i) \rangle = 10^{am_i}$, where A and a are the coefficients of the productivity law in Eq. (3.4). The temporal $(t - t_i)$ and spatial (x_{ij}) distance from the mother event, of the first-order generation aftershocks, is randomly extracted with probability $(t-t_i+c)^p$ and $G(x_{ij})$, respectively. Isotropic aftershock distribution is also assumed and magnitudes are always assigned according to the GR law. Once all first-generation aftershocks have been triggered, the process is iterated at the subsequent step, considering as the mother event the first-order aftershocks. The process is then iterated until no further aftershocks are triggered. A final sorting procedure is necessary to temporally order events. The effect of STAI can be included in the ETAS model under the assumption that each event is followed by a blind time Δt removing from the original ETAS catalogue all events occurring at a temporal distance less than Δt after a larger event. In this way we generate ETASI catalogues. The blind time can be considered constant or dependent on the magnitude difference between each couple of events.

Concerning the choice of parameters, A and a in Eq. (3.3) can be fixed imposing that the simulated catalogue has the same after-to-mainshock ratio as the considered instrumental catalogue. More precisely, the first observation (Lippiello et al., 2017) is that for any choice of a, the ratio n_a/n_M for events identified as aftershocks, follows the productivity law $n_a/n_M \simeq K_a 10^{\alpha_a m_M}$ with $\alpha_a \simeq a$. Therefore, this fixes a to the value α_a measured in the instrumental catalogue. The value of A is then fixed in order to have the same value of K_a in the synthetic and the reference instrumental catalogue. The choices of other parameters p and c are quite arbitrary and usually one chooses $p \gtrsim 1$ and a small c value. Finally, the number of independent background events is chosen in order to have a similar number of events in the synthetic and instrumental catalogues and their spatial distribution is obtained from the smoothed distribution of instrumental earthquakes.

Concerning the spatial kernel $G(x_{ij})$, this is chosen implementing in the numerical code the experimental scaling relationship [Eq. (3.29)]. In particular, we fit the functional form of $F(x)$ from the following procedure. Starting from the aftershock spatial distribution $\rho_a(\Delta r, m_M)$ we evaluate $h(\Delta r) = \int_0^{\Delta r} d\Delta r' \rho_a(\Delta r', m_M)$ in the instrumental catalogue and extract a random number $x*$ uniformly distributed in $[0, 1]$. After numerically inverting the relationship $h(\Delta r*) = x*$ we obtain a random number $\Delta r*$ distributed according to $\rho_a(\Delta r*, m_M)$. Finally, assuming Eq. (3.29), the epicentral distance from a triggering earthquake of magnitude m_i is given by $\Delta r = \Delta r * 10^{\gamma(m_i - m_M)}$. We have considered $m_M = 4$ as a good compromise between an extended aftershock zone and good statistics.

References

Baiesi, M., Paczuski, M., 2005. Complex networks of earthquakes and aftershocks. Nonlinear Processes Geophys. 12 (1), 1−11. Available from: https://doi.org/10.5194/npg-12-1-2005. URL http://www.nonlin-processes-geophys.net/12/1/2005/.

Beeler, N.M., Lockner, D.A., 2003. Why earthquakes correlate weakly with the solid earth tides: effects of periodic stress on the rate and probability of earthquake occurrence. J. Geophys. Res.: Solid Earth 108 (B8), 2391. Available from: https://doi.org/10.1029/2001JB001518. URL https://doi.org/10.1029/2001JB001518.

Bottiglieri, M., Lippiello, E., Godano, C., de Arcangelis, L., 2009. Identification and spatiotemporal organization of aftershocks. J. Geophys. Res.: Solid Earth 114 (B3), B03303. Available from: https://doi.org/10.1029/2008JB005941. URL https://doi.org/10.1029/2008JB005941.

Bottiglieri, M., Lippiello, E., Godano, C., de Arcangelis, L., 2011. Comparison of branching models for seismicity and likelihood maximization through simulated annealing. J. Geophys. Res.: Solid Earth 116 (B2), n/a−n/a, b02303. https://doi.org/10.1029/2009JB007060. URL https://doi.org/10.1029/2009JB007060.

Bouchon, M., Marsan, D., 2015. Reply to artificial seismic acceleration. Nat. Geosci. 8, 83.

Bouchon, M., Durand, V., Marsan, D., Karabulut, H., Schmittbuhl, J., 2013. The long precursory phase of most large interplate earthquakes. Nat. Geosci. 6, 299302.

Brodsky, E.E., 2011. The spatial density of foreshocks. Geophys. Res. Lett. 38 (10), L10305, l10305. https://doi.org/10.1029/2011GL047253. URL https://doi.org/10.1029/2011GL047253.

Brodsky, E.E., Lay, T., 2014. Recognizing foreshocks from the 1 April 2014 Chile earthquake. Science 344 (6185), 700−702.

Bufe, C.G., Varnes, D.J., 1993. Predictive modeling of the seismic cycle of the greater san Francisco bay region. J. Geophys. Res.: Solid Earth 98 (B6), 9871−9883. Available from: https://doi.org/10.1029/93JB00357. URL https://doi.org/10.1029/93JB00357.

Chris, M., 1998. Laboratory-derived friction laws and their application to seismic faulting. Annu. Rev. Earth Planet. Sci. 26, 643−696. Available from: https://doi.org/10.1146/annurev.earth.26.1.643.

de Arcangelis, L., Godano, C., Grasso, J.R., Lippiello, E., 2016. Statistical physics approach to earthquake occurrence and forecasting. Phys. Rep. 628, 1−91. statistical physics approach to earthquake occurrence and forecasting. https://doi.org/10.1016/j.physrep.2016.03.002. URL www.sciencedirect.com/science/article/pii/S0370157316300011.

Dieterich, J.H., 1972. Time-dependent friction as a possible mechanism for aftershocks. J. Geophys. Res. 77 (20), 3771−3781. Available from: https://doi.org/10.1029/JB077i020p03771. URL https://doi.org/10.1029/JB077i020p03771.

Dieterich, J.H., 1994. A constitutive law for rate of earthquake production and its application to earthquake clustering. J. Geophys. Res.: Solid Earth 99 (B2), 2601−2618. Available from: https://doi.org/10.1029/93JB02581. URL https://doi.org/10.1029/93JB02581.

Dodge, D.A., Beroza, G.C., Ellsworth, W.L., 1996. Detailed observations of California foreshock sequences: implications for the earthquake initiation process. J. Geophys. Res.: Solid Earth 101 (B10), 22371−22392. Available from: https://doi.org/10.1029/96JB02269. URL https://doi.org/10.1029/96JB02269.

Enescu, B., Mori, J., Miyazawa, M., 2007. Quantifying early aftershock activity of the 2004 mid-Niigata prefecture earthquake (mw6.6). J. Geophys. Res.: Solid Earth 112 (B4), B04310. Available from: https://doi.org/10.1029/2006JB004629. URL https://doi.org/10.1029/2006JB004629.

Felzer, K.R., Brodsky, E.E., 2006. Decay of aftershock density with distance indicates triggering by dynamic stress. Nature 441, 735–738.

Felzer, K.R., Abercrombie, R.E., Ekström, G., 2003. Secondary aftershocks and their importance for aftershock forecasting. Bull. Seismol. Soc. Am. 93 (4), 1433–1448. arXiv:http://bssa.geoscienceworld.org/content/93/4/1433.full.pdf, https://doi.org/10.1785/0120020229. URL http://bssa.geoscienceworld.org/content/93/4/1433.

Felzer, K.R., Abercrombie, R.E., Ekstrm, G., 2004. A common origin for aftershocks, foreshocks, and multiplets. Bull. Seismol. Soc. Am. 94 (1), 88–98. Available from: https://doi.org/10.1785/0120030069.

Felzer, K.R., Page, M.T., Michael, A.J., 2015. Artificial seismic acceleration. Nat. Geosci. 8, 82–83.

Gu, C., Schumann, A.Y., Baiesi, M., Davidsen, J., 2013. Triggering cascades and statistical properties of aftershocks. J. Geophys. Res.: Solid Earth 118 (8), 4278–4295. Available from: https://doi.org/10.1002/jgrb.50306. URL https://doi.org/10.1002/jgrb.50306.

Hainzl, S., 2013. Comment on self-similar earthquake triggering, Båth's Law, and foreshock/aftershock magnitudes: simulations, theory, and results for southern California by P. M. Shearer. J. Geophys. Res.: Solid Earth 118 (3), 1188–1191. Available from: https://doi.org/10.1002/jgrb.50132. URL https://doi.org/10.1002/jgrb.50132.

Hainzl, S., 2016a. Apparent triggering function of aftershocks resulting from rate-dependent incompleteness of earthquake catalogs. J. Geophys. Res.: Solid Earth 121 (9), 6499–6509. 2016JB013319. https://doi.org/10.1002/2016JB013319. URL https://doi.org/10.1002/2016JB013319.

Hainzl, S., 2016b. Rate dependent incompleteness of earthquake catalogs. Seismol. Res. Lett. 87 (2A), 337–344.

Hainzl, S., Zller, G., Kurths, J., 1999. Similar power laws for foreshock and aftershock sequences in a spring-block model for earthquakes. J. Geophys. Res.: Solid Earth 104 (B4), 7243–7253. Available from: https://doi.org/10.1029/1998JB900122. URL https://doi.org/10.1029/1998JB900122.

Hainzl, S., Christophersen, A., Enescu, B., 2008. Impact of earthquake rupture extensions on parameter estimations of point-process models. Bull. Seismol. Soc. Am. 98 (4), 2066–2072. arXiv:http://www.bssaonline.org/content/98/4/2066.full.pdf + html, https://doi.org/10.1785/0120070256. URL http://www.bssaonline.org/content/98/4/2066.abstract.

Hardebeck, J.L., Felzer, K.R., Michael, A.J., 2008. Improved tests reveal that the accelerating moment release hypothesis is statistically insignificant. J. Geophys. Res.: Solid Earth 113 (B8), B08310. Available from: https://doi.org/10.1029/2007JB005410. URL https://doi.org/10.1029/2007JB005410.

Hauksson, E., Shearer, P., Yang, W., 2012. Waveform relocated earthquake catalog for southern California (1981 to June 2011). Bull. Seismol. Soc. Am. 102 (5), 2239–2244. Available from: https://doi.org/10.1785/0120120010.

Helmstetter, A., 2003. Is earthquake triggering driven by small earthquakes? Phys. Rev. Lett. 91, 058501. Available from: https://doi.org/10.1103/PhysRevLett.91.058501. URL http://link.aps.org/doi/10.1103/PhysRevLett.91.058501.

Helmstetter, A., Shaw, B.E., 2009. Afterslip and aftershocks in the rate-and-state friction law. J. Geophys. Res.: Solid Earth 114 (B1), B01308. Available from: https://doi.org/10.1029/2007JB005077. URL https://doi.org/10.1029/2007JB005077.

Helmstetter, A., Sornette, D., 2003. Foreshocks explained by cascades of triggered seismicity. J. Geophys. Res.: Solid Earth 108 (B10), 2457. Available from: https://doi.org/10.1029/2003JB002409. URL https://doi.org/10.1029/2003JB002409.

Helmstetter, A., Sornette, D., Grasso, J.-R., 2003. Mainshocks are aftershocks of conditional foreshocks: how do foreshock statistical properties emerge from aftershock laws. J. Geophys. Res.: Solid Earth 108 (B1), 2046. https://doi.org/10.1029/2002JB001991. URL https://doi.org/10.1029/2002JB001991.

Helmstetter, A., Kagan, Y.Y., Jackson, D.D., 2005. Importance of small earthquakes for stress transfers and earthquake triggering. J. Geophys. Res.: Solid Earth 110 (B5), B05S08. Available from: https://doi.org/10.1029/2004JB003286.

Helmstetter, A., Kagan, Y.Y., Jackson, D.D., 2006. Comparison of short-term and time-independent earthquake forecast models for southern California. Bull. Seismol. Soc. Am. 96 (1), 90−106. arXiv:http://www.bssaonline.org/content/96/1/90.full.pdf + html, https://doi.org/10.1785/0120050067. URL http://www.bssaonline.org/content/96/1/90.abstract.

Hough, S., 2009. Predicting the Unpredictable: The Tumultuous Science of Earthquake Prediction. Princeton University Press, Princeton, NJ.

Kagan, Y.Y., 2004. Short-term properties of earthquake catalogs and models of earthquake source. Bull. Seismol. Soc. Am. 94 (4), 1207−1228.

Kagan, Y.Y., Jackson, D.D., 1995. New seismic gap hypothesis: five years after. J. Geophys. Res.: Solid Earth 100 (B3), 3943−3959. Available from: https://doi.org/10.1029/94JB03014.

Landes, F.P., Lippiello, E., Scaling laws in earthquake occurrence: disorder, viscosity, finite size effects in Olami−Feder−Christensen models. Phys. Rev. E 93 (2016) 051001. https://doi.org/10.1103/PhysRevE.93.051001.

Lippiello, E., Godano, C., de Arcangelis, L., 2007a. Dynamical scaling in branching models for seismicity. Phys. Rev. Lett. 98, 098501. Available from: https://doi.org/10.1103/PhysRevLett.98.098501. URL http://link.aps.org/doi/10.1103/PhysRevLett.98.098501.

Lippiello, E., Bottiglieri, M., Godano, C., de Arcangelis, L., 2007b. Dynamical scaling and generalized Omori law. Geophys. Res. Lett. 34 (23), L23301. Available from: https://doi.org/10.1029/2007GL030963. URL https://doi.org/10.1029/2007GL030963.

Lippiello, E., de Arcangelis, L., Godano, C., 2008. Influence of time and space correlations on earthquake magnitude. Phys. Rev. Lett. 100, 038501. Available from: https://doi.org/10.1103/PhysRevLett.100.038501. URL http://link.aps.org/doi/10.1103/PhysRevLett.100.038501.

Lippiello, E., de Arcangelis, L., Godano, C., 2009a. Role of static stress diffusion in the spatiotemporal organization of aftershocks. Phys. Rev. Lett. 103, 038501. Available from: https://doi.org/10.1103/PhysRevLett.103.038501. URL http://link.aps.org/doi/10.1103/PhysRevLett.103.038501.

Lippiello, E., de Arcangelis, L., Godano, C., 2009b. Time, space and magnitude correlations in earthquake occurrence. Int. J. Mod. Phys. B 23 (28n29), 5583−5596. arXiv:http://www.worldscientific.com/doi/pdf/10.1142/S0217979209063870, https://doi.org/10.1142/S0217979209063870. URL http://www.worldscientific.com/doi/abs/10.1142/S0217979209063870.

Lippiello, E., Marzocchi, W., de Arcangelis, L., Godano, C., 2012a. Spatial organization of foreshocks as a tool to forecast large earthquakes. Sci. Rep. 2, 1−6. URL https://doi.org/10.1038/srep00846.

Lippiello, E., Godano, C., de Arcangelis, L., 2012b. The earthquake magnitude is influenced by previous seismicity. Geophys. Res. Lett. 39 (5), L05309. Available from: https://doi.org/10.1029/2012GL051083. URL https://doi.org/10.1029/2012GL051083.

Lippiello, E., Giacco, F., Arcangelis, L.D., Marzocchi, W., Godano, C., 2014. Parameter estimation in the ETAS model: approximations and novel methods. Bull. Seismol. Soc. Am. 104 (2), 985−994. arXiv:http://www.bssaonline.org/content/104/2/985.full.pdf + html, https://doi.org/10.1785/0120130148. URL http://www.bssaonline.org/content/104/2/985.abstract.

Lippiello, E., Giacco, F., Marzocchi, W., Godano, C., de Arcangelis, L., 2015. Mechanical origin of aftershocks. Sci. Rep. 5, 1−6.

Lippiello, E., Cirillo, A., Godano, G., Papadimitriou, E., Karakostas, V., 2016. Real-time forecast of aftershocks from a single seismic station signal. Geophys. Res. Lett. 43 (12), 6252−6258. 2016GL069748. https://doi.org/10.1002/2016GL069748. URL https://doi.org/10.1002/2016GL069748.

Lippiello, E., Giacco, F., Marzocchi, W., Godano, C., Arcangelis, L.D., 2017. Statistical features of foreshocks in instrumental and etas catalogs. Pure Appl. Geophys. 1−19. Available from: https://doi.org/10.1007/s00024-017-1502-5. URL https://doi.org/10.1007/s00024-017-1502-5.

Lippiello, E., Godano, C., de Arcangelis, L., 2018. The intrinsic incompleteness of earthquake catalogs and aftershock forecasting, Submitted to Geophys. Res. Lett.

Mignan, A., 2014. The debate on the prognostic value of earthquake foreshocks: a meta-analysis. Sci. Rep. 4, 4099−4103.

Moradpour, J., Hainzl, S., Davidsen, J., 2014. Nontrivial decay of aftershock density with distance in southern California. J. Geophys. Res.: Solid Earth 119 (7), 5518−5535. Available from: https://doi.org/10.1002/2014JB010940. URL https://doi.org/10.1002/2014JB010940.

Nakanishi, H., 1990. Cellular-automaton model of earthquakes with deterministic dynamics. Phys. Rev. A 41, 7086−7089. Available from: https://doi.org/10.1103/PhysRevA.41.7086. URL http://link.aps.org/doi/10.1103/PhysRevA.41.7086.

Nakanishi, H., 1991. Statistical properties of the cellular-automaton model for earthquakes. Phys. Rev. A 43, 6613−6621. Available from: https://doi.org/10.1103/PhysRevA.43.6613. URL http://link.aps.org/doi/10.1103/PhysRevA.43.6613.

Nakanishi, H., 1992. Earthquake dynamics driven by a viscous fluid. Phys. Rev. A 46, 4689−4692. Available from: https://doi.org/10.1103/PhysRevA.46.4689. URL http://link.aps.org/doi/10.1103/PhysRevA.46.4689.

Ogata, Y., 1988. Statistical models for earthquake occurrences and residual analysis for point processes. J. Am. Stat. Assoc. 83, 9−27.

Ogata, Y., 1998. Space-time point-process models for earthquake occurrences. Ann. Inst. Stat. Math. 50, 379402.

Ogata, Y., Katsura, K., 2014. Comparing foreshock characteristics and foreshock forecasting in observed and simulated earthquake catalogs. J. Geophys. Res.: Solid Earth 119 (11), 8457−8477. 2014JB011250. https://doi.org/10.1002/2014JB011250. URL https://doi.org/10.1002/2014JB011250.

Ohnaka, M., 1992. Earthquake source nucleation: a physical model for short-term precursors. Tectonophysics 211 (14), 149−178.

Ohnaka, M., 1993. Critical size of the nucleation zone of earthquake rupture inferred from immediate foreshock activity. J. Phys. Earth 41 (1), 45−56. Available from: https://doi.org/10.4294/jpe1952.41.45.

Omi, T., Ogata, Y., Hirata, Y., Aihara, K., 2013. Forecasting large aftershocks within one day after the main shock. Sci. Rep. 3, 2218.

Omi, T., Ogata, Y., Shiomi, K., Enescu, B., Sawazaki, K., Aihara, K., 2016. Automatic aftershock forecasting: A test using realtime seismicity data in Japan automatic aftershock forecasting: A test using realtime seismicity data in Japan. Bull. Seismol. Soc. Am. 106 (6), 2450. Available from: https://doi.org/10.1785/0120160100.

Omori, F., 1894. On the after-shocks of earthquakes. J. Coll. Sci., Imp. Univ. Tokyo 7, 111−200.

Parsons, T., Velasco, A.A., 2009. On near-source earthquake triggering. J. Geophys. Res.: Solid Earth (1978−2012) 114 (B10), B10307.

Pelletier, J., 2000. Spring-block models of seismicity: review and analysis of a structurally heterogeneous model coupled to a viscous asthenosphere. In: Rundle, J.B., Turcotte, D.L., Klein, W. (Eds.), GeoComplexity and the Physics of Earthquakes Geophysical Monograph. American Geophysical Union, Washington, DC, pp. 120−128.

Peng, Z., Zhao, P., 2009. Migration of early aftershocks following the 2004 Parkfield earthquake. Nat. Geosci. 2 (12), 877−881.

Peng, Z., Vidale, J.E., Houston, H., 2006. Anomalous early aftershock decay rate of the 2004 Mw6.0 Parkfield, California, earthquake. Geophys. Res. Lett. 33 (17), L17307, l17307. https://doi.org/10.1029/2006GL026744. URL https://doi.org/10.1029/2006GL026744.

Peng, Z., Vidale, J.E., Ishii, M., Helmstetter, A., 2007. Seismicity rate immediately before and after main shock rupture from high-frequency waveforms in Japan. J. Geophys. Res.: Solid Earth 112 (B3), n/a–n/a, b03306. https://doi.org/10.1029/2006JB004386. URL https://doi.org/10.1029/2006JB004386.

Perfettini, H., Avouac, J.-P., 2004. Postseismic relaxation driven by brittle creep: a possible mechanism to reconcile geodetic measurements and the decay rate of aftershocks, application to the Chi-Chi earthquake, Taiwan. J. Geophys. Res.: Solid Earth 109 (B2), B02304. Available from: https://doi.org/10.1029/2003JB002488. URL https://doi.org/10.1029/2003JB002488.

Rabinowicz, E., 1951. The nature of the static and kinetic coefficients of friction. J. Appl. Phys. 22, 1373–1379.

Rabinowicz, E., 1956. Autocorrelation analysis of the sliding process. J. Appl. Phys. 27, 131–135.

Rabinowicz, E., 1958. The intrinsic variables affecting the stick-slip process. Proc. Phys. Soc. 71, 668–675.

Reasenberg, P.A., Jones, L.M., 1989. Earthquake hazard after a mainshock in California. Science 243 (4895), 1173–1176. arXiv:http://www.sciencemag.org/content/243/4895/1173.full.pdf, https://doi.org/10.1126/science.243.4895.1173. URL http://www.sciencemag.org/content/243/4895/1173.abstract.

Richards-Dinger, K., Stein, R., Toda, S., 2010. Decay of aftershock density with distance does not indicate triggering by dynamic stress. Nature 467, 0402.

Ruina, A., 1983. Slip instability and state variable friction laws. J. Geophys. Res.: Solid Earth 88 (B12), 10359–10370. Available from: https://doi.org/10.1029/JB088iB12p10359. URL https://doi.org/10.1029/JB088iB12p10359.

Schaff, D.P., Beroza, G.C., Shaw, B.E., 1998. Postseismic response of repeating aftershocks. Geophys. Res. Lett. 25 (24), 4549–4552. Available from: https://doi.org/10.1029/1998GL900192. URL https://doi.org/10.1029/1998GL900192.

Schoenberg, F.P., 2013. Facilitated estimation of ETAS. Bull. Seismol. Soc. Am. 103 (1), 601–605. arXiv:http://www.bssaonline.org/content/103/1/601.full.pdf + html, https://doi.org/10.1785/0120120146. URL http://www.bssaonline.org/content/103/1/601.abstract.

Scholz, C.H., 2002. The Mechanics of Earthquakes and Faulting. Cambridge University Press, New York, NY.

Shcherbakov, R., Turcotte, D.L., Rundle, J.B., 2004. A generalized Omori's law for earthquake aftershock decay. Geophys. Res. Lett. 31 (11), L11613. Available from: https://doi.org/10.1029/2004GL019808. URL https://doi.org/10.1029/2004GL019808.

Shearer, P.M., 2013. Reply to comment by S. Hainzl on self-similar earthquake triggering, Båth's Law, and foreshock/aftershock magnitudes: simulations, theory and results for southern California. J. Geophys. Res.: Solid Earth 118 (3), 1192–1192. https://doi.org/10.1002/jgrb.50133. URL https://doi.org/10.1002/jgrb.50133.

Shearer, P.M., 2012a. Self-similar earthquake triggering, Bath's Law, and foreshock/aftershock magnitudes: simulations, theory, and results for southern California, J. Geophys. Res.—Solid Earth 117, n/a. https://doi.org/10.1029/2011jb008957. URL Go to ISI://WOS:000305638400001.

Shearer, P.M., 2012b. Space-time clustering of seismicity in California and the distance dependence of earthquake triggering, J. Geophys. Res.—Solid Earth 117, n/a.

Stein, R.S., 1999. The role of stress transfer in earthquake occurrence. Nature 402 (6762), 605–609.

Turcotte, D.L., Schubert, G., 2002. Geodynamics. Cambridge University Press, Cambridge.

Utsu, T., 1970. Aftershocks and earthquake statistics (ii)—further investigation of aftershocks and other earthquake sequences based on a new classification of earthquake sequences. J. Fac. Sci. Hokkaido Univ., Ser. VII 3, 197–266.

Utsu, T., Seki, A., 1954. A relation between the area of aftershock region and the energy of mainshock. J. Seismol. Soc. Jpn 7, 233–240.

Utsu, T., Ogata, Y., Matsu'ura, R.S., 1995. The centenary of the Omori formula for a decay law of aftershock activity. J. Phys. Earth 43 (1), 1–33. Available from: https://doi.org/10.4294/jpe1952.43.1.

Werner, M.J., Helmstetter, A., Jackson, D.D., Kagan, Y.Y., 2011. High-resolution long-term and short-term earthquake forecasts for California. Bull. Seismol. Soc. Am. 101 (4), 1630−1648. arXiv:http://www.bssaonline.org/content/101/4/1630.full.pdf + html, https://doi.org/10.1785/0120090340. URL http://www.bssaonline.org/content/101/4/1630.abstract.

Zhuang, J., Ogata, Y., Vere-Jones, D., 2002. Stochastic declustering of space-time earthquake occurrences. J. Am. Stat. Assoc. 97, 369380.

Zhuang, J., Ogata, Y., Vere-Jones, D., 2004. Analyzing earthquake clustering features by using stochastic reconstruction. J. Geophys. Res.: Solid Earth 109 (B5), B05301. Available from: https://doi.org/10.1029/2003JB002879. URL https://doi.org/10.1029/2003JB002879.

Fractal, Informational and Topological Methods for the Analysis of Discrete and Continuous Seismic Time Series: An Overview

Luciano Telesca

NATIONAL RESEARCH COUNCIL, INSTITUTE OF METHODOLOGIES FOR ENVIRONMENTAL ANALYSIS, TITO (PZ), ITALY

CHAPTER OUTLINE

4.1 Introduction

Observational time series are the physical means by which information about a natural process is conveyed. On the basis of time structure, time series can be time-continuous or point process. A time-continuous series is a generally regularly sampled signal measuring a specific property (electrical, chemical, magnetic, hydrologic, etc.) of the natural process; for this type of series the information is stored in the amplitude, the variation of which with time indicates the changing dynamics underlying the mechanism governing that process. Point processes are discrete processes of events randomly occurring in time; for these processes the information is stored not only in the amplitude but also in the occurrence time.

Examples of natural point processes are earthquakes, rain events, eruptions, lightning, etc. They are mathematically expressed by a sum of Dirac's delta centred on the occurrence time with amplitude proportional to the event intensity that, for instance, is the magnitude for earthquakes, the volcanic explosivity index for eruptions, or the water column for rain.

Time series can be classified on the basis of their correlation properties. A purely random signal, which is a realization of a white-noise process, is characterized by independence of any sample and absence of any memory phenomena. In contrast, a correlated series can be persistent or antipersistent, where persistence indicates that the increments of the series tend to maintain the same sign, depicting the time series as smoothly varying; while antipersistence indicates that the increments of the series tend to alternate between positive and negative signs, therefore depicting the series as rapidly fluctuating.

Another classification of observational series is based on their topological organization. Let's consider the annual sunspot time series from 1830 to 1880 (Fig. 4−1); if we connect each value with any other (i.e. visible to the previous one) we get a network of links or a graph. Most of the temporal points of the decreasing phase of one solar cycle are connected to points of the increasing phase of the next cycle. Therefore, the graph is clustered into communities, each of which mainly consists of the temporal information of two subsequent solar cycles. When the sunspot number reaches a larger but more infrequent extreme maximum, inter-community connections appear, since these extreme maxima have better visibility contact with more neighbours than other time points, forming hubs in the graph (Zou et al., 2014).

Time series can, furthermore, be classified on the basis of their structural organization: they can be more or less organized or ordered. For instance, a series drawn by a normally distributed distribution can be considered more organized and ordered than that drawn by a uniformly distributed process. In fact, the normal distribution process is more peaked than the uniform one, thus implying that all the states of the process are not equally probable, but there is a tendency for the process to be in the average state more frequently.

These three different classifications are reflected into three different types of statistical methodologies for the investigation of observational time series: fractal, informational and visibility graph (VG)-based methods linked with correlation, structural and topological classifications of time series, respectively.

In this chapter, we will present an overview of the more advanced statistical methodologies relying on these three different classifications of time series that have been applied for the investigation of time dynamics of earthquake processes and earthquake-related signals.

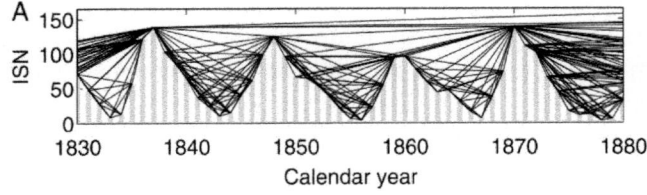

FIGURE 4–1 Time series of the annual sunspot number (grey bars) (from Zou et al., 2014); the black lines connects two time points fulfilling the visibility condition (see Section 4.4).

Within the context of this chapter, for seismic time series we mean sequences of earthquakes and seismograms. If seismograms are obviously time-continuous time series, earthquake sequences are temporal stochastic point processes marked by their magnitude. However, a discrete-time earthquake process can be derived from the stochastic earthquake point process in two equivalent ways: (1) using the interevent time (IET) series, or (2) forming its relative counting process. In the first representation a discrete-time series is formed by the rule $\tau_i = t_{i+1} - t_i$, where t_i indicates the time of the earthquake numbered by the index i. In the second representation, the time axis is divided into equally spaced contiguous counting windows of duration T to produce a sequence of counts $\{N_k(T)\}$, where $N_k(T)$ represents the number of earthquakes falling into the kth window of duration T. The duration T of the window is called the counting time or timescale. The latter approach considers earthquakes as the events of interest, assuming the existence of an objective clock for the timing of the events. The first approach emphasizes the interevent intervals using the event number as an index of the time. Both representations permit different statistical techniques to be performed in order to investigate the temporal fluctuations of a seismic process.

4.2 Fractal Methods

The quantitative characterization of earthquake dynamics is performed using methodologies able to extract robustly features that are hidden in the complexity of their time fluctuations. Fractality is one of the aspects of such complexity. A fractal is an object whose sample path included within some radius scales with the size of the radius. This definition configures a fractal process as characterized by a scaling behaviour, and this implies that statistics employed to describe a fractal have to be a power-law. In fact, if $f(x)$ is the statistics describing the fractal, and if the changing the scale x by a factor a effectively scales the statistics $f(x)$ by a factor $g(a)$, $f(ax) = g(a)f(x)$, then $f(x) = bg(x)$, $g(x) = x^c$, for some constants b and c (Thurner et al., 1997, and references therein). Therefore, power-law statistics and fractals are very closely related concepts.

Second-order fractal measures convey information about the correlation properties of a time series; therefore they are well suited to investigate the temporal fluctuations of a temporal process. Thus, in terms of temporal processes, fractality can be considered as a very close concept to correlation.

The standard manner to investigate the fractality of a time series is the well-known method of the power spectral density. The power spectral density represents the frequency distribution of the power of the time series. Since it is calculated by Fourier transforming the series, it permits the identification of periodic, multiperiodic or no-periodic behaviours. If the time series is fractal, thus characterized by scaling, the power spectral density behaves as a power-law function of the frequency f, $S(f) \propto f^{-\beta}$, with the spectral exponent β indicating the type of the temporal fluctuations and quantifying the strength of the time-correlation structures intrinsic in the time series (Kantelhardt et al., 2002). If $\beta = 0$ the fluctuations can be considered purely random, typical of signals completely uncorrelated, whose samples are independent of each other. If $\beta > 0$, the fluctuations are persistent, meaning that positive (negative) increments are very likely followed by positive (negative) increments; this is typical of systems governed by positive feedback mechanisms. If $\beta < 0$, the fluctuations are antipersistent, meaning that positive (negative) increments are very likely followed by negative (positive) increments; this is typical of systems governed by negative feedback mechanisms.

In a point process, the concept of fractal is equivalent to that of time-clustering, which is 'opposite' to time behaviour displayed by a homogeneous Poissonian process. Clusterization in a point process leads to a power-law (fractal) behaviour of some of the statistics, used to describe its properties (Thurner et al., 1997), and allows to estimate the so-called fractal exponent α (Lowen and Teich, 1995; Teich et al., 1996), which quantifies the strength of clusterization (Lowen and Teich, 1993). If the point process is Poissonian, the earthquake occurrence times are uncorrelated; for this memoryless process $\alpha \approx 0$. On the other side, $\alpha \neq 0$ is typical of earthquake processes with self-similar behaviour; self-similar meaning that parts of the whole can be made to fit to the whole in some way by scaling (Mandelbrot, 1983). Thus, we can understand that estimating the fractal exponent α for a seismic sequence is very important in the general characterization of the mechanism underlying the seismic phenomenon.

Depending on the time structure of the representation of a seismic sequence (IETs or counting process), various statistical quantities can be defined and used for the time dynamics characterization of a sequence of earthquakes. In this overview, we will describe the global coefficient of variation (CV), the local coefficient of variation (LV), detrended fluctuation analysis (DFA), and multifractal detrended fluctuation analysis (MF-DFA) for seismic IETs or continuous seismic signals, and the Allan factor (AF) for seismic counting processes.

4.2.1 Coefficient of Variation

A simple and fast method to evaluate the time-clustering (and possibly fractality) in a seismic sequence is the global CV (Kagan and Jackson, 1991), defined as

$$C_V = \frac{\sigma_T}{<T>}, \tag{4.1}$$

where $<T>$ is the mean of the IETs and σ_τ is the standard deviation: a Poisson seismic process (memoryless, completely random) has $C_V = 1$, a periodic or regular process has $C_V < 1$, but a clusterized seismic process has $C_V > 1$. This coefficient permits obtaining very simply a first idea about the clustering properties of a seismic sequence, although it does not give any information about the timescale ranges where the process can be reliably characterized as a clusterized process (this aspect is discussed in detail later in this chapter). This could be a limit of this measure, since the deep comprehension of a complex process, like earthquakes, is possible only if the different timescale ranges governing its dynamics are well identified and understood.

Recently, Telesca et al. (2016a) investigated the time-clustering properties of the volcanic seismicity at El Hierro, Canary Islands (Spain), by employing the local CV (LV), defined by Shinomoto et al. (2005):

$$L_v = \frac{1}{N-1} \sum_{i=1}^{N-1} 3 \frac{(T_i - T_{i+1})^2}{(T_i + T_{i+1})^2} \tag{4.2}$$

where N is the number of IETs. Like the C_v, L_v is larger, equal or smaller than 1 if the process is clusterized, Poissonian or regular, respectively. But, if C_v is a global index of temporal variability of seismic sequences, L_v describes locally the variability of the IETs. Fig. 4−2 can help in understanding the difference between the two measures: the point process shown in Fig. 4−2 is a combination of two periodic point processes; if one calculates the global CV, it can be seen that $C_v \gg 1$ because globally the process appears strongly clusterized, but $L_v \sim 0$, because at local scales the process is periodic.

Telesca et al. (2016b) described an experiment using a microseismic monitoring array consisting of 12 three-component geophones placed in a nearly vertical borehole able to record seismic waves of microseismic events induced by hydraulic fracturing of a long horizontal well, separated into sections, called stages, with each section separately consecutively fractured. The collected induced microseisms were analysed using the CV and LV (Fig. 4−3). If the LV indicates locally Poissonian time dynamics of the induced microseismic events (the value of LV is within the 95% Poisson confidence band), the CV reveals different time dynamics between the stages, and, in particular, would suggest interaction with a preexisting natural fracture, where sequences of mainshock-aftershocks are expected. In fact, for stages 9, 11 and 12, where natural preexisting fractures were suspected based on the locations of the

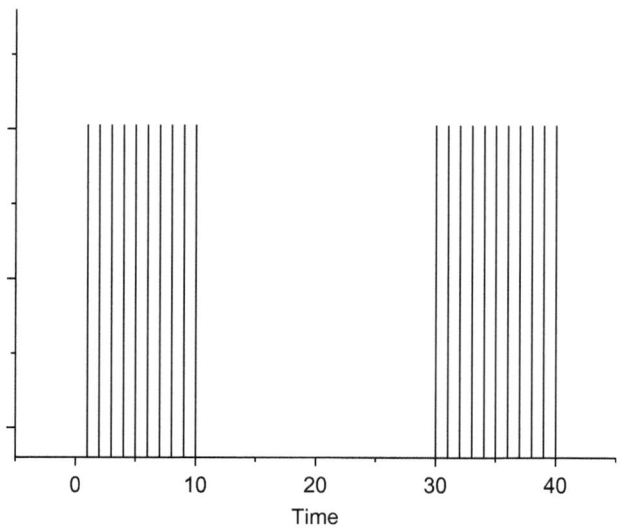

FIGURE 4–2 Superposition of two periodic point processes (Telesca et al., 2016b).

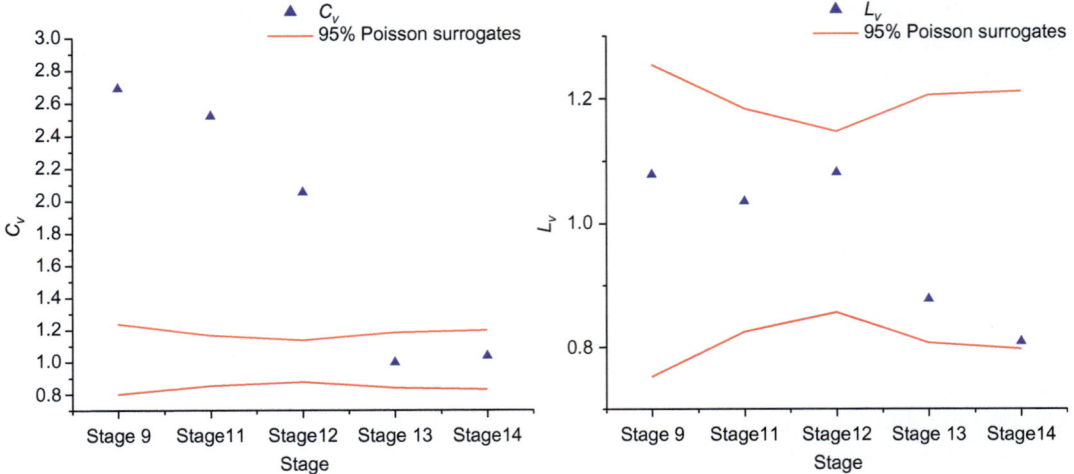

FIGURE 4–3 CV and LV for microseism induced by hydraulic fracturing (Telesca et al., 2016b). *CV*, coefficient of variation.

microseismic events, the CV is beyond the 95% confidence band, while for stages 13 and 14, in which natural fractures with significant stress release were not stimulated and hydrofracturing probably stimulated new volumes, are characterized by a significantly Poissonian behaviour. The CV calculation helped, in this case, to discriminate between volumes with a preexisting natural fracture system and those without, where the fluid injection contributed to the opening of new fractures (Telesca et al., 2016b).

4.2.2 Detrended Fluctuation Analysis

DFA was developed by Peng et al. (1994) to investigate the power-law correlations of nonstationary signals; in fact, trends in nonstationary signals would produce artificial scaling behaviour, thus their effective removal represents a crucial task if one wants to reveal intrinsic fluctuations, especially if the time series is observational or experimental, and is often affected by possible trends with unknown source and shape. The method operates on the time series $x(i)$, where $i = 1,2,\ldots,N$ where N is the length of the series. With x_{ave} we indicate the mean value

$$x_{ave} = \frac{1}{N}\sum_{k=1}^{N} x(k). \tag{4.3}$$

The signal is first integrated

$$y(k) = \sum_{i=1}^{k} [x(i) - x_{ave}]. \tag{4.4}$$

Next, the integrated time series is divided into boxes of equal length, n. In each box a least-squares line is fit to the data, representing the trend in that box. The y coordinate of the straight line segments is denoted by $y_n(k)$. Next we detrend the integrated time series $y(k)$ by subtracting the local trend $y_n(k)$ in each box. The root-mean-square fluctuation of this integrated and detrended time series is calculated by

$$F(n) = \sqrt{\frac{1}{N} \sum_{k=1}^{N} \left[y(k) - y_n(k) \right]^2}. \tag{4.5}$$

Repeating this calculation over all box sizes, we obtain a relationship between $F(n)$, that represents the average fluctuation as a function of box size, and the box size n. If $F(n)$ behaves as a power-law function of n, data present scaling:

$$F(n) \propto n^d, \tag{4.6}$$

where d is the fractal exponent.

Under these conditions the fluctuations can be described by the scaling exponent d, representing the slope of the line fitting $log[F(n)]$ to $log(n)$. For a white noise process, $d = 0.5$. If there are only short-range correlations, the initial slope may be different from 0.5 but will approach 0.5 for large window sizes. $0.5 < d < 1.0$ indicates the presence of persistent long-range correlations, meaning that a large (compared to the average) value is more likely to be followed by a large value and vice versa. $0 < d < 0.5$ indicates the presence of antipersistent long-range correlations, meaning that a large (compared to the average) value is more likely to be followed by a small value and vice versa. $d = 1$ indicates flicker-noise dynamics, typical of systems in a self-organized critical state. $d = 1.5$ characterizes processes with Brownian-like dynamics.

The DFA of seismic IETs was applied to the seismicity of Italy from 1986 to 2001 (Fig. 4−4) by Telesca et al. (2003); as a result, the areas with a significant persistent long-range behaviour were identified. Fig. 4−5 shows a map of Italy with cells characterized by uncorrelated (grey) and correlated (black) for both the full (Fig. 4−5A) and aftershock-depleted (Fig. 4−5B) catalogues. Almost the same cells in both catalogues are featured by correlated temporal behaviour, suggesting that there are areas with a strong degree of time-clustering that is not only due to aftershock generation after a large earthquake, but relies on the underlying time structure of the seismicity of those areas.

The scaling behaviour of the two-dimensional sequence $(\Delta s, \Delta t)$ of the 1981−98 southern California (SC) seismicity (where Δs is the distance between two consecutive earthquakes and Δt is their interevent interval) was investigated using 2-D DFA, which is a simple extension of DFA to a two-dimensional space (Telesca et al., 2007). The estimated scaling exponent α, that gives information on the scaling behaviour of the joint time series, indicates the presence of persistent long-range correlations in the seismic two-dimensional sequence (Fig. 4−6A), showing a twofold pattern with the threshold magnitude: (1) between 1.5 and 3.0, α is rather constant; (2) α decreases for magnitudes

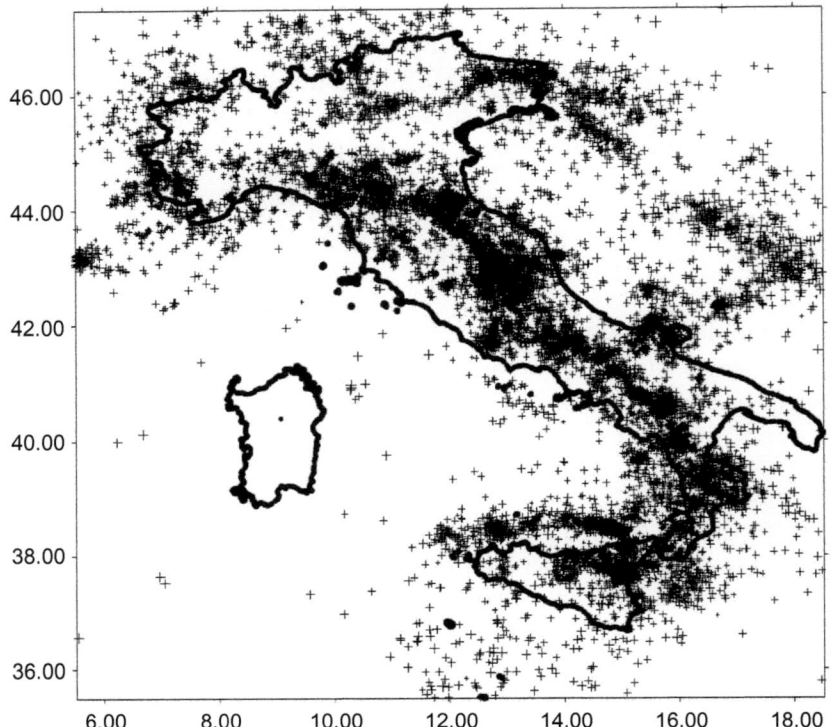

FIGURE 4–4 Epicentral distribution of Italian seismicity that occurred from 1986 to 2001 (Telesca et al., 2003).

larger than 3.0 revealing a tendency toward a two-dimensional space—time Poisson process for large events.

Space-magnitude-dependent scaling behaviours in seismic IET series were identified using DFA applied to the seismicity of central Italy from 1981 to 2007 (Telesca et al., 2008) in a circular area centred on the epicentre of the strongest event, which occurred on 26 September 1997 (MD = 5.8) at a changing radius from 40 to 120 km (Fig. 4–7). The analysis revealed the variability of the DFA exponent that is characterized by three different patterns within three different magnitude ranges. From 2.1 to 2.4, the DFA scaling exponents seem to not depend on threshold magnitude, distance from the centre and the number of events, varying between 0.78 and 0.8; at these magnitudes, the dynamics of the seismic process is mainly governed by small events. MD = 2.4 could also be viewed as a sort of magnitude cut-off, since for larger magnitudes the scaling behaviour significantly changes with threshold magnitude and distance from the centre. From 2.5 to 2.9, the DFA scaling exponents are clearly separated in two disjoint sets depending on the distance from the centre: from 40 to 80 km from the centre, and from 90 to 120 km from the centre. In particular, the scaling exponent calculated for distances between 40 and 80 km is larger than that calculated for distances between 90 and 120 km, probably due to a smoothing effect induced by the events located far

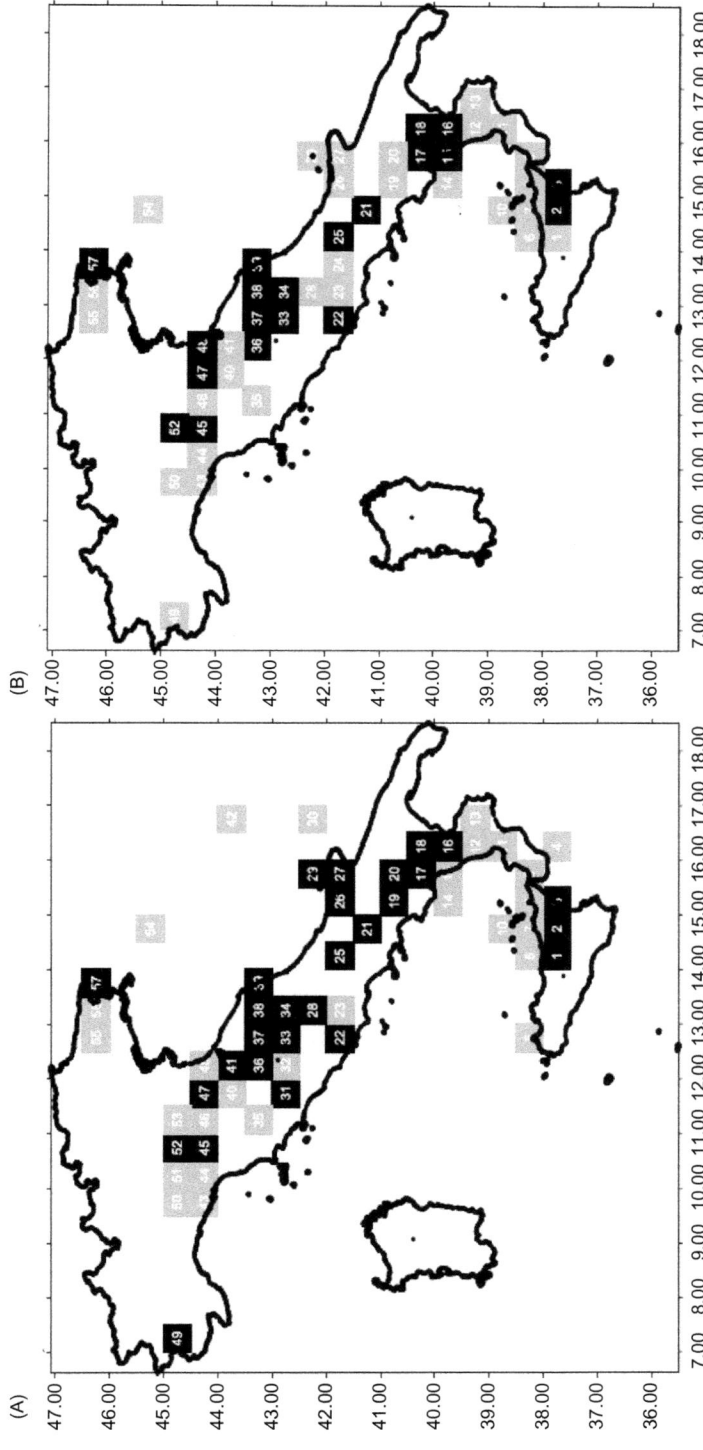

FIGURE 4–5 Map of Italy with cells characterized by uncorrelated (grey) and correlated (black) for both the full (A) and aftershock-depleted (B) catalogues (Telesca et al., 2003).

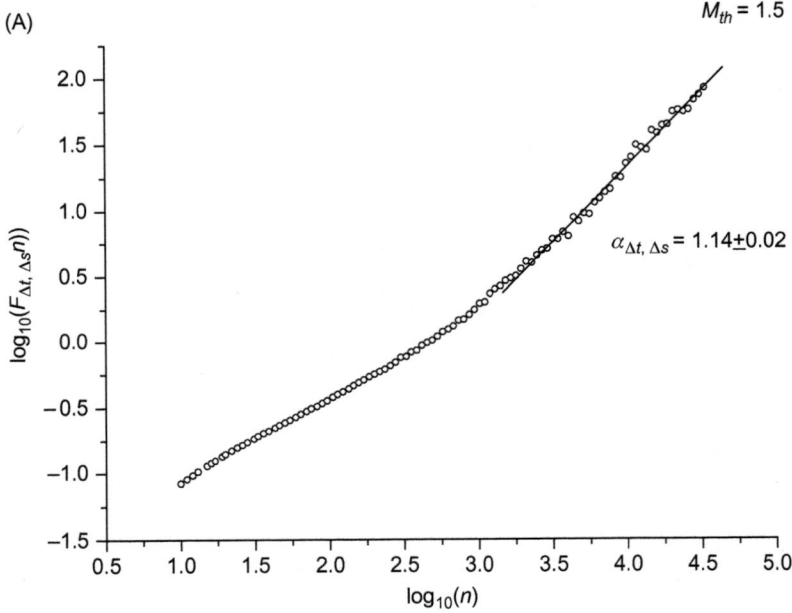

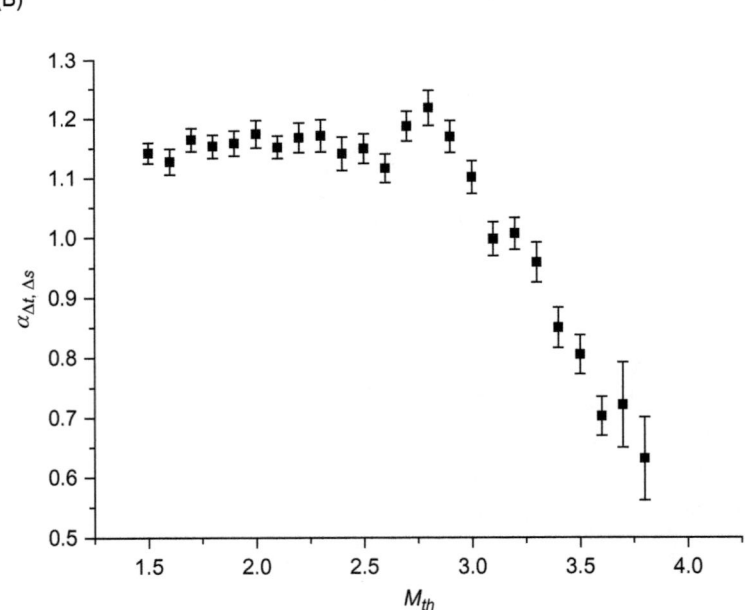

FIGURE 4–6 (A) Fluctuation function for the 2-D sequence (Δt, Δs); (B) variation of the scaling exponent α with the threshold magnitude (Telesca et al., 2007).

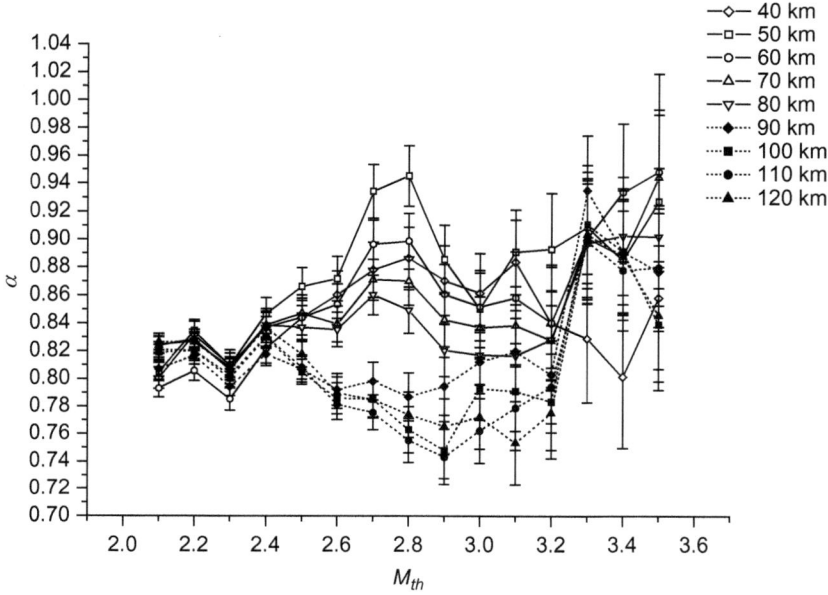

FIGURE 4–7 Space–magnitude variation of the scaling exponent α, estimated using the DFA for the 1981–2007 Umbria–Marche seismicity (Telesca et al., 2008). *DFA*, detrended fluctuation analysis.

from the centre, which contribute to lower the time clustering. Even the concavity of the magnitude-dependent curves of the DFA scaling exponent changes with the distance from the centre: the events located not farther than 80 km contribute to the increase in the time clustering (the scaling exponent increases), but those located between 90 and 120 km from the centre cause a general decrease in the time clustering. From 3.0 to 3.5, the scaling exponents behave quite irregularly and no clear separation between the different distance ranges is visible, very probably due to a finite size effect, especially at larger threshold magnitudes.

The instability index β is another parameter, which helps to identify and quantify deviations from uniform power-law scaling, and thus, indicates changing dynamics in the seismic process. The instability index is calculated as the standard deviation of the local slopes of the DFA curve, and signals deviation from uniform power-law; therefore, high values of β suggest a large deviation from uniform power-law. In the case of the 1981–2007 seismicity of Umbria–Marche (central Italy) significant deviations from uniform power-law scaling were found using the instability index, apparently linked with the occurrence of rather large earthquakes or seismic clusters (Fig. 4–8) (Telesca and Lovallo, 2009).

4.2.3 Multifractal Detrended Fluctuation Analysis

Multifractals are characterized by high time variability on a wide range of timescales, related to intermittent fluctuations and long-range power-law correlations. Many phenomena show multifractality: the signal, which explains the dynamics of such phenomena, is featured by

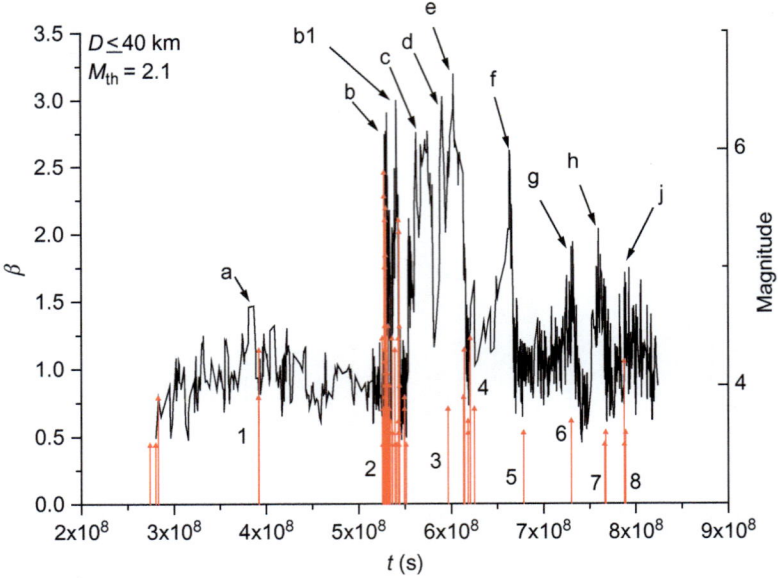

FIGURE 4–8 Time variation of DFA instability for the seismicity in central Italy from 1981 to 2007. The red vertical arrows labelled with numbers from 1 to 8 indicate the seismic events with magnitude $M \geq 3.5$. The oblique black arrows labelled with letters from a to j indicate the anomalous increases of the instability index (Telesca and Lovallo, 2009). *DFA*, detrended fluctuation analysis.

sudden bursts of high-frequency fluctuations, indicating that different scaling behaviours for different intensities of fluctuations coexist in its time variability (Currenti et al., 2005; Telesca et al., 2001). MF-DFA extends DFA to consider an infinity of moment order q (DFA can be viewed as a particular case of MF-DFA for $q = 2$). The fluctuation function in the context of the MF-DFA becomes

$$F_q(s) = \left\{ \frac{1}{N_S} \sum_{\nu=1}^{N_S} \left[F^2(s, \nu) \right]^{q/2} \right\}^{1/q} \tag{4.7}$$

where $\nu = 1, \ldots, N_S$, N_S is the number of nonoverlapping windows of duration s and the index variable q represents the moment order, whose meaning is detailed below. Varying the time-scale s, $F_q(s)$ will increase with increasing s. Then analysing log–log plots $F_q(s)$ versus s for each value of q, we determine the scaling behaviour of the fluctuation functions

$$F_q(s) \propto s^{h(q)}. \tag{4.8}$$

The exponent $h(q)$, called the generalized Hurst exponent (Kantelhardt et al., 2002), is an indicator of multifractality, being constant for monofractals but decreasing with increases in q for multifractals. If $q > 0$, then segments ν with large variance (i.e., a large deviation from the corresponding fit) dominate the average $F_q(s)$. Thus, positive q describes the scaling of

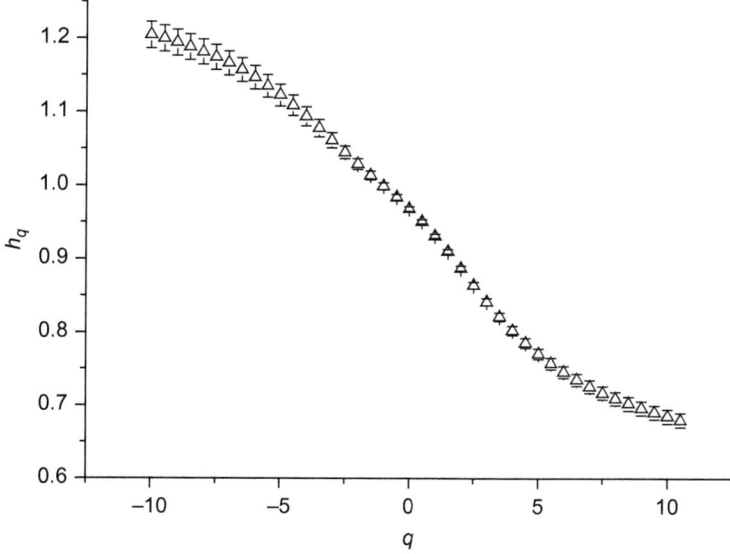

FIGURE 4–9 Generalized Hurst exponents h_q versus q ranging between -10 and 10 (step of 0.5) (Telesca and Lapenna, 2006).

the large fluctuations, generally leading to smaller scaling exponents $h(q)$. If $q < 0$, then segments ν with small variance dominate the average $F_q(s)$, thus describing the scaling behaviour of the small fluctuations, usually characterized by a larger scaling exponent.

Studying the 1983–2003 seismicity of central Italy, struck by a strong event ($M_D = 5.8$) on 26 September 1997, Telesca and Lapenna (2006) applied MF-DFA to the IET series of earthquakes located within a 100-km radius circular area centred on the epicentre of the strongest event. The slope h_q of the fluctuation functions $F_q(s)$, plotted in log–log scales, for scales s ranging from 10 events to $N/4$, where N is the total length of the series, and for $q \in [-10, 10]$, with a 0.5 step, shows the typical multifractal q-dependence, monotonically decreasing with the increase in q (Fig. 4–9).

The Legendre transform is used to calculate the multifractal spectrum that is another way to quantify the characteristics of a multifractal. The generalized Hurst exponents $h(q)$ defined in Eq. (4.8) can be put in a relationship with the scaling exponents $\tau(q)$ defined by the standard partition function multifractal formalism (Kantelhardt et al., 2002)

$$\tau(q) = qh(q) - 1. \tag{4.9}$$

Monofractal series with long-range correlations are characterized by linearly dependent q-order exponents $\tau(q)$, i.e., the exponents $\tau(q)$ of different moments q are linearly dependent on q

$$\tau(q) = Hq - 1, \tag{4.10}$$

because $h(q)$ is independent of q for monofractal series. Long-range correlated multifractal signals have a multiple Hurst exponent, i.e., the generalized Hurst exponent $h(q)$,

$$h(q) = \frac{d\tau}{dq} \neq \text{const}, \qquad (4.11)$$

where $\tau(q)$ depends nonlinearly on q.

The multifractal spectrum $f(\alpha)$ can be derived by $\tau(q)$ using the Legendre transform,

$$\alpha = \frac{d\tau}{dq} \qquad (4.12)$$

$$f(\alpha) = q\alpha - \tau(q), \qquad (4.13)$$

where α is the Hölder exponent and $f(\alpha)$ indicates the dimension of the subset of the series that is characterized by α. The multifractal spectrum explains how important the various fractal exponents present in the series are. Its width, in particular, indicates the range of exponents present, being wider for time series with a higher degree of multifractality. To quantify multifractality, the multifractal spectrum can be fitted to a quadratic function (Shimizu et al., 2002) around the position of its maximum at α_0, i.e. $f(\alpha) = A(\alpha - \alpha_0)^2 + B$ $(\alpha - \alpha_0) + C$: the coefficients are obtained by an ordinary least-squares procedure. The parameter B serves as an asymmetry parameter, which is zero for symmetric shapes, and positive or negative for a left- or right-skewed (centred) shape, respectively. The asymmetry parameter B captures the dominance of low or high fractal exponents with respect to the other; a right-skewed spectrum indicates relatively strongly weighted high fractal exponents, and low ones for left-skewed shapes (Shimizu et al., 2002). The width of the spectrum, calculated as the distance between the two zero-crossings of the fitted curve, is a measure of the degree of multifractality (Ashkenazy et al., 2003); the wider the range of possible fractal exponents, the 'richer' the process in structure. The maximum α_0, which is the value of α where $f(\alpha)$ assumes its maximum value, gives an indication of whether the underlying process is regular in appearance; furthermore, the larger α_0, the more regular the process. Fig. 4–10 shows the multifractal spectrum of the IETs of the 1983–2003 seismicity of central Italy and the quadratic fit. The position of its maximum is at $\alpha_0 = 0$. The best fit, obtained by a least square method, furnished the following values of the parameter $A = -7.51$, $B = -0.05$ and $C = 1.02$. The width of the multifractal spectrum is $W = 0.74$. From these parameter estimations we can deduce that the seismic process is rather multifractal, with more weighted low-fractal exponents. Much more interesting is the time-dependent analysis of multifractality, to detect significant changes in the variation of the three multifractal parameters signalling significant changes in the dynamics of the process. By using the concept of a sliding window (Gamero et al., 1997; Martin et al., 2000) one can calculate the temporal evolution of the multifractality. With a sliding window of 10^3 events and a shift between two successive windows of 50 events, Fig. 4–11 shows the time variation of the maximum α_0, asymmetry B and width W of the 1983–2003 seismicity of central Italy (Telesca and Lapenna, 2006). Strong

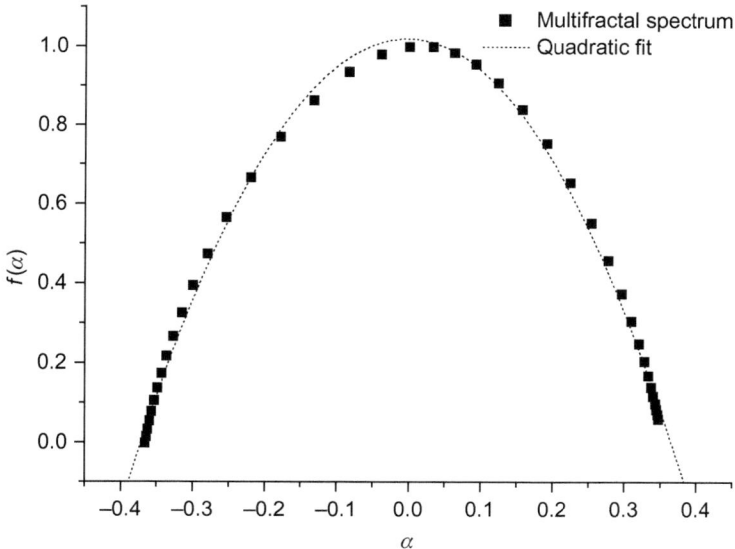

FIGURE 4–10 Multifractal spectrum of the interevent time series of 1983–2003 seismicity of central Italy. The dotted curve represents the quadratic fit (Telesca and Lapenna, 2006).

variability in the multifractality is revealed by the strong variability of the three multifractal parameters, especially during the aftershocks, following an M5.8 event. The maximum α_0 shows a sudden increase, indicating that most clusters possess a higher local fractal dimension than before the mainshock. This means that areas with high values of the measure increase, while almost void areas expand, which in turn implies a rising energy concentration. The asymmetry B decreases, changing from positive to negative values, thus from a left-skewed to a right-skewed shape, suggesting a relative dominance of high fractal exponents. The width W suddenly decreases, indicating a change from a heterogeneous to a more homogeneous dynamics. Fig. 4–12 shows a 3-D visualization of the time variation of the generalized Hurst exponents $h_q(t)$ and the projection on the plane $h_q - t$, where the upper and the lower curves are the envelopes of the highest ($q = -10$) and the lowest ($q = 10$) Hurst exponents, respectively. The variability range of the exponents is approximately constant up to the occurrence of the largest earthquake of the series, lowering during the aftershock activation. Fig. 4–13 shows a greyscale map of the variability of the Hurst exponents h_q as a function of the event number and the parameter q. During aftershocks (the red vertical line indicates the mainshock) the values of h_q reduce their variability. The multifractal method is well suited to explaining the complex geophysical phenomena underlying earthquakes. The multifractality of a seismic phenomenon relies on the different scalings for long and short interevent intervals. During the aftershock activation the loss of multifractality indicates the transition from a heterogeneous state to a homogeneous one. In particular, the mechanisms of generating aftershocks can be of two different types (Godano et al., 1999): (1) the mainshock does not release the stress completely, and some patches of the fault remain unruptured or the amount of slip is less than on the patch where the main event

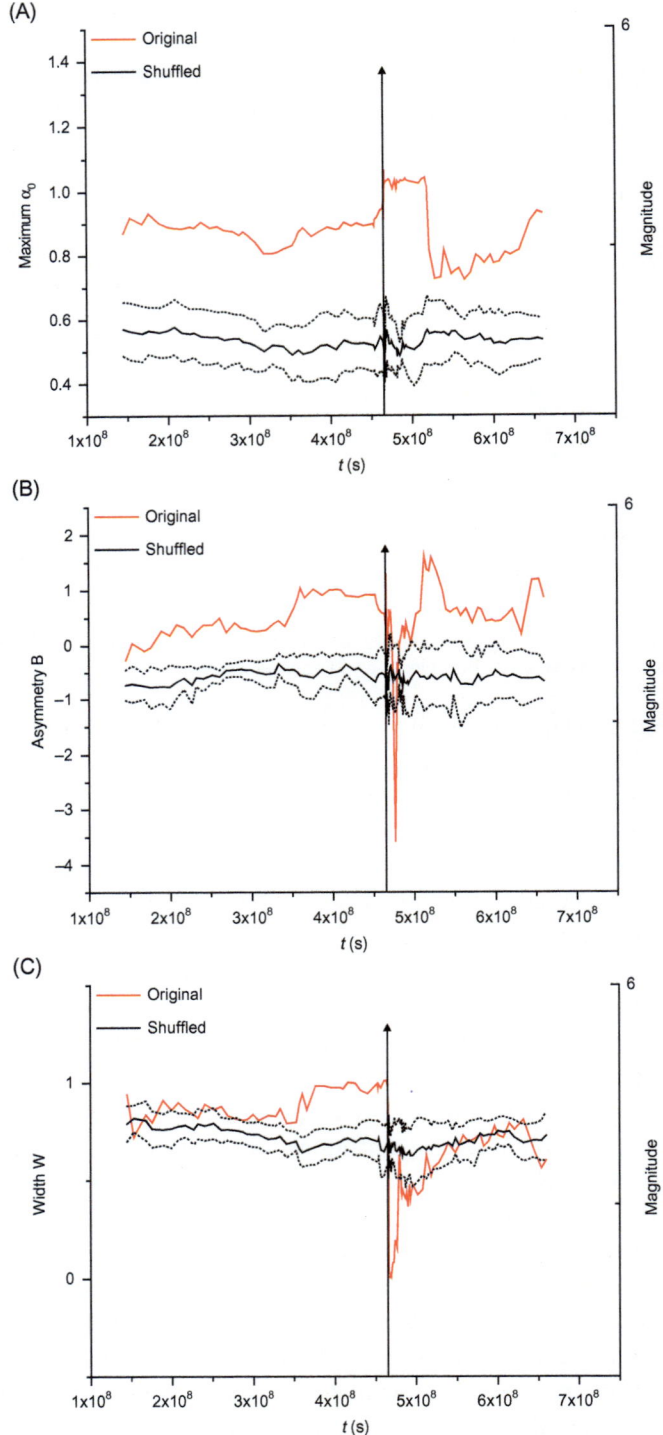

FIGURE 4–11 Time variation of maximum (A), asymmetry (B) and width (C); the average of parameters obtained by 10 randomly shuffled versions of the original series (black curves) within their $1 - \sigma$ range (dotted lines) are also plotted (Telesca and Lapenna, 2006).

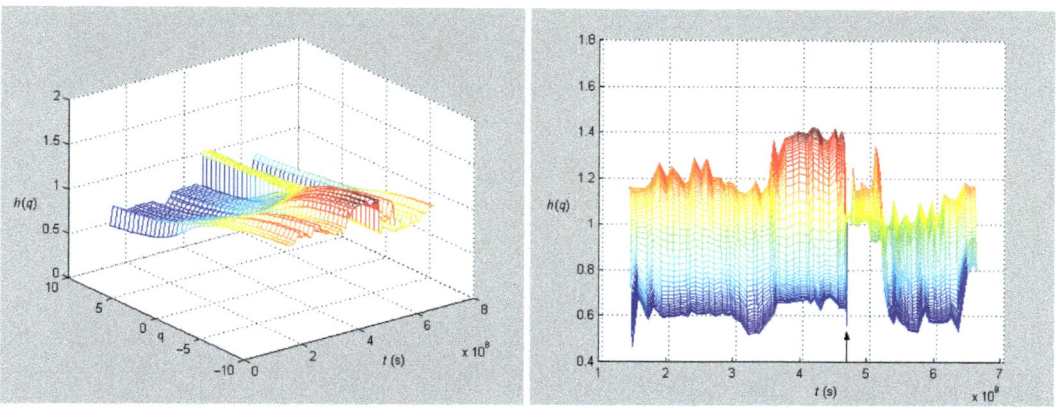

FIGURE 4–12 Three-dimensional visualization of the time variation of the generalized Hurst exponents $h_q(t)$ and the projection on the plane $h_q - t$ (Telesca et al., 2005).

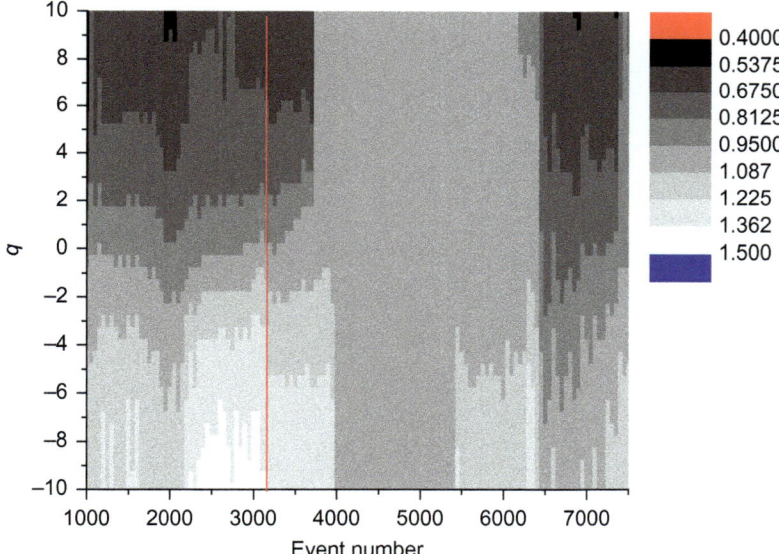

FIGURE 4–13 A greyscale map of the variability of the Hurst exponents h_q as a function of the event number and the parameter q (Telesca et al., 2005).

occurs; or (2) the main event modifies the stress field in the volume surrounding the fault (Dietrich, 1991). Both mechanisms are diffusive. After the main event, which took place on the main fault, many surrounding small faults are activated, leading to a diffusion of the stress; a homogeneous diffusion of the stress could explain the monofractal distribution of the aftershock occurrences. Similar behaviours have been reported in Goltz (1998).

Multifractality in earthquake magnitude series was investigated by Aggarwal et al. (2015), who analysed the seismicity of the Kachchh region, one of the most seismically active areas in India. The magnitude sequence of the aftershock-depleted catalogue is more multifractal than that of the entire catalogue (Fig. 4−14), indicating a higher heterogeneity of the magnitude sequence of the aftershock-depleted catalogue than that of the whole catalogue. Since multifractality depends on the different long-range correlations between large-magnitude

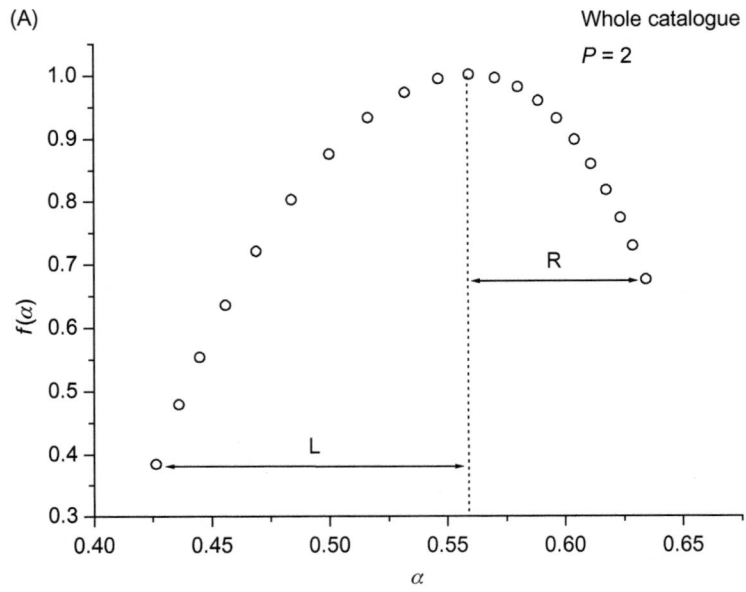

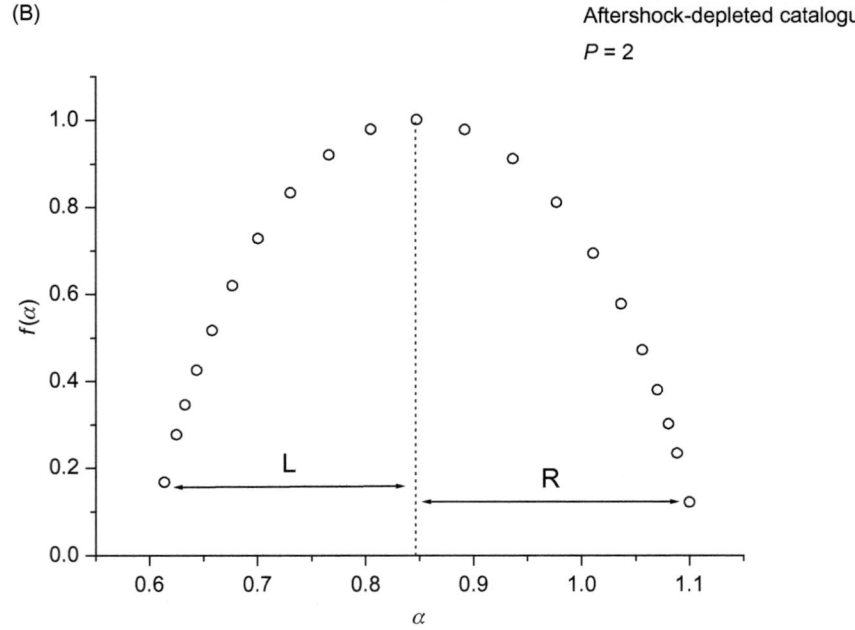

FIGURE 4−14 Multifractal spectra (two-degree detrending polynomial) of the magnitude sequence of Kachchh region (India) (Aggarwal et al., 2015).

and small-magnitude fluctuations, this result suggests that the intensity of such a difference is larger for the aftershock-depleted catalogue than for the whole catalogue. Removing aftershocks implies depleting all those events that are closely dependent on the mainshock, but swarms and other smaller earthquakes not directly dependent on the mainshock remain, leading to lower homogeneity. The magnitude sequence of the whole catalogue is characterized by a left-skewed multifractal spectrum, while that of the magnitude series of the aftershock-depleted one is quasisymmetric, indicating the dependence of the multifractality of the magnitudes of the whole catalogue, especially on the large-magnitude fluctuations arising from the difference in magnitude between earthquakes. Reasonably, aftershocks, being smaller events depending on the largest shock, contribute to enhance such a difference in magnitude between events. In fact, removing smaller aftershock magnitudes, the source of such a difference in magnitude is weakened, along with the dominance of the large-magnitude fluctuations, and a more symmetric multifractal spectrum is generated in the aftershock-depleted catalogue.

The multifractal spectrum of the magnitude sequence of the Pannonian region (central Europe) of deep earthquakes is wider than that of the shallow events (Fig. 4–15) and is right-skewed, while that of shallow earthquakes is quasisymmetric, indicating more heterogeneous dynamics of magnitudes at greater depths and a dominance of small-magnitude fluctuations that arise from the difference in magnitude between earthquakes; the maximum magnitude in both shallow and deep catalogues is 5.8, but the minimum magnitude of the deep catalogue is 3.6 (while it is 2.5 in the shallow catalogue), suggesting that the smaller magnitude variations are more dominant in the deep series than in shallow ones. The

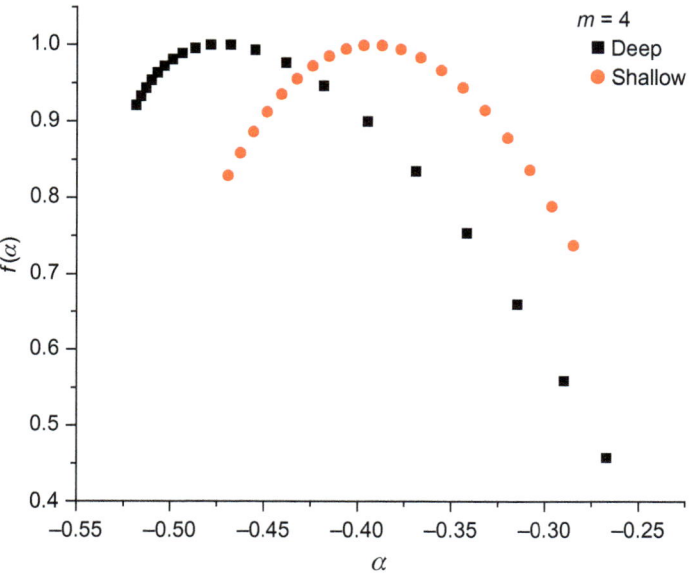

FIGURE 4–15 Multifractal spectrum of deep (black squares) and shallow (red circles) Pannonian magnitude sequences (Telesca and Toth, 2016).

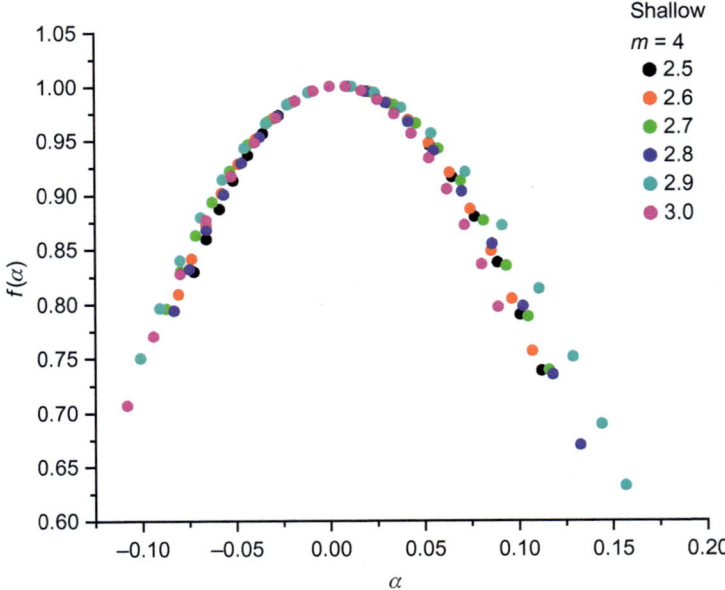

FIGURE 4–16 Comparison among the multifractal spectra of the shallow catalogue centred on the corresponding maxima (Telesca and Toth, 2016).

multifractal spectra of the shallow catalogue calculated for different minimum magnitudes overlap, and this indicates a magnitude-invariant multifractality (Fig. 4–16); this effect would suggest a possible link between multifractality (quantified by the width of the Legendre spectrum) and the *b*-value of the Gutenberg–Richter law, since neither (multifractality and *b*-value) change with the increase in the minimum threshold magnitude (Telesca and Toth, 2016). Of course, more investigations are necessary to assess the reliability of such a postulated link.

MF-DFA was useful to evidence dynamic changes in the continuous seismic signal RSAM (Fig. 4–17) recorded at El Hierro volcano (Canary Islands), where a submarine monogenetic eruption occurred in October 2011. The analysis of the three frames of the signal, measured before the onset of eruption and after it, during two distinct eruptive episodes, showed a striking difference in the width of the multifractal spectrum (Fig. 4–18): during the eruptive episodes the multifractal spectra, almost identical, are much narrower than those of the signal frame measured before the onset of the eruption. The unrest was mainly characterized by seismicity generated by the fracturing of the host rock under magma overpressure, and by variation in the epicentral location with time, leading to higher heterogeneity of the seismic signal (López et al., 2012). A similar behaviour has been recently observed at the Bardarbunga Volcano in Iceland (Gudmundsson et al., 2014). During the eruption the seismic signal reflects the flow of magma and associated fluids through the upper part of the eruption conduit (tremor) and the mechanical (gravitational) readjustment of the plumbing system (tectonic seismicity) following decompression in magma reservoirs during eruption. The additional

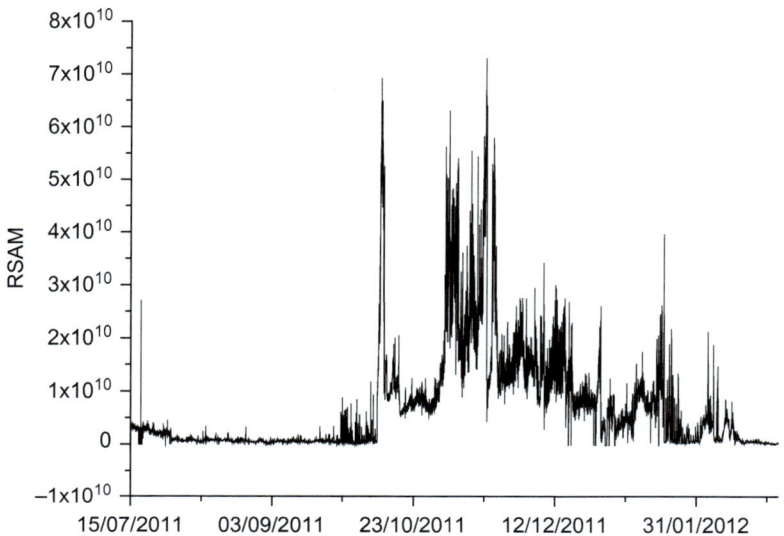

FIGURE 4–17 Continuous seismic signal RSAM recorded at El Hierro volcano (Canary Islands), where a submarine monogenetic eruption occurred in October 2011 (Telesca et al., 2015a).

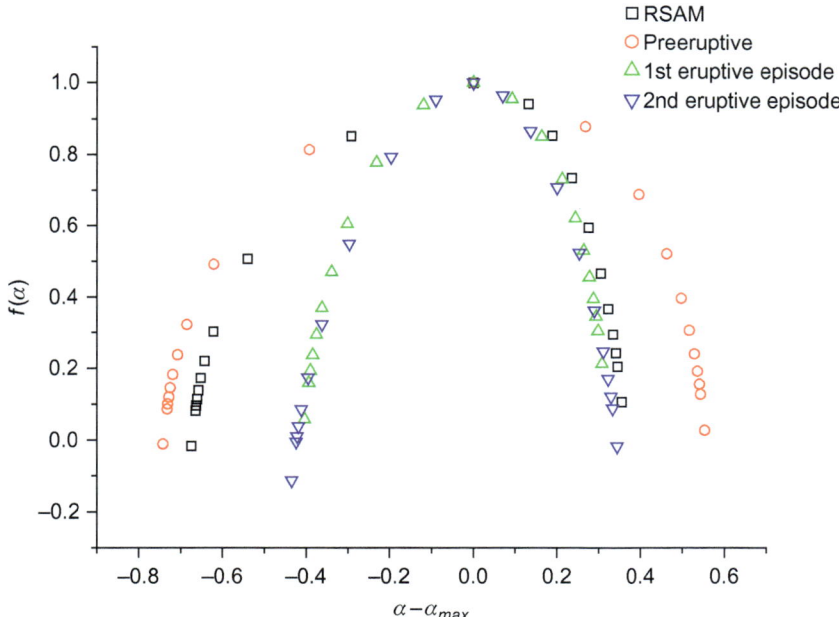

FIGURE 4–18 Multifractal spectra of three frames of the RSAM measured before the onset of eruption, and after, during two distinct eruptive episodes (Telesca et al., 2015a).

tremor during the eruption, along with the shear movements during the gravitational read-justment that accompanied the eruption, could have caused such a difference in multifractal-ity between unrest and eruption, clearly indicating different behaviours of the plumbing systems under overpressure or decompression conditions. The very similar shape of the two multifractal spectra during the two eruptions would suggest that the sources of the seismic signal during both episodes are similar, with the dominance of the tremor associated with the movement of pressured fluids and associated host-rock deformation (Telesca et al., 2015a). The similarity between the two episodes was also identified in geophysical (Martí et al., 2013a; Tárraga et al., 2014; López et al., 2014) and petrological data (Martí et al., 2013b).

4.2.4 Allan Factor

The AF (Allan, 1966) is a measure, related to the variability of successive counts, that is use-ful to detect event clustering in a point process. After fixing a counting time T and dividing the entire period in successive nonoverlapping windows of duration T, the counting process of the seismic sequence $\{N_k(T)\}$ is constructed by counting how many events fall in each kth window. For such a specific counting time T, called the timescale, the AF is defined as the variance of successive counts divided by twice the mean number of events in that counting time

$$\text{AF}(T) = \frac{<(N_{k+1}(T) - N_k(T))^2>}{2 <N_k(T)>}.$$ (4.14)

Repeating the same procedure for all available timescales (generally, the highest time-scale is given by $P/10$, where P is the duration of the whole investigation period), a function $\text{AF} \sim T$ is obtained, whose shape can disclose the characteristics of the time dynamics of the earthquake sequence. This measure reduces the effect of possible nonstationarity of the point process, because it is defined in terms of the difference in successive counts (Viswanathan et al., 1997). For a homogeneous Poissonian process AF is near unity; for a nonhomogeneous Poisson or clusterized process, AF is larger than unity. A homogeneous Poisson process models earthquake sequences where each event is independent of the others, while in a nonhomogeneous Poisson or clusterized process, the events are not inde-pendent of each other and a sort of memory can feature earthquake sequences.

In particular, if the time-clustering is self-similar or fractal, the AF scales with the time-scale T as a power-law:

$$\text{AF}(T) = 1 + \left(\frac{T}{T_1}\right)^{\alpha}$$ (4.15)

with $0 < \alpha < 3$ over a large range of timescales T (Lowen and Teich, 1995); T_1 is the fractal onset time for the AF and is estimated as the crossover timescale between homogeneous Poissonian and scaling behaviours.

The AF represents the most used measure of the time-clustering of earthquake sequences and was employed in many study-cases worldwide. Telesca et al. (2012b) analysed the Caucasus seismicity performing the AF analysis on the whole (Fig. 4−19A) and aftershock-depleted (Fig. 4−19B) catalogues. The AF analysis revealed several characteristics in the time

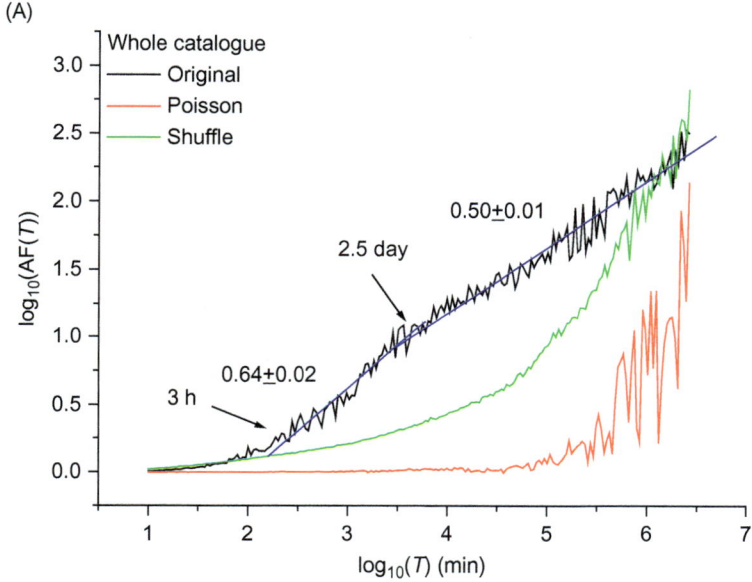

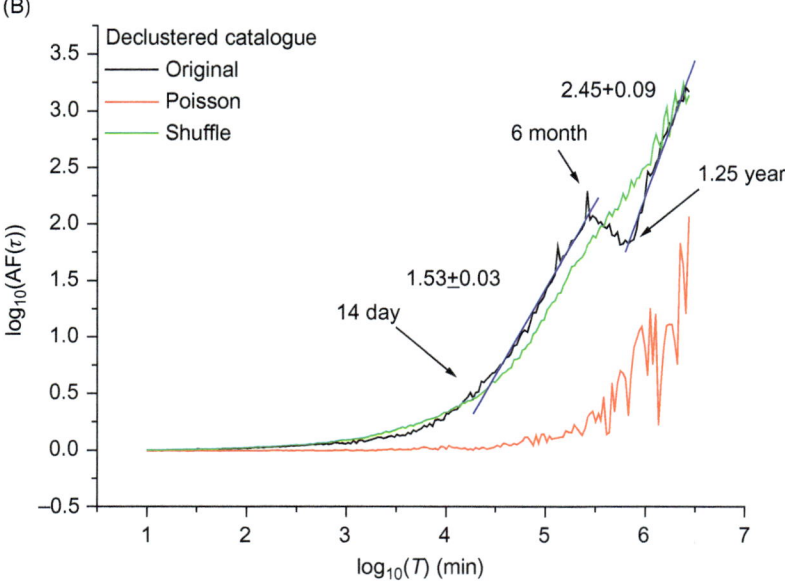

FIGURE 4–19 Allan factor curve for the whole (A) and aftershock-depleted catalogue (B) of the Caucasus (Telesca et al., 2012a).

dynamics of the whole seismicity: the seismic process is not a homogeneous Poissonian process because the AF curve is not constant but manifests more than one scaling regime; a cutoff timescale of about 3 hours can be estimated as the lowest timescale below which a Poissonian behaviour is visible; intermediate timescales between approximately 3 hours and 2.5 days revealed with a scaling exponent of ~ 0.64; another scaling region involving high timescales between about 2.5 days and 5 years is evidenced with a scaling exponent of ~ 0.5; the coexistence of the two scaling regions suggests the coexistence of two different time dynamics in the seismic process. The intermediate scaling region, whose exponent is larger than that of the region at higher timescales, could likely describe aftershock activation following the largest earthquakes of the sequence. The significance of the results can be checked by using two surrogate methods: by generating Poissonian sequences with identical number and mean intervent time of the original sequence, and shuffled sequences with identical interevent interval probability density function as the original one. Generating many surrogates, one applies the AF method to each surrogate (Poissonian or shuffled) and for each timescale calculates the 95th percentile among the AF values of the surrogates for a specific timescale. The 95% confidence AF curve is obtained enveloping the 95th percentiles for each timescale. Fig. 4−18A shows the 95% confidence AF curves for Poissonian (red) and shuffled (green) surrogates: the AF curve of the original sequence is significantly above both the 95% confidence AF curves, indicating that the found time-clustering is significant and depends on the ordering of the interevent intervals. Similarly to the whole sequence, the AF of the aftershock-depleted catalogue of Caucasus also shows several features: the AF is not flat for any timescale, but is also significantly different from the 95% confidence curve obtained for Poissonian surrogates (red); two scaling regions are detectable, involving the first timescales between about 14 days and 6 months with a scaling exponent of ~ 1.53, and the second timescales larger than 1.25 years with a larger scaling exponent of ~ 2.45; the AF curve drops down at about 1.25 years, and this suggests the presence of a cycle in the time dynamics of the seismic sequence with a period close to 1.25 years. It is possible that quasi-periodic variations of the water level of the Enguri dam could influence the time dynamics of the local seismicity in the western Caucasus (Matcharashvili et al., 2008), modulating its variability with a cycle of around 1.25 years. However, on regional scales, 1-year cycles in seismicity could also be explained by seasonal variations of surface loading (snow melting and precipitation) in the mountainous Caucasian region. The high similarity of the AF of the original aftershock-depleted sequence with the 95% confidence curve for the shuffles (red) indicates that the time-clustering of the aftershock-depleted seismicity is mainly to the shape of the probability density function of the IETs and would not depend on their ordering, in contrast to the whole sequence.

The AF can also be used to assess the performance of time declustering of earthquake catalogues. Telesca et al. (2016c) have evaluated the performance of two earthquake declustering methods, the Gardner and Knopoff (with Grünthal and Uhmhammer window) (Gardner and Knopoff, 1974) and the Reasenberg (with different setting parameters) (Reasenberg, 1985) on two seismic areas: SC and Switzerland. Both methods [described in van Stiphout et al. (2012)] are implemented in the software ZMAP (Wiemer, 2001); in particular the parameters of the Grünthal window, optimized for central Europe, are used in Giardini et al. (2004). Fig. 4−19 shows the AF results for the SC declustered catalogue. The red curve in each plot is the AF of the original declustered catalogue, while the black curves are the AFs of 100 surrogates obtained by shuffling the IETs of the original sequence. The comparison with surrogates is performed to evaluate the significance. In particular, shuffling the IETs does not alter their probability density function, but destroys all the correlation structures, producing, then, uncorrelated sequences. If the AF of the declustered catalogue is beyond the region depicted by the AFs of the surrogates in a certain timescale range, this means that in that range the AF is significantly different from that of the surrogates. Fig. 4−20 shows the AF curve of the SC seismicity declustered using the methods of Gardner and Knopoff with the Uhmhammer window compared with shuffled surrogates (Fig. 4−20A), Gardner and Knopoff with Grünthal window (Fig. 4−20B) and Reasenberg (Fig. 4−20C) compared with Poissonian surrogates. In all these cases, the declustered catalogue is time-uncorrelated at low and intermediate timescales (up to a maximum of 6 months), while it is correlated at larger timescales. Fig. 4−21 shows a comparison of all the AF curves obtained for the Swiss whole and declustered catalogues: the method of Gardner and Knopoff with the Grünthal window produces a declustered catalogue with the lowest AF curve, indicating less overdispersion with respect to the catalogues declustered with the other methods, and, therefore a more efficient declustering.

4.3 Informational Methods

The Fisher information measure (FIM) and Shannon entropy are important tools in elucidating quantitative information about the level of organization/order and complexity of a natural process. The FIM allows the investigation of complex and nonstationary time series through the quantification of their degree of order or organization, while Shannon entropy permits measurement of their degree of disorder. The FIM was introduced by Fisher (1925) in the context of statistical estimation (Fisher, 1925), and later Frieden (1990) employed it as a versatile tool to describe the evolution laws of physical systems. The FIM accurately describes the behaviour of dynamic systems and the complex signals generated by these systems (Vignat and Bercher, 2003). Martin et al. (1999) used FIM to characterize the dynamics of electroencephalogram signals, and to detect significant changes in the behaviour of nonlinear dynamic systems, thus enhancing the usefulness of FIM in studying many theoretical and observational features of natural phenomena (Martin et al., 2001). The FIM was also used for precursory ability of critical events (Telesca et al., 2009, 2010).

Shannon entropy is normally used to define the degree of uncertainty involved in predicting the output of a probabilistic event (Shannon, 1948; Hilborn, 1994; de Araujo et al., 2003; Fuhrman et al., 2000). For discrete distributions, if one predicts the outcome exactly before it happens, the probability assumes the maximum value; in this case the Shannon entropy assumes the minimum value. Therefore, for an absolutely predictable event, the Shannon entropy will be zero. This is not the case for distributions (probability densities) on a

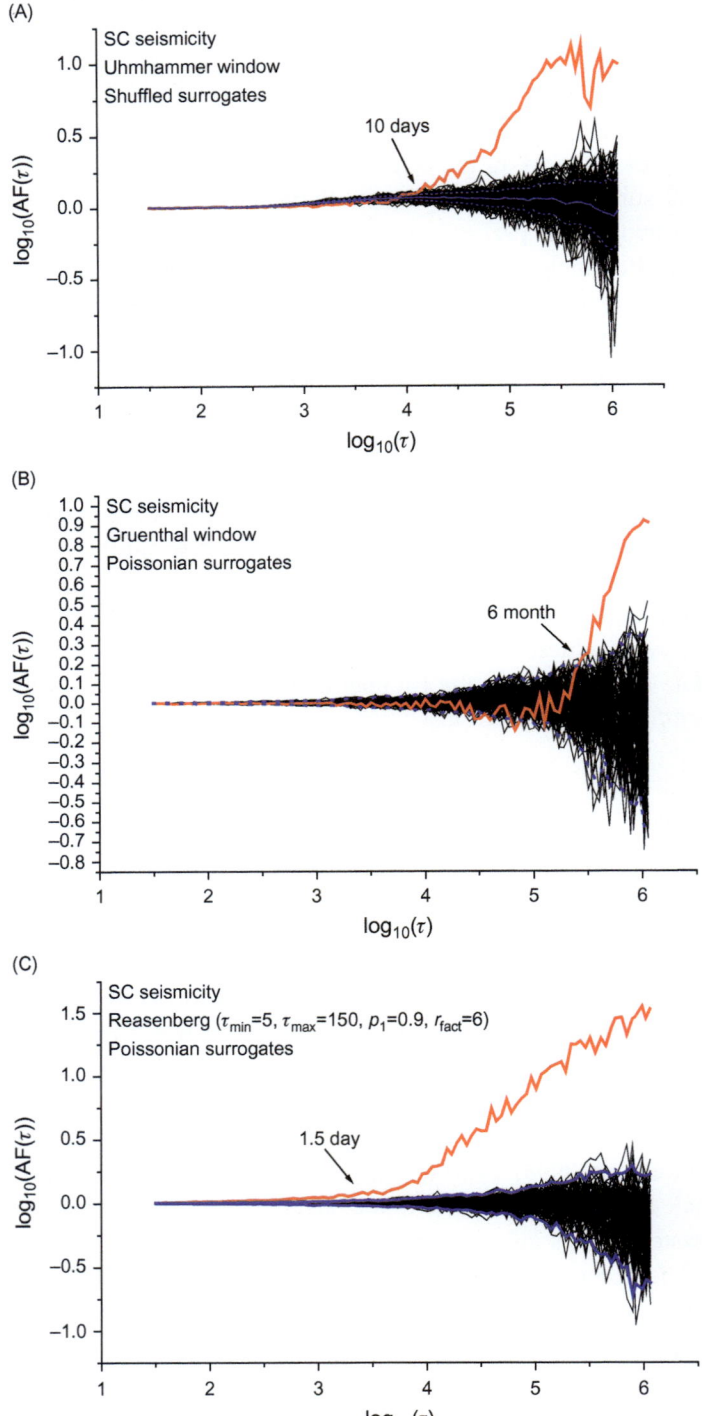

FIGURE 4–20 AF curve of the southern California seismicity declustered (red) by using the methods of Gardner and Knopoff with the Uhmhammer window compared with shuffled surrogates (A), Gardner and Knopoff with Grünthal window (B) and Reasenberg (C) compared with Poissonian surrogates (black). The blue curves limit the 95% confidence level of the Poissonian surrogates (Telesca et al., 2016c). *AF*, Allan factor.

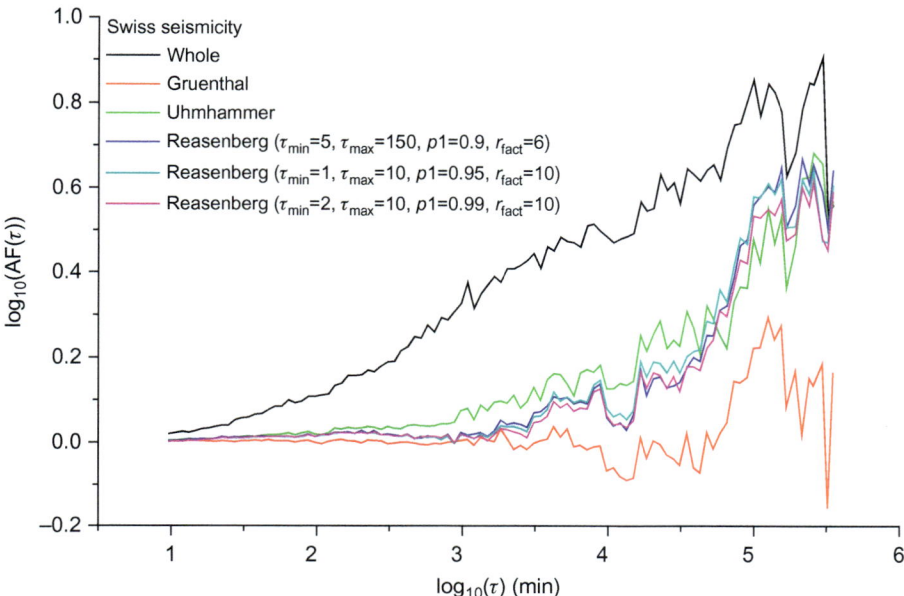

FIGURE 4–21 Comparison of all the AF curves obtained for the Swiss whole and declustered catalogues (Telesca et al., 2016c). *AF*, Allan factor.

continuous variable, ranging, e.g., over the real line. In this case, the Shannon entropy can reach any arbitrary value, positive or negative. Therefore, the use of the Shannon power entropy avoids the difficulty of dealing with negative information measures. The use of Shannon entropy power instead of Shannon entropy arises from the so-called 'isoperimetric inequality' (Dembo et al., 1991; Romera and Dehesa, 2004; Angulo et al., 2008; Esquivel et al., 2010), a lower bound to the product between the FIM and the Shannon entropy power is 1, for one-dimensional systems. The 'isoperimetric inequality' suggests that the FIM and Shannon entropy are intrinsically linked, so that the dynamic characterization of signals should be improved when analysing them in the so-called Fisher−Shannon (FS) information plane (Martin et al., 1999; Vignat and Bercher, 2003). Several further applications have shown the potential of this tool to gain an insight into the dynamics of processes (Lovallo and Telesca, 2011; Telesca and Lovallo, 2011; Telesca et al., 2011).

If $f(x)$ indicates the probability density function of a time series x, the FIM is defined as

$$I = \int_{-\infty}^{+\infty} \left(\frac{\partial}{\partial x} f(x) \right)^2 \frac{dx}{f(x)}, \tag{4.16}$$

while the Shannon entropy is given by the following formula (Martin et al., 1999):

$$H_X = - \int_{-\infty}^{+\infty} f_X(x) \log f_X(x) dx. \tag{4.17}$$

As stated above, it is more convenient to use the alternative notion of Shannon entropy power (Dembo et al., 1991) N_X

$$N_X = \frac{1}{2\pi e} e^{2H_X}.$$ (4.18)

Telesca et al. (2012b) investigated the IET and interevent-distance (IED) series of seismic events that occurred in the Aswan area (Egypt) from 2004 to 2010 in the FS information plane, varying depth and magnitude thresholds. Fig. 4−22 shows the FIM and N_X varying the threshold magnitude and depth for the IET series. The FIM is larger for lower threshold magnitudes and for deeper events; while N_X is larger for larger magnitudes mainly occurring in the shallower strata. Such results suggest a higher level of organization and order for sequences of earthquakes with lower threshold magnitudes and larger threshold depths. Fig. 4−23 shows the FIM and N_X varying the threshold magnitude and depth for the IED series; the FIM increases with a decrease in threshold magnitude and an increase in the threshold depth, as for the IET series; N_X is rather independent of the threshold magnitude, but increases with a decrease in the threshold depth. These findings indicate that more intense and shallower events are less organized and less predictable, being characterized by lower FIM and higher Shannon entropy power of the IET series; this can also feature these events as more 'Poissonian'. In fact, several papers have shown that increasing the magnitude threshold of seismicity, earthquakes tends to be a Poissonian process, characterized by loss of memory and, thus, loss of predictability (Telesca et al., 2002). The level of uncertainty of the IED series does not change significantly if one considers relatively high-magnitude events or also lower-magnitude events; and this suggests that the IED series shows the same degree of predictability even when varying the threshold magnitude. This is in agreement with De Santis et al. (2011), in which a relationship between the Shannon entropy of seismicity and the b-value of the Gutenberg−Richter law (Gutenberg and Richter, 1944) was derived; in fact, for a given depth threshold, the earthquake series constitutes a complete earthquake series, with a single b-value for any magnitude, and so a single Shannon entropy (Telesca et al., 2012b).

The FIM was used to study the properties of seismograms of nine tectonic earthquakes that occurred in Vrancea (Romania) and were recorded by three seismic stations located in Moldova: two (MILM and LEOM) within an area with high seismic hazard, and one (SORM) in a less hazardous region (Telesca et al., 2014a). The site of the seismic station SORM is characterized by the presence of a single soft soil 5 m thick layer overlying the geological bedrock. The site of the seismic station MILM consisted of Sarmatian limestone, characterized by the absence of soft soil. The seismic station LEOM is located on a site with soft soils represented by alternating layers of sandy loam, clay and water-saturated sands underlying dense clay, with a total thickness of soft soils of 12 m. The three sites have different ground conditions; however, the soft sedimentary soil is absent in MILM and is not thick in SORM and LEOM. A clear discrimination on the base of the informational properties of the recorded seismograms corresponding to the same earthquakes was found between the two

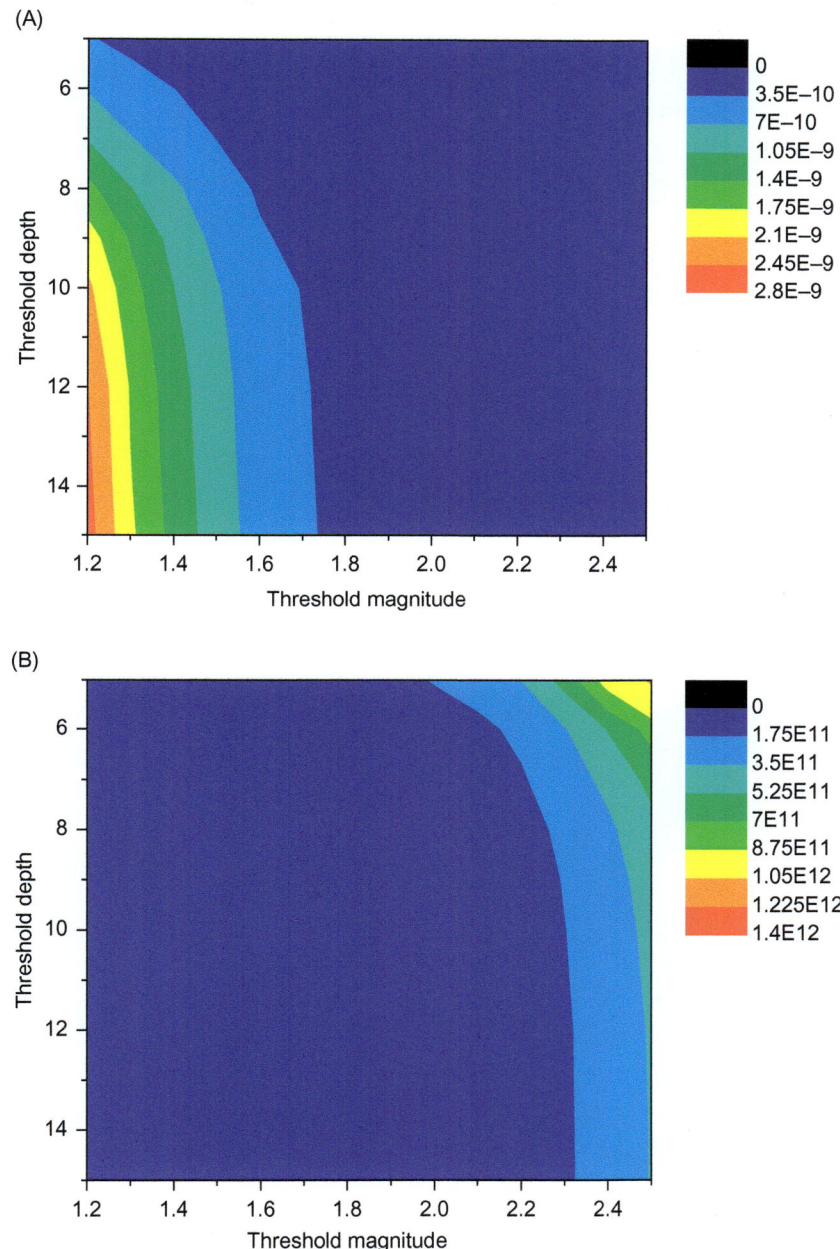

FIGURE 4–22 FIM (A) and N_x (B) varying the threshold magnitude and depth for the interevent times of Aswan seismicity (Telesca et al., 2012b). *FIM*, Fisher information measure.

types of stations. For each earthquake, three seismic components were analysed: East−West (EW), North−South (NS) and vertical (Z). Fig. 4−24 shows an example of a seismogram, recorded by the seismic station SORM, containing both p- and S-waves. Fig. 4−25 shows the

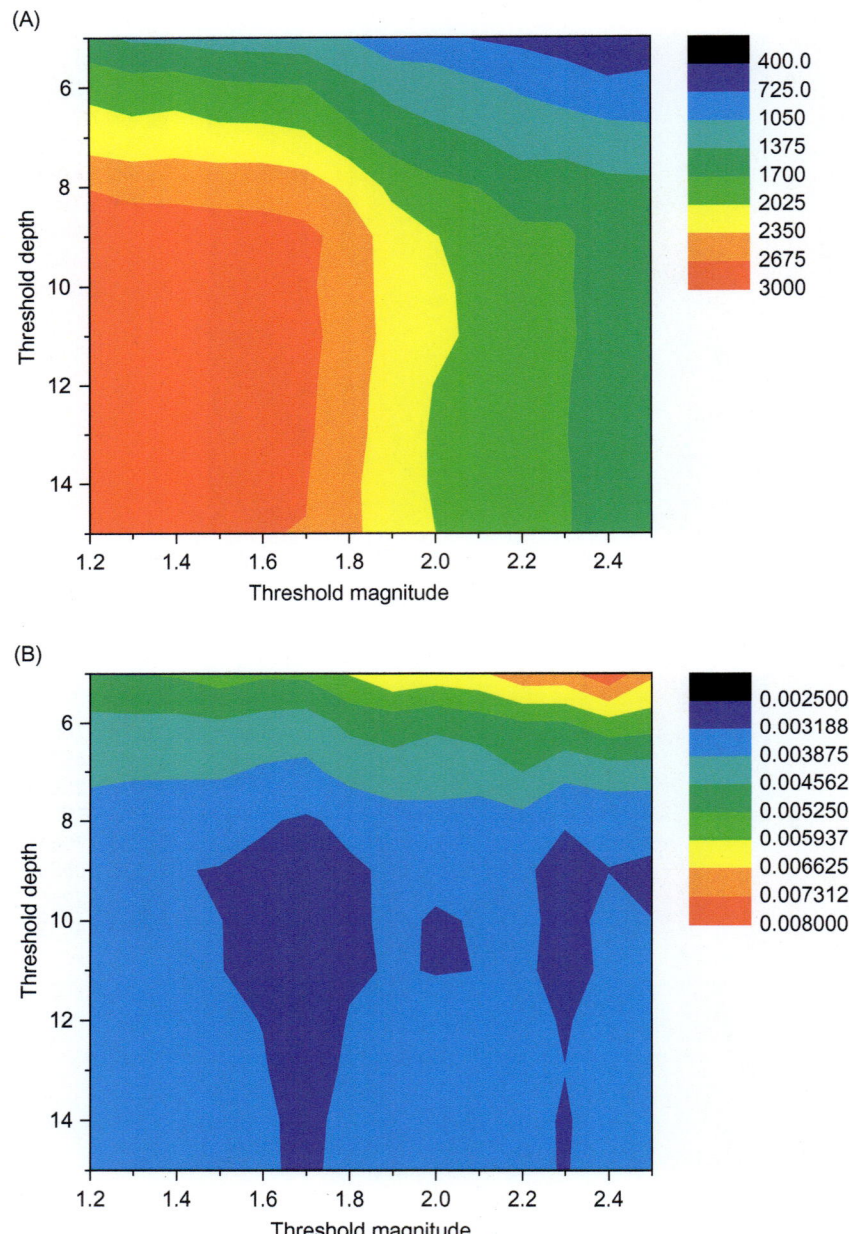

FIGURE 4–23 FIM (A) and N_X (B) varying the threshold magnitude and depth for the interdistance times of Aswan seismicity (Telesca et al., 2012b). *FIM*, Fisher information measure.

FIM of each component of each station versus the azimuth and the distance from the epicentre of the corresponding earthquake. A quite clear separation between seismograms registered by SORM and those recorded by LEOM and MILM is visible for each component.

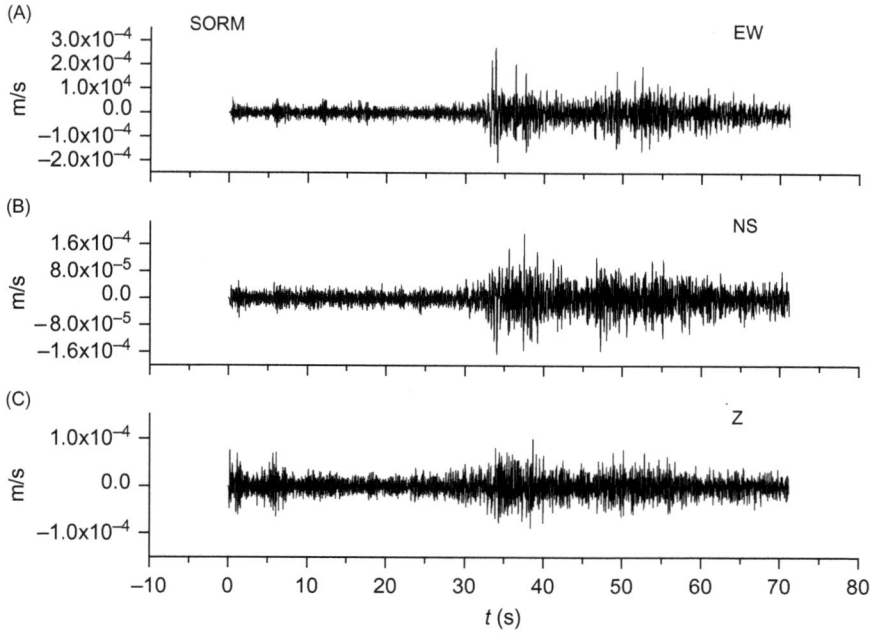

FIGURE 4–24 Example of seismogram recorded by SORM seismic station on 1May 2011. (A) EW component, (b) NS component, and (c) Z component (Telesca et al., 2014a). *EW*, East–West; *NS*, North–South; *Z*, vertical.

Fig. 4–26 shows the relationship between the mean FIM with mean earthquake azimuth and the mean distance from the seismic station; a greater distance and lower azimuth characterize seismograms with lower FIM, and, thus, with lower organization and higher disorder in seismograms recorded by SORM. The dependence of FIM on the azimuth could be explained by the azimuthal anisotropy of Vrancea seismic radiation, and by the structural anisotropy of the seismic energy spread due to the inhomogeneity of the propagation medium, in agreement with Radulian et al. (2006) who showed that the frequency-dependent attenuation of the seismic waves travelling from Vrancea subcrustal sources has an azimuth-dependent attenuation. The dependence of FIM on the distance from the earthquake epicentre could be explained by the increase in attenuation of the seismic wave with the distance from the earthquake focus; the seismic stations closer to the epicentre detect less attenuated waves; and this implies that their seismograms are more structured and organized. Therefore, the seismograms recorded by MILM and LEOM, that are closer to Vrancea earthquake epicentres, detect the seismic wave with a larger amount of information; while SORM, that is more distant, detected a more dispersed and more random seismogram for the same seismic wave that reached MILM and LEOM. A relationship can be found between the informational properties of the seismograms and the seismic hazard, through the wave attenuation; in fact, the three Moldovan seismic stations are located, two (MILM and LEOM) in an area with medium seismic hazard (0.8–2.4 m/s^2, PGA), while the third (SORM) is located in a less hazardous region (0.2–0.8 m/s^2, PGA) (Alkaz, 2005).

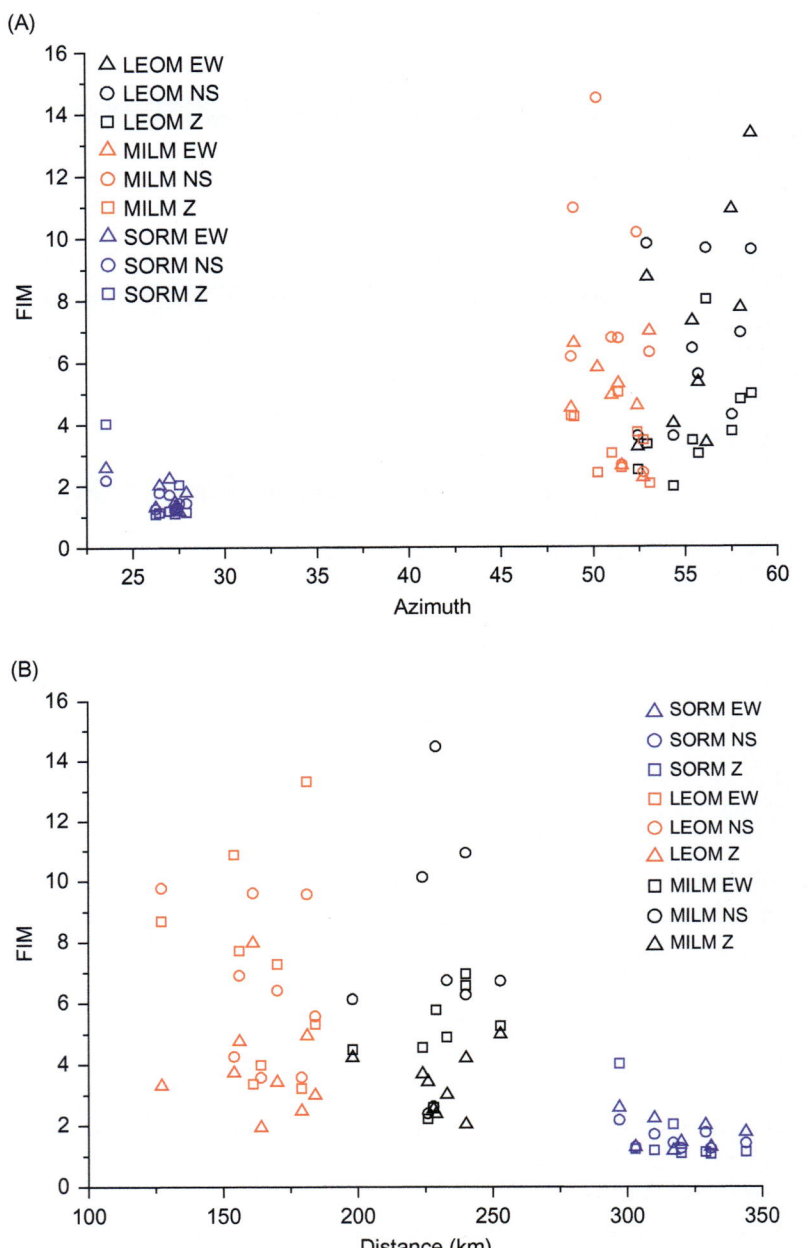

FIGURE 4–25 FIM versus azimuth (A) and epicentral distance (B) for the seismograms recorded by the three seismic stations (Telesca et al., 2014a). *FIM*, Fisher information measure.

A study performed by Telesca et al. (2015b) showed the usefulness of informational measures in discriminating between seismograms of tsunamigenic (TS) and nontsunamigenic (NTS) strong earthquakes. Telesca et al. (2015b) calculated the Shannon entropy and the

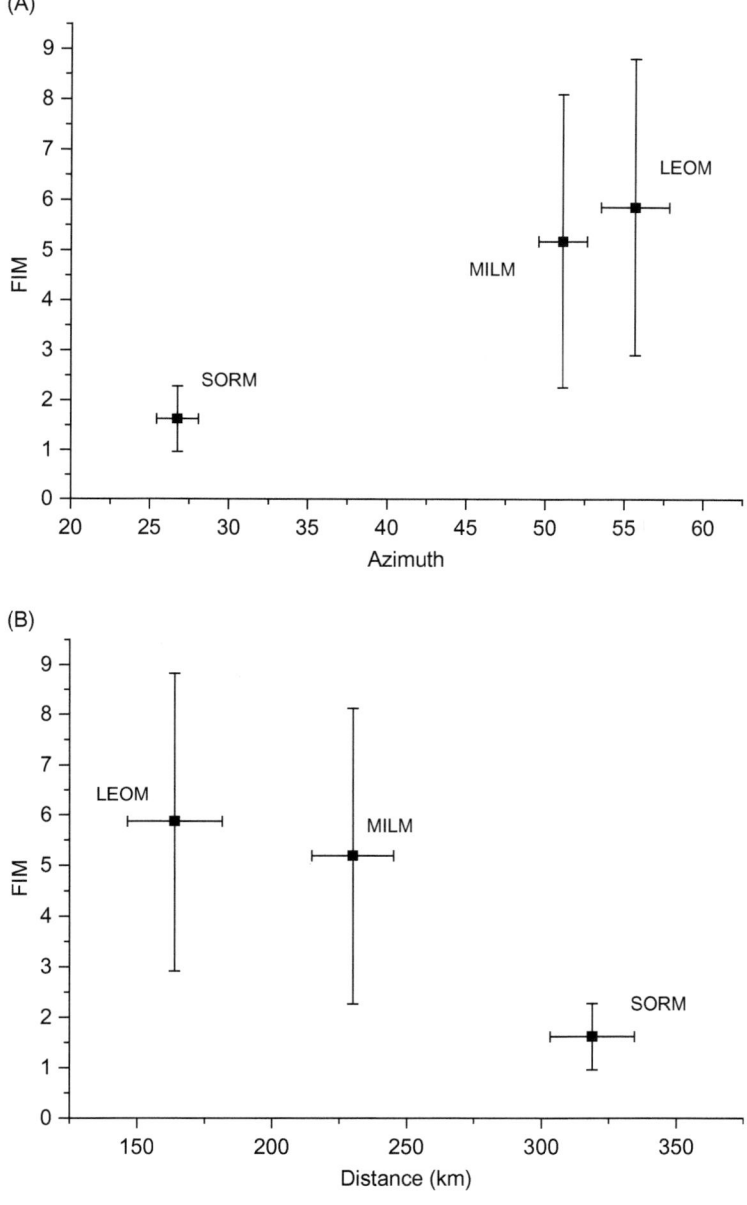

FIGURE 4–26 Mean values of FIM versus mean azimuth and mean epicentral distance (Telesca et al., 2014a). *FIM*, Fisher information measure.

FIM for normalized P-waves of 26 NTS and 17 TS earthquakes. They used the receiver operating characteristic (ROC) analysis to check the performance of the informational parameters. The ROC curve relates the true positive rate, called sensitivity (SE), which is the ratio between the number of correctly classified positives and the total number of positives, versus the false positive rate, indicated as 1-specificity (1-SP), which is the ratio between the number of negatives incorrectly classified and the total number of negatives; this is a well-known method to evaluate the performance of binary classifiers (Kharin and Zwiers, 2003). The position of the classifier in the ROC space determines its goodness or not. The point (0, 1) represents perfect classification; and one point in ROC space is better than another if it is to the northwest of the first. The diagonal line $y = x$ represents the strategy of randomly guessing a class. A classifier is better than another if its ROC graph is located above that of the other, or if the area under the ROC curve (AUC) is larger. Therefore, the performance of the classifier can be judged from the value of the AUC. Fig. 4–27 shows the ROC curve of the Shannon entropy (Fig. 4–27A), FIM (Fig. 4–27B) and complexity (Fig. 4–27C), which is simply the product of the FIM and the Shannon entropy power. The ROC curve of the Shannon entropy is almost overlapping that corresponding to the random guess ($y = x$) and its AUC is about 0.45; this indicates that the Shannon entropy should not be used as a classifier to discriminate between NTS and TS seismograms. The ROC curve of the FIM is a little above the diagonal $y = x$ and its AUC is about 0.58. The ROC curve of the complexity is well above the diagonal $y = x$ and its AUC is about 0.66, indicating that the informational parameter complexity can be considered as a rather good classifier of seismograms of TS and NTS earthquakes. This could have clear implications in tsunami warning schemes.

4.4 Topological Methods

Recently, a new class of methods of investigation of dynamics of time series has become more popular, based on the topological properties of time series. The seminal paper by Lacasa et al. (2008) stimulated several studies in different scientific fields for its potential in describing the dynamic features of time series using the VG method. This method maps time series into networks that reflect several properties of the time series. In turn, investigating networks derived from time series through the VG contribute to disclosing nontrivial information about the time series.

The visibility criterion is defined as follows: two arbitrary values of the time series (t_a, y_a) and (t_b, y_b) are visible to each other if any other data (t_c, y_c) placed between them fulfils the following constraint:

$$y_c < y_b + (y_a - y_b)\frac{t_b - t_c}{t_b - t_a}. \tag{4.19}$$

A sketch of the method applied to seismicity of Italy is illustrated in Fig. 4–28. The VG method always holds: (1) Connection: each node is visible at least by its nearest neighbours (left and right); (2) Absence of directionality: no direction is defined in the links; and (3) Invariance under affine transformations (rescaling of both axes and horizontal and vertical translations) of the time series (Lacasa et al., 2008).

The VG-based network or graph of the time series saves its main properties: periodic time series are converted into regular graphs, random time series into random graphs, and

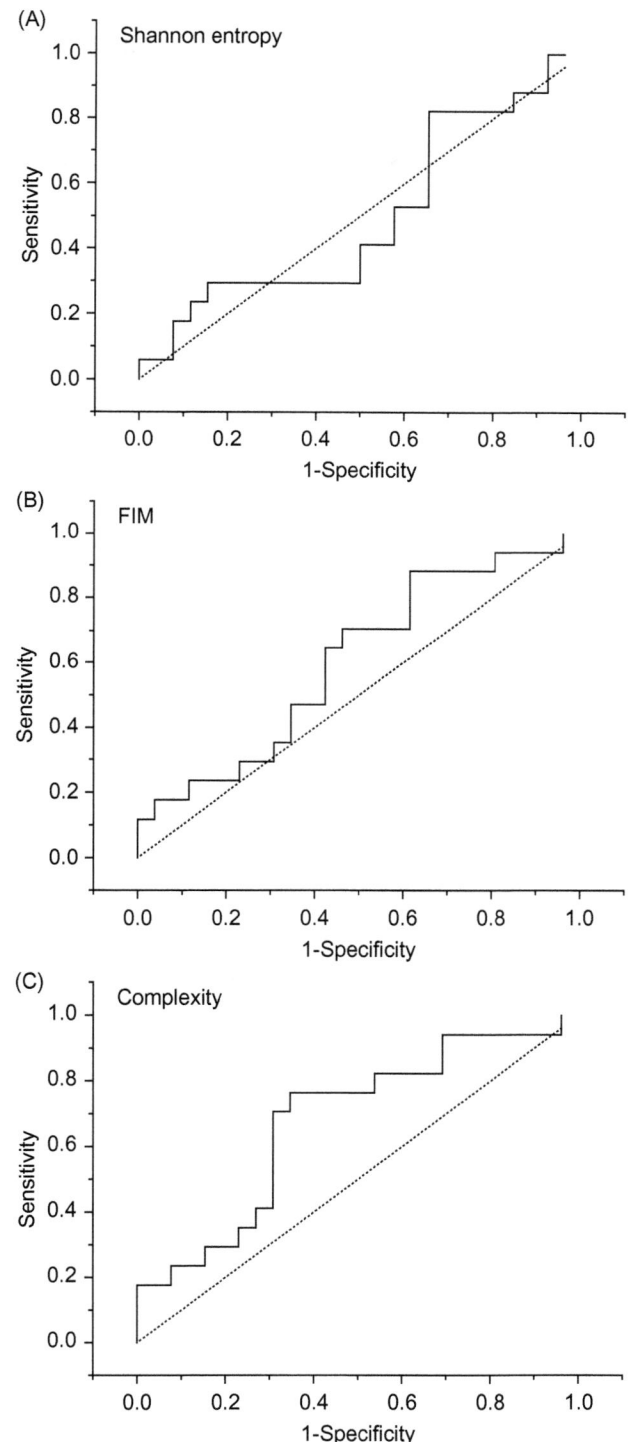

FIGURE 4–27 ROC curves for the Fisher–Shannon parameters of seismograms of tsunamigenic and nontsunamigenic earthquakes (Telesca et al., 2015b). *ROC*, receiver operating characteristic.

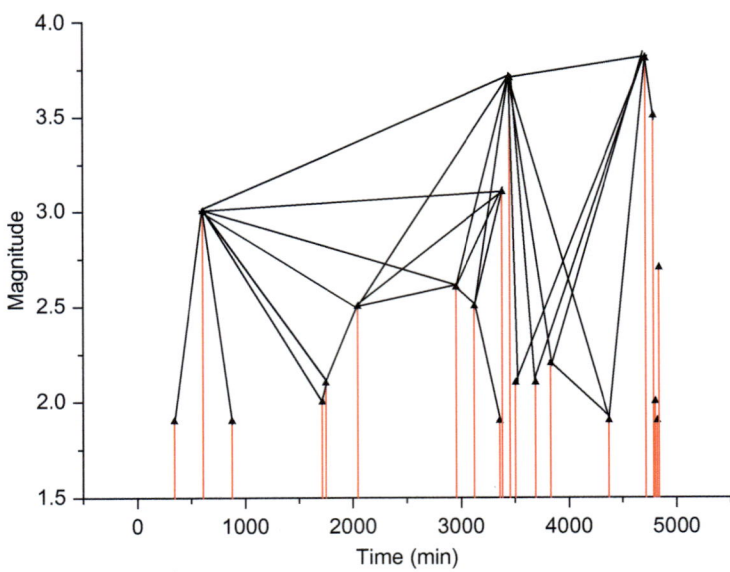

FIGURE 4–28 The first 20 magnitude data (red vertical arrows) of the Italian seismic sequence from 16 April 2005 to 31 December 2010 connected by the visibility rays (black lines) (Telesca and Lovallo, 2012).

fractal time series into scale-free networks (Lacasa et al., 2008; Campanharo et al., 2011; Donner et al., 2011). Lacasa et al. (2009) have shown that the VG derived from a fractional Brownian motion (fBm) is scale-invariant, with a power-law distribution of the degree distribution, $P(k) \sim k^{-\gamma}$, where k is the degree of a specified node, and γ is the power-law exponent linearly related to the Hurst exponent H of the fBm, $\gamma = 3 - 2H$. Through the relationship between H and the spectral exponent β of a $1/f^{\beta}$ noise, the following relationship holds, $\gamma = 4 - \beta$ (Lacasa et al., 2009).

Generally, the VG is applied to continuous time series. Telesca and Lovallo (2012) first applied it to a sequence of earthquakes that is a point process. Fig. 4–27 illustrates visually the VG graph for a portion of the earthquake magnitudes that occurred in Italy from 16 April 2005 to 31 December 2010 (Telesca and Lovallo, 2012); the magnitudes are indicated by vertical arrows that represent the nodes of the graph; two nodes are connected if visibility exists between them; that is, there exists a straight line connecting the two magnitudes, but not intersecting any intermediate one. One interesting result of this study was the collapsing phenomenon of all the degree distributions calculated for increasing the threshold magnitude from 1.9 to 3.5 (Fig. 4–29), suggesting the existence of a 'universal' scaling law permitting a unified description of the shape of the degree distribution, in terms of a power-law distribution.

A relationship between the seismological parameter b-value of the Gutenberg–Richter law and the topological parameter k–M slope, defined by Telesca et al. (2013) suggests the appropriateness of the VG method in investigating seismic sequences. The k–M slope is the

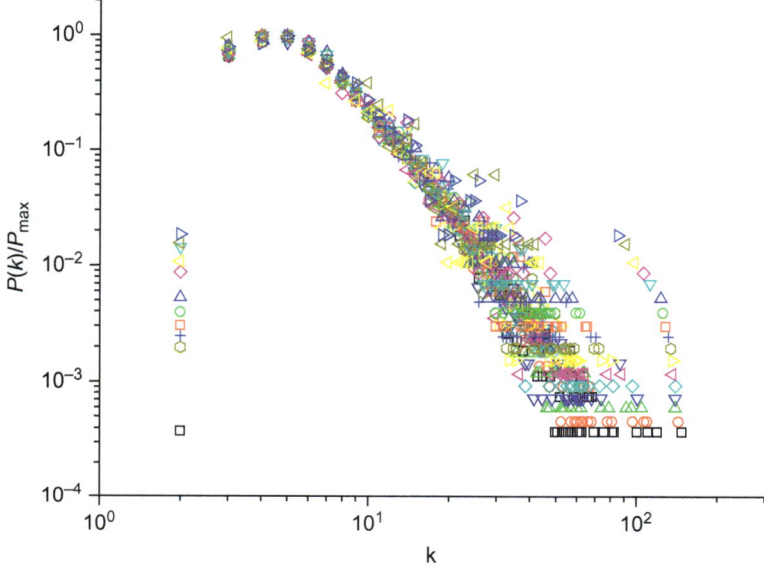

FIGURE 4–29 Connectivity degree distributions for magnitude series with threshold magnitudes ranging from 1.9 to 3.5, normalized to their maximum (Telesca and Lovallo, 2012).

slope of the least square fitting line of the $k-M$ scatterplot (the relationship between the magnitude of each event and its connectivity degree). Studying five seismic regions of the Mexican subduction zone, Telesca et al. (2013) found that all the $k-M$ plots were characterized by an increasing trend of the connectivity degree with an increase in the magnitude, explained by the property of hub typical of the higher magnitudes; furthermore, they observed a positive correlation between the $k-M$ slope and the b-value.

The positive correlation between the $k-M$ slope and the b-value is clearly assessed in Telesca et al. (2014b), who studied the time variation of both parameters for the shallow and deep seismicity that occurred at Pannonia (central Europe) (Fig. 4–30), thus reinforcing the link between the seismological and topological characteristics of the seismic sequences.

The universal character of this relationship was better disclosed by Telesca et al. (2014c), who studied the topological properties of synthetic seismicity generated by a simple stick–slip system with asperities. The experimental setup is based on the interaction between a block and a table, both covered by sandpapers of certain degrees of roughness, thus quantifying the status of asperity. The ageing of the fault is simulated by implementing successive runs of the same block without changing the covering sandpaper [for details on the experiment, please refer to Telesca et al. (2014c)]. Fig. 4–31 shows the relationship ($R^2 \sim 0.97$) between the b-value and the $k-M$ slope for two synthetic sequences and for the real data studied in Telesca et al. (2013). The results strengthen the suggestion that the VG method could represent a way to analyse the earthquake magnitude distribution in a

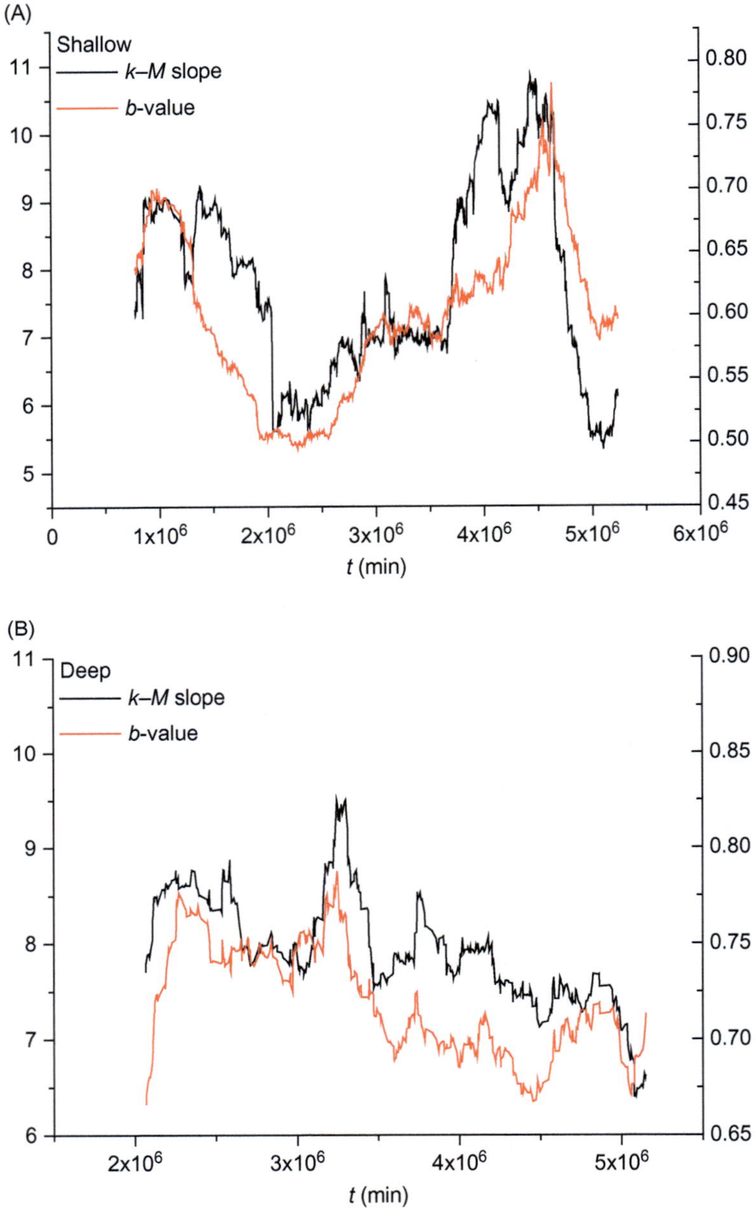

FIGURE 4–30 Time variation of the $k-M$ slope (black) and b-value (red) for the shallow (A) and deep (B) Pannonian seismicity (Telesca et al., 2014b).

more general way than the Gutenberg–Richter law, because it takes into account not only the magnitudes but also the occurrence times of the events, due to the connectivity law [Eq. (4.19)] by which the seismic events are linked with each other; thus suggesting that the

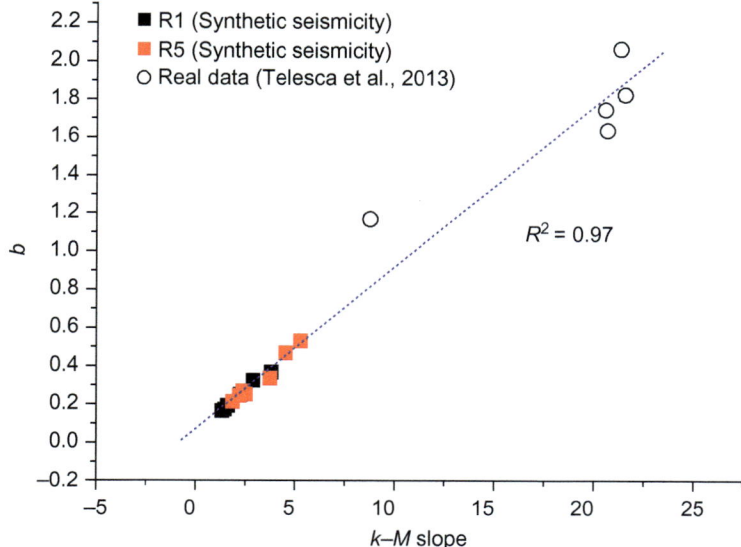

FIGURE 4–31 Relationship between the $k-M$ slope and the b-value for the synthetic seismicity (black and red squares) compared with that for the real seismicity data (white circles) (Telesca et al., 2014c).

classical Gutenberg−Richter law could be considered as a particular case of the relationship derived by the VG method.

The VG method has been used also for proposing possible candidates in seismic forecasting. Telesca et al. (2015c) analysed the time variation of the $k-M$ slope of the seismicty of Kachchh (western India) from 2003 to 2012, where the largest earthquake occurred on 7 March 2006 ($M = 5.7$). The occurrence of M4.5+ events is associated with peaks of the $k-M$ slope (Fig. 4−32); and in particular the largest shock is preceded by a sharp increase and decrease in the $k-M$ slope. The increase and sharp decrease in the $k-M$ slope before the occurrence of a large event is very similar to the b-value that also shows such an increase/decrease before a large earthquake occurs. Such similarity is strengthened by the positive correlation between the $k-M$ slope and the b-value shown above.

Similarly, the topological parameter defined by Telesca et al. (2016d), the *window mean interval connectivity time* $<T_c>$, which informs about the mean linkage time between earthquakes, has revealed its potential to be a good candidate as an earthquake precursor. Considering a sliding window of N events, for each event in the window the *interval connectivity time* t_c is calculated as the time interval between two visible events [events that satisfy the visibility condition of Eq. (4.19)]; then for each event the *mean interval connectivity time* $\langle t_c \rangle$ is computed, averaging over all the intervals t_c for that event; finally, for each window averaging among all the $<t_c>$ *window mean interval connectivity time* $<T_c>$ is obtained. Telesca et al. (2016d) showed that the time variation of $<T_c>$ in the 2003−12 aftershock depleted catalogue of the Kachchh Gujarat (Western India) seismicity appeared to significantly decrease before 7 March 2006 event (Fig. 4−33).

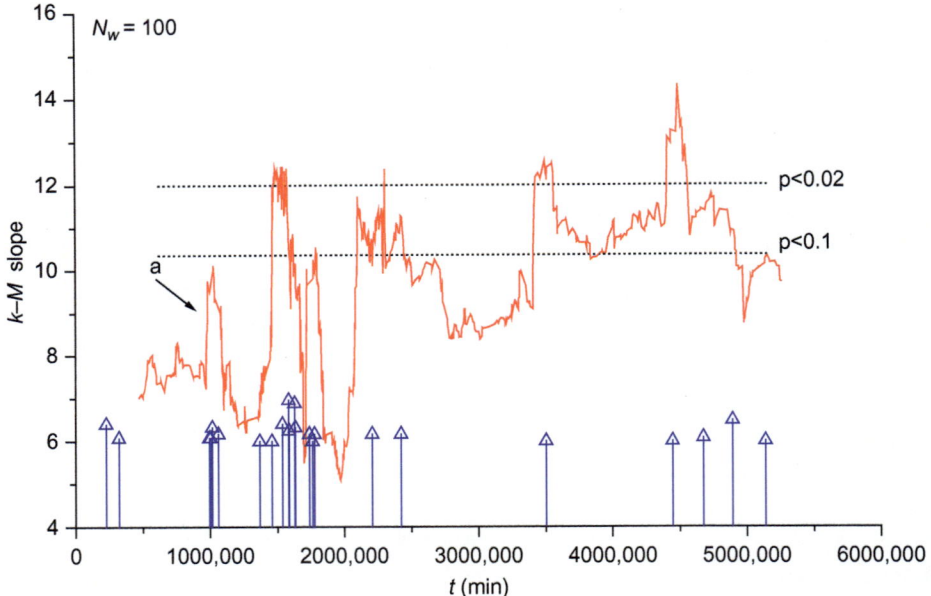

FIGURE 4–32 Time variation of the $k-M$ slope of the seismicity of Kachchh (India), where M4.5 + earthquakes are indicated by vertical blue arrows. The largest earthquake occurred on 7 March 2006 ($M = 5.7$), and was preceded by a sharp increase and decrease of the parameter. The black arrow labelled 'a' indicates the increase of the $k-M$ slope with a P-value larger than 0.1 (Telesca et al., 2015c).

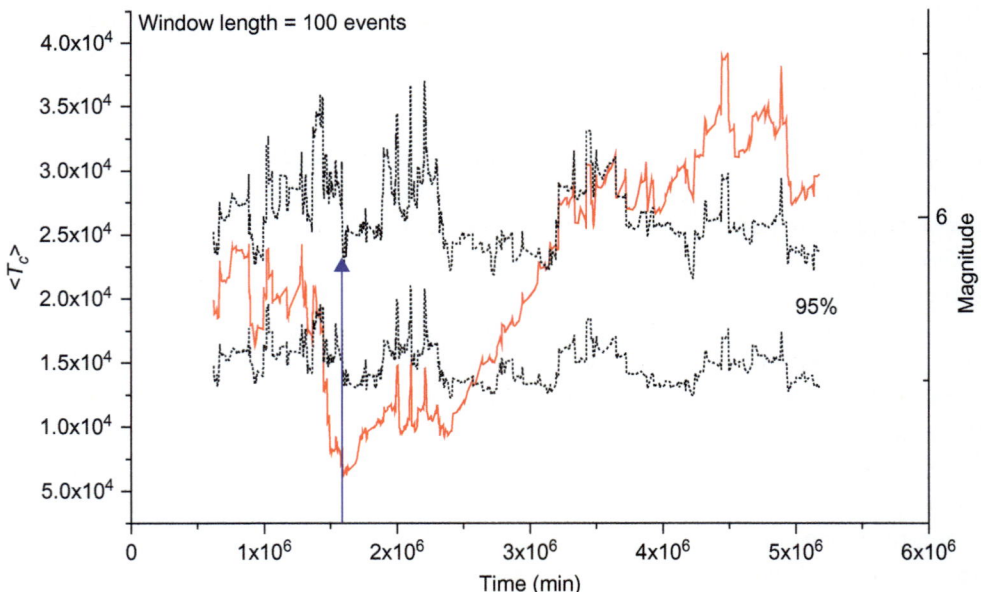

FIGURE 4–33 Time variation of $\langle T_c \rangle$ (red) and the 95% confidence band (black dotted), using a window length of 100 events. The largest earthquakes occurred on 7 March 2006 ($M = 5.7$) (indicated by the blue vertical arrow) (Telesca et al., 2016d).

4.5 Conclusions

In this chapter an exhaustive overview of the most commonly used methods of investigation of seismic time series is shown. The three different types of statistical methodologies (fractal, informational and topological) allow evidence to be gathered for different characteristics and to study different properties of the same seismic phenomena. The fractal methods are based on a self-similar view of the seismic series and are used to identify and quantify correlation properties; the informational methods allow evidence to be collected for organization/order degree in a seismic time series linked with the properties of its probability density function; and the VG method permits converting a seismic sequence in a network with several topological properties that are useful not only for a nonstandard description of the seismic phenomenon but also for identifying candidate parameters in the context of seismic forecasting.

References

Aggarwal, S.K., Lovallo, M., Khan, P.K., Rastogi, B.K., Telesca, L., 2015. Multifractal detrended fluctuation analysis of magnitude series of seismicity of Kachchh region, Western India. Phys. A 426, 56−62.

Alkaz, V., 2005. Earthquake hazard assessment for the Territory Republic of Moldova. Bull. Inst. Geol. Seismol. ASM 1, 5−11.

Allan, D.W., 1966. Statistics of atomic frequency standards. Proc. IEEE 54, 221−230.

Angulo, J.C., Antolin, J., Sen, K.D., 2008. Fisher−Shannon plane and statistical complexity of atoms. Phys. Lett. A 372, 670−674.

Ashkenazy, Y., Baker, D.R., Gildor, H., Havlin, S., 2003. Nonlinearity and multifractality of climate change in the past 420,000 years. Geophys. Res. Lett. 30, 2146−2149.

Campanharo, A.S.L.O., Sirer, M.I., Malgren, R.D., Ramos, F.M., Amaral, L.A.N., 2011. Duality between time series and network. PLoS ONE 6, e23378.

Currenti, G., Del Negro, C., Lapenna, V., Telesca, L., 2005. Multifractality in local geomagnetic field at Etna volcano, Sicily (southern Italy). Nat. Hazards Earth Sys. Sci. 5, 555−559.

de Araujo, D.B., Tedeschi, W., Santos, A.C., Elias Jr., J., Neves, U.P.C., Baffa, O., 2003. Shannon entropy applied to the analysis of event-related fMRI time series. NeuroImage 20, 311−317.

Dembo, A., Cover, T.A., Thomas, J.A., 1991. Information theoretic inequalities. IEEE Trans. Inf. Theory 37, 1501−1518.

De Santis, A., Cianchini, G., Favali, P., Beranzoli, L., Boschi, E., 2011. The Gutenberg−Richter law and entropy of earthquakes: two case studies in Central Italy. Bull. Seismol. Soc. Am. 101, 1386−1395.

Dietrich, J., 1991. A constitutive law for rate of earthquake production and its application to earthquake clustering. J. Geophys. Res. 99, 2601−2618.

Donner, R.V., Small, M., Donges, J.F., Marwan, N., Zou, Y., Xiang, R., et al., 2011. Int. J. Bifurcation Chaos 21, 1019−1048.

Esquivel, R.O., Angulo, J.C., Antolin, J., Dehesa, J.S., Lopez-Rosa, S., Flores-Gallegos, N., 2010. Analysis of complexity measures and information planes of selected molecules in position and momentum spaces. Phys. Chem. Chem. Phys. 12, 7108−7116.

Fisher, R.A., 1925. Theory of statistical estimation. Proc. Cambridge Philos. Soc. 22, 700−725.

Frieden, B.R., 1990. Fisher information, disorder, and the equilibrium distributions of physics. Phys. Rev. A 41, 4265−4276.

Fuhrman, S., Cunningham, M.J., Wen, X., Zweiger, G., Seilhamer, J.J., Somogyi, R., 2000. The application of Shannon entropy in the identification of putative drug targets. Biosystems 55, 5—14.

Gamero, L., Plastino, A., Torres, M.E., 1997. Wavelet analysis and nonlinear dynamics in a nonextensive setting. Phys. A 246, 487—509.

Gardner, J.K., Knopoff, L., 1974. Is the sequence of earthquakes in southern California, with aftershocks removed, Poissonian? Bull. Seismol. Soc. Am. 64, 1363—1367.

Giardini, D., Wiemer, S., Fäh, D., Deichmann, N., Sellami, S., Jenny, S., et al., 2004. Seismic Hazard Assessment of Switzerland. Swiss Seismological Service, ETH Zurich.

Godano, C., Tosi, P., De Rubeis, V., Augliera, P., 1999. Scaling properties of the spatio-temporal distribution of earthquakes: a multifractal approach applied to a Californian catalogue. Geophys. J. Int. 136, 99—108.

Goltz, B., 1998. Fractals and Chaotic Properties of Earthquakes. Springer, New York.

Gudmundsson, A., Lecoeur, N., Mohajeri, N., Thordarson, T., 2014. Dike emplacement at Bardarbunga, Iceland, induces unusual stress changes, caldera deformation, and earthquakes. Bull. Volcanol. 76, 869. Available from: https://doi.org/10.1007/s00445-014-0869-8.

Gutenberg, B., Richter, C.F., 1944. Frequency of earthquakes in California. Bull. Seismol. Soc. Am. 34, 185—188.

Hilborn, R.C., 1994. Chaos and Nonlinear Dynamics. Oxford University Press, Oxford.

Kagan, Y.Y., Jackson, D.D., 1991. Long-term earthquake clustering. Geophys. J. Int. 104, 117—133.

Kantelhardt, J.W., Zschiegner, S.A., Koscielny-Bunde, E., Havlin, S., Bunde, A., Stanley, H.E., 2002. Multifractal detrended fluctuation analysis of nonstationary time series. Phys. A 316, 87—114.

Kharin, V.V., Zwiers, F.V., 2003. On the ROC score of probability forecasts. J. Clim. 16, 4145—4150.

Lacasa, L., Luque, B., Ballesteros, F., Luque, J., Nuño, J.C., 2008. From time series to complex networks: the visibility graph. PNAS 105, 4972—4975.

Lacasa, L., Luque, B., Luque, J., Nuño, J.C., 2009. The visibility graph: a new method for estimating the Hurst exponent of fractional Brownian motion. Europhys. Lett. 86, 30001.

López, C., Blanco, M.J., Abella, R., Brenes, B., CabreraRodríguez, V.M., Casas, B., et al., 2012. Monitoring the unrest of El Hierro (Canary Islands) before the onset of the 2011 Submarine Eruption. Geophys. Res. Lett. 39, L13303.

López, C., Martí, J., Abella, R., Tárraga, M., 2014. Applying fractal dimensions and energy-budget analysis to characterize fracturing processes during magma migration and eruption: 2011—2012 El Hierro (Canary Islands) submarine eruption. Surv. Geophys. 35, 1023—1044.

Lovallo, M., Telesca, L., 2011. Complexity measures and information planes of X-ray astrophysical sources. J. Stat. Mech. P03029.

Lowen, S.B., Teich, M.C., 1993. Estimating the dimension of a fractal point process. In: 64/SPIE Chaos in Biology and Medicine. Vol. 2036.

Lowen, S.B., Teich, M.C., 1995. Estimation and simulation of fractal stochastic point processes. Fractals 3, 183—210.

Mandelbrot, B.B., 1983. The Fractal Geometry of Nature. W. H. Freeman, New York.

Martí, J., Pinel, V., López, C., Geyer, A., Blanco, M.J., Abella, R., et al., 2013a. Causes and mechanisms of the 2011—2012 El Hierro (Canary Islands) submarine eruption. J. Geophys. Res. 118, 823—839.

Martí, J., Castro, A., Rodríguez, C., Costa, F., Carrasquilla, S., Pedreira, R., et al., 2013b. Correlation of magma evolution and geophysical monitoring during the 2011—2012 El Hierro (Canary Islands) submarine eruption. J. Petrol. 54, 1349—1373.

Martin, M.T., Pennini, F., Plastino, A., 1999. Fisher's information and the analysis of complex signals. Phys. Lett. A 256, 173—180.

Martin, M.T., Plastino, A.R., Plastino, A., 2000. Tsallis-like information measures and the analysis of complex signals. Phys. A 275, 262−271.

Martin, M.T., Perez, J., Plastino, A., 2001. Fisher information and nonlinear dynamics. Phys. A 291, 523−532.

Matcharashvili, T., Chelidze, T., Peinke, J., 2008. Increase of order in seismic processes around large reservoir induced by water level periodic variation. Nonlinear Dyn. 51, 399−407.

Peng, C.K., Buldyrev, S.V., Havlin, S., Simons, M., Stanley, H.E., Goldberger, A.L., 1994. Mosaic organization of DNA nucleotides. Phys. Rev. E 49 (2), 1685−1689.

Radulian, M., Panza, G.F., Popa, M., Grecu, B., 2006. Seismic wave attenuation for Vrancea events revisited. J. Earthquake Eng. 10, 411−427.

Reasenberg, P., 1985. Second-order moment of central California seismicity, 1969−82. J. Geophys. Res. 90, 5479−5495.

Romera, E., Dehesa, J.S., 2004. The Fisher−Shannon information plane, an electron correlation tool. J. Chem. Phys. 120, 8906−8912.

Shannon, C.E., 1948. A mathematical theory of communication. Bell Syst. Tech. J. 27, 379−423.

Shimizu, Y., Thurner, S., Ehrenberger, K., 2002. Multifractal spectra as a measure of complexity in human posture. Fractals 10, 103−116.

Shinomoto, S., Miura, K., Koyama, S., 2005. A measure of local variation of inter-spike intervals. BioSystems 79, 67−72.

Tárraga, M., Martí, J., López, C., Carniel, R., Abella, R., 2014. Volcanic tremors: good indicators of change in plumbing systems during volcanic eruptions. J. Volcanol. Geother. Res. 273, 33−40.

Teich, M.C., Heneghan, C., Lowen, S.B., Turcott, R.G., 1996. Estimating the fractal exponent of point processes in biological systems using wavelet- and Fourier-transform methods. In: Aldroubi, A., Unser, M. (Eds.), Wavelets in Medicine and Biology. CRC Press, Boca Raton, FL, pp. 383−412.

Telesca, L., Lapenna, V., 2006. Measuring multifractality in seismic sequences. Tectonophysics 423, 115−123.

Telesca, L., Lovallo, M., 2009. Non-uniform scaling features in central Italy seismicity: a non-linear approach in investigating seismic patterns and detection of possible earthquake precursors. Geophys. Res. Lett. 36, L01308.

Telesca, L., Lovallo, M., 2011. Analysis of time dynamics in wind records by means of multifractal detrended fluctuation analysis and Fisher-Shannon information plane. J. Stat. Mech. P07001.

Telesca, L., Lovallo, M., 2012. Analysis of seismic sequences by using the method of visibility graph. Europhys. Lett. 97, 50002.

Telesca, L., Toth, L., 2016. Multifractal detrended fluctuation analysis of Pannonian earthquake magnitude series. Phys. A 448, 21−29.

Telesca, L., Cuomo, V., Lapenna, V., Macchiato, M., 2001. Intermittent-type temporal fluctuations in seismicity of the Irpinia (southern Italy) region. Geophys. Res. Lett. 28, 3765−3768.

Telesca, L., Cuomo, V., Lapenna, V., Macchiato, M., 2002. 1/fα fluctuations of seismic sequences. Fluctuation Noise Lett. 2, L357−L367.

Telesca, L., Lapenna, V., Macchiato, M., 2003. Spatial variability of time-correlated behaviour in Italian seismicity. Earth Planet. Sci. Lett. 212, 279−290.

Telesca, L., Lapenna, V., Macchiato, M., 2005. Multifractal fluctuations in seismic interspike series. Phys. A 354, 629−640.

Telesca, L., Lovallo, M., Lapenna, V., Macchiato, M., 2007. Long-range correlations in 2-dimensional spatiotemporal seismic fluctuations. Phys. A 377, 279−284.

Telesca, L., Lovallo, M., Lapenna, V., Macchiato, M., 2008. Space-magnitude dependent scaling behaviour in seismic interevent series revealed by detrended fluctuation analysis. Phys. A 387, 3655−3659.

Telesca, L., Lovallo, M., Ramirez-Rojas, A., Angulo-Brown, F., 2009. A nonlinear strategy to reveal seismic precursory signatures in earthquake-related self-potential signals. Phys. A 388, 2036—2040.

Telesca, L., Lovallo, M., Carniel, R., 2010. Time-dependent Fisher Information Measure of volcanic tremor before 5 April 2003 paroxysm at Stromboli volcano, Italy. J. Volcanol. Geoterm. Res. 195, 78—82.

Telesca, L., Lovallo, M., Hsu, H.-L., Chen, C.-C., 2011. Analysis of dynamics in magnetotelluric data by using the Fisher-Shannon method. Phys. A 390, 1350—1355.

Telesca, L., Matcharashvili, T., Chelidze, T., 2012a. Investigation of the temporal fluctuations of the 1960—2010 seismicity of Caucasus. Nat. Hazards Earth Syst. Sci. 12, 1905—1909.

Telesca, L., Lovallo, M., Mohamed, A.E.-E.A., ElGabry, M., El-hady, S., Abou Elenean, K.M., et al., 2012b. Informational analysis of seismic sequences by applying the Fisher Information Measure and the Shannon entropy: An application to the 2004—2010 seismicity of Aswan area (Egypt). Phys. A 391, 2889—2897.

Telesca, L., Lovallo, M., Ramirez-Rojas, A., Flores-Marquez, L., 2013. Investigating the time dynamics of seismicity by using the visibility graph approach: application to seismicity of Mexican subduction zone. Phys. A 392, 6571—6577.

Telesca, L., Lovallo, M., Alcaz, V., Ilies, I., 2014a. Investigating the inner time properties of seismograms by using the Fisher Information Measure. Phys. A 409, 154—161.

Telesca, L., Lovallo, M., Toth, L., 2014b. Visibility graph analysis of 2002—2011 Pannonian seismicity. Phys. A 416, 219—224.

Telesca, L., Lovallo, M., Ramirez-Rojas, A., Flores-Marquez, L., 2014c. Relationship between the frequency magnitude distribution and the visibility graph in the synthetic seismicity generated by a simple stick—slip system with asperities. PLoS ONE 6, e106233.

Telesca, L., Lovallo, M., Martì Molist, J., López Moreno, C., Abella Meléndez, R., 2015a. Multifractal investigation of continuous seismic signal recorded at El Hierro volcano (Canary Islands) during the 2011—2012 pre- and eruptive phases. Tectonophysics 642, 71—77.

Telesca, L., Chamoli, A., Lovallo, M., Stabile, T.A., 2015b. Investigating the tsunamigenic potential of earthquakes from the analysis of the informational and multifractal properties of seismograms. Pageoph 172, 1933—1943.

Telesca, L., Lovallo, M., Aggarwal, S.K., Khan, P.K., 2015c. Precursory signatures in the visibility graph analysis of seismicity: an application to the Kachchh (Western India) seismicity. Phys. Chem. Earth 85—86, 195—200.

Telesca, L., Lovallo, M., Lopez, C., Martì Molist, J., 2016a. Multiparametric statistical investigation of seismicity occurred at El Hierro (Canary Islands) from 2011 to 2014. Tectonophysics 672—673, 121—128.

Telesca, L., Eisner, L., Stabile, T.A., Vlček, J., 2016b. Investigating the time clustering of induced microseismicity generated by hydraulic fracturing. Europhys. Lett. 116, 59002.

Telesca, L., Lovallo, M., Golay, J., Kanevski, M., 2016c. Comparing seismicity declustering techniques by means of the joint use of Allan Factor and Morisita index. Stoch. Environ. Res. Risk Analysis 30, 77—90.

Telesca, L., Lovallo, M., Aggarwal, S.K., Khan, P.K., Rastogi, B.K., 2016d. Visibility graph analysis of 2003—2012 earthquake sequence in Kachchh region, Western India. Pageoph 173, 125—132.

Thurner, S., Lowen, S.B., Feurstein, M., Heneghan, C., Feichtinger, H.G., Teich, M.C., 1997. Analysis, synthesis, and estimation of fractal-rate stochastic point processes. Fractals 5, 565—595.

van Stiphout, T., Zhuang, J., Marsan, D., 2012. Seismicity Declustering, Community Online Resource for Statistical Seismicity Analysis. https://doi.org/10.5078/corssa-52382934. Available at http://www.corssa.org.

Vignat, C., Bercher, J.F., 2003. Analysis of signals in the Fisher−Shannon information plane. Phys. Lett. A 312, 27−33.

Viswanathan, G.M., Peng, C.-K., Stanley, H.E., Goldberger, A.L., 1997. Deviations from uniform power law scaling in nonstationary time series. Phys. Rev. E 55, 845−849.

Wiemer, S., 2001. A software package to analyze seismicity: ZMAP. Seismol. Res. Lett. 72, 373−382.

Zou, Y., Small, M., Liu, Z., Kurths, J., 2014. Complex network approach to characterize the statistical features of the sunspot series. New J. Phys. 16, 013051.

5

Modelling of Persistent Time Series by the Nonlinear Langevin Equation

Zbigniew Czechowski

POLISH ACADEMY OF SCIENCES, WARSAW, POLAND

CHAPTER OUTLINE

5.1 Introduction

Geophysical processes belong to the group of complex processes. Due to their specific properties, direct observations of these processes are impossible in many cases. Therefore, we must confine ourselves to monitoring one or a few parameters which characterize the phenomenon. Quite often, we have at our disposal only a scalar time series which contains hidden and partial information about the process under investigation. The data reflect the complexity of the phenomenon and its important features, e.g., power law distributions, nonlinear dynamics or long-range correlations. The typical nonregularity of geophysical data induces us to accept the assumption about stochastic basis of geophysical time series, so the registered data can be treated as realizations of stochastic processes. Therefore, stochastic models were constructed and tested in order to explain the origins of these characteristic features. In the case of power law distributions it was shown that the mechanism of privilege introduced to models led to long-tail solutions (Czechowski, 1993, 2001, 2003, 2005, 2010, 2015; Czechowski and Rozmarynowska, 2008).

Complexity of Seismic Time Series. DOI: https://doi.org/10.1016/B978-0-12-813138-1.00005-5

Long-range persistent time series have been documented and discussed for many processes in the Earth sciences. Examples include river run-off and precipitation (Hurst, 1951; Mandelbrot and van Ness, 1968; Montanari et al., 1996; Kantelhardt et al., 2003; Mudelsee, 2007; Khaliq et al., 2009), atmospheric variability (Govindan et al., 2002), temperatures over short to very long timescales (Pelletier and Turcotte, 1999; Fraedrich and Blender, 2003), fluctuations of the North-Atlantic Oscillation index (Collette and Ausloos, 2004), surface wind speeds (Govindan and Kantz, 2004), the geomagnetic auroral electrojet index (Chapman et al., 2005), geomagnetic variability (Anh et al., 2007) and ozone records (Kiss et al., 2007).

Complex features of time series require using nonlinear stochastic models. However, nonlinear methods are much less understood than classical methods for linear cases. The Langevin equation is a widely accepted nonlinear stochastic model of time series, because the procedure of reconstruction of the equation from data was elaborated (see review paper by Friedrich et al., 2011). The Langevin equation introduces nonlinearity in drift and diffusion terms and leads to a wide class of distributions. Moreover, it can be applied as a generator of multifractal time series (Telesca et al., 2015; Czechowski et al., 2016). Therefore, using the Langevin equation may provide some progress in constructing nonlinear models from data. The Langevin model was applied in geophysics (Yin and Ranalli, 1995; Rundle et al., 1999; Klein et al., 2000; Matthews et al., 2002; Lind et al., 2005; Czechowski and Telesca, 2011; Telesca and Czechowski, 2012; Matcharashvili et al., 2013).

However, the standard Langevin equation describes Markov processes only. The aim of this work is to generalize the approach for some class of non-Markov processes, namely, persistent/antipersistent processes. Therefore, first of all, a novel generalized Langevin-like equation and the associated Fokker−Planck equation are derived. Then, new procedures of reconstruction of the modified Langevin equation from the time series are proposed and tested. It should be underlined that procedures (Siegert et al., 1998; Gottschall and Peinke, 2008) for the standard Langevin equation fail in non-Markov cases, because they lead to the wrong reconstruction of the drift term. Corrections due to the finite sampling rate or due to large magnitudes of parameters in drift and diffusion terms are taken into account in our procedures. Having the drift and diffusion functions enables derivation of short-time transition probability, which can be useful in forecasting.

5.2 Modified Langevin Equation

The standard Langevin equation for the one-dimensional stochastic process $y(t)$ has the form

$$\frac{dy(t)}{dt} = a(y(t)) + \sqrt{b(y(t))}\zeta(t), \tag{5.1}$$

where the random force $\zeta(t)$ is the Gaussian variable with zero mean and δ correlation function (white noise). In this work we are using the following discrete version (according to the forward Euler scheme; Grasman and van Herwaarden, 1999) of the Langevin equation

$$y(t + \Delta t) = y(t) + a(y(t))\Delta t + \sqrt{b(y(t))}\sqrt{\Delta t}\,\xi_t\,, \tag{5.2}$$

where ξ_t is an independent random variable with normal density. The Langevin equation can describe an evolution of Markov processes only. For non-Markov processes, nonlocal effects or memory must be taken into account.

A direct method of generalization of the Langevin equation for non-Markov processes consists of substitution of the white noise by the fractional noise. However, this method has its weaknesses. Due to the construction procedure of the fractional noise the correlations there are infinitely long, which is not a feature of natural time series.

Here we propose a simple generalization of the Langevin equation for persistent/antipersistent processes. We follow some ideas and results of our earlier papers (Czechowski, 2016; Czechowski and Telesca, 2013).

We assume that the next state, $y(t + \Delta t)$, of the process is dependent not only on the present state, $y(t)$, but also on signs s_{tk} of p previous jumps $\Delta y = y(t-(k-1)\Delta t) - y(t-k\Delta t)$, where $k = 1, 2, \ldots, p$, and

$$s_{tk} \equiv S(y(t - (k - 1)\Delta t) - y(t - k\Delta t)) \equiv \begin{cases} 1 & \text{if } \Delta y \geq 0 \\ -1 & \text{if } \Delta y < 0 \end{cases}. \tag{5.3}$$

To this aim, Eq. (5.2) is modified by introducing a new random function $c(s_t, r_t; \boldsymbol{d})$ which determines the sign of the diffusion term, i.e.,

$$y(t + \Delta t) = y(t) + a(y(t))\Delta t + c(s_t, r_t; \boldsymbol{d})\sqrt{b(y(t))}\sqrt{\Delta t}|\xi_t| \tag{5.4}$$

The function $c(s_t, r_t; \boldsymbol{d})$ depends on vector random variable $\boldsymbol{s_t} = [s_{t1}, s_{t2}, \ldots, s_{tp}]$, random scalar variable r_t and on vector parameter $\boldsymbol{d} = [d_1, d_2, \ldots, d_2{}^p]$. The function can be equal to 1 or -1 randomly, according to the following rules (for clarity, the case $p = 3$ is presented)

$$\begin{aligned} + + + &\Rightarrow P(c = 1) = p_1\,, \\ - - - &\Rightarrow P(c = -1) = p_2\,, \\ - + + &\Rightarrow P(c = 1) = p_3\,, \\ + - - &\Rightarrow P(c = -1) = p_4\,, \\ - - + &\Rightarrow P(c = 1) = p_5\,, \\ + + - &\Rightarrow P(c = -1) = p_6\,, \\ + - + &\Rightarrow P(c = 1) = p_7\,, \\ - + - &\Rightarrow P(c = -1) = p_8\,, \end{aligned} \tag{5.5}$$

where the notation "$- + +$" denotes: $s_{t3} = -1$, $s_{t2} = +1$, $s_{t1} = +1$, $p_i = 1 - d_i$, "c" is the abbreviation for $c(s_t, r_t; \boldsymbol{d})$ and the arrow represents the implication relation. Probability P corresponds to random variable r_t drawn with the uniform distribution in $(0, 1)$.

For example, for $p = 1$ (with $p_1 = p_2$) the function $c(s_t, r_t; d)$ can be given by the clear mathematical formula (see Czechowski, 2016).

$$c(s_t, r_t; d) = \begin{cases} 1 & if \quad (s_t = 1 \wedge r_t \geq d) \vee (s_t = -1 \wedge r_t < d) \\ -1 & if \quad (s_t = -1 \wedge r_t \geq d) \vee (s_t = 1 \wedge r_t < d) \end{cases} \tag{5.6}$$

where $d = 1 - p_1 = 1 - p_2$.

The sign of function $c(s_t, r_t; d)$ is calculated before every step at discrete times $t + \Delta t$ (i.e., with the same rate as a magnitude of the noise ξ_t is drawn). For persistent processes of order p the function $c(s_t, r_t; d)$ keeps the tendency to increase/decrease $y(t)$ in the next step according to given probabilities p_i, where $i = 1, 2, \ldots, 2^p$. When all $p_i = 1/2$ then Eq. (5.4) reduces to the standard Langevin Eq. (5.2) without the modification.

5.3 Reconstruction Procedures

The main task in time series modelling, besides choosing the proper mathematical model, is to propose a reconstruction procedure of the model from data.

Following the well-known correspondence of the Langevin and Fokker–Planck equations, Siegert et al. (1998) introduced a numerical procedure for reconstruction of the standard Langevin equation, which was based on numerical estimations of joint distribution function (i.e., histograms). The method, called the standard reconstruction procedure (SRP), was then developed in other papers (Sura and Barsugli, 2002; Kleinhaus et al., 2005; Kleinhaus and Friedrich, 2007; Gottschall and Peinke, 2008; Anteneodo and Riera, 2009; Lamauroux and Lehnertz, 2009; Hindriks et al., 2011), however it works well only for standard Langevin Eq. (5.1), which describes the Markov processes. Unfortunately, for persistent processes the procedure leads to a wrong reconstruction of the drift function $a(y)$. Fig. 5–1 shows a comparison of reconstructed and input functions for the example of time series generated by modified Langevin Eq. (5.4) with definition (Eq. 5.6) and with: $a(y) = -y^2/2 + y$, $b(y) = y^2$, $d = 0.2$ and time increment $\Delta t = 0.001$. We can observe an essential deviation $E(y) = a_1(y) - a(y)$ of the reconstructed $a_1(y)$ (represented by points) from the input function. However, the estimation of the diffusion function $b(y)$ is good.

Such a deviation was also noted by Stratonovich (1963). He derived the Fokker–Planck equation for non-Markov processes in the form

$$\frac{\partial}{\partial t} p(y, t) = -\frac{\partial}{\partial y} \left[(a(y) + E(y)) p(y, t) \right] + \frac{1}{2} \frac{\partial^2}{\partial y^2} \left[b(y) p(y, t) \right]. \tag{5.7}$$

where the function $E(y)$, which modifies the drift term, contains a contribution of memory effects. However, in Stratonovich (1963), $E(y)$ was given in a formal form (i.e., by an integral of an averaged expression) only.

We should also note that for a finite time increment Δt or/and for large magnitudes of parameters in drift and diffusion functions, big deviations in reconstruction of these

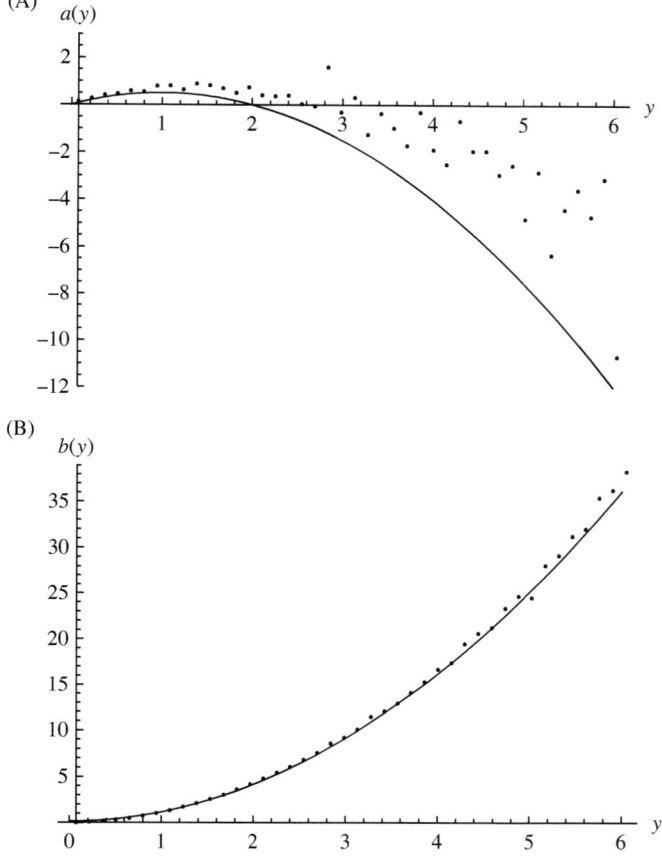

FIGURE 5–1 Comparison of SRP results (points) with input functions, $a(y) = -y^2/2 + y$ and $b(y) = y^2$ (lines), for persistent time series with $d = 0.2$, generated by Eq. (5.4) with $c(s_t, r_t; d)$ given by Eq. (5.6).

functions by SRP can be observed. Therefore, the reconstruction of $b(y)$ can be affected only by the deviation $\beta^\tau(y)$ due to the finite sampling rate, while $a(y)$ can be affected by two sources of deviations in reconstruction procedure: $\alpha^\tau(y)$ due to the finite sampling rate and $\alpha^c(y)$ due to correlations introduced by function $c(s_t, r_t; d)$, i.e.,

$$
\begin{aligned}
a_1(y) &= a(y) + \alpha^\tau(y) + \alpha^c(y), \\
b_1(y) &= b(y) + \beta^\tau(y),
\end{aligned}
\tag{5.8}
$$

where $a_1(y)$ and $b_1(y)$ are reconstructed functions using SRP. Hence, in order to reconstruct the input functions $a(y)$ and $b(y)$ we need to estimate the deviations $\alpha^c(y)$, $\alpha^\tau(y)$ and $\beta^\tau(y)$.

Here we introduce two novel reconstruction procedures of the modified Langevin Eq. (5.4), a purely numerical one and a semianalytical one, which can estimate these deviations.

5.3.1 Modified Numerical Reconstruction Procedure (MNRP)

The algorithm can be summarized in three steps as follows:

Step 1. Reconstruction of $b(y)$.

The first use of SRP to the input persistent time series leads to the first reconstruction $a_1(y)$ of function $a(y)$ and $b_1(y)$ of function $b(y)$:

$$\begin{aligned} a_1(y) &= a(y) + \alpha_1^{c\tau}(y), \\ b_1(y) &= b(y) + \beta_1^{\tau}(y). \end{aligned} \tag{5.9}$$

We generate new time series using the standard Langevin Eq. (5.2) with functions $a_1(y)$ and $b_1(y)$. The second use of SRP to this time series leads to reconstruction $a_2(y)$ of function $a_1(y)$ and $b_2(y)$ of function $b_1(y)$. Therefore, for the diffusion function we have:

$$b_2(y) = b_1(y) + \beta_2^{\tau}(y) = b(y) + \beta_1^{\tau}(y) + \beta_2^{\tau}(y). \tag{5.10}$$

Under the assumption that deviation $|\beta^{\tau}(y)| << |b(y)|$ and then $\beta_1^{\tau}(y) \approx \beta_2^{\tau}(y)$ we obtain

$$b(y) = b_1(y) - \beta_1^{\tau}(y) \approx b_1(y) - \beta_2^{\tau}(y) = b_1(y) - [b_2(y) - b_1(y)] = 2b_1(y) - b_2(y) \equiv b_R(y). \tag{5.11}$$

Step 2. Estimation of parameter ***d***.

A direct method of estimation of parameters p_i from data is based on histograms P^H of joint probability $P(s_{t3}, s_{t2}, s_{t1}, c)$. Then, e.g., according to rules (Eq. 6.5), the estimator for p_1 is given by $p_1^E = P(1, 1, 1, 1)$.

Step 3. Reconstruction of $a(y)$.

Time series generated by the modified Langevin Eq. (5.4) with parameter ***d*** estimated in step 2 and reconstructed functions $a_1(y)$ and $b_R(y)$ is treated as the input to the third use of SRP. At the result, functions $a_3(y)$ and $b_3(y)$ are reconstructed, where

$$a_3(y) = a_1(y) + \alpha_3^{c\tau}(y). \tag{5.12}$$

Assuming that deviation $|\alpha_1^{c\tau}(y)| << |a(y)|$ and then $\alpha_1^{c\tau}(y) \approx \alpha_3^{c\tau}(y)$ we obtain

$$a(y) = a_1(y) - \alpha_1^{c\tau}(y) \approx a_1(y) - \alpha_3^{c\tau}(y) = a_1(y) - [a_3(y) - a_1(y)] = 2a_1(y) - a_3(y) \equiv a_R(y). \tag{5.13}$$

The reconstructed modified Langevin equation has a form of Eq. (5.4) with $a(y) = a_R(y)$, $b(y) = b_R(y)$ and parameter ***d*** estimated in step 2.

5.3.2 Modified Semianalytical Reconstruction Procedure (MsARP)

For $p = 1$ (with $p_1 = p_2$), profiting from the simplicity of function $c(s_t, r_t; d)$ (see Eq. (5.6)), we can calculate analytically deviations $\alpha^c(y)$, $\alpha^{\tau}(y)$ and $\beta^{\tau}(y)$ in Eq. (5.8). To this end the

Fokker–Planck equation associated with the modified Langevin equation was derived (see Appendix):

$$\frac{\partial p(y)}{\partial t} = -\frac{\partial}{\partial y}\left[\left(a(y) - \frac{1-2d}{\pi}b(y)\frac{p_s'(y)}{p_s(y)}\right)p(y)\right]$$
$$+ \frac{1}{2}\frac{\partial^2}{\partial y^2}\left[\left(b(y) + \Delta t\, a^2(y) - \Delta t\frac{2-4d}{\pi}a(y)b(y)\frac{p_s'(y)}{p_s(y)}\right)p(y)\right],$$

(5.14)

where $p_s(y)$ is the stationary distribution function

$$p_s(y) = \frac{c}{b_1(y)}\exp\left(2\int_\varepsilon^y \frac{a_1(x)}{b_1(x)}dx\right)$$

(5.15)

and $p'(y)$ denotes the derivative dp/dy. Therefore,

$$\frac{p_s'(y)}{p_s(y)} = \frac{1}{b_1(y)}\left[2a_1(y) - b_1'(y)\right]$$

(5.16)

and

$$b_1(y) = b(y) + \Delta t\, a^2(y) - \Delta t\frac{2-4d}{\pi}\frac{a(y)b(y)}{b_1(y)}\left[2a_1(y) - b_1'(y)\right],$$
$$a_1(y) = a(y) - \frac{1-2d}{\pi}\frac{b(y)}{b_1(y)}\left[2a_1(y) - b_1'(y)\right].$$

(5.17)

For the aim of reconstruction we revert the system (Eq. (5.17)) to find the following explicit formulas for reconstructed functions $a(y)$ and $b(y)$:

$$a_R(y) = \frac{1 + 2\Delta t\dfrac{a_1(y)z(y)}{b_1(y)} - \sqrt{D(y)}}{2\Delta t\dfrac{z(y)}{b_1(y)}},$$

(5.18)

$$b_R(y) = b_1(y) - \Delta t\, a(y)[2a_1(y) - a(y)],$$

where

$$z(y) \equiv \frac{1-2d}{\pi}\left[2a_1(y) - b_1'(y)\right]$$

(5.19)

and

$$D(y) \equiv \left(1 + 2\Delta t\frac{a_1(y)z(y)}{b_1(y)}\right)^2 - 4\Delta t\frac{z(y)}{b_1(y)} - 4\Delta t\frac{z^2(y)}{b_1(y)}.$$

(5.20)

The formulas depend on the persistence parameter d and on first reconstructions $a_1(y)$ and $b_1(y)$ given by SRP. Without the persistence (i.e., for $d = 1/2$ and for $\Delta t\text{-} > 0$ we have $a_R(y) = a_1(y)$ and $b_R(y) = b_1(y)$.

Therefore, the semianalytical procedure of reconstruction of discrete modified Langevin Eq. (5.4) from time series with memory of order $p = 1$ can be summarized in three steps:

Step 1. First reconstruction of functions $a_1(y)$ and $b_1(y)$ by the procedure SRP.
Step 2. Estimation of the persistence parameter d by the method presented in the purely numerical procedure MNRP.
Step 3. Calculation of final reconstructions of drift and diffusion functions $a_R(y)$ and $b_R(y)$ according to Eqs. (5.18–5.20).

5.4 Testing of the Reconstruction Procedures

In order to test the two procedures we generate time series by using the modified Langevin Eq. (5.4) with different functions $a(y)$ and $b(y)$, different values of the parameter d, different time increments Δt and considering different time series lengths. This enables comparison to be made of the input parameters and functions to the reconstructed ones. We present two examples of efficiency of each procedure.

The MsARP was derived only for persistent processes of the order 1, so the persistence parameter d is a scalar.

In the first example we put $a(y) = -20y$, $b(y) = 500 + 20y^2$ and $d = 0.7$ (antipersistent). The length of the generated time series is $N = 100,000$, and time increment $\Delta t = 0.01$. Fig. 5–2 shows big deviations in the drift and diffusion functions $a_1(y)$ and $b_1(y)$ reconstructed by SRP (*long-dashed lines*) from the input functions (*continuous lines*). The deviation in diffusion function $b(y)$ is due to large magnitudes of the parameter in drift function $a(y) = -20y$. The estimated parameter $d = 0.668$ is close to the input value. Step 3 of the MsARP leads to very good reconstructions of functions $a(y)$ and $b(y)$ (*dashed lines*).

In the second example we assume $a(y) = -y^2 + 1/2$, $b(y) = y$, $d = 0.2$ (persistent). The length of the generated time series is much longer, $N = 1,500,000$. Time increment $\Delta t = 0.1$ is 10 times greater than in the previous case. Fig. 5–3 also shows a large deviation in the drift and diffusion functions reconstructed by SRP, however, here, the deviation in $b(y)$ is mainly due to the large time increment. The estimated parameter $d = 0.223$. We can observe small deflections in the final MsARP reconstruction of the drift function (this is due to some errors in accuracy of least-square fit in SRP) but the final reconstruction of the diffusion function is good.

The purely numerical procedure MNRP is not restricted to time series which have an order of persistency equal to 1 only. We present two examples of synthetic data with persistence of order 3.

In the first case we generate the persistent time series with parameters: $d_1 = d_2 = 0.2$, $d_3 = d_4 = 0.25$, $d_5 = d_6 = 0.3$, $d_7 = d_8 = 0.4$, and assumed functions $a(y) = -\frac{1}{2}\text{Log}(y)$,

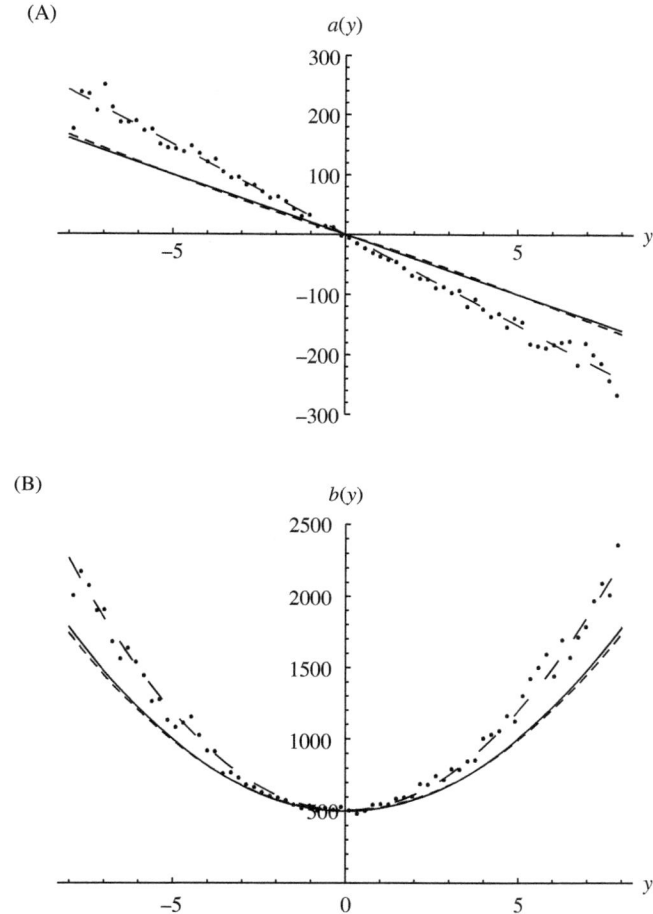

FIGURE 5–2 Comparison of MsARP reconstructions of drift and diffusion functions (*dashed lines*) with input functions (*continuous lines*), for time series generated by Eq. (5.4) with $a(y) = -30y$, $b(y) = 500 + 10y^2$ and $d = 0.7$. Points represent results of SRP and long-dashed lines are least-square fits which give first reconstructions $a_1(y)$ and $b_1(y)$.

$b(y) = y$ ($N = 1,000,000$, $\Delta t = 0.01$). In the first step of MNRP we apply SRP in order to find first reconstructions $a_1(y)$ and $b_1(y)$. These are represented by least-square fits to clouds of points in Fig. 5–4A and Fig. 5–5B, respectively. Because $b_1(y)$ is a good estimation of the input diffusion function we can omit continuation of step 1 here. In the second step of the procedure the estimation for persistence parameters gives: $d_1 = d_2 = 0.205$, $d_3 = d_4 = 0.254$, $d_5 = d_6 = 0.301$, $d_7 = d_8 = 0.406$. Next, in a frame of the third step, we generate a new time series using the modified Langevin Eq. (5.4) with reconstructed functions $a_1(y)$ and $b_R(y) = b_1(y)$ and estimated parameter **d**. We again apply SRP to obtain $a_3(y)$ (see Fig. 5–4B). Relation (5.13) leads to the final reconstruction $a_R(y)$ (*continuous line* in Fig. 5–5A) which is a little deviated from the input function (thick *continuous line*). The

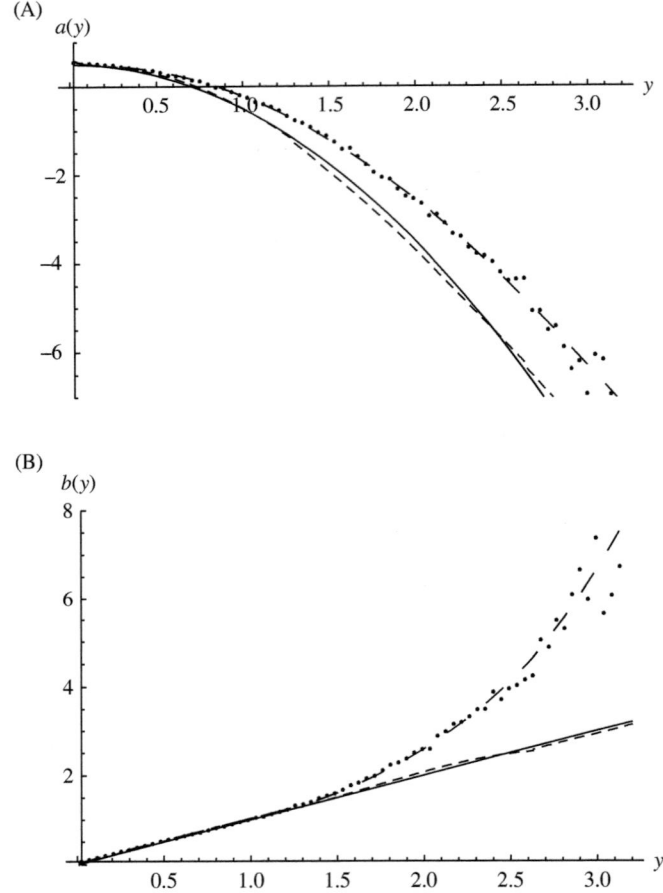

FIGURE 5–3 Comparison of MsARP reconstructions of drift and diffusion functions (*dashed lines*) with input functions (*continuous lines*), for time series generated by Eq. (5.4) with $a(y) = -y^2 + 1/2$, $b(y) = y$ and $d = 0.2$. Points represent results of SRP and long-dashed lines are least-square fits which give first reconstructions $a_1(y)$ and $b_1(y)$.

origin of the deviation corresponds to the assumption in step 3 that $|\alpha_1^{cr}(y)| << |a(y)|$, which cannot be fulfilled in this case.

For the second test of numerical procedure MNRP we generate the antipersistant time series using the modified Langevin Eq. (5.4) with: $a(y) = \frac{1}{2}(1 - 2y)$, $b(y) = y^2$, $d_1 = d_2 = 0.75$, $d_3 = d_4 = 0.65$, $d_5 = d_6 = 0.6$, $d_7 = d_8 = 0.5$, $N = 1,000,000$, $\Delta t = 0.01$. The results of MNRP are presented in Figs. 5–6 and 5–7 (description is the same as in the previous example). Estimation of the persistence parameters gives: $d_1 = d_2 = 0.745$, $d_3 = d_4 = 0.647$, $d_5 = d_6 = 0.598$, $d_7 = d_8 = 0.494$.

We can see that the final reconstruction of the drift function (*continuous line*) is close to the input function $a(y)$ (thick *continuous line*). Also in this case, the first SRP reconstruction $b_1(y)$ of the diffusion function fits very well to the input function $b(y)$. This is due to the small time step $\Delta t = 0.01$ and small magnitudes of the drift function.

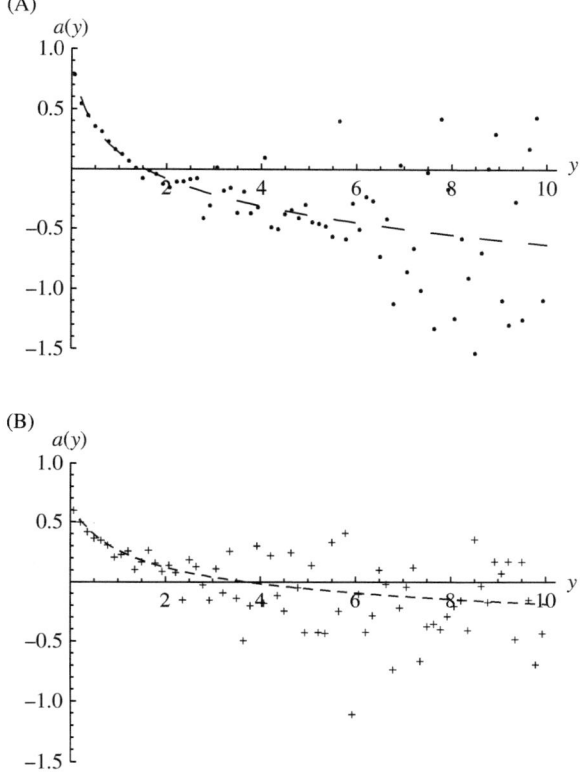

FIGURE 5–4 Intermediate reconstructions of drift function $a(y) = -\frac{1}{2}\log(y)$ according to MNRP.
(A)   First reconstruction $a_1(y)$ (top),
(B)   third reconstruction $a_3(y)$ (bottom).
Points represent results of SRP and long-dashed lines are least-square fits.

5.5 Conclusions

The standard Langevin equation was modified for description of some non-Markov processes – persistent and antipersistent. For this goal we introduced an additional random factor to the diffusion term, i.e., a function $c(s_t, r_t; d)$, which determines a sign of the term and is dependent on signs of p previous jumps. For the modified Langevin equation SRPs fail. Therefore, we introduced two novel procedures: one purely numerical and another semianalytical. The semianalytical one requires an explicit form of the Fokker–Planck equation associated with the modified Langevin equation. We could derive the Fokker–Planck equation starting from the Ito formula and then applying an averaging procedure of some terms over correlated random variables.

Working of the procedures was tested on a synthetic time series generated by different modified Langevin equations with memory. The results showed good efficiency of the two novel procedures. The semianalytical procedure (MsARP) leads to exact formulas for

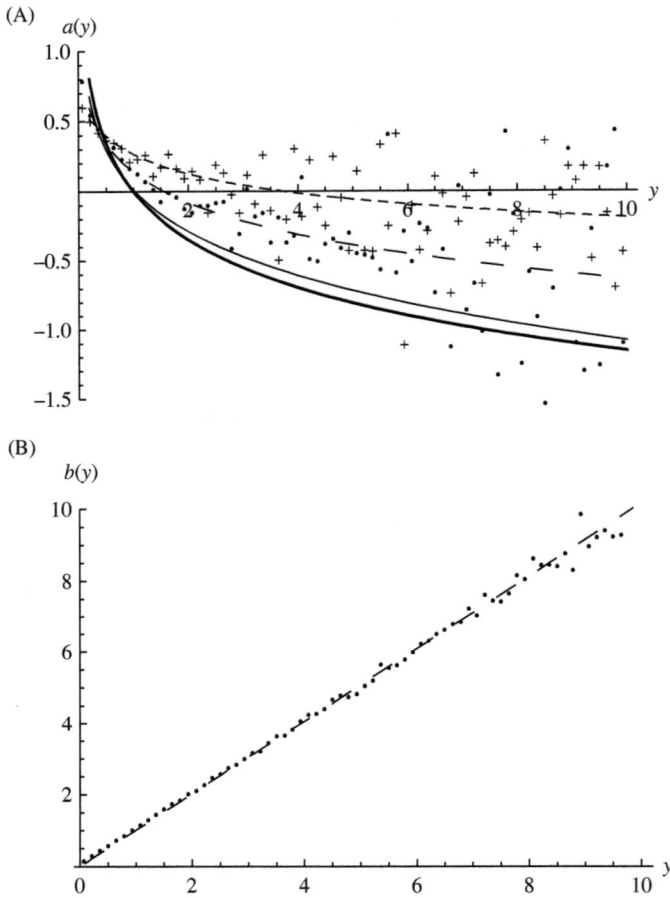

FIGURE 5–5 Final results of MNRP:
(A) For drift function (top). The final reconstruction $a_R(y)$ (*continuous line*) is a little deviated from the input function $a(y) = -\frac{1}{2}\log(y)$ (thick *continuous line*),
(B) for diffusion function (bottom). The final reconstruction $b_R(y) = b_1(y)$ (long-*dashed line*) agrees with the input function $b(y) = y$.

deviations $\alpha^c(y)$, $\alpha^\tau(y)$ and $\beta^\tau(y)$, and, therefore, its efficiency is dependent only on good fitting of first reconstruction of functions $a_1(y)$ and $b_1(y)$ to clouds of points given by SRP. The numerical method (MNRP) can suffer because of deviations due to assumptions in step 1 and step 3 and due to the triple execution of the procedure SRP. However, unlike MsARP, it can be applied to time series with persistence of order p greater than 1.

The results form a novel tool for modelling long-correlated nonlinear time series. Having the explicit form of the Langevin-type or Fokker–Planck equation makes it possible to determine a role for drift and diffusion terms in the behaviour of fluctuations given by time series. This can be applied to physical interpretation of the phenomenon under investigation. Moreover, it is possible to find the short-time transition probability which enables forecasting.

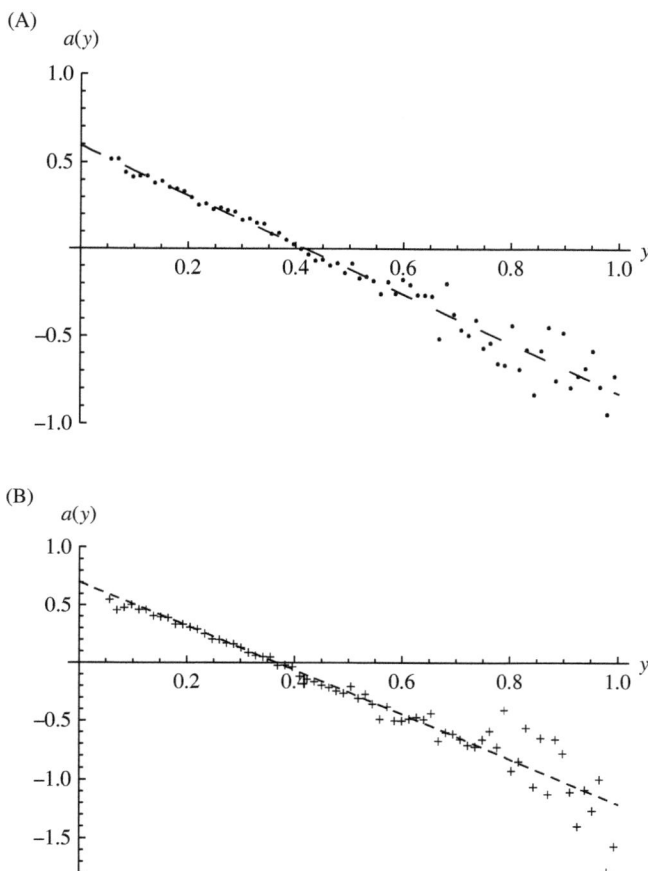

FIGURE 5–6 Intermediate reconstructions of drift function $a(y) = \frac{1}{2}(1 - 2y)$ according to MNRP.
(A) First reconstruction $a_1(y)$ (top),
(B) third reconstruction $a_3(y)$ (bottom).
Points represent results of SRP and long-dashed lines are least-square fits.

Appendix: Derivation of the Fokker–Planck Equation Associated With the Modified Langevin Equation

Following the method presented in Grasman and van Herwaarden (1999) the Fokker–Planck equation associated with the modified Langevin Eq. (5.5) is derived here. Let $f(y)$ be an arbitrary twice continuously differentiable function, then we have

$$df(y) = f'(y)dy + \frac{1}{2}f''(y)(dy)^2 + O(dy^3) \tag{5.A1}$$

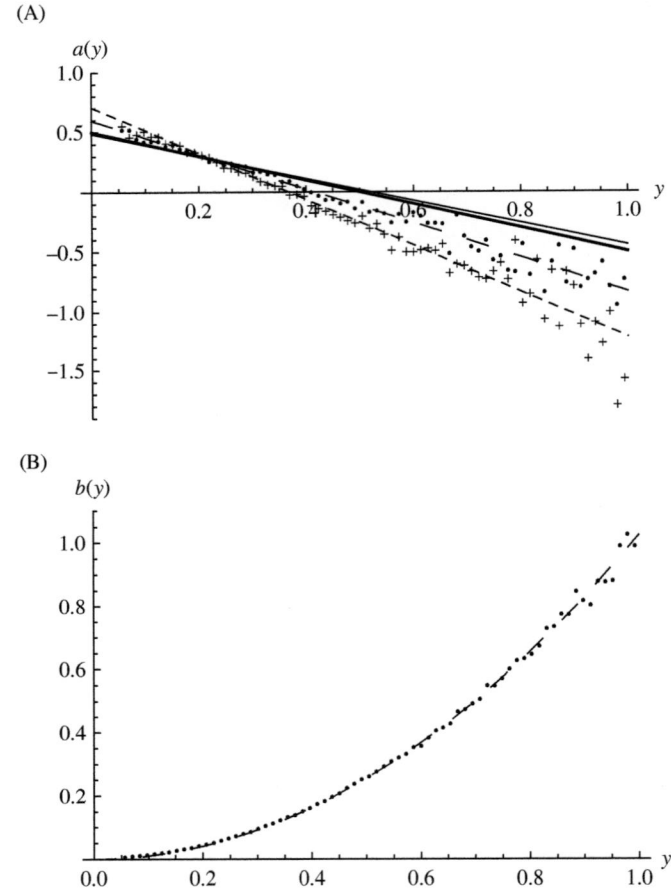

FIGURE 5–7 Final results of MNRP:

(A) For drift function (top). The final reconstruction $a_R(y)$ (*continuous line*) is a little deviated from the input function $a(y) = \frac{1}{2}(1 - 2y)$ (thick *continuous line*),

(B) for diffusion function (bottom). The final reconstruction $b_R(y) = b_1(y)$ (long-*dashed line*) agrees with the input function $b(y) = y^2$.

up to the second order in dy. Using the modified Langevin equation

$$dy = a(y)dt + c(s_t, r_t; d)\sqrt{b(y)}|dW_t| \tag{5.A2}$$

and the relation $dW_t^2 = dt$ we obtain

$$
\begin{aligned}
(dy)^2 &= a^2(y)(dt)^2 + c^2(s_t, r_t; d)b(y)|dW_t|^2 + 2a(y)c(s_t, r_t; d)\sqrt{b(y)}|dW_t|dt \\
&= a^2(y)(dt)^2 + b(y)dt + 2a(y)c(s_t, r_t; d)\sqrt{b(y)}|dW_t|dt.
\end{aligned}
\tag{5.A3}
$$

up to terms of order $(dt)^2$. Then, from Eqs (5.A1), (5.A2) and (5.A3):

$$df(y) = f'(y)[a(y)dt + c(s_t, r_t; d)\sqrt{b(y)}|dW_t|]$$

$$+ \frac{1}{2}f''(y)\left[b(y)dt + a^2(y)(dt)^2 + 2a(y)c(s_t, r_t; d)\sqrt{b(y)}|dW_t|dt\right]$$

$$= dt\, f'(y)\left[a(y) + c(s_t, r_t; d)\sqrt{b(y)}\frac{|dW_t|}{dt}\right] \tag{5.A4}$$

$$+ dt\, \frac{1}{2}f''(y)\left[b(y) + a^2(y)dt + 2a(y)c(s_t, r_t; d)\sqrt{b(y)}\frac{|dW_t|}{dt}dt\right].$$

There are four random variables: y, s_t, r_t and ω_t in Eq. (5.A4) where

$$\omega_t \equiv dW_t \quad p(\omega_t) = \frac{1}{\sqrt{2\pi\, dt}}\exp\left(-\frac{1}{2}\frac{\omega_t^2}{dt}\right). \tag{5.A5}$$

Only two of them, y and s_t, are dependent, the others are independent. Hence, the joint distribution function can be presented by the product

$$p(y, s_t, r_t, \omega_t) = p(y, s_t)p(r_t)p(\omega_t) \tag{5.A6}$$

or, neglecting the subindex t, by

$$p(y, s, r, \omega) = p(y, s)p(r)p(\omega). \tag{5.A7}$$

Taking the expected value of Eq. (5.A4) and using the fact that the expected value of df/dt equals the time derivative of the expected value of f we obtain:

$$\frac{\partial}{\partial t}\int f(y)p(y)dy = \int \left[f'(y)a(y) + \frac{1}{2}f''(y)\left[b(y) + a^2(y)dt\right]\right]p(y)dy +$$

$$+ \int f'(y)c(s, r; d)\sqrt{b(y)}\frac{|\omega|}{dt}p(y, s)p(r)p(\omega)dydsdrd\omega + \tag{5.A8}$$

$$+ \int f''(y)a(y)c(s, r; d)\sqrt{b(y)}\frac{|\omega|}{dt}dt\, p(y, s)p(r)p(\omega)dydsdrd\omega$$

Integration by parts at the right and interchanging the order of differentiation and integration at the left yields

$$\int f(y)\frac{\partial p(y)}{\partial t}dy = \int f(y)\left\{-\frac{\partial}{\partial y}[a(y)p(y)] + \frac{1}{2}\frac{\partial^2}{\partial y^2}\left[(b(y) + a^2(y)dt)p(y)\right]\right\}dy +$$

$$+ \int [f'(y) + f''(y)a(y)dt]c(s, r; d)\sqrt{b(y)}\frac{|\omega|}{dt}p(y, s)p(r)p(\omega)dydsdrd\omega. \tag{5.A9}$$

We neglected the surface integrals because we have chosen $f(y)$ to be arbitrary but vanishing on the surface (see Gardiner, 1985, p. 51). Our main task is to calculate the second integral on the right-hand side. We denote this by the symbol J.

Integration over ω gives the following factor:

$$\frac{1}{dt}\int_{-\infty}^{\infty}|\omega|\frac{1}{\sqrt{2\pi\,dt}}\exp\left(-\frac{1}{2}\frac{\omega_t^2}{dt}\right)d\omega = \frac{1}{dt}\sqrt{\frac{2}{\pi}}\sqrt{dt} = \sqrt{\frac{2}{\pi\,dt}}. \qquad (5.A10)$$

Next, integration of the function $c(s,r;d)$:

$$c(s,r;d) = \begin{cases} 1 & if & (s=1 \wedge r \geq d) \vee (s=-1 \wedge r<d) \\ -1 & if & (s=-1 \wedge r \geq d) \vee (s=1 \wedge r<d) \end{cases} \qquad (5.A11)$$

over the random variable r, which is uniformly distributed in $(0, 1)$, leads to

$$\int_0^1 c(s,r;d)dr = \int_0^d \left(\begin{cases} 1 & if & s=-1 \\ -1 & if & s=1 \end{cases}\right)dr + \int_d^1 \left(\begin{cases} 1 & if & s=1 \\ -1 & if & s=-1 \end{cases}\right)dr = \left(\begin{cases} d & if & s=-1 \\ -d & if & s=1 \end{cases}\right)$$

$$+ \left(\begin{cases} 1-d & if & s=-1 \\ -(1-d) & if & s=1 \end{cases}\right) = \begin{cases} 1-2d & if & s=1 \\ -(1-2d) & if & s=-1 \end{cases} = (1-2d)\begin{cases} 1 & if & s=1 \\ -1 & if & s=-1 \end{cases}.$$

$$\qquad (5.A12)$$

Therefore, the integral J has the following form:

$$J = (1-2d)\sqrt{\frac{2}{\pi\,dt}}\int [f'(y)+f''(y)a(y)dt]\sqrt{b(y)}\left(\begin{cases} 1 & if & s=1 \\ -1 & if & s=-1 \end{cases}\right)p(y,s)ds\,dy. \qquad (5.A13)$$

The last integration we must do over the random variable s, which is dependent on y. However, the joint distribution function $p(y, s)$ is a discrete-continuous distribution and has a form of two marginal distributions, $p(y)/2$ and $p(y+\Delta)/2$, shifted by $\Delta = <\Delta y>$. Therefore, the integral in Eq. (5.A13) can be reduced to the form

$$\int [f'(y)+f''(y)a(y)dt]\sqrt{b(y)}\left(\begin{cases} 1 & if & s=1 \\ -1 & if & s=-1 \end{cases}\right)p(y,s)ds$$

$$= [f'(y)+f''(y)a(y)dt]\sqrt{b(y)}\int\left(\begin{cases} 1 & if & s=1 \\ -1 & if & s=-1 \end{cases}\right)p(y,s)ds$$

$$= [f'(y)+f''(y)a(y)dt]\sqrt{b(y)}\,[p(y,\,1)-p(y,\,-1)]$$

$$= \frac{1}{2}[f'(y)+f''(y)a(y)dt]\sqrt{b(y)}\,[p(y)-p(y+\langle|\Delta y|\rangle)] \qquad (5.A14)$$

$$= -\frac{1}{2}[f'(y)+f''(y)a(y)dt]\sqrt{b(y)}\,\frac{dp(y)}{dy}\langle|\Delta y|\rangle$$

$$= -\frac{1}{2}[f'(y)+f''(y)a(y)dt]p(y)\langle|\Delta y|\rangle\sqrt{b(y)}\,\frac{p'(y)}{p(y)}.$$

Here, we need an estimation of the average absolute value of the jump Δy in state y. Because

$$|\Delta y| = \sqrt{\Delta t}|a(y)\sqrt{\Delta t} + c(s_t, r_t; d)\sqrt{b(y)}|\xi|| \approx \sqrt{\Delta t}|c(s_t, r_t; d)\sqrt{b(y)}|\xi|| = \sqrt{\Delta t}\sqrt{b(y)}|\xi| \qquad (5.A15)$$

then

$$\langle|\Delta y|\rangle \approx \sqrt{\Delta t}\sqrt{b(y)} \int_{-\infty}^{\infty} |\xi| \frac{1}{\sqrt{2\pi}} \exp\left(-\frac{1}{2}\xi^2\right) d\xi = \sqrt{\frac{2}{\pi}}\sqrt{\Delta t}\sqrt{b(y)}. \qquad (5.A16)$$

Using Eqs. (5.A14) and (5.A16) we can present the integral J in the form:

$$
\begin{aligned}
J &\equiv \int [f'(y) + f''(y)a(y)dt]c(s, r; d)\sqrt{b(y)} \frac{|\omega|}{dt} p(y, s)p(r)p(\omega)dy ds dr d\omega \\
&= \sqrt{\frac{2}{\pi}}\frac{1 - 2d}{\sqrt{\Delta t}}\left(-\frac{1}{2}\right) \int [f'(y) + f''(y)a(y)dt]p(y)\langle|\Delta y|\rangle\sqrt{b(y)} \frac{p'(y)}{p(y)} dy \\
&= -\frac{1 - 2d}{\pi} \int [f'(y) + f''(y)a(y)dt]p(y)b(y) \frac{p'(y)}{p(y)} dy.
\end{aligned}
\qquad (5.A17)
$$

Then, Eq. (5.A9) reduces to

$$\int f(y) \frac{\partial p(y)}{\partial t} dy =$$
$$\int f(y) \left\{ -\frac{\partial}{\partial y}\left[\left(a(y) - \frac{1 - 2d}{\pi}b(y)\frac{p'(y)}{p(y)}\right)p(y)\right] + \frac{1}{2}\frac{\partial^2}{\partial y^2}\left[\left(b(y) + \Delta t\, a^2(y) - \Delta t\frac{2 - 4d}{\pi}a(y)b(y)\frac{p'(y)}{p(y)}\right)p(y)\right] \right\} dy. \qquad (5.A18)$$

Hence, since $f(y)$ is arbitrary, the Fokker–Planck equation associated with the modified Langevin Eq. (5.5) has the following form:

$$
\begin{aligned}
\frac{\partial p(y)}{\partial t} &= -\frac{\partial}{\partial y}\left[\left(a(y) - \frac{1 - 2d}{\pi}b(y)\frac{p'(y)}{p(y)}\right)p(y)\right] \\
&+ \frac{1}{2}\frac{\partial^2}{\partial y^2}\left[\left(b(y) + \Delta t\, a^2(y) - \Delta t\frac{2 - 4d}{\pi}a(y)b(y)\frac{p'(y)}{p(y)}\right)p(y)\right].
\end{aligned}
\qquad (5.A19)
$$

Acknowledgements

This work was partially financed by a project of the National Science Centre (contract No. 2016/21/B/ST10/02998), by statutory activities No 3841/E-41/S/2017 of the Ministry of Science and Higher Education of Poland, and from funds from the Leading National Research Centre (KNOW) received by the Centre for Polar Studies for the period 2014–2018.

References

Anh, V., Yu, Z.-G., Wanliss, J.A., 2007. Analysis of global geomagnetic variability. Nonlinear Process. Geophys. 14, 701–708.

Anteneodo, C., Riera, R., 2009. Arbitrary-order corrections for finite-time drift and diffusion coefficients. Phys. Rev. E 80, 031103.

Chapman, S.C., Hnat, B., Rowlands, G., Watkins, N.W., 2005. Scaling collapse and structure functions: identifying self-affinity in finite length time series. Nonlinear Process. Geophys. 12, 767–774.

Collette, C., Ausloos, M., 2004. Scaling analysis and evolution equation of the North Atlantic Oscillation index fluctuations. Int. J. Mod. Phys. C 15, 1353–1366.

Czechowski, Z., 1993. A kinetic model of nucleation, propagation and fusion of cracks. J. Phys. Earth 41, 127–137.

Czechowski, Z., 2001. Transformation of random distributions into power-like distributions due to non-linearities: application to geophysical phenomena. Geophys. J. Int. 144, 197–205.

Czechowski, Z., 2003. The privilege as the cause of the power distributions in geophysics. Geophys. J. Int. 154, 754–766.

Czechowski, Z., 2005. The importance of the privilege in resource redistribution models for appearance of inverse-power solutions. Physica A 345, 92–106.

Czechowski, Z., 2010. The importance of privilege for appearance of long-tail distributions, Chapter 7. In: De Rubeis, V., Czechowski, Z., Teisseyre, R. (Eds.), Synchronization and Triggering: From Fracture to Earthquake Processes. Springer, Berlin, pp. 97–121.

Czechowski, Z., 2015. On microscopic mechanisms which elongate the tail of cluster size distributions: an example of Random Domino Automaton. Pure Appl. Geoph. 172, 2075–2082.

Czechowski, Z., 2016. Reconstruction of the modified discrete Langevin equation from persistent time series. CHAOS 26, 053109.

Czechowski, Z., Rozmarynowska, A., 2008. The importance of the privilege for appearance of inverse-power solutions in Ito equations. Physica A 387, 5403–5416.

Czechowski, Z., Telesca, L., 2011. Construction of Ito model for geoelectrical signals. Physica A 390, 2511–2519.

Czechowski, Z., Telesca, L., 2013. Construction of a Langevin model from time series with a periodical correlation function: application to wind speed data. Physica A 392, 5592–5603.

Czechowski, Z., Lovallo, M., Telesca, L., 2016. Multifractal analysis of visibility graph-based Ito-related connectivity time series. CHAOS 26, 023118.

Fraedrich, K., Blender, R., 2003. Scaling of atmosphere and ocean temperature correlations in observations and climate models. Phys. Rev. Lett. 90, 108501.

Friedrich, R., Peinke, J., Sahimi, M., Reza Rahimi, Tabar, M., 2011. Approaching complexity by stochastic methods: from biological systems to turbulence. Phys. Rep. 506, 87–162.

Gardiner, C.W., 1985. Handbook of Stochastic Methods for Physics, Chemistry and the Natural Sciences. Springer-Verlag, Berlin, Heidelberg, New York.

Gottschall, J., Peinke, J., 2008. On the definition and handling of different drift and diffusion estimates. New J. Phys. 10, 083034.

Govindan, R.B., Kantz, H., 2004. Long-term correlations and multifractality in surface wind speed. Europhys. Lett. 68, 184–190.

Govindan, R.B., Vyushin, D., Bunde, A., Brenner, S., Havlin, S., Schellnhuber, H.-J., 2002. Global climate models violate scaling of the observed atmospheric variability. Phys. Rev. Lett. 89, 028501.

Grasman, J., van Herwaarden, O.A., 1999. Asymptotic Methods for the Fokker-Planck Equation and the Exit Problem in Applications. Springer-Verlag, Berlin.

Hindriks, R., Jansen, R., Bijama, F., Mansvelder, H.D., de Gunst, M.C.M., van der Vaart, A.W., 2011. Unbiased estimation of Langevin dynamics from time series with application to hippocampal field potentials in vitro. Phys. Rev. E 84, 021133-1—-13.

Hurst, H.E., 1951. Long-term storage capacity of reservoirs. Trans. Am. Soc. Civil Eng. 116, 770—799.

Kantelhardt, J.W., Rybski, D., Zschiegner, S.A., Braun, P., Koscielny-Bunde, E., Livina, V., et al., 2003. Multifractality of river runoff and precipitation: comparison of fluctuation analysis and wavelet methods. Physica A 330, 240—245.

Khaliq, M.N., Ouarda, T.B.M.J., Gachon, P., 2009. Identification of temporal trends in annual and seasonal low flows occurring in Canadian rivers: the effect of short- and long-term persistence. J. Hydrol. 369, 183—197.

Kiss, P., Muller, R., Janosi, I.M., 2007. Long-range correlations of extrapolar total ozone are determined by the global atmospheric circulation. Nonlinear Process. Geophys. 14, 435—442.

Klein, W., Anghel, M., Ferguson, C.D., Rundle, J.B., Sa Martins, J.S., 2000. In: Rundle, J.B., Turcotte, D.L., Klein, W. (Eds.), GeoComplexity and the Physiscs of Earthquakes. Am. Geophys. Union, Washington, DC, pp. 43—71.

Kleinhaus, D., Friedrich, R., 2007. Maximum likelihood estimation of drift and diffusion functions. Phys. Lett. A 368, 194—198.

Kleinhaus, D., Friedrich, R., Nawroth, A., Peinke, J., 2005. An iterative procedure for the estimation of drift and diffusion coefficients of Langevin processes. Phys. Lett. A 346, 42.

Lamauroux, D., Lehnertz, K., 2009. Kernel-based regression of drift and diffusion coefficients of stochastic processes. Phys. Lett. A 373, 3507—3512.

Lind, P.G., Mora, A., Gallas, J.A.C., Haase, M., 2005. Reducing stochasticity in the North Atlantic Oscillation index with coupled Langevin equations. Phys. Rev. E 72, 056706.

Mandelbrot, B.B., van Ness, J.W., 1968. Fractional Brownian motions, fractional noises and applications. SIAM Rev. 10, 422—437.

Matcharashvili, T., Chelidze, T., Javakhishvili, Z., Zhukova, N., Jorjiashvili, N., Shengelia, I., 2013. Discrimination between stochastic dynamics patterns of ambient noises (Case study for Oni seismic station). Acta Geophys. 61, 1659—1676.

Matthews, M.V., Ellsworth, W.L., Reasenberg, P.A., 2002. A Brownian model for recurrent earthquakes. Bull. Seism. Soc. Am 92, 2233—2250.

Montanari, A., Rosso, R., Taqqu, M.S., 1996. Some long-run properties of rainfall records in Italy. J. Geophys. Res. D21, 431—438.

Mudelsee, M., 2007. Long memory of rivers from spatial aggregation. Water Resour. Res. 43, W01202.

Pelletier, J.D., Turcotte, D.L., 1999. Self-affine time series: II. applications and models. Adv. Geophys. 40, 91—166.

Rundle, J.B., Klein, W., Gross, S., 1999. Physical basis for statistical patterns in complex earthquake populations: models, predictions and tests. Pure Appl. Geophys. 155, 575—607.

Siegert, S., Friedrich, R., Peinke, J., 1998. Analysis of data sets of stochastic systems. Phys. Lett. A 243, 275—280.

Stratonovich, R.L., 1963. Topics in the Theory of Random Noise, Vol.1. Gordon and Breach, New York.

Sura, P., Barsugli, J., 2002. A note on estimating drift and diffusion parameters from timeseries. Phys. Lett. A 305, 304—311.

Telesca, L., Czechowski, Z., 2012. Discriminating geoelectrical signals measured in seismic and aseismic areas by using Ito models. Physica A 391, 809–818.

Telesca, L., Czechowski, Z., Lovallo, M., 2015. Multifractal analysis of time series generated by discrete Ito equations. CHAOS 25, 063113.

Yin, Z.-M., Ranalli, G., 1995. Modelling of earthquake rupturing as a stochastic process and estimation of its distribution function from earthquake observations. Geophys. J. Int. 123, 838–848.

6

Synchronization of Geophysical Field Fluctuations

Alexey Lyubushin

INSTITUTE OF PHYSICS OF THE EARTH, RUSSIAN ACADEMY OF SCIENCES, MOSCOW, RUSSIA

CHAPTER OUTLINE

6.1 Introduction

Random fluctuations in geophysical fields carry important information about the processes occurring in the Earth's crust, including for preparation for major geological disasters. The synchronization of noise measurements obtained in different points of the monitoring network is an important indicator of the approximation of complex systems to a drastic change in its properties by virtue of its own dynamics. Therefore, the development and testing of methods of analysis of monitoring data, with the purpose of allocating time intervals of increased coherence and to determine the characteristic frequencies of the synchronization of a large number of time series of geophysical monitoring systems is a very urgent geosciences task. This chapter outlines the methods of analysis of multivariate time series, allowing exploration of the dynamics of change in the degree of coherence of the noise in the data stream from the monitoring systems.

One of the fundamental problems with geophysical monitoring is the 'complexation' of different observations and measurements. This term means the joint analysis of observations of different geophysical fields or observations within the same field, but at different measurement points, or both simultaneously. The idea of complexation is based on the hypothesis that using of a large number of monitored parameters can help to extract a weak common signal, which 'drowns' in the strong noise when individual measurements are considered separately. The main feature of such a common signal must be its coherence (correlation) in a variety of observations, the use of which allows one to detect the very existence of the common components, despite the fact that the frequency content of the common signal may coincide with the frequency content of the strong local noise.

The identification of precursors of earthquakes or other geological catastrophes is among the most challenging problems of complexing measurements. In such a problem, a weak common signal is related to the earthquake preparation processes, such as consolidation of the matter in the Earth's crust in a future earthquake focus and around it (Rice, 1980). The search for precursors of catastrophes as the occurrence of synchronous components in a variety of observations is a general idea about increasing the correlation radius of the random fluctuations of the parameters of a complex system, when it approaches a sharp change in its properties as a result of its own dynamics (Gilmore, 1981; Nicolis and Prigogine, 1989).

The idea of complexation requires using methods of analysis of multivariate time series. In this chapter a set of algorithms for the analysis of multivariate time series, obtained by monitoring systems, will be presented; these are intended for discovering the hidden relations between processes, including those with different natures and structures. An important part of the developed algorithms is a preliminary analysis of time series with different scales for the purpose of extracting the dimensionless characteristics within adjacent short time intervals. These characteristics were chosen in dimensionless forms which are independent of the physical meaning of the initial time series. Such transformation of initial high-frequency (HF) time series into low-frequency (LF) series is an important tool, which helps to detect hidden coherence effects, that may not be found when the initial data are processed. Analysis of this noise is often neglected, although statistical regularities in the noise structure are an important source of hidden information about upcoming sharp changes in the properties of the objects under consideration. The methods are based on the analysis of canonical coherences, multidimensional spectral matrices and canonical correlation coefficients of the wavelet decomposition of signals in moving time windows. The purpose of these algorithms is the detection of very weak nonstationary signals of common origin, having both harmonic oscillation behaviour or sharply nonstationary, wavelet character, and finding the time intervals and frequency bands of these common signals.

6.2 Wavelet-Based Robust Coherence Measure

The robust wavelet-based measure of coherence is a modification of the approach to the analysis of multidimensional time series proposed in Lyubushin (2000) and Lyubushin and Kopylova (2004). In the paper by Lyubushin (2015) this measure was applied in the investigation of global seismic noise coherence. Let $Z(t) = (Z_1(t), \ldots, Z_q(t))^T$ be the multidimensional time series of the dimensionality $q \geq 2$, $t = 0, 1, \ldots$ is a discrete time index which numerates successive samples. The scale-dependent measure of coherent behaviour in a moving time window of a given length of N samples is constructed. Let's consider that time index t within each time window varies from 0 up to $(N - 1)$. The analysis is performed independently for each position of the time window (moved to the right by one sample). Before the wavelet decomposition of the analysed time series fragments, which is presented in the current time window, the following sequence of operations is applied to each fragment:

1. The general linear trend within the time window is removed;
2. A sample estimate of the standard deviation is obtained, and each value is divided by this estimate;
3. A tapering operation is performed within each time window;
4. The window fragment is padded by zeros to the full length of $M = \min\{2^m: 2^m \geq N\}$ samples.

Operation (1) removes the strongest LF variations in signals, which are not statistically representative within the window. The division of each signal within the window by its standard deviation mutually adjusts different time series by reducing the total energy of their variations to the same value. Tapering in point (3) is a usual preliminary operation in spectral and wavelet analysis before applying discrete Fourier or wavelet transform and consists of multiplying the samples within the current time window by the positive function, which is tending to zero, when samples approach the left and right ends of the window. We use a cosine tapering function which equals $(1 - \cos(\pi t / L))/2$ for $0 \leq t \leq L$ and $(1 - \cos(\pi(t - (N - 1))/L))/2$ for $(N - 1) - L \leq t \leq (N - 1)$. Here L is the length of time intervals at the beginning and at the end of the time window, where the tapering operation is performed. We have used the value $L = N/8$. Tapering operation (3) is necessary to reduce the negative 'wrap-around' effect of the finite discrete wavelet transform (Press et al., 1996). Finally, the last operation (4) is necessary for the subsequent application of the fast discrete wavelet transform.

Let τ be the position of the right-hand end of a moving time window N samples wide. After performing discrete orthogonal wavelet transform within the window after preliminary operations (1)–(4) we have a set of tables of wavelet-coefficients:

$$c_j^{(\beta,\tau)}(k), \quad j = 1, \ldots, q; \quad k = 1, \ldots, M_\beta = 2^{(m-\beta)}; \quad \beta = 1, \ldots, m \tag{6.1}$$

We chose the Haar wavelet from the family of orthogonal wavelets as the most compact and suitable for analysis of the most abrupt variations in signals. Here β is the number of detail levels of wavelet decomposition, m is a power of 2 in the presentation $M = 2^m$ of minimum integer which is not less than N. The general number of wavelet coefficients at the detail level β equals $M_\beta = 2^{(m-\beta)}$. The index j corresponds to different scalar components of the multidimensional time series $Z(t)$, whereas index k successively enumerates the coefficients belonging to the level $\beta = 1, \ldots, m$. However, some of these coefficients may correspond to the zero padding of the sample in point (4) of the preliminary transformations. Therefore, the actual number of coefficients at the level β, reflecting the behaviour of the signal inside the window, is equal to $L_\beta(N) = 2^{(m-\beta)}(N/M) = 2^{-\beta}N$. Each coefficient $c_j^{(\beta,\tau)}(k)$ reflects the signal behaviour in the frequency band with approximate bounds $[\Omega_{min}^{(\beta)}, \Omega_{max}^{(\beta)}] = [1/(2^{(\beta+1)}\Delta s), 1/(2^\beta \Delta s)]$, where Δs is the length of the sampling interval, in the neighbourhood of the sample with the number $\tau_k^{(\beta)} = k \, 2^\beta$, $k = 1, \ldots, M_\beta$, measured from the position of the left end of the time window. The width of this neighbourhood (the temporal 'zone of responsibility' of the coefficient) is equal to $\Delta s \, 2^\beta$.

The number of wavelet coefficients is reduced exponentially with increasing detail level β. That is why $L_\beta(N)$ decreases with the same rate and starts from some detail level number β the number of wavelet coefficients $L_\beta(N)$ could be equal to zero. Therefore, it is natural to introduce the parameter of statistical significance L_{min} as the minimum possible value of the number of wavelet coefficients $L_\beta(N)$ that allows one to perform statistical estimates using wavelet coefficients at the βth detail level. It is possible to determine the maximum possible detail level β_{max} from the formula $\beta_{max} = \max\{\beta\colon L_\beta(N) \geq L_{min}\}$. Thus, the length of the window N and the significance threshold L_{min} together set the maximum possible detail level β_{max}, the wavelet coefficients of which can be included into the analysis.

Now we address a scalar time series j_0 from multiple series $Z(t)$ and construct the measure describing the relationship between this series and all other scalar signals within the current time window. Naturally, this relationship depends on the scale of the variations in question and, therefore, should be sought at various detail levels between wavelet expansion coefficients. The problem to be solved for this purpose is

$$\sum_{k=1}^{L_\beta} \left| c_{j_0}^{(\beta,\tau)}(k) - d_{j_0}^{(\beta,\tau)}(k|\gamma) \right| \to \min_{\gamma_j} \tag{6.2}$$

where $d_{j_0}^{(\beta,\tau)}(k|\gamma)$ is a linear combination of wavelet coefficients at the detail level β from all other scalar signals except a signal with number j_0 with unknown coefficients γ_j:

$$d_{j_0}^{(\beta,\tau)}(k|\gamma) = \sum_{j=1, j\neq j_0}^{q} \gamma_j \cdot c_{j_0}^{(\beta,\tau)}(k) \tag{6.3}$$

We should emphasize that the sum in Eq. (6.3) is a linear combination of expansion coefficients of time series except the chosen series j_0. The problem (6.2) is solved by the method of generalized gradient (Clarke, 1975). Finding the vector γ from the solution of problem (6.2), we obtain certain values of $d_{j_0}^{(\beta,\tau)}(k|\gamma)$. Now we can find the correlation coefficient between samples of values of $c_{j_0}^{(\beta,\tau)}(k)$ and $d_{j_0}^{(\beta,\tau)}(k|\gamma)$ for $k = 1, \ldots, L_\beta$; however, instead of the classic Pearson formula for calculating the sample value of the correlation coefficient, we use its robust modification (Huber and Ronchetti, 2009), according to which the correlation coefficient between samples $x(k)$ and $y(k)$, $k = 1, \ldots, n$, can be calculated by the formula

$$\rho(x,y) = \frac{S\left(\hat{z}^2\right) - S(\check{z}^2)}{S\left(\hat{z}^2\right) + S(\check{z}^2)} \tag{6.4}$$

where

$$\widehat{z}(r) = a \cdot x(r) + b \cdot y(r), \quad \check{z}(r) = a \cdot x(r) - b \cdot y(r),$$
$$a = \frac{1}{S(x)}, \quad b = \frac{1}{S(y)}, \quad S(x) = \text{med}\left|x - \text{med}(x)\right| \tag{6.5}$$

where $\text{med}(x)$ is the median value of the sample $x(k)$, and $S(x)$ is its absolute median deviation.

Substituting $x(k)$ for $c_{j_0}^{(\beta,\tau)}(k)$, $y(k)$ for $d_{j_0}^{(\beta,\tau)}(k|\gamma)$, and n for $L_\beta(N)$, we obtain the robust value $\nu_{j_0}(\beta,\tau)$ of the correlation coefficient describing the degree of connection of the process j_0 with all other signals from multiple time series $Z(t)$. If we replace in Eq. (6.2) the sum of the moduli of deviations by the sum of their squares, the problem can be reduced to the classic Hotelling problem of canonical correlations (Hotelling, 1936; Rao, 1965). Therefore, the quantity $\nu_{j_0}(\beta,\tau)$ is here referred to as the robust canonical correlation of the time series j_0. The need to replace the classic scheme of the calculation of canonical correlations by its robust variant is dictated by the strong instability of the result of the classic calculations with respect to outliers in wavelet coefficients. The presence of such outliers is due to the well-known fact that the wavelet decomposition is capable of accumulating maximum information about the signal behaviour in a relatively small number of wavelet coefficients. We should emphasize that the method is robust in two procedures: the solution of minimization problem (6.2) by the method of least moduli rather than by least squares and the calculation of the correlation coefficient by formula (6.4).

Since, with an increase in the number of the detail level, the number of wavelet coefficients involved in the estimation of $\nu_k(\beta,\tau)$ exponentially decreases, we reduce statistical fluctuations in estimates by introducing additional averaging over a certain number of coefficients obtained within preceding windows:

$$\bar{\nu}_k(\beta,\tau) = \sum_{s=1}^{m_\beta} \nu_k(\beta,\tau - s + 1)/m_\beta, \quad m_\beta = 2^\beta \tag{6.6}$$

The higher the detail level, the deeper the averaging [Eq. (6.6)] over the past time windows; this fact considerably decreases the dependence of the variance of statistical fluctuations in estimation (6.6) on the detail level number and makes this variance nearly the same for different values of β. According to formula (6.6), the effective width of the time window becomes scale-dependent and equal to $(N + 2^\beta - 1)$.

We define the robust wavelet-based measure of coherence by the formula

$$\rho(\tau,\beta) = \prod_{k=1}^{q} \left|\bar{\nu}_k(\tau,\beta)\right| \tag{6.7}$$

The values of measure (6.7) range from 0 to 1. The larger the value of Eq. (6.7), the stronger the overall connection between all analysed processes on scales corresponding to the number β. We should emphasize that the value of Eq. (6.7) is the product of q nonnegative values with moduli less than unity. Therefore, the greater the number q of the series analysed, the lower the absolute values of $\rho(\tau, \beta)$. As a consequence, the absolute values of statistic (6.7) can be compared only for the same number of series q. Most interesting are not the absolute values of measure (6.7) but its relative values for different values of τ. Thus, with a fixed Haar wavelet in use, the method has two free parameters: the time window length N and the significance threshold L_{min}. Later we will use the threshold $L_{min} = 16$.

It should be noticed that the value in Eq. (6.7) could be calculated without wavelet decomposition of the signals within a moving time window as well. At this case it is not dependent on the level β and it is not necessary to make additional smoothing [Eq. (6.6)] using values within the preceding time windows. Thus, the by-component robust correlations ν_k become level-independent and the value [Eq. (6.7)] becomes the formula for robust multiple correlation coefficient:

$$\rho(\tau) = \prod_{k=1}^{q} |\nu_k(\tau)| \tag{6.8}$$

If $q = 2$ then the value [Eq. (6.8)] presents a robust estimate of the squared correlation coefficient between two time series.

6.3 Multiple Spectral Coherence Measure

The multiple spectral coherence is analogous to wavelet-based multiple coherence but it is based on classic Fourier basis functions instead of orthogonal wavelets. It uses frequency-dependent canonical coherences, which are similar to level-dependent wavelet-based canonical correlations.

Canonical coherences are the generalization of the usual squared coherence spectrum between two scalar time series for the case when two vector time series are considered: m-dimensional time series $X(t)$ and n-dimensional time series $Y(t)$. Here t is an integer time index. Without loss of generality let us suppose that $m \leq n$. Squared maximum canonical coherence $\mu^2(\omega)$ between multiple time series $X(t)$ and $Y(t)$ is computed as the maximum eigenvalue of the following frequency-dependent matrix (Brillinger, 1975; Hannan, 1970):

$$U(\omega) = S_{xx}^{-1} S_{xy} S_{yy}^{-1} S_{yx} \tag{6.9}$$

where ω is the frequency, $S_{xx}(\omega)$ is spectral matrix of the size $m \times m$ of time series $X(t)$, $S_{xy}(\omega)$ is a cross-spectrum matrix of size $m \times n$ between time series $X(t)$ and $Y(t)$, $S_{yx}(\omega) = S_{xy}^H(\omega)$, H is the sign of Hermitian conjunctions (i.e., transposition of the matrix and complex conjugation), $S_{yy}(\omega)$ is the spectral matrix of size $n \times n$ of time series $Y(t)$. The value of $\mu^2(\omega)$ is used instead of the usual squared coherence spectrum when two scalar time series are regarded, i.e., when $m = n = 1$.

If we take $X(t)$ as the scalar ith component of q-dimensional time series $Z(t)$ and $Y(t)$ as the $(q-1)$-dimensional time series composed of all other scalar components of $Z(t)$, then function (6.9) became scalar and could be named the by-component canonical coherence $v_i^2(\omega)$.

The value $v_i^2(\omega)$ is the measure of connection of variations of the ith component of q-dimensional time series $Z(t)$ with variations of all other scalar components of $Z(t)$ at frequency ω. The inequality $0 \le |v_i(\omega)| \le 1$ is fulfilled, and the closer the value of $|v_i(\omega)|$ to unity, the stronger the linear relation of variations at the frequency ω of the ith scalar series to analogous variations in all other series. Now we can define the multiple spectral coherence measure by the following formula:

$$\lambda(\omega) = \prod_{i=1}^{q} |v_i(\omega)| \tag{6.10}$$

The value (6.10) provides a frequency-dependent measure of linear joint synchronization of variations of all scalar components of time series $Z(t)$ at frequency ω. Because the dimensionality of series $X(t)$ in formula (6.9) equals 1, $m = 1$, the matrix $U(\omega)$ in fact is a scalar. Thus, its 'maximum eigenvalue' is the value of the following quadratic form divided by the power spectrum of the ith component:

$$v_i^2(\omega) = \frac{S_i^H(\omega)\left(S_{ZZ}^{(i)}(\omega)\right)^{-1} S_i(\omega)}{P_i(\omega)} \tag{6.11}$$

where $S_{ZZ}^{(i)}(\omega)$ is a Hermitian matrix of the size $(q-1) \times (q-1)$, which is obtained from the full spectral matrix $S_{ZZ}(\omega)$ of the size $q \times q$ of multiple time series $Z(t)$ by removing its ith column and ith row, $S_i(\omega)$ is a $(q-1)$-dimensional vector consisting of cross-spectra between the ith component of $Z(t)$ with all its other scalar components. It is evident that vector $S_i(\omega)$ is composed of elements of spectral matrix $S_{ZZ}(\omega)$ from the ith column except the elements in the ith row. Finally $P_i(\omega)$ is a power spectrum of the ith component of $Z(t)$, i.e., the ith element on the main diagonal of the matrix $S_{ZZ}(\omega)$. The matrix $S_{ZZ}^{(i)}(\omega)$ is Hermitian and positively defined — that is why quadratic form $S_i^H(\omega)(S_{ZZ}^{(i)}(\omega))^{-1}S_i(\omega)$ is real and positive.

For calculating the measure (6.10) using values (6.11) it is necessary to estimate the spectral matrix $S_{ZZ}(\omega)$ of the size $q \times q$. For this we use a vector autoregression model (Marple, 1987):

$$Z(t) + \sum_{k=1}^{p} A_k \cdot Z(t-k) = e(t) \tag{6.12}$$

where p is an autoregression order, A_k are matrices of autoregression coefficients of size $q \times q$, $e(t)$ is a q-dimensional residual signal with zero mean and covariance matrix $\Phi = M\{e(t)e^T(t)\}$ of size $q \times q$. Matrices A_k and Φ are defined using the Durbin–Levinson procedure and the spectral matrix is calculated using formula:

$$S_{ZZ}(\omega) = F^{-1}(\omega) \cdot \Phi \cdot F^{-H}(\omega), \quad F(\omega) = E + \sum_{k=1}^{p} A_k \cdot \exp(-i\omega k) \tag{6.13}$$

where E is a unit matrix of size $q \times q$.

When $q = 2$ the value (6.10) equals the usual squared coherence spectrum:

$$\lambda(\omega) = \frac{|S_{12}(\omega)|^2}{(S_{11}(\omega) \cdot S_{22}(\omega))} \tag{6.14}$$

where $S_{11}(\omega)$ and $S_{22}(\omega)$ are diagonal elements of the matrix (6.13), i.e., parametric estimates of the power spectra of two signals, and $S_{12}(\omega)$ is their mutual cross-spectrum.

Let us consider moving a time window of the certain length and let τ be the time coordinate of the right-hand end of the moving time window. If the function (6.10) is estimated within each time window independently then we will have a time–frequency function:

$$\lambda(\tau, \omega) = \prod_{i=1}^{q} |\nu_i(\tau, \omega)| \tag{6.15}$$

The value (6.15) presents the evolution of linear synchronization measure for multiple time series $Z(t)$. It was important that, before calculating the spectral matrix, each scalar component of the multidimensional time series was subjected (independently in each time window) to the following preliminary operations. First of all, the general linear trend was eliminated and optional conversion to the increments was carried out. Then, the obtained data were winsorized (Huber and Ronchetti, 2009): the sample mean and standard deviation σ were iteratively calculated, the mean was subtracted from the sample, after which the counts were divided by σ and all the values that fell beyond the limits of $\pm 3\sigma$ were replaced by their limiting values. The iterations were repeated until σ stopped changing. These procedures ensure the robustness of the estimate of the coherence measure to the outliers (extreme values).

Besides the frequency–time dependence $\lambda(\tau, \omega)$, the pure time-dependent measures of the maximum and mean coherence in the current time window with coordinate τ were also used:

$$\lambda_{\max}(\tau) = \max_{\omega_{\min} \leq \omega \leq \omega_{\max}} \lambda(\tau, \omega), \quad \lambda_{\mean}(\tau) = \sum_{\omega_{\min} \leq \omega \leq \omega_{\max}} \lambda(\tau, \omega)/m(\omega_{\min}, \omega_{\max}) \tag{6.16}$$

where $m(\omega_{\min}, \omega_{\max})$ is the number of discrete frequency values within frequency band $[\omega_{\min}, \omega_{\max}]$. We note that quantities (6.16) are certain analogues of the coefficient of multiple correlation $\rho(\tau)$ from Eq. (6.8) calculated in the moving time window. However, since the maximum and mean values in formula (6.16) are taken over the frequencies, these coefficients allow accounting for the time shifts between the scalar components of multidimensional time series within the current time window.

Multiple spectral coherence measure in the form of Eqs. (6.10) and (6.11) was suggested in Lyubushin (1998) for multidimensional time series processing in the problems of geophysical monitoring. In the following papers (Lyubushin, 1999, 2008b, 2009, 2010a,b, 2014a, 2016a,b; Lyubushin et al., 2003, 2004; Lyubushin and Sobolev, 2006; Lyubushin and Klyashtorin, 2012) this spectral measure was applied to different problems of multidimensional time series analysis in geophysics, meteorology, hydrology and climate sciences.

6.4 Statistics of Time Fragments

For characterizing changing of statistical properties of analysed geophysical time series estimated in moving time windows we have chosen three parameters: entropy of distribution of squared orthogonal wavelet coefficients *En* and two multifractal parameters — singularity spectrum support width $\Delta\alpha$ and generalized Hurst exponent α^*.

Minimum normalized entropy En of squared wavelet coefficients. Let $x(t)$ be the finite sample of the signal $t = 1,\ldots,N$ — index, numerating the counts. The normalized entropy is defined by the formula:

$$En = - \sum_{k=1}^{N} p_k \cdot \frac{\log(p_k)}{\log(N)}, \quad p_k = \frac{c_k^2}{\sum_{j=1}^{N} c_j^2}, \quad 0 \le En \le 1 \qquad (6.17)$$

where c_k, $k = 1, N$ are the orthogonal wavelet coefficients which were found from minimization of the value (6.17). We try 17 orthogonal wavelets (Mallat, 1998): 10 usual wavelets of Daubechies (number of vanishing moments equals integer numbers from 1 up to 10) and seven so-called symlets with numbers of vanishing moments varying from 4 up to 10. For geophysical monitoring time series the parameter *En* was estimated within adjacent 'short' time windows of the certain length after removing the trend by polynomial of some order. This operation provides one of the possible transformations of a HF initial time series to the LF series of its properties.

Minimum normalized entropy *En* was suggested in Lyubushin (2012) and was used for investigating seismic noise properties in Lyubushin (2013a,b) and Lyubushin et al. (2014). This entropy measure has some common features with multiscale entropy which was introduced in Costa et al. (2003, 2005) for analysis of time series. In particular, the orthogonal wavelet transform of the signal, which is used in Eq. (6.17), is multiscale as well because it provides decomposition into discrete dyadic time-frequency 'atoms' with energy that is equal to c_k^2.

Multifractal parameters $\Delta\alpha$ and α^*. Let $x(t)$ be a random signal. Let us define its measure of variability $\mu_X(t,\delta)$ on the time interval $[t, t+\delta]$ as the difference between maximum and minimum values $\mu_x(t,\delta) = \max_{t \le u \le t+\delta} x(u) - \min_{t \le u \le t+\delta} x(u)$ and calculate the mean value of its power degree q: $M(\delta, q) = M[(\mu_x(t, \delta))^q]$. A random signal is scale-invariant (Taqqu, 1988) if $M(\delta, q) \sim \delta^{\rho(q)}$ when $\delta \to 0$, that is, the following limit exists:

$$\rho(q) = \lim_{\delta \to 0} \left(\frac{\ln M(\delta, q)}{\ln \delta} \right) \tag{6.18}$$

If $\rho(q)$ is a linear function $\rho(q) = Hq$, where $H = const$, $0 < H < 1$, the process is called monofractal. In the case where $\rho(q)$ is a nonlinear concave function of q, the signal is called multifractal. To estimate the value of $\rho(q)$ using a finite sample $x(t)$, $t = 0, 1, \ldots, N - 1$ we used a method based on the approach of detrended fluctuation analysis (DFA) (Kantelhardt et al., 2002). Let us split the entire time series into nonoverlapping intervals of length s:

$$I_k^{(s)} = \left\{ t: \ 1 + (k - 1)s \le t \le ks, \quad k = 1, \ldots, [N/s] \right\} \tag{6.19}$$

and let

$$y_k^{(s)}(t) = x((k - 1)s + t), \quad t = 1, \ldots, s \tag{6.20}$$

be a part of the signal $x(t)$, corresponding to interval $I_k^{(s)}$. Let $p^{(s, m)}{}_k(t)$ be a polynomial of the order m, best fitted to the signal $y_k^{(s)}(t)$. Let us consider the deflections from the local trend:

$$\Delta y_k^{(s,m)}(t) = y_k^{(s)}(t) - p^{(s, m)}{}_k(t), \quad t = 1, \ldots, s \tag{6.21}$$

and calculate the values

$$Z^{(m)}(q, s) = \left(\frac{\sum_{k=1}^{[N/s]} \left(\max_{1 \le t \le s} \Delta y_k^{(s,m)}(t) - \min_{1 \le t \le s} \Delta y_k^{(s,m)}(t) \right)^q}{[N/s]} \right)^{1/q} \tag{6.22}$$

that can be regarded as the estimate of $(M(\delta_s, q))^{1/q}$. Let us define the function $h(q)$ as a coefficient of linear regression between $\ln(Z^{(m)}(q, s))$ and $\ln(s)$: $Z^{(m)}(q, s) \sim s^{h(q)}$ fitted for a scales range of $s_{\min} \le s \le s_{\max}$. It is evident that $\rho(q) = qh(q)$ and, for a monofractal signal, $h(q) = H = const$. The multifractal singularity spectrum $F(\alpha)$ is equal to the fractal dimensionality of the set of time moments t for which the Hölder−Lipschitz exponent is equal to α, i.e., for which $|x(t + \delta) - x(t)| \sim |\delta|^\alpha$, $\delta \to 0$ (Feder, 1988). The singularity spectrum can be estimated using the standard multifractal formalism, which consists of calculating the Gibbs sum:

$$W(q, s) = \sum_{k=1}^{[N/s]} \left(\max_{1 \le t \le s} \Delta y_k^{(s,m)}(t) - \min_{1 \le t \le s} \Delta y_k^{(s,m)}(t) \right)^q \tag{6.23}$$

and estimating the mass exponent $\tau(q)$ from the condition $W(q, s) \sim s^{\tau(q)}$. From Eq. (6.22) it follows that $\tau(q) = \rho(q) - 1 = qh(q) - 1$. In the next step, the spectrum $F(\alpha)$ is calculated with the Legendre transform:

$$F(\alpha) = \max \left\{ \min_q (\alpha q - \tau(q)), 0 \right\} \tag{6.24}$$

If the singularity spectrum $F(\alpha)$ is estimated in a moving window, its evolution can give useful information on the variations in the structure of the 'chaotic' pulsations of the series. In particular, the position and width of the support of the spectrum $F(\alpha)$, i.e., the values $\alpha_{\min}$, $\alpha_{\max}$, $\Delta\alpha = \alpha_{\max} - \alpha_{\min}$, and α^*, such that $F(\alpha^*) = \max_\alpha F(\alpha)$, are characteristics of the noisy signal. The value α^* can be called a generalized Hurst exponent and it gives the most typical value of the Lipschitz−Holder exponent. Parameter $\Delta\alpha$, singularity spectrum support width, could be regarded as a measure of the variety of stochastic behaviour. In the case of a monofractal signal, the quantity $\Delta\alpha$ should vanish and $\alpha^* = H$. Usually $F(\alpha^*) = 1$, but there exist time windows for which $F(\alpha^*) < 1$. Estimates of the minimum Lipschitz−Holder exponent $\alpha_{\min}$ are mainly positive. Nevertheless, negative values of $\alpha_{\min}$ are quite possible as well (Currenti et al., 2005; Telesca et al., 2005; Telesca and Lovallo, 2011; Chandrasekhar et al., 2016) for time fragments which are characterized by high-amplitude spikes and steps.

In the calculation of $\Delta\alpha$ and α^* we were guided by the following considerations. The exponent q in the formula (6.23) varied within the interval $q \in [-Q, +Q]$, where Q is a certain sufficiently large number, for example $Q = 10$. For each probe value of α within interval $\alpha \in [A_{\min}, A_{\max}]$ where $A_{\min} = \min_{q \in [-Q, +Q]} d\tau(q)/dq$ and $A_{\max} = \max_{q \in [-Q, +Q]} d\tau(q)/dq$ we calculated the value $\tilde{F}(\alpha) = \min_{q \in [-Q, +Q]}(\alpha q - \tau(q))$. If the value of α is close to $A_{\min}$ then $\tilde{F}(\alpha) < 0$, and this value is unsuitable as an estimate of the singularity spectrum, which must be nonnegative. However, beginning from some certain α, the value of $\tilde{F}(\alpha)$ becomes nonnegative, and this condition defines the $\alpha_{\min}$ value. At a further α increase, the value $\tilde{F}(\alpha)$ increases, reaches its maximum when $\alpha = \alpha^*$, then begins to decrease and, finally, attains a certain value $\alpha_{\max} < A_{\max}$, at which $\tilde{F}(\alpha)$ again becomes negative for $\alpha > \alpha_{\max}$. Thus, the condition $\tilde{F}(\alpha) \geq 0$ determines the interval of singularity spectrum support $\alpha \in [\alpha_{\min}, \alpha_{\max}]$, where $F(\alpha) = \tilde{F}(\alpha)$. The derivative $d\tau(q)/dq$ is calculated numerically from the values of $\tau(q)$, $q \in [-Q, +Q]$, and the accuracy of its calculation is of little significance, because this derivative is used for a rough determination of an a priori interval of possible values of exponent q. The minimum value of scale $s_{\min}$ within formulae (6.22) and (6.23) was chosen from 20 samples, and the maximum scale equals $s_{\max} = N/5$.

Multifractal analysis is a rather popular tool in geophysical studies (Ramirez-Rojas et al., 2004; Currenti et al., 2005; Ida et al., 2005; Telesca et al., 2005; Lyubushin et al., 2012, 2014; Chandrasekhar et al., 2016). In the paper by Lyubushin et al. (2012) the multifractal analysis of geomechanical monitoring time series was applied for its fragmentation into intervals with different behaviours. In Lyubushin (2008b, 2009, 2010a,c, 2011a,b, 2012, 2013a,b, 2014a,b) estimates of multifractal properties $\Delta\alpha$, α^* and $\alpha_{\min}$ of LF seismic noise were used for the purposes of earthquake prediction and dynamic estimation of seismic danger.

6.5 First Principal Component

There is a necessity to aggregate the used sequences of properties of time series such as $(En, \Delta\alpha, \alpha^*)$ into time series of scalar characteristics, which carries the most common properties from the set of initial properties. We have here used the most popular approach of principal components (Jolliffe, 1986). Let $P(t) = (P_1(t), \ldots, P_m(t))^T$, $t = 0, 1, \ldots$ be a multiple time series of dimensionality m. Let L be a number of samples within the time window, which is moving from left to right with minimum mutual shift 1, which we will name a 'window of adaptation'. Let s be the number of the sample corresponding to the right-hand end of the moving time window. This means that the time window contains samples with time indexes t which obey the condition $s - L + 1 \le t \le s$. Let's calculate a correlation matrix $\Phi(s)$ of size $m \times m$ within each time window after normalization of multiple time series components:

$$\Phi(s) = \left(\varphi^{(s)}_{ab} \right), \quad \varphi^{(s)}_{ab} = \sum_{t=s-L+1}^{s} q^{(s)}_a(t) q^{(s)}_b(t) / L, \qquad a, b = 1, \ldots, m \tag{6.25}$$

where

$$q^{(s)}_a(t) = (P_a(t) - \overline{P}^{(s)}_a)/\sigma^{(s)}_a, \quad \overline{P}^{(s)}_a = \sum_{t=s-L+1}^{s} P_a(t)/L,$$

$$(\sigma^{(s)}_a)^2 = \sum_{t=s-L+1}^{s} (P_a(t) - \overline{P}^{(s)}_a)^2/(L-1), \quad a = 1, \ldots, m \tag{6.26}$$

The first principal component $\psi^{(s)}(t)$ is calculated using

$$\psi^{(s)}(t) = \sum_{\alpha=1}^{m} \theta^{(s)}_a \cdot q^{(s)}_a(t) \tag{6.27}$$

where m-dimensional vector $\theta^{(s)} = (\theta^{(s)}_1, \ldots, \theta^{(s)}_m)^T$ is an eigenvector of the correlation matrix $\Phi(s)$ corresponding to its maximum eigenvalue. Let us define a scalar time series of first principal components $\psi(t)$ within the window of adaptation of the length L samples by the rule:

$$\psi(t) = \begin{cases} \psi^{(L-1)}(t) & \text{for } 0 \le t \le (L-1) \\ \psi^{(t)}(t) & \text{for } t \ge L \end{cases} \tag{6.28}$$

Thus, within the first time window of the adaptation time series $\psi(t)$ is composed of values calculated by Eq. (6.27), whereas for all further time indexes $\psi(t)$ equals to the value (6.27) corresponding to the right-hand-most end of the time window, i.e., outside the first window of adaptation $\psi(t)$ depends on past values of $P(t)$ only.

6.6 Properties of Global Low-Frequency Seismic Noise

Microseismic oscillations in a wide frequency range are one of the most frequently investigated topics of geophysical studies. This is due to their accessibility, the presence of numerous regional and global seismic networks, and the well-developed practice of seismic observations. Even an approximate review of the literature devoted to analysis of microseisms apparently cannot be made. This is particularly true of the analysis of HF microseisms (from 0.01 to 100 Hz and higher, up to seismoacoustic waves). The widespread occurrence of HF microseismic observations is due to the relative simplicity and mobility of instrumentation free from rigid requirements on long-term stability of sensors that can by no means be neglected in problems of LF geophysical monitoring. In the paper by McNamara and Buland (2004) the results of detailed research into the microseismic background of natural and industrial origins in the frequency band 0.01–16 Hz were presented, including the construction of estimators for the temporal (diurnal and seasonal) and spatial distribution of power spectrum properties. More recent studies on the composition of short-period microseisms are presented in Koper and de Foy (2008) and Koper et al. (2010). With an increase in the period of microseismic background oscillations studied, the role of atmospheric and oceanic waves as major sources of microseisms becomes predominant. Berger et al. (2004) presented a review of the use of IRIS broadband seismic stations for the study of background microseisms. Microseismic oscillations in the period range 5–40 seconds were studied in Stehly et al. (2006), where their oceanic origin was established. Continuously observed microseismic oscillations at periods of 100–500 seconds were examined in Friedrich et al. (1998), Kobayashi and Nishida (1998), Tanimoto (2001, 2005) and Ardhuin et al. (2011). These oscillations are generated both by weak earthquakes and by processes in the atmosphere and ocean. In the following papers (Grevemeyer et al., 2000; Aster et al., 2008; Kedar et al., 2008; Schimmel et al., 2011) variability in the field of microseisms due to climate change and ocean processes was studied.

In order for earthquakes to be a source of continuously present microseismic oscillations, at least one earthquake with a magnitude of 6 should occur daily to maintain the observed intensity of such oscillations. The cumulative effect of all weak earthquakes estimated from the Gutenberg–Richter recurrence law yields an energy contribution one to two orders smaller than the observed value. The effect of atmospheric processes (movement of cyclones) and oceanic waves generated by them, as well as the impact of waves on the shelf and coasts, contributes most to the energy of the LF microseismic background. The origin of an LF seismic hum with a predominant period of 4 minutes was studied in Rhie and Romanowicz (2004, 2006). A significant correlation was established between the intensity of these oscillations and the storm wave height in the oceans, and it was shown that the hum intensity is independent of the Earth's seismic activity: the authors presented an example of a seismically quiet time interval (31 January–3 February 2000) characterized, however, by anomalously high amplitudes of microseismic background in the vicinity of the 4-minute period. As a possible mechanism of excitation of such oscillations, they proposed the perturbation of the gravitational field by high waves resulting in the excitation of LF seismic waves on the sea floor. The main regions of excitation of these oscillations are suggested to be the north Pacific Ocean in winter and the southern Atlantic Ocean in summer. This frequency range of the ambient seismic noise ('seismic hum') has been investigated (Nishida et al., 2008, 2009; Fukao et al., 2010).

In spite of the fact that the main source of energy for LF microseisms is an external one with respect to the Earth's crust, and the latter is merely the propagation medium, the conditions in the Earth's crust affect the statistical characteristics and the specific features in the behaviour of LF microseismic vibrations. Consequently, if we study the time variations of the characteristics of seismic noise, this study will hopefully yield important information concerning the changes in the Earth's crust, including those linked with the seismic process and with the preparation of strong earthquakes.

This basically simple idea of the use of LF microseismic oscillations for monitoring the lithosphere, nevertheless, cannot be realized simply. The main difficulty consists in a strong influence of numerous uncorrelated sources of data. These sources are often diffusely distributed over the Earth's surface. Therefore, it is impossible in this case to investigate the transmitting properties of the lithosphere by controlling input actions and responses. Additionally, the division into 'a signal' and 'noise,' which is typical of the traditional methods used for data analysis, loses its sense, when microseismic oscillations are processed. Only tidal variations in the amplitude of microseisms, as well as the arrivals and coda from the well-known strong earthquakes, can be related to 'signals'. These signals have been traditionally used in geophysics for a long time. All other microseism variations relate to 'noise'.

In this section the main tool for overcoming the influence of uncorrelated random sources is using $(En, \Delta\alpha, \alpha^*)$ statistics calculated within adjacent 'short' time fragments. Thus, seismic noise records are transformed into time series with a 'big' sampling time step: 1 day for instance. These time series are much more correlated and are more suitable for investigating synchronization effects.

The seismic records were taken by requests to the Incorporated Research Institutions for Seismology (IRIS) database at the address http://www.iris.edu/forms/webrequest/ from 229 seismic stations of three global broadband seismic networks:

- Global Seismographic Network: http://www.iris.edu/mda/_GSN;
- GEOSCOPE: http://www.iris.edu/mda/G;
- GEOFON: http://www.iris.edu/mda/GE.

Vertical components with a sampling rate of 1 Hz (LHZ-records) were downloaded for 20 years of observation from 1 January 1997 to 31 December 2016. The initial LHZ-records were transformed to a sampling time step of 1 minute by calculating mean values within successive time intervals of 60 seconds. A further analysis is based on estimating statistical properties of LF seismic noise waveforms (periods exceeding 2 minutes) within successive daily time intervals of the 1440 samples with time steps of 1 minute.

Fig. 6−1 presents the positions of 229 broadband seismic stations all over the world and splits them into eight groups of stations. Each group has a three-letter identification code and the number of stations within each group is given in brackets. The names of the groups have the following abbreviation definitions: the first letter is 'N' or 'S' indicating North or

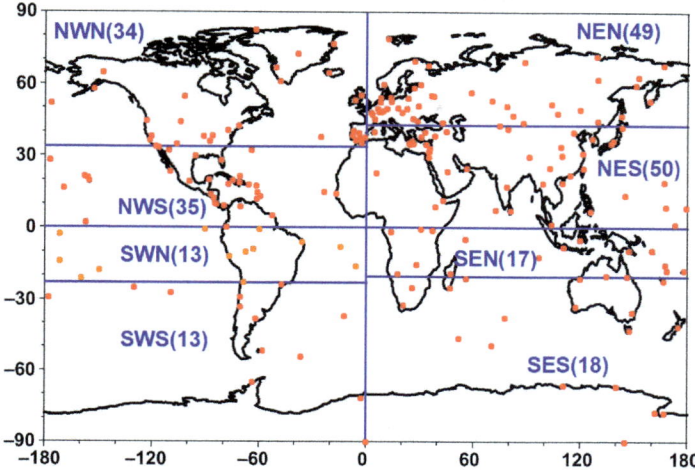

FIGURE 6–1 Positions of 229 broadband seismic stations split into eight groups, with the number of stations in each group in brackets.

South, respectively. The second letter is 'E' or 'W' indicating East or West, respectively. Thus, initially all stations were divided into four groups by splitting them into North-East, North-West, South-East and South-West quarter-spheres. Finally, each of these four parts was split into North and South parts (the third letter is 'N' or 'S') by the rule that the number of stations within each part must be approximately equal each other.

The seismic records from each station after reaching the 1 minute sampling time step were split into adjacent time fragments of 1 day (1440 samples) and for each fragment three parameters of LF daily seismic noise waveforms were calculated. Two of these are multifractal parameters: generalized Hurst exponent α^* and singularity spectrum support width $\Delta\alpha$. For removing scale-dependent trends (which are mostly caused by tidal variations) in the method of singularity spectrums estimates a local polynomials of the eighth order were used in the Eq. (6.21). The other seismic noise parameter is minimum normalized entropy En of the squared orthogonal wavelet coefficients. Before computing entropy En in each daily time window polynomial trends of the eighth order were removed from seismic noise waveforms. Thus, time series of α^*, $\Delta\alpha$ and En values with a sampling time step of 1 day were obtained from each of the 229 seismic stations presented in Fig. 6–1. Fig. 6–2 illustrates the sequence of data transform operations.

Fig. 6–3 presents graphics of daily median values of multifractal singularity spectrum support width $\Delta\alpha$, generalized Hurst exponent α^* and minimum normalized entropy En of squared orthogonal wavelet coefficients for four regions NWN (North-West-North), NEN (North-East-North), NWS (North-West-South) and NES (North-East-South) in the Northern hemisphere and first principal components calculated using Eq. (6.28) with length of adaptation of 365 days from the daily time series of ($En, \Delta\alpha, \alpha^*$) for all eight regions both from the Northern and Southern hemispheres. From the plots of moving averages in a window of 57 days it can be seen that a strong annual periodicity is characterized for all regions,

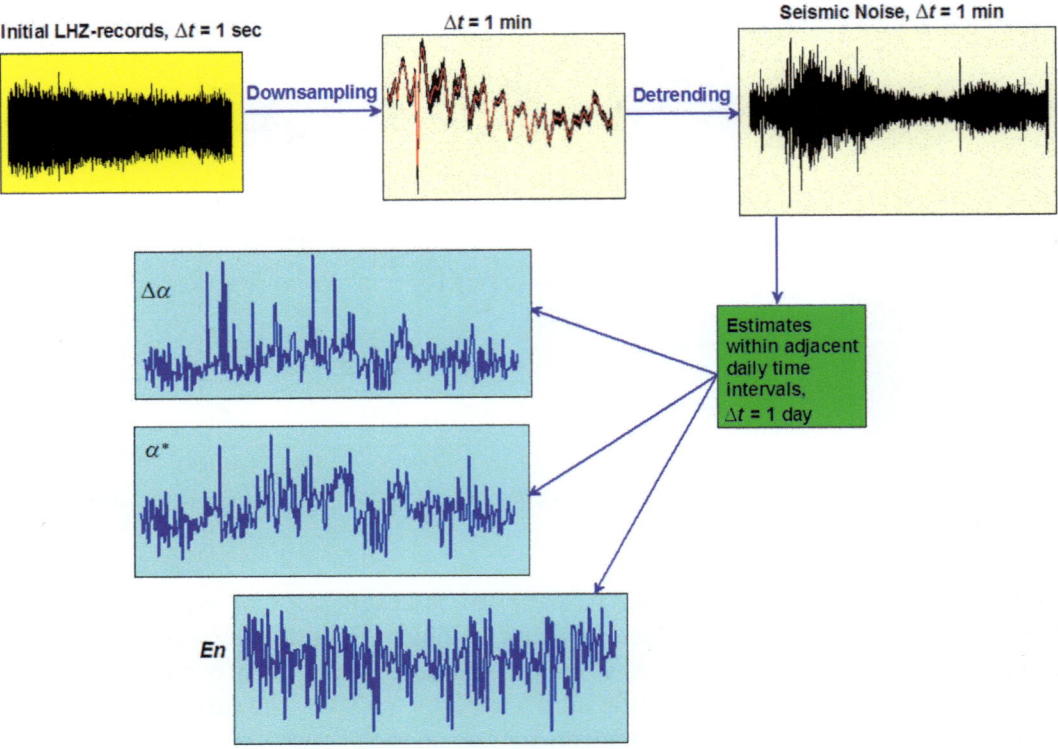

FIGURE 6–2 Scheme of data transformation from initial seismic records with a sampling rate of 1 Hz to time series of multifractal and entropy properties with a sampling time step of 1 day.

especially for NWN and NEN. We believe that this periodicity is caused by the influence of periodic seasons of strong oceanic storms, which are known to be an important source of energy for permanent seismic noise (Rhie and Romanowicz, 2004, 2006; Nishida et al., 2008, 2009; Fukao et al., 2010).

First principal component time series from eight parts of the world compose multiple time series, which is the object of applying wavelet-based and spectral coherence measures with the purpose of detecting coherence effects in global seismic noise properties. The length of the moving time window of 365 days is quite natural for such an application. Fig. 6–4 presents the results of estimating measures of coherence. Fig. 6–4A1–A4 demonstrate evolution of the wavelet-based measure $\rho(\tau, \beta)$ [Eq. (6.7)] with the use of Haar wavelets for four detail levels with a range of scales 2–4, 4–8, 8–16 and 16–32 days, respectively. The occurrence of the first four detail levels of wavelet decomposition is a consequence of the length of 365 samples of time window and the significance threshold $L_{min} = 16$. The graphs in Fig. 6–4A1–A4 are plotted versus time of the right-hand end of the moving time window. The main peculiarity of these graphs is the increasing coherence, which is observed starting from the window position at 2007–08. This increase is not gradual and has rather strong fluctuations with a timescale of 2–3 years at the background of the general positive trend.

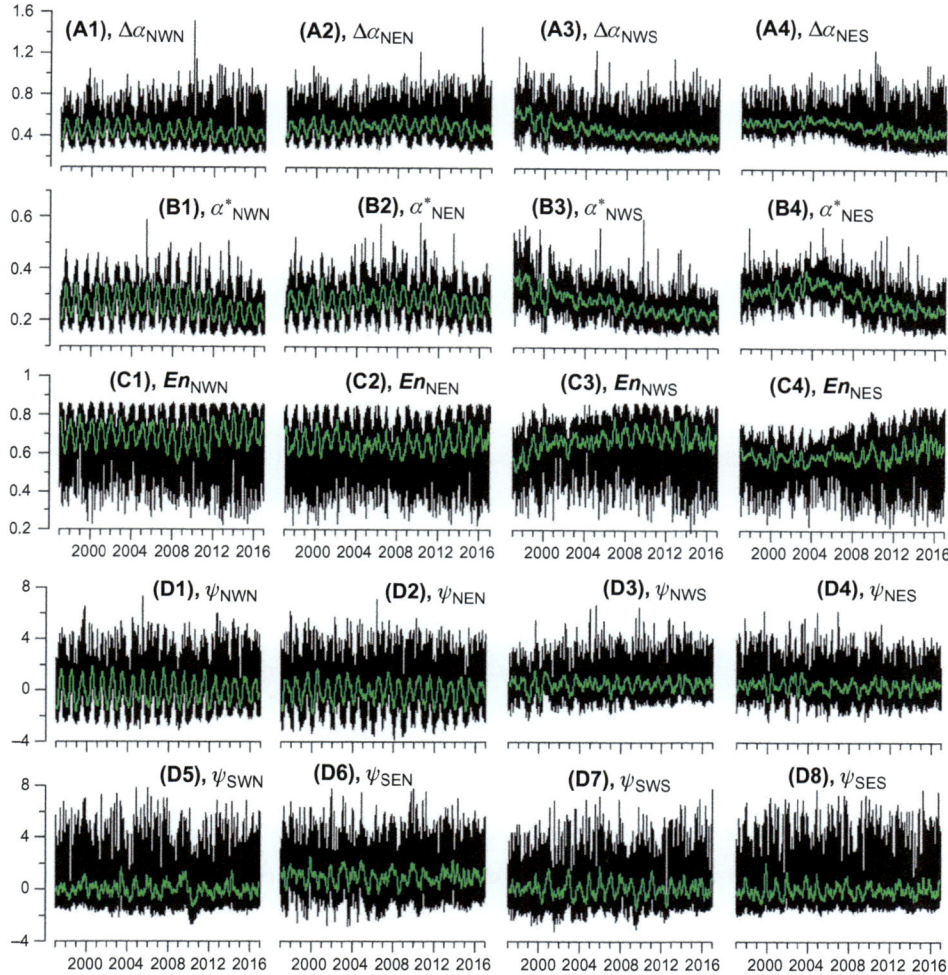

FIGURE 6–3 (A1)–(A4), black lines - plots of daily median values of $\Delta\alpha$ for four regions (NWN, NEN, NWS and NES) in the Northern hemisphere (Fig. 6–1); (B1)–(B4), black lines - plots of daily median values of α^* for regions NWN, NEN, NWS and NES; (C1)–(C4), black lines - plots of daily median values of *En* for regions NWN, NEN, NWS and NES; (D1)–(D4), black lines - plots of values of first principal component ψ, calculated for daily median values of $(\Delta\alpha, \ \alpha^*, \ En)$ for regions NWN, NEN, NWS and NES respectively within the length of adaptation of 365 days; (D5)–(D8) as for (D1)–(D4) but calculated for four regions in the Southern hemisphere SWN (South-West-North), SEN (South-East-North), SWS (South-West-South) and SES (South-East-South) (Fig. 6–1). Bold green lines represent the running average within the time window of 57 days for each curve.

The time−frequency diagram of spectral measure of coherence $\lambda(\tau,\omega)$ [Eq. (6.15)] in Fig. 6−4B confirms this conclusion. Fig. 6−4B was obtained by estimating $\lambda(\tau,\omega)$ within the moving time window of the same length of 365 samples using a vector autoregression model [Eq. (6.12)] of the fifth order. Thus, using both wavelets (compact basic functions) and a Fourier spectral approach independently extracts the same effect of coherence increases for variations of seismic noise properties from stations all around the world.

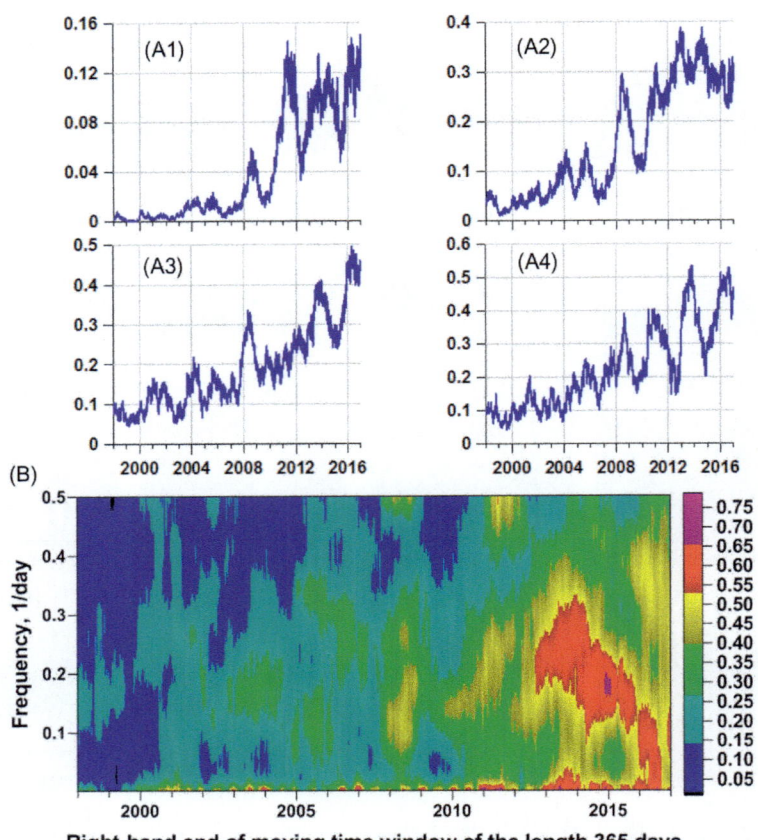

FIGURE 6–4 (A1)–(A4) Plots of multiple wavelet-based coherence measures for eight-dimensional time series of first principal components (Fig. 6–3D1–D8) for detail levels 1–4 depending on the right-hand end of a moving time window of 365 days; (B) time–frequency diagram of multiple spectral measures of coherence for the same eight-dimensional time series estimated within a moving time window of 365 days.

Table 6–1 presents information about 20 of the strongest earthquakes that occurred from the beginning of the 20th century.

According to the information in Table 6–1, of the 20 strongest earthquakes that have occurred since the beginning of the 20th century, seven events have taken place since the middle of 2001, with four events happening after September 2007. It turns out that the increasing coherence of global seismic noise properties coincides with a dramatic increase in the rate of the strongest earthquakes, which has been observed starting from the Sumatra megaearthquake on 26 December 2004, and especially since 2007. Taking into account that we have investigated a range of periods from 2 minutes up to 500 minutes, this coherence increase could not be the direct consequence of aftershocks of the strongest earthquakes. Our hypothesis is that slow movements of small Earth's crust blocks are synchronized in regions preparing for huge earthquakes (Lyubushin, 2009, 2010b, 2011a,b, 2012, 2013a,b)

Table 6–1 Strongest Earthquakes, $M \geq 8.4$, From the Beginning of the 20th Century

Date	Magnitude	Latitude	Longitude	Date	Magnitude	Latitude	Longitude
31 January 1906	8.8	0.955	−79.369	13 October 1963	8.5	44.872	149.483
11 November 1922	8.5	−28.293	−69.852	28 March 1964	9.2	60.908	−147.339
3 February 1923	8.4	54.486	160.472	4 February 1965	8.7	51.251	178.715
2 March 1933	8.4	39.209	144.59	23 June 2001	8.4	−16.265	−73.641
1 February 1938	8.5	−5.045	131.614	26 December 2004	9.1	3.295	95.982
1 April 1946	8.6	53.492	−162.832	28 March 2005	8.6	2.085	97.108
15 August 1950	8.6	28.363	96.445	12 September 2007	8.4	−4.438	101.367
4 November 1952	9	52.623	159.779	27 February 2010	8.8	−36.122	−72.898
9 March 1957	8.6	51.499	−175.626	11 March 20111	9.1	38.297	142.373
22 May 1960	9.5	-38.143	−73.407	11 April 2012	8.6	2.327	93.063

Source: https://earthquake.usgs.gov/earthquakes/search/

and we see that this synchronization is a global phenomenon starting from the beginning of the 2000s (Lyubushin, 2014a, 2015). The increase in seismic noise synchronization that has been observed could be a precursor of the strongest earthquakes in the near future.

6.7 Low-Frequency Seismic Noise at Japan Islands

The attention of this section is primarily focused on the processing of seismic noise data from the network on Japan islands. This peculiarity follows from the fact that one of the strongest megaearthquakes with a magnitude of 9.1 in the latest period of instrumental seismology happened on 11 March 2011 in Japan, which is a region with an extremely dense network of geophysical observations which are available through open access via the Internet. Such a combination gives the unique possibility to test different hypotheses about the ways in which processes of seismic catastrophe preparation influence the statistical properties of seismic noise in the active region.

For this analysis, vertical broadband seismic oscillation components with 1-second sampling time steps (LHZ-records) from the broadband seismic network F-net stations in Japan were downloaded from the following Internet address http://www.fnet.bosai.go.jp starting from the beginning of 1997 up to 31 August 2017. The whole list of F-net seismic stations includes 84 positions. We considered the stations, which are located northward from 30°N and, thereby excluded the data from six solitary stations located on remote small islands. The locations of 78 stations, which were chosen for analysis, are shown in Fig. 6−5 with the epicentres of two of the strongest earthquakes which occurred during the observations: near Hokkaido on 25 September 2003 with magnitude 8.3 and Tohoku megaearthquake on 11 March 11 2011 with magnitude 9.1. These 78 stations were split into five clusters, which are presented in Fig. 6−5 by numbers of clusters with different colours.

The F-net seismic data were analysed after transforming them to a sampling time step of 1 minute by calculating mean values in adjacent time windows of 60 seconds and further

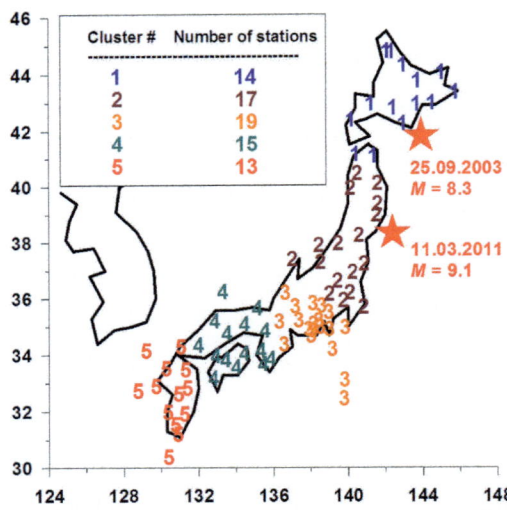

FIGURE 6–5 Positions of 78 broadband seismic stations of the network F-net in Japan split into five clusters. Positions of stations belonging to the same cluster are indicated by the same number (from 1 up to 5). The number of stations within each cluster is given in the frame. Red stars show the epicentres of two of the strongest earthquakes since the beginning of 1997.

calculating values of $(En, \Delta\alpha, \alpha^*)$ within adjacent time windows of 1440 samples with 1-minute sampling time steps, i.e., of 1 day (see Fig. 6–2).

Let us analyse in detail the time series of $\Delta\alpha$. For each time window we calculate median values of $\Delta\alpha$ estimates from all operable stations and thus, scalar time series of median values of $\Delta\alpha^{(m)}$ will be obtained as integral characteristics of seismic noise from the whole network. Our special attention to $\Delta\alpha^{(m)}$ follows from the fact that reducing the singularity spectrum support width is known to be an indicator for transforming the complex system to a critical state with a more simple structure of the ambient noise (less multifractal) preceding some catastrophic changes.

It should be noted that using multifractal singularity spectrum support width has a rather long history in the investigation of nonlinear system behaviour. Particularly the 'loss of multifractality,' i.e., decreasing the singularity spectrum support width, is a well-known effect before the abrupt change of different system properties. This effect has mainly been investigated in biological and medicine systems (Ivanov et al., 1999; Humeaua et al., 2008; Dutta et al., 2013), but in Pavlov and Anishchenko (2007) it was shown that it has a rather universal character and is also observed in physical systems. The analogy between the effect of singularity spectrum support narrowing of seismic noise waveforms and the loss of multifractality in the behaviour of other nonlinear systems gave the author the impetus to make a hypothesis about using Japanese islands for seismic catastrophe work (Lyubushin, 2008a).

Before analysing the peculiarities of $\Delta\alpha^{(m)}$ let us compute Gaussian kernel smoothing $\xi(t)$ (Hardle, 1990) of the signal $\Delta\alpha^{(m)}(t)$ by the formula

$$\xi(t|h) = \int \Delta\alpha^{(m)}(t + h\,s) \cdot \psi(s)\,ds, \quad \psi(s) = \exp(-s^2/2)/\sqrt{2\pi} \tag{6.29}$$

where $h > 0$ is a smoothing parameter which could be called the averaging radius of Gaussian smoothing. The smoothed value of $\Delta\alpha^{(m)}$ with a radius averaging 13 days is represented in Fig. 6–6A by a black line.

The two red vertical lines indicate time moments of the strongest earthquakes and they split the history of $\xi(t|h)$ into a sequence of three fragments. The first time fragment belongs to the time interval before the Hokkaido earthquake on 25 September 2003. It can be seen that the mean value of $\xi(t)$ for this fragment is greater than the mean value for the other time interval before the Tohoku earthquake on 11 March 2011. Let us find time point t_C of

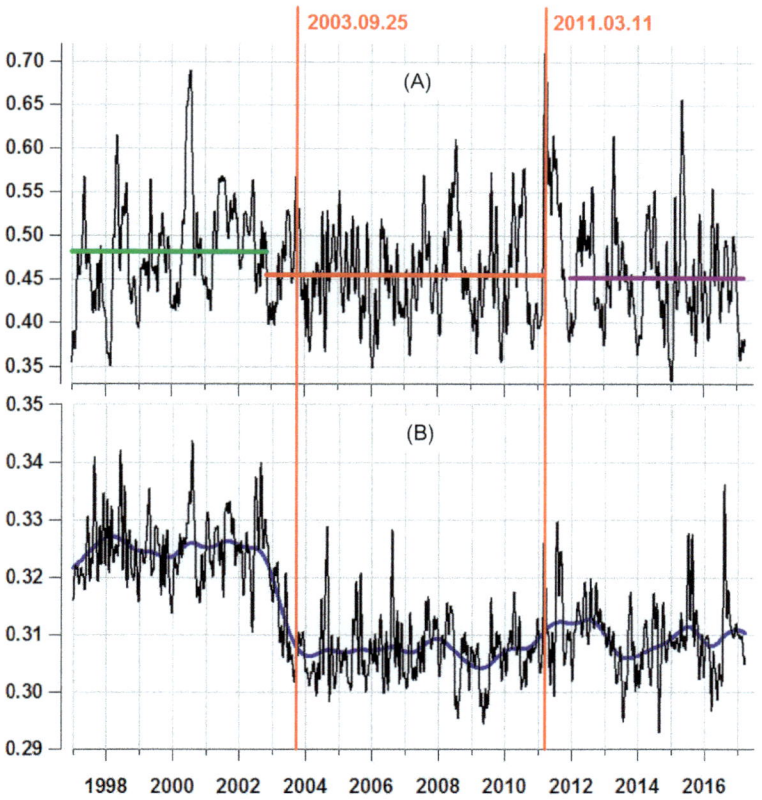

FIGURE 6–6 (A) Plot of Gaussian kernel smoothing with a radius averaging 13 days of median values of singularity spectrum support width $\Delta\alpha^{(m)}$ from network F-net after coming to a sampling time step 1 minute within the adjacent time windows of 1 day. Vertical red lines indicate time moments of earthquakes on 25 September 2003, $M = 8.3$ and on 11 March 2011, $M = 9.1$. Parallel bold lines of green, red and purple show mean values for time intervals 1 January 1997–8 November 2002, 9 November 2002–10 March 2011 and 1 January 2012–31 August 2017. (B) Black line — plot of Gaussian kernel smoothing with averaging radius 13 days of median of $\Delta\alpha$ from all F-net networks for vertical seismic noise waveforms with a sampling time step of 1 s within adjacent time windows of 30 min; (A), bold blue line — as in (B) with an averaging radius of 0.5 years.

maximum change of mean values of $\xi(t)$ at the vicinity of the time moment of Hokkaido earthquake by using Fisher criterion from analysis of variance (ANOVA) (Rao, 1965). Let us calculate the general mean value of $\xi(t)$ by all analysed time intervals $\overline{\xi}_0 = \sum_{t=1}^{N} \xi(t)/N$ where N is the general number of samples and mean values from the left and right sides of the probe time moment t_C: $\overline{\xi}_1 = \sum_{t=1}^{tc} \xi(t)/t_C$ and $\overline{\xi}_2 = \sum_{t=t_C+1}^{N} \xi(t)/(N - t_C)$. Change point t_C is found from the condition

$$F(t_C) = \frac{S_1^2(t_C)}{S_2^2(t_C)} \to \max_{tc} \tag{6.30}$$

where

$$S_1^2(t_C) = t_C \cdot \left(\overline{\xi}_1 - \overline{\xi}_0\right)^2 + (N - t_C) \cdot \left(\overline{\xi}_2 - \overline{\xi}_0\right)^2,$$
$$S_2^2(t_C) = \frac{\left(\sum_{t=1}^{tc} \left(\xi(t) - \overline{\xi}_1\right)^2 + \sum_{t=t_C+1}^{N} \left(\xi(t) - \overline{\xi}_2\right)^2\right)}{(N - 2)} \tag{6.31}$$

The using of criterion (6.30) detects 8 November 2002 as the changing point t_C with mean values 0.482 and 0.454 for $\overline{\xi}_1$ and $\overline{\xi}_2$. These mean values are shown in Fig. 6−6 by bold green and red parallel lines. This drop in $\Delta\alpha^{(m)}$ detects some transient geodynamic process to an unstable state, which began with preparation of the Hokkaido earthquake on 25 September 2003. After the Tohoku earthquake, the smoothed value of $\Delta\alpha^{(m)}$ shows a rapid increase and this is a reaction to processes of stress relaxation in the Earth's crust immediately after the megaearthquake. However, starting from the beginning of 2012 the mean smoothed value of $\Delta\alpha^{(m)}$ returns approximately to its previous level of 0.452. This means that the Tohoku megaearthquake on 11 March 2011 did not return the region to its 'normal' state as it was before the Hokkaido earthquake and the situation of high seismic danger continues currently.

The difference between mean values of $\Delta\alpha$ before and after the Hokkaido earthquake on 25 September 2003 becomes more explicit if they are estimated for initial vertical seismic records with 1 second sampling time step in the adjacent time windows of 30 minutes (1800 samples). In this case, multifractal singularity spectra were calculated using DFA by removing local trends by polynomials of the fourth order. In Fig. 6−6B the black line represents the result of Gaussian smoothing of these $\Delta\alpha$ values with an averaging radius of 13 days, whereas the bold blue line has an averaging radius of 0.5 years. Fig. 6−6B shows the same drop in $\Delta\alpha$ before the earthquake on 25 September 2003 as shown in Fig. 6−6A.

Let us consider the five groups of stations that are presented in Fig. 6−5. We are interested in time-dependent evolution of coherence effects between median values of daily seismic noise properties calculated from these five parts of the network. For this purpose let us compute first principal components $\psi_k(t)$, $k = 1, ..., 5$, from median values of parameters $(En, \Delta\alpha, \alpha^*)$ estimated daily from all operational stations within each cluster of stations in Fig. 6−5. These first principal components were computed within the moving window of adaptation of 365 days using Eq. (6.28). Plots in Fig. 6−7A1−A5 present graphs of these first

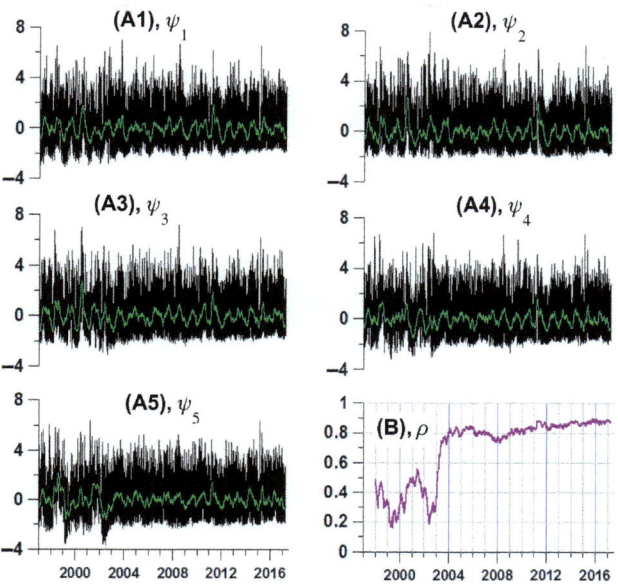

FIGURE 6–7 (A1)–(A5) Plots of values of first principal components ψ_k, $k = 1, \ldots, 5$, calculated for daily median values of multifractal singularity spectrum support width $\Delta\alpha$, generalized Hurst exponent α^* and minimum normalized entropy *En* of squared orthogonal wavelet coefficients for five clusters of stations in Japan (Fig. 6–5) after coming to sampling time step 1 minute (see Fig. 6–2) correspondingly within the length of adaptation of 365 days; green lines represent the running average within the time window of 57 days for each curve. Plot (B) presents values of multiple robust correlation coefficient ρ which was calculated within moving time window of the length 365 days. The values of ρ are plotted in dependence on time moments corresponding to right-hand end of moving time window.

principal components. The simplest measure of coherence is the squared robust multiple correlation coefficient ρ [Eq. (6.8)], which is shown in Fig. 6–7B, estimated in the moving time window of the same length of 365 days. We can see that the value of ρ demonstrates a rapid increase for time windows lying in the interval 2002–04, i.e., in the vicinity of time of the Hokkaido earthquake on 25 September 2003. This peculiarity of seismic noise correlation independently confirms the conclusion, which was made from the analysis of Fig. 6–6: 2002 was a year of rapid change in the structure of seismic noise on Japan islands, which is an indicator of preparing for the future seismic catastrophe on 11 March 2011. In this sense, the Hokkaido earthquake of 25 September 2003 could be interpreted as a foreshock for the Tohoku earthquake on 11 March 2011, despite its strength. If we continue comparing Fig. 6–6 and 6–7 we can notice one more conformity: the Tohoku megaearthquake has a short-term response both in the median value of $\Delta\alpha$ and in the value of multiple correlation ρ. This means that the Tohoku megaearthquake has not changed the correlation structure of seismic noise, and moreover, this correlation becomes stronger because the value of ρ has a slight positive trend after 2012.

It is interesting to obtain independent confirmation about the conclusion on the growth of seismic noise correlations using another approach − estimation of coherence with the

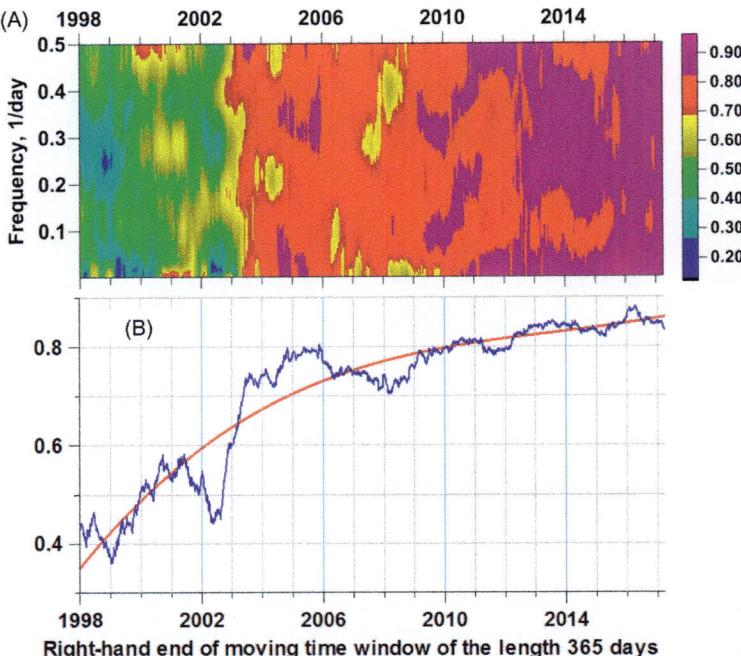

FIGURE 6−8 (A) Time−frequency diagram of the multiple spectral measure of coherence for the five-dimensional time series of first principal components ψ_k, $k = 1,\ldots,5$ presented in Fig. 6−7A1−A5 estimated within a moving time window of 365 days; (B) plot of mean values $\lambda_{mean}(\tau)$ of multiple spectral coherence calculated by averaging all frequency-dependent values within each time window; the bold red line represents the best-fitted polynomial of the third order.

help of multiple spectral measure [Eq. (6.10)]. For this purpose, let us consider a five-dimensional time series of first principal components $\psi_k(t)$, $k = 1,\ldots,5$ and calculate the multiple spectral coherence measure [Eq. (6.10)] within the moving time window of 365 days using a vector autoregression model [Eq. (6.12)] of fifth order. The time−frequency 2D diagram in Fig. 6−8A presents the result of such an estimate, whereas Fig. 6−8B shows a graph of time-dependent averaging $\lambda_{mean}(\tau)$ of the coherence measure [Eq. (6.10)] using all frequency values within each time window [Eq. (6.16)].

The results of the coherence estimates presented in Fig. 6−8 confirm the conclusion, which was made from the analysis of the behaviour of multiple correlation ρ in Fig. 6−7B: seismic noise coherence effects became stronger after the transient process in 2002−04 with a slightly positive trend which is still continuing currently despite the occurrence of the Tohoku megaearthquake on 11 March 2011. Using the time−frequency diagram in Fig. 6−8A gives a possibility to visualize frequency decomposition of coherence growth and we can see that this increase is practically independent of the frequency band.

One of the seismic noise parameters used is the normalized entropy of squared wavelet coefficients *En*. This quantity is a kind of antipode to $\Delta\alpha$. Fig. 6−9 presents examples of four

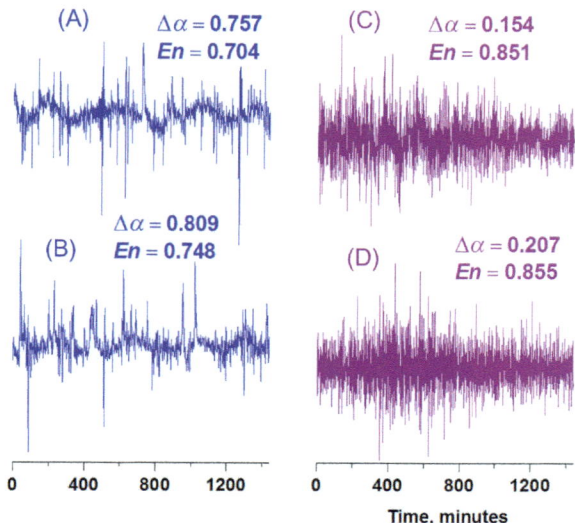

FIGURE 6–9 Two types of daily low-frequency seismic noise waveforms after removing tidal trends by eighth-order polynomial: (A and B) with relatively large values of singularity spectrum support width $\Delta\alpha$ and high values of normalized entropy *En* and (C, D) with relatively low values of $\Delta\alpha$ and *En*.

daily noise waveforms with different values of $\Delta\alpha$ and *En*: the left-hand panels of graphics (Fig. 6–9A and B) present noise waveforms with a high value of $\Delta\alpha$ and low values of *En*, whereas the right-hand panels (Fig. 6–9C and D) correspond to two noise waveforms with low values of $\Delta\alpha$ and high values of *En*. The difference in waveform peculiarities between Fig. 6–9A, B and C, D is evident: high values of $\Delta\alpha$ and low values of *En* occur because of the existence of irregular high-amplitude spikes, which are intermittent with intervals with stationary behaviour. This is a typical multifractal: different types of stochastic behaviour are observed. Low values of $\Delta\alpha$ correspond to much more stationary behaviour: the noise structure is more simple and less multifractal.

Our hypothesis consists of a correlation between low values of $\Delta\alpha$ and the growth of seismic danger. Thus, the increasing entropy *En* could also be connected to the increasing seismic danger. A possible physical interpretation of the ability of low values of $\Delta\alpha$ and high values of *En* to extract seismically dangerous regions was given in Lyubushin (2012, 2013a,b). This is the consequence of the consolidation of small blocks of the Earth's crust into a large one before the strong earthquake. Consolidation implies that seismic noise does not include spikes, which are connected to mutual movements of small blocks. The absence of irregular spikes in the noise follows a decrease in $\Delta\alpha$ and an increase in entropy *En*.

Having the values of $\Delta\alpha$ and *En* from all seismic stations, it is possible to create maps of spatial distribution of these seismic noise statistics. For this we consider the regular grid of 30×30 nodes covering the rectangular domain with latitudes between 30 and 46°N and longitudes between 128 and 148°E (see Fig. 6–5). For each node of this grid the corresponding daily values of $\Delta\alpha$ and *En* are found, and are calculated as the median for the values of

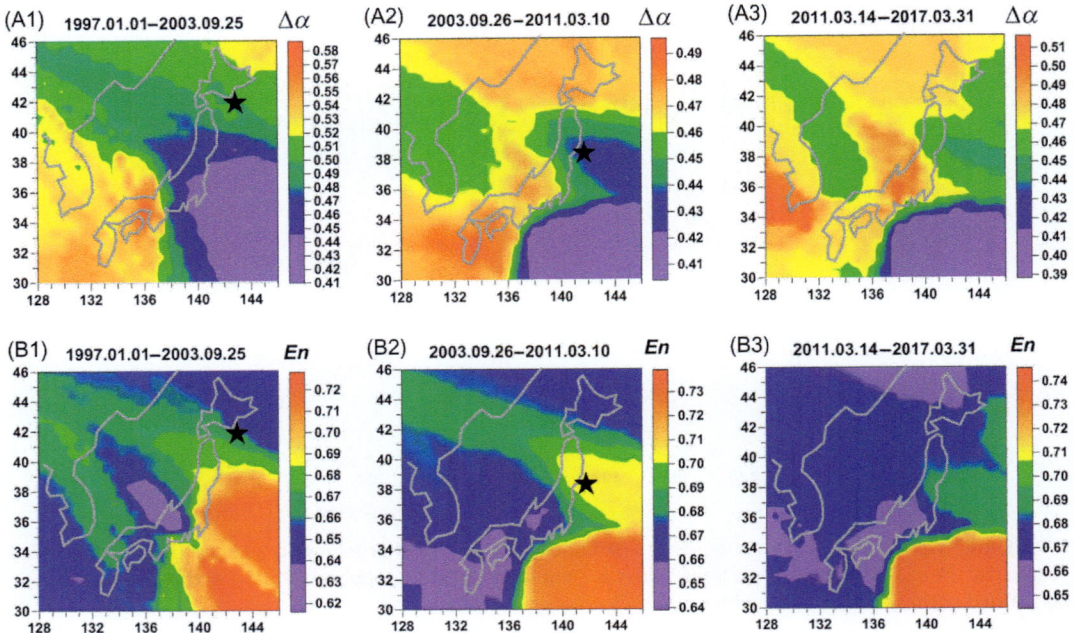

FIGURE 6–10 Averaged maps of multifractal singularity spectrums support width $\Delta\alpha$ ((A1)–(A3)) and minimum normalized entropy *En* of squared orthogonal wavelet coefficients ((B1)–(B3)) for three time intervals: 1 January 1997–25 September 2003 ((A1) and (B1)), 26 September 2003–10 March 2011 ((A2) and (B2)) and 14 March 2011–31 August 2017 ((C1) and (C2)). Stars indicate epicentres of the earthquakes on 25 September 2003, $M = 8.3$ ((A1) and (B1)) and 11 March 2011, $M = 9.1$ ((A2) and (B2)).

the five nearest to the node operable seismic stations. This simple procedure provides a sequence of daily maps for all parameters. The averaged maps are created by averaging daily maps for all days between two given dates. Taking into account that almost all stations of the F-net are located on large Japanese islands, these maps in ocean regions are less accurate than on islands. However, we had to work with those data that we had at our disposal. The method of nearest neighbours, which is used in this chapter, provides a rather natural extrapolation of the used values into domains, which have no observation points.

Fig. 6–10 presents averaged maps of $\Delta\alpha$ and *En* for three adjacent time fragments: from the beginning of 1997 up to 25 September 2003, the day of the earthquake with magnitude 8.3 near Hokkaido; from 26 September 2003 up to 10 March 2011, the day before the Tohoku megaearthquake of 11 March with magnitude 9.1 and from 14 March 2011 up to 31 August 2017. Three days (11, 12 and 13 March 2011) are excluded from the analysis because during these days a lot of seismic stations of F-net were not working properly after the seismic shock of 11 March 2011.

Fig. 6–10A2 shows that plotting averaged maps of spatial distribution of singularity spectrum support width $\Delta\alpha$ could extract the place of future catastrophe as the regions with relatively low values of $\Delta\alpha$. Fig. 6–10A1 presents a map where an area of relatively low $\Delta\alpha$

can be seen which includes the place of the future megaearthquake and it is not split into north and south parts. In Fig. 6−10A2 we can see that after the event on 25 September 2003 this area was split into north and south parts and the north part turned out to be the area with aftershocks from the Great Japan earthquake on 11 March 11 2011, whereas the south part remains a region of relatively low values of singularity spectrum support width $\Delta\alpha$ before and after 11 March 2011 (Fig. 6−10A3). According to the interpretation of regions with low values of $\Delta\alpha$ as being seismically dangerous, we could propose a hypothesis that during the Tohoku earthquake only a part of the accumulated seismic energy was dropped and that the above-mentioned south region (the north part of the Philippine plate, Nankai Through) could be the area for a future megaearthquake. Such an hypothesis explains why the coherence of seismic noise remains high after 11 March 2011 (Figs 6−7B and 6−8).

Fig. 6−10B1−B3 confirms the conclusion, which was made from the analysis of graphs in Fig. 6−9, that minimum normalized entropy En is an 'antipode' to the parameter $\Delta\alpha$ and almost everything that was written above about the properties of $\Delta\alpha$ could be repeated for En by changing 'minimum' to 'maximum': relatively maximum values of normalized entropy locate seismically dangerous domains before earthquakes.

Let us call the regions identified with low values of $\Delta\alpha$ and high values of En as 'spots of seismic danger' (SSD). Mean values of $\Delta\alpha$ and En are strongly anticorrelated − that is why statistics $\Delta\alpha$ and En obtain approximately the same SSD. Nevertheless, their mutual consideration is expedient because these parameters are based on different approaches. The maps of $\Delta\alpha$ and En could be plotted as the sequence in the moving time window − such estimates provide the possibility to visualize the origin and evolution of SSDs.

The problem of predicting the strongest earthquakes in Japan at the region of the Nankai Trough has been a great traditional problem for seismologists in Japan (Rikitake, 1999; Mogi, 2004). In Rikitake (1999) the probability of an earthquake with magnitude more than 8.5 at the Tokai-Nankai zone, the region where the Philippine Sea plate is approaching central Japan, was estimated as 0.35−0.45 'for a 10-year period following the year 2000'. In Simons et al. (2011) the seismic danger for Japan was estimated immediately after the Tohoku earthquake based on the analysis of GPS data and the conclusion was that 'estimates … suggest the need to consider the potential for a future large earthquake just south of this event.' In the paper by Kagan and Jackson (2013) the problem of why the Tohoku earthquake was a surprise for the scientific community is discussed. One of the conclusions given is 'A magnitude 9 earthquake off Tohoku should not have been a surprise'. This conclusion was made by retrospective analysis of seismic catalogues. Nevertheless, the Tohoku event was a great surprise for all traditional methods of earthquake prediction. Another conclusion in Zoller et al. (2014) is that even a magnitude 10 earthquake is quite possible for the Japan Trench.

Maps of correlations between pairs of parameters $(\alpha^*, \Delta\alpha)$ and (α^*, En) possess interesting prognostic properties. Similar to the maps presented in Fig. 6−10 we can plot averaged maps of the correlation between increments of $(\alpha^*, \Delta\alpha)$. For this purpose, let's estimate evolution of the correlation coefficient between $(\alpha^*, \Delta\alpha)$ for each station a within moving time window of 365 days. For each position of a 1-year moving time window, we can plot a map by calculating the median of correlation coefficients for the five operable seismic stations

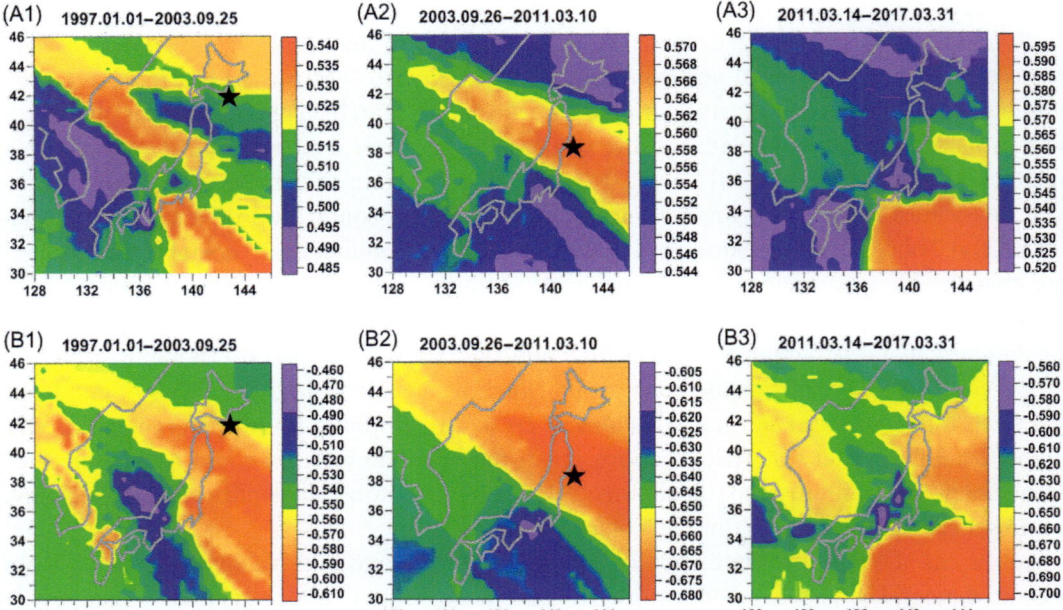

FIGURE 6–11 (A1)–(A3) Averaged maps of the correlation coefficient between increments of generalized Hurst exponent α^* and multifractal singularity spectrums support width $\Delta\alpha$ for three time intervals: 1 January 1997–25 September 2003 (A1), 26 September 2003–10 March 2011 (A2) and 14 March 2011–31 August 2017 (A3). Stars indicate epicentres of the earthquakes on 25 September 2003, $M = 8.3$ (A1) and 11 March 2011, $M = 9.1$ (A2). Plots ((B1)–(B3)) present similar maps for correlation coefficients between α^* and minimum entropy of wavelet coefficients, En.

which are nearest to each node of the regular grid. The averaged maps are created by averaging maps corresponding to all 1-year time fragments, which lie entirely between and including two given dates. In a similar way maps of correlation coefficient could be plotted for pair (α^*, En); such maps are presented in Fig. 6–11. It is interesting to note that in Fig. 6–11A2 and B2 the region of the future Tohoku earthquake is extracted by relatively high absolute values of correlations. Note that correlations in Fig. 6–11B1–B3 are negative. For the period after the Tohoku earthquake the region of SSD according to Fig. 6–10A3 and B3 coincides with the region of maximum absolute values of correlations – Fig. 6–11A3 and B3.

Let us add one more independent multifractal parameter and consider three median values of $(\Delta\alpha, \alpha^*, \alpha_{\min})$ where $\alpha_{\min}$ is the minimum Hölder–Lipschitz exponent. Fig. 6–12A1–A3 presents plots of these values. Considering clustering properties of the clouds of the daily sequence of these 3D vectors within the moving time window gives the possibility of estimating natural fluctuations in seismic danger and even discovering its periodical structure.

Let's consider a moving time window of $L = 365$ days and let $\vec{\xi}^{(t)} = (\Delta\alpha, \alpha^*, \alpha_{\min})^{(t)}$ be 3D vector within the current time window, $t = 1, \ldots, L$, t is time index, numerating vectors. Our

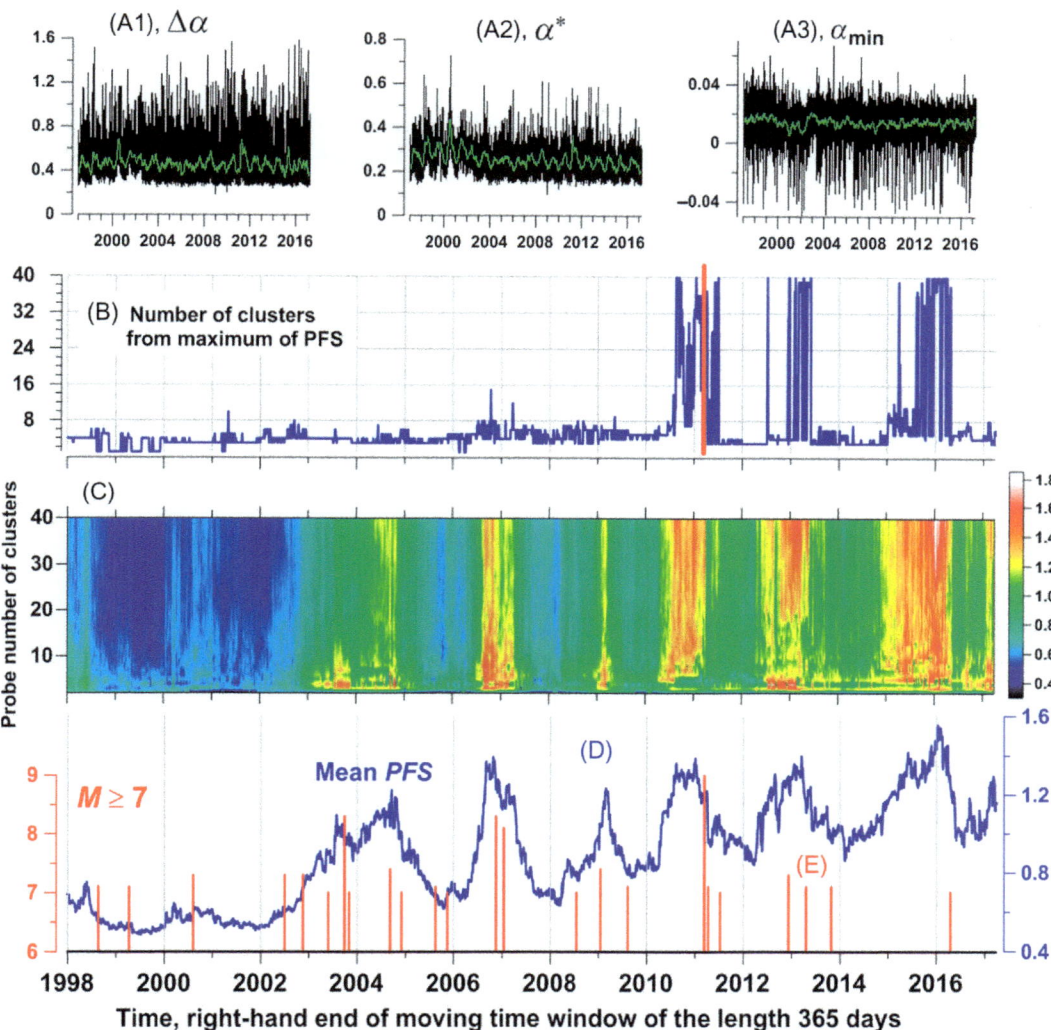

FIGURE 6–12 (A1)–(A3) Plots of daily median values of multifractal singularity spectrum support width $\Delta\alpha$, generalized Hurst exponent α^* and minimum Hölder–Lipschitz exponent α_{min} from all 78 stations of broadband seismic network F-net in Japan; green lines represent the running average within time windows of 57 days; (B) plot of the best numbers of clusters for the sequence of clouds consisting of 365 daily 3D vectors ($\Delta\alpha$, α^*, α_{min}) from the moving time window of 365 days with a mutual shift of 3 days. The best number of clusters is defined from the maximum of pseudo-F-statistics. The vertical red line indicates the time of the Tohoku megaearthquake on 11 March 2011, $M = 9.1$. A two-dimensional diagram (C) presents dependence of pseudo-F-statistics on the probe number of clusters, which varies from 2 up to 40 within each time window. Plot (D) presents the mean value of the pseudo-F-statistics averaged by all probe numbers of clusters in dependence on the right-hand end of the moving time window of 365 days. Plot (E) presents the sequence of time moments of strong earthquakes $M \geq 7$ in the rectangular domain with coordinates latitude: 28–48°N; longitude 128–156°E, which is a rather broad area of the Japanese islands.

purpose is to investigate clustering properties of clouds of 3D vectors $\vec{\xi}^{(t)}$ with each 1-year time window. In particular, we are interested in what the 'best' number of clusters is.

Before carrying out the cluster procedure, a preliminary operation of normalizing and iterative clipping of outliers (Huber and Ronchetti, 2009) was performed within each time window for each scalar component $\vec{\xi}^{(t)}_k$ of vectors $\vec{\xi}^{(t)}$. Here $k = 1, 2, 3$ is the index numerating scalar components of the vector $\vec{\xi}^{(t)}$. Let $\bar{\xi}_k = 1/L \sum_{t=1}^{L} \xi^{(t)}_k$, $\sigma^2_k = 1/(L-1) \sum_{t=1}^{L} (\xi^{(t)}_k - \bar{\xi})^2$ be sample estimates of mean values and variance of scalar components of the 3D vector $\vec{\xi}^{(t)}$. Let's perform iterations which consist of coming to values $\zeta^{(t)}_k = (\xi^{(t)}_k - \bar{\xi}_k)/\sigma_k$ and clipping values $\zeta^{(t)}_k$ exceeding thresholds $\pm 3\sigma_k$. These iterations are stopped when the values $\bar{\xi}_k$ and σ_k become stable and equal to the following values: $\bar{\xi}_k = 0$, $\sigma_k = 1$. After this preliminary operation at each current time window we have a cloud consisting of L 3D vectors $\vec{\zeta}^{(t)}$.

Let's split some cloud into a given probe number q of clusters using a standard k-means cluster procedure (Duda et al., 2000). Let Γ_r, $r = 1, \ldots, q$ be clusters, $\vec{z}_r = \sum_{\vec{\zeta} \in \Gamma_r} \vec{\zeta}/n_r$ – vector of the centre of cluster Γ_r, n_r be a number of vectors $\vec{\zeta}^{(t)}$ within cluster Γ_r, $\sum_{r=1}^{q} n_r = L$. Vector $\vec{\zeta}^{(t)} \in \Gamma_r$ if the distance $|\vec{\zeta}^{(t)} - \vec{z}_r|$ is the minimum among all positions of cluster centres. K-means procedure minimizes the sum

$$S\left(\vec{z}_1, \ldots, \vec{z}_q\right) = \sum_{r=1}^{q} \sum_{\vec{\zeta} \in \Gamma_r} \left|\vec{\zeta} - \vec{z}_r\right|^2 \to \min_{\vec{z}_1, \ldots, \vec{z}_q} \tag{6.32}$$

with respect to positions of cluster centres $\vec{z}_r$. Let $J(q) = \min_{\vec{z}_1, \ldots, \vec{z}_q} S(\vec{z}_1, \ldots, \vec{z}_q)$. We try a probe number of clusters within the range $2 \le q \le 40$. The problem of selecting the best number of clusters $q*$ was solved from maximum of pseudo-F-statistics (Vogel, Wong, 1979), which is similar to F-criterion [Eq. (6.30)] from ANOVA:

$$\text{PFS}(q) = \frac{\sigma^2_1(q)}{\sigma^2_0(q)} \to \max_{2 \le q \le 40} \tag{6.33}$$

where

$$\sigma^2_0(q) = \frac{J(q)}{(L-q)}, \quad \sigma^2_1(q) = \sum_{r=1}^{q} \nu_r \left|\vec{z}_r - \vec{z}_0\right|^2$$

$$\nu_r = \frac{n_r}{L}, \quad \vec{z}_0 = \sum_{t=1}^{L} \frac{\vec{\zeta}^{(t)}}{L} \tag{6.34}$$

The PFS $\to$ max rule does not working if we try to distinguish cases $q* = 1$ and $q* = 2$ because the value $\sigma^2_1(q)$ is not defined for $q = 1$. These cases could be distinguished by the existing break point of the monotonous function $J(q)$ at the argument $q = 2$ (Lyubushin, 2011a). The value $\sigma^2_0(q)$ monotonically increases as q decreases, and usually the dependence of $\log(\sigma^2_0(q))$ on $\log(q)$ is close to linear, that is, it scales as $q^{-\mu}$. As is known, the optimal number of clusters can also be determined from the break point of the monotonic dependence $\sigma^2_0(q)$ for $q = q*$: as q decreases, the function $\sigma^2_0(q)$ increases faster at $q < q*$ than at

$q > q*$. This rule for determining the number of clusters is known as the 'elbow method' (Ketchen, Jr and Shook, 1996). This criterion of identification of $q = q*$ is more susceptible to noise and exhibits a poorer performance compared to the technique $q* = \text{argmaxPFS}(q)$ but this is the only possibility to discern the case $q* = 1$ from the case $q* = 2$. Let $\delta(q)$ denote the deviation of $\log(\sigma_0^2(q))$ from the best fit straight line approximating the dependence on $\log(q)$, i.e., be defined from formula $\delta(q) = \log(\sigma_0^2(q)) - (a\log(q) + b)$, where coefficients (a, b) are found by least squares: $\sum_{q=1}^{40} \delta^2(q) \to \min_{a,b}$. Then, we assume that the point $q = 2$ is the break point of the dependency $\sigma_0^2(q)$ if $\delta(1)$ exceeds all values of $\delta(q)$ for $q \geq 2$. Let $q_0 = \arg \max_{2 \leq q \leq 40} \text{PFS}(q)$. Thus, we define the optimal number $q*$ of clusters according to the following rule:

If $q_0 > 2$ then $q* = q_0$. Else, if $\delta(1) / \max_{2 \leq q \leq 40} \delta(q) \leq 1$ then $q* = 1$, else $q* = 2$.

Fig. 6−12B presents the evolution of the estimates of the best number of clusters $q*$ in dependence on the right-hand end of the moving time window of 1 year.

The values [Eq. (6.33)] computed within the moving time window for all probe number q of clusters are dependent on the position of the time window. Thus, pseudo-F-statistics could be presented as a 2D map as dependence on q and the right-hand end of the moving time window. This 2D map is shown in Fig. 6−12C. From Fig. 6−12C it can be seen that the $q*$ before the Tohoku megaearthquake on 11 March 2011, has an extremely chaotic regime with jumps from minimum up to maximum values in the 1 year before the event and this time interval was characterized by high PFS(q) values. Let's consider the sequence of mean values $\overline{P} = \sum_{q=2}^{40} \text{PFS}(q)/39$ of pseudo-F-statistics in each time window in dependence on time moments of the right-hand end of the window. This dependence is presented in Fig. 6−12D, which demonstrates a positive trend and strong periodicity for almost 2 years, which was established after the Hokkaido earthquake on 25 September 2003.

Our hypothesis assumes that maps of pseudo-F-statistics, which are built for multifractal properties of seismic noise, similar to those presented in Fig. 6−12C and graphs of its mean values, similar to Fig. 6−12D, could be useful for the visualization of natural fluctuations of seismic danger in some rather large region. The basis for this hypothesis can be seen in comparison of Fig. 6−12D with the sequence of strong earthquakes $M \geq 7$ within the rectangular domain with coordinates for latitude 28−48°N and longitude 128−156°E, which is presented in Fig. 6−12E. In particular starting from 2006, all strong events belong to time intervals with large values of mean PFS(q) and the last time interval of large values of $q*$ and PFS(q) in Fig. 6−12B and C precedes the Kumamoto earthquake $M = 7$ on 15 April 2016 with the hypocentre at latitude 32.78°N and longitude 130.72°E in south Japan. Mean values of pseudo-F-statistis in Japan beginning from 2004 have fluctuations with approximate period 2 years.

6.8 Results for Surrogate Time Series

Let us check the stability of conclusions about multiple correlation, which is presented in Fig. 6−7, by calculating the same measure for surrogate time series obtained by simple shuffling of samples. To this end, in order to preserve the visible LF structure of the graphs, let

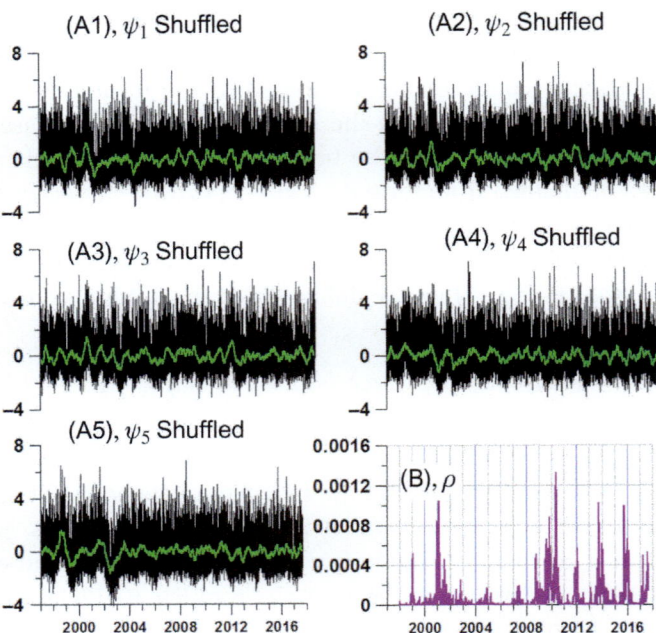

FIGURE 6–13 (A1)–(A5) black lines - plots of values of shuffled first principal components ψ_k, $k = 1, \ldots, 5$ (low-frequency trends were preserved), calculated for daily median values of multifractal singularity spectrum support width $\Delta\alpha$, generalized Hurst exponent α^* and minimum normalized entropy En of squared orthogonal wavelet coefficients for five clusters of stations in Japan (Fig. 6–5) for seismic noise waveforms after coming to sampling time step 1 min from initial waveforms with sampling rate 1 Hz (see Fig. 6–2), respectively, within a length of adaptation of 365 days; green lines represent running average within the time window of 57 days for each curve. Plot (B) presents values of squared multiple robust correlation coefficient ρ estimated within the moving time window of 365 days for shuffled time series depending on the right-hand end of the moving time window.

us compute trends for each principal component by applying Gaussian kernel smoothing [Eq. (6.29)] with averaging radius h of 90 days. These Gaussian trends were subtracted from principal components and the residuals were shuffled. After this the surrogate time series were constructed as the sum of trends and shuffled residuals. The result of the computing multiple correlation is similar to the case presented in Fig. 6–7 when seen in Fig. 6–13.

Comparing Figs 6–7 and 6–13 shows that shuffling of HF components of time series (HF components with periods less than 180 days) destroys high correlation which is presented at the Fig. 6–7.

6.9 Conclusion

In studies of such a complex multicomponent system as the Earth's crust, it is highly challenging to identify a set of the main deterministic reasons that would define all the features of the global seismic regime, particularly those which control long-term changes in the intensity of potential seismic events. Solving this problem may be facilitated by the phenomenological approach based on the use of coherent noise generated by the system in the course of its evolution. For the Earth's crust, the ambient noise is a product of its 'life'. Coherence (or synchronization) of the behaviour of characteristics of a complex system, described by data of different nature and structure, is an important feature for assessments of its approach to rapid changes in the condition, which are often referred to as a 'catastrophe'. Searching for precursors of catastrophes, which may be manifested by the occurrence of synchronous components in a variety of observations, is a general idea for increasing the correlation radius of random fluctuations of parameters of a complex system as it approaches a sharp change in its properties, resulting from its own dynamics (Gilmore, 1981; Nicolis, Prigogine, 1989). This property of the coherence of ambient noise of the Earth is investigated in this chapter.

The analysis of coherence between properties of global seismic noise, measured at the network of 229 broadband stations all over the world since the beginning of 1997 until now, extracts the effect of progressively increasing synchronization after the Sumatra megaearthquake on 26 December 2004, which could be a precursor to a further rise in the intensity of the strongest seismic events (Lyubushin, 2014a, 2015). A particular case of this effect is a strong increase in the coherence between the behaviour of different parameters of LF seismic noise in Japan and California before the Tohoku megaearthquake on 11 March 2011 in Japan, detected by the analysis of seismic noise waveforms from regional broadband seismic networks (Lyubushin, 2016a,b).

Plotting the maps of different properties of LF seismic noise (multifractal singularity spectrum support width and minimum normalized entropy of squared orthogonal wavelet coefficients) within the moving time window could present a new method for dynamic seismic hazard estimation. It raises the possibility of inspecting the origin and evolution of the 'spots of seismic danger'. The analysis of seismic noise on Japanese islands from broadband seismic network F-net gave the possibility for the prediction of the Great Japan earthquake of 11 March 2011. This prediction was published in a number of scientific papers and abstracts at international conferences in advance of the seismic catastrophe (Lyubushin, 2008a, 2009, 2010a,c, 2011a,b,c). According to the analysis of seismic noise after 11 March 2011 the next megaearthquake with magnitude 8.5−9.0 could occur at the region of the Nankai Trough (Lyubushin, 2012, 2013a,b, 2014b). To estimate the time interval for the occurrence of this seismic event the periodic structure of seismic danger natural fluctuations with a period of about 2 years (Fig. 6−12D) could be used.

Acknowledgement

This work was supported by the Russian Foundation for Basic Research (project no. 18-05-00133).

References

Ardhuin, F., Stutzmann, E., Schimmel, M., Mangeney, A., 2011. Ocean wave sources of seismic noise. J. Geophys. Res. 116, C09004.

Aster, R., McNamara, D., Bromirski, P., 2008. Multidecadal climate induced variability in microseisms. Seismol. Res. Lett. 79, 194–202.

Berger, J., Davis, P., Ekstrom, G., 2004. Ambient earth noise: a survey of the global seismographic network. J. Geophys. Res. 2004 109, B11307.

Brillinger, D.R., 1975. Time Series. Data Analysis and Theory. Holt, Rinehart and Winston, Inc, New York, Chicago, San Francisco, p. 540.

Chandrasekhar, E., Sanjana, S.P., Gopi, K.S., Nayana, S., 2016. Multifractal detrended fluctuation analysis of ionospheric total electron content data during solar minimum and maximum. J. Atmos. Sol.—Terr. Phys. 149 (2016), 31–39. Available from: https://doi.org/10.1016/j.jastp.2016.09.007.

Clarke, E., 1975. Generalized gradients and applications. Trans. Am. Math. Soc. 1975 205 (2), 247–262.

Costa, M., Peng, C.-K., Goldberger, A.L., Hausdorf, J.M., 2003. Multiscale entropy analysis of human gait dynamics. Physica A 330 (2003), 53–60.

Costa, M., Goldberger, A.L., Peng, C.-K., 2005. Multiscale entropy analysis of biological signals. Phys. Rev. E 71 (2005), 021906.

Currenti, G., del Negro, C., Lapenna, V., Telesca, L., 2005. Multifractality in local geomagnetic field at Etna volcano, Sicily (southern Italy). Nat. Hazard. Earth Syst. Sci. 5 (555–559).

Duda, R.O., Hart, P.E., Stork, D.G., 2000. Pattern Classification. Wiley-Interscience Publication, New York, Chichester, Brisbane, Singapore, Toronto, p. 680.

Dutta, S., Ghosh, D., Chatterjee, S., 2013. Multifractal detrended fluctuation analysis of human gait diseases. Front. Physiol., 2013 v.4. Available from: https://doi.org/10.3389/fphys.2013.00274.

Feder, J., 1988. Fractals. Plenum Press, New York, London, p. 284.

Friedrich, A., Krüger, F., Klinge, K., 1998. Ocean-generated microseismic noise located with the Gräfenberg array. J. Seismol. 2 (1), 47–64.

Fukao, Y.K., Nishida, K., Kobayashi, N., 2010. Seafloor topography, ocean infragravity waves, and background Love and Rayleigh waves. J. Geophys. Res. 115, B04302.

Gilmore, R., 1981. Catastrophe Theory for Scientists and Engineers. John Wiley and Sons, Inc, New York, NY, p. 666.

Grevemeyer, I., Herber, R., Essen, H.-H., 2000. Microseismological evidence for a changing wave climate in the northeast Atlantic Ocean. Nature 408, 349–352.

Hannan, E.J., 1970. Multiple Time Series. John Wiley and Sons, Inc, New York, London, Sydney, Toronto.

Hardle, W., 1990. Applied Nonparametric Regression. (Biometric Society Monographs No. 19.). Cambridge University Press, Cambridge, p. 333.

Hotelling, H., 1936. Relations between two sets of variates. Biometrika. 28, 321–377.

Huber, P.J., Ronchetti, E.M., 2009. Robust Statistics, second ed John Wiley & Sons, Inc, p. 354. Available from: https://doi.org/10.1002/9780470434697.ch1.

Humeaua, A., Chapeau-Blondeau, F., Rousseau, D., Rousseau, P., Trzepizur, W., Abraham, P., 2008. Multifractality, sample entropy, and wavelet analyses for age-related changes in the peripheral cardiovascular system: preliminary results. Med. Phys., Am. Assoc. Phys. Med. 35 (.2), 717–727. February 2008.

Ida, Y., Hayakawa, M., Adalev, A., Gotoh, K., 2005. Multifractal analysis for the ULF geomagnetic data during the 1993 Guam earthquake. Nonlinear Processes Geophys. vol.12, 157–162.

Ivanov, P. Ch, Amaral, L.A.N., Goldberger, A.L., Havlin, S., Rosenblum, M.B., Struzik, Z., et al., 1999. Multifractality in healthy heartbeat dynamics. Nature 399, 461–465.

Jolliffe, I.T., 1986. Principal Component Analysis. Springer-Verlag, p. 487. Available from: https://doi.org/10.1007/b98835.

Kagan, Y.Y., Jackson, D.D., 2013. Tohoku earthquake: a surprise? Bull. Seismol. Soc. Am. 103 (.2B), 1181–1194. Available from: https://doi.org/10.1785/0120120110.

Kantelhardt, J.W., Zschiegner, S.A., Konscienly-Bunde, E., Havlin, S., Bunde, A., Stanley, H.E., 2002. Multifractal detrended fluctuation analysis of nonstationary time series. Physica A 316 (1−4), 87−114. Available from: https://doi.org/10.1016/S0378-4371(02)01383-3.

Kedar, S., Longuet-Higgins, M., Webb, F., Graham, N., Clayton, R., Jones, C., 2008. The origin of deep ocean microseisms in the North Atlantic Ocean. Proc. R. Soc. A 464, 777−793.

Ketchen Jr, D.J., Shook, C.L., 1996. The application of cluster analysis in strategic management research: an analysis and critique. Strategic Manage. J. 17 (6), 441−458.

Kobayashi, N., Nishida, K., 1998. Continuous excitation of planetary free oscillations by atmospheric disturbances. Nature. 395, 357−360.

Koper, K.D., de Foy, B., 2008. Seasonal anisotropy in short-period seismic noise recorded in South Asia. Bull. Seismol. Soc. Am. 98, 3033−3045.

Koper, K.D., Seats, K., Benz, H., 2010. On the composition of Earth's short-period seismic noise field. Bull. Seismol. Soc. Am. April 2010 100 (2), 606−617.

Lyubushin, A., 2012. Prognostic properties of low-frequency seismic noise. Nat. Sci. 4 (8A), 659−666. Available from: https://doi.org/10.4236/ns.2012.428087.

Lyubushin, A., 2013b. How soon would the next mega-earthquake occur in Japan?. Nat. Sci. 5 (8A1), 1−7. Available from: https://doi.org/10.4236/ns.2013.58A1001.

Lyubushin, A.A., 1998. Analysis of canonical coherences in the problems of geophysical monitoring. Izv. Phys. Solid Earth 34 (1), 52−58.

Lyubushin, A.A., 1999. Analysis of multidimensional geophysical monitoring time series for earthquake prediction. Ann. Geofis. 42 (5), 927−937. Available from: https://doi.org/10.4401/ag-3757.

Lyubushin, A.A., 2000. Wavelet-aggregated signal and synchronous peaked fluctuations in problems of geophysical monitoring and earthquake prediction. Izv. Phys. Solid Earth 36 (2000), 204−213.

Lyubushin, A.A., 2008a. Multifractal properties of low-frequency microseismic noise in Japan, 1997−2008. In: Book of Abstracts of Seventh General Assembly of the Asian Seismological Commission and Japan Seismological Society, 2008 Fall Meeting, Tsukuba, Japan, 24−27 November 2008, 92.

Lyubushin, A.A., 2008b. Microseismic noise in the low frequency range (periods of 1−300 min): properties and possible prognostic features. Izv. Phys. Solid Earth 44 (4), 275−290. Available from: https://doi.org/10.1134/s11486-008-4002-6.

Lyubushin, A.A., 2009. Synchronization trends and rhythms of multifractal parameters of the field of low-frequency microseisms. Izv. Phys. Solid Earth 45 (5), 381−394.

Lyubushin, A.A., 2010a. The statistics of the time segments of low-frequency microseisms: trends and synchronization. Izv. Phys. Solid Earth 46 (6), 544−554. Available from: https://doi.org/10.1134/S1069351310060091.

Lyubushin A.A., 2010c. Synchronization of multifractal parameters of regional and global low-frequency microseisms. In: European Geosciences Union General Assembly 2010, Vienna, 02−07 May 2010. Geophys. Res. Abstr. V. 12. EGU2010-696.

Lyubushin, A.A., 2011a. Cluster analysis of low-frequency microseismic noise. Izv. Phys. Solid Earth 47 (6), 488−495. Available from: https://doi.org/10.1134/S1069351311040057.

Lyubushin, A.A., 2011b. Seismic catastrophe in Japan on March 11, 2011: long-term prediction on the basis of low-frequency microseisms. Izv. Atmos. Oceanic Phys. 46 (8), 904−921. Available from: https://doi.org/10.1134/S0001433811080056. 2011.

Lyubushin, A.A., 2013a. Mapping the properties of low-frequency microseisms for seismic hazard assessment. Izv. Phys. Solid Earth 49 (1), 9−18. Available from: https://doi.org/10.1134%2FS1069351313010084.

Lyubushin, A.A., 2014a. Analysis of coherence in global seismic noise for 1997−2012. Izv. Phys. Solid Earth 50 (3), 325−333. Available from: https://doi.org/10.1134/S1069351314030069.

Lyubushin, A.A., 2014b. Dynamic estimate of seismic danger based on multifractal properties of low-frequency seismic noise. Nat. Hazard. 70 (1), 471–483. Available from: https://doi.org/10.1007/s11069-013-0823-7.

Lyubushin, A.A., 2015. Wavelet-based coherence measures of global seismic noise properties. J. Seismol. 19 (2), 329–340. Available from: https://doi.org/10.1007/s10950-014-9468-6.

Lyubushin, A.A., 2016a. Long-range coherence between seismic noise properties in Japan and California before and after Tohoku mega-earthquake. Acta Geodaetica Geophys . Available from: https://doi.org/10.1007/s40328-016-0181-5.

Lyubushin, A.A., 2016b. Coherence between the fields of low-frequency seismic noise in Japan and California. Izv. Phys. Solid Earth 52 (6), 810–820. Available from: https://doi.org/10.1134/S1069351316050086. 2016.

Lyubushin, A.A., Klyashtorin, L.B., 2012. Short term global dT prediction using (60–70)-years periodicity. Energy Environ. 23 (.1), 75–85. Available from: https://doi.org/10.1260/0958-305X.23.1.75. 2012.

Lyubushin, A.A., Kopylova, G.N., 2004. Multidimensional wavelet analysis of time series of electrotelluric observations in Kamchatka. Izv. Phys. Solid Earth 40 (.2), 163–175. 2004.

Lyubushin, A.A., Sobolev, G.A., 2006. Multifractal measures of synchronization of microseismic oscillations in a minute range of periods. Izv. Phys. Solid Earth 42 (.9), 734–744.

Lyubushin, A.A., Pisarenko, V.F., Bolgov, M.V., Rukavishnikova, T.A., 2003. Study of general effects of rivers runoff variation. Russ. Meteorol. Hydrol. (.7), 59–68.

Lyubushin, A.A., Pisarenko, V.F., Bolgov, M.V., Rodkin, M.V., Rukavishnikova, T.A., 2004. Synchronous variations in the Caspian sea level from coastal observations in 1977–1991. Atmos. Oceanic Phys. 40 (.6), 737–746. 2004.

Lyubushin, A.A., 2010b. Multifractal parameters of low-frequency microseismsIn: de Rubeis, V., et al., (Eds.), Synchronization and Triggering: from Fracture to Earthquake Processes, GeoPlanet. Earth and Planetary Sciences, Springer, Verlag Berlin Heidelberg, pp. 253–272. 2010, 388 p., Chapter 15. Available from: https://doi.org/10.1007/978-3-642-12300-9_15.

Lyubushin, A.A., Kaláb, Z., Lednická, M., 2012. Geomechanical time series and its singularity spectrum analysis. Acta Geodaetica Geophys Hungarica 47 (1), 69–77. March.

Lyubushin, A.A., Kaláb, Z., Lednická, M., 2014. Statistical properties of seismic noise measured in underground spaces during seismic swarm. Acta Geodaetica Geophys 49 (2), 209–224.

Lyubushin, A.A., Bobrovskiy, V.S., Shopin, S.A., 2016. Experience of complexation of global geophysical observations. Geodyn. Tectonophys. 7 (1), 1–21. Available from: https://doi.org/10.5800/GT-2016-7-1-0194.

Mallat, S., 1998. A Wavelet Tour of Signal Processing. Academic Press, San Diego, London, Boston, NY, Sydney, Tokyo, Toronto, 577 p.

Marple (Jr.), S.L., 1987. Digital Spectral Analysis with Applications. Prentice-Hall, Inc., Englewood Cliffs, NJ.

McNamara, D.E., Buland, R.P., 2004. Ambient noise levels in the continental United States. Bull. Seismol. Soc. Am. 2004 (94), 1517–1527.

Mogi, K., 2004. Two grave issues concerning the expected Tokai Earthquake. Earth Planets Space 56, li-lxvi.

Nicolis, G., Prigogine, I., 1989. Exploring Complexity, An Introduction. W.H. Freedman and Co., New York, NY, p. 328.

Nishida, K., Kawakatsu, H., Fukao, Y., Obara, K., 2008. Background Love and Rayleigh waves simultaneously generated at the Pacific Ocean floors. Geophys. Res. Lett. 35, L16307.

Nishida, K., Montagner, J., Kawakatsu, H., 2009. Global surface wave tomography using seismic hum. Science 326 (5949), 112.

Pavlov, A.N., Anishchenko, V.S., 2007. Multifractal analysis of complex signals. Physics – Uspekhi Fizicheskikh Nauk. Russ. Acad. Sci. 50 (.8), 819–834. Available from: https://doi.org/10.1070/PU2007v050n08ABEH006116.

Press, W.H., Flannery, B.P., Teukolsky, S.A., Vettering, W.T., 1996. Numerical Recipes. second ed., Chapter 13, Wavelet Transforms, Cambridge Univ. Press, Cambridge.

Ramirez-Rojas, A., Munoz-Diosdado, A., Pavia-Miller, C.G., Angulo-Brown, F., 2004. Spectral and multifractal study of electroseismic time series associated to the Mw = 6.5 earthquake of 24 October 1993 in Mexico. Nat. Hazard. Earth Syst. Sci. 4, 703–709.

Rao, C.R., 1965. Linear Statistical Inference and Its Applications. John Wiley and Sons, New York, London, Sydney, 1965.

Rhie, J., Romanowicz, B., 2004. Excitation of Earth's continuous free oscillations by atmosphere-ocean-seafloor coupling. Nature 2004 (431), 552–554.

Rhie, J., Romanowicz, B., 2006. A study of the relation between ocean storms and the Earth's hum. Geochem. Geophys. Geosyst. 7 (10).

Rice, J., 1980. The mechanics of earthquake rupture. In: Dziewonski, A.M., Boschi, E. (Eds.), Physics of the Earth's Interior. Proceedings of the International School of Physics "Enrico Fermi", Course 78, 1979. Italian Physical Society, Amsterdam, North-Holland, pp. 555–649. , 1980.

Rikitake, T., 1999. Probability of a great earthquake to recur in the Tokai district, Japan: reevaluation based on newly-developed paleoseismology, plate tectonics, tsunami study, micro-seismicity and geodetic measurements. Earth Planets Space 51, 147–157.

Schimmel, M., Stutzmann, E., Ardhuin, F., Gallart, J., 2011. Polarized Earth's ambient microseismic noise. Geochem. Geophys. Geosyst. 12, Q07014.

Simons, M., Minson, S.E., Sladen, A., Ortega, F., Jiang, J., Owen, S.E., et al., 2011. The 2011 Magnitude 9.0 Tohoku-Oki earthquake: mosaicking the megathrust from seconds to centuries. Science 332 (6032), 911. Available from: https://doi.org/10.1126/science.332.6032.911.

Stehly, L., Campillo, M., Shapiro, N.M., 2006. A study of the seismic noise from its long-range correlation properties. J. Geophys. Res. 111, B10306.

Tanimoto, T., 2001. Continuous free oscillations: atmosphere-solid earth coupling. Annu. Rev. Earth Planet. Sci. 29, 563–584.

Tanimoto, T., 2005. The oceanic excitation hypothesis for the continuous oscillations of the Earth. Geophys. J. Int. 160, 276–288.

Taqqu, M.S., 1988. Self-similar processes, Encyclopedia of Statistical Sciences, vol. 8. Wiley, New York, NY, pp. 352–357.

Telesca, L., Lovallo, M., 2011. Analysis of the time dynamics in wind records by means of multifractal detrended fluctuation analysis and the Fisher–Shannon information plane. J. Stat. Mech.: Theory Exp. . Available from: https://doi.org/10.1088/1742-5468/2011/07/P070012011.

Telesca, L., Colangelo, G., Lapenna, V., 2005. Multifractal variability in geoelectrical signals and correlations with seismicity: a study case in southern Italy. Nat. Hazard. Earth Syst. Sci. 5, 673–677.

Vogel, M.A., Wong, A.K.C., 1979. PFS clustering method. IEEE Trans. Pattern Anal. Mach. Intell 1 (3), 237–245. Available from: https://doi.org/10.1109/TPAMI.1979.4766919.

Zoller, G., Holschneider, M., Hainzl, S., Zhuang, J., 2014. The largest expected earthquake magnitudes in Japan: the statistical perspective. Bull. Seismol. Soc. Am. 104 (2), 769–779. Available from: https://doi.org/10.1785/0120130103. April 2014.

Further Reading

Schreiber, T., Schmitz, A., 2000. Surrogate time series. Physica D 142 (2000), 346–382.

Natural Time Analysis of Seismic Time Series

Nicholas V. Sarlis, Efthimios S. Skordas, Panayiotis A. Varotsos

NATIONAL AND KAPODISTRIAN UNIVERSITY OF ATHENS, ATHENS, GREECE

CHAPTER OUTLINE

Complexity of Seismic Time Series. DOI: https://doi.org/10.1016/B978-0-12-813138-1.00007-9

7.1 Introduction

In general, when studying phase transitions (critical phenomena), which take a central place in statistical physics because of their crucial importance in applications in a variety of fields, we meet two major difficulties (Varotsos et al., 2011d). First, it is the choice of an order parameter which, according to Sethna (1992), 'is an art, since usually it's a new phase which we do not understand yet, and guessing the order parameter is a piece of figuring out what's going on'. Assuming that we have overcome this difficulty (thus having chosen the order parameter), we meet the second one: the order parameter of the system in the critical state is expected to undergo non-Gaussian fluctuations (note that critical phenomena are by no means limited to the order parameter, e.g., see Stanley (1999)). However, almost nothing is known — with the remarkable exception (Botet, 2011) of the site-percolation on the Bethe lattice — about the mathematical form of the possible probability distributions of the order parameter. Any effort to understand which kind of fluctuations the order parameter can experience at criticality is considered to be of chief importance.

Here we focus on seismicity, which exhibits complex correlations in time, space and magnitude (M) (Bak et al., 2002; Baiesi and Paczuski, 2004; Corral, 2004; Tirnakli and Abe, 2004; Davidsen and Paczuski, 2005; Saichev and Sornette, 2005; Livina et al., 2005; Eichner et al., 2007; Lippiello et al., 2007, 2012; Tiampo et al., 2007; Lennartz et al., 2008, 2011; Zaliapin et al., 2008; Bunde and Lennartz, 2012; Tiampo and Shcherbakov, 2012; Zaliapin and Ben-Zion, 2013a,b; Batac and Kantz, 2014; Zaliapin and Ben-Zion, 2015) and the observed earthquake scaling laws (Turcotte, 1997; Scholz, 2002) widely accepted to indicate the existence of phenomena closely associated with the proximity of the system to a critical point (Carlson et al., 1994; Rundle et al., 1996; Sornette, 2000; Holliday et al., 2006; Xia et al., 2008). Specifically, in the vast majority of the proposals (e.g., Carlson et al., 1994; Fisher et al., 1997; Sornette, 2000) earthquake dynamics is associated with a critical point (second-order phase transition), or with a mean field spinodal (Rundle et al., 1996) that can be understood as a line of critical points. Alternative models based on first-order phase transitions have also been forwarded (e.g., see Rundle et al., 2003, and references therein).

In this chapter we review the following topic which has been developed during the last few years (Sarlis et al., 2010b; Varotsos et al., 2011b, 2012b,a, 2015b; Sarlis et al., 2013, 2015b, 2015c; Skordas and Sarlis, 2014; Varotsos et al., 2014b; Huang, 2015): taking the view that earthquakes are (nonequilibrium) critical phenomena, the analysis of the earthquake catalogue in the new time domain, termed natural time χ (Varotsos et al., 2001, 2002a) (see below), in the frame of which an order parameter for seismicity has been introduced (Varotsos et al., 2005b), reveals the following. Restricting ourselves to regional earthquakes in California, Greece and Japan, a study of the order parameter in seismic time series leads to an estimation of the occurrence time of the impending mainshock with an accuracy from a few days to 1 week or so if geolectric data are also available. In addition, using a seismological catalogue, the fluctuations of this order parameter enable the identification of certain characteristic features that are precursory to very strong earthquakes in both regional and global scales.

This chapter is organized as follows. In Section 7.2 we give a brief overview of the analysis in natural time. In Section 7.3, the order parameter for seismicity is introduced and we present its main applications. The main features of the fluctuations of this order parameter deduced by using a natural time window which has either a varying length or a fixed length sliding through the earthquake catalogue are presented in Sections 7.4 and 7.5, respectively. Finally, in Section 7.6 we give our main conclusions.

We clarify in advance that, for the reader's convenience and in order to conform with the original publications, the symbol l, which stands in general for the natural time window length (number of successive events), turns to W in Sections 7.4 and 7.5 to specify that it refers to excerpts taken from a seismic catalogue.

7.2 A Brief Overview of Natural Time Analysis

7.2.1 Background of Natural Time Analysis

Natural time analysis, introduced early in the 21st century (Varotsos et al., 2001, 2002a, 2003a,b), has found useful applications in a variety of fields: statistical physics (e.g., Varotsos et al., 2006b, 2011c; Vallianatos et al., 2013; Flores-Márquez et al., 2014; Christopoulos and Sarlis, 2014, 2015; Tsuji and Katsuragi, 2015), cardiology (e.g., Varotsos et al., 2004a, 2005a, 2007; Papasimakis and Pallikari, 2010; Sarlis et al., 2015a, 2009a), geophysics (e.g., Varotsos et al., 2002a, 2009; Ramírez-Rojas et al., 2011; Potirakis et al., 2013, 2015, 2016a; Skordas and Sarlis, 2014; Hayakawa et al., 2015a,b;Vargas et al., 2015), atmospheric sciences (e.g., Varotsos and Tzanis, 2012b,a; Varotsos et al., 2016), seismology (e.g., Tanaka et al., 2004; Sarlis et al., 2009b, 2010a; Lennartz et al., 2011; Sarlis, 2011; Sarlis and Christopoulos, 2012; Papadopoulou et al., 2016), physics of earthquakes (e.g., Sarlis et al., 2010b, 2013; Varotsos et al., 2013b, 2011b; Ramírez-Rojas and Flores-Márquez, 2013), earthquake prediction (e.g., Varotsos et al., 2001; Uyeda et al., 2009; Sarlis, 2013; Vallianatos et al., 2014, 2015; Huang, 2015; Sarlis et al., 2015c,b; Potirakis et al., 2016b; Christopoulos and Sarlis, 2017) etc. (for a recent review see Varotsos et al., 2011d). For example, in geophysics, the distinction between seismic electric signal (SES) activities − which are a series of low-frequency electric signals that precede earthquakes (Varotsos and Alexopoulos, 1984a,b; Varotsos et al., 1986, 1988, 2006a; Varotsos and Lazaridou, 1991; Varotsos et al., 2003c) − from signals due to manmade sources (Varotsos et al., 2003b,a, 2009, 2011a, 2005c; Skordas et al., 2010) is possible − beyond other techniques (Varotsos and Lazaridou, 1991) − by employing natural time analysis. SES activities are presumably generated from a cooperative orientation of electric dipoles formed due to defects (e.g., Varotsos, 2008) when − before an earthquake − the gradually increasing stress in the focal area reaches a *critical* value (Varotsos et al., 1982; Varotsos and Alexopoulos, 1986; Varotsos, 2005).

Focusing on the case of earthquakes, in a time series comprising N earthquakes (see Fig. 7−1), the natural time $\chi_k = k/N$ serves as an index for the occurrence of the k-th earthquake. It is the combination of this index with the energy Q_k released during the k-th

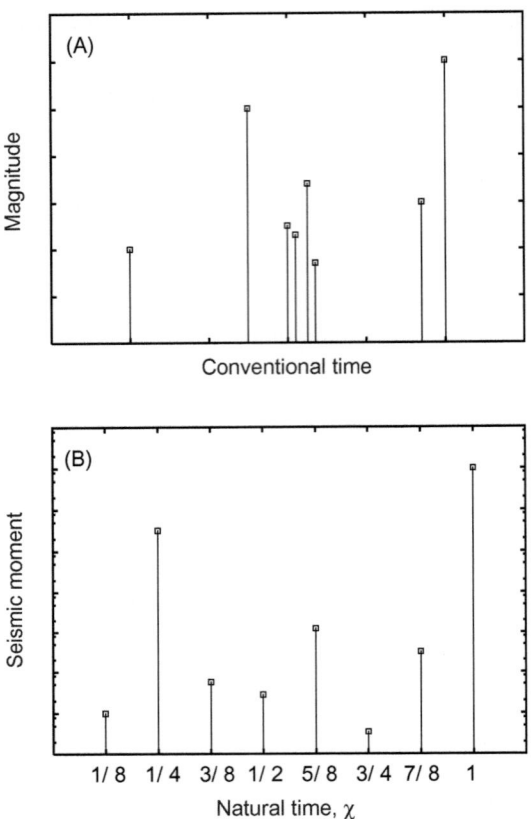

FIGURE 7–1 How a series of eight earthquakes is represented in conventional time (A) and in natural time (B).

earthquake of magnitude M_k, i.e., the pair (χ_k, Q_k), which is studied in natural time analysis. Alternatively, one can study the pair (χ_k, p_k), where

$$p_k = Q_k / \sum_{n=1}^{N} Q_n \tag{7.1}$$

denotes the normalized energy released during the k-th earthquake. In natural time analysis it has been found (Varotsos et al., 2001, 2002b, 2003a,b, 2005b, 2011d) that the variance of χ weighted for p_k, designated by κ_1, which is given by

$$\kappa_1 = \sum_{k=1}^{N} p_k (\chi_k)^2 - \left(\sum_{k=1}^{N} p_k \chi_k \right)^2, \tag{7.2}$$

plays a prominent role. This quantity becomes equal to 1/12 for a 'uniform' (u) distribution, e.g., when all p_k are equal or Q_k are positive independent and identically distributed random variables of finite variance (Varotsos et al., 2004a). In this case, κ_1 is designated by $\kappa_u (= 1/12)$.

Note that Q_k, and hence p_k, for earthquakes is estimated through the usual relation (Kanamori, 1978)

$$Q_k \propto 10^{1.5M_k}, \tag{7.3}$$

where M_k denotes the (moment) magnitude of the k-th earthquake.

7.2.2 The Seismic Data Used

For California the following two earthquake catalogues have been used (e.g., Sarlis et al., 2010b; Varotsos et al., 2012b): The Southern California Earthquake Catalog (SCEC) available from www.data.scec.org/eq-catalogs and the United States Geological Survey Northern California Seismic Network catalogue available from the Northern California Earthquake Data Center (NCEDC): www.ncedc.org/ncedc/catalog-search.html. All these seismic data have been analyzed in natural time in a similar fashion to that shown in Fig. 7−1. Following Eq. (7.3), the energy emitted in each earthquake was obtained through the relation $Q_k \propto 10^{1.5M_w}$ by assuming $M_w = M$, where M is the magnitude reported in these two catalogues.

For Japan (e.g., Sarlis et al., 2013, 2015c; Skordas and Sarlis, 2014; Varotsos et al., 2014b), the Japan Meteorological Agency (JMA) seismic catalogue has been employed and the energy of each earthquake was obtained from M_{JMA} after converting to the moment magnitude M_w using the formulae given by Tanaka et al. (2004).

For worldwide seismicity (e.g., Sarlis et al., 2015b; Papadopoulou et al., 2016), the events in the Global Centroid Moment Tensor (CMT) catalogue (Dziewonski et al., 1981; Ekström et al., 2012) available online at http://www.ldeo.columbia.edu/~gcmt/projects/CMT/catalog/ have been used. Additionally, the Centennial Earthquake Catalog (called CEC hereafter) (Engdahl and Villasenor, 2002) available from https://earthquake.usgs.gov/data/centennial/ has been also studied.

The most important earthquakes that occurred in California during the 25-year period 1979−2003, in Japan during the 27-year period from 1 January 1984 to 11 March 2011, the day of the $M9$ Tohoku earthquake, and globally during the 40-year period 1 January 1976 to 1 January 2015 are given in Tables 7−1, 7−2, and 7−3, respectively.

7.2.3 The Two Origins of Self-Similarity and Their Distinction in the Case of Seismicity

A large variety of natural systems exhibit irregular and complex behaviours, which at first look seem to be erratic, but in fact possess scale-invariant structure, e.g., see Peng et al. (1995a) and Kalisky et al. (2005). A stochastic process $X(t)$ is called self-similar with index $H > 0$ if it has the property (Lamperti, 1962)

$$X(\lambda t) \underset{=}{d} \lambda^H X(t) \quad \forall \quad \lambda > 0. \tag{7.4}$$

Table 7–1 All Major Earthquakes in California With $M \geq 7.0$ Within $N_{31.7}^{45.7} W_{127.5}^{112.1}$ During the Period 1979–2003

Earthquake Date	Lat.	Long.	M (NCDC)	Earthquake NAME
8 November 1980	41.08	− 124.62	7.2	Eureka
18 October 1989	37.04	− 121.88	7.0	Loma Prieta
28 June 1992	34.19	− 116.46	7.4	Landers
17 January 1994	*34.23*	*− 118.55*	*6.9*	*Northridge*
1 September 1994	40.41	− 126.30	7.0	Mendocino
16 October 1999	34.60	− 116.34	7.0	Hector Mine

In addition, the *M*6.9 Northridge earthquake is included in italics.

Table 7–2 All Shallow Earthquakes in Japan With $M_{JMA} \geq 7.6$ Within $N_{25}^{46} E_{125}^{148}$ Since 1 January 1984 Until the M9 Tohoku Earthquake on 11 March 2011

Earthquake Date	Lat.	Long.	M_{JMA}	Earthquake Name
12 July 1993	42.78	139.18	7.8	Southwest-Off Hokkaido
4 October 1994	43.38	147.67	8.2	East-Off Hokkaido
28 December 1994	40.43	143.75	7.6	Far-Off Sanriku
26 September 2003	41.78	144.08	8.0	Off Tokachi
22 December 2010	27.05	143.93	7.8	Near Chichi-jima
11 Marc 2011	38.10	142.86	9.0	Tohoku

Table 7–3 All Earthquakes With $M \geq 8.5$ During the Period 1 January 1976 to 1 January 2015 According to the CMT Catalogue

earthquake Date	LAT	LON	M_w(CMT)	Earthquake Name
26 December 2004	3.30	95.78	9.0	Sumatra-Andaman
28 March 2005	2.09	97.11	8.6	Sumatra-Nias
12 September 2007	− 4.44	101.37	8.5	Sumatra, Indonesia
27 February 2010	− 35.85	− 72.71	8.8	Chile
11 March 2011	38.32	142.37	9.1	Tohoku, Japan
11 April 2012	2.33	93.06	8.6	Indian Ocean

where the equality concerns the finite-dimensional distributions of the process $X(t)$ on the right- and the left-hand side of the equation (*not* the values of the process).

In several systems, this nontrivial structure points to long-range temporal correlations which alternatively means that self-similarity results from *process memory* only (but we stress that long-range temporal correlations do not automatically imply self-similarity of a process, e.g., Varotsos et al. (2006b)). This is the case, e.g., of fractional Brownian motion (fBm) or of SES activities. Alternatively, the self-similarity may solely result from the *process increments infinite variance* (heavy tails in the distribution). This is the case, e.g., with Levy stable motion. Note that Levy stable distributions, which are followed by many natural processes (e.g., see Tsallis et al., 1995, 1996), have heavy tails and their variance is infinite (Scafetta and West, 2005; Weron et al., 2005; Ausloos and Lambiotte, 2006). In general, the distinction of these two origins of self-similarity, i.e., process memory and process increments infinite variance, which may coexist, is a difficult task. This has been attempted in Varotsos et al. (2006b) by employing natural time analysis and further investigated in Sarlis et al. (2009b).

According to Varotsos et al. (2006b), the use of natural time analysis can lead to the identification of the origin of self-similarity as follows: First, if self-similarity results from the process memory *only*, the κ_1 value should change to $\kappa_u = 1/12$ for the (randomly) shuffled data. Second, if the self-similarity *exclusively* results from process increments 'infinite' variance, the $\kappa_{1,p}$ value, at which the probability distribution function (PDF) $P(\kappa_1)$ (see Section 7.3.1) maximizes, should be the same (but different from κ_u) for the original and the randomly shuffled data. This procedure answers, e.g., to the fundamental problem of distinguishing between stochastic models characterized by different statistics, e.g., between fractal Gaussian intermittent noise and Levy-walk intermittent noise, that may equally well reproduce some patterns of a time series (Scafetta and West, 2004, 2005). When *both* sources of self-similarity are present in the time series, quantitative conclusions on their relative strength can be obtained on the basis of Eqs. (7.12) and (7.13) of Varotsos et al. (2006b) that will now be explained.

These two equations relate either the expectation value $E(\kappa_1)$ of κ_1 in the actually observed time series, or the expectation value $E(\kappa_{1,shuf})$ in a randomly shuffled time series, when a (natural) time window of length l is sliding through the time series $Q_k \geq 0, k = 1, 2, \ldots N$. For such a window, starting at $k = k_0$, the quantities $p_j = Q_{k_0+j-1} / \sum_{m=1}^{l} Q_{k_0+m-1}$ in natural time are defined and Varotsos et al. (2006b) find that $E(\kappa_1)$ in the actually observed time series is given by

$$E(\kappa_1) = \kappa_{1,\mathbb{M}} + \sum_{\text{all pairs}} \frac{(j-m)^2}{l^2} \text{Cov}(p_j, p_m), \tag{7.5}$$

where $\kappa_{1,\mathbb{M}}$ is the value of κ_1 corresponding to the time series of the averages $\mu_j \equiv E(p_j)$ of p_j, i.e.,

$$\kappa_{1,\mathbb{M}} = \sum_{j=1}^{l} (j/l)^2 \mu_j - \left(\sum_{j=1}^{l} \mu_j j/l \right)^2, \tag{7.6}$$

and $\text{Cov}(p_j, p_m)$ stands for the covariance of p_j and p_m defined as

$$\text{Cov}(p_j, p_m) \equiv E[(p_j - \mu_j)(p_m - \mu_m)], \tag{7.7}$$

while the variance of p_j is given by

$$\text{Var}(p_j) = E\left[\left(p_j - \mu_j \right)^2 \right]. \tag{7.8}$$

The symbol $\sum_{\text{all pairs}}$ stands for $\sum_{j=1}^{l-1}\sum_{m=j+1}^{l}$. Eq. (7.5) reveals that $E(\kappa_1)$ is determined by two factors that involve: (1) the correlation of the data as reflected in the averages μ_j, e.g., due to a decrease in the magnitude of aftershocks in an earthquake time series, and (2) the covariance term which sums up the correlations between all natural time lags up to $l-1$.

On the other hand, $E(\kappa_{1,shuf})$ obtained upon randomly shuffling the original time series is found (Varotsos et al., 2006b) to be

$$E(\kappa_{1,shuf}) = \kappa_u\left(1 - \frac{1}{l^2}\right) - \kappa_u(l+1)\text{Var}(p) \tag{7.9}$$

(note that for the shuffled data $\text{Var}(p_j)$ is independent of j, and hence we merely write $\text{Var}(p) \equiv \text{Var}(p_j)$). If Q_k do not exhibit heavy tails and have finite variance, Eq. (7.9) rapidly converges to $E(\kappa_{1,shuf}) = \kappa_u$. Otherwise, $E(\kappa_{1,shuf})$ differs from κ_u, and the difference

$$\Delta E(\kappa_{1,shuf}) \equiv \kappa_u\left(1 - \frac{1}{l^2}\right) - E(\kappa_{1,shuf}) = \kappa_u(l+1)\text{Var}(p) \tag{7.10}$$

provides a measure of the process increments 'infinite' variance. Comparing the results obtained from Eqs. (7.5), (7.9) and (7.10), we can draw quantitative conclusions on the existence of temporal correlations in a real time series even if the process increments exhibit 'infinite' variance. This procedure was followed by Sarlis et al. (2009b) demonstrating the existence of temporal correlations between earthquake magnitudes in the seismicity of Southern California. The same method was later employed (Sarlis, 2011; Sarlis and Christopoulos, 2012) for the identification of the correlations between successive magnitudes in global seismicity.

Calculating the κ_1 value by means of a window $l = 6$ to 40 consecutive events sliding through either the original earthquake catalogue or a shuffled one (see Section 7.3.1), Varotsos et al. (2006b) obtained the following results for the SCEC (Southern California) as well as for the JMA earthquake catalogue (Japan): In both catalogues, the most probable values of κ_1, denoted by $\kappa_{1,p}$, are found to be $\kappa_{1,p} \approx 0.066$ for the original data, while they are $\kappa_{1,p} \approx 0.064$ for the surrogate data. Both these $\kappa_{1,p}$ values differ markedly from the value $\kappa_u = 1/12$ of the 'uniform' distribution. This could be interpreted as reflecting that the self-similarity mainly originates from the process increments infinite variance. In addition, since the $\kappa_{1,p}$ value of the original earthquake data does not differ considerably from the value $\kappa_1 \approx 0.070$ identified (Varotsos et al., 2002a, 2003a,b, 2011c,d) in infinitely ranged temporal correlations, this indicates the importance of temporal correlations rather than their absence in the earthquake catalogues. In other words, the temporal correlations between earthquake magnitudes are responsible for the difference between the value of $\kappa_{1,p} \approx 0.064$ of the surrogate data from the value of $\kappa_{1,p} \approx 0.066$ of the original data (both these values have a plausible uncertainty of ± 0.001; Sarlis et al., 2009b). At this point, it is worthwhile mentioning that by means of natural time analysis of a generalized Cantor set (multiplicative cascade), a theoretical interrelation between $\kappa_{1,p}$ of the (randomly) shuffled earthquake data and the parameter b of the Gutenberg–Richter law can be found (Sarlis et al., 2009b)

$$\kappa_{1,p} = \frac{2^{\frac{3}{2b}}}{3\left(1+2^{\frac{3}{2b}}\right)^2}.$$ (7.11)

Recall that according to Gutenberg and Richter (1954) the (cumulative) number of earthquakes with magnitude greater than (or equal to) M, $N(\geq M)$, occurring in a specified area and time, is given by

$$N(\geq M) = 10^{a-bM},$$ (7.12)

where the constant a gives the logarithm of the number of earthquakes with magnitude greater than zero (e.g., Shcherbakov et al., 2004). Eq. (7.11), if we just adopt a reasonable value of b, i.e., $b \approx 1$, leads to a value of $\kappa_{1,p}$ that is very close to 0.064, thus agreeing with the aforementioned $\kappa_{1,p}$ value deduced by Varotsos et al. (2006b) from the natural time analysis of the shuffled experimental data of SCEC and Japan.

7.2.4 Temporal Correlations Between Earthquake Magnitudes by Means of Detrended Fluctuation Analysis

The problem of detection of temporal correlations in the earthquake magnitude time series M_k can be treated, as already mentioned in Section 7.2.3, on the basis of natural time analysis. Among other approaches (Lippiello et al., 2007, 2008, 2012; Lennartz et al., 2008; Sarlis et al., 2010a; Davidsen and Green, 2011) to solve the same problem, independent studies (e.g., Lennartz et al., 2008) have also reported the detection of long-range temporal correlations in M_k by means of the detrended fluctuation analysis (DFA) (Peng et al., 1994; Taqqu et al., 1995; Goldberger et al., 2000).

DFA has been established as a robust method suitable for detecting long-range power-law correlations embedded in nonstationary signals. It has been applied to diverse fields such as DNA (Stanley et al., 1999), heart dynamics (Peng et al., 1995b; Ashkenazy et al., 2000), circadian rhythms (Hu et al., 2004; Ivanov, 2007), economics (Liu et al., 1997; Vandewalle and Ausloos, 1998; Ausloos et al., 1999; Ivanov et al., 2004), etc. For recent applications see Bashan et al. (2008), Ma et al. (2010), Xu et al. (2011), Carretero-Campos et al. (2012) and Varotsos et al. (2013a, 2014a, 2015a).

Lennartz et al. (2008) showed that, in the regimes of stationary seismic activity (i.e., during periods at which large aftershock sequences are missing) in California, long-range correlations exist between successive earthquake magnitudes. In this study, Lennartz et al. (2008), as well as Lennartz et al. (2011), used the sequence index k (i.e., the sequential order in which an earthquake had occurred, which is just the natural time χ_k multiplied by the number N of the events) for the detection of the long-range correlations and found the DFA exponent $\alpha = 0.59$ (if $\alpha > 0.5$, the signal is long-range correlated). This study, as explained in Section 7.4.1, was extended (Sarlis et al., 2010a,b; Varotsos et al., 2014b) to include the nonstationary periods of seismicity.

7.3 The Order Parameter of Seismicity in Natural Time

7.3.1 Definition of the Order Parameter and the Construction of Its Probability Density Function

Varotsos et al. (2005b) argued in detail (see also pp. 249–253 of Varotsos et al., 2011d) that the quantity κ_1 given by Eq. (7.2) – or the normalized power spectrum in natural time $\Pi(\omega)$ as defined by Varotsos et al. (2001, 2002a) for natural cycling frequency $\omega \to 0$ – can be considered as an order parameter for seismicity since its value changes abruptly when a mainshock (the new phase) occurs, and, in addition, the statistical properties of its fluctuations resemble those in other nonequilibrium critical systems and in equilibrium critical phenomena, as explained in Section 7.3.2.

In a seismic catalogue comprising a number of events, the procedure to construct the PDF $P(\kappa_1)$ is, briefly, the following: Starting from the first earthquake, we calculate the κ_1 values taking natural time windows of length from $l = 6$ to 40 consecutive events (including the first one). We then proceed to the second earthquake, and repeat the calculation of κ_1 and so on. Thus, after sliding event by event (see Fig. 7–2) through the earthquake catalogue, the calculated κ_1 values enable the construction of the PDF $P(\kappa_1)$.

7.3.2 Universal Curve of Seismicity

The properties of the $P(\kappa_1)$ versus κ_1 curve for the *long-term seismicity* have been studied by Varotsos et al. (2005b) after constructing this PDF by means of the procedure described in Section 7.3.1. The following data from different areas, i.e., the San Andreas fault system and Japan, have been analysed in natural time: First, the earthquakes that occurred during the

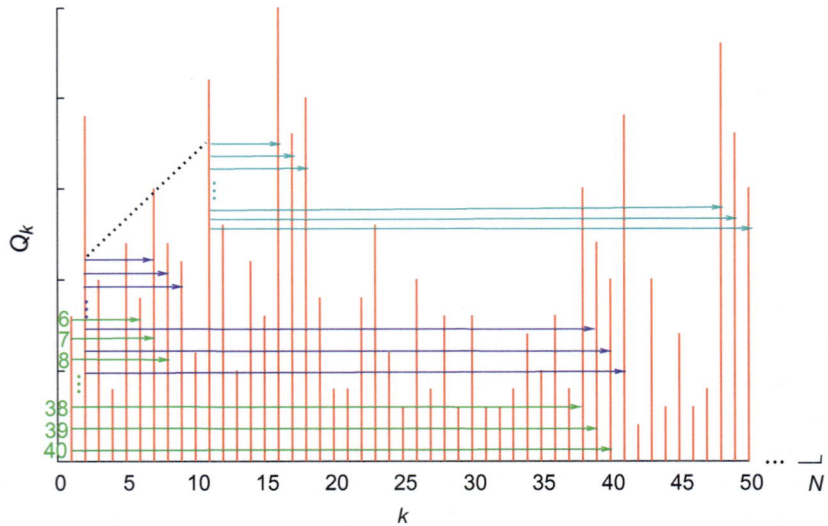

FIGURE 7–2 Schematic diagram for the construction of the probability density function $P(\kappa_1)$ versus k in a seismic catalog by using a sliding natural time window of length $l = 6$–40 events.

period 1981−2003 within the area $N_{32}^{37}W_{114}^{122}$ using the SCEC catalogue (see Sarlis, 2011). Second, the earthquakes within $N_{25}^{46}E_{125}^{146}$ for the period 1967−2003 using the JMA catalogue (see Varotsos et al., 2005b). The magnitude thresholds $M \geq 2.0$ and $M_{JMA} \geq 3.5$ have been considered for SCEC and JMA, respectively, for the sake of data completeness. Additionally, the following two global catalogues were later analysed in natural time (Sarlis, 2011; Sarlis and Christopoulos, 2012): the aforementioned CMT catalogue during the period 1 January 1977 to 30 September 2009 and the CEC for the period since 1900 until 30 September 2007, with magnitude thresholds $M_w \geq 5.0$ and $M \geq 7.0$, respectively. All these data from the aforementioned four earthquake catalogues resulted in the four $P(\kappa_1)$ versus κ_1 curves depicted in Fig. 7−3A.

We now study the order parameter fluctuations relative to the standard deviation of its distribution. Thus, we plot in Fig. 7−3B the quantity $P(y) \equiv \sigma_{\kappa_1}P(\kappa_1)$ versus $y \equiv (E(\kappa_1) - \kappa_1)/\sigma_{\kappa_1}$, where $E(\kappa_1)$ and σ_{κ_1} refer to the expectation value and the standard deviation of κ_1. We find that the results shown in Fig. 7−3B for all four catalogues $P(y)$ almost fall on the *same* curve, which clearly consists of two segments: the segment to the left shows a decrease of $P(y)$ almost by four orders of magnitude, while the upper right segment has an almost constant $P(y)$ value. The feature of this plot is strikingly reminiscent of the one obtained by Bak et al. (2002) (see their Fig. 7−4) on different grounds, using earthquakes in California only. More precisely, they measured $P_{S,l}(T)$, the distribution of waiting times T, between earthquakes occurring within range l whose magnitudes are greater than $M(\equiv \log_{10}S)$. Then, they plotted $T^{\alpha_0}P_{S,l}(T)$ versus $TS^{-b}l^d$ and found that, for a suitable choice of the exponent α_0 (i.e., $\alpha_0 = 1$), the Gutenberg−Richter law exponent b (i.e., $b = 1$) and the spatial dimension d (i.e., fractal dimension $d = 1.2$), all the data collapse onto a single curve which is similar to that of Fig. 7−3B.

After a further inspection of Fig. 7−3B, the following points have been clarified (for their compilation see pp. 255−256 of Varotsos et al. (2011d)).

First, the rapidly decaying part (i.e., the left segment), which is consistent with an almost exponentially decaying function over around four orders of magnitude, remains practically unchanged, upon *randomly* shuffling the data (Varotsos et al., 2004b).

Second, the feature of the plot of Fig. 7−3B is not altered upon changing the seismic region. As an example, we mention that three different regions in Japan, as well as the whole of Japan, result in almost identical plots (Varotsos et al., 2005b).

Third, the 'upturn branch' in the upper right part of Fig. 7−3B arises from the presence of aftershocks. It disappears when, in Japan, e.g., we delete the earthquakes with $M_{JMA} \lesssim 5.7$ (and hence drastically reduce the number of aftershocks), but it does not when deleting earthquakes with smaller threshold, i.e., $M_{JMA} < 4.0$ (note that this threshold still allows the presence of a reasonable number of aftershocks).

Fourth, when considering the relevant results for the worldwide seismicity by taking a large magnitude threshold (so that the data are complete; Varotsos et al. 2005b), we find that they fall onto the same curve with the results of both Japan and SCEC, as described.

To summarize, if we analyse the long-term seismicity in natural time and study the order parameter fluctuations relative to the standard deviation of its distribution, we find without making use of *any* adjustable parameter that the scaled distributions of different seismic areas (as well as that of the worldwide seismicity) fall on the *same* (*universal*) curve.

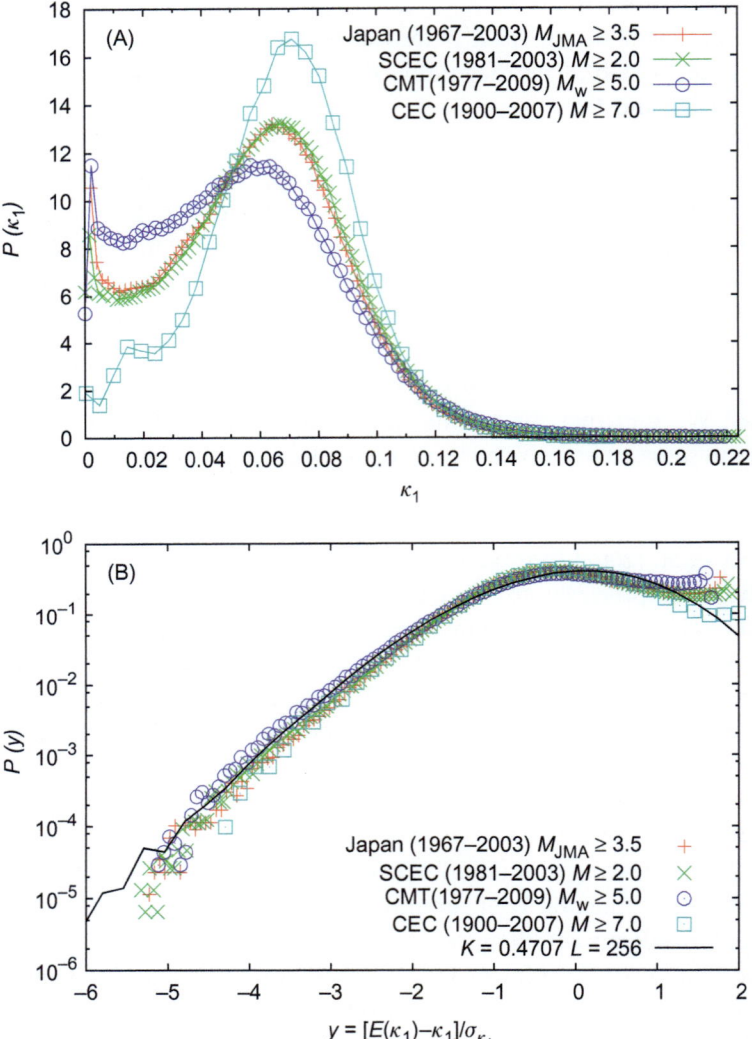

FIGURE 7–3 (A) The probability density function $P(\kappa_1)$ versus κ_1 for the four earthquake catalogues: Japan, SCEC, CMT and CEC. (B) Universality of the log-linear plot of $P(y) \equiv \sigma_{\kappa_1} P(\kappa_1)$ versus $y = [E(\kappa_1) - \kappa_1]/\sigma_{\kappa_1}$. The points correspond to Japan, SCEC, CMT and CEC with magnitude thresholds written in the insets. The corresponding results (Varotsos et al., 2005b) for the 2D Ising model for inverse temperature parameter $K = 0.4707$ for $L = 256$ are also depicted.

7.3.3 Similarity of Fluctuations in Correlated Systems Including Seismicity

Great interest has been focused on the fluctuations of correlated systems in general and of critical systems in particular (Bramwell et al., 1998, 2000, 2001b; Bramwell et al., 2001a; Zheng and Trimper, 2001; Bramwell et al., 2002; Watkins et al., 2002; Zheng, 2003; Clusel et al., 2004).

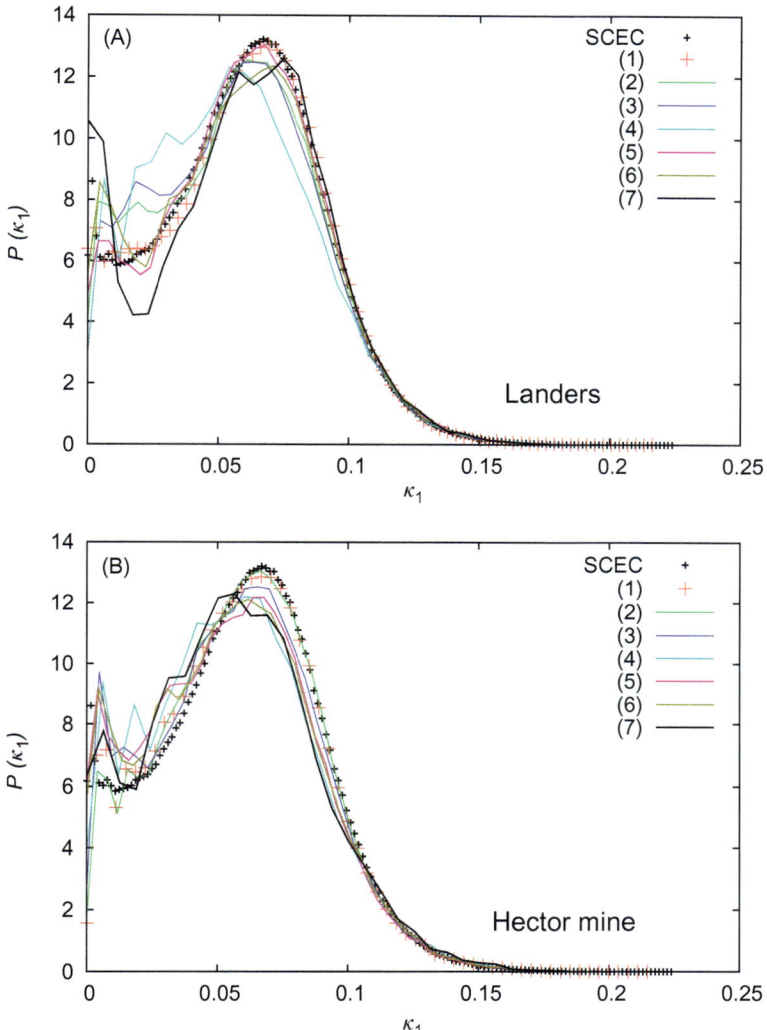

FIGURE 7–4 The probability density function $P(\kappa_1)$ versus κ_1 for SCEC (black plus, all panels) along with those resulting from (1) the aftershock sequence as for the Landers earthquake (A) and the Hector Mine earthquake (B). The results (2)–(7) depict $P(\kappa_1)$ for $W = 5000, 3000, 1000$ earthquakes immediately after and $W = 5000, 3000, 1000$ earthquakes immediately before the Landers earthquake (A) and the Hector Mine earthquake (B). In panel (C), we depict the results for: (1) 1000 earthquakes immediately before the Landers earthquake, (2) 1000 earthquakes immediately after the Landers earthquake, (3) 5000 earthquakes immediately before the Hector Mine earthquake, and (4) 5000 earthquakes immediately after the Hector Mine earthquake. *Taken from Sarlis, N.V., Skordas, E.S., Varotsos, P.A., 2010b. Order parameter fluctuations of seismicity in natural time before and after mainshocks. EPL 91, 59001.*

In particular, Bramwell et al. (1998), in an experiment on a closed turbulent flow, found that a normalized form of the PDF of the power fluctuations has the same functional form as that of the magnetization (M) of the finite-size 2D (two-dimensional) XY equilibrium model in the

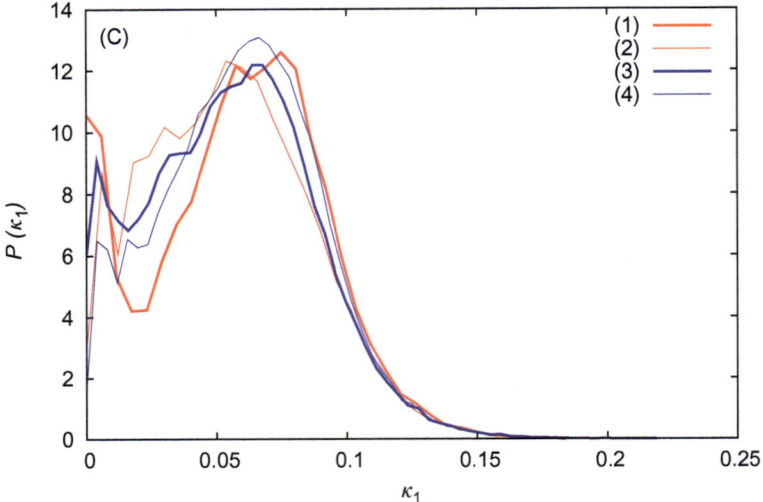

FIGURE 7–4 (Continued).

critical region below the Kosterlitz-Thouless transition temperature (magnetic ordering is then described by the order parameter M). The 'normalized' form of the PDF, denoted by P(m), is defined by introducing the reduced magnetization (Bramwell et al., 1998) $m = (M - \langle M \rangle)/\sigma$, where $\langle M \rangle$ denotes the mean and σ the standard deviation. For both systems, Bramwell et al. (1998) found that while the high end $(m > 0)$ of the distribution has a Gaussian shape the asymptote of which was later clarified (Bramwell et al., 2001b) to have a double exponential form, a distinctive exponential tail appears towards the low end $(m < 0)$ of the distribution. The latter tail, which will be hereafter simply called, for the sake of convenience, the 'exponential tail', provides the main region of interest. This is because such a tail shows that the probability for a rare fluctuation, e.g., greater than six standard deviations from the mean, is almost five orders of magnitude higher than in the Gaussian case. Subsequent independent simulations (Bramwell et al., 2000, 2001b; Zheng and Trimper, 2001; Zheng, 2003; Clusel et al., 2004) showed that a variety of highly correlated (nonequilibrium as well as equilibrium) systems, under certain conditions, exhibit approximately the 'exponential tail'. Varotsos et al. (2005b) found that this 'tail' is identified in seismicity only *if* we analyze the time series of earthquakes in natural time. In addition Varotsos et al. (2005b) found that the 'universal' curve deduced for seismicity in Fig. 7−3B, exhibits an 'exponential tail' form similar to that observed in certain nonequilibrium systems, e.g., 3D turbulent flow, as well as in several equilibrium critical phenomena, e.g., 2D Ising, 3D Ising, and 2D XY.

7.3.4 Identifying the Occurrence Time of a Mainshock Using the Value of the Order Parameter Itself

It has been empirically found (Varotsos et al., 2001, 2006c,b, 2011d, 2015b; Sarlis et al., 2008; Sarlis, 2013) that a mainshock occurs in a few days to 1 week after the κ_1 value is recognized to have approached 0.070 in the natural time analysis of the seismicity (in the candidate area to suffer the mainshock) subsequent to the initiation of an SES activity. This is justified by the fact that, according to the SES generation mechanism, as mentioned in Section 7.2.1, the initiation of an SES activity marks when the system enters the critical stage and in addition that the epicentral area of the impending mainshock is estimated on the basis of SES data (e.g., Varotsos and Lazaridou, 1991). As for the validity of the condition $\kappa_1 = 0.070$ at the critical state, it has been demonstrated for various dynamical models by Varotsos et al. (2011c) (see also Table 8.1 in p. 343 of Varotsos et al. (2011d) in which 14 cases obeying this condition have been compiled). This condition has been ascertained in a number of major mainshocks that occurred in various seismic-prone areas including Greece, Japan and California. In particular, in Greece all mainshocks with $M_w \geq 6.4$ during the decade 2001−11, the identification of the occurrence time of the impending earthquake was documented well in advance, as described in Section 7.2 of Varotsos et al. (2011d). In Japan, Uyeda et al. (2009), by analysing in natural time the seismic data subsequent to the relevant SES activity (that started on 26 April 2000) before the volcanic-seismic swarm activity in 2000 in the Izu Island region of Japan, found that the condition $\kappa_1 = 0.070$ was fulfilled on 27 June 2000, i.e., a few days before the magnitude 6.0 earthquake that occurred on 1 July 2000. This swarm was then characterized by JMA as the largest earthquake swarm ever recorded (Japan Meteorological Agency, 2000). Finally, concerning California, the condition $\kappa_1 = 0.070$ has been ascertained (Varotsos et al., 2010) a few days before the occurrence of the Loma Prieta earthquake on 18 October 1989 after starting the computation of the κ_1-values from the initiation of magnetic field variations reported by Fraser-Smith and coworkers (Fraser-Smith et al., 1990; Bernardi et.al., 1991), which were strikingly similar to those that accompanied the SES activities recorded in Greece (Varotsos et al., 2003c).

7.3.5 The Two Types of Complexity Measures Involving Order Parameter Fluctuations and Their Complementarity

When employing natural time analysis to study the approach of a complex system to the critical point by means of the order parameter fluctuations, we use two types of complexity measures that capture different aspects of dynamics (see Varotsos et al., 2004a, 2005a; see also Varotsos et al., 2011b; as well as Sections 3.6 and 9.2.6 of Varotsos et al., 2011d): First, complexity measures that take into account order parameter fluctuations on different natural time window length scales (we call them here, for the sake of convenience, type I measures), and second, complexity measures that employ fluctuations on a *fixed* natural time window length scale W (i.e., the number of successive events) sliding event by event through the time series (type II measures).

Our study has shown that a *complementarity* holds in the following sense (Varotsos et al. (2005a, 2007; see also pages 410−413 of Varotsos et al., 2011d)): If in the frame of the one type an ambiguity emerges in identifying the approach to the critical point, the other type of complexity measure gives a clear answer. (This is so because, as mentioned above, these two types of complexity measure are focused on different aspects of dynamics.) We emphasize that *both* types of complexity measure must be envisaged in order to get the optimum performance when dealing with a heterogeneous database like the case of the analysis of electrocardiograms to identify the sudden cardiac death risk or the case to identify the occurrence time of earthquakes. Thus, we discuss here both types of complexity measures for seismicity. In particular, in Section 7.4, the type I complexity measures and in Section 7.5 the type II complexity measures, are discussed.

7.4 Order Parameter Fluctuations Upon Varying the Natural Time Window Length

7.4.1 The Probability Density Function of the Order Parameter Fluctuations Before a Mainshock

The natural time analysis of the time series with $W = 5000,\ 3000$ and 1000 earthquakes just *before* and just *after* Landers and Hector Mine earthquake has led (Sarlis et al., 2010b) to the results shown in Fig. 7−4A and B for each of these two earthquakes, respectively. We observe that the PDF $P(\kappa_1)$ versus κ_1 curves differ in general from the corresponding curve obtained from the whole SCEC. In other words, we see that either just *before* or just *after* a significant earthquake, the behaviour of seismicity in natural time deviates from its mean behaviour in natural time (compare these curves with the one for the SCEC for long-term seismicity plotted in Fig. 7−3A). For example, in the case of Landers earthquake Sarlis et al. (2010b), plotted $P(\kappa_1)$ versus κ_1 for $W = 1000$ earthquakes before and after this earthquake (see Fig. 7−4C). These two curves were found to be markedly different in the following respect: Before this mainshock a *significant bimodal feature* appears in the $P(\kappa_1)$ vs. κ_1 plot. This finding, which emerged from natural time analysis alone, is of profound importance since it is reminiscent of the bimodal feature observed in the PDF of the order parameter when approaching (from below) T_c in equilibrium critical phenomena as reported by Varotsos et al. (2006b). Recalling that κ_1 is the order parameter for seismicity it is reasonable to expect a similar behaviour before every mainshock.

To further elucidate the aforementioned conclusion, Varotsos et al. (2012a) extended the study to smaller W values, i.e., $W = 500$, $W = 300$ and $W = 100$ earthquakes *before* the mainshocks in California. The PDFs $P(\kappa_1)$ versus κ_1 obtained after considering all events with $M \geq 2.0$ reported by SCEC within the area $N_{32}^{37} W_{114}^{122}$ (which is marked with a rectangle in Fig. 7−11A) are shown in Fig. 7−5A, B, and C for $W = 10{,}000,\ 7000,\ 5000,\ 1000,\ 500,\ 300$ and 100 events *before* the Landers, Northridge and Hector Mine earthquakes, respectively. A detailed inspection of these figures sheds more light on the changes in the feature of the $P(\kappa_1)$ curve upon decreasing W from 10,000 down to 100 events before the corresponding mainshock. Let us consider, e.g., Landers earthquake: at $W = 10{,}000$, the curve is practically unimodal since there exists a main peak of amplitude around 13.4 at a κ_1-value close to 0.066 (in agreement with the mean behaviour reported in the last but one paragraph of Section 7.2.3, recall also the $P(\kappa_1)$ curve for SCEC in Fig. 7−3A) and a weaker peak of

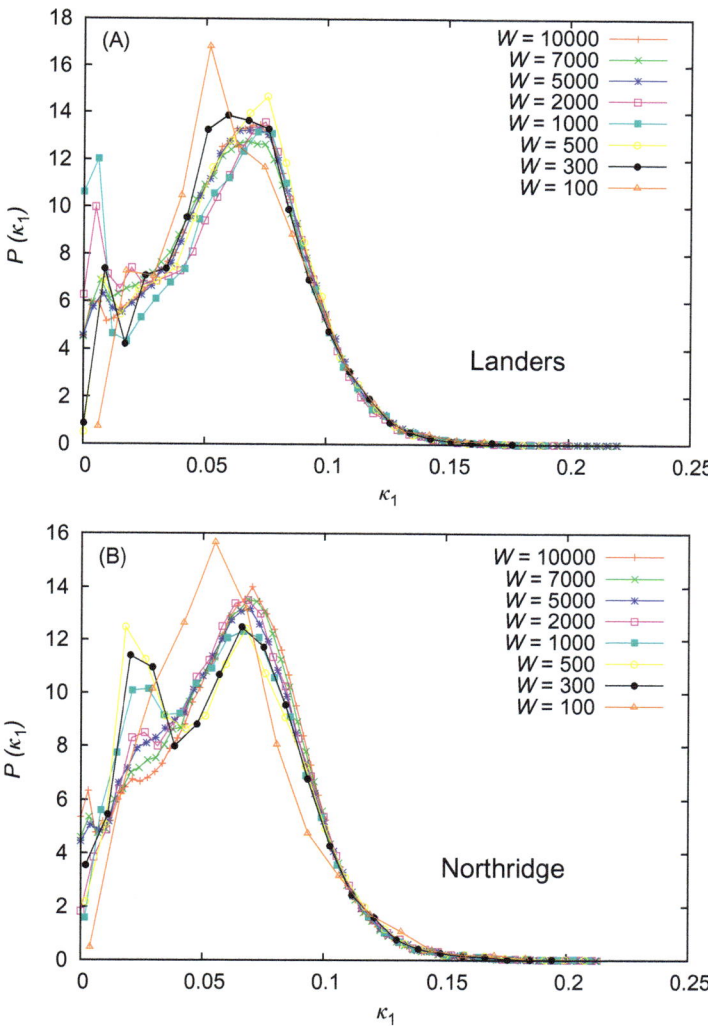

FIGURE 7–5 The probability density function $P(\kappa_1)$ versus κ_1 for SCEC with $M \geq 2.0$ for the following W values W = 10,000 (red pluses), 7000 (green crosses), 5000 (blue asterisks), 1000 (cyan squares), 500 (yellow circles), 300 (black solid circles) and 100 (orange triangles) *before* the following earthquakes: (A) Landers, (B) Northridge and (C) Hector Mine. *Taken from Varotsos, P., Sarlis, N., Skordas, E., 2012a. Remarkable changes in the distribution of the order parameter of seismicity before mainshocks. EPL 100, 39002.*

amplitude around 6.0 at an appreciably smaller κ_1 value close to $\kappa_1 \approx 0$. As W decreases from W = 10,000 to W = 1000, the part of the $P(\kappa_1)$ curve at smaller values around $\kappa_1 \approx 0$ becomes higher for smaller W, reaching the largest amplitude around 12.0 at W = 1000 which is more or less comparable with the corresponding amplitude (around 13.2 or so) of the initial main peak at $\kappa_1 \approx 0.066$. This is why the $P(\kappa_1)$ curve could then be termed *bimodal*. In other words, upon decreasing the W from W = 10,000 to W = 1000, the $P(\kappa_1)$

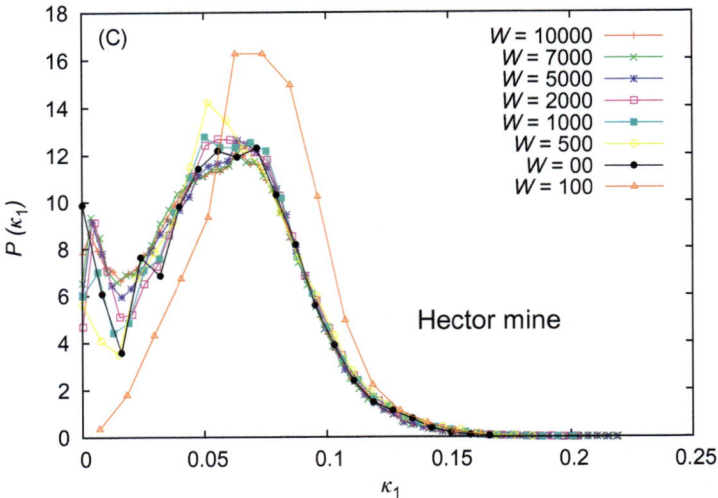

FIGURE 7–5 (Continued).

curve is initially practically unimodal and gradually becomes bimodal, thus confirming the findings of Sarlis et al. (2010b). Upon focusing on shorter W values, however, the opposite trend is observed: for the Landers earthquake, by comparing the $P(\kappa_1)$ curves for $W = 1000$, $W = 500$, $W = 300$ and $W = 100$, we see that the aforementioned bimodal curve for $W = 1000$ gradually changes, and finally returns practically to a unimodal feature for $W = 100$. Concerning the Northridge earthquake, for $W = 10,000$, the $P(\kappa_1)$ curve is unimodal (in a similar sense to that described above for the Landers case) maximizing at around $\kappa_1 \approx 0.07$ with amplitude 14.0 and then gradually becomes bimodal for $W = 500$, where the two peaks have comparable amplitudes, i.e., around 12.4 at the larger κ_1 value ≈ 0.068 and an amplitude of around 12.5 at a smaller κ_1 value of around $\kappa_1 \approx 0.018$. For even shorter W values, i.e., $W = 100$, the $P(\kappa_1)$ curve returns to a unimodal feature, having a prominent peak with enhanced amplitude. Finally, in the case of the Hector Mine earthquake, upon decreasing the W value from $W = 10,000$ to around $W = 2000$ the change of the feature of the $P(\kappa_1)$ curve is not significant. For shorter W values, however, by comparing, e.g., for $W = 1000$ with that for $W = 300$, the change becomes remarkable. At the latter case, i.e., $W = 300$ the feature of $P(\kappa_1)$ could be characterized as bimodal since its amplitude at the larger κ_1 value (≈ 0.065) is around 12.2 and at the smaller κ_1 value (≈ 0) the amplitude is around 9.8. This 'bimodal' feature, however, returns to a unimodal feature for $W = 100$, having a prominent peak with enhanced amplitude around 16.3 (compared to the amplitude of around 12.1 for $W = 10,000$) being located at $\kappa_1 \approx 0.068$.

Let us now summarize the aforementioned changes in the feature of the $P(\kappa_1)$ curve observed upon decreasing the W value, which means that we gradually approach the main-shock: well before the mainshock, i.e., for large W values, the $P(\kappa_1)$ curve is almost unimodal in the sense that its main part maximizes at a κ_1 value around $\kappa_1 = 0.066$. This main part lowers only slightly upon decreasing the W value, but another part of the curve develops maximizing at a smaller κ_1 value close to $\kappa_1 \approx 0$, reaching for W values around a few hundreds to one thousand, a height comparable to that of the initial main part, *thus the $P(\kappa_1)$ curve becomes almost bimodal.* Finally, for even shorter W values, e.g., $W = 100$ (meaning that the mainshock is imminent) the curve *regains its unimodal feature, which, however, is drastically different to the initial one*, i.e., having a significantly enhanced height and a different shape compared to that of the $P(\kappa_1)$ curve well before the mainshock.

7.4.2 The Variability of the Order Parameter Upon Gradually Approaching a Mainshock

The challenging finding in Section 7.4.1 is that upon approaching a mainshock the feature of the probability distribution function $P(\kappa_1)$ versus κ_1 exhibits remarkable changes. To quantify these changes, we employ the *variability* of κ_1, defined by Sarlis et al. (2010b) which is just the ratio

$$\beta \equiv \sigma(\kappa_1)/\mu(\kappa_1) \tag{7.13}$$

where $\sigma(\kappa_1)$ and $\mu(\kappa_1)$ stand for the standard deviation and the mean value of κ_1. The aforementioned appearance of the bimodal feature reflects that upon approaching the mainshock with the number W of the earthquakes before mainshock decreasing, the variability of κ_1 should increase. In other words, this means that *upon considering various natural time window lengths ending at a given mainshock, we must observe a considerable increase in the fluctuations of κ_1 before the mainshock.* This has been clearly observed before the Tohoku $M9$ earthquake that occurred in Japan on 11 March 2011 and before the $M8.3$ Tokachi-Oki earthquake in Japan on 25 September 2003 (see Figs. A.4 and A.5 on pp. 213 and 214 of Lazaridou-Varotsos (2013)).

The above phenomenon has been studied in more detail in California by Varotsos et al. (2012a). Fig. 7−6 shows the values of the variability of κ_1 versus the conventional time that have been deduced from excerpts of the SCEC catalogue with $M \geq 2.0$ comprising W earthquakes *before* each of the mainshocks, i.e., Landers (red circles), Northridge (green circles) and Hector Mine (blue circles). Recall that the data from these mainshocks themselves were *not* included in the calculation. In this figure, the points are calculated at every hundred W interval and the W values for the closest points to each mainshock are 100. An inspection of Fig. 7−6 shows that the variability of κ_1 well before these mainshocks exhibited small changes lying more or less on the same level, but it markedly changes upon approaching the mainshock. A similar behaviour for the variability of κ_1 is observed before all these three mainshocks, as follows.

An increase in the κ_1 variability has been clearly observed before the mainshocks, i.e., on 1 May 1992 before Landers, on 24 November 1993 before Northridge and on 29 August 1999 before Hector Mine (see Fig. 7−3A, B and C of Varotsos et al., 2012a, respectively) *followed by a decrease before the mainshock occurrence.*

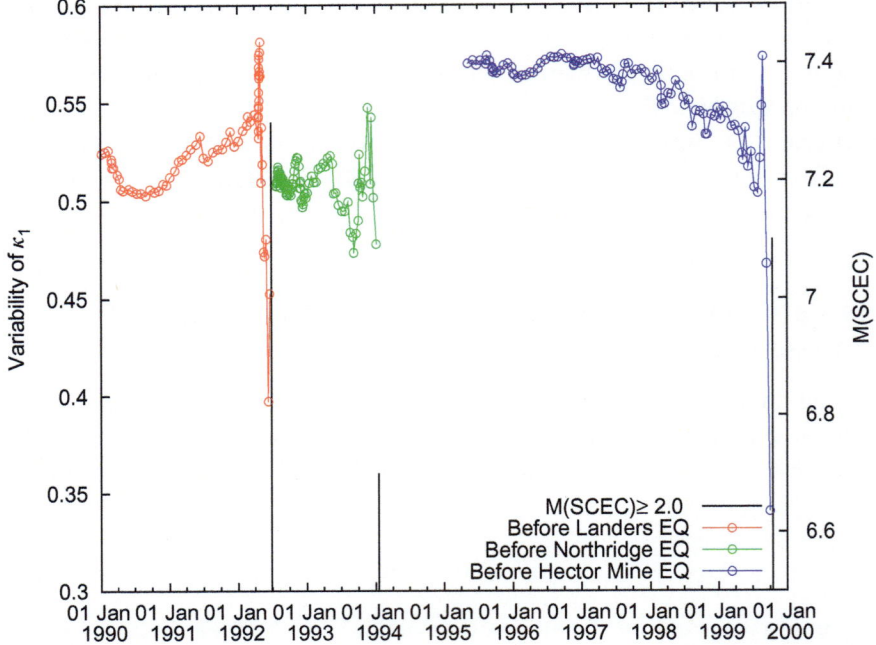

FIGURE 7–6 The values of the variability of κ_1 plotted versus the conventional time (UT). These values have been deduced from excerpts of the SCEC catalogue comprising W earthquakes with $M \geq 2.0$, before each of the three mainshocks: Landers (red), Northridge (green) and Hector Mine (blue). The points are calculated at every hundred W interval and the W values for the closest points to the mainshock are 100. The earthquakes are depicted with black bars and their magnitudes are shown in the right scale. *Taken from Varotsos, P., Sarlis, N., Skordas, E., 2012a. Remarkable changes in the distribution of the order parameter of seismicity before mainshocks. EPL 100, 39002.*

This may be understood from a further inspection of the PDFs $P(\kappa_1)$ versus κ_1 in Fig. 7−5 in the following context: well before all these three mainshocks, the main part of their $P(\kappa_1)$ curves for large W values, e.g., $W = 5000$ to around 2000, is almost unimodal maximizing at almost the same κ_1 value, i.e., $\kappa_1 \approx 0.066$. Upon decreasing W to a value around some hundred events the $P(\kappa_1)$ curves become bimodal, thus reflecting an increase in the κ_1 variability. At even shorter W values, however, e.g., $W = 100$, the $P(\kappa_1)$ curves regain practically a unimodal feature, thus causing a decrease in the κ_1 variability.

We now give in Fig. 7−7 the results for the variability of κ_1 in the same fashion as in Fig. 7−6, but after selecting a higher magnitude threshold in the SCEC catalogue, i.e., $M \geq 2.5$. The relevant excerpts of Fig. 7−7 are given in Fig. 7−8A, B, C for the Landers (red), Northridge (green) and Hector Mine (blue) earthquakes, respectively. We see that before the strongest mainshock, i.e., the Landers earthquake, the precursory increase in the κ_1 variability is again visible very clearly, apart from a small shift on the date of its

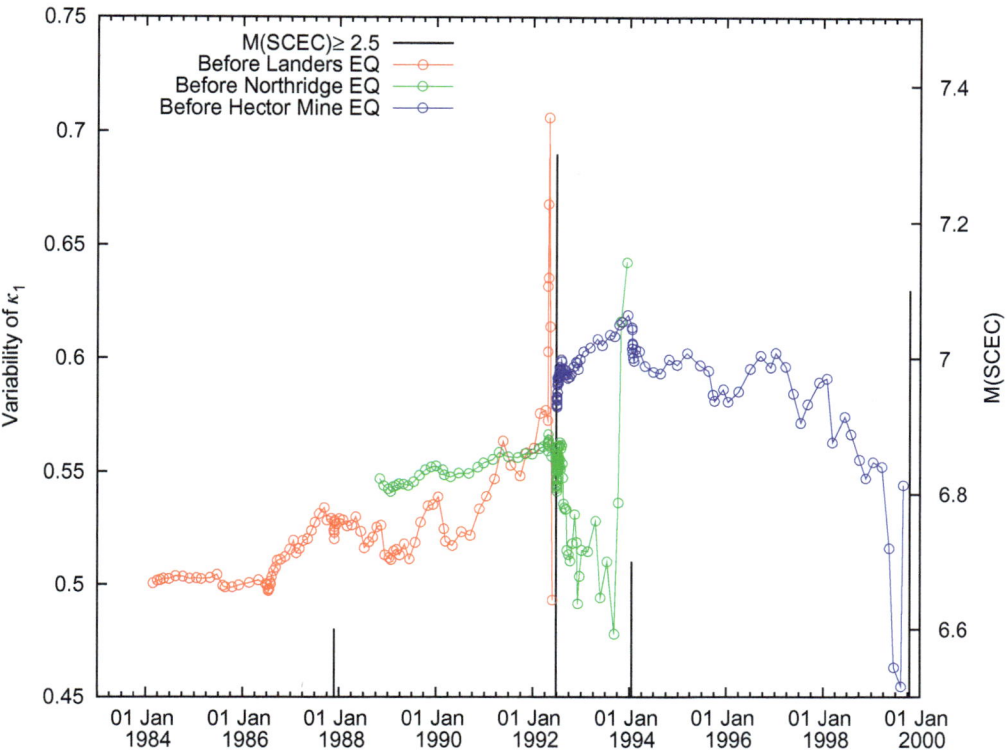

FIGURE 7–7 As Fig. 7–6, but when using the SCEC catalogue with $M \geq 2.5$.

observation. In particular, it is observed on 30 April 1992 when increasing the magnitude threshold to $M \geq 2.5$ in Fig. 7–8A, while on 1 May 1992 when using $M \geq 2.0$. In the case of the Northridge earthquake, the precursory increase of the κ_1 variability appears just before the mainshock after a transient decrease and seems to be more visible in Fig. 7–8B for $M \geq 2.5$ than what it was for $M \geq 2.0$, beyond again a shift on the date of its observation, i.e., 24 November 1993 but 8 December 1993 in Fig. 7–8B. In the case of the Hector Mine earthquake, the κ_1 variability in Fig. 7–8C for $M \geq 2.5$ exhibits first an evident decrease and then a sharp increase just before the mainshock on 28 August 1999 compared to 29 August 1999 for $M \geq 2.0$.

The SCEC catalogue does not contain all major mainshocks that occurred in California during the period 1979–2003. In particular, the M7.2 Eureka earthquake on 8 November 1980, the M7.0 Loma Prieta earthquake on 18 October 1989 and the Mendocino earthquake on 1 September 1994 are not included in SCEC. In order to also investigate these mainshocks, we now consider the NCEDC catalogue with $M \geq 2.5$ which extends to an appreciably wider area and reports all the major earthquakes shown in Table 7–1 with $M \geq 7.0$ that occurred in California during this period. Repeating the

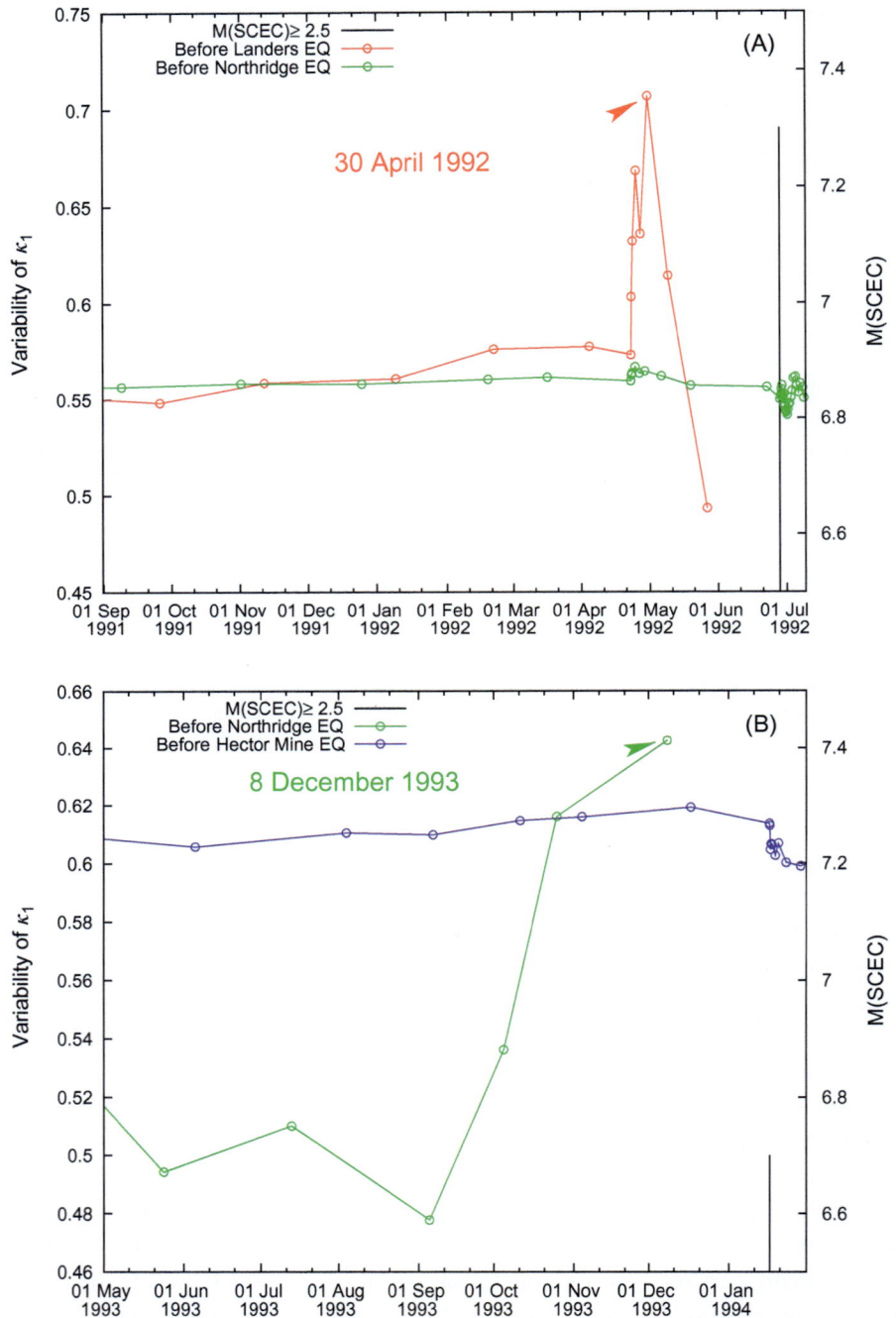

FIGURE 7–8 Excerpts of Fig. 7–7 plotted in expanded timescale showing what happened before the following mainshocks: (A) Landers (red), (B) Northridge (green) and (C) Hector Mine (blue). The arrows indicate the date at which the κ_1 variability has been maximized.

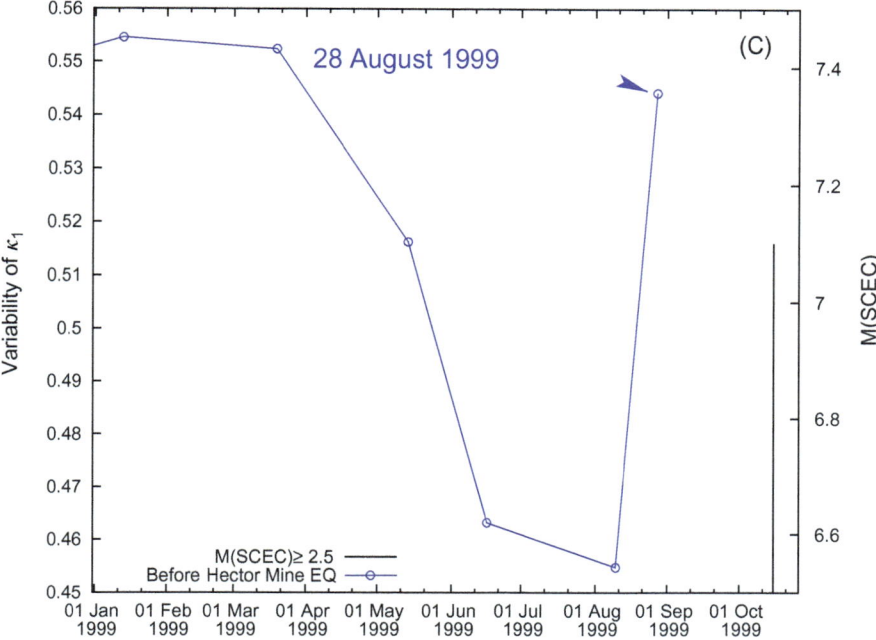

FIGURE 7–8 (Continued).

calculation for the variability of κ_1, as described above, we find the results depicted in Fig. 7−9. They show an evident increase in the κ_1 variability only before the strongest earthquake, i.e., the Landers earthquake (and a somewhat weaker effect before the Hector Mine earthquake as well as some changes before the Loma Prieta earthquake). Recall that in Fig. 7−6, when using the SCEC catalogue with low threshold $M \geq 2.0$ the precursory effect in the κ_1 variability of seismicity was clearly seen before *all* three cases of the major mainshocks (Landers, Northridge and Hector Mine). The same seems more or less to hold in Fig. 7−7 -as well as in Fig. 7−8A, B and C − when again using the SCEC catalogue with $M \geq 2.5$. In contrast, in Fig. 7−9 upon using the NCEDC catalogue with $M \geq 2.5$, such a precursory effect in the κ_1 variability is clearly seen only in one case (or in three) out of the six major mainshocks in California listed in Table 7−1. *This is important since it dictates that we may lose useful information due to coarse graining and hence the precursory increase of the κ_1 variability could be lost or hardly seen.* This coarse graining can be realized if we take into account that during the 25-year period (1979−2003) investigated, while SCEC contains 86,284 earthquakes with $M \geq 2.0$, and 30,163 earthquakes with $M \geq 2.5$, NCEDC reports 31,832 earthquakes with $M \geq 2.5$, but we have to consider also that the latter catalogue extends to an appreciably wider area, i.e., $N_{31.7}^{45.7} W_{112.1}^{127.5}$, compared to the former, i.e., $N_{32}^{37} W_{114}^{122}$.

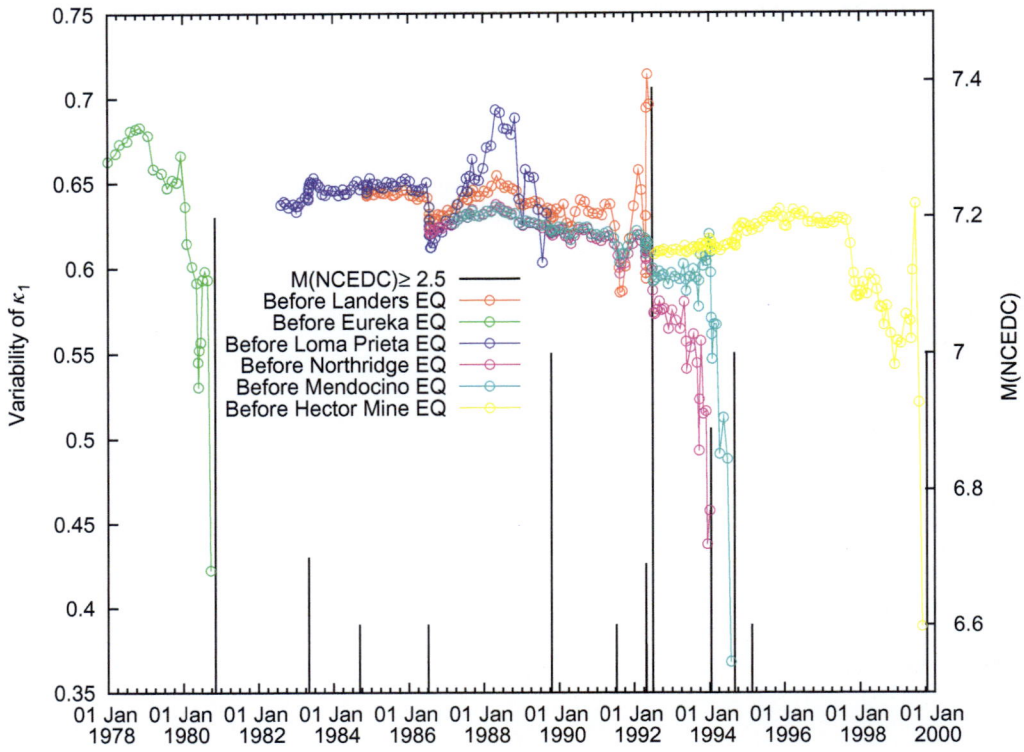

FIGURE 7–9 The values of the variability of κ_1 plotted versus the conventional time (UT). These values have been deduced from excerpts of the NCEDC catalogue comprising W earthquakes with $M \geq 2.5$ before each of the six major mainshocks that occurred in California (during the 25-year period 1979–2003) given in Table 7–1. The points are calculated at every hundred W interval and the W values for the closest points to the mainshock are 100. All earthquakes that occurred during the same period with $M(NCEDC) \geq 6.5$ (right scale) are also shown with black vertical bars.

7.5 Order Parameter Fluctuations Upon Sliding a Natural Time Window of Fixed Length: The Global Minimum of the Variability of the Order Parameter Before the Strongest Mainshock

In this section, we consider a natural time window of fixed length W (that means comprising a constant number W of consecutive seismic events) sliding through the seismic catalogue. In particular, we find here that the results become exciting upon using a *crucial* scale, i.e., when these W consecutive events extend to a time period comparable to the lead time (Varotsos and Lazaridou, 1991; Varotsos et al., 1993, 2011d) of the precursory SES activities. The motivation to focus on such a scale becomes clear if we consider the following points.

First, according to the SES generation mechanism mentioned in Section 7.2.1 SES emission occurs when the stress in the future focal region of an earthquake reaches a *critical* value. The validity of this SES generation mechanism is strengthened by the fact that the up to date experimental data of SES activities (along with their associated magnetic field variations) have been shown to exhibit infinitely ranged temporal correlations (Varotsos et al., 2009), thus being in accord with the conjecture of *critical* dynamics. Thus, the observation of SES marks when the system enters a critical stage.

Second, observations of SES activities in Japan in the 2000s (Uyeda et al., 2002, 2009) as well as in Mexico (see Ramírez-Rojas et al., 2011, and references therein) and in California (see Bernardi et al., 1991; Fraser-Smith et al., 1990), where magnetic field variations similar to those associated with the SES activities in Greece have been reported — as already mentioned in Section 7.3.4 — have shown that their lead time is of the order of a few months, in agreement with earlier observations in Greece (Varotsos and Lazaridou, 1991; Varotsos et al., 1993).

Hence, the SES observations in various countries reveal that before the occurrence of major earthquakes there is a crucial timescale of around a few months or so, in which long-range correlations are seriously affected. In other words, the SES observations dictate that the critical stress σ_{cr} is attained a few months or so before a mainshock which may reflect that changes in the correlation properties of other associated physical quantities like seismicity may become detectable at this timescale if studying the corresponding order parameter fluctuations in natural time (Varotsos et al., 2011b).

Let us now recall the procedure explained in Fig. 7–3, in which we take a natural time window of fixed length comprising W consecutive events. Starting from the first earthquake, we calculate the κ_1 values using say $N = 6$ to 40 consecutive events. We next turn to the second earthquake, and repeat the calculation of κ_1. After sliding, event by event, through the whole natural time window, the computed values enable the calculation of the average value $\mu(\kappa_1)$ and the standard deviation $\sigma(\kappa_1)$ that correspond to this natural time window of length W. We then determine the variability of κ_1, i.e., the quantity β defined in Eq. (7.13). Here, however, we employ the following change: the calculation of the variability β of κ_1 for various windows W, for each earthquake e_i in the seismic catalogue, is made (Varotsos et al., 2011b) by calculating the κ_1 values resulting when using the *previous* 6–40 consecutive earthquakes. Then, the hitherto obtained κ_1 values for the earthquakes e_{i-W+1} to e_i were considered for the estimation of the variability β for a natural time window length W. The resulting β value, labelled β_i, was attributed to e_i, the data of which were obviously *not included* in the β_i estimation.

Hereafter, we shall use the symbol β_W where the subscript W designates the natural time window length from which the β value under discussion was obtained. In addition, a second subscript 'min' will be needed later so that the symbol $\beta_{W,min}$ will designate the corresponding minimum of the β_W value.

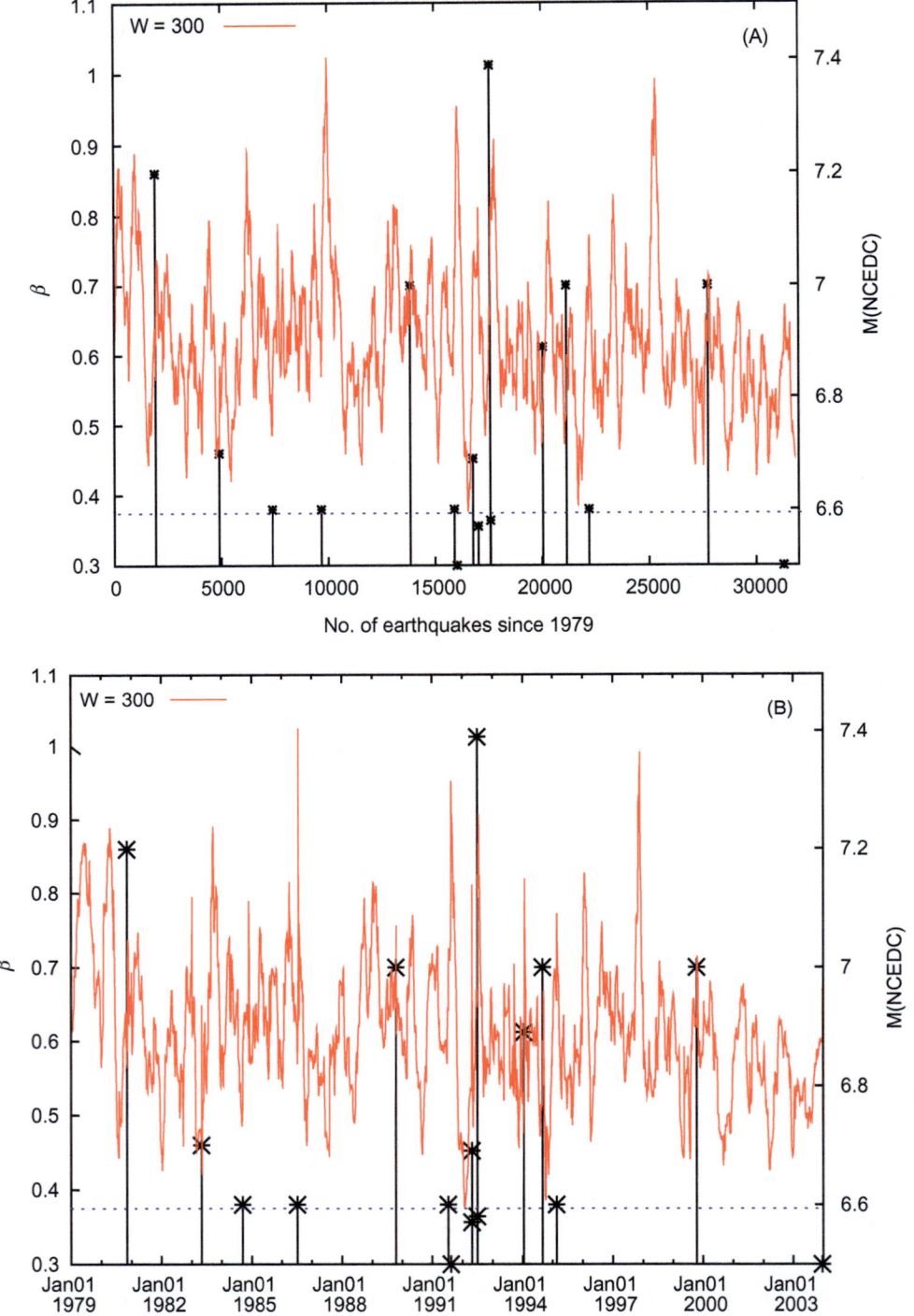

FIGURE 7–10 (A) The variability of κ_1 versus the number of events (earthquakes) when a natural time window of fixed length $W = 300$ events is sliding through the NCEDC catalogue for the seismicity ($M \geq 2.5$) within the area $N_{31.7}^{45.7}W_{127.5}^{112.1}$ during the 25-year period 1 January 1979 to 31 December 2003. The earthquakes that occurred are shown in black (with magnitudes labelled $M(NCEDC)$ in the right scale). (B) The same as in (A) but here the variability of κ_1 is plotted versus the conventional time (UT). (C) An excerpt of (B) showing the variability of κ_1 versus the conventional time during the almost 14-month period from 00:00 UT on 1 May 1991 until the occurrence of the Landers earthquake on 28 June 1992. The horizontal dotted (blue) lines were drawn as a guide to the eye indicating the minimum β value. *Taken from Varotsos, P., Sarlis, N., Skordas, E., 2011b. Scale-specific order parameter fluctuations of seismicity in natural time before mainshocks. EPL 96, 59002.*

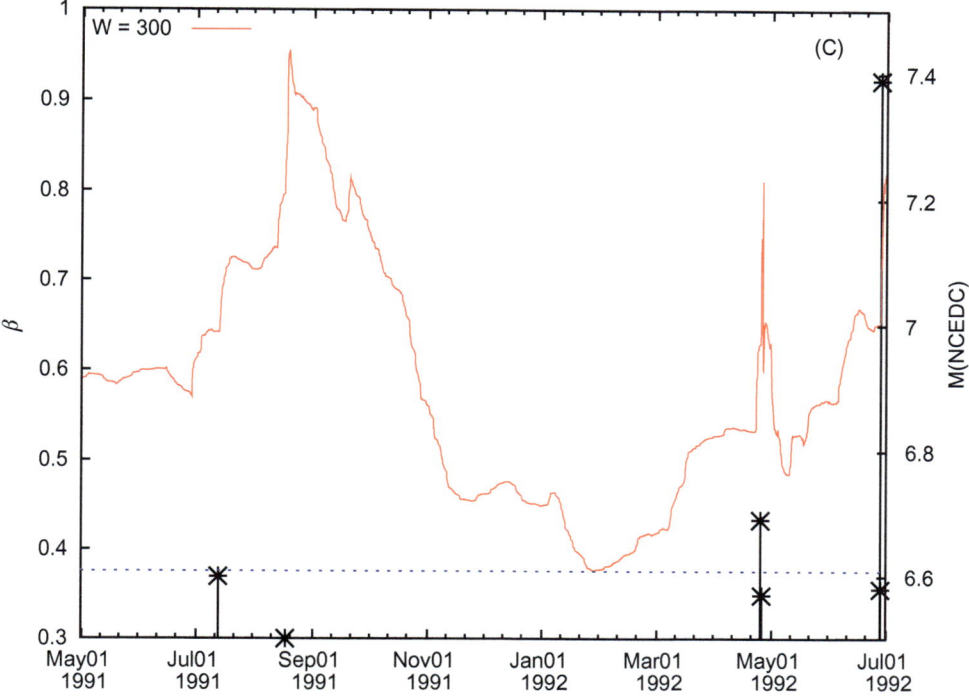

FIGURE 7–10 (Continued).

Varotsos et al. (2011b), for the case of California, considered all earthquakes reported by NCEDC within the area $N_{31.7}^{45.7}W_{127.5}^{112.1}$ during the 25-year period from 1 January 1979 to 1 January 2004 as in Section 7.4.2. Considering that 31,832 earthquakes occurred with $M \geq 2.5$, as mentioned, we have on average $\sim 10^2$ earthquakes per month. Thus, the lead time of SES activities, which is around a few months, say 2 months, with an upper limit of around 5 months (Varotsos et al., 2011d), corresponds to W values lying in the range from $W = 200$ to $W = 500$ events. Hence, Varotsos et al. (2011b) focused on such window lengths, i.e., $W = 200$ to 500. For example, Fig. 7–10A depicts the results for $W = 300$ of the variability of κ_1 versus the number of events (earthquakes) during the aforementioned 25-year period. The same results for the variability of κ_1 are plotted in Fig. 7–10B versus conventional time. An inspection of Fig. 7–10A, B reveals that before the strongest earthquake, which is the Landers earthquake that occurred on 28 June 1992, with $M = 7.4$ (see Table 7–1), a transient change of the κ_1 variability is observed, which exhibits the *lowest value* (around 0.38) during the 25-year period investigated. We emphasize that *such a minimum value solely stems from earlier earthquakes* in view of the procedure followed in our computation. To better visualize what happened before this strongest earthquake, Fig. 7–10C shows in expanded timescale, the variability of κ_1 during the 14-month period from 1 May 1991 to 1 July 1992. A close inspection of this figure shows that the lowest value of the κ_1 variability, i.e., $\beta_{300,min}$ was observed around the period from the last days of January to the first days of February 1992. The $M7.4$ Landers mainshock occurred somewhat less than 5 months later.

Varotsos et al. (2011b) also reported a similar investigation of the seismicity in Japan from 1 January 1984 to 11 March 2011, the day of the $M9$ Tohoku earthquake by setting a magnitude threshold $M_{JMA} = 3.5$ in the JMA catalogue to ensure data completeness. This reflects on the average $\sim 10^2$ earthquakes per month, thus Varotsos et al. (2011b) chose values $W = 200$ and 300, which would cover on average a period of around a few months. They found that the variability of κ_1 exhibited a very clear global minimum during the first week of January 2011, i.e., almost 2 months before the mainshock occurrence on 11 March 2011. Sarlis et al. (2013) extended this study and found that the fluctuations of the order parameter κ_1 of seismicity exhibited distinct minima a few months before all six shallow earthquakes of magnitude 7.6 or larger that occurred during this 27-year period in the Japanese area (Fig. 7–11B). Among these minima, the minimum before the $M9$ Tohoku earthquake was the deepest. In a later study, Varotsos et al. (2014b), by also applying DFA to the earthquake magnitude time series found that each of these minima is preceded as well as followed by characteristic changes to temporal correlations between earthquake magnitudes.

7.6 Conclusions

Natural time analysis of seismic times series enables the introduction of an order parameter κ_1 for seismicity and uncovers dynamic features hidden behind these time series as follows.

Concerning the value of the order parameter of seismicity κ_1 in the region prone to suffer the strong earthquake if SES data are available, it may lead to the estimation of the occurrence time of the impending earthquake within a week or so.

Upon decreasing the value of the natural time window length W, which means that we gradually approach the mainshock, an increase in the κ_1 variability is observed before the mainshock followed by a decrease before the mainshock occurrence.

Finally, when considering a sliding natural time window of *fixed* length comprising a number of earthquakes that would occur in a few months (which corresponds to the average lead time of the SES activities), we find that the fluctuations of the order parameter of seismicity exhibit a distinct minimum a few months before major mainshocks.

Acknowledgements

We would like to express our deepest thanks to the eminent figure of contemporary geophysics, Professor Seiya Uyeda for long discussions on natural time analysis of seismicity.

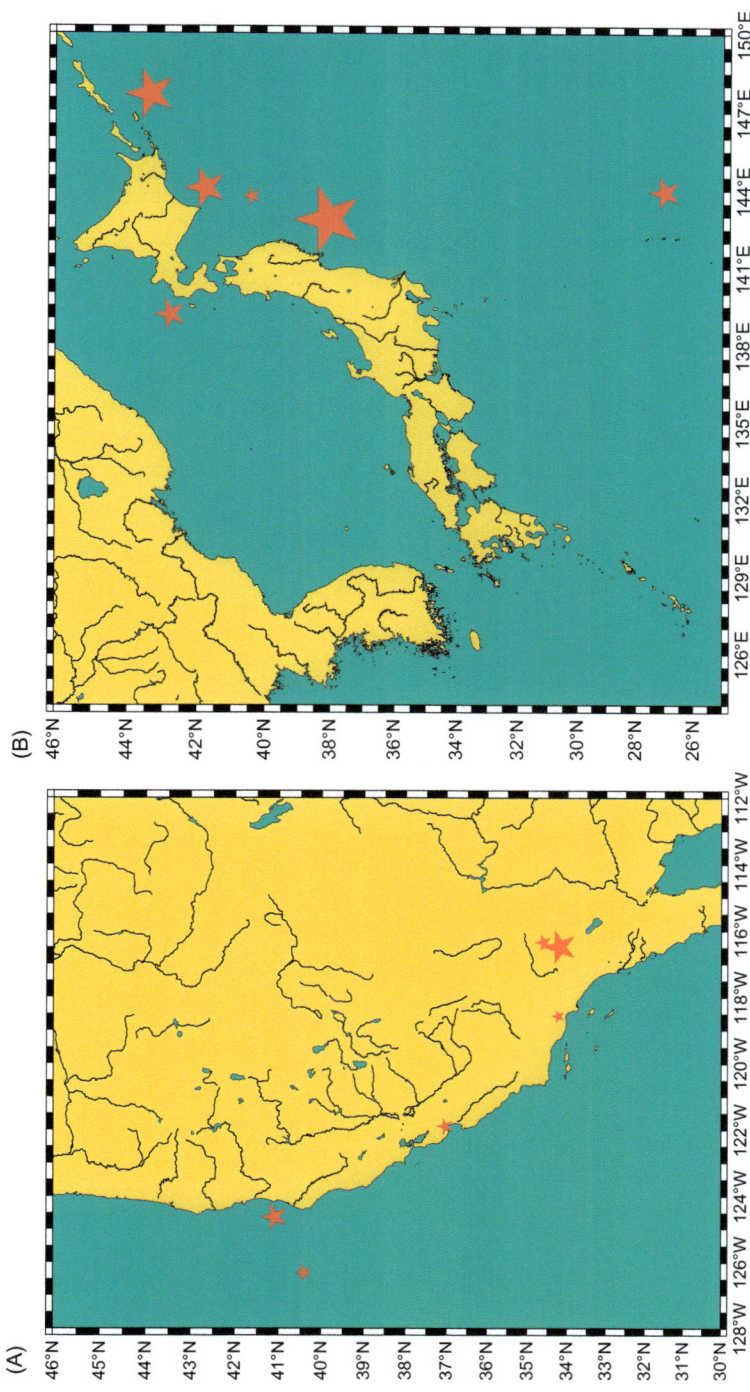

FIGURE 7–11 Maps of the areas in California (A) and Japan (B) the seismicities of which are analysed in natural time. The epicentres of the major earthquakes reported in Tables 7–1 and 7–2 are also marked.

References

Ashkenazy, Y., Ivanov, P.C., Havlin, S., Peng, C.K., Yamamoto, Y., Goldberger, A.L., et al., 2000. Decomposition of heartbeat time series: scaling analysis of the sign sequence. Comput. Cardiol. 27, 139–143.

Ausloos, M., Lambiotte, R., 2006. Brownian particle having a fluctuating mass. Phys. Rev. E 73, 011105.

Ausloos, M., Vandewalle, N., Boveroux, P., Minguet, A., Ivanova, K., 1999. Applications of statistical physics to economic and financial topics. Physica A 274, 229–240.

Baiesi, M., Paczuski, M., 2004. Scale-free networks of earthquakes and aftershocks. Phys. Rev. E 69, 066106.

Bak, P., Christensen, K., Danon, L., Scanlon, T., 2002. Unified scaling law for earthquakes. Phys. Rev. Lett. 88, 178501.

Bashan, A., Bartsch, R., Kantelhardt, J.W., Havlin, S., 2008. Comparison of detrending methods for fluctuation analysis. Physica A 387, 5080–5090.

Batac, R., Kantz, H., 2014. Observing spatio-temporal clustering and separation using interevent distributions of regional earthquakes. Nonlin. Processes Geophys. 21, 735–744.

Bernardi, A., Fraser-Smith, A.C., McGill, P.R., Villard, O.G., 1991. ULF magnetic field measurements near the epicenter of the Ms 7.1 Loma Prieta earthquake. Phys. Earth Planet. Inter. 68, 45–63.

Botet, R., 2011. Order parameter fluctuations at a critical point an exact result about percolation. J. Phys. Conf. Ser. 297, 012005.

Bramwell, S.T., Holdsworth, P.C.W., Pinton, J.F., 1998. Universality of rare fluctuations in turbulence and critical phenomena. Nature (London) 396, 552–554.

Bramwell, S.T., Christensen, K., Fortin, J.Y., Holdsworth, P.C.W., Jensen, H.J., Lise, S., et al., 2000. Universal fluctuations in correlated systems. Phys. Rev. Lett. 84, 3744–3747.

Bramwell, S.T., Christensen, K., Fortin, J.Y., Holdsworth, P.C.W., Jensen, H.J., Lise, S., et al., 2001a. Phys. Rev. Lett. 87, 188902.

Bramwell, S.T., Fortin, J.Y., Holdsworth, P.C.W., Peysson, S., Pinton, J.F., Portelli, B., et al., 2001b. Magnetic fluctuations in the classical xy model: the origin of an exponential tail in a complex system. Phys. Rev. E 63, 041106.

Bramwell, S.T., Christensen, K., Fortin, J.Y., Holdsworth, P.C.W., Jensen, H.J., Lise, S., et al., 2002. Phys. Rev. Lett. 89, 208902.

Bunde, A., Lennartz, S., 2012. Long-term correlations in earth sciences. Acta Geophysica 60, 562–588.

Carlson, J.M., Langer, J.S., Shaw, B.E., 1994. Dynamics of earthquake faults. Rev. Mod. Phys. 66, 657–670.

Carretero-Campos, C., Bernaola-Galván, P., Ivanov, P.C., Carpena, P., 2012. Phase transitions in the first-passage time of scale-invariant correlated processes. Phys. Rev. E 85, 011139.

Christopoulos, S.R., Sarlis, N., 2014. q-exponential relaxation of the expected avalanche size in the coherent noise model. Physica A 407, 216–225.

Christopoulos, S.R., Sarlis, N., 2015. Corrigendum to "q-exponential relaxation of the expected avalanche size in the coherent noise model" [physica a 407 (2014) 216225]. Physica A 438, 667.

Christopoulos, S.R.G., Sarlis, N.V., 2017. An Application of the Coherent Noise Model for the Prediction of Aftershock Magnitude Time Series. Complexity 2017.

Clusel, M., Fortin, J.Y., Holdsworth, P.C.W., 2004. Criterion for universality-class-independent critical fluctuations: example of the two-dimensional ising model. Phys. Rev. E 70, 046112.

Corral, A., 2004. Long-term clustering, scaling, and universality in the temporal occurrence of earthquakes. Phys. Rev. Lett. 92, 108501.

Davidsen, J., Green, A., 2011. Are earthquake magnitudes clustered? Phys. Rev. Lett. 106, 108502.

Davidsen, J., Paczuski, M., 2005. Analysis of the spatial distribution between successive earthquakes. Phys. Rev. Lett. 94, 048501.

Dziewonski, A., Chou, T.A., Woodhouse, J., 1981. Determination of earthquake source parameters from waveform data for studies of global and regional seismicity. J. Geophys. Res. Solid Earth 86, 2825–2852.

Eichner, J.F., Kantelhardt, J.W., Bunde, A., Havlin, S., 2007. Statistics of return intervals in long-term correlated records. Phys. Rev. E 75, 011128.

Ekström, G., Nettles, M., Dziewoński, A., 2012. The global CMT project 2004–2010: centroid-moment tensors for 13,017 earthquakes. Phys. Earth Planet. Inter. 200, 1–9.

Engdahl, E.R., Villasenor, A., 2002. Global seismicity: 1900–1999. In: Lee, W.H.K., Kanamori, H., Jennings, P. C., Kisslinger, C. (Eds.), International Handbook of Earthquake and Engineering Seismology, Part A, Chapter 41. Academic Press, London, pp. 665–690.

Fisher, D.S., Dahmen, K., Ramanathan, S., Ben-Zion, Y., 1997. Statistics of earthquakes in simple models of heterogeneous faults. Phys. Rev. Lett. 78, 4885–4888.

Flores-Márquez, E., Vargas, C., Telesca, L., Ramírez-Rojas, A., 2014. Analysis of the distribution of the order parameter of synthetic seismicity generated by a simple springblock system with asperities. Physica A 393, 508–512.

Fraser-Smith, A.C., Bernardi, A., McGill, P.R., Ladd, M.E., Helliwell, R.A., Villard, O.G., 1990. Low-frequency magnetic-field measurements near the epicenter of the Ms-7.1 Loma-Prieta earthquake. Geophys. Res. Lett. 17, 1465.

Goldberger, A.L., Amaral, L.A.N., Glass, L., Hausdorff, J.M., Ivanov, P.C., Mark, R.G., et al., 2000. Physiobank, physiotoolkit, and physionet - components of a new research resource for complex physiologic signals. Circulation 101, E215 (see also www.physionet.org).

Gutenberg, B., Richter, C.F., 1954. Seismicity of the earth and associated phenomena. Princeton Univ. Press, Princeton, New York.

Hayakawa, M., Schekotov, A., Potirakis, S., Eftaxias, K., 2015a. Criticality features in ULF magnetic fields prior to the 2011 Tohoku earthquake. Proc. Jpn Acad. Ser. B Phys. Biol. Sci. 91, 25–30.

Hayakawa, M., Schekotov, A., Potirakis, S., Eftaxias, K., Li, Q., Asano, T., 2015b. An integrated study of ULF magnetic field variations in association with the 2008 sichuan earthquake, on the basis of statistical and critical analyses. Open J. Earthquake Res. 4, 85–93.

Holliday, J.R., Rundle, J.B., Turcotte, D.L., Klein, W., Tiampo, K.F., Donnellan, A., 2006. Space-time clustering and correlations of major earthquakes. Phys. Rev. Lett. 97, 238501.

Hu, K., Ivanov, P.C., Chen, Z., Hilton, M.F., Stanley, H.E., Shea, S.A., 2004. Non-random fluctuations and multi-scale dynamics regulation of human activity. Physica A 337, 307.

Huang, Q., 2015. Forecasting the epicenter of a future major earthquake. Proc. Natl. Acad. Sci. USA 112, 944–945.

Ivanov, P.C., 2007. Scale-invariant aspects of cardiac dynamics - observing sleep stages and circadian phases. IEEE Eng. Med. Biol. 26, 33.

Ivanov, P.C., Yuen, A., Podobnik, B., Lee, Y., 2004. Common scaling patterns in intertrade times of U. S. stocks. Phys. Rev. E 69, 056107.

Japan Meteorological Agency, 2000. Recent seismic activity in the Miyakejima and Niijima-Kozushima region, Japan -the largest earthquake swarm ever recorded-. Earth Planets and Space 52, i–viii.

Kalisky, T., Ashkenazy, Y., Havlin, S., 2005. Volatility of linear and nonlinear time series. Phys. Rev. E 72, 011913.

Kanamori, H., 1978. Quantification of earthquakes. Nature 271, 411–414.

Lamperti, J.W., 1962. Semi-stable stochastic processes. Trans. Am. Math. Soc. 104, 62.

Lazaridou-Varotsos, M.S., 2013. Earthquake Prediction by Seismic Electric Signals. The Success of the VAN Method Over Thirty Years. Springer-Verlag, Berlin Heidelberg.

Lennartz, S., Bunde, A., Turcotte, D.L., 2011. Modelling seismic catalogues by cascade models: do we need long-term magnitude correlations?. Geophys. J. Int. 184, 1214−1222.

Lennartz, S., Livina, V.N., Bunde, A., Havlin, S., 2008. Long-term memory in earthquakes and the distribution of interoccurrence times. EPL 81, 69001.

Lippiello, E., Godano, C., de Arcangelis, L., 2007. Dynamical scaling in branching models for seismicity. Phys. Rev. Lett. 98, 098501.

Lippiello, E., de Arcangelis, L., Godano, C., 2008. Influence of time and space correlations on earthquake magnitude. Phys. Rev. Lett. 100, 038501.

Lippiello, E., Godano, C., de Arcangelis, L., 2012. The earthquake magnitude is influenced by previous seismicity. Geophys. Res. Lett. 39, L05309.

Liu, Y.H., Cizeau, P., Meyer, M., Peng, C.K., Stanley, H.E., 1997. Correlations in economic time series. Physica A 245, 437−440.

Livina, V.N., Havlin, S., Bunde, A., 2005. Memory in the occurrence of earthquakes. Phys. Rev. Lett. 95, 208501.

Ma, Q.D.Y., Bartsch, R.P., Bernaola-Galván, P., Yoneyama, M., Ivanov, P.C., 2010. Effect of extreme data loss on long- range correlated and anticorrelated signals quantified by detrended fluctuation analysis. Phys. Rev. E 81, 031101.

Papadopoulou, K.A., Skordas, E.S., Sarlis, N.V., 2016. A tentative model for the explanation of Bth law using the order parameter of seismicity in natural time. Earthquake Sci. 29, 311−319.

Papasimakis, N., Pallikari, F., 2010. Correlated and uncorrelated heart rate fluctuations during relaxing visualization. EPL 90, 48003.

Peng, C.K., Buldyrev, S.V., Havlin, S., Simons, M., Stanley, H.E., Goldberger, A.L., 1994. Mosaic organization of DNA nucleotides. Phys. Rev. E 49, 1685−1689.

Peng, C.K., Buldyrev, S.V., Goldberger, A.L., Havlin, S., Mantegna, R.N., Simons, M., et al., 1995a. Statistical properties of dna sequences. Physica A 221, 180.

Peng, C.K., Havlin, S., Stanley, H.E., Goldberger, A.L., 1995b. Quantification of scaling exponents and crossover phenomena in nonstationary heartbeat time series. Chaos 5, 82−87.

Potirakis, S.M., Karadimitrakis, A., Eftaxias, K., 2013. Natural time analysis of critical phenomena: the case of pre- fracture electromagnetic emissions. Chaos 23, 023117.

Potirakis, S.M., Contoyiannis, Y., Eftaxias, K., Koulouras, G., Nomicos, C., 2015. Recent Field Observations Indicating an Earth System in Critical Condition Before the Occurrence of a Significant Earthquake. IEEE Geosci. Remote Sensing Letters 12, 631−635.

Potirakis, S., Eftaxias, K., Schekotov, A., Yamaguchi, H., Hayakawa, M., 2016a. Criticality features in ultra-low frequency magnetic fields prior to the 2013 M6.3 Kobe earthquake. Ann. Geophys. 59, S0317.

Potirakis, S.M., Contoyiannis, Y., Melis, N.S., Kopanas, J., Antonopoulos, G., Balasis, G., et al., 2016b. Recent seismic activity at Cephalonia (Greece):a study through candidate electromagnetic precursors in terms of non-linear dynamics. Nonlin. Processes Geophys. 23, 223−240.

Ramírez-Rojas, A., Flores-Márquez, E.A., 2013. Order parameter analysis of seismicity of the Mexican Pacific coast. Physica A 392, 2507−2512.

Ramírez-Rojas, A., Telesca, L., Angulo-Brown, F., 2011. Entropy of geoelectrical time series in the natural time domain. Nat. Hazards Earth Syst. Sci. 11, 219−225.

Rundle, J.B., Klein, W., Gross, S., 1996. Dynamics of a traveling density wave model for earthquakes. Phys. Rev. Lett. 76, 4285−4288.

Rundle, J.B., Turcotte, D.L., Shcherbakov, R., Klein, W., Sammis, C., 2003. Statistical physics approach to understanding the multiscale dynamics of earthquake fault systems. Rev. Geophys. 41, 1019.

Saichev, A., Sornette, D., 2005. Vere-jones self-similar branching model. Phys. Rev. E 72, 056122.

Sarlis, N.V., 2011. Magnitude correlations in global seismicity. Phys. Rev. E 84, 022101.

Sarlis, N.V., 2013. On the recent seismic activity in North-Eastern Aegean Sea including the Mw 5.8 earthquake on 8 January 2013. Proc. Jpn. Acad. Ser. B Phys. Biol. Sci. 89, 438–445.

Sarlis, N.V., Christopoulos, S.R.G., 2012. Natural time analysis of the Centennial Earthquake Catalog. Chaos 22, 023123.

Sarlis, N.V., Skordas, E.S., Lazaridou, M.S., Varotsos, P.A., 2008. Investigation of seismicity after the initiation of a Seismic Electric Signal activity until the main shock. Proc. Jpn. Acad. Ser. B Phys. Biol. Sci. 84, 331–343.

Sarlis, N.V., Skordas, E.S., Varotsos, P.A., 2009a. Heart rate variability in natural time and 1/f "noise". EPL 87, 18003.

Sarlis, N.V., Skordas, E.S., Varotsos, P.A., 2009b. Multiplicative cascades and seismicity in natural time. Phys. Rev. E 80, 022102.

Sarlis, N.V., Skordas, E.S., Varotsos, P.A., 2010a. Nonextensivity and natural time: the case of seismicity. Phys. Rev. E 82, 021110.

Sarlis, N.V., Skordas, E.S., Varotsos, P.A., 2010b. Order parameter fluctuations of seismicity in natural time before and after mainshocks. EPL 91, 59001.

Sarlis, N.V., Skordas, E.S., Varotsos, P.A., Nagao, T., Kamogawa, M., Tanaka, H., et al., 2013. Minimum of the order parameter fluctuations of seismicity before major earthquakes in Japan. Proc. Natl. Acad. Sci. USA 110, 13734–13738.

Sarlis, N.V., Christopoulos, S.R.G., Bemplidaki, M.M., 2015a. Change ΔS of the entropy in natural time under time reversal: complexity measures upon change of scale. EPL 109, 18002.

Sarlis, N.V., Christopoulos, S.R.G., Skordas, E.S., 2015b. Minima of the fluctuations of the order parameter of global seismicity. Chaos 25, 063110.

Sarlis, N.V., Skordas, E.S., Varotsos, P.A., Nagao, T., Kamogawa, M., Uyeda, S., 2015c. Spatiotemporal variations of seismicity before major earthquakes in the Japanese area and their relation with the epicentral locations. Proc. Natl. Acad. Sci. USA 112, 986–989.

Scafetta, N., West, B.J., 2004. Multiscaling comparative analysis of time series and a discussion on "earthquake conversations" in california. Phys. Rev. Lett. 92, 138501.

Scafetta, N., West, B.J., 2005. Multiscaling compapative analysis of time seples and geophysical phenomena. Complexity 10 (4), 51–56.

Scholz, C.H., 2002. The Mechanics of Earthquakes and Faulting, second ed Cambridge University Press, Cambridge, UK.

Sethna, J.P., 1992. Order parameters, broken symmetry, and topology. In: Nagel, L., Stein, D. (Eds.), 1991 Lectures in Complex Systems, Santa Fe Institute Studies in the Sciences of Complexity. Addison-Wesley, New York, Proc. Vol. XV.

Shcherbakov, R., Turcotte, D.L., Rundle, J.B., 2004. A generalized Omori's law for earthquake aftershock decay. Geophys. Res. Lett. 31, L11613.

Skordas, E., Sarlis, N., 2014. On the anomalous changes of seismicity and geomagnetic field prior to the 2011 9.0 Tohoku earthquake. J. Asian Earth Sci. 80, 161–164.

Skordas, E.S., Sarlis, N.V., Varotsos, P.A., 2010. Effect of significant data loss on identifying electric signals that precede rupture estimated by detrended fluctuation analysis in natural time. Chaos 20, 033111.

Sornette, D., 2000. Critical Phenomena in the Natural Sciences: Chaos, Fractals, Selforganization, and Disorder: Concepts and Tools. Springer-Verlag, Berlin Heidelberg.

Stanley, H.E., 1999. Scaling, universality, and renormalization: three pillars of modern critical phenomena. Rev. Mod. Phys. 71, S358–S366.

Stanley, H.E., Buldyrev, S.V., Goldberger, A.L., Havlin, S., Peng, C.K., Simons, M., 1999. Scaling features of noncoding DNA. Physica A 273, 1–18.

Tanaka, H.K., Varotsos, P.A., Sarlis, N.V., Skordas, E.S., 2004. A plausible universal behaviour of earthquakes in the natural time-domain. Proc. Jpn. Acad. Ser. B Phys. Biol. Sci. 80, 283–289.

Taqqu, M.S., Teverovsky, V., Willinger, W., 1995. Estimators for long-range dependence: an empirical study. Fractals 3, 785–798.

Tiampo, K.F., Shcherbakov, R., 2012. Seismicity-based earthquake forecasting techniques: ten years of progress. Tectonophysics 522–523, 89–121.

Tiampo, K.F., Rundle, J.B., Klein, W., Holliday, J., Martins, J.S.S., Ferguson, C.D., 2007. Ergodicity in natural earthquake fault networks. Phys. Rev. E 75, 066107.

Tirnakli, U., Abe, S., 2004. Aging in coherent noise models and natural time. Phys. Rev. E 70, 056120.

Tsallis, C., Levy, S.V.F., Souza, A.M.C., Maynard, R., 1995. Statistical-mechanical foundation of the ubiquity of lévy distributions in nature. Phys. Rev. Lett. 75, 3589–3593.

Tsallis, C., Levy, S.V.F., Souza, A.M.C., Maynard, R., 1996. Statistical-mechanical foundation of the ubiquity of the lévy distributions in nature. Phys. Rev. Lett. 77, 5442.

Tsuji, D., Katsuragi, H., 2015. Temporal analysis of acoustic emission from a plunged granular bed. Phys. Rev. E 92, 042201.

Turcotte, D.L., 1997. Fractals and Chaos in Geology and Geophysics, second ed Cambridge University Press, Cambridge.

Uyeda, S., Hayakawa, M., Nagao, T., Molchanov, O., Hattori, K., Orihara, Y., et al., 2002. Electric and magnetic phenomena observed before the volcano-seismic activity in 2000 in the Izu Island Region, Japan. Proc. Natl. Acad. Sci. USA 99, 7352–7355.

Uyeda, S., Kamogawa, M., Tanaka, H., 2009. Analysis of electrical activity and seismicity in the natural time domain for the volcanic-seismic swarm activity in 2000 in the Izu Island region, Japan. J. Geophys. Res. 114, B02310.

Vallianatos, F., Michas, G., Benson, P., Sammonds, P., 2013. Natural time analysis of critical phenomena: the case of acoustic emissions in triaxially deformed Etna basalt. Physica A 392, 5172–5178.

Vallianatos, F., Michas, G., Papadakis, G., 2014. Non-extensive and natural time analysis of seismicity before the Mw6.4, October 12, 2013 earthquake in the South West segment of the Hellenic Arc. Physica A 414, 163–173.

Vallianatos, F., Michas, G., Hloupis, G., 2015. Multiresolution wavelets and natural time analysis before the January-February 2014 Cephalonia (Mw6.1 - 6.0) sequence of strong earthquake events. Phys. Chem. Earth, Parts A/B/C 85–86, 201–209.

Vandewalle, N., Ausloos, M., 1998. Crossing of two mobile averages: a method for measuring the roughness exponent. Phys. Rev. E 58, 6832–6834.

Vargas, C., Flores-Marquez, E., Ramírez-Rojas, A., Telesca, L., 2015. Analysis of natural time domain entropy fluctuations of synthetic seismicity generated by a simple stick-slip system with asperities. Physica A 419, 23–28.

Varotsos, C., Tzanis, C., 2012a. A new El Niño-Southern Oscillation forecasting tool based on Southern Oscillation Index. Atmos. Chem. Phys. Discuss. 12, 17443–17463.

Varotsos, C., Tzanis, C., 2012b. A new tool for the study of the ozone hole dynamics over Antarctica. Atmos. Environ. 47, 428–434.

Varotsos, C.A., Efstathiou, M.N., Cracknell, A.P., 2013a. On the scaling effect in global surface air temperature anomalies. Atmosp. Chem. Phys. 13, 5243–5253.

Varotsos, P.A., Sarlis, N.V., Skordas, E.S., Lazaridou, M.S., 2013b. Seismic electric signals: an additional fact showing their physical interconnection with seismicity. Tectonophysics 589, 116−125.

Varotsos, C.A., Franzke, C.L.E., Efstathiou, M.N., Degermendzhi, A.G., 2014a. Evidence for two abrupt warming events of sst in the last century. Theor. Appl. Climatol. 116, 51−60.

Varotsos, P.A., Sarlis, N.V., Skordas, E.S., 2014b. Study of the temporal correlations in the magnitude time series before major earthquakes in Japan. J. Geophys. Res.: Space Physics 119, 9192−9206.

Varotsos, C.A., Lovejoy, S., Sarlis, N.V., Tzanis, C.G., Efstathiou, M.N., 2015a. On the scaling of the solar incident flux. Atmos. Chem. Phys. 15, 7301−7306.

Varotsos, P.A., Sarlis, N.V., Skordas, E.S., Christopoulos, S.R.G., Lazaridou-Varotsos, M.S., 2015b. Identifying the occurrence time of an impending mainshock: a very recent case. Earthquake Science 28, 215−222.

Varotsos, C.A., Tzanis, C.G., Sarlis, N.V., 2016. On the progress of the 2015−2016 El Niño event. Atmos. Chem. Phys. 16, 2007−2011.

Varotsos, P., 2005. The Physics of Seismic Electric Signals. TERRAPUB, Tokyo.

Varotsos, P., 2008. Point defect parameters in β-PbF$_2$ revisited. Solid State Ionics 179, 438−441.

Varotsos, P., Alexopoulos, K., 1984a. Physical Properties of the variations of the electric field of the earth preceding earthquakes, I. Tectonophysics 110, 73−98.

Varotsos, P., Alexopoulos, K., 1984b. Physical Properties of the variations of the electric field of the earth preceding earthquakes, II. Tectonophysics 110, 99−125.

Varotsos, P., Alexopoulos, K., 1986. Thermodynamics of Point Defects and their Relation with Bulk Properties. North Holland, Amsterdam.

Varotsos, P., Lazaridou, M., 1991. Latest aspects of earthquake prediction in Greece based on Seismic Electric Signals. Tectonophysics 188, 321−347.

Varotsos, P., Alexopoulos, K., Nomicos, K., 1982. Comments on the pressure variation of the Gibbs energy for bound and unbound defects. Phys. Status Solidi B 111, 581.

Varotsos, P., Alexopoulos, K., Lazaridou, M., 1993. Latest aspects of earthquake prediction in Greece based on Seismic Electric Signals, II. Tectonophysics 224, 1−37.

Varotsos, P., Alexopoulos, K., Nomicos, K., Lazaridou, M., 1986. Earthquake prediction and electric signals. Nature (London) 322, 120.

Varotsos, P., Alexopoulos, K., Nomicos, K., Lazaridou, M., 1988. Official earthquake prediction procedure in Greece. Tectonophysics 152, 193−196.

Varotsos, P., Sarlis, N., Skordas, E., 2011a. Identifying long-range correlated signals upon significant periodic data loss. Tectonophysics 503, 189−194.

Varotsos, P., Sarlis, N., Skordas, E., 2011b. Scale-specific order parameter fluctuations of seismicity in natural time before mainshocks. EPL 96, 59002.

Varotsos, P., Sarlis, N.V., Skordas, E.S., Uyeda, S., Kamogawa, M., 2011c. Natural time analysis of critical phenomena. Proc. Natl. Acad. Sci. USA 108, 11361−11364.

Varotsos, P.A., Sarlis, N.V., Skordas, E.S., 2011d. Natural TimeAnalysis: The New View of Time. Precursory Seismic Electric Signals, Earthquakes and other Complex Time-Series. Springer-Verlag, Berlin Heidelberg.

Varotsos, P., Sarlis, N., Skordas, E., 2012a. Remarkable changes in the distribution of the order parameter of seismicity before mainshocks. EPL 100, 39002.

Varotsos, P., Sarlis, N., Skordas, E., 2012b. Scale-specific order parameter fluctuations of seismicity before mainshocks: natural time and detrended fluctuation analysis. EPL 99, 59001.

Varotsos, P.A., Sarlis, N.V., Skordas, E.S., 2001. Spatio-temporal complexity aspects on the interrelation between seismic electric signals and seismicity. Practica of Athens Academy 76, 294−321.

Varotsos, P.A., Sarlis, N.V., Skordas, E.S., 2002a. Long-range correlations in the electric signals that precede rupture. Phys. Rev. E 66, 011902.

Varotsos, P.A., Sarlis, N.V., Skordas, E.S., 2002b. Seismic electric signals and seismicity: on a tentative interrelation between their spectral content. Acta Geophys. Pol. 50, 337–354.

Varotsos, P.A., Sarlis, N.V., Skordas, E.S., 2003a. Attempt to distinguish electric signals of a dichotomous nature. Phys. Rev. E 68, 031106.

Varotsos, P.A., Sarlis, N.V., Skordas, E.S., 2003b. Long-range correlations in the electric signals the precede rupture: further investigations. Phys. Rev. E 67, 021109.

Varotsos, P.V., Sarlis, N.V., Skordas, E.S., 2003c. Electric fields that "arrive" before the time derivative of the magnetic field prior to major earthquakes. Phys. Rev. Lett. 91, 148501.

Varotsos, P.A., Sarlis, N.V., Skordas, E.S., Lazaridou, M.S., 2004a. Entropy in natural time domain. Phys. Rev. E 70, 011106.

Varotsos, P.A., Sarlis, N.V., Skordas, E.S., Tanaka, H.K., 2004b. A plausible explanation of the b-value in the gutenberg-richter law from first principles. Proc. Jpn. Acad. Ser. B Phys. Biol. Sci. 80, 429–434.

Varotsos, P.A., Sarlis, N.V., Skordas, E.S., Lazaridou, M.S., 2005a. Natural entropy fluctuations discriminate similar-looking electric signals emitted from systems of different dynamics. Phys. Rev. E 71, 011110.

Varotsos, P.A., Sarlis, N.V., Tanaka, H.K., Skordas, E.S., 2005b. Similarity of fluctuations in correlated systems: the case of seismicity. Phys. Rev. E 72, 041103.

Varotsos, P.A., Sarlis, N.V., Tanaka, H.K., Skordas, E.S., 2005c. Some properties of the entropy in the natural time. Phys. Rev. E 71, 032102.

Varotsos, P.A., Sarlis, N.V., Skordas, E.S., 2006a. On the recent advances in the study of seismic electric signals (van method). Phys. Chem. Earth 31, 189.

Varotsos, P.A., Sarlis, N.V., Skordas, E.S., Tanaka, H.K., Lazaridou, M.S., 2006b. Attempt to distinguish long-range temporal correlations from the statistics of the increments by natural time analysis. Phys. Rev. E 74, 021123.

Varotsos, P.A., Sarlis, N.V., Skordas, E.S., Tanaka, H.K., Lazaridou, M.S., 2006c. Entropy of seismic electric signals: analysis in the natural time undertime reversal. Phys. Rev. E 73, 031114.

Varotsos, P.A., Sarlis, N.V., Skordas, E.S., Lazaridou, M.S., 2007. Identifying sudden cardiac death risk and specifying its occurrence time by analyzing electrocardiograms in natural time. Appl. Phys. Lett. 91, 064106.

Varotsos, P.A., Sarlis, N.V., Skordas, E.S., 2009. Detrended fluctuation analysis of the magnetic and electric field variations that precede rupture. Chaos 19, 023114.

Varotsos, P.A., Sarlis, N.V., Skordas, E.S., Uyeda, S., Kamogawa, M., 2010. Natural time analysis of critical phenomena. the case of seismicity. EPL 92, 29002.

Watkins, N.W., Chapman, S.C., Rowlands, G., 2002. Comment on "universal fluctuations in correlated systems". Phys. Rev. Lett. 89, 208901.

Weron, A., Burnecki, K., Mercik, S., Weron, K., 2005. Complete description of all self-similar models driven by levy stable noise. Phys. Rev. E 71, 016113.

Xia, J., Gould, H., Klein, W., Rundle, J.B., 2008. Near-mean-field behavior in the generalized Burridge-Knopoff earthquake model with variable-range stress transfer. Phys. Rev. E 77, 031132.

Xu, Y., Ma, Q.D., Schmitt, D.T., Bernaola-Galván, P., Ivanov, P.C., 2011. Effects of coarse-graining on the scaling behavior of long-range correlated and anti-correlated signals. Physica A 390, 4057–4072.

Zaliapin, I., Ben-Zion, Y., 2013a. Earthquake clusters in southern California I: identification and stability. J. Geophys. Res. Solid Earth 118, 2847–2864.

Zaliapin, I., Ben-Zion, Y., 2013b. Earthquake clusters in southern California II: classification and relation to physical properties of the crust. J. Geophys. Res. Solid Earth 118, 2865–2877.

Zaliapin, I., Ben-Zion, Y., 2015. Artefacts of earthquake location errors and short-term incompleteness on seismicity clusters in southern California. Geophys. J. Int. 202, 1949–1968.

Zaliapin, I., Gabrielov, A., Keilis-Borok, V., Wong, H., 2008. Clustering analysis of seismicity and aftershock identification. Phys. Rev. Lett. 101.

Zheng, B., 2003. Generic features of fluctuations in critical systems. Phys. Rev. E 67, 026114.

Zheng, B., Trimper, S., 2001. Comment on "Universal Fluctuations in Correlated Systems". Phys. Rev. Lett. 87, 188901.

Complexity of Time Series of Stick-Slip (Models of Seismic Process)

8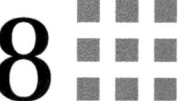

Complexity in Laboratory Seismology: From Electrical and Acoustic Emissions to Fracture

Vassilis Saltas[1], Filippos Vallianatos[1], Dimos Triantis[2], Ilias Stavrakas[2]

[1]*TECHNOLOGICAL EDUCATIONAL INSTITUTE OF CRETE AND UNESCO CHAIR ON SOLID EARTH PHYSICS AND GEOHAZARDS RISK REDUCTION, CHANIA, CRETE, GREECE*
[2]*UNIVERSITY OF WEST ATTICA, ATHENS, GREECE*

CHAPTER OUTLINE

8.1 Introduction

It is well-established that the application of mechanical stress in brittle materials such as rocks is accompanied by electromagnetic and acoustic emission (AE), infrared radiation and other fracto-emission phenomena (emission of electrons, atoms, molecules, etc.). All these fracture-induced phenomena are of great importance within a wide range of applications such as the search for precursor signals used for health monitoring of materials and underground mines as well as in earthquake prediction. Thus, the different underlying physical processes have been widely investigated both experimentally and from the theoretical point of view.

Here, we focus on the investigation of the emission of transient weak electric currents and elastic waves which constitute the basis of the PSCs (pressure-stimulated currents) technique and the well-known AE monitoring technique, respectively (Stavrakas et al., 2003, 2004). Microcrack development in a quasi-brittle, nonmetallic material produces electric charges due to the formation of electric dipoles that constitute, in general, a complicated charged system (Varotsos et al., 2002; Vallianatos et al., 2004). Such electric dipoles produce an electric potential across the crack, leading to the ability of a current to flow. PSCs are detected using a sensitive electrometer and a pair of electrodes which is attached at proper locations on the surface of a specimen that is subjected to mechanical stress tests.

The detection of such electric currents may be useful as a precursor of large-scale fracture and they have been measured at both laboratory (Vallianatos et al., 2004; Triantis et al., 2007; Frid et al., 2009) and geodynamic scales (Nomikos and Vallianatos, 1997; Nomikos et al., 1997; Varotsos, 2005). Specifically, the emission of weak electric currents prior to and during the failure of rocks subjected to mechanical loading has been extensively studied during recent decades and numerous experimental results on both dry and wet rock specimens, such as marble and amphibolites, have been published (Stavrakas et al., 2004; Triantis et al., 2006, 2008, 2012). The scientific challenge in these studies is the applicability of the laboratory findings to the understanding of mega-scale phenomena such as field observations of electric earthquake precursors (Varotsos, 2005).

Fracture-induced phenomena exhibit long-range interactions, fractality and memory effects, and in this sense they can be described by scale-invariant laws. Qualitatively, the similarity of the earthquake mechanism and the formation/propagation of cracks inside a material is quite obvious (Vallianatos and Sammonds, 2010, 2011; Vallianatos et al., 2012a,b, 2013a). The similarities between fracture experiments and fundamental laws of statistical seismology (Gutenberg–Richter, Omori, aftershock productivity and the waiting-time scaling law) have been recently investigated by means of AE monitoring during the uniaxial compression of a mesoporous silica ceramics with 40% porosity (Baró et al., 2013).

In recent years, it has been reported that the collective properties of earthquake populations in global, regional and local scales can be described quite well using nonextensive statistical physics (NESP). The same type of statistical physics has been also demonstrated to be suitable for describing similar phenomena that are observed in the laboratory scale, such as generated microcracks and emitted pressure-induced weak electric currents, during the application of mechanical stress to rocks (Vallianatos et al., 2012a,b,c, 2013a,b; Vallianatos and Triantis, 2012). In this context, the pressure-induced emissions of acoustic waves and weak electric currents will be investigated in the following sections.

8.2 AE and PSC Experimental Studies: Laboratory and Field Observations

8.2.1 Fundamentals of Experimental Techniques

The generation and propagation of pressure-induced microcracks inside rock samples can be investigated by means of the well-established nondestructive technique of AEs. In the AE technique, piezoelectric sensors are properly attached to the surface of the rock specimen in order to record the elastic waves that are produced due to crack formation and propagation. The recorded signals are analysed in terms of various AE parameters such as the hit rate or the cumulative number of hits (or events), their amplitude (in dB) and absolute energy, the rising time and rising angle, etc. (see Fig. 8–1). The number of AE events reflects the number of growing microcracks, while their amplitudes are related to the length of crack growth increments. The AE signals can be also analysed and investigated in combination with the emitted PSC, during the accumulation of damage in rock specimens under stress. The correlation of both techniques will give us a deeper insight to the origin of the related physical mechanisms which are responsible for the observed precursory signals. This correlation will be highlighted in detail in the next section.

The AE and PSC experiments are carried out using loading frames of high capacities (>300 kN) under quasistatic loading or displacement control mode, depending on the conducted experiment. Preliminary tests were always realized in order to estimate the strength of the specific batch of specimens.

Concerning the PSC technique, the measuring system consists of ultrasensitive programmable electrometers (Keithley 6517A) combined with multichannel electrical current cards (Keithley 6521 scanner card) with recorded currents ranging from 0.1 fA to 20 mA in several ranges. The data of the electrometer are stored in a computer using a GPIB interface. The

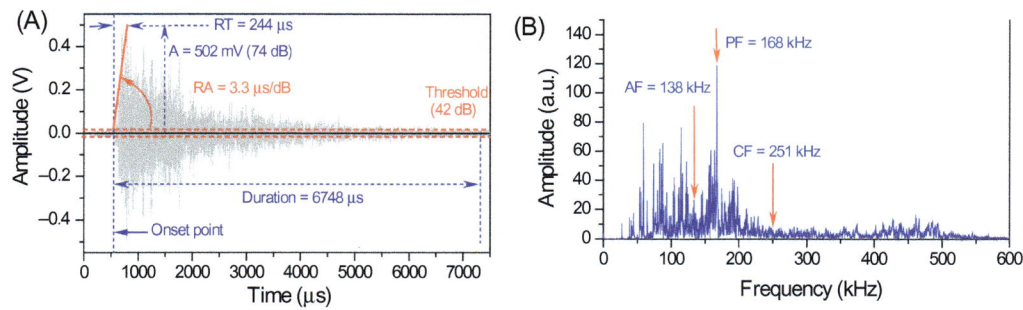

FIGURE 8–1 (A) A typical waveform signal (hit) recorded in an AE sensor during uniaxial loading of a marble specimen from which various AE parameters have been determined. The threshold was set to 42 dB (12.6 mV), the amplitude (*A*) is 502 mV (74 dB), the duration is 6748 μs, the RT is 244 μs and the RA is 3.3 μs/dB. (B) The waveform in the frequency domain (amplitude of the Fast Fourier Transform spectrum) where its main characteristic parameters are indicated (AF, CF and PF). *AE*, acoustic emission; *AF*, average frequency; *CF*, centroid frequency; *PF*, peak frequency; *RA*, rising angle; *RT*, rising time.

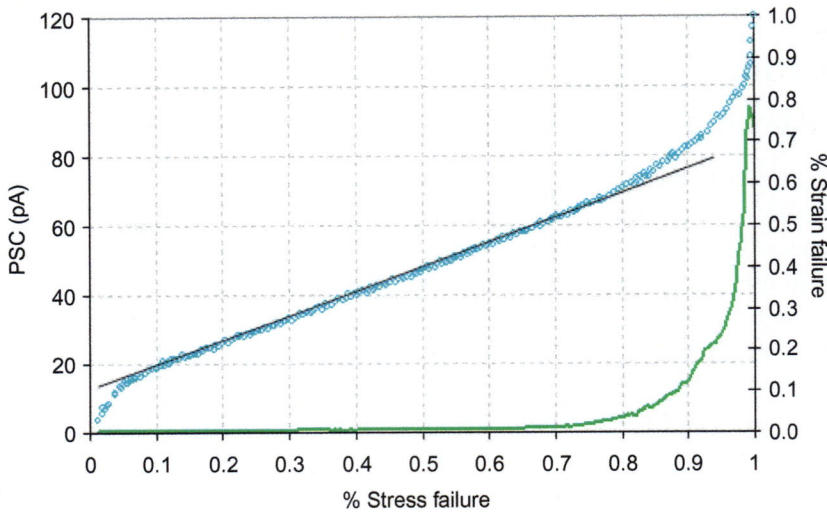

FIGURE 8–2 The evolution of the PSC emission (green line) with respect to the normalized mechanical stress and the corresponding behaviour of the axial strain, during the uniaxial loading of a marble specimen. Considerable PSC is recorded when the specimen exits its elastic behaviour, i.e., ~75% of the stress failure. *PSC*, pressure-stimulated current.

sensing system consists of pairs of electrodes installed along the loading axis, enabling the collection of electric emissions as close as possible to any potential source of electric current (or equivalently to any point where damage occurs). The distribution of the electrode grids on the surface of specimens depends on the loading protocol, the size and the shape of the specimens. An indicative recording of a PSC signal with respect to the normalized stress combined with the axial strain during uniaxial loading of a marble specimen is presented in Fig. 8–2. In Fig. 8–3, the PSC recordings are shown in stepped-like loading experiments of marble specimens up to their ultimate failure, with sufficient time between steps to relax the PSC (see Fig. 8–3). Qualitative and quantitative descriptions of PSCs are given in following sections.

For the detection of AE activity, two-channel AE cards have been used (PCI-2 of Physical Acoustics Corp.). The AE sensors were either the R15a (150 kHz resonant frequency) or the broadband picosensors and they were placed at the appropriate positions in order to enable 3D location of the AE events, as well as the estimation of the characteristics of the AE sources (amplitude, energy and the mode of cracking). The AE sensors are attached to the specimens by means of vacuum grease or silicon glue. Preamplifiers with 40 dB gain were used and the threshold of the AE amplitudes was usually set to 40 dB.

In addition to the electrodes and the AE sensors, electric strain gauges are glued at the mid-height of the specimens, in order to measure their axial and/or lateral strain, during compressive loading. Kyowa strain gauges have been used together with the Microlink 770 resistor bridge. Both the force exerted and the strain developed were digitized and sent to a computer using the KUSB-3100 (manufactured by Keithley) A/D data acquisition module.

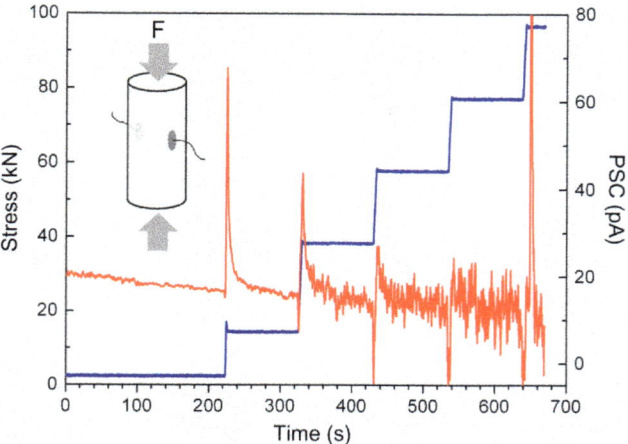

FIGURE 8–3 Typical PSCs recorded in a cylindrical marble specimen during a stepped-like uniaxial loading, up to the fracture. The weak current is measured at a direction perpendicular to the uniaxial applied stress. The drastic increase of PSC at the last step corresponds to the failure of the marble specimen. *PSC*, pressure-stimulated current.

8.2.2 Characteristics of PSCs and AE During Stress-Induced Rock Fracturing

Different electrification mechanisms have been proposed to explain the emission of the weak transient electric currents that are observed when rock specimens under mechanical stress approach failure and are associated with the opening and propagation of microcracks in the material (Freund, 2002; Lavrov, 2005; Bleier et al., 2010). These microfracturing-induced electrification mechanisms include (1) electrokinetic effects due to the water flow in permeable rocks where gravity and the crustal strain are the driving forces, (2) piezoelectricity in quartz-bearing rocks, (3) the activation of positive holes in quartz-free rocks, (4) the motion of charged dislocations (MCDs), (5) flowing gases and (6) the motion of conductive earth materials forced by acoustic waves.

All these kinds of mechanisms should be consistent with the observed electric recordings in the field prior to or concurrently with earthquake events and they could contribute collectively to the overall appearance of spontaneous charge production. However, the efficiency of each possible mechanism and their (constructive or destructive) interaction is still a matter of investigation.

The MCD electrification mechanism, which was first proposed by Slifkin (1993) and developed further by Vallianatos and Tzanis (1998, 1999, 2003) and Tzanis and Vallianatos (2002) is quite attractive as it is associated with brittle fracture in both piezoelectric and nonpiezoelectric rocks. According to this model, the electric current density, which is defined as the rate of change in polarization, is expressed through the following relation:

$$J = \frac{\partial P}{\partial t} = \sqrt{2} \cdot \frac{\Lambda^+ - \Lambda^-}{\Lambda^+ + \Lambda^-} \cdot \frac{q_l}{b} \cdot \frac{d\varepsilon}{dt} \tag{8.1}$$

where Λ^+ is the density of edge dislocations of the type required to accommodate the uniaxial compression, Λ^- is the corresponding density of the opposite type, q_l is the charge per unit length on the dislocation, b is the Burger's vector and $d\varepsilon/dt$ is the strain rate. Eq. (8.1) implies that the recorded transient electric current is related to the nonstationary accumulation of deformation. In the case of elastic deformation of the sample, $\sigma = Y_o \cdot \varepsilon$ where Y_o is the Young's modulus of the material and the stress rate is proportional to the strain rate. Thus, according to Eq. (8.1), the PSC should be proportional to the strain rate, i.e., $J \propto \partial \sigma / \partial t$. The latter implies that when the rock specimen is subjected to stress with a constant rate, we do not expect to observe any transient PSC effect in the elastic region. However, when the stress enters the plastic deformation region where microcracks are formed and propagate, Young's modulus does not remain constant and an effective Young's modulus, Y_{eff} should be considered ($\sigma = Y_{eff} \cdot \varepsilon$). In this case, the strain rate increases and consequently PSC also increases.

The previous considerations of the MCD model have been verified by experiments that were conducted in dried marble specimens subjected to uniaxial loading with different stress modes (constant stress rate, stepped-like stress and cycling load) (Stavrakas et al., 2004; Vallianatos et al., 2004). As is evident in Fig. 8–4, the emitted PSC (curve b) varies in

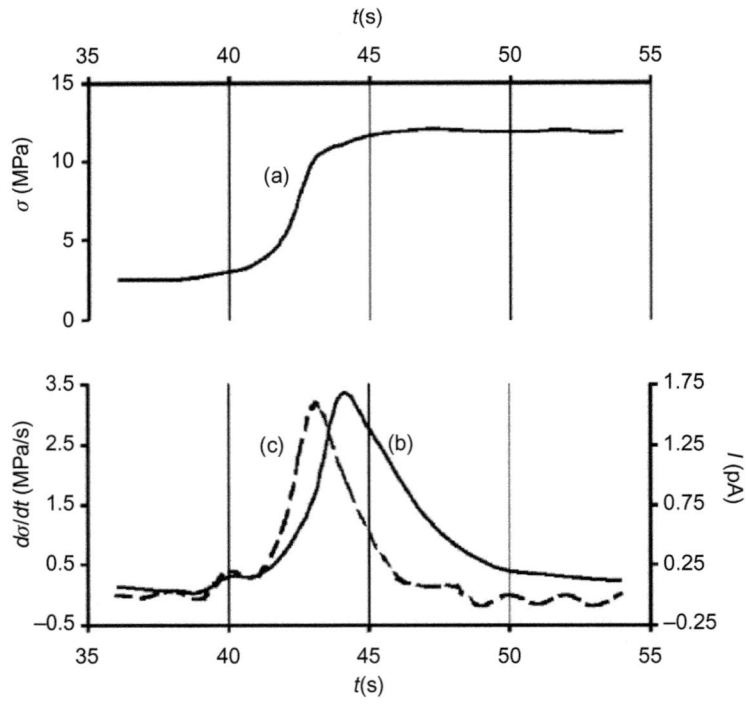

FIGURE 8–4 Time evolution of stress and PSC during the uniaxial loading of Dionysos marbles in the elastic deformation range. Curves (a) and (c) represent the applied stress and the corresponding stress rate, while curve (b) shows the emitted PSC. PSC, pressure-stimulated current. *From Vallianatos, F., Triantis, D., Tzanis, A., Anastasiadis, C., Stavrakas, I., 2004. Electric earthquake precursors: from laboratory results to field observations. Phys. Chem. Earth 29, 339–351.*

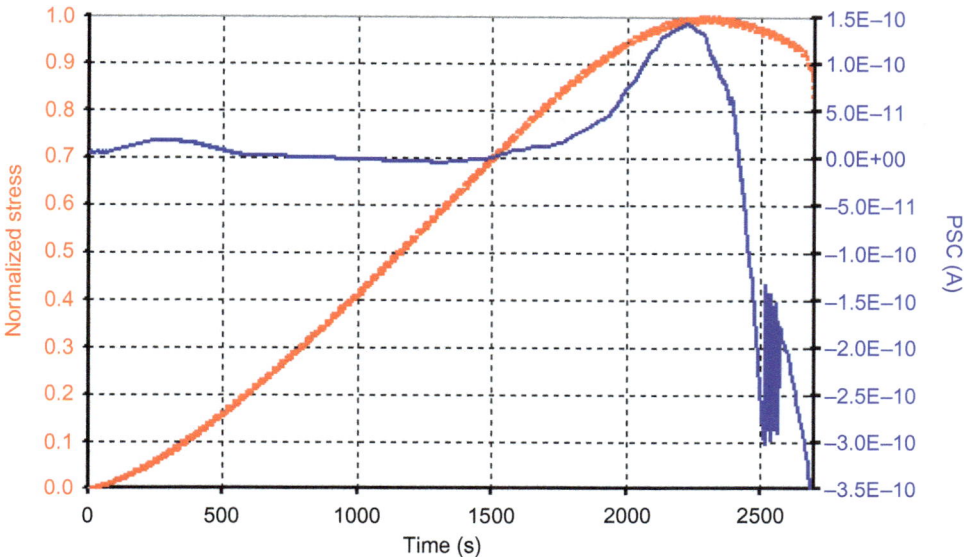

FIGURE 8–5 Time evolution of the recorded PSC signals during the uniaxial compression test of a marble specimen under a constant deformation rate. The normalized stress is also shown (red line). *PSC*, pressure-stimulated current.

accordance with the stress rate (curve c), during a stepped-like uniaxial loading of a Dionysos marble specimen, when the stress is in the elastic region. In this case, the maximum value of the recorded PSC is proportional to the corresponding peak stress rate. For marble specimens that are subjected to a constant stress rate up to failure, PSCs are observed only when stress enters the plastic deformation range (Triantis et al., 2006; Pasiou and Triantis, 2017).

A characteristic variation of the PSC signal recorded from a marble specimen that was subjected to uniaxial compressive stress under constant deformation rate $(5 \cdot 10^{-4}\,\text{mm/s})$ is depicted in Fig. 8–5. Four distinct regions may be distinguished when observing this diagram. First, a minor PSC excitation at low stress values is observed $(t < 500\ \text{s})$ which may be related to the pore-closing processes. The Young's modulus increases continuously and remains constant when the material enters the linear stress-deformation region. Subsequently, the PSC obtains very low values with low variation $(500 < t < 1500\ \text{s})$. In this case, the material remains in the elastic region regarding its mechanical behaviour, while the Young's modulus remains constant. Next, a significant increase in the PSC is observed when the material enters the plastic deformation region $(1500 < t < 2200\ \text{s})$. The PSC reaches its maximum in the vicinity of the maximum stress value. Finally, the PSC obtains reverse polarity after the maximum stress value is reached and shows intense variations. During this range, sequential and continuous macrocracks develop.

Another property of the PSCs regarding the total electric charge, $Q(t) = \int I_{\text{PSC}}(t)dt$, released during experimental tests where loading increases up to failure, seems to be independent of the applied stress rate. Experiments that were carried out on marble specimens confirmed this fact, while this observation was also verified theoretically (Triantis et al., 2006). The temporal development of the electric charge release in marble samples has been examined and it was found that a linear relationship between the electric charge and the deformation exists. This linear relation is valid when the applied stress becomes higher than the value of the yield stress and remains valid until the specimen approaches failure (Triantis et al., 2008). The experimental verification of this linear relationship is in accordance with the theoretical predictions of the MCD model. It must be noted that according to the MCD model an analogy exists between the electric current density and the strain rate.

The previous experimental observations of PSC in the laboratory scale that are interpreted by the MCD model could in principle be applied to the mega-scale of earthquake events which produce electric precursory signals. If so, then the recorded electrical precursors of numerous earthquake events could be attributed to the collective action of microcrack formation and propagation, which accumulates damage on a larger scale (Vallianatos and Tzanis, 1998, 1999, 2003; Tzanis and Vallianatos, 2002).

It is worthwhile mentioning that, when a rock specimen has already suffered damage in previous stress cycles, a subsequent loading cycle will produce PSC with amplitudes smaller than that produced in the previous load (see Fig. 5 in Vallianatos et al., 2004). This constitutes another characteristic property, similar to that which has been examined and deals with the AE, and is known as the Kaiser effect (memory effect).

A recent publication (Stavrakas, 2017) describes the correlation of the PSC signals and the AE emissions during sequential loading cycles. According to the Kaiser effect, an absence of AE activity is detected while the applied mechanical stress is lower than any previously applied stress on the specimen under test. Such an observation clearly shows that the discontinuities in the specimen's bulk do not grow further when the applied mechanical stress is not higher than any previously applied stress. Contrary to the Kaiser effect, a quantitative estimation of the specimen mechanical quality is known as the Felicity effect. A specimen that shows the Felicity effect obtains AE activity even at lower stress values than any previously applied stress value (see Fig. 8−6). This may be qualitatively estimated using the Felicity ratio (FR), which is described as the ratio of the applied mechanical load when the AE emissions initiate over the maximum value of the previously applied mechanical load. Thus, the Kaiser effect obtains a FR equal to 1. The lower the FR becomes, the more critical the mechanical status of a specimen under test may be considered. Regarding the PSC signals, a similar ratio may be introduced: R_{PSC}. Specifically, the R_{PSC} may be defined as the stress value where the PSC starts to grow beyond a background value (PSC_b), over the previously applied maximum stress value. Fig. 8−7 demonstrates the recorded PSCs during a compressive test on a Dionysos marble specimen consisting of three consecutive loading−unloading cycles. We clearly observe that the PSC decreases significantly in each repetitive cycle, while strong PSC emissions are detected when the applied load exceeds 60 MPa and the specimen approaches failure. During this specific experiment the R_{PSC} ratio for the second loading equals 0.78, while during the third loading it reaches the value of 0.94.

It is noteworthy that the PSC technique has been adopted by several researchers while others have used similar techniques (Cartwright-Taylor et al., 2014; Li et al., 2015;

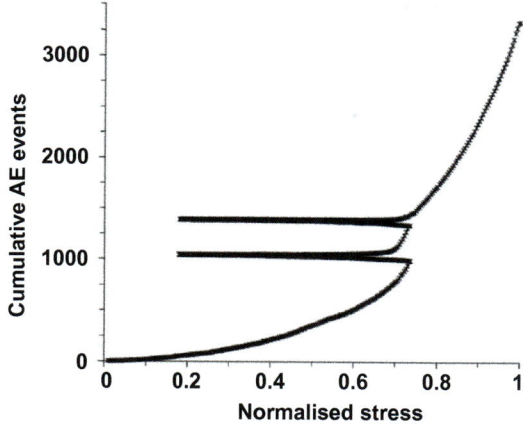

FIGURE 8–6 The cumulative number of AE events as a function of the normalized mechanical stress applied on a Dionysos marble specimen, during three sequential loading cycles. The felicity effect is clearly observed in each loading cycle, as an AE activity recorded at lower stress values than the maximum stress of the previous loading cycle. *AE*, acoustic emission. *Modified from Stavrakas, I., 2017. Acoustic emissions and pressure stimulated currents experimental techniques used to verify Kaiser effect during compression tests of Dionysos marble. Fract. Struct. Integrity 40, 32−40.*

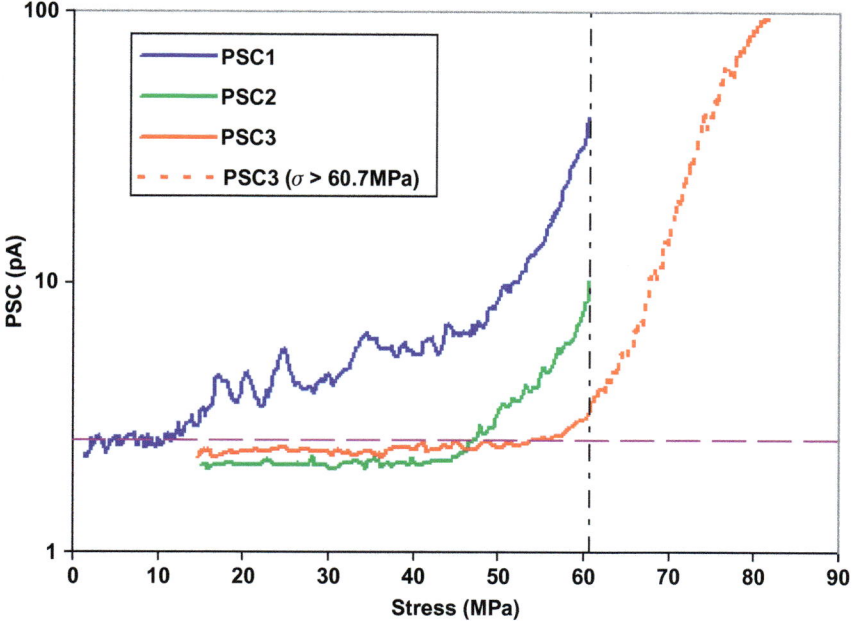

FIGURE 8–7 The recorded PSCs during a compressive test with three loading−unloading cycles of a marble specimen, as a function of the applied stress. Note the logarithmic scale of the PSC axis. *PSC*, pressure-stimulated current. *Modified from Stavrakas, I., 2017. Acoustic emissions and pressure stimulated currents experimental techniques used to verify Kaiser effect during compression tests of Dionysos marble. Fract. Struct. Integrity 40, 32−40.*

Archer et al., 2016; Fursa et al., 2016, 2017). For example, the PSC technique was applied to coal and rock in rock burst coal mine samples extracted from the Changgouyu coal mine in Beijing (Li et al., 2015). According to the findings from the coal samples, when the applied loading lays in the elastic deformation stage, a weak variation of the electrical current is observed. When the coal samples show plastic deformation, the electric current increases rapidly, along with a loading increase. Once the peak stress is reached, the samples enter into an unstable failure stage and the electric current shows a significant reverse. This observation constitutes a signature characteristic of the PSC signal and has been confirmed and reported several times.

8.2.3 A Qualitative Correlation of AE and PSC Features

The correlation of the PSC with the AE constitutes a field of research that has attracted considerable scientific interest (see, for example, Stavrakas et al., 2015 and references therein). Specifically, it is aimed at studying the similarities between the AE and PSC recordings as AE is a direct method that provides information regarding the mechanical status of a specimen when it is subjected to mechanical loading. It must be noted that the AE measurements were adopted as a nondestructive testing method a long time ago, in order to examine the mechanical characteristics (i.e., stress vector orientation, elastic wave speed, shear—tensile or mixed mode fracture) of the microfracture processes that occur in the bulk of a specimen.

Simultaneous measurements of AE activity and PSC emissions have been carried out in porous sandstone samples subjected to uniaxial loading, in order to investigate the role of water content and the pore fluid (Saltas et al., 2015). The PSCs were recorded in different directions, perpendicular to the applied stress and a set of AE picosensors was used for the simultaneous recordings of the AE events. The experimental setup is described in Fig. 8—8. During the uniaxial loading of the dried sandstone specimen, weak electric currents (less than 2 nA) are detected at the final stage of the failure (after 385 s), as is evident in Fig. 8—9A. However, considerable AE activity initiates much earlier (at 340 s), at around 75% of the fracture load, as indicated by the increased hit (or energy) rate in Fig. 8—9B and C. This corresponds roughly to the transition from the elastic range to the brittle deformation through the generation of microcracks. The observed delay in PSC emission with respect to the preceding AE activity could be attributed to the existence of a critical concentration of microcracks, which enables the electrical transport of the redistributed charges inside the sandstone specimen. Different intensities of PSCs with different polarities are recorded in each channel, indicating that the weak currents due to crack formation and propagation are emitted unevenly in the bulk, with a preferable direction that is associated with the formation of oriented shear planes inside the bulk prior to failure.

In the case of the water-saturated sandstone (Fig. 8—10), the intensity of the PSCs is almost three orders of magnitude greater than that observed in the specimen with low water content, i.e., the dried one which remained in air for 3 days (Fig. 8—9). In the latter case, the absence of free or loosely bound water should inhibit the generation and propagation of

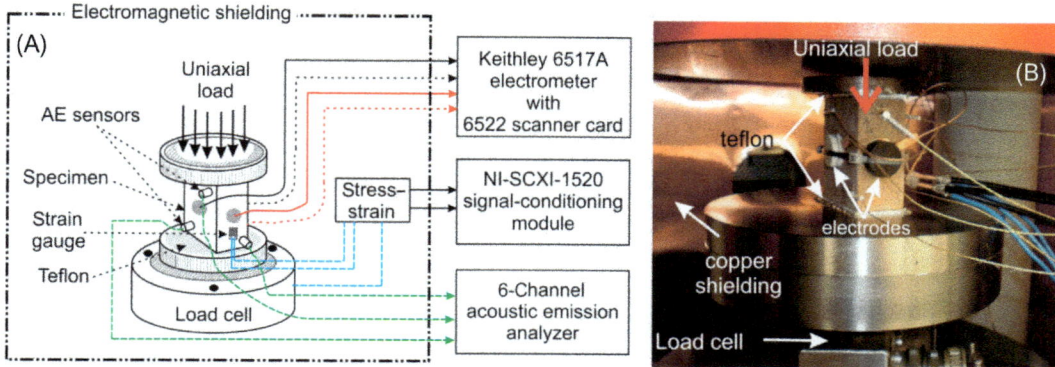

FIGURE 8–8 (A) Schematic view of the experimental setup for the combined recordings of AE and PSCs during the uniaxial loading of a porous prismatic sandstone specimen (7 cm × 7 cm × 3.5 cm, approximately). A sensitive electrometer (Keithley 6517A) equipped with a multichannel scanner card (Keithley 6522) is used to record the pressure-stimulated electrical currents in the pA region. Silver paste ensured good electric contact between the current electrodes and the surface of the specimen. Thin Teflon plates are used for the electrical isolation of the rock specimen. A strain gauge mounted on the lateral surface of the specimen and a load cell placed at the bottom side of the bearing plate are used for continuous monitoring of stress–strain. Miniature piezoelectric sensors (PICO sensors, 200 kHz–1 MHz, MISTRAS Group) have been placed on the side surfaces of the specimen in order to measure the acoustic activity, through a PCI-2 card-based AE system by Physical Acoustics Corporation. The threshold of AE detection was set to 40 dB to avoid background noise. (B) Photograph of the specimen inside the loading machine (ALPHA S-3000 by Form + Test, GmbH), before the compression test. The load frame is covered with copper sheets for electromagnetic shielding. (For more information on the experimental setup, refer to Saltas et al., 2015.). *AE*, acoustic emission; *PSC*, pressure-stimulated current.

weak electric currents. However, in water-saturated sandstone the presence of free water in the pore volume should affect considerably the conduction paths, resulting in higher values of the emitted currents. Finally, in the case of the brine-saturated sandstone, the excess of ionic charge results in even higher emitted currents (around one order of magnitude) than that recorded in water-saturated sandstone. Irrespective of the water content of the samples, the AE activity exhibits similar features in all cases.

In an attempt to investigate the pressure-induced variation of the electrical properties of rocks during mechanical stress, Saltas et al. (2014) conducted combined complex electrical impedance and AE measurements in limestone samples subjected to linear and stepped-like uniaxial loading. Notably, they found that the DC conductivity of both, grain interiors and grain boundaries, increases with increasing uniaxial loading, indicating negative values of activation volumes at elevated uniaxial stresses, ranging from −10 to −61 cm^3/mol. The latter finding, in combination with other reported negative activation volumes in hydrated rocks such as, leukolite, granodiorite and amphibolites (Shankland et al., 1997; Papathanassiou et al., 2010, 2011, 2012), are crucial for the proper documentation of the physical mechanism of seismic electric signals generated prior to and/or during seismic events (Varotsos and Alexopoulos, 1984, 1987; Varotsos, 2005). Furthermore, they reported that AC conductivity during linear loading of limestone is strongly correlated with AE activity,

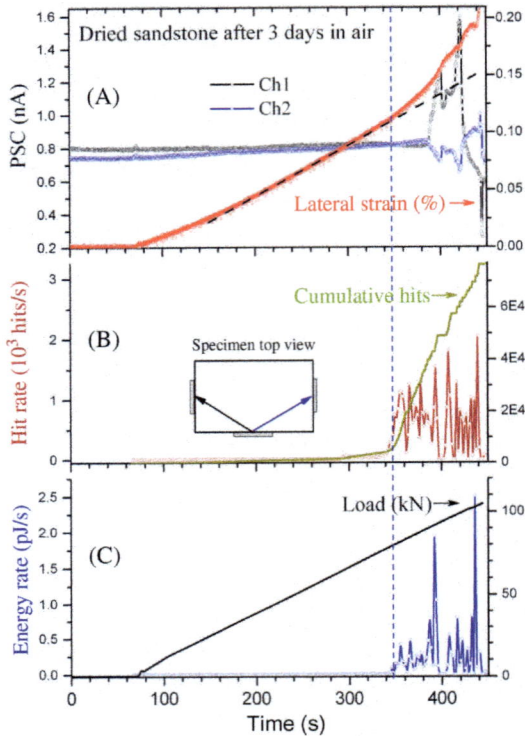

FIGURE 8–9 PSC recordings (A) and AE activity (B, C) from a porous sandstone specimen subjected to linear loading up to the ultimate failure. The specimen was dried at 105°C for a sufficiently long time and after that it remained in air for 3 days to achieve equilibrium conditions in regards to the absorbed water vapour from the air. The linear loading rate is 0.3 kN/s. The emitted weak electric currents were measured at two different directions, perpendicular to the applied loading, as can be seen in (B). The dotted line indicates the initiation of considerable AE activity. *AE*, acoustic emission; *PSC*, pressure-stimulated current. *Modified from Saltas, V., Fitilis, I., Makris, J.P., Vallianatos, F., 2015. Acoustic and electrical emissions from sandstone under uniaxial compression. In: International Conference 'Science in Technology' SCinTE 2015.*

obeying the general self-similar law of critical phenomena for energy release before material fracture. This strong correlation of AE activity and electrical AC conductivity has been also observed in other types of uniaxial loading, such as stepped and sawtooth, as shown in Fig. 8–11.

In each case, the cumulative number of hits from the recording channels exhibits the same characteristics as the AC conductivity. However, in the longer term experiment of stepped-like loading (Fig. 8–11A), a gradual decrease in conductivity is clearly observed during the time periods where the load remains constant. This relaxation phenomenon of AC conductivity is more pronounced when AE activity is low and is not observed in the short-term experiment of sawtooth uniaxial loading (Fig. 8–11B).

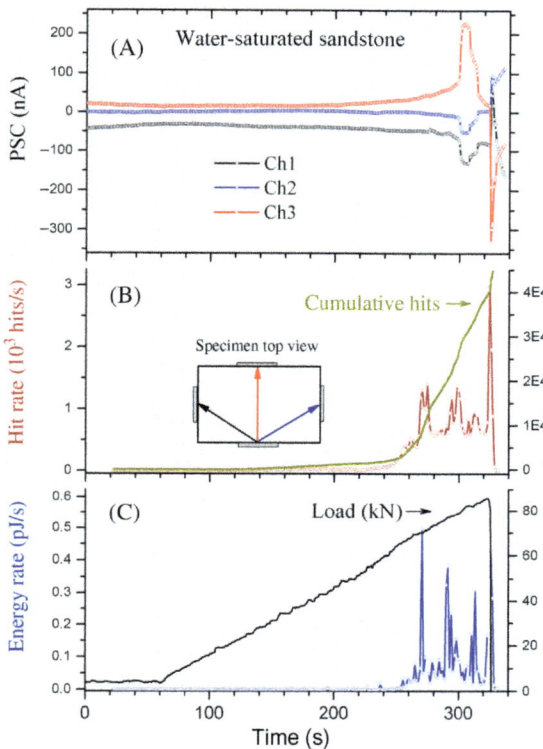

FIGURE 8–10 PSCs (A) and AE recordings (B, C) during uniaxial loading of water-saturated sandstone up to the fracture. The specimen was saturated with water in a chamber under vacuum. PSCs were measured in three different directions perpendicular to the applied load [see the inset of (B)]. *AE*, acoustic emission; *PSC*, pressure-stimulated current. *Modified from Saltas, V., Fitilis, I., Makris, J.P., Vallianatos, F., 2015. Acoustic and electrical emissions from sandstone under uniaxial compression. In: International Conference 'Science in Technology' SCinTE 2015.*

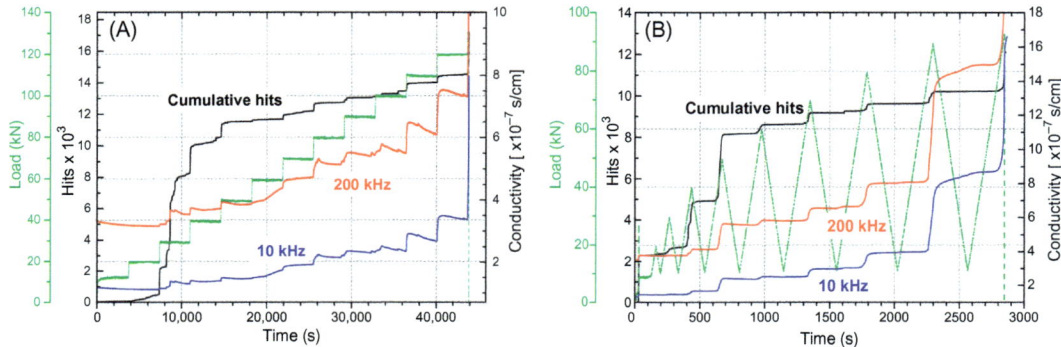

FIGURE 8–11 Time-series of AC conductivity at selected frequencies (10 and 200 kHz) and the corresponding AE activity (expressed as cumulative hits) during (A) stepped-like and (B) sawtooth uniaxial compression of limestone samples. *AE*, acoustic emission.

8.2.4 A Quantitative Correlation of AE and PSCs

Different approaches may be used in order to correlate the experimental data of the PSC and AE techniques. As a consequence of the similarities between the occurrence of earthquake events and the AEs recorded during fracture processes in solids, the b-value that has been introduced from the field of seismology may also be used in the analysis of AE events, according to the following modified Gutenberg–Richter law (Colombo et al., 2003):

$$\log_{10} N = a - b\left(\frac{A_{dB}}{20}\right) \tag{8.2}$$

where N is the number of AE events (or hits) with amplitude higher than A_{dB}, a is a constant and b is the AE base b-value calculated from the gradient of the plotted curve of Eq. (8.2).

Alternatively, the so-called improved b-value, I_b introduced by Shiotani et al. (1994) may be used instead of Eq. (8.2) to describe the stages of fracture in a solid from the amplitude distribution in AE analysis. The improved b-value is defined by utilizing statistical values of the amplitude distribution, i.e., the mean value μ and the standard deviation σ, according to the following relation (Shiotani et al., 1994):

$$I_b = \frac{\log N(\mu - \alpha_1 \sigma) - \log N(\mu + \alpha_2 \sigma)}{(\alpha_1 + \alpha_2)\sigma} \tag{8.3}$$

where $N(\mu - \alpha_1 \sigma)$ and $N(\mu + \alpha_2 \sigma)$ represent the cumulative number of events with amplitudes greater than $\mu - \alpha_1 \sigma$ and $\mu + \alpha_2 \sigma$, respectively. The empirical constants a_1 and a_2 are user-defined and usually set equal to unity (Shiotani et al., 1994). The characteristic trend of the variation of I_b may serve as a qualitative index of damage in the material under test. Values of I_b lower that 1 are indicative of the generation and propagation of microcracks and thus are related to the upcoming failure of the material.

The correlation of the I_b values and the PSCs during the uniaxial loading of marble specimens has been reported by Pasiou and Triantis (2017) and is illustrated in Fig. 8–12A. The initial decrease in the I_b-value to 30% of the maximum stress has been attributed to the closure of the preexisting microcracks. The improved b-value remains almost constant up to the deviation of stress–strain from linearity ($\sim$80% of the maximum stress), due to the low generation of microcracks, but then it decreases significantly to values close to 1.0 (point A_2' in Fig. 8–12A) due to the formation of a large number of cracks.

In their work, Pasiou and Triantis also related the energies of the AE events and the emitted PSCs during the compression of various brittle materials. In the case of the PSCs, the released energy is calculated from the relation:

$$E_{PSC} = \int\limits_{t_i}^{t_i + \Delta t} (I_{PSC}(t))^2 dt \tag{8.4}$$

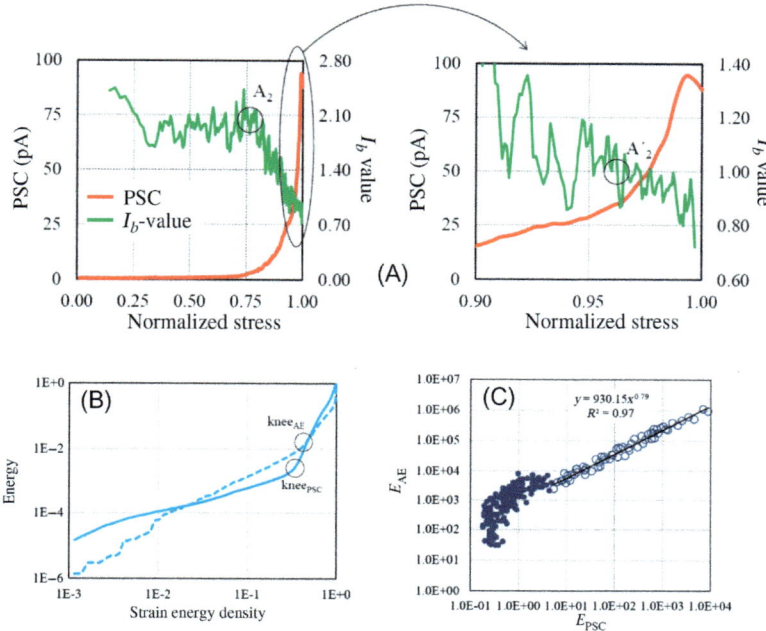

FIGURE 8–12 (A) PSCs and calculated I_b values as a function of normalized stress during uniaxial loading of a marble specimen. (B) The energies of PSCs and AE versus the strain energy density. All the quantities are normalized with respect to their maximum values. (C) The correlation of the AE energy and the PSC energy. Solid circles correspond to weak electric currents with negligible variation, while the empty symbols represent the upward trend of the PSC, until it reaches its maximum value. *AE*, acoustic emission; *PSC*, pressure-stimulated current. *From Pasiou, E.D., Triantis, D., 2017. Correlation between the electric and acoustic signals emitted during the compression of brittle materials, Frattura ed Integrità Strutturale 40, 41–51.*

where the integral is calculated over time intervals Δt ($\Delta t = 1$ s) and $t_i = 0, 1, \ldots, n-1$ with n the duration of the experiment. The absolute energy of an AE hit (measured in aJ) is derived from the integral of the squared voltage signal divided by the $10\,\text{k}\Omega$ reference resistance, over the duration of the AE waveform packet.

The calculated energies of PSCs and AE events with respect to the strain energy density (SED) for a marble specimen are shown in Fig. 8–12B. The SED was calculated from the area below the recorded stress–strain curves during the tests. Both curves comprise two almost straight segments of different slopes, while a knee point is observed in each case. Notably, the knee point of the PSC energy is observed prior to the corresponding knee of the AE, indicating that the PSC technique is more sensitive to the detection of the internal damage of the marble specimen. The correlation between the AE and PSC energies in logarithmic scale is depicted in Fig. 8–12C. Ignoring the range where the energy of PSC remains roughly constant (solid circles), a power law applies to the region where the emitted energy increases considerably (open circles), i.e., $E_{\text{AE}} \propto (E_{\text{PSC}})^m$ with $m \sim 0.8$.

The above correlation of AE and PSC energies is in accordance with a previous study by Vallianatos and Triantis (2008), who conducted uniaxial stress experiments in marble and cement mortar specimens and performed a statistical analysis of the recorded PSC interevent times, χ (i.e., time between successive PSC events). They showed that the associated cumulative probability density function (PDF), $P(\chi)$, is independent of the type of fracture experiments (constant stress or constant stress rate) and the kind of material (marble of cement mortar). They suggested that PDF is determined by its mean waiting time and can be approximated by a universal scaling function which has the form of a gamma distribution, i.e., $P(\chi) \propto \chi^{-(1-\gamma)} \exp(-\chi/B)$, with $\gamma \approx 0.8$ and $B \approx 0.7$. They concluded that this function is very similar to that observed for AE and earthquake data, which implies a possible universality for rock fracture and a self-similarity over at least 3–4 orders of magnitude for the experiments that have been conducted. Based on irreversible thermodynamics, they also described the cumulative energy of PSC events with a time-to-failure power law as the specimen approaches failure, similar to that observed in AE data and seismicity catalogues (Kawada and Nagahama, 2006; Kawada et al., 2007).

In the framework of the application of PSC and electromagnetic emission technologies to field measurements, the technique of pressure-stimulated voltages (PSVs) has recently been introduced (Archer et al., 2016). This technique is based on the use of an electric potential sensor (EPS) developed and patented by the University of Sussex (Prance et al., 2000). It was verified experimentally that the detection of PSVs in various rock lithologies originates from the rocks and not from the equipment that was used or the local environment (Aydin et al., 2009). In the same work, a uniaxial loading experiment was carried out to investigate the correlation between AE events and PSV activity in quartz-rich syenogranite and nonquartzose halite-stone. The experimental results showed that a very strong positive correlation exists between AE and PSV. Changes in PSV were recorded either concurrently or after a minor delay to the corresponding AE recordings. The authors supported that the changes in PSV that precede AE (and therefore cracking) may be due to stress accumulation and be occurring at the atomic scale. This would be consistent with the aforementioned moving charge dislocation mechanism.

8.3 Complexity of Rock Fracture and Similarity With Seismicity

8.3.1 Fundamentals of Tsallis Entropy: A Statistical Mechanics Approach to Laboratory Seismology

According to the NESP proposed by Tsallis (1988), the generalized entropy, S_q (Tsallis entropy) of a system in terms of the probability distribution p for the discrete case is defined as:

$$S_q = k_B \frac{1 - \sum_{i=1}^{W} p_i^q}{q-1}, \quad q \in R \text{ and } \sum_{i=1}^{W} p_i = 1 \tag{8.5}$$

where k_B is Boltzmann's constant and the sum refers to the set of probabilities p_i of the possible configurations of the system, with a total number W. The entropic index, q, represents a measure of the nonextensivity of the system. For $q \to 1$, the Tsallis entropy reduces to Boltzmann–Gibbs (BG) entropy of classical statistical physics.

For a system consisting of two statistically independent subsystems A and B, Tsallis entropy, S_q, violates the additivity property obeyed by BG entropy, according to the following expression:

$$S_q(A, B) = S_q(A) + S_q(B) + \frac{1 - q}{k_B} S_q(A) S_q(B) \tag{8.6}$$

The last term on the right-hand side of Eq. (8.6) reflects the nonadditivity due to long-range interactions in the physical system under consideration. Different values of the index q, i.e., $q > 1$, $q = 1$ and $q < 1$ correspond to subadditivity, additivity and superadditivity, respectively.

In the case of a continuous variable X with probability distribution $p(X)$, Eq. (8.5) is modified to:

$$S_q = k_B \frac{1 - \int_0^\infty p^q(X) dX}{q - 1} \tag{8.7}$$

The X variable could be a fundamental seismic parameter such as seismic moment, interevent time or interevent distance between two successive events, such as earthquakes or microcracks formed inside a rock under stress.

The distribution $p(X)$ that optimizes the Tsallis entropy, S_q, is subjected to the natural constraint, i.e., the normalization condition,

$$\int_0^\infty p(X) dX = 1 \tag{8.8}$$

and the condition concerning the generalized expectation value X_q (q-expectation value) which is defined as:

$$X_q = \langle X_q \rangle = \int_0^\infty X P_q(X) dX = 1 \tag{8.9}$$

where $P_q(X)$ is the escort probability that is given from the following expression (Tsallis, 2009):

$$P_q(X) = \frac{p^q(X)}{\int_0^\infty p^q(X) dX} \tag{8.10}$$

By means of Lagrange multipliers, we finally obtain the physical probability:

$$p(X) = \frac{\left[1-(1-q)\beta_q X\right]^{1/(1-q)}}{Z_q} \tag{8.11}$$

In the above equation, the nominator represents the q-exponential function, $\exp_q(-\beta_q X)$, which is defined as:

$$\exp_q(X) = \begin{cases} \left[1+(1-q)X\right]^{1/(1-q)} & \text{for} \quad 1+(1-q)X \geq 0 \\ 0 & \text{for} \quad 1+(1-q)X < 0 \end{cases} \tag{8.12}$$

The denominator in Eq. (8.11) is the q-partition function

$$Z_q = \int_0^{X_{max}} \exp_q\left(-\beta_q X\right) dX \tag{8.13}$$

where the term β_q is related to the Lagrange multiplier β^* as follows:

$$\beta_q = \frac{\beta^*}{c_q + (1-q)\beta X_q} \quad \text{and} \quad c_q = \int_0^{X_{max}} p^q(X) dX$$

The q-exponential function [Eq. (8.12)] for different values of q has been plotted in Fig. 8–13, in lin–lin and log–log scales. For $q = 1$, the distribution reduces to the standard exponential distribution. For $q > 1$, it has an asymptotic slope, $-1/(q-1)$ while, for $q < 1$, it has a vertical asymptote at $x = 1/(1-q)$.

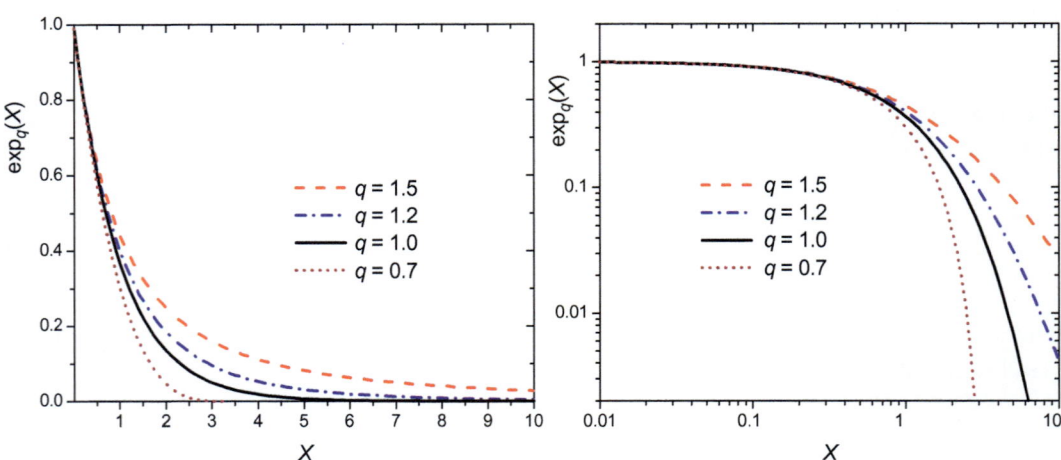

FIGURE 8–13 The q-exponential function for various values of q, in linear and logarithmic scales.

The previous standard representation of NESP can be further generalized by considering a distribution of q-indices instead of a single q-value, as has been proposed by Tsallis (1988) and Tsekouras and Tsallis (2005). In this case, a generalized differential equation is introduced to describe the crossover to another type of behaviour that is observed at larger values of the variable X, as follows:

$$\frac{dp}{dX} = -\beta_r p^r - (\beta_q - \beta_r)p^q \tag{8.14}$$

In trivial cases of $r = 0$ or $r = q$, Eq. (8.14) reduces to the nonlinear differential equation, $dp/dX = -\beta_q p_i^q$, whose solution is the generalized probability $p(X)$ given by Eq. (8.11). For $1 \le r < q$, the general solution of Eq. (8.14) is given as follows:

$$X = \int_p^1 \frac{dx}{\beta_r x^r + (\beta_q - \beta_r)x^q} = \frac{1}{\beta_r} \int_p^1 dx \left[\frac{1}{x^r} - \frac{((\beta_q/\beta_r) - 1)x^{q-2r}}{1 + ((\beta_q/\beta_r) - 1)x^{q-r}} \right]$$

which finally leads to:

$$X = \frac{1}{\beta_r} \left\{ \frac{p^{1-r} - 1}{r - 1} - \frac{(\beta_q/\beta_r) - 1}{1 + q - 2r} \left[H\left(1; q - 2r; q - r; \frac{\beta_q}{\beta_r} - 1\right) - H\left(p; q - 2r; q - r; \frac{\beta_q}{\beta_r} - 1\right) \right] \right\} \tag{8.15}$$

where

$$H(\xi; a; b; c) = \xi^{1+a} F\left(\frac{1+a}{b}, 1; \frac{1+a+b}{c}; -\xi^b c\right)$$

and F is the hypergeometric function.

When applying NESP in various systems, the question that arises is which distribution we shall compare with the distribution of the system under consideration. If the escort distribution is used, instead of the physical probability $p(X)$ given in Eq. (8.11), then we obtain the following cumulative distribution function (CDF):

$$P_{\text{cum}}(>X) = \int_{X_{\min}}^{\infty} P_q^{esc}(X)dX = \exp_q(-X/X_0) \tag{8.16}$$

Alternatively, by integrating the physical probability $p(X)$, the following cumulative probability emerges after proper transformations (Michas et al., 2013):

$$P(>X) = \left[1 - (1-q)\frac{X}{X_0} \right]^{(2-q)/(1-q)} \tag{8.17}$$

A generalized form of the well-known Gaussian distribution may be derived by optimizing Tsallis entropy S_q for the squared variable X^2. We thus obtain the following, q-Gaussian distribution:

$$p(X) = p_0 \left[1 - (1 - q) \left(\frac{X}{X_0} \right)^2 \right]^{1/(1-q)} \tag{8.18}$$

For $q \to 1$, the normal Gaussian distribution is obtained while for $q > 1$, the tails of q-Gaussian follow a power law, enhancing the probability of the high values of X (Fig. 8–14).

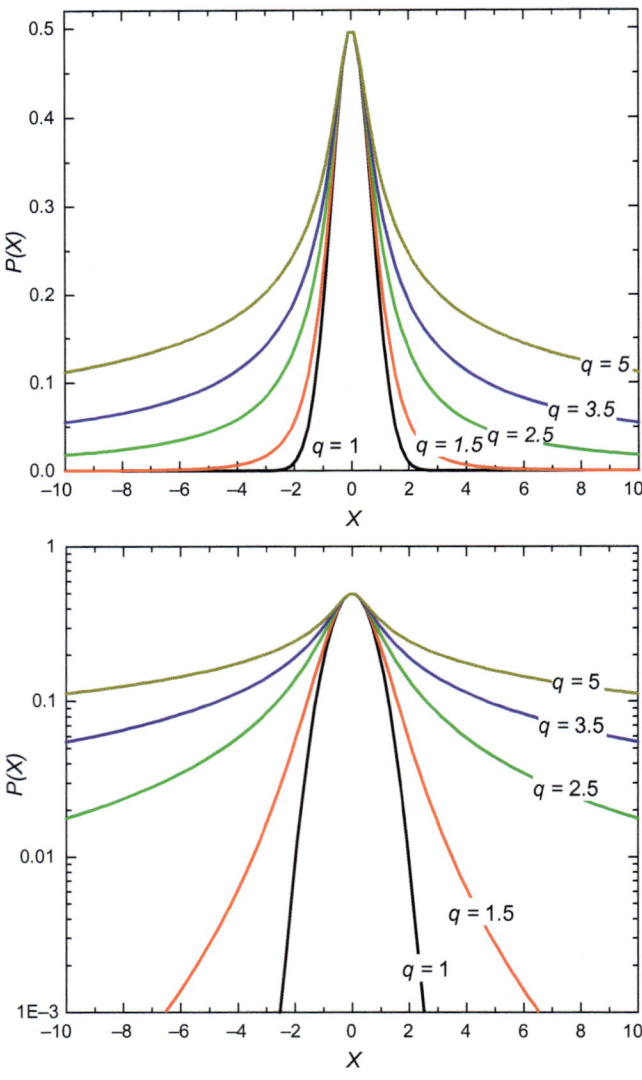

FIGURE 8–14 The q-Gaussian distribution for various values of q, according to Eq. (8.16), in linear and log-linear scales. The values of p_0 and X_0 are set at 0.5 and 1, respectively. The distribution for $q = 1$ is the well-known Gaussian distribution.

8.3.2 Applications to Laboratory Experiments

Recently, several studies have been carried out in order to investigate the applicability of NESP in laboratory-scale experiments of AE and PSCs emitted from rocks under stress (Vallianatos et al., 2011, 2012a,b,c, 2013a,b; Vallianatos and Triantis, 2012, 2013; Cartwright-Taylor et al., 2014). Notably, the concept of NESP has been also extended to large-scale systems such as plate tectonics and global seismicity (Vallianatos and Sammonds, 2010, 2011, 2013).

The relaxation of the recorded PSCs when fine-grained amphibolite rocks from KTB drilling were subjected to high-rate stress steps has been studied from a phenomenological point of view in the frame of NESM by Vallianatos and Triantis (2012). In their experiments, each rock specimen was subjected to a step-like uniaxial stress with a high stress rate (~ 3 MPa/s), reaching a different high constant stress value, σ_H. In each case, the PSCs were recorded for a sufficiently long time, in order to observe the relaxation signal after the maximum current that was emitted, when the stress reached its high value at the specific step. They observed that the quantity $\xi(t)$, defined as the PSC relaxation signal, $I(t)$, divided by the maximum current, I_o ($= I(t = 0)$), recorded in each step ($\xi(t) = I(t)/I_o$), decreases monotonically with time, following an exponential law at low values of σ_H, i.e., $\xi(t) = \exp(-\beta_1 t)$. However, as σ_H increased beyond the yielding stress, a deviation to a power law was observed at intermediate time range, especially for σ_H close to the fracture stress. This deviation, which is related to a fractal picture of the recorded relaxation currents, was described with the generalized q-exponential function, $\xi(t) \propto \exp_q(-\beta_q t)$, defined in Eq. (8.12). The fitting curves of the experimental data according to the q-exponential function are illustrated in Fig. 8–15 in green for the cases where $\sigma_H > \sigma_Y$. However, by considering the long time departures of the experimental data from the power law behaviour at high applied stress, Vallianatos and Triantis suggested the following differential equation which describes a crossover to another type of behaviour at very long times:

$$\frac{d\xi(t)}{dt} = -\beta_1 \xi - (\beta_q - \beta_1)\xi^q \tag{8.19}$$

This equation is a specific case of the generalized Eq. (8.14), for $r = 1$ with the following solution:

$$\xi(t) = \left[1 - \frac{\beta_q}{\beta_1} + \frac{\beta_q}{\beta_1}\exp\left[(q-1)\beta_1 t\right]\right]^{-(1/(q-1))} \tag{8.20}$$

As can be observed in Fig. 8–15, the fittings of the pressure-stimulated relaxation currents using Eq. (8.20) exhibit very good correlation over the entire time range.

The same analysis in the framework of NESP has been also applied to a modified set of the previous experiments by the same authors (Vallianatos and Triantis, 2013) who investigated the emitted PSCs in marble and amphibolite samples under sequential loading–unloading cycles. They observed, in accordance with their previous results, that the normalized

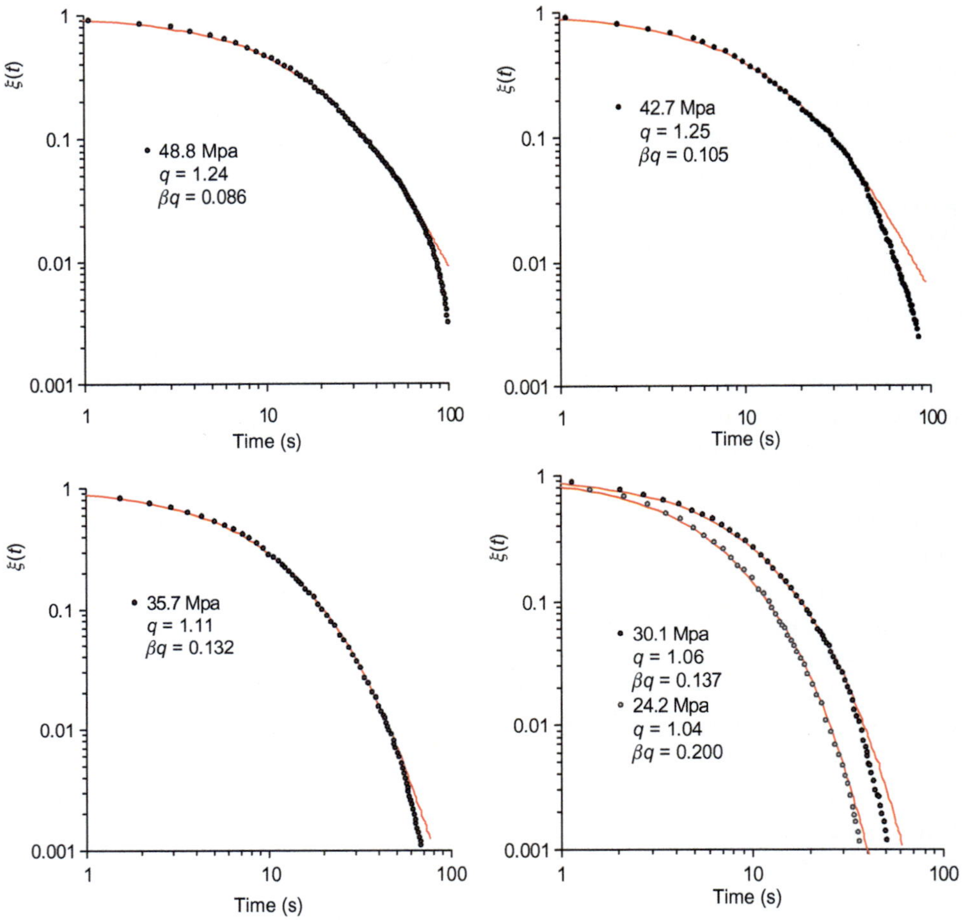

FIGURE 8–15 The experimental data and the corresponding fitting curves of the normalized pressure-stimulated relaxation currents, $\xi = I(t)/I(t = 0)$ vs time at various uniaxial stresses. *Modified from Vallianatos, F., Triantis, D., 2012. Is pressure stimulated current relaxation in amphibolite a case of non-extensivity? EPL 99, 18006.*

PSC relaxation signals, $\xi(t)$, in both marble and amphibolite specimens, deviate from the exponential behaviour, exhibiting power law characteristics of the generalized q-exponential function, $\exp_q(-\beta_q t)$, in the large time range (see Fig. 2 from Vallianatos and Triantis, 2013). In each loading—unloading cycle of the five successive cycles, the Tsallis q-parameter, which is related to the PSC relaxation, decreases, gradually approaching the value $q = 1$, as the number of sequential cycles gets higher (Fig. 8–16). The latter implies that the distribution which describes the network of fractures leads to the ordinary exponential by repeating the loading—unloading cycle, suggesting a subadditive process of fracturing with hierarchically constrained dynamics. They concluded that their description of the PSC relaxation function in the framework of NESP pointed to a fractal picture originating from the fractures' network that drives the emitted relaxation currents.

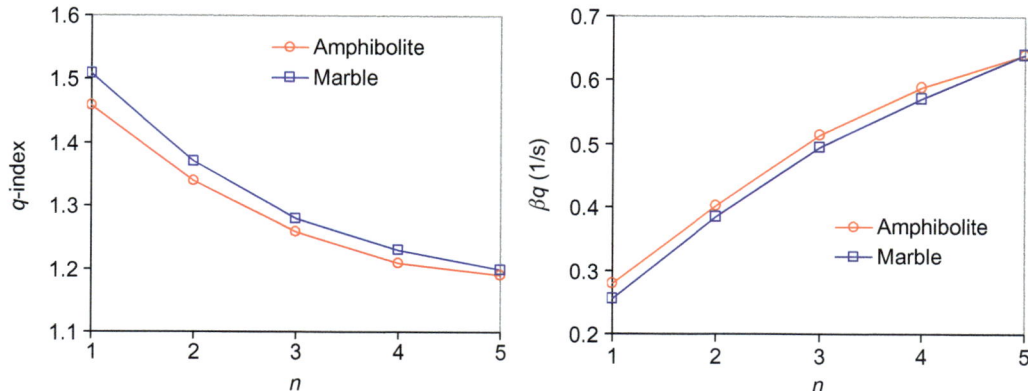

FIGURE 8–16 The parameters q and β_q as a function of the number of loading–unloading cycles in amphibolite and marble samples. *Modified from Vallianatos, F., Triantis, D., 2013. A non-extensive view of the pressure stimulated current relaxation during repeated abrupt uniaxial load-unload in rock samples. EPL 104, 68002-p6.*

In their work, they also used a complementary approach, so-called superstatistics, to investigate the dynamic reason for the NESP applied to the macroscopic behaviour of PSC relaxation. In this approach, a simple model consisting of a large number of relaxed subdomains is considered and the local relaxation is expressed by a Debye exponential relaxation, $\xi(t/\beta) = \beta e^{-\beta t} = (1/\tau)e^{-t/\tau}$. For a certain value of the local relaxation parameter β, the relaxation of a subdomain is given by $\xi(t/\beta)$. The observed PSC relaxation signal is the macroscopic result of the relaxation of these subdomains formed by a complex network of fractures and, thus, the intensive parameter β of the system's relaxation should fluctuate over a large range. This assumption leads to a superstatistical view of the exponential model. They considered a χ^2-distribution for the PDF of β, as follows

$$f(\beta) = \left(\frac{z}{2\beta_0}\right)^{z/2} \cdot \beta^{(z/2)-1} \cdot \exp\left(-\frac{z\beta}{2\beta_0}\right) \tag{8.21}$$

that led to the relaxation function

$$\xi(t) = \int_0^\infty f(\beta)\xi\left(\frac{t}{\beta}\right)d\beta = \int_0^\infty f(\beta)\beta e^{-\beta t}d\beta = \left[1 + B(q-1)t\right]^{1/(1-q)} \tag{8.22}$$

which is exactly the result obtained in the framework of NESP, with $q = 1 + \left(2/(z+2)\right)$ and $B = 2\beta_0/(2-q) = \beta_q$.

A superstatistical view of stress-induced electrical current fluctuations in triaxially deformed Carrara marble specimens has been recently reported by Cartwright-Taylor et al. (2014). The nonextensive behaviour of the failure process in their deformation experiments was investigated for the PSC fluctuations as described by the variable $u_i = (F_i - \langle F \rangle)/\sigma_F$, where $F_i = I(t_{i+1}) - I(t_i)$ denotes the incremental PSC fluctuations and $\langle F \rangle$, σ_F stand for the

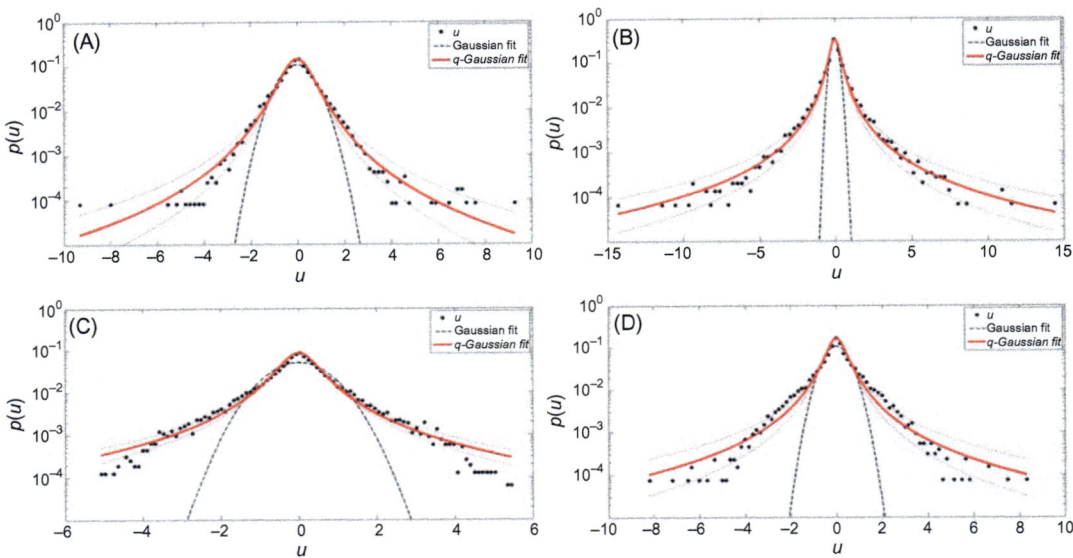

FIGURE 8–17 The probability density function of the normalized PSC fluctuations, $p(u)$ for four different confining pressures: (A) 10, (B) 20, (C) 30 and (D) 100 MPa. The corresponding q-Gaussian and the well-known Gaussian fits are also shown. *PSC*, pressure-stimulated current. *From Cartwright-Taylor, A., Vallianatos, F., Sammonds, P., 2014. Superstatistical view of stress-induced electric currents fluctuations in rocks. Physica A 414, 368–377.*

average value and the standard deviation, respectively. The PDF, $p(u)$ of the normalized increments, $u(t)$, for different confining pressures is presented in Fig. 8–17. At all confining pressures, the distributions were described by a nonextensive q-Gaussian function, with q-parameter ranging from 1.53 (at 30 MPa) to 1.78 (at 100 MPa). The authors interpreted the observed variation of the q-value with an increase of long-range interactions due to the transition from a localized brittle fracture ($P < 30$ MPa) to a cataclastic flow mechanism ($P > 30$ MPa) with homogeneously distributed microcracks.

The application of the superstatistics concept on the previous PSC fluctuations led to the appearance of a q-Gaussian distribution due to the varying β parameter [see Eqs (8.21) and (8.22)] of the PDF, $p(u|\beta)$ which describes the system at short timescales. When the timescale of variation in β is much larger than that of $u(t)$, the PDF of $u(t)$ is obtained as a superposition of temporary equilibria. In local equilibrium where β is taken as constant, the PDF of $u(t)$ is approximated as

$$p(u|\beta) = \sqrt{\frac{\beta}{2\pi}} \; \exp\left(-\frac{\beta}{2}u^2\right) \tag{8.23}$$

while, if β fluctuates over a longer timescale, the distribution of $u(t)$ is given by

$$p(u) = \int p(u|\beta)f(\beta)d\beta \tag{8.24}$$

The authors considered a Γ-distribution to describe the stationary distribution of β,

$$f(\beta) = \frac{1}{\Gamma(a)} \left(\frac{a}{\beta_0}\right)^\alpha \beta^{\alpha-1} \exp\left(-\frac{a\beta}{\beta_0}\right)\beta_0 \tag{8.25}$$

leading to a PDF, $p(u) \propto (1+B(q-1)u^2)^{-1/(q-1)}$, which has the form of a q-Gaussian distribution.

The concept of NESP has also been applied in AEs data obtained from triaxially deformed specimens of Etna basalt (Vallianatos et al., 2012a,b,c). The laboratory experiments were carried out on highly fractured samples of a heterogeneous brittle material (basalts from Mount Etna, Italy), with a constant axial strain rate at various confined pressures. The spatial and temporal evolution of AE activity were monitored continuously and the associated location pattern of growing cracks was estimated with high accuracy. In their work, they studied the CDFs of the AE scalar moment M, $P(>M)$, the AE interevent times, $P(>T)$ and the AE interevent Euclidean distances, $P(>D)$. In each case, the CDF (see Eq. 8.16) should obey the relation:

$$\frac{P^{1-q}(>X)-1}{1-q} = -BX \tag{8.26}$$

where the left side of the equation is the q-logarithmic function, $\ln(P(>X))$. According to Eq. (8.26), the q-index of the scalar moment M was estimated to be $q_M = 1.82$, consistent with other estimations of q for earthquakes and other natural hazards. Accordingly, the q-indices of AE interevent times and distances were estimated as $q_T = 1.34$ and $q_D = 0.65$, respectively. The latter suggests that the AE activity described by interevent time and distance exhibits a nonextensive spatiotemporal duality ($q_T + q_D \approx 2$), similar to that observed in Earth seismicity (Abe and Suzuki, 2003, 2005).

8.3.3 Natural Time Analysis

In order to study the correlation of AE and PSC signals with fracture processes in rocks and to estimate the load level of fracture from early loading stages, complexity techniques could be promising (Hloupis et al., 2016). Recently, the natural time analysis was introduced to describe the dynamic evolution of a complex system and its entry to a critical state (Varotsos et al., 2001, 2002, 2011a,b; Vallianatos et al., 2012a,b,c, 2013a,b; Hloupis et al., 2015).

In a time series of N recorded AE events, the natural time, χ is defined as $\chi_k = k/N$, and it serves as an index for the occurrence of the kth event (Fig. 8–18A and B) (Varotsos et al., 2011a). In the analysis of AEs, where we consider the seismic moment, M_k released during the kth event, the continuous function $F(\omega)$ is defined as (Varotsos et al., 2011a,b):

$$F(\omega) = \sum_{k=1}^{N} M_k \exp\left(i\omega\frac{k}{N}\right) \tag{8.27}$$

where $\omega = 2\pi\varphi$, and φ corresponds to the natural frequency.

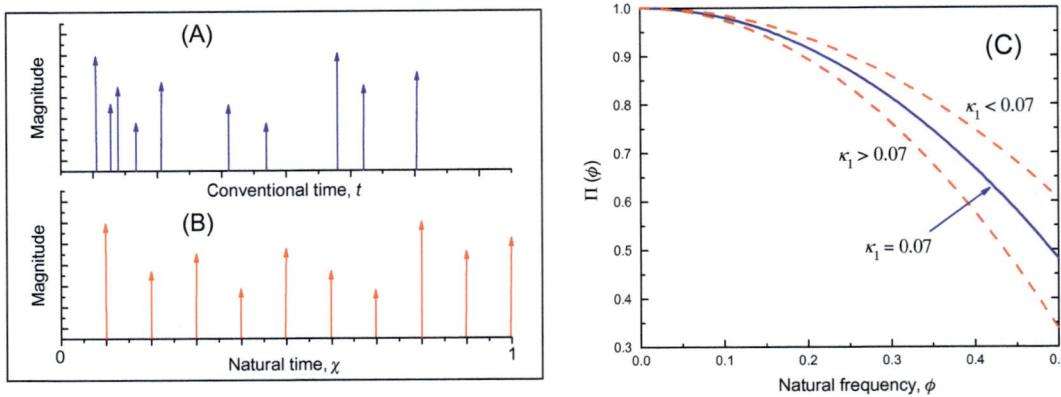

FIGURE 8–18 A time series of AE events in (A) time domain and (B) natural time domain. (C) The normalized power spectrum at the critical value, $\kappa_1 = 0.07$ (black line) and two other values of κ_1 for noncritical stages. AE, acoustic emission. *Modified from Vallianatos, F., Michas, G., Papadakis, G., 2014. Non-extensive and natural time analysis of seismicity before the Mw6.4, October 12, 2013 earthquake in the South West segment of the Hellenic arc. Physica A 414, 163–173.*

The normalized function, $\Phi(\omega)$ is derived by dividing $F(\omega)$ with $F(0)$, as follows:

$$\Phi(\omega) = \frac{\sum_{k=1}^{N} M_k \exp\left(i\omega\left(k/N\right)\right)}{\sum_{n=1}^{N} M_n} = \sum_{k=1}^{N} p_k \exp\left(i\omega\frac{k}{N}\right) \tag{8.28}$$

where $p_k(= M_k/\sum_{n=1}^{N} M_n)$ corresponds to the probability of observing the kth AE event at natural time, χ_k. Subsequently, the normalized power spectrum of Eq. (8.28), defined as $\Pi(\omega) = |\Phi(\omega)|^2$, reduces to the following relation, for $\varphi < 0.5$:

$$\Pi(\omega) = \frac{18}{5\omega^2} - \frac{6\cos\omega}{\omega^2} - \frac{12\sin\omega}{5\omega^3} \tag{8.29}$$

According to probability theory, the distribution function can be approximately determined if its behaviour is known around zero. Thus, for $\omega \to 0$, Eq. (8.29) leads to

$$\Pi(\omega) \approx 1 - \kappa_1 \omega^2 \tag{8.30}$$

where the coefficient κ_1 is the variance of the natural time, expressed as:

$$\kappa_1 = \langle \chi^2 \rangle - \langle \chi \rangle^2 = \sum_{k=1}^{N} p_k \chi_k^2 - \left(\sum_{k=1}^{N} p_k \chi_k\right)^2 \tag{8.31}$$

It has been shown that when a complex system approaches a critical state, the parameter κ_1 should approach the value $\kappa_1 = 0.07$ (Varotsos et al., 2011b). The normalized power spectrum, $\Pi(\varphi)$, for different values of the parameter κ_1 has been plotted in Fig. 8–18C.

Vallianatos et al. (2013a,b) have studied, in the context of natural time, the AE activity in triaxially deformed specimens of Etna basalt (Fig. 8–19). The AE magnitude as a function of

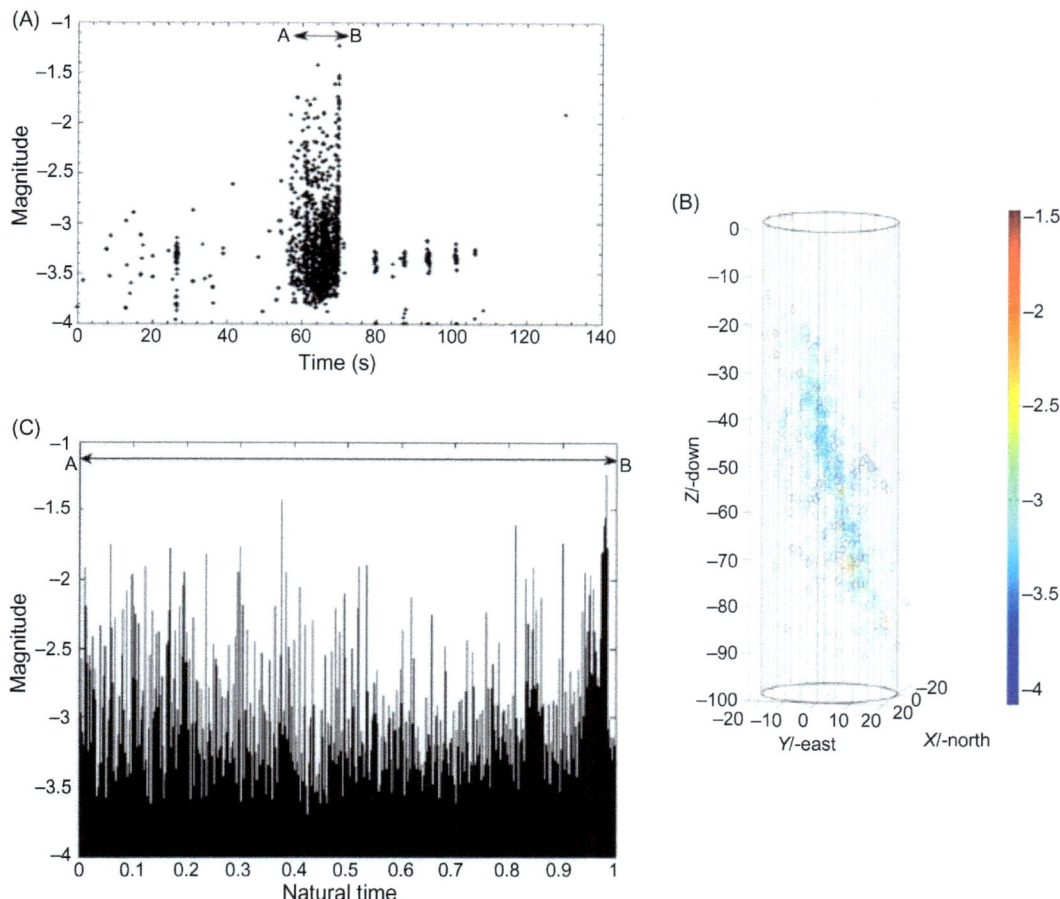

FIGURE 8–19 (A) The AE magnitude, M_{AE} as a function of time, in triaxially deformed Etna basalt. M_{AE} has been calculated according to a similar expression as for earthquakes, $M_{AE} = 2/3\log M − 10.73$, where M is the scalar moment (in dyne cm). The event location for the time period A−B is illustrated in (B). During this period, microcracks start to grow in the lower right-hand part of the specimen and then propagate diagonally across the entire specimen. (C) The AE magnitude of the period A−B as presented in the natural time domain. *AE*, acoustic emission. *From Vallianatos, F., Michas, G., Benson, P., Sammonds, P., 2013a. Natural time analysis of critical phenomena: the case of acoustic emissions in triaxially deformed Etna basalt. Physica A 392, 5172−5178.*

time is shown in Fig. 8−19A. Considerable AE activity was observed during the final stage of deformation, which is denoted as region A−B in Fig. 8−19A. The AE data in this time period were analysed in the natural time domain (refer to Fig. 8−19C). A sliding natural time window (of length l and starting point k_0) for the W events of the AE catalogue was used, in order to calculate the average values of probabilities p_j, according to the relation:

$$\mu_j = \frac{1}{W-l+1} \sum_{k_0=1}^{W-l+1} \frac{M_{k_0+j-1}}{\sum_{m=1}^{l} M_{k_0+m-1}} \tag{8.32}$$

Subsequently, the expectation value of the parameter κ_1 obtained from the $W - l + 1$ time windows of the AE catalogue is expressed as:

$$E(\kappa_1) = \kappa_{1,M} + \sum_{j=1}^{l-1} \sum_{i=j+1}^{l} \frac{(j-l)^2}{l^2} \text{Cov}(p_j, p_i) \tag{8.33}$$

where $\kappa_{1,M}$ is obtained from Eq. (8.31) by substituting μ_k for p_k and $\text{Cov}(p_j, p_i)$ is the covariance of p_j. In order to identify the presence of AE magnitude correlations in a natural time basis, similarly to field scale seismicity, they compared the expectation values of κ_1, $E(\kappa_1)$ of the original AE time series with the distribution obtained for $E(\kappa_{1, \text{shuf}})$, given by:

$$E(\kappa_{1, \text{shuf}}) = \kappa_u\left(1 - \frac{1}{l^2}\right) - \kappa_u(l+1)\text{Var}(p) \tag{8.34}$$

when many randomly shuffled copies of the original AE data set are being used. In the above equation, κ_u corresponds to a uniform distribution and $\text{Var}(p)$ is the variance in the shuffled catalogue. They found that $E(\kappa_{1, \text{shuf}})$ follows a normal distribution ($\mu_s = 0.0573$, $\sigma = 0.00097$), while the original AE catalogue results in an average κ_1 value ($\mu_0 = 0.0614$), which is unlikely to come from this distribution. This was attributed to the fact that μ_0 is related to the presence of long-range correlations in the original AE catalogue.

Remarkably, based on the statistical properties of the order parameter κ_1 in natural time, they also observed a similarity between their AE catalogue with the Centennial Earthquake Catalogue of the period 1900−2007, suggesting a universality of AEs and global seismicity. They suggested that the evolution of a shear zone formed in compressed rock specimens at the laboratory scale is quite similar to the evolution of the shear process in the macroscopic scale, such as fault zones.

Recently, an experimental protocol was applied involving AE and PSC measurements on double-edged notched tensile (DENT) specimens made of marble quarried from Dionysos Mountain in Attica, Greece (Kourkoulis et al., 2017). The mechanical response and fracture of the specimens were monitored simultaneously using the AE and the PSC techniques. In addition, the displacement field developed was determined using the digital image correlation (DIC) technique, while the onset and propagation of the crack were captured with the aid of an ultrahigh-speed camera. In parallel, clip gauges were used to measure the notch mouth opening displacement, mainly for comparison and calibration purposes. The main advantage of the complex sensing systems that were used is that they provide data, both from the surface and also from the interior of the specimens tested. The aim of that work was to approach the problem of early detection of upcoming catastrophic fracture of structural elements made of marble. The experimental data were studied under the framework of the hypothesis that unique and hidden dynamic features of complex systems could be detected using the time series of measurements and analysed in terms of natural time (Varotsos et al., 2011a,b; Vallianatos et al., 2013a,b, 2014; Hloupis et al., 2015). In that first approach, the results of the mechanical characteristics that indicated the upcoming criticality were put in contrast to the corresponding results after applying the natural time analysis. In a previously described published work (Hloupis et al., 2015) it was shown that the analysis of AE provides information regarding the criticality approach. Specifically, the quantity RA (Rise_Time/Amplitude) was used as input to the natural time field. The striking observation was that on all the tested experiments, the NT analysis was able to reveal a critical point (entrance point to 'critical stage'). Specifically, the κ_1 parameter was studied using as input the RA quantity calculated using the AE hits from the sensors that they were placed near the fracture zone (Varotsos et al., 2011a,b; Vallianatos et al., 2013b, 2014; Hloupis et al., 2015). Natural time analysis revealed indicators that dictate entrance to the 'critical stage' (see Fig. 8a–d in Hloupis et al., 2015). An interesting observation was that the time when the critical point was approached (criticality initiation time) was not the same for all the AE recording channels. Detailed justification of this can be found in Hloupis et al. (2015). The promising results after this study justified the robustness of the NT method since the number N (of separate events in NT terminology — of recorded AE hits in this study) didn't affect the method's applicability. As during the above experiment a main issue was to define the crack initiation location, the entrance to criticality was examined separately for each group of AE sensors. It must be noted that a pair of AE sensors was placed near the notch of the left side of the specimen and another pair of AE sensors was placed near the right notch of the specimen. Following that, the relative time where the criticality conditions were fulfilled for the left and the right sides of the specimen is provided allowing one to obtain the criticality initiation order. Besides the existence of critical points introducing the entrance to a critical stage, another interesting pattern was systematically observed for all specimens tested: the fracture initiation zone of the specimen (i.e., the notch from which fracture crack propagation started) was the opposite from the one from which the earliest criticality indicator was received. Although this observation seems to contradict the expected order, it could be explained by interpreting the term 'critical stage' in case of a specimen made of brittle material subjected to tension. It is generally accepted that before crack initiation a 'process zone' is formed around the tip of the preexisting notch. This zone is characterized by strong microcracking, void coalescence and thus material within this zone is badly damaged and cannot be characterized by the properties of the initial virgin material. The new properties are considerably undergraded. As a result the stress field in this zone is redistributed, the singularity order becomes lower and the material appears to be less brittle. In this context entrance to criticality designates an increase of the 'apparent' ductility, making crack initiation more difficult from the specific tip, thus favouring sudden crack initiation from the opposite notch. Finally, all the above observations of this case regarding the detection of criticality were also verified by the analysis of the PSC emissions. Specifically, a slow PSC increase was detected well before the time of fracture. This PSC increase becomes significantly more intense before the fracture reaches a maximum value (Kourkoulis et al., 2017).

The study of prefailure indicators for DENT marble specimens is also a very interesting topic from the engineering point of view, taking into account the complexity of the sequence of failure mechanisms activated during fracture of brittle rock-like materials.

8.4 Summary and Open Questions

Earthquakes occur due to instabilities in the deformation of rocks in the Earth's crust. At shallower depths, referred to as the seismogenic zone, rocks resist plate motion and remain locked, until the force becomes large enough that the material fails and the fault slips rapidly — of the order of metres/second in an earthquake. A primary goal in seismology is to identify constraints arising from the small-scale physics of fracture that can provide bounds on seismic hazard and ground motion at the fault scale. It is obvious that an understanding of the laws of this emerging interdisciplinary field of earthquake physics, demands a connection with geophysics, material science, laboratory studies and seismic observations to reduce uncertainties in seismic hazards.

Over the last decade, great progress has been made in scientifically understanding earthquake physics from the point of view of rock physics and complexity theory. However, the resolution of seismological data observed in the field is not sufficient, not only to strictly formulate the fundamental laws, but also to fully elucidate the physical nature of a scale-dependent earthquake rupture generation process from its nucleation to the subsequent dynamic propagation on a heterogeneous fault. In contrast, in laboratory experiments, the experimental method can be properly organized and high temporal and spatial resolution measurements can be made.

In order to view earthquake physics in terms of rock physics and complexity theory, it is critically important to formulate the fundamental law based on positive facts elucidated by high-resolution laboratory experiments. In light of rock physics and complexity theory, it is possible to rationally formulate the governing law for earthquake rupture and earthquake physics, based on fundamental research on the physics of rocks.

Recently, the use of electrical and acoustic signal emissions as fracture precursors when brittle materials like rocks are subjected to mechanical stress has been investigated extensively, offering information on the complex physical processes into the earthquake focus. This chapter has given a detailed review and description within the framework of a complexity approach performed on laboratory experiments that have been conducted and show up the generation and behaviour of electrical and acoustic signal emissions, mainly when the Earth material is subjected to mechanical stress. Specifically, the temporal variation of the electrical and AEs, in combination with the adopted mechanical stress application protocol and the type of stress test (compression, bending, etc.), is discussed. Furthermore, modelling that involves statistical physics, which is currently used for the analysis of electrical and AEs, is described and the latest findings are reported. The similarities with the observations associated with fracture are viewed in relation to the electrical and acoustic signals and discussed in view of seismicity patterns.

The experimental results of PSC and AE techniques have been examined within the framework of their capacity to provide information regarding the initial stages of microcrack generation, propagation and coalescence, with the aim of using these as fracture precursors.

From an applied point of view, the PSC technique has also been applied in complex systems such as real models that are used for the ancient monument restoration processes or in standardized tests according to ASTM (three-point bending and direct tensional tests), highlighting the need for further standardization of this technique. It should be emphasized that the proposed technique of PSV using EPSs could potentially find applications in monitoring landslides as well as in the structural health monitoring of tunnels, caverns and basements.

However, in addition to the potential applicability of PSC in materials science, fundamental questions should be addressed, such as the size-scale effect, in order to extend the application of the PSC technique into field measurements, i.e., in a scale similar to that of earthquake events. The influence of the fracture mode (tensile, shear or mixed mode) of samples to the corresponding recordings of PSC and AE signals should also be investigated thoroughly. Additionally, the PSC technique should be further developed in order to estimate in detail the spatial distribution of the internal damage when a specimen is subjected to an external mechanical load. Thus, efforts should be made towards the standardization of the PSC technique not only as a laboratory testing method but also as a field method.

In the context of NESP, the study of the distribution of interevent times (or distances) in AE data could give insights into the critical state of a complex system, by means of the variation of the entropic index, q. In the self-organized growth of correlated microcracks in rocks under mechanical stress high q-values have been observed, in accordance with corresponding q-values before the occurrence of high-magnitude seismic events. Further studies are necessary to correlate the variation of the q-index with the entrance of a complex system to a critical state.

Acknowledgements

This work of VS and FV has been partly supported by the project 'HELPOS. Hellenic System for Lithosphere Monitoring' (MIS 5002697) of the Operational Programme NSRF 2014-20, co-financed by Greece and the European Union (European Regional Development Fund).

References

Abe, S., Suzuki, N., 2003. Law for the distance between successive earthquakes. J. Geophys. Res. 108, 19-4.

Abe, S., Suzuki, N., 2005. Scale-free statistics of time interval between successive earthquakes. Physica A 350, 588−596.

Archer, J.W., Dobbs, M.R., Aydin, A., Reeves, H.J., Prance, R.J., 2016. Measurement and correlation of acoustic emissions and pressure stimulated voltages in rock using an electric potential sensor. Int. J. Rock Mech. Min. Sci. 89, 26−33.

Aydin, A., Prance, R.J., Prance, H., Harland, C.J., 2009. Observation of pressure stimulated voltages in rocks using an electric potential sensor. Appl. Phys. Lett. 95, 124102-3.

Baró, J., Corral, Á., Illa, X., Planes, A., Salje, E.K.H., Schranz, W., et al., 2013. Statistical similarity between the compression of a porous material and earthquakes. Phys. Rev. Lett. 110, 088702-5.

Bleier, T., Dunson, C., Alvarez, C., Freund, F., Dahlgren, R., 2010. Correlation of pre-earthquake electromagnetic signals with laboratory and field rock experiments. Nat. Hazard. Earth Syst. Sci. 10, 1965−1975.

Cartwright-Taylor, A., Vallianatos, F., Sammonds, P., 2014. Superstatistical view of stress-induced electric currents fluctuations in rocks. Physica A 414, 368−377.

Colombo, I.S., Main, I.G., Ford, M.C., 2003. Assessing damage of reinforced concrete beam using b-value analysis of acoustic emission signal. J. Mat. Civ. Eng. 15, 280−286.

Freund, F., 2002. Charge generation and propagation in igneous rocks. J. Geodyn. 33, 543−570.

Frid, V., Goldbaum, J., Rabinovitch, A., Bahat, D., 2009. Electric polarization induced by mechanical loading of Solnhofen limestone. Philos. Mag. Lett. 89 (7), 453−463.

Fursa, T.V., Dann, D.D., Osipov, K.Yu, 2016. Evaluation of freeze−thaw damage in concrete by the parameters of electric response under impact excitation. Constr. Build. Mater. 102, 182−189.

Fursa, T.V., Dann, D.D., Petrov, M.V., Lykov, A.E., 2017. Evaluation of damage in concrete under uniaxial compression by measuring electric response to mechanical impact. J. Nondestr. Eval. 36 (2), 30.

Hloupis, G., Stavrakas, I., Pasiou, E.D., Triantis, D., Kourkoulis, S.K., 2015. Natural time analysis of acoustic emissions in double edge notched tension (DENT) marble specimens. Procedia Eng. 109, 248−256.

Hloupis, G., Stavrakas, I., Vallianatos, F., Triantis, D., 2016. A preliminary study for prefailure indicators in acoustic emissions using wavelets and natural time analysis. J. Mater. Des. Appl. 230 (3), 780−788.

Kawada, Y., Nagahama, H., 2006. Cumulative Benioff strain-release, modified Omori's law and transient behaviour of rocks. Tectonophysics 424, 157−166.

Kawada, Y., Nagahama, H., Nakamura, N., 2007. Time-scale invariances in preseismic electromagnetic radiation, magnetization and damage evolution of rocks. Nat. Hazard. Earth Syst. Sci. 7, 599−606.

Kourkoulis, S.K., Triantis, D., Stavrakas, I., Pasiou, E.D., Dakanali, I., 2017. Recording the mechanical response and fracture of marble DENT specimens using modern sensing techniques. Procedia Struct. Integrity 3, 326−333.

Lavrov, A., 2005. Fracture-induced physical phenomena and memory effects in rocks: a review. Strain 41, 135−149.

Li, Z., Wang, E., He, M., 2015. Laboratory studies of electric current generated during fracture of coal and rock in rock burst coal mine. J. Min. 2015, . Available from: https://doi.org/10.1155/2015/235636. Article ID 235636.

Michas, G., Vallianatos, F., Sammonds, P., 2013. Non-extensivity and long-range correlations in the earthquake activity at the West Corinth rift (Greece). Nonlinear Processes Geophys. 20, 713−724.

Nomikos, K., Vallianatos, F., 1997. Transient electric variations associated with large intermediate-depth earthquakes in South Aegean. Tectonophysics 269, 171−177.

Nomikos, K., Vallianatos, F., Kaliakatsos, J., Sideris, S., Bakatsakis, M., 1997. Latest aspects of telluric and electromagnetic variations associated with shallow and intermediate depth earthquakes in South Aegean. Ann. Geofis. XL (2), 361−374.

Papathanassiou, A.N., Sakellis, I., Grammatikakis, J., 2010. Negative activation volume for dielectric relaxation in hydrated rocks. Tectonophysics 490, 307−309.

Papathanassiou, A.N., Sakellis, I., Grammatikakis, J., 2011. Dielectric properties of granodiorite partially saturated with water and its correlation to the detection of seismic electric signals. Tectonophysics 511, 148−151.

Papathanassiou, A.N., Sakellis, I., Grammatikakis, J., 2012. Dielectric relaxation under pressure in granular dielectrics containing water: compensation rule for the activation parameters. Solid State Ionics 209-210, 1−4.

Pasiou, E.D., Triantis, D., 2017. Correlation between the electric and acoustic signals emitted during the compression of brittle materials. Frattura ed Integrità Strutturale (40), 41−51.

Prance, R.J., Debray, A., Clark, T.D., Prance, H., Nock, M., Harland, C.J., et al., 2000. An ultra-low-noise electrical-potential probe for human-body scanning. Meas. Sci. Technol. 11, 291−297.

Saltas, V., Fitilis, I., Vallianatos, F., 2014. A combined complex electrical impedance and acoustic emission study in limestone samples under uniaxial loading. Tectonophysics 637, 198–206.

Saltas, V., Fitilis, I., Makris, J.P., Vallianatos, F., 2015. Acoustic and electrical emissions from sandstone under uniaxial compression. In: International Conference 'Science in Technology' SCinTE 2015.

Shankland, T.J., Duba, A.G., Mathez, E.A., Peach, C.L., 1997. Increase of electrical conductivity with pressure as an indicator of conduction through a solid phase in midcrustal rocks. J. Geophys. Res. 102, 14741–14750.

Shiotani, T., Fujii, K., Aoki, T., Amou, K., 1994. Evaluation of progressive failure using AE sources and improved b-value on slope model tests. Prog. Acoust. Emiss. VII 7, 529–534.

Slifkin, L., 1993. Seismic electric signals from displacement of charged dislocations. Tectonophysics 224, 149–152.

Stavrakas, I., 2017. Acoustic emissions and pressure stimulated currents experimental techniques used to verify Kaiser effect during compression tests of Dionysos marble. Fract. Struct. Integrity 40, 32–40.

Stavrakas, I., Anastasiadis, C., Triantis, D., Vallianatos, F., 2003. Piezo stimulated currents in marble samples: precursory and concurrent − with − failure signals. Nat. Hazard. Earth Syst. Sci. 3, 243–247.

Stavrakas, I., Triantis, D., Agioutantis, Z., Maurigiannakis, S., Saltas, V., Vallianatos, F., et al., 2004. Pressure stimulated currents in rocks and their correlation with mechanical properties. Nat. Hazard. Earth Syst. Sci. 4, 563–567.

Stavrakas, I., Triantis, D., Kyriazopoulos, A., Ninos, K., 2015. Pressure stimulated currents and acoustic emission combined recordings for the detection of compressive strength of dionysos marble. In: Proceedings of the International Conference Science in Technology, SCinTE-2015, vol. 1 pp. 19–22.

Triantis, D., Stavrakas, I., Anastasiadis, C., Kyriazopoulos, A., Vallianatos, F., 2006. An analysis of pressure stimulated currents (PSC) in marble samples under mechanical stress. Phys. Chem. Earth 31, 234–239.

Triantis, D., Anastasiadis, C., Vallianatos, F., Kyriazis, P., Nover, G., 2007. Electric signal emissions during repeated abrupt uniaxial compressional stress steps in amphibolite from KTB drilling. Nat. Hazard. Earth Syst. Sci. 7, 149–154.

Triantis, D., Anastasiadis, C., Stavrakas, I., 2008. The correlation of electrical charge with strain on stressed rock samples. Nat. Hazard. Earth Syst. Sci. 8, 1243–1248.

Triantis, D., Vallianatos, F., Stavrakas, I., Hloupis, G., 2012. Relaxation phenomena of electrical signal emissions from rock following application of abrupt mechanical stress. Ann. Geophys. 55 (1), 207–212.

Tsallis, C., 1988. Possible generalization Boltzmann−Gibbs statistics. J. Stat. Phys. 52, 479–487.

Tsallis, C., 2009. Introduction to Non-extensive Statistical Mechanics: Approaching a Complex World. Springer, Berlin.

Tsekouras, G.A., Tsallis, C., 2005. Generalized entropy arising from a distribution of q indices. Phys. Rev. E71, 046144-8.

Tzanis, A., Vallianatos, F., 2002. A physical model of electrical earthquake precursors due to crack propagation and the motion of charged edge dislocations. Seismo Electromagnetics (Lithosphere-Atmosphere-Ionosphere Coupling). Terra Scientific Publishing Co., Tokyo.

Vallianatos, F., Sammonds, P., 2010. Is plate tectonics a case of non-extensive thermodynamics? Physica A 389, 4989–4993.

Vallianatos, F., Sammonds, P., 2011. A non-extensive statistics of the fault-population of the Valles Marineris extensional province, Mars. Tectonophysics 509, 50–54.

Vallianatos, F., Sammonds, P., 2013. Evidence of non-extensive statistical physics of the lithospheric instability approaching the 2004 Sumatran-Andaman and 2011 Honshu mega-earthquakes. Tectonophysics 590, 52–58.

Vallianatos, F., Triantis, D., 2008. Scaling in pressure stimulated currents related with rock fracture. Physica A 387, 4940–4946.

Vallianatos, F., Triantis, D., 2012. Is pressure stimulated current relaxation in amphibolite a case of non-extensivity? EPL 99, 18006.

Vallianatos, F., Triantis, D., 2013. A non-extensive view of the Pressure Stimulated Current relaxation during repeated abrupt uniaxial load-unload in rock samples. EPL 104, 68002-p6.

Vallianatos, F., Tzanis, A., 1998. Electric current generation associated with the deformation range of a solid: preseismic and coseismic signals. Phys. Chem. Earth 239, 933–938.

Vallianatos, F., Tzanis, A., 1999. A model for the generation of precursory electric and magnetic fields associated with the deformation rate of the earthquake focus. In: Hayakawa, M. (Ed.), Seismic Atmospheric & Ionospheric electromagnetic Phenomena. Terra Scientific Publishing Co., Tokyo.

Vallianatos, F., Tzanis, A., 2003. On the nature, scaling and spectral properties of pre-seismic ULF signals. Nat. Hazard. Earth Syst. Sci. 3, 237–242.

Vallianatos, F., Triantis, D., Tzanis, A., Anastasiadis, C., Stavrakas, I., 2004. Electric earthquake precursors: From laboratory results to field observations. Phys. Chem. Earth 29, 339–351.

Vallianatos, F., Triantis, D., Sammonds, P., 2011. Non-extensivity of the isothermal depolarization relaxation currents in uniaxial compressed rocks. EPL 94, 68008.

Vallianatos, F., Benson, P., Meredith, P., Sammonds, P., 2012a. Experimental evidence of a non-extensive statistical physics behaviour of fracture in triaxially deformed Etna basalt using acoustic emissions. EPL 97, 58002-6.

Vallianatos, F., Michas, G., Papadakis, G., Sammonds, P., 2012b. A non-extensive statistical physics view to the spatiotemporal properties of the June 1995, Aigion earthquake (M6.2) aftershock sequence (West Corinth rift, Greece). Acta Geophysica 60, 758–768.

Vallianatos, F., Nardi, A., Carcuccio, R., Chiappini, M., 2012c. Experimental evidence of a non-extensive statistical physics behavior of electromagnetic signals emitted from rocks under stress up to fracture. Preliminary results. Acta Geophys. 60 (3), 894–909.

Vallianatos, F., Michas, G., Benson, P., Sammonds, P., 2013a. Natural time analysis of critical phenomena: the case of acoustic emissions in triaxially deformed Etna basalt. Physica A 392, 5172–5178.

Vallianatos, F., Michas, G., Papadakis, G., Tzanis, A., 2013b. Evidence of non-extensivity in the seismicity observed during the 2011–2012 unrest at the Santorini volcanic complex, Greece. Nat. Hazard. Earth Syst. Sci. 13, 177–185.

Vallianatos, F., Michas, G., Papadakis, G., 2014. Non-extensive and natural time analysis of seismicity before the Mw6.4, October 12, 2013 earthquake in the South West segment of the Hellenic arc. Physica A 414, 163–173.

Varotsos, P., Alexopoulos, K., 1987. Physical properties of the variations in the electric field of the earth preceding earthquakes, III. Tectonophysics 136, 335–339.

Varotsos, P.A., 2005. The Physics of Seismic Electric Signals. TERRAPUB, Tokyo.

Varotsos, P.A., Alexopoulos, K., 1984. Physical properties of the variations of the electric field of the earth preceding earthquakes, I. Tectonophysics 110, 73–98.

Varotsos, P.A., Sarlis, N.V., Skordas, E.S., 2001. Spatio-temporal complexity aspects on the interrelation between seismic electric signals and seismicity. Pract. Athens Acad. 76, 294–321.

Varotsos, P.A., Sarlis, N.V., Skordas, E.S., 2002. Long-range correlations in the electric signals that precede rupture. Phys. Rev. E 66, 011902.

Varotsos, P.A., Sarlis, N.V., Skordas, E.S., 2011a. Natural time analysis: the new view of time. Precursory Seismic Electric Signals, Earthquakes and Other Complex Time Series. Springer-Verlag, Berlin, Heidelberg.

Varotsos, P.A., Sarlis, N.V., Skordas, E.S., Uyeda, S., Kamogawa, M., 2011b. Natural time analysis of critical phenomena. PNAS 108, 11361–11364.

Further Reading

Anastasiadis, C., Stavrakas, I., Triantis, D., Vallianatos, F., 2007. Correlation of pressure stimulated currents in rocks with the damage variable. Ann. Geophys. 50, 1–6.

Dakanali, I., Stavrakas, I., Triantis, D., Kourkoulis, S.K., 2016. Pull-out of threaded reinforcing bars from marble blocks. Procedia Struct. Integrity 2, 2865–2872.

Kourkoulis, S.K., Pasiou, E.D., Triantis, D., Stavrakas, I., Hloupis, G., 2015. Innovative experimental techniques in the service of restoration of stone monuments – Part I: The experimental set up. Procedia Eng. 109, 268–275.

Kyriazopoulos, A., Anastasiadis, C., Triantis, D., Brown, J.C., 2011. Non-destructive evaluation of cement-based materials from pressure-stimulated electrical emission – preliminary results. Constr. Build. Mater. 25, 1980–1990.

Pasiou, E.D., Stavrakas, I., Hloupis, G., Kourkoulis, S.K., Triantis, D., 2015. Electrical and acoustic emissions during three point bending tests of pre-notched marble specimens. In: Proceedings of the International Conference, "SCience in TEchnology – SCinTE – 2015, vol. 1, pp. 86–90.

Stavrakas, I., Triantis, D., Kourkoulis, S., Pasiou, E., Dakanali, I., 2016. Acoustic emission analysis when cement mortar specimens are subjected to three point bending repetitive cycles. Lat. Am. J. Solids Struct. 13, 2283–2297.

Triantis, D., Stavrakas, I., Pasiou, E.D., Hloupis, G., Kourkoulis, S.K., 2015. Innovative experimental techniques in the service of restoration of stone monuments – Part II: Marble epistyles under shear. Procedia Eng. 109, 276–284.

9

Complexity and Synchronization Analysis in Natural and Dynamically Forced Stick—Slip: A Review

Tamaz Chelidze, Temur Matcharashvili, Nodar Varamashvili,
Ekaterine Mepharidze, Dimitri Tephnadze, Zurab Chelidze
IVANE JAVAKHISHVILI TBILISI STATE UNIVERSITY, TBILISI, GEORGIA

CHAPTER OUTLINE

Complexity of Seismic Time Series. DOI: https://doi.org/10.1016/B978-0-12-813138-1.00009-2

9.1 Fracture or Friction

The basis of earlier models of the seismic process was the linear fracture mechanics of solids, rooted in the Griffiths–Irvin crack model (Griffiths, 1921; Irwin, 1957): as this approach is focused on isolated crack growth under external stress, it mainly can be applied to the analysis of the main rupture development. One of the main deficiencies of mechanical fracture models is a lack of cyclicity (recurrence) property, characteristic for the seismic process. These fracture models work well for representation of just a single act of destruction: namely, they predict some precursory effects of the final rupture of the sample under stress or geological fault under tectonic force, as a result of approaching the critical state, disregarding the complicated dynamics of the earthquake (EQ) source area before incipient main rupture formation. The full model of seismic process should include, besides the final rupture pattern, the description of microcrack nucleation, their multiplication and clustering, the appearance of the incipient main rupture and, most importantly, the phenomenon of recurrence of main ruptures.

Brace and Byerlee (1966) suggested the relevant physical mechanism – unstable friction or stick–slip – which explains practically all the main features of the natural seismic process.

9.2 Natural Stick–Slip: Basics

9.2.1 From Static to Dynamic Friction

The science of friction actually begins with formulation of Amonton's law of friction, inspired by the ideas of Leonardo da Vinci:

$$\tau = \mu\sigma_n, \tag{9.1}$$

where τ is shear stress, μ is the coefficient of friction, and σ_n is normal (nominal) stress. Coulomb took into consideration the forces of adhesion between contacting surfaces:

$$\tau = c + \mu\sigma_n, \tag{9.2}$$

where c is the coefficient of adhesion between surfaces.

Important adjustments were introduced to the initial Amonton formulation by Hubbert and Rubbey (1959), who elucidated an important role of pore fluid pressure P_p, which can change significantly friction resistance:

$$\tau = c + \mu(\sigma_n - P_p) = c + \mu\sigma_{\mathit{eff}}, \tag{9.3}$$

where $\sigma_{\mathit{eff}} = (\sigma_n - P_p)$ is the effective normal stress.

The first three above-mentioned friction law formulations are independent of the time and correspond to the stable friction state. At the same time, from the ancient epoch it was known that it is easier to slide a moving body than to move a body at rest. In addition, Coulomb found that static friction depends on the time of repose of sliding surfaces. Thus the next development of friction equation is connected with the names of Brace and Byerlee (1966), Burridge and Knopoff (1967), Dieterich (1972,1979) and Ruina (1983). Accordingly, the modern rate-and-state friction law takes into account the effect of the transition from static to kinetic friction and the duration of repose (stick) time on the dynamics of friction:

$$\tau = \left[\mu\sigma_n + a\ln(V/V_0) + b\ln(V_0\Theta/L]\sigma_{\mathit{eff}}, \tag{9.4}$$

$$d\Theta/dt = (V\Theta/L) \tag{9.5}$$

where a and b are constants, V is sliding velocity, V_0 is reference velocity, Θ is state variable, and L is critical slip length.

This so-called rate-and-state-dependent law of friction for shear stress describes almost all the main features of stick–slip, obtained in numerous spring-slider experiments: it shows that the frictional force is not a constant, but is time-dependent and undergoes complex evolution during the slip event. The equation is nonlinear: consequently, the slip process should manifest such properties as high sensitivity to weak external forcing, hysteresis effect, etc.

9.2.2 Laboratory Experiments

Experimental investigations into friction regularities are carried out on three main systems: traditional slider-spring arrangement (Nasuno, 1998), biaxial (Marone, 1998), three-axial shearing apparatus (Beeler and Lockner, 2003), and rotary systems. Depending on conditions (spring stiffness k, velocity of drag V, normal stress σ_n, slip surface state Θ), three main types of friction are observed by displacement recording — stick–slip, inertial regime and quasistable regime. The stick–slip regime is observed at relatively low velocities V and low stiffness. At higher V the transition to inertial periodic oscillations occurs; at still higher V we have stable sliding with small fluctuations.

The mean period of stick–slip recurrence T depends on the drive velocity V and spring stiffness K_s, (Nasuno, 1998); according to Karner and Marone (2000) the recurrence period as a function of V follows the power law:

$$T \sim aV^b$$

where a and b are coefficients.

It is important that the lateral movement along the slip surface is always preceded by small vertical displacements, which are necessary to override asperities; its value depends on the roughness of the surface and the normal stress. Bowden and Tabor (1950) suggested the physical model, which takes into account the role of contact roughness on the friction coefficient. As a rule, the experimental value of μ for rock samples at small/moderate normal stress is less than 1. Bowden and Tabor (1950) explain this effect by the roughness (fractality) of contacting surfaces, what means that the real contact area, $A_r = n\langle a \rangle$, where n is the number of contacts and $\langle a \rangle$ is the average area of contact, is much less than the nominal contact area A. It is evident that $A > A_r$. In the elastic regime A_r grows with the normal stress σ_n as: $A_r \propto \sigma_n$, as both n and $\langle a \rangle$ increase with load. Accordingly, the real normal stress on the contact $\sigma_r = \sigma_n(A/A_r)$ is larger than nominal, σ_n and correspondingly, the measured friction coefficient will be equal to 1 only if $A = A_r$, i.e., at the very high load, leading to plastic deformation of the contact area.

It seems that the percolation theory, namely, the model of percolation for the tangential shift of contacting fractal surfaces, may explain the transition of friction coefficient from the static to kinetic value at attaining some critical value of contact points' number n_c of shearing fractal surfaces. This suggestion is confirmed by experimental evidence of small vertical displacements preceding the slip event, which means that the number of contact points n decreases to some threshold value n_c, due to the uplift, facilitating the macroscopic tangential displacement. Ben-David et al. (2010) showed experimentally that the contact area before a slip drops suddenly by 20%, which confirms the existence of the threshold value of the contact area. The mathematical formalism, similar to that of the percolation model of fracture (Chelidze, 1982), could be developed for the transition of static to kinetic friction at initiation of the slip process, when the number of local contacts between two contacting fractal surfaces decreases to the critical percolation threshold value n_c due to a vertical uplift of the sliding plate before the slip phase.

Until recently, one of the main problems with friction experiments was inaccessibility of the microscale processes to the contact area before and during sliding. Now, due to the development of innovative technology, we can obtain a very detailed picture of the real contact area and its transformation before and during frictional sliding, at least for optically transparent substances like PMMA blocks (Rubinstein et al., 2004, 2009; Ben-David et al., 2010; Capozza et al., 2011, 2012). The laser sheet illuminates the contact area at an angle, larger than the total incident reflection angle: this allows revealing of the net of contact points of the rough interface, i.e., measuring the real contact area A_r. The direct experiments of Rubinstein et al. (2004) showed that the net contact area of the interface is proportional to the normal load and is linearly proportional to the applied force, validating the Bowden and Tabor model. The fast camera fixed the transmitted light on the time intervals in the submillisecond range, which records in detail transitions in A_r during the nucleation and slip phase. Ben-David et al. (2010) distinguished four successive phases of process evolution: (1) sudden reduction of all local contact area (percolation threshold n_c) within several microseconds; (2) rapid slip phase with some characteristic time; (3) transition to an order of magnitude slower slip phase; and (4) commencement of the stick phase, when the ageing process restores the preexisting contact area during some characteristic time. The sophisticated methods of recording/analysing acoustic emission (AE), which accompanies stick–slip, add important information on the microphysics of the process (Johnson et al., 2013). Probably, the main experimental data on friction microphysics, obtained over the last few decades due to using new technologies, is the discovery of three/four displacement–time scales in friction dynamics: in addition to early recognized slow (accumulation) and fast (discharge) phases, the so-called precursory or ultra-fast microscale displacements and corresponding microacoustic signals were discovered (Johnson et al., 2013). AE amplitudes for the fast and precursory events are accordingly 5×10^{-7} and 5×10^{-9} strains and the corresponding waiting times of 10^{-1} and 5×10^{-4} s (approximately) according to Johnson et al.'s (2013) experiments. The precursory events appear when the system is close to the critical (discharge) state and disappear immediately after discharge.

From the dynamic point of view stick–slip is one of typical representatives of complex integrate-and-fire systems, where the slow nucleation (integrate, stick, stress accumulation) phase terminates necessarily by a slip (fire, stress drop) phase (Pikovsky et al., 2003). The mathematical expressions for the shear stress τ evolution, formulated by Dieterich (1972, 1979) and Ruina (1983) are in agreement with the majority of the observed data on stick–slip. It has been shown that for some critical stiffness k_c the system undergoes Hopf bifurcation, leading finally to instability. The solution of the rate-and-state-dependent equations' system of stick-slip motion (9.4 and 9.5) demonstrates all details, characteristic for low-dimentional chaotic nonlinear systems: strange attractors, bifurcations, etc. (Becker, 2000).

Since the seminal paper of Brace and Byerly (1966) and several following main works (Burrige and Knopoff, 1967; Dieterich, 1972, 1979; Ruina, 1983), the stick–slip has been considered as a main mechanism for explaining the EQ process and the existence of seismic cycles (Scholz, 1998, 2002). Here we have to note that a half-century before Brace and Byerly (1966), Reid (1910) formulated the elastic-rebound theory, where he considered seismic cycles as the sequence of recurrent (sudden) releases of elastic strain energy, which slowly accumulated in the preceding period, i.e., conceptually, as the integrate-and-fire process.

9.3 Forced Stick–Slip

9.3.1 Elementary Concepts and Geophysical Consequences

The simplified mathematical model of a relaxation oscillator, an example of which is the slider-spring system, can be represented as a superposition of linearly increasing shear stress $\tau_s(t) = b(t - t_0)$ and of a weak periodic stress with amplitude, frequency and initial phase, a, ω, φ, respectively. The resulting summary stress τ_s is:

$$\tau_s(t) = \tau(t) + a\sin(\omega t + \varphi) \tag{9.6}$$

When the summary stress mounts to some critical value τ_c, the friction resistance is overcome and the slip event, accompanied with a stress drop, occurs:

$$\tau_s(t) = \tau(t) + a\sin(\omega t + \varphi) = \tau_c \tag{9.7}$$

If we assume that at t_0 the stress is equal to τ_{0s} and after that the stress increases linearly as $\tau_s(t) = b(t - t_0)$, then:

$$\tau_{0s} + b(t - t_0) = \tau_s(t) + a\sin(\omega t + \varphi) = \tau_{cs}$$

or finally

$$b(t - t_0) = \tau_{cs} - \tau_{0s} - a\cos(\omega t + \varphi) \tag{9.8}$$

Solution of Eq. (9.8) allows composing of the distribution of slip moments versus certain phases of the forcing period. The synchronization effect reveals itself as an intensive peak in the distribution within a certain range of forcing period phases.

The generalized approach to a problem of synchronization or phase locking of an autonomous oscillator (here the spring-slider system) by a weak forcing is presented in Rosenblum et al. (1996, 1997) and Pikovsky et al. (2003). According to this, the natural frequency ω_0 of the autonomous oscillator subjected to a weak forcing of frequency ω and intensity I changes its frequency ω_0 to a different value Ω. Note that the natural frequency of the autonomous oscillator changes under forcing, which means that the observed synchronization frequency Ω can be both lower or higher than ω_0. The difference $(\omega - \omega_0)$ is called detuning: on the phase space plot of I versus ω the synchronization area of autonomous oscillator forms an inverse bell-shaped area, so-called Arnolds' tongue (Arnold, 1983), with a minimum at $(\omega - \omega_0) = 0$, i.e., at the point where forcing frequency is close to the natural frequency of the oscillator. As the detuning increases, stronger forcing is needed for phase locking and at very large detuning, synchronization becomes impossible. Sometimes, besides 1:1 phase locking at $\omega = \omega_0$, the phenomenon of high-order synchronization (HOS) can be observed at multiples of natural frequency $m\omega_o$ (Pikovsky et al., 2003; Chelidze et al., 2009, 2010). It is necessary to discriminate between synchronization (the result of weak forcing) and modulation, which is observed at strong forcing. In the case of modulation the period of oscillation (here, the dominant waiting time of AE bursts) coincides exactly with the forcing period at any intensity and frequency of forcing, i.e., the phase locking is very rigid. Synchronization means that: (1) both the autonomous oscillator and forcing have their own rhythms, (2) weak forcing is enough to adjust the rhythms of these two systems, and (3) the synchronization of rhythms occurs only in a certain range of detuning (Pikovsky et al., 2003), i.e., when the difference between the frequency of forcing ω and the natural frequency of the autonomous oscillator ω_0 is not too large and the forcing intensity is not too low.

The full solution of the rate-and-state-dependent equation of friction with additional forcing is quite complicated (Bureau et al., 2000; Chelidze and Varamashvili, 2010; Putelat et al., 2007; Matsukawa and Saito, 2007) to interpret experimental data, so we will use in the following the aforementioned simplified approach.

The issue of mechanically forced (triggered) stick–slip is closely related to the problem of induced or triggered EQs, which are generated by a weak natural or anthropogenic impact. The oldest example is an attempt to study the correlation of seismic activity with Earth tides, which leads to controversial results. In the last few decades it was discovered that the wave trains of remote strong EQs can trigger small to moderate events (dynamic tremors) on very large epicentral distances. For example, dynamic tremors can be triggered by long period wave pressures of the order of 5 kPa (Brodsky and Prejean, 2005). The interest in seismic activity increase, caused by human activity, had also been growing significantly during the last few decades, due to the social and economic impacts of hydrocarbon production, reservoir impoundment, mining operations, geothermal energy production, underground gas and fluid storages, etc. The additional stresses, connected with all these forcings, are much smaller than the main tectonic stresses, which points to the closeness of some volumes of the Earth's crust to the critical state (Scholz, 1990, 1998), where even forcing as small as a few kPa can trigger seismic events.

Triggering and synchronization are actually two faces of the same coin, with the important difference that in the synchronization approach we can correlate phases of dozens and hundreds of events with their source (forcing), which makes synchronization analysis a reliable tool for the assessment of the forcing effect.

The problem of mechanically forced stick–slip has been considered in several papers, published in leading geophysical journals (Beeler and Lockner, 2003; Savage and Marone, 2007; Bartlow et al., 2012) for the analysis of tidal forcing of EQs.

9.3.2 Laboratory Experiments on Forced Stick–Slip

For laboratory study of forced stick–slip, the same experimental schemes are used as in natural stick–slip studies, with the addition of a weak forcing source. Despite versatile stress application schemes, realized in the two- and three-axial devices, the simple uniaxial system also has merit, because it is free from massive inertial loading frames, which can complicate the interpretation of results. Fig. 9–1 presents the scheme of the simplest system, used in the series of experiments on mechanical and electromagnetic (EM) forcing of stick–slip (Chelidze et al., 2002, 2009; Chelidze and Lursmanashvili, 2003; Chelidze and Matcharashvili, 2007).

Mechanical forcing. Laboratory set up in synchronization experiments represents a system of two horizontally oriented saw-cut basalt plates (Fig. 9–1). The height of surface asperities was in the range 0.1–0.2 mm. A constant pulling force F_p of the order of several N was applied to the upper (sliding) plate; in addition, the same plate was subjected to periodic mechanical perturbations (forcing) from a vibrator. The normal load was constant and equal to the weight of the sliding sample (700 g). In our experiments the following quantities were varied: (1) the frequency ω of superimposed periodical mechanical perturbation; (2) the amplitude I of the external mechanical excitation (forcing); and (3) the stiffness of the spring, $K_s(78 \text{ N/m} < K_s < 1700 \text{ N/m})$. The forcing with a variable frequency (from 1 to 120 Hz) and amplitude was applied normally to the slip plane. Mechanical pull from the forcing was much weaker (of the order of $10^3 - 10^{-4}$N) compared to the pulling force F_p of the spring ($F_p >$ several N), and this indicates that experiments were performed in the synchronization regime and not in the modulation regime (Pikovsky et al., 2003). The mechanical forcing

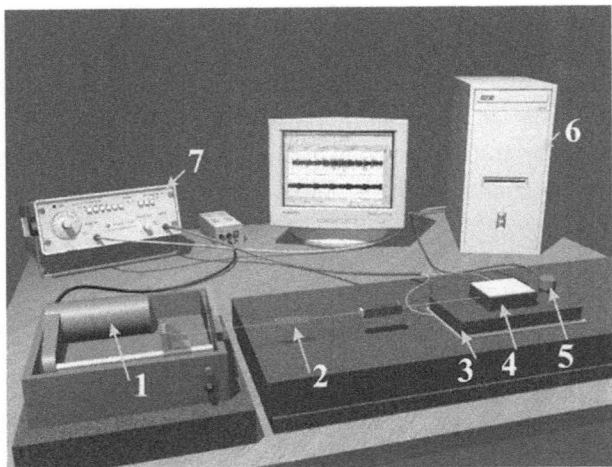

FIGURE 9–1 Scheme of the laboratory system for studying forced stick–slip: 1. driving block, 2. spring, 3. fixed plate, 4. sliding plate, 5. vibrator, 6. computer and monitor with two-channel recording of forcing signal and AE, 7. generator for varying intensity and frequency of forcing.

strength I was varied by applying a voltage from 0 to 5 V to a mechanical vibrator. Slip events in synchronization experiments were registered as acoustic bursts by the sound card of PC (Chelidze et al., 2002; Chelidze and Lursmanashvili, 2003).

Electromagnetic forcing. The spring-block system was the same as in the mechanical forcing experiments, described above. The only difference was that forcing, given by periodic electric perturbations, was applied directly to electrodes, glued to the external surfaces of sliding and fixed blocks. The mechanical equivalent to the effect of the EM field on the sliding surface was much weaker (of the order of 1 N) compared to the driving force. The electric field lines were normal to the sliding plane that generates (ponderomotive) attraction forces between fixed and slipping plates. Slip events were recorded as AE bursts as in case of mechanical forcing. Details of the setup and technique are given in Chelidze and Lursmanashvili (2003).

9.4 Forced Stick–Slip Results: Mechanical Forcing

We investigated the synchronization of the same spring-slider system under weak periodic mechanical forcing (Chelidze and Matcharashvili 2007; Chelidze et al., 2010b; Varamashvili et al., 2008; Varamashvili and Chelidze, 2004). We conducted experiments for two modes of periodic mechanical forcing: normal to the slip surface and parallel to the slip. F_{mech} was in the range $10^{-3} - 5 \times 10^{-5}$N, which means that forcing was always much less than the minimal driving force $F = 4$ N. The driving velocity was generally 0.47 mm/s.

Fig. 9–2 presents an experimental record of a single acoustic pulse (upper channel) and corresponding tangential mechanical forcing (lower channel) at mechanically forced stick−slip: the mechanical forcing is in the range $(5 \times 10^{-5} - 2 \times 10^{-3})$ N, which corresponds to voltages of 1−4 V, applied to the mechanical vibrator.

In Fig. 9−2 the arrows mark the onsets and termination moments of a single acoustic pulse during the slip event (upper panel of Fig. 9−2). The arrows in the lower panel show maxima of forcing signal, relative to which we count off the phases of onsets and terminations. In Fig. 9−3A we show distributions of the AE burst onsets relative to the phase (in decimals) of the mechanical forcing period for normal forcing. At low voltages (up to 1 V) the

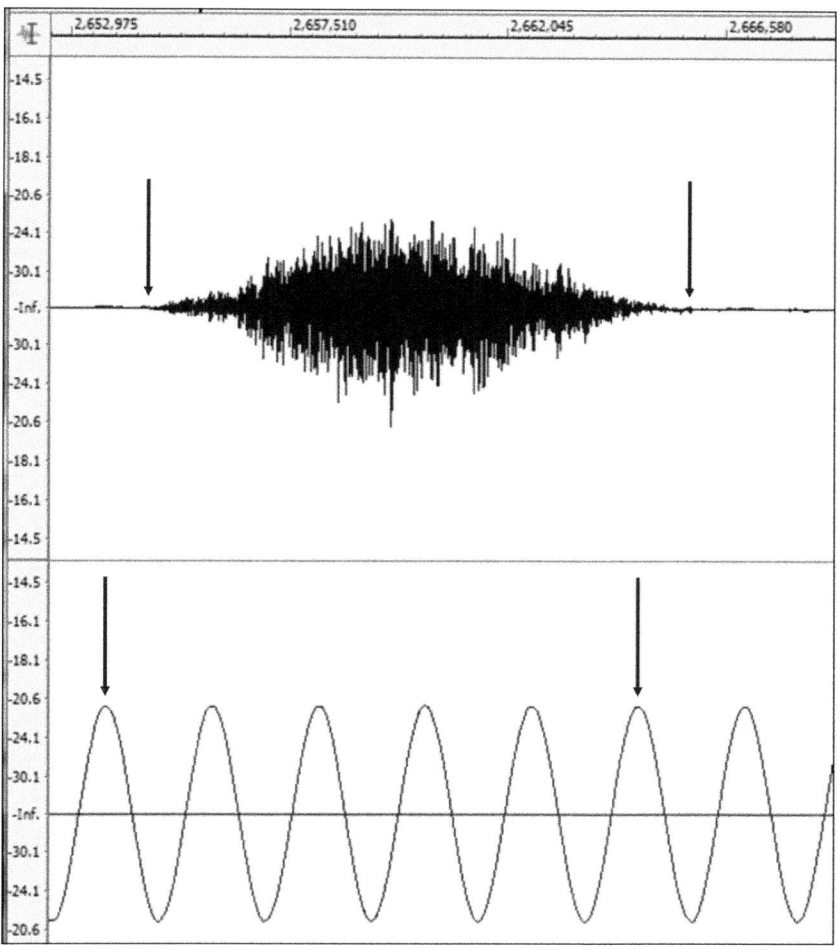

FIGURE 9–2 Print screen of the recording of a single acoustic pulse (upper channel) and corresponding tangential mechanical forcing (lower channel) at mechanically forced stick−slip: the *Y*-axis shows amplitude of acoustic emission (AE, upper panel) or amplitude of forcing vibration intensity (lower channel) in decibels at application of 4 V to the vibrator; *X*-axis is time in milliseconds. The initial small deviation from the background line is considered as the onset of the AE pulse relative to the previous forcing maximum; the moment of pulse termination is also determined relative to the previous forcing maximum; these moments are marked by arrows.

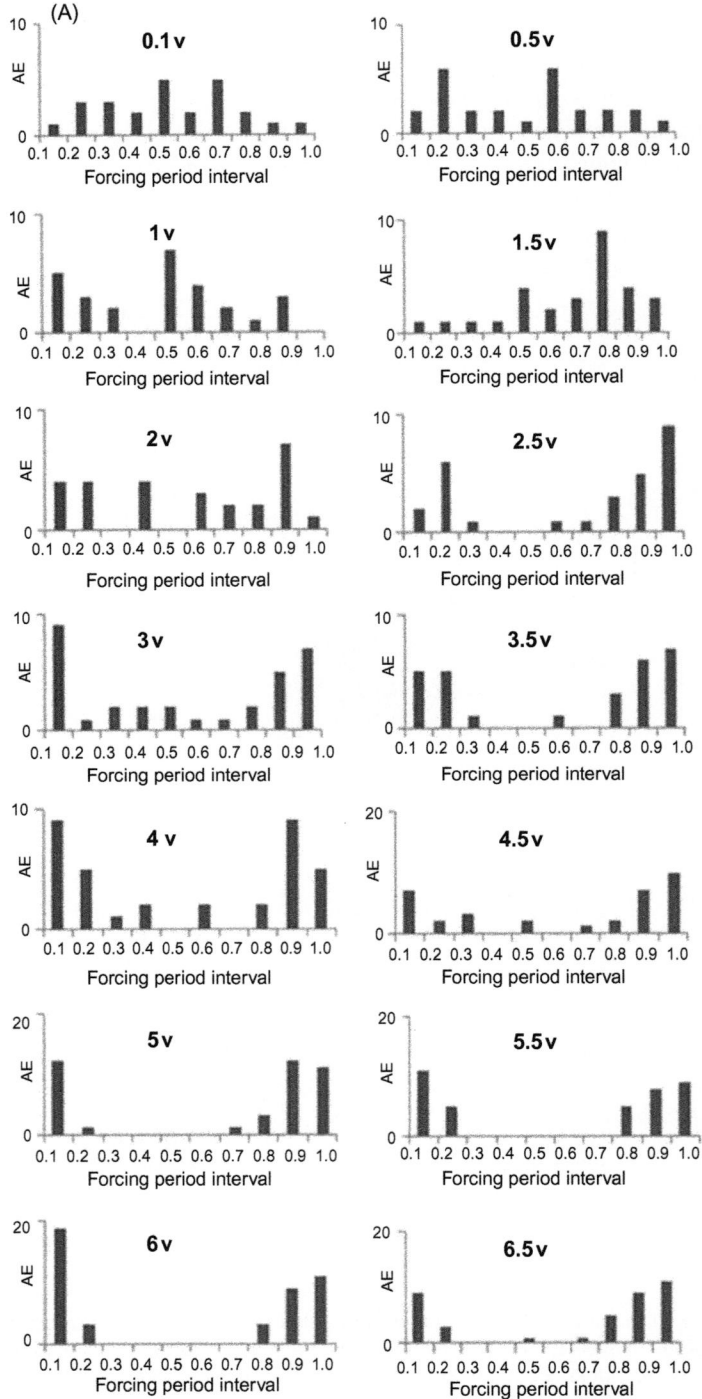

FIGURE 9–3 (A) Distribution of acoustic emission (AE) onset number relative to mechanical forcing period phases (in decimals) for different intensities of normal forcing. Forcing frequency − 20 Hz; (B) Distribution of AE onsets (the left column) and terminations (the right column) relative to the (mechanical) forcing period phase (in 12 of the forcing period) for different intensities of tangential forcing. Spring stiffness = K_s = 223.7 N/m, drive velocity V = 0.47 mm/s. Forcing frequency − 80 Hz.

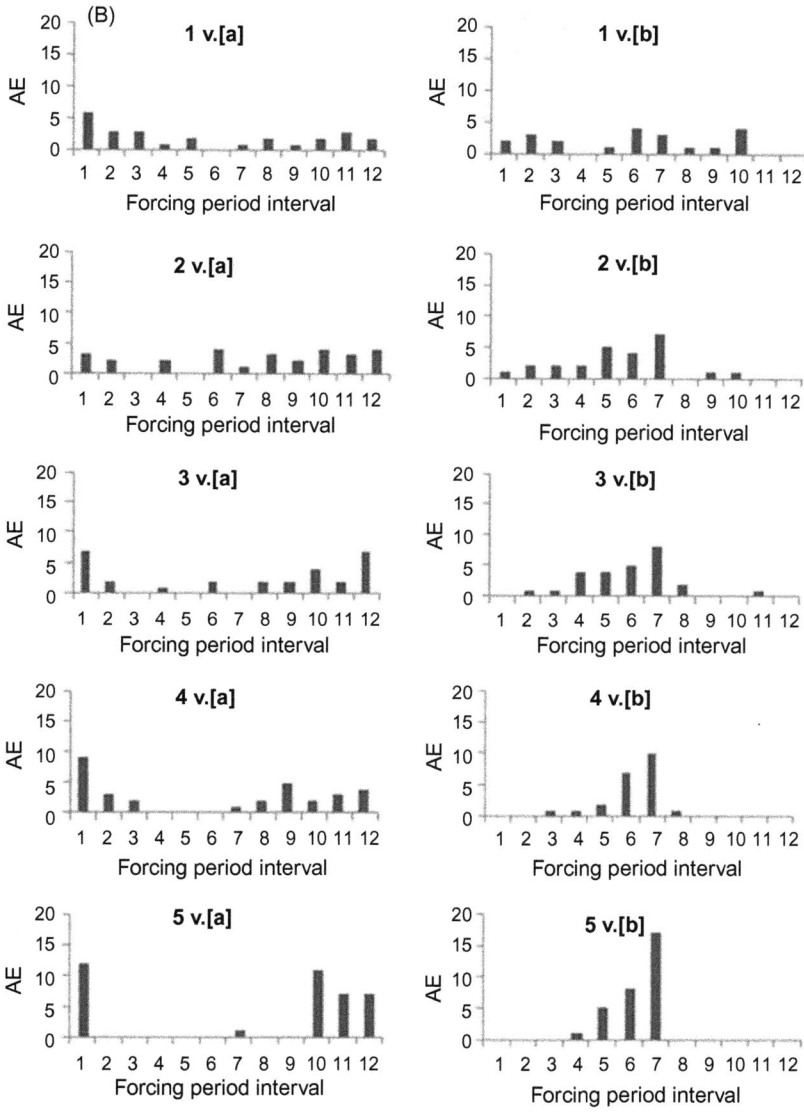

FIGURE 9–3 (Continued)

onsets are more or less randomly distributed in the decimals of the forcing period. A voltage increase results in concentration of the offsets at a definite part of the forcing period, namely in the first and the last decimals of the forcing phase. Voltage increase results in concentration of the offsets at a definite part of the forcing period, namely in the first and the last decimals of the forcing period for normal forcing. Evidently, increasing the voltage applied to the mechanical vibrator promotes synchronization of AE offsets with external forcing. Similar behaviour is observed for tangential mechanical forcing (Fig. 9.3B; Chelidze et al., 2010).

It was a surprise to discover that a weak mechanical forcing synchronizes not only AE onsets (Fig. 9–3B, left column) but also the terminal parts of the signal (Fig. 9–3B, right column), by forcing at the same strength as for onsets. We conclude that the forcing can affect not only the phases of onsets but also the phases of terminations of AE bursts: their influence is different at different phases of forcing. Tangential forcing triggers slips (AE onsets) around the minimum phase and suppressed at the maximum phases of the forcing period.

An increase of the forcing intensity, besides giving better synchronization of onsets and terminations, also brings on regular shortening of duration of AE bursts (Fig. 9–4), i.e., duration of the slip phase.

Fig. 9–5A shows the distribution of waiting times between slips T (AE onsets) for natural T_0 and forced stick–slip T_{obs} for $K_s = 223.7 \text{N/m}$, drive velocity $V = 0.47 \text{ mm/s}$: note the decrease in the dominant waiting times from 6–7 s for natural to 2 s for forced stick–slip, as well as a strong decrease in the PDF half-width in the case of forcing. Fig. 9–5B presents PDFs of phase shifts of slips $\Delta\varphi$ between the forcing signal and AE onset for forced and natural stick–slip; of course, in the latter case the phase shift was calculated between the AE (slips) and fictitious sinusoid phase. It is evident that phase shifts of the forced process tend to some dominant value around 30 ms (i.e., synchronization is present), whereas natural stick–slip demonstrates flat distribution of AE phase shifts relative to the imaginary sinusoid phase.

9.5 Measuring Complexity/Ordering of Natural Processes: Nonlinear Dynamics Tools

Natural processes are complex (Rundle et al., 2009), mainly due to their nonlinearity. Complexity incorporates phenomena with a very broad diversity of dynamic features. Generally, this diversity manifests itself in a certain kind of hierarchy of dynamic patterns ranging from strict determinism to total randomness. Between these extremes there are many intermediate states that reveal different degrees of orderliness, e.g., periodicity, quasiperiodicity, deterministic chaos, low- and high-dimensional dynamics, hyperchaos, etc. (Kantz and Schreiber, 1997). The fundamental problem is how to measure the complexity

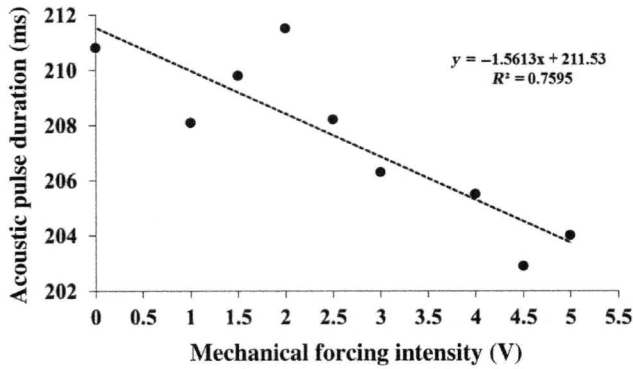

FIGURE 9–4 Mean duration of stick–slip-generated acoustic pulses for different intensities of normal mechanical forcing, with a trend line.

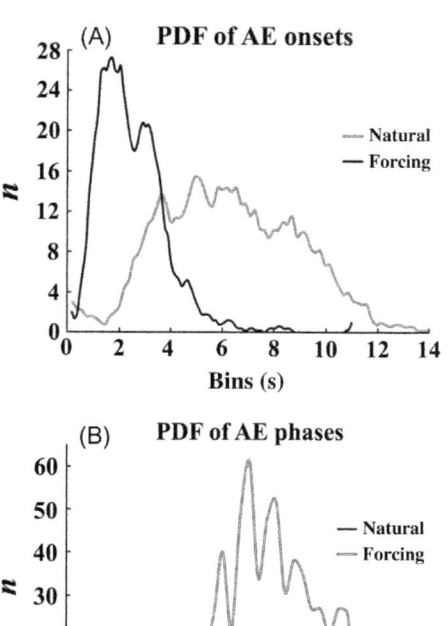

FIGURE 9–5 (A) Acoustic emission (AE) waiting time distributions for natural (grey) and forced (black) stick–slip in seconds. (B) Distribution of AE phase shifts $\Delta\varphi$ relative to the (imaginary) forcing sinusoid phase for natural (black) and forced (grey) stick–slip in seconds. In both plots on the Y-axis is shown the number (n) of AE bursts in a bin; spring stiffness K_s = 223.7 N/m, drive velocity V = 0.47 mm/s.

from the observed time series (Matcharashvili and Chelidze, 2010). One of the most often used methods is the time delay method for phase space reconstruction of original time series (Takens, 1981; Sprott, 2003). After phase space reconstruction, a phase space structure (attractor) can be analysed using both qualitative and quantitative methods (e.g., Packard et al., 1980; Takens, 1981; Abarbanel and Tsimring, 1993; Kantz and Schreiber, 1997; Sprott, 2003).

The evolution of phase space trajectories can be analysed by calculation of the spectrum of Lyapunov exponents or simply by calculation of the maximal Lyapunov exponent $\lambda_{\max}$ (Rosenstein, 1993). Calculation of fractal dimension measures of phase space point sets reveals the unknown dynamics of the process by analysis of self-similar properties of phase space objects. These measures are correlation dimension (d_c), information dimension (d_i), Hausdorf dimension d_H, etc. (Grassberger and Procaccia, 1983; Abarbanel and Tsimring, 1993; Abarbanel et al., 1996; Kantz and Schreiber, 1997; Sprott, 2003; Sornette, 2000; Strogatz, 2000).

Since linear correlations lead to many errors in nonlinear time series analyses, it is important to verify obtained results using the so-called surrogate data approach (Theiler and Farmer, 1992). The surrogates are constructed from the original time series using different null hypotheses. The most commonly used significance measure of the difference between the original time series and the surrogate data is the criterion: $S = |\langle M_{surr} \rangle - M_{orig}|/\sigma_{surr}$, where σ_{surr} denotes standard deviation of M_{surr} (Theiler and Farmer, 1992).

In the last few decades new techniques for the analysis of not too long and noisy time series have been developed, like the method of recurrence plots (RPs) (Eckmann et al., 1987). Generally, RPs are designed to locate hidden recurring patterns and structures in time series and are defined as $N \times N$ symmetric matrix:

$$R_{i,j} = \Theta(\varepsilon_i - \|\vec{x}_i - \vec{x}_j\|), \quad i,j = 1, \ldots, N, \tag{9.9}$$

where $\vec{x}_{i,j}$ are phase space vectors reconstructed using Taken's time delay method, and $\Theta(x)$ is the Heaviside function. RP visualizes the distance matrix, which represents autocorrelation in the series at all possible time (distance) scales. As far as distances are computed for all possible pairs, on the RP plots elements near the diagonal correspond to short-range correlation, whereas the points distant from the diagonal reveal the long-range correlations. Hence, if the analysed time series dynamics is deterministic (ordered, regular), then the RP shows line segments parallel to the main upward diagonal. At the same time if dynamics is purely random, the RP will not present any structure at all.

Zbilut and Webber (1992) and Webber and Marwan (2015) developed a tool, which quantifies the complexity level of structures in RPs, namely, the recurrence quantitative analysis (RQA). They define several measures using the recurrence point density, the length of diagonal and vertical structures in the % *recurrence* (% REC), is simply the percentage of those pairs of points, whose spacing is below ε_i, the predefined cutoff distance: the more periodic is the signal dynamics, the higher is the % REC value. The second RQA measure is called % *determinism* (%DET); this measures the percentage of recurrent points in an RP that are contained in lines parallel to the main diagonal. Again, the more ordered is the signal dynamics, the higher is the % DET: in addition, since diagonals represent points close to each other successively forward in time, %DET also contains the information about the duration of a stable phase: the longer the interactions, the higher the %DET value. The third most used RQA measure, the *entropy*, is closely related to %DET. Thus, whereas % DET accounts for the number of the diagonals, %ENT quantifies the distribution of the diagonal line lengths. The more different are the lengths of the diagonals, the more complex is the deterministic structure of the RP. One can find detailed descriptions of the RQA method and its applications to different processes in Zbilut and Webber (1992) and Webber and Marwan (2015).

Besides phase space testing, there are also other tests for complex data analysis, namely, information statistics methods. For example, Shannon's information criterion allows derivation of the information content (or lack of it) in a message, i.e., assessing underlying regularity in a data set. Well-known information complexity tests include the following: iterated function systems (IFS) (Jeffrey, 1992), Lempel–Ziv complexity (LZC) measure (Lempel and Ziv, 1976), mutual information (MI, Sprott, 2003), Shannon and Tsallis entropies (Tsallis, 1988; Vallianatos et al., 2016), etc.

Synchronization is one of manifestations of the nonlinearity/complexity of a system (Rosenblum et al., 1996, 1997; Pikovsky et al., 2003; Meyers, 2009). That is why the tools of nonlinear dynamics are the main methods for revealing synchronization and measuring its strength. It is reasonable to assume that the existence of some determinism (i.e., recurrence of definite states) in the structure of forced system phase space, dependent on the forcing intensity/frequency, should be a signature of synchronization. That is why we used tools of nonlinear dynamics for quantitative assessment of both complexity of system and the strength of synchronization.

Several new methods prove their effectiveness for quantitative analysis of synchronization strength between forced process and forcing signal. The methods are: compilation of the phase space plot of forcing intensity versus forcing frequency (Arnold's plot), calculation of the generalized phase difference, MI, Shannon and Tsallis entropies, conditional probability of phases, flatness of stripes of synchrograms (or the stroboscopic approach, founded on Poincare section technique), and coefficients of phase diffusion (Rosenblum et al., 1996, 1997; Pikovsky et al., 2003). These methods give good results in analysis of relaxation (integrate-and-fire type) processes, when the signal is composed of well-defined marker events (slips, AE bursts, EQs).

To reveal hidden periodicities, related to synchronization, in short and noisy time series, the methods of singular spectral analysis (SSA) and detrended fluctuation analysis (DFA) have been suggested (Broomhead and King, 1986; Telesca et al., 2012). Lastly, in the case of very restricted statistics (tens of events), less demanding methods such as the Schuster test (Schuster, 1897) can be used.

9.6 Complexity Analysis of Mechanically Forced Stick—Slip

9.6.1 On the Patterns of the Synchronization Area in the Phase Space Plot

According to the general theory of synchronization of integrate-and-fire processes (Pikovsky et al., 2003) the phase space plots (the Arnold's tongue plot) for synchronized AE sequences independent of the processing method should have an inverse triangle (bell-curve) form, namely, they should have both low-frequency (LF) and high-frequency (HF) branches. Note that according to this general model: (1) the observed frequency at synchronization Ω can be lower or higher than ω_0 due to coupling of the forcing with autonomous oscillator (actually, this leads to Arnold's tongue formation); and (2) in addition to 1:1 phase locking at $\omega = \omega_0$, the phenomenon of HOS can be observed at multiples of natural frequency $m\omega_o$ as is shown in Fig. 9—6A (Pikovsky et al., 2003; Chelidze et al., 2009, 2010). The parameter m is called a winding number for the ratio of frequencies.

In recent years the 'nucleation' model has been suggested, where it is supposed, that if the forcing's period is less than some nucleation phase duration, it cannot lead to synchronization: as a result the synchronization area plot reveals only the low-frequency branch (Beeler and Lockner, 2003; Bartlow et al., 2012; Savage, 2007; Scholz, 2003).

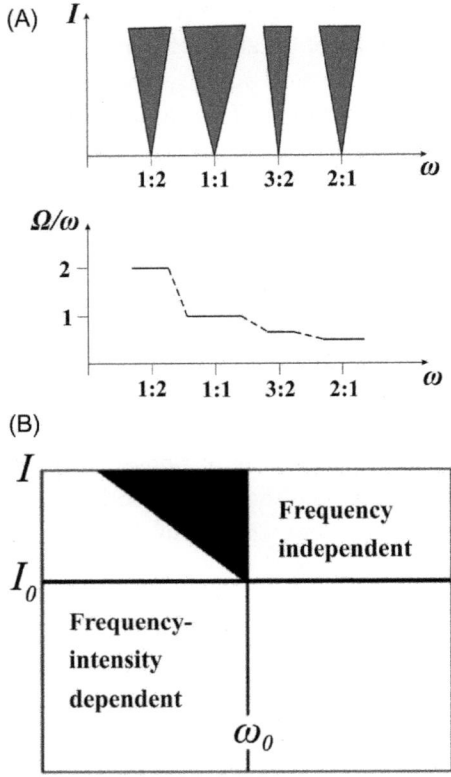

FIGURE 9–6 (A, B) Different patterns of synchronization area: (A) (upper panel) Conventional Arnold's Plot or CAP model of synchronization areas in the forcing frequency–intensity phase space plot (multiple Arnold's tongues versus forcing frequency ω), modified from http://www.scholarpedia.org/article/Synchronization. When the frequency of forcing ω is close to the natural frequency of autonomous oscillator ω_0, the 1:1 synchronization regime is established. In addition, the high-order synchronization (HOS) regime establishes at multiples of ω_0 (1/2, 3/2, 2/1), marked on the ω axis. (A) (lower panel) Horizontal plateaus mark Ω/ω rational frequency ratios, where synchronization is possible, versus forcing frequency ω; (B) TAP model: On the plot of I versus ω the synchronization area is similar to Arnolds' tongue only at low forcing frequencies $\omega < \omega_0$ and at high frequencies the correlation of slips with forcing frequency vanishes. The HOS phenomenon is not considered in the TAP model.

In the following, we will describe the conventional Arnold's plot (CAP) and the nucleation model (truncated Arnold's plot – TAP). The main parameter in both these models is the natural frequency of slip events ω_0 (or waiting times of slips T_0, which constrains the range of synchronizing forcing frequencies ω). Namely, the frequency detuning ($\omega - \omega_0$) should be small enough to ensure phase locking. Fig. 9–6A,B illustrates the pattern of synchronization area according to both these models in the phase space plot of forcing frequency versus forcing intensity. Note that the TAP model (Fig. 9–6B) does not envisage the possibility of synchronization at frequencies higher than ω_0, when the CAP model (Fig. 9–6A) predicts an inverse bell-curve form of synchronization area with a minimum at ω_0 as well as the existence of multiple synchronization areas on both sides of ω_0 (multiple Arnold's tongues). The fractal curve Ω/ω in Fig. 9–6A is a so-called devil's staircase curve of synchronization areas versus forcing frequency. Horizontal plateaus mark Ω/ω rational frequency ratios, where synchronization is possible, including HOS. The possibility of the HOS phenomenon is not considered in the TAP model.

Here we have to clarify the concept of slip nucleation time T_0 (or T_{obs} in the case of forced slip). In accordance with the general theory of synchronization of integrate-and-fire systems, we define the nucleation time as a mean waiting time between slips in contrast to some other definitions, where the nucleation time is just a part of T_0 (or T_{obs}) immediately before the slip, where one observes some precursory phenomena (pr-slips, microacoustic events).

Our experiments show that the conventional Arnold's tongue model works well, though stick−slip is definitely a process with a nucleation phase: the nucleation phase corresponds to a whole stick phase, when the stress accumulates to overcome friction resistance. In the general model of integrate-and-fire oscillators (Pikovsky et al., 2003) in each cycle of the recurrent process the driving force slowly accumulates (integrates) and at some threshold value instantly relaxes (fires). This picture fully describes the stick−slip process: the stick phase is a slow phase, when the stress accumulates and the fire phase is an instant slip, which we identify with acoustic bursts.

The absence of a HF branch, declared in Beeler and Lockner (2003), Bartlow et al. (2012) and Savage and Marone (2007) and interpreted in terms of the TAP model (Fig. 9−6B), can be explained by some experimental restrictions (e.g., by a small number of experimental points in the synchronization area, high inertia of the loading frame, etc.). It should be noted that in some experiments by these cited authors the incipient HF branch of Arnold's plot seems to be present (see Figs 5a and 7a in Beeler and Lockner (2003) and Fig. 3a in Bartlow et al. (2012)). We present our interpretation of the above-mentioned experimental data (Fig. 5a in Beeler and Lockner, 2003) in modified Fig. 9−7, where the trend-line is a possible inverse bell-curve configuration of the synchronization area with a minimum at the forcing period of approximately 100 s.

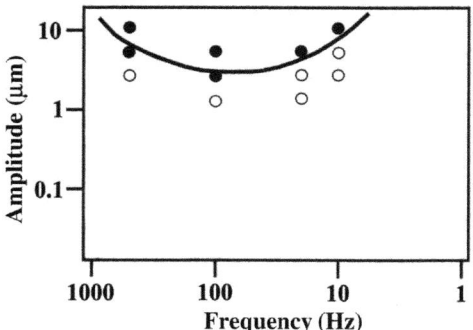

FIGURE 9–7 Reinterpretation of the experimental data of Beeler and Lockner (2003). The *Y*-axis shows the amplitude and the *X*-axis shows the period of mechanical forcing. Solid circles correspond to the high degree of synchronization of slips with the phase of oscillating forcing (Schuster probability >99.5%), and open circles represent a lower synchronization strength. Loading rate 0.1 microns/s. Each symbol represents a mean for 20 slip events. The solid line is our version of a possible trend-line, corresponding to the Arnold's tongue mode with a minimum at the forcing period approximately 100s.

More detailed experiments with detailed filling of the intensity–frequency phase space plot could reveal the full configuration of Arnold's plot with both LF and HF branches (including possible high-order Arnold's tongues).

At the same time, we cannot totally exclude the possibility of the truncated pattern shown in Fig. 9–6B. For example, in the case of the noise-induced van der Pole oscillator with additional noise, phases can slip, constraining the region of entrainment and, in this case, the Arnold's tongue plot has practically no HF branch (Mitarai et al., 2013). Similar results are obtained for a rotator under harmonic forcing (García-Álvarez et al., 2008). The TAP-like configuration of the synchronization area in the cited publications is due to the specific influence of additional noise and not to the restricting effect of nucleation phase duration.

9.6.2 Phase Space Plot of the Mechanical Synchronization Area (Arnold Tongue Plot)

In the previous sections, we have shown that weak mechanical forcing synchronizes the slip recurrence moments with the phase of forcing (Fig. 9–3). In order to have the full pattern of the process, it is necessary to delineate the synchronization area in the phase plot of applied forcing intensities I versus forcing frequencies f (Pikovsky et al., 2003). We have carried out a corresponding series of experiments for different values of stiffness K_S of spring (235, 555, and 1700 N/m) in the frequency range 0.5–120 Hz for different voltages V, applied to a vibrator (which in our case is a proxy of mechanical forcing intensity I). As the absolute values of mechanical forcing intensity I are not decisive for our task of phase plot construction, in the following we present our results as phase plots of synchronization strength, which depends on the frequency $f = \omega/2\pi$ of superimposed periodical perturbation and voltage V, applied to the mechanical vibrator. In other words, we will use interchangeably the terms voltage V and the mechanical intensity variable I. For each frequency and intensity of forcing three sets of experiments were carried out and then merged into one file in order to have a sufficient data set for statistical analysis. The experimental time series of phase differences between forcing signal and slip initiation (AE start) $\Delta\varphi$ were picked up in seconds, and were then converted into 2π(Rad) units and corresponding histograms were compiled.

In experiments with a spring of stiffness $K_s = 555\ N/m$ and drive velocity 8.47 mm/s we were able to observe at least two synchronization areas: the general pattern resembles the scheme presented in Fig. 9–6A. The recurrence period of slips during the natural stick–slip process was 0.37 s; correspondingly, the recurrence frequency was 2.7 Hz (Fig. 9–8A). The 1:1 synchronization was fixed at the forcing frequencies, close to the natural slip frequency, in the range 2–3 Hz (Fig. 9–8B). The low-frequency mode of HOS is observed at 1 Hz (Fig. 9–8C). In this case the periods' winding number n is: $\frac{T_{obs}}{T} = n = 2$.

The corresponding histograms of phase differences $\Delta\varphi$ between forcing and slip phases are presented in Fig. 9–9. In the histograms in Fig. 9–9 the X-axes represent phase differences $\Delta\varphi$ in units of 2π(Rad) and Y-axes represent the number of events with certain $\Delta\varphi$ in a given bin; a bin is taken as being equal to 1/10th of the forcing period.

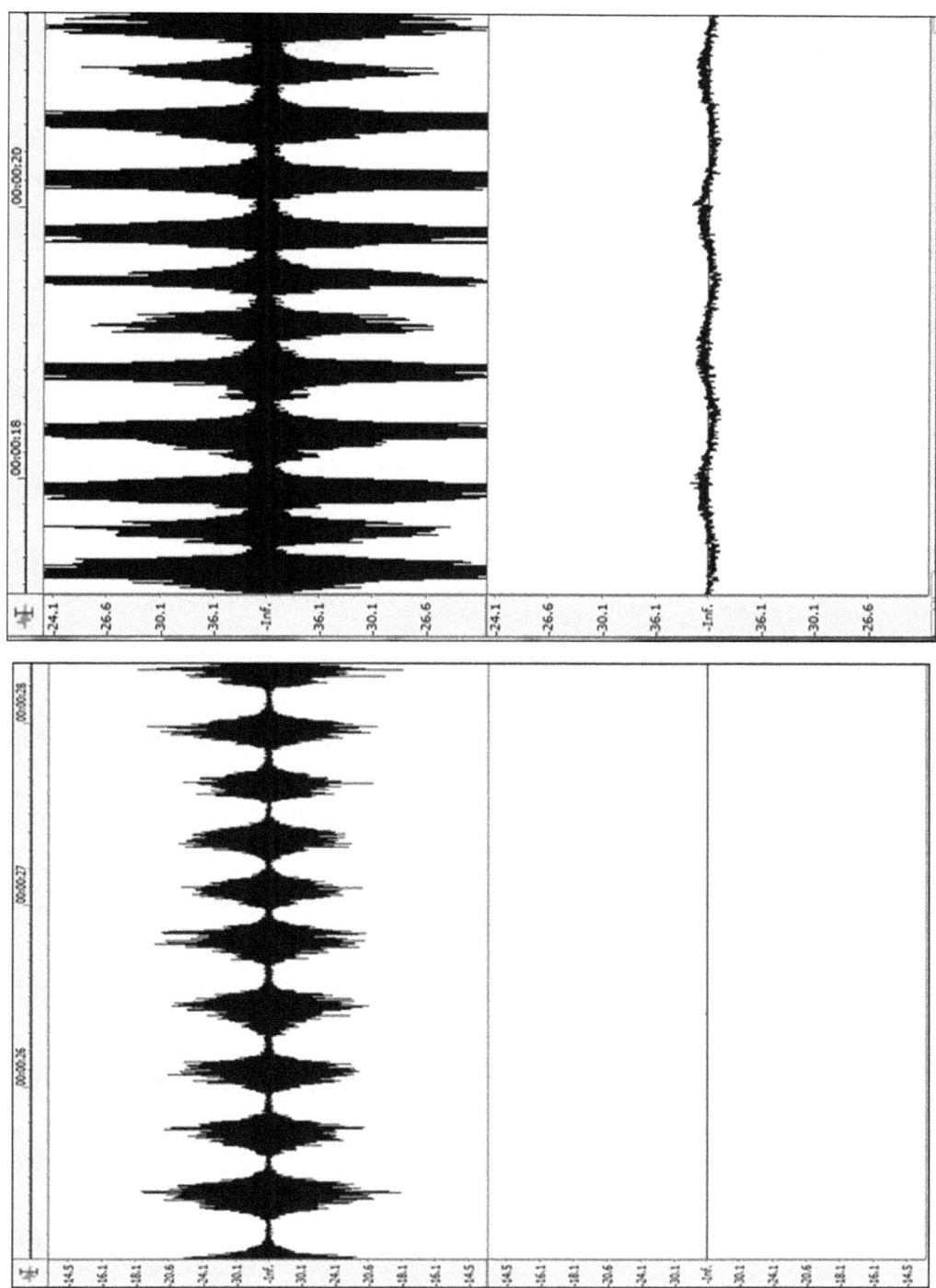

FIGURE 9–8 Slip recurrence modes for the spring of stiffness $K_s = 555\ N/m$ and drive velocity = 8.47 mm/s. Upper channel: acoustic bursts generated by slips; lower channel: forcing signal. (A) Natural stick–slip, mean frequency of slips: 2.7 Hz. (B) Forced stick–slip at forcing frequency and intensity: 1 Hz and 2 V. (C) Forced stick–slip at forcing frequency and intensity: 2 Hz and 2 V. Note transition from 1:1 synchronization to HOS mode at 1 Hz forcing.

FIGURE 9–8 (Continued)

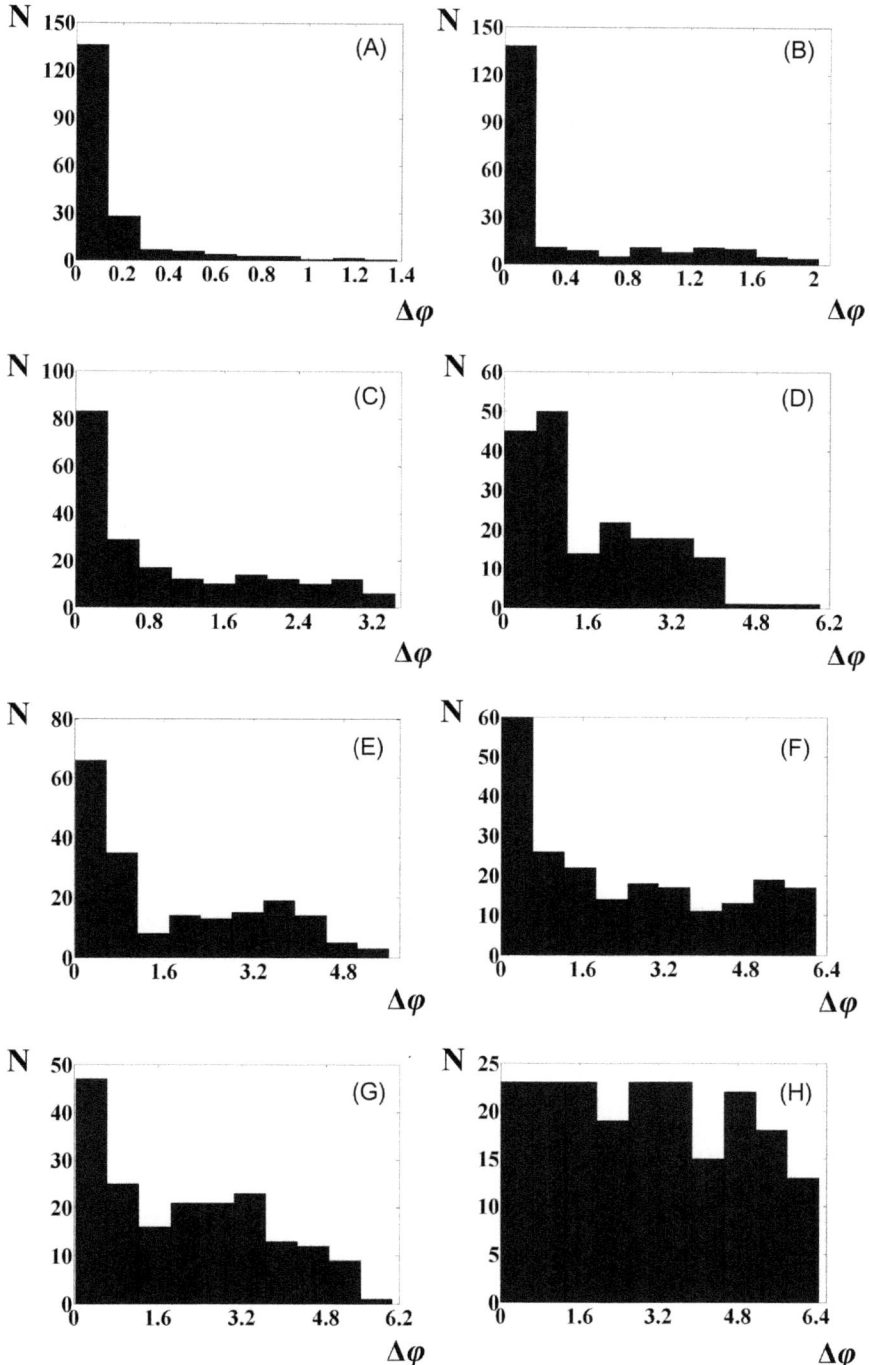

FIGURE 9–9 Histograms of number N of AE signals in a bin, normalized to the total number of the AE events for a given frequency $\sum N$, with certain phase differences $\Delta\varphi$ between forcing signal maximum and AE onset versus phases of the forcing period, expressed in the units of 2π(Rad), in 1/10th bins of the forcing period. In this set of experiments: spring stiffness $K_s = 555\ N/m$ and drive velocity = 8.47 mm/s; (A) and (B) show histograms for 1:2 HOS area at forcing frequency of 1 Hz and forcing intensities of 0.5 V and 2 V, respectively. (C,E,G) show histograms for 1:1 synchronization area at forcing frequencies of: (C) 2 Hz, (E) 3 Hz, (G) 4 Hz, and a forcing intensity of 0.5 V. (D,F,H) show histograms for 1:1 synchronization area at forcing frequencies of: (D) 2 Hz; (F) 3 Hz; (H) 4 Hz, and a forcing intensity of 2.0 V.

Evidently, the distribution of phase differences in the presented histograms is quite different. Some (Fig. 9–9A–G) manifest preferred values of $\Delta\varphi$, which is evidence of synchronization and some correspond to a flat distribution (i.e., to the absence of phase locking).

Fig. 9–9A,B show histograms for 1:2 HOS area at a forcing frequency of 1 Hz and forcing intensities of 0.5 V and 2 V, respectively. Fig. 9–9C,E,G show histograms for 1:1 synchronization area at forcing frequencies of: (C) 2 Hz, (E) 3 Hz, (G) 4 Hz and a forcing intensity of 0.5 V. Fig. 9–9D,F,H show histograms for 1:1 synchronization area at forcing frequencies of: (D) 2 Hz; (F) 3 Hz; (H) 4 Hz and a forcing intensity of 2.0 V.

Here we want to stress that similarity to uniform distribution in Fig. 9–9G,H does not mean that the recurrence is absent in the nonsynchronized (natural) stick–slip. This nonforced (or too weakly forced) stick–slip process is still quasiperiodic with a dominant waiting time between slips, equal to T_0, but the phases of slips are not synchronized with the forcing frequency. Thus, the quasiuniform $\Delta\varphi$ distribution in Fig. 9–9G,H means only that the forcing intensity I is too weak to change the natural rhythm of stick–slip (to synchronize it with forcing frequency). As a result, phase differences $\Delta\varphi$ between forcing signal and slips are distributed almost uniformly.

Histograms allow for assessment of the synchronization strength visually: the almost uniform distribution of phase differences in the whole period of forcing means that synchronization is absent. On the other hand, the appearance of close to unimodal histograms with a well-expressed maximum at a certain phase of the forcing period means that the onsets of AE bursts are strongly correlated with forcing phases, i.e., there is some synchronization (see Figs. 9–9A–F). The better expressed the maximum in the histogram plot (or the smaller histogram half-width), the higher the synchronization strength is.

The distribution of bins in the histograms (Fig. 9–9) is asymmetric: the first bins contain most of the events. We assessed the strength of synchronization by a full-width at half-maximum (fwhm) of histogram distribution in the same way as in the analysis of nearly symmetric histogram distributions. Namely, we identify the height of the highest bin of the histogram and eliminate bins to the right side with heights less than 50% of the value of the highest bin.

After this we compiled Arnold's plots in the range of forcing frequencies 1–4 Hz (Fig. 9–10), assuming that a full-width at half-maximum or fwhm less than 2 corresponds to strong synchronization.

In the phase space plot (Fig. 9–10) the mechanical forcing amplitude is substituted by the forcing intensity I (voltage) at the vibrator and ω by the cyclic frequency $f = 1/T$ in Hz. Note that our Arnold's plot does not touch the frequency axis, this happens only in the ideal case in the absence of any noise (Pikovsky et al., 2003).

We conclude that the pattern of phase space plots in Fig. 9–10 is close to that predicted by the general theory of synchronization of integrate-and-fire processes (Pikovsky et al., 2003) as in addition to 1:1 phase locking between 2 and 3 Hz, we also obtained 1:2 mode at frequency around 1 Hz (compare Figs. 9–10 and 9–6A).

In another series of experiments, we tried to study in detail HOS at frequencies higher than the area of 1:1 synchronization, for which we used less stiff spring. In experiments with a less stiff spring, $K_s = 235.2\,N/m$ in the range of frequencies 0.5–120 Hz we observe only an HOS effect. The histograms (Fig. 9–11A–H) show the number N of AE burst onsets with

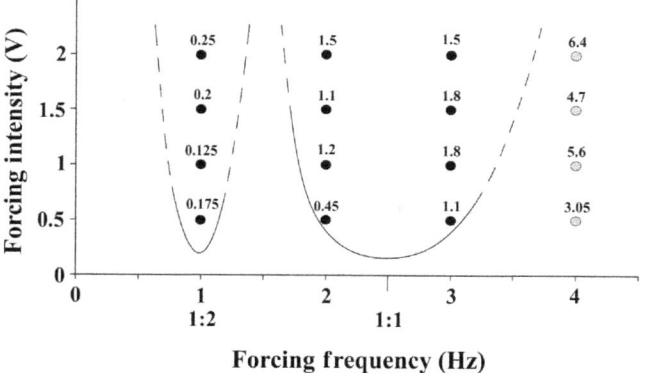

FIGURE 9–10 Phase space plot of stick–slip synchronization strength for spring stiffness $K_s = 555$ *N/m* and drive velocity = 8.47 mm/s at various forcing intensities and frequencies (Arnold's tongue). The synchronization strength is expressed as a full width at half-maximum (fwhm) of the histogram at 50% of the highest bin. Black dots correspond to good synchronization (fwhm < 2) and grey dots to absence of good synchronization (fwhm > 2). The full line delineates a conditional border of the synchronization area; the dashed line is its interpolation to the experimentally less examined phase space. Note the existence of two areas of synchronization: 1:1 and 1:2.

a phase difference $\Delta\varphi$ relative to the phase of the previous maxima of the forcing signal in the certain intervals (bins) of the forcing signal period. The histograms manifest clear differences in distribution at varying intensities I and frequencies f of forcing.

At a certain intensity and frequency of forcing, the AE onsets begin to concentrate in the definite phases of forcing signal (Fig. 9–11B–F). The qualitative assessment of the synchronization strength is possible by calculation of the full width at half-maximum (fwhm) of the histogram: for this, we used the simple procedure from http://cdn.teledynelecroy.com/files/manuals/dda-ref-e09.pdf (accessed January 18, 2017). Namely, we identify the height of the highest bin of the histogram and eliminate bins to the right and left if it with heights less than 50% of the value of the highest bin. The length of the line (in units of *x*-axis) connecting the centrepoints of the remaining marginal bins, normalized to the full length of the *x*-axis is the absolute value of fwhm. In other words, the fwhm is the population of *n* bins of the truncated histogram, normalized to the full population of 15 bins of the histogram.

To compile the Arnold's tongue, a fwhm of histograms were calculated for a limited interval of phase differences $\Delta\varphi$, in the range $0 < \Delta\varphi < 6.4$. In order to delineate the synchronization area more or less accurately, we accept the following rules: good synchronization (fwhm < 0.22), moderate ($0.24 \leq$ fwhm ≤ 0.253) and absence of synchronization (fwhm > 0.253): at larger values of fwhm the PDF of $\Delta\varphi$ is practically uniform. The resulting Arnold plot, compiled according to these rules, is presented in Fig. 9–12.

Besides direct analysis of the histograms' half-width, we also approximated histograms by some distribution function. For this the versatile Wakeby distribution function was used, as it can represent complicated histograms, calculate the half-widths of corresponding PDFs and compile the resulting Arnold's plot. Finally, we used the polynomial of the second degree (parabola) to approximate the phase distributions of events and obtained exactly the same Arnold plot. As all these plots are quite similar to Fig. 9–12, they are not shown here.

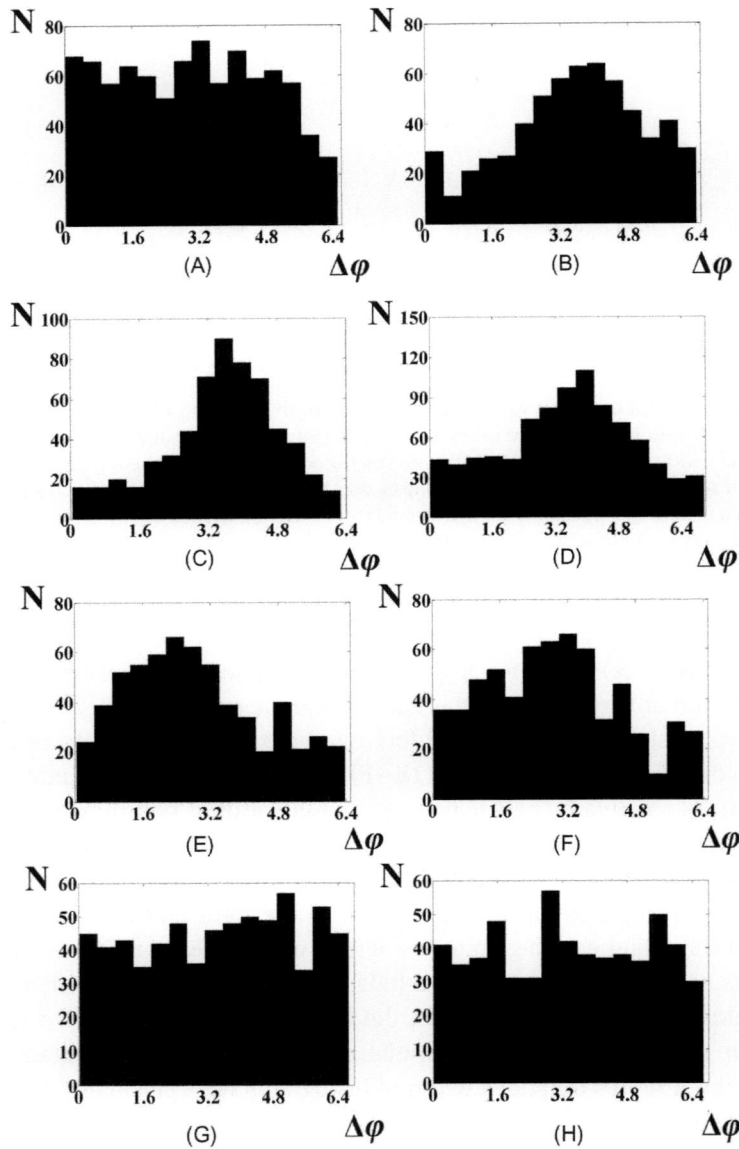

FIGURE 9–11 Histograms of number N of acoustic emission (AE) signals in a bin, normalized to the total number of AE events for a given frequency $\sum N$, with certain phase differences $\Delta\varphi$ between forcing signal maximum and AE onset versus phases of the forcing period, expressed in the units of 2π(Rad), in 1/15th bins of the forcing period. In this set of experiments: spring stiffness $K_s = 235.2$ N/m and drive velocity $V = 0.69$ mm/s; vibrator forcing intensity (voltage applied to mechanical vibrator) is 1 V; forcing frequencies are: (A) 5 Hz; (B) 10 Hz; (C) 20 Hz; (D) 30 Hz; (E) 40 Hz; (F) 50 Hz; (G) 80 Hz; (H) 120 Hz.

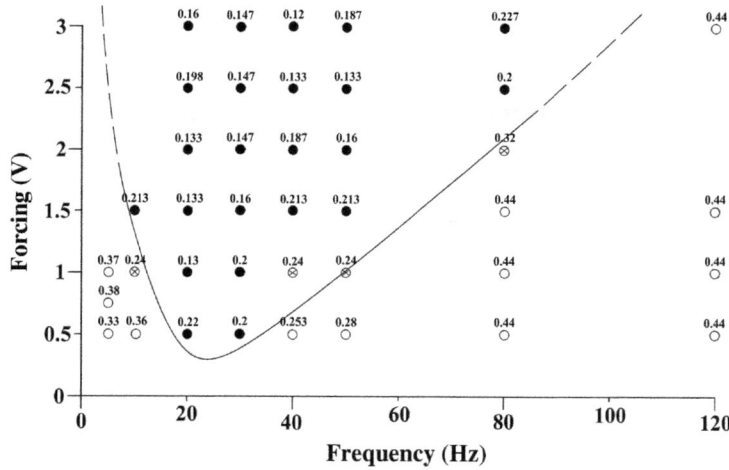

FIGURE 9–12 Phase space plot of synchronization strength for spring stiffness $K_s = 235.2 \, N/m$ and drive velocity $V = 0.69$ mm/s at various forcing intensities and frequencies (Arnold's tongue). The synchronization strength is expressed as a full width at half-maximum (fwhm) of the histogram at the 50% of the highest bin. Hollow dots mean absence of (fwhm > 0.253), dots with crosses moderate (0.24 ≤ fwhm ≤ 0.253) and filled dots good synchronization (fwhm < 0.22). The full line delineates a conditional border of the synchronization area; the dashed line is its interpolation to the experimentally less examined phase space.

9.6.3 Analysis of High-Order Synchronization in the High-Frequency Mode

Here we analyse the HF mode of HOS in the stick–slip process (Fig. 9–12), which is incompatible with the TAP model. In our case, the minimum mechanical forcing frequency needed for synchronization at the minimal voltage, applied to vibrator ≈ 0.5 *V*, corresponds to the range 20–40 Hz (Fig. 9–12). At the same time, the natural period T_0 of the slips in our experiments with a spring stiffness $K_s = 235.2 \, N/m$ was close to 3 s. This means that in the case of 1:1 synchronization the forcing period should be close to a natural period T_0 of the slips, i.e., it should be in the range 2–3 s (or in the range of frequencies 0.33–0.5 Hz). According to the TAP model, at frequencies higher than 0.33 Hz the forcing should not affect the phase of autonomous oscillator (stick–slip) at all, but experiments show that at much higher frequencies there is a whole new synchronization area. This phenomenon can be explained only by the general theory of integrate-and-fire oscillators (Pikovsky et al., 2003), which predicts that besides the 1:1 phase locking, it is also possible to observe HF synchronization areas (Arnold's plots) at frequencies multiples of ω_0, i.e., at frequencies $m\omega_0$. In terms of periods, the HOS is possible at $T_{\text{obs}} = nT$ (or at $n = \frac{T_{\text{obs}}}{T}$). In the following we will define the forcing winding number N_f as a number of forcing periods N within one observed period of autonomous oscillator $N_f = 1/N$. The value of N_f can be either larger or smaller than 1, depending to the type of synchronization.

In our experiments N_f varied in a wide range, depending on the experimental conditions: in the case of experimental data, shown in Fig. 9–13, the winding number N_f equals 1:44 for the forcing frequency 10 Hz at intensity 0.5 V and 1:310 for 120 Hz and 0.5 V. In other words, there are many forcing periods during one period (nucleation phase) of the oscillator. We counted the number of periods N between the onsets of successive AE bursts (slips) at various forcing frequencies ω and found that N increases systematically with a rise in ω at all intensities (Fig. 9−14A,B) with a small kink at 30−40 Hz, which is discussed later.

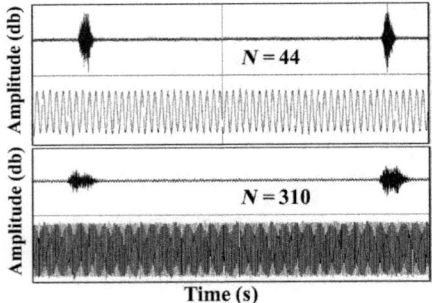

FIGURE 9–13 Schematic plot of variation of the number of periods between slips at two forcing frequencies. The upper channels in both panels show slips recorded as acoustic pulses, and lower channels in both panels show the corresponding mechanical forcing (periodic oscillations of mechanical vibrator). The number of periods N between successive slips (acoustic pulses) at two frequencies are, for the forcing frequency 10 Hz and intensity 0.5 V, $N = 44$ (upper panel), for 120 Hz and 0.5 V, $N = 310$ (lower panel). Waiting times T at both forcing frequencies 10 Hz and 120 Hz are in the range 3.5−1.5 s. In both experiments $K_s = 235.2\ N/m$ and drive velocity $V = 0.69\ mm/s$.

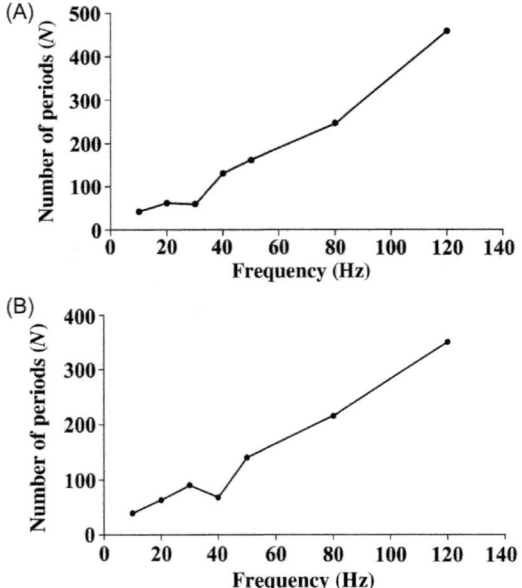

FIGURE 9–14 (A) The average number of forcing periods N between onsets of successive acoustic emission (AE) bursts (slips) versus forcing frequencies (Hz) at the input voltage on vibrator 0.5 V; (B) The same as in Fig. 9–13A at the input voltage on vibrator 3.0 V. In both experiments $K_s = 235.2\ N/m$ and drive velocity $V = 0.69\ mm/s$.

The tests show that the number of periods N between onsets of successive AE bursts increases greatly with forcing frequency at any intensity. The observed winding ratio N_f at mechanical forcing decreases with forcing frequency applied, from 1:50 = 0.02 at 10 Hz to 1:350 = 0.003 at 120 Hz (for intensity 3 V), from 1:50 = 0.02 at 10 Hz to 1:400 = 0.0025 at 120 Hz (for intensity 1 V) and from 1:50 = 0.02 to 1:450 = 0.002 at 120 Hz (for intensity 0.5 V).

Next, we verified the relationship between the increase in the forcing periods' number N between successive slips (Fig. 9−14) and the change of waiting times T_{obs} between the slips. In Fig. 9−15A−C we plot the observed waiting times T_{obs} between slip onsets versus forcing frequency. It is evident that the T_{obs} is relatively constant (3−4 s), except a drop to 1.5−2 s at

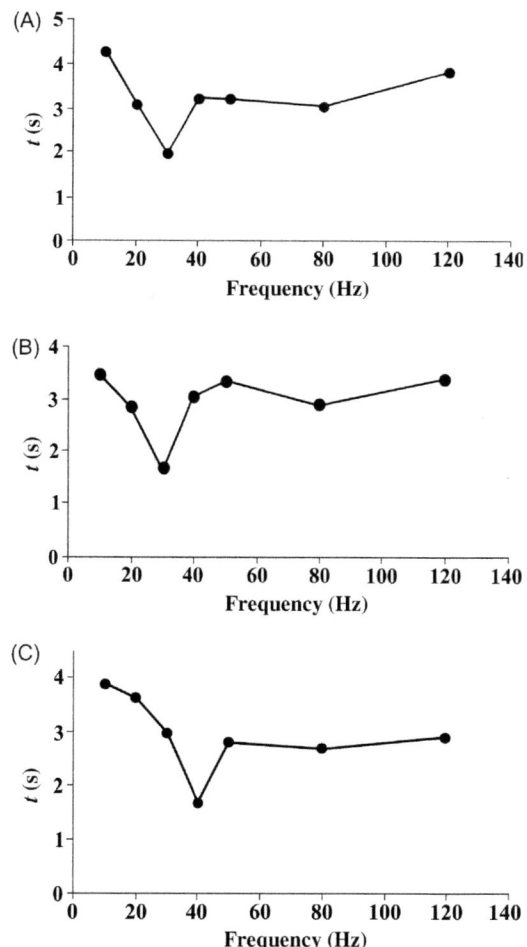

FIGURE 9–15 The waiting time t (s) between successive acoustic emission (AE) bursts versus forcing frequency (Hz): (A) at the input voltage on the vibrator of 0.5 V; (B) the same for the input voltage on the vibrator of 1.0 V; (C) the same for the input voltage on the vibrator of 3.0 V. In all experiments K_s = 235.2 N/m and drive velocity V = 0.69 mm/s.

forcing frequencies 30–40 Hz. We observe an identical pattern at all used forcing frequencies and intensities. The drop of T_{obs} from 3–4 s to 1.5–2 s at frequencies of 30–40 Hz corresponds to the minimum of our 'high-order' Arnold's tongue, i.e., to the area of maximal coupling strength between the autonomous oscillator and the forcing system in the HOS area (Fig. 9–12). The dominant waiting times of 3–4 s (Fig. 9–15) are observed mostly outside the HOS area, and correspond to a natural period T_0 of nonforced (or too weakly forced) stick–slip.

The systematic drop of the slip waiting times T_{obs} at forcing frequencies 30–40 Hz (or in other words, acceleration of the stick–slip process) could be explained within the framework of the 'synchronization by variation of the threshold' concept (Pikovsky et al., 2003). The significance is that superimposed periodic forcing increases or decreases the threshold u_t of the natural integrate-and-fire oscillator (the threshold, at which oscillator fires), by the value $u = u_t - \varepsilon(\sin\omega t)$. Periodic forcing regularly decreases (or increases) the threshold of firing by the value of the forcing amplitude at the corresponding phase of the autonomous oscillator. Generally, if the forcing period is shorter than that of the oscillator, the firing threshold is reached earlier; correspondingly, the waiting time T_{obs} in the HOS area shortens (Fig. 9–15).

Note that the forcing frequency/intensity combination, characteristic for the synchronization area minimum, provides optimal conditions of friction stabilization. This is an important issue in tribology (Bureau et al., 2000; Capozza et al., 2012) and it is of interest for geophysics, as the synchronization effect can make destabilization of a tectonic fault easier at the appropriate forcing frequency and shorten/prolong its slip recurrence time. Friction stabilization by weak periodic forcing does not imply complete elimination of the stick–slip phenomenon: experimental data (Bureau et al., 2000) show that even when 'stabilized' by the appropriate forcing stick–slip process, there are still small-scale stick–slip events, synchronized with the forcing frequency. It seems that instead of the term friction stabilization, the term stick–slip suppression (minimization) by external forcing would be more appropriate.

9.6.4 Additional Tools for Complexity/Synchronization Analysis

An effective tool for revealing synchronization is the phase stroboscope technique (Pikovsky et al., 2003), when the spikes of the fastest signal (here AE bursts) are compared to phases of slow signal (here forcing). If AE burst onset positions within one forcing period are ordered relative to the chosen phase of forcing we can conclude that forcing strongly affects the autonomous oscillator regime, leading to synchronization. We applied the stroboscopic technique to the experimental data on the very low-frequency (0.01 Hz) mechanically forced stick–slip presented in Fig. 2.6 in Savage (2007). The obtained synchrogram (Fig. 9–16) shows the regular timings of successive slips within two separate periods of slow forcing. The exact timing of these two series of slips in both forcing periods reflects a strong coupling between an autonomous oscillator (spring-slider system) and applied mechanical forcing, leading to HOS with the winding number $N_f = 1/12$. This is a reverse

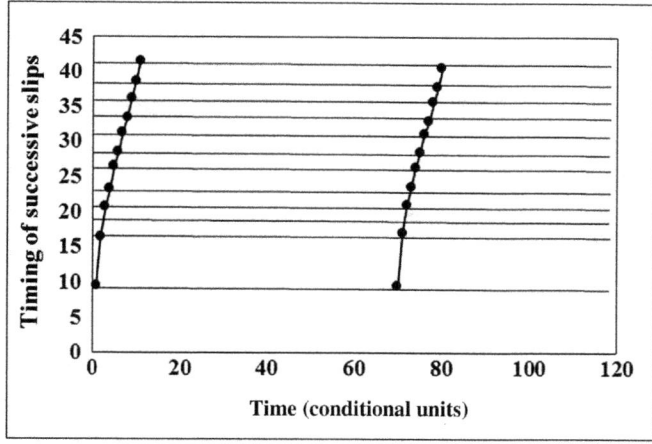

FIGURE 9–16 Synchrogram, showing the timings of successive slips in conditional units versus time also in conditional units within two periods of low-frequency (0.01 Hz) forcing. We use for synchrogram plotting Fig. 2.6 from Savage (2007). The exact coincidence of the timings from these two series of slips, which occurred within each period of forcing, reflects the strong coupling between autonomous oscillator (spring-slider system) and applied mechanical forcing, leading to strong high-order synchronization.

case of HOS, which we present in Fig. 9−12 and where the winding number N_f was less than one: $N_f \ll 1$. The waiting times of slips within one forcing period are of the order of several seconds (Fig. 9−16).

For qualitative and quantitative analysis of stick−slip time series complexity in addition to Arnold's plot some other nonlinear tests were performed, namely recurrence plots (RP) and RQA (Kononov, 2006; Marwan et al., 2007; Webber and Marwan, 2015)

RP plots for original data contain horizontal and vertical lines/clusters, which means that during both natural and driven stick−slip some states are 'laminar', i.e., they change slowly or do not change at all: we attribute laminar states of AE waiting times to the stick periods. RPs for various frequencies (at $K_s = 235.2$N/m and constant forcing intensity 1.5 V) show clear recurrence structures at frequencies of 20−40 Hz (Fig. 9−17), which are fading at higher frequencies (the plots became similar to RP of random sequence) in accordance with the Arnold's plot (compare with Fig. 9−12).

In order to characterize the recurrence pattern for the same case of HOS ($K_s = 235.2 \, N/m$ and constant forcing intensity 1.5 V) in a quantitative manner, we calculated the percent of determinism (%DET) at various frequencies and intensities of mechanical forcing using RQA (Marwan, 2003; Marwan et al., 2007). The results, presented in Fig. 9−18, show that the % DET parameter, which reflects the level of ordering (synchronization strength here) in the slip time series, is practically absent at all frequencies for a forcing intensity of 0.5 V and at frequencies of 5, 80 and 120 Hz for all intensities. %DET significantly increases in the frequency range from 20 to 60 Hz and at intensities of forcing larger than 0.5 V, in accordance with Fig. 9−12.

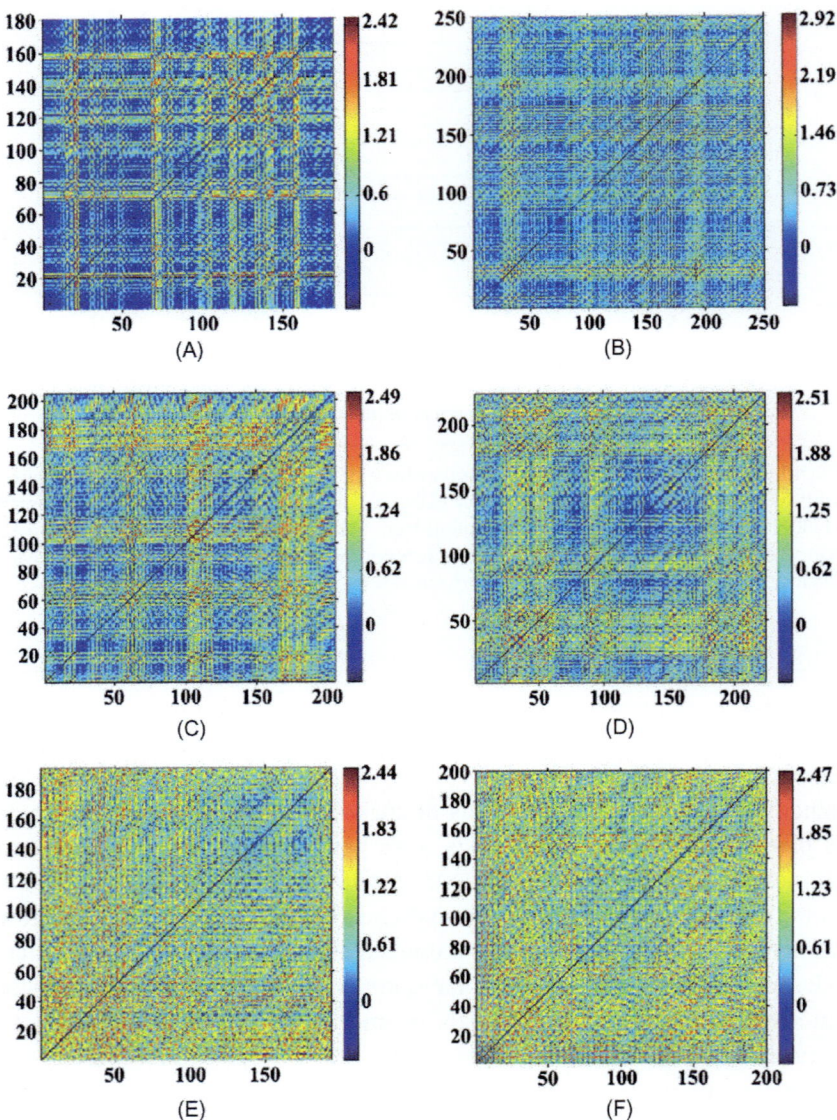

FIGURE 9–17 Recurrence plots (RPs) of slip-generated acoustic emission (AE) signals (original sequence of events) for the stiffness of the spring $K_s = 235.2\ N/m$ and forcing 1.5 V (natural frequency of sensor 20 Hz). The synchronization strength decreases from maximal at forcing frequencies: 20 Hz (A); 30 Hz (B); 40 Hz (C); 50 Hz (D), to minimal at 80 Hz (E). At 120 Hz (F) synchronization is practically absent.

The results obtained shows that the application of nonlinear dynamics methods can be very helpful in the quantitative analysis of the forced stick–slip process, namely for the localization of the synchronization area.

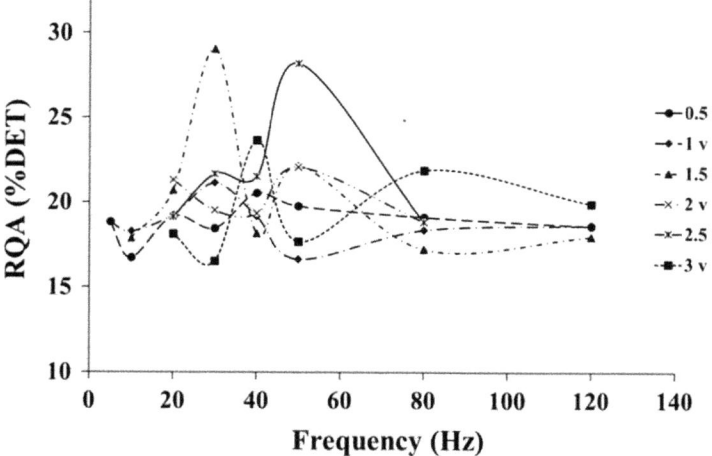

FIGURE 9–18 Recurrence quantitative analysis (RQA) %DET versus frequency of forcing at various intensities of mechanical forcing from 0.5 to 3 V. %DET significantly increases in the frequency range from 20 to 60 Hz and at intensities of forcing larger than 0.5 V; synchronization is practically absent at frequencies less than 20 Hz and larger than 100 Hz in accordance with the phase space plot in Fig. 9–12.

One of the most interesting methods for revealing synchronization in relaxation-type processes is a phase-locked loop technique, which compares the frequency of a local reference oscillator to that of a received signal (here, the AE time series) and uses a feedback scheme to lock the local oscillator's frequency to the incoming signal. It is a well-known technique in the analysis of radio technical systems (Blekhman, 1988). Its use for stick–slip generated AE time series is described in detail in Lursmanashvili et al. (2010) as a 'gap'the technique. The method is founded on the effect of concentration of events (slips and associated seismic/acoustic events) in the definite phases of forcing period similar to Figs 9–9 and 9–11. Naturally, this means that the occurrence of dynamic events in the remaining phases of forcing is less probable (prohibited). The width of the prohibited zone is larger for stronger synchronization. We tested the method on the data, obtained on a laboratory spring-slider model with mechanical forcing. In these experiments, the forcing frequency (30 Hz) was known beforehand and the objective was to retrieve it from the observed data as if it was unknown. Therefore, the known forcing frequency was used only for validation of the gap method. The mechanical forcing was realized by application of 5 V to the vibrator. From the analysis of the gap distribution (Fig. 9–19a) and the corresponding periods' spectra (Fig. 9–19b), we can conclude that synchronization is present at the forcing frequency (30 Hz) and the genuine forcing period $T_g = 0.03323$ can be extracted with no less than 0.01 Hz accuracy. This means that the forcing frequency can be determined accurately from the observed synchronized slip recurrence spectra.

The above approach was also tested on the Catalogue of Caucasian EQs of M. Nodia Institute of Geophysics and some significant gaps related to tidal effects were revealed (Lursmanashvili et al., 2010).

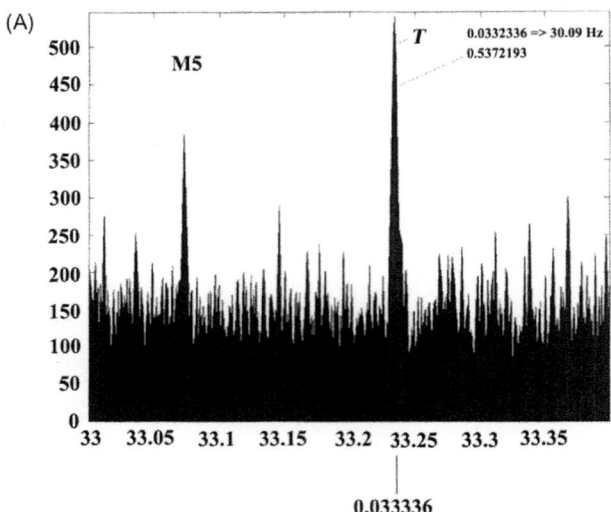

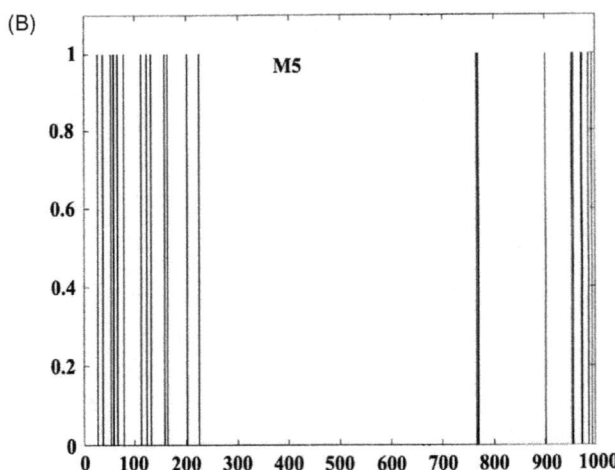

FIGURE 9–19 The spectrum of gap widths and slip periods T (s) (X-axis) for vibrator exciting voltage of 5 V: (A) the spectrum of gaps, Y-axis shows the local maxima of the gap widths: X-axis shows trial periods T_i· 1000; (B) the distribution of slip moments inside the genuine forcing period T_g, divided into 1000 intervals.

9.7 Forced Stick—Slip Results: Electromagnetic Forcing

9.7.1 Experimental Set Up for Electromagnetic Forcing

The same simple spring-slider system (Fig. 9—1) was used to study the EM forcing of stick—slip (Chelidze et al., 2002, 2005, 2009, 2010; Chelidze and Matcharashvili, 2003; Chelidze and Lursmanashvili, 2003). In this case the voltage was applied directly to electrodes, glued to external surfaces of fixed and pulled blocks, in contrast with the mechanical forcing case described above, when the voltage V was used only to activate the mechanical vibrator. In our experiments the following parameters were varied: (1) the stiffness of the spring K_s; (2) the frequency, f, of the superimposed periodical perturbation; (3) the amplitude of the excitation (applied voltage V_a) and (4) the direction of the applied electrical field.

9.7.2 Electromagnetic Synchronization: Results

EM synchronization was observed only at some definite values for the set of parameters (K_s, f, V_a). In addition, with a variation of the forcing frequency we observed different patterns of synchronization, with the winding number N_f equal to, larger or smaller than 1. Fig. 9−20a shows the transition from 1:1 to a HOS with an increasing period of EM forcing from 0.5 to 4.5 s and Fig. 9−20b for the transition from 1:2 to 1:1 synchronization at simultaneous action of direct $V(0)$ and periodic $V(p)$ voltages after $V(0)$ became larger than $V(p)$.

9.8 Complexity Analysis of Electromagnetically Forced Stick−Slip

9.8.1 Phase Space Plot of the Electromagnetic Synchronization Area: Arnold Tongue Plot

The quantitative analysis of stick−slip synchronization strength under EM forcing using several different methods of nonlinear dynamics is presented in Chelidze and Matcharashvili (2003, 2013, 2015), Chelidze et al. (2002, 2005, 2009, 2010) and Chelidze and Lursmanashvili (2003). In particular, in the case of EM synchronization, the phase diagram for the variables frequency and intensity also has a typical inverse bell-form (Fig. 9−21).

Fig. 9−22 presents the evolution of phase difference $D = \Delta\Phi$, obtained from the Hilbert transform of waiting times' series when increasing smoothly the forcing intensity from 0 (at $t = 12$ s) to 1000 V (at $t = 42$ s) and then decreasing the intensity to zero at $t = 72$ s. The constant value of D correspond to situation, when slips are phase-synchronized with the phase of forcing. Thus, well-defined horizontal part of D versus t represents the interval, during which the AE (slips) become phase-synchronized with the external sinusoidal EM forcing. Phases are synchronized in a wide range of forcing amplitudes, from approximately 500 V (at $t = 22$ s) to 1000 V (at $t = 62$ s).

Frequency locking, expressed as a minimum of the phase diffusion coefficient D, is also a quantitative measure of the phase synchronization (Fig. 9−23):

$$D = d/dt \left[<\Delta\varphi^2> - <\Delta\varphi>^2 \right], \tag{9.10}$$

where $\Delta\varphi$ is the phase difference between forcing and AE (or slip) at time t. The results are very similar to Fig. 9−22: synchronization is strong in the area of forcing intensity from 500 to 1000 V (or from 22−62 s).

(A)

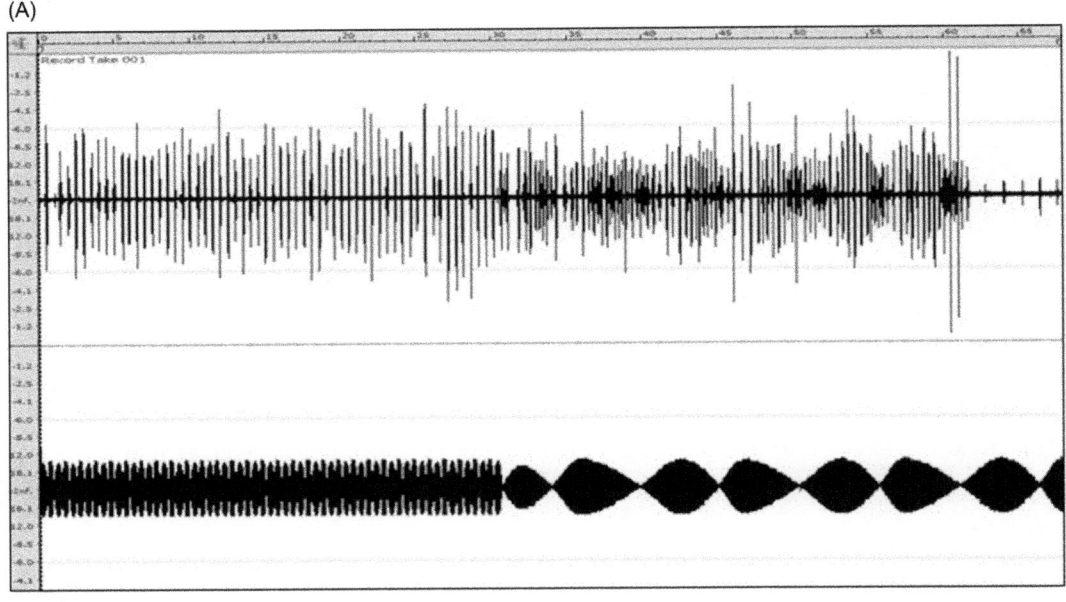

(B)

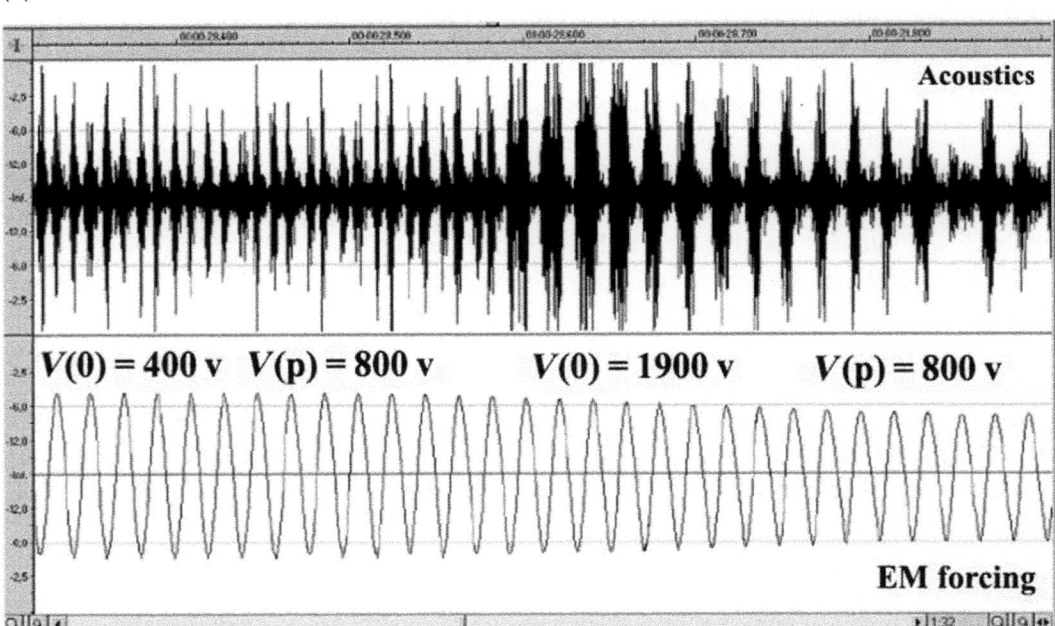

FIGURE 9–20 (A) Transition (bifurcation) from 1:1 synchronization of stick–slip to high-order synchronization when increasing the period of electromagnetic (EM) forcing from 0.5 to 4.5 s. Note that in Figs 9–19 and 9–24 the LF forcing signal was filled by high-frequency (HF) oscillations in order to visualize the low-frequency forcing on the computer screen. The HF signal was applied only to the computer and not to the rock plates. (B) Transition (bifurcation) in stick–slip from HOS (period doubling) to 1:1 synchronization with simultaneous action of the direct $V(0)$ and periodic $V(p)$ voltages; transition occurred at $V(0) > V(p)$.

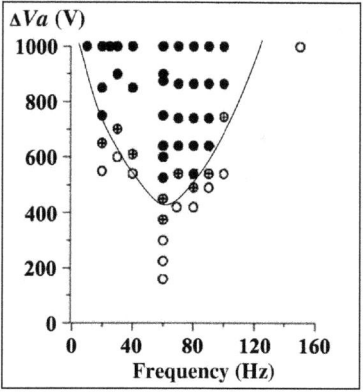

FIGURE 9–21 Synchronization area of forcing phase and acoustic emission (AE) starts (Arnold's tongue) for various intensities (V_a) and frequencies (*f*) of the external electromagnetic (EM) periodic forcing. Filled circles are perfect, circles with crosses are intermittent and empty circles indicate an absence of synchronization. *Modified from Chelidze and Matcharashvili, 2013*

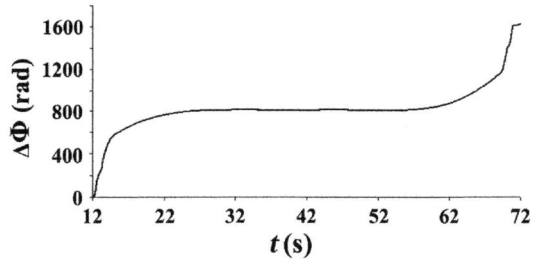

FIGURE 9–22 Phase difference $\Delta\Phi$ between maximums of acoustic emission bursts and forcing phase at smooth variation of the forcing strength: rising of the external EM sinusoidal forcing intensity from 0 V at 12 s to 1000 V at 42 s and decreasing to 0 at the 72 s mark. Note the strong phase locking for the forcing intensities at 500–1000 V (from 22 to 62 s).

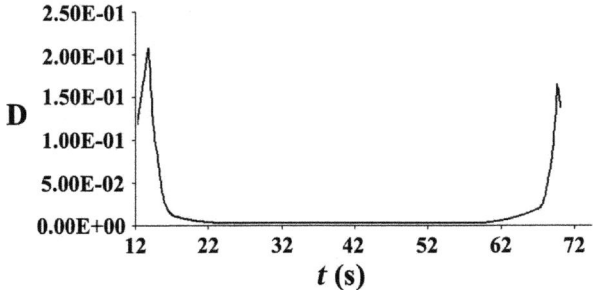

FIGURE 9–23 Variation of phase diffusion coefficient *D* of phase differences $\Delta\varphi(t)$, between AE maximums and forcing phase calculated for consecutive sliding windows, containing 500 events. Synchronization is strong in the area of forcing intensity from 500 to 1000 V (from 22 to 62 s), where the phase diffusion is negligible.

A clear increase in Mutual Information (MI) value *MI* indicates that the dynamics of AE becomes much more regular in the synchronized part of the AE data set (Fig. 9–24).

We also applied a phase stroboscope technique (Pikovsky et al., 2003) to analyse HOS, due to EM forcing (Fig. 9–25). The timings of spikes of the fastest signal (here AE bursts) were compared to phases of the slow signal (here EM forcing). It is evident that EM forcing by very long EM pulses invokes very ordered swarms of AE bursts within one EM pulse, i.e., we observe HOS with the winding number N_f 1:40 = 0.025. Fig. 9–25 presents a stroboscopic diagram (synchrogram) for the first 31 AE bursts in the swarms, generated by 11 forcing EM pulses. Note the horizontal striped structure of the synchrogram, which shows that the phase shift between EM forcing and onsets of AE bursts in the successive swarms is almost constant for the first 10 slips. Only later slips show small phase diffusion. This is a sign of strong synchronization of the initial AE burst timings (slips) within the forcing period.

9.9 Implications of the Forced Stick–Slip Model for Geophysical Phenomena

Since the 1970s stick–slip has been considered as the main model for EQ cycling or the Brace–Byerly–Dietrich–Ruina (BBDR) rate-and-state friction model to the natural seismic process. In the following paragraphs, we consider shortly implications of our results on synchronization of stick–slip to some natural and anthropogenic processes using modern tools of nonlinear dynamics theory (Kantz and Schreiber, 1997; Strogatz, 2000; Webber and Marwan, 2015). As it is practically impossible to control experimental conditions (forcing frequency and intensity) in nature in order to compile the full synchronization plot, similar to Figs. 9–10 and 9–12, we can only assess the possibility of synchronization of induced/triggered seismic activity under weak external forcing. We consider such natural forcings as earth tides, periodic loading by water reservoir, and dynamic triggering of so-called nonvolcanic tremors (NVTs) by teleseismic waves from strong remote EQs (Hill and Prejean, 2009), where the frequency and magnitude of external forcing can be assessed more or less accurately.

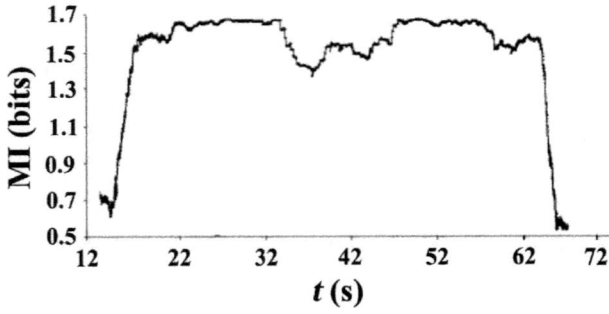

FIGURE 9–24 Variation of Mutual Information *MI* of phase differences, calculated for consecutive sliding windows, containing 500 events. Synchronization is strong in the area of forcing intensity from 500 to 1000 V (from 22 to 62 s).

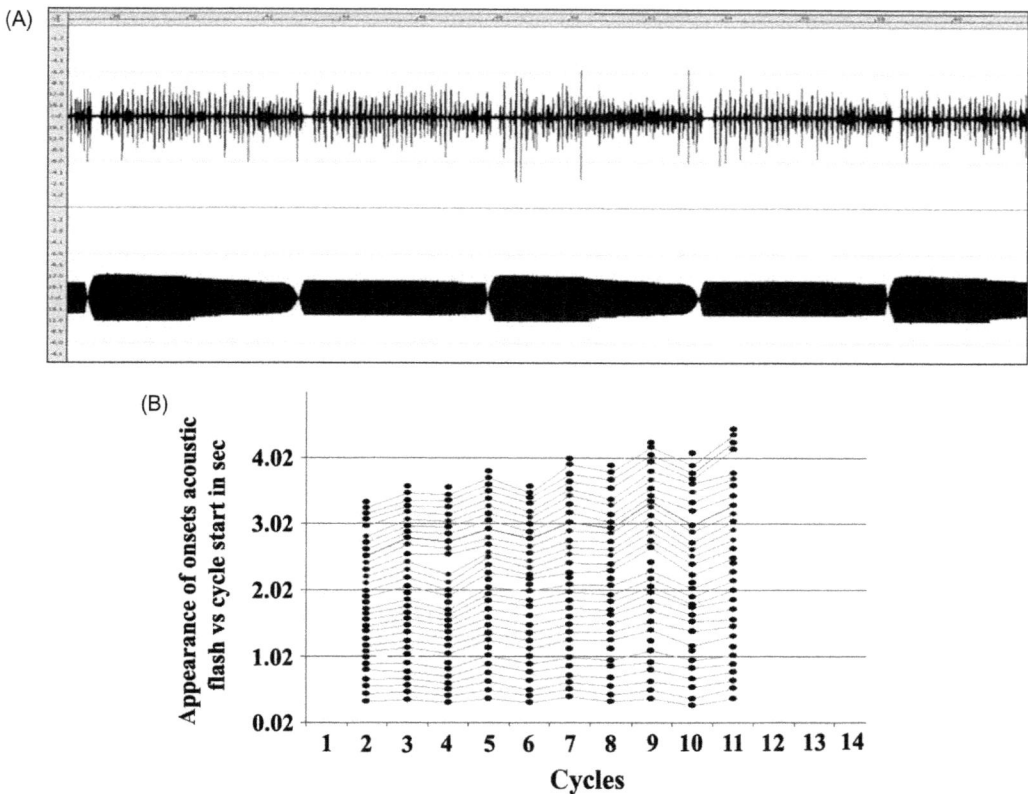

FIGURE 9–25 High-order synchronization with winding number 1:(1/40) = 40: (A) swarms of acoustic emission (AE) pulses, generated during forcing by very long electromagnetic (EM) pulses; (B) stroboscopic diagram (synchrogram) for the first 31 AE bursts in the swarms, generated by 11 forcing EM pulses. The *Y*-axis shows the phase shift of spikes of the fastest signal (here AE bursts) in seconds within the period of the slow signal (here EM forcing) for 11 consecutive forcing signals. Note the striped structure of the synchrogram, which shows that the phase shift between EM forcing and onsets of AE pulses in the swarm is almost constant for the first 10 slips.

9.9.1 Tides

The correlation between the Earth, oceanic tides and seismicity has been totally rejected by some authors (Vidale et al., 1998) but also claimed to be very significant by others (Nikolaev, 2003; Scholz, 2003; Métivier, 2009; Iwata, 2002). In principle, weak periodic mechanical tidal forcing, applied to the nonlinear tectonic stick–slip (accumulation-and-fire) process can lead to synchronization of seismicity with tides. At the same time, the EQ nucleation time T_0 is important for the location of the most appropriate frequency range of the synchronization area. It is accepted that EQs' nucleation time T_0 is a function of the magnitude of the impending event. According to Kasahara (1981) the nucleation (stress build up, preparation, predictor) time depends on the magnitude M of the impending event as: $\lg T_0 = 0.79M - 1.88$, where T_0 is nucleation time, and M is the magnitude of the impending event. This empirical power law equation was later confirmed (Bak, 2002; Corral, 2004; De Arcangelis, 2016) by revealing the scaling behaviour of the distribution D of interevent times Δt between EQs with magnitude m larger than m_c: $D(\Delta t, m_c) = m_c^{2/3b} G(\Delta t, m_c^{2/3b})$, where b is the Gutenberg–Richter coefficient and G is the scaling function.

The condition of 1:1 synchronization for tidal impact means that due to relatively small periods of strong tidal components (M2, O2), we should expect the 1:1 synchronization of seismic events with relatively short T_0, i.e., for small EQs and for AE in underground cavities (Iwata, 2002). Rikunov et al. (1980) supported this conclusion when studying seismoAEs in the Earth.

At the same time, as has been proved in hundreds of works and as we show in this chapter, synchronization can appear even when the accumulation phase is much longer than the forcing period – this is so-called HOS. Therefore, seismic HOS can appear at multiples of tidal periods, which can be much larger or smaller than the EQ preparation (nucleation) time T_0 and, in addition, the tidal forcing can affect the dominant EQ recurrence time (see Section 9.1).

9.9.2 Reservoir Load–Unload

One interesting example of EQ (tectonic stick–slip) synchronization by a weak external force is correlation of the local seismicity rate with the large reservoir load–unload regime. It is well known that the filling of a large reservoir activates a local seismic process such as reservoir-triggered seismicity or RTS (Gupta, 2002; Chen and Talwani, 1998; Assumpção et al., 2002; Telesca, 2010). Here we use the data acquired by the monitoring system at the high Enguri dam in Georgia. After a transition period (months or years) the quasiperiodic regime of water level (WL) variations will set up. In this period, the level of RTS decreases and the regular loading–unloading of the reservoir entrains the seismic regime. We suggest calling this effect reservoir-induced synchronization of seismicity (RISS) (Chelidze et al., 2010; Matcharashvili et al., 2010). To assess the synchronization strength in a quantitative way, several tools of nonlinear dynamics were applied to a long-term series of WL and seismicity. The nonlinear characteristics of seismicity were calculated in both the transitional (reference) period (before and during water irregular filling, 1974–87) as well as during quasiperiodic WL change (regular period, 1987–2013). We calculated changes in the mean effective phase diffusion coefficient D (Pikovsky et al., 2003) and carried out RQA (Marwan et al., 2007; Webber and Marwan, 2015), as well as singular spectrum analysis (SSA) (Ghil and Vautard, 1991; Ghil et al., 2002), in order to investigate the relationship between phases of the local seismic activity and reservoir WL variations (Matcharashvili et al., 2010; Chelidze and Matcharashvili, 2015).

RQA calculations indicate that the extent of regularity in both magnitude and waiting time sequences of EQs, evaluated as the percent of determinism (RQA %DET), increases approximately five times in the period of regular load-unload of reservoir in comparison to a reference one (Fig. 9–26B). Note, that randomized/shuffled data do not show any change in %DET in the whole period of observations (Fig. 9–26A).

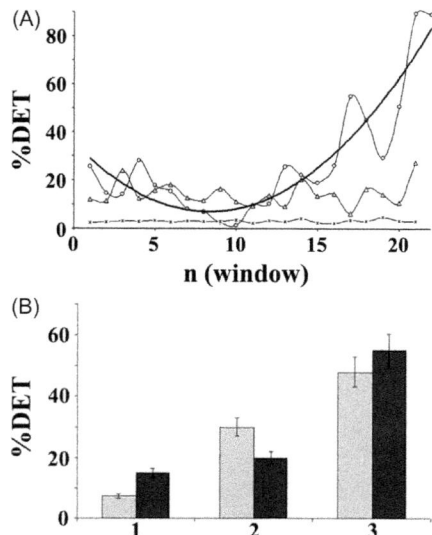

FIGURE 9–26 (A) Recurrence quantitative analysis (RQA) %DET of the natural daily number of earthquakes (EQs) calculated for consecutive nonoverlapping 1-year sliding windows (circles) and the average (heavy line). Averaged results of RQA %DET for 20 shuffled (asterisks) and phase randomized (triangles) surrogates of the daily number of EQs in consecutive 1-year sliding windows: (1) before impoundment – windows from 0 to 5; (2) during flooding and reservoir filling – windows from 5 to 10; (3) periodic change of water level (WL) in reservoir – windows from 10 to 20; (B) RQA %DET of magnitudes (black columns) and waiting times (grey columns) of local EQ time series around the high Enguri dam: (1) before impoundment, (2) during flooding and reservoir filling, and (3) during periodic change to the WL in the reservoir (Chelidze et al., 2006; Matcharashvili et al., 2010).

We found that the phase diffusion coefficient, $D = d/dt\left[<\Delta\varphi^2> - <\Delta\varphi>^2\right]$, where $\Delta\varphi$ is the phase difference between daily released seismic energy and WL daily variations, is large in the reference period ($D \approx 300$) and decreased to $D \approx 100$ in the regular load–unload period, which means that local seismicity became phase-synchronized with WL variation (Chelidze et al., 2010; Matcharashvili et al., 2010).

In addition to RQA, we used the SSA technique in order to investigate the correlation between reservoir water variations and local seismicity. The time series of the monthly number of EQs was compared to the monthly mean of the WL and several main periodic components of both processes were reconstructed using SSA methodology (Telesca et al., 2012). The power spectrum of the second and third reconstructed components revealed a clear annual cycle in the regular period. In contrast, the main periodicities in the reference period were fixed at 15, 9, 8 and 7 months, but there was no evidence of an annual cycle. This clearly indicates synchronization of the seismicity with WL variations around Enguri dam. What is interesting, in addition to annual cyclicity, the fourth component in seismicity shows the presence of a 4-month cycle, possibly, due to HOS effects in the seismic process, connected with WL variation (Telesca et al., 2012).

9.9.3 Nonvolcanic Tremors

In recent years, a new phenomenon, dynamic triggering of so-called NVTs by teleseismic waves from strong remote EQs (Prejean and Hill, 2009) has been discovered. Some observations of Rubinstein et al. (2009, 2010), Hill and Prejean (2009) and Peng et al. (2010) show that weak forcing by Rayleigh (R) or Love (L) surface waves from remote strong EQs can not only trigger, but also synchronize tremors with surface wave peaks. This is illustrated by a high cross-correlation coefficient between R (or L) wave phases and NVT events (see, e.g., Fig. 5.13 in Chao et al., 2012). These characteristics of NVTs allow for the hypothesis that the observed 1:1 synchronization of dynamically triggered events with teleseismic surface waves is the result of an instantaneous elastic response of the fault state to $L - R$ wave passage. These waves induce small synchronous quasiperiodic reversible displacements of the fault faces in the potential wells, formed by asperities and notches of contacting surfaces. These vibrations (abortive EQs), due to friction resistance, could generate secondary seismic waves (tremors), synchronized with $L - R$ teleseisms.

9.10 Future Developments

In order to properly understand the forced stick–slip phenomenon, it is necessary to carry out an enormous amount of experimental and theoretical work. Experimenters will need to cover a very wide range of variables (drag velocities, normal stresses, classes of surfaces and gouges, intensities and frequencies of forcing, etc.) in order to reveal all possible synchronization regimes.

Theoretical solutions for a rate-and-state friction model with weak external forcing need to be obtained and reasonable simplifications found for comparisons of theory with experimental data.

The possible influences of natural and anthropogenic forcings on recurrent geophysical phenomena (seismicity, displacements, energy discharges) should be reconsidered taking into account complications introduced by HOS and natural waiting time changes under forcing.

9.11 Conclusions

We have presented the results of laboratory experiments on the mechanical and EM periodic forcing of stick–slip of a slider-spring system, which prove that a weak external force evokes the phase synchronization of the stick–slip process. AE bursts were used as the markers of slip events.

Our experiments confirm predictions of the general theory of synchronization of integrate-and-fire processes in the particular case of the stick–slip process. We show that by varying the forcing frequency and drag velocity it is possible to reveal all predicted theory regimes of phase locking from 1:1 synchronization to HOS at multiples of the natural stick–slip frequency. We found that not only the onsets/maxima of AE signals are synchronized with forcing, but also AE wave train terminations.

The data allow delineation of the synchronization regions (Arnold's tongues) in the phase space plot of forcing intensity versus forcing frequency. Our experiments with mechanical periodic forcing, as well as our earlier experiments with EM forcing on the same slider-spring system, support the view that the forced stick–slip phenomenon belongs to the general class of integrate-and-fire systems and, consequently, phase space plot of synchronization area has a configuration, close to the inverse bell-curve form with both LF and HF branches, known as the Arnold's tongue. In the phase space plot of the synchronization area of forced stick–slip (as in general for integrate-and-fire systems), both low- and HF branches of the Arnold's plot are frequency-dependent, in contrast to the 'truncated' model, where only the low-frequency branch is present.

The systematic drop in the slip waiting times T_{obs} at (mechanical) forcing frequencies of 30–40 Hz (or in other words, acceleration of the stick–slip process) in the HOS regime can be explained within the framework of synchronization by variation of the threshold. According to this hypothesis, periodic forcing regularly decreases (or increases) the threshold of firing by the value of the forcing amplitude at the corresponding phase of the autonomous oscillator. If, as in our case, the forcing period is shorter than that of the oscillator, then the firing threshold is reached earlier: correspondingly, the waiting time T_{obs} in the HOS area shortens significantly, from 3–4 to 1–2 s.

For quantitative measuring of the complexity level (here the synchronization strength) several modern nonlinear dynamics tools were used, including the mean effective phase diffusion coefficient, Mutual Information, RPs and RQA, namely, the percent of determinism (%DET), phase space plot construction for revealing the synchronization area, etc. It is evident that these complexity measurement tools are very useful for revealing hidden structures in analysed time series.

Finally, we applied the tools of nonlinear dynamics to reveal complexity/synchronization patterns in EQ time series under weak external periodic forcing, such as WL variation in a large reservoir. Our analysis revealed the appearance of the annual component as well as the 4-month harmonic in the local seismicity during the periodic loading–unloading of the reservoir, which are absent in the preceding reference period. The 4-month harmonic could be an example of HOS related to the annual cycle.

We find that weak forcing can change the synchronization regime of stick-slip, namely high-order synchronization can appear in phase space plots. In turn, this points to the possibility of the existence of seismic tidal effects at frequencies, different from (multiples of) the classic values of tide frequencies. Besides, tidal forcing can distort dominant EQ recurrence (waiting) time due to effect of decreasing the firing threshold. As we show in stick-slip experiments, decrease of firing threshold lead to shortening of the EQ recurrence time.

The results, obtained by complexity analysis of the forced stick–slip in laboratory and field data, are important both in seismology and in stick–slip tribology, as the minimum of the Arnold's tongue corresponds to the condition of the optimal friction control (stick–slip minimization/suppression) under the application of minimal external mechanical/EM forcing.

Further investigations in this direction should reveal new details of frictional motion complexity and, in particular, the areas of validity of different synchronization area models for periodically forced stick–slip.

Acknowledgements

The authors acknowledge grants from the Shota Rustaveli National Science Foundation of Georgia [Projects 31/31-"FR/567/9—140/12" and #216732], which have greatly supported this research.

The authors provide credit to Nova Science Publishers for granting the nonexclusive permission to Elsevier for use of the following figures (Fig. 1; Fig. 2, Fig. 3; Fig. 4; Fig. 5; Fig. 11; Fig. 15; Fig. 16; Fig. 17; Fig. 18; Fig. 24; Fig. 26; Fig. 32; Fig. 33) from: T. Chelidze, T. Matcharashvili "Triggering and Synchronization of Seismicity: Laboratory and Field Data — A Review", in the e-book: Triggers, Environmental Impact and Potential Hazards, Ed. K. Konstantinou, Nova Science Publishers, 2013, pp. 165—231 free of charge for use in this chapter.

References

Abarbanel, H., Tsimring, L.S., 1993. The analysis of observed chaotic data in physical systems. Rev. Mod. Phys. 65, 1331—1392.

Abarbanel, H., Rabinovich, M., Sushchik, M., 1996. Introduction to nonlinear dynamics for physicists. World Scientific, Singapur.

Arnold, V., 1983. Notes on the perturbation theory for the problems of Matieu type. Russ. Mat. Surveys 38, 215—233.

Assumpção, M., Marza, V., Barros, L., Chimpliganond, C., Soares, J.E., Carvalho, J., et al., 2002. Reservoir-Induced Seismicity in Brazil. Pure Appl. Geophys. 159 (1—3), 597—617.

Bak, P., Christensen, K., Danon, L., Scanlon, T., 2002. Unified scaling law for earthquakes. Phys. Rev. Lett. 88, 178501. Available from: https://doi.org/10.1103/.

Bartlow, N.M., Lockner, D.A., Beeler, N.M., 2012. Laboratory triggering of stick-slip events by oscillatory loading in the presence of pore fluid with implications for physics of tectonic tremor. J. Geophys. Res. 117 (B11411). Available from: https://doi.org/10.1029/2012JB009452.

Becker, T., 2000. Deterministic chaos in two state-variable friction sliders and the effect of elastic interactions. In: Rundle. J., Turcotte, D., Klein, W. (Eds.). GeoComplexity and Physics of Earthquakes. American Geophysical Union, Washington, D.C. pp. 5—27.

Beeler, N.M., Lockner, D.A., 2003. Why earthquakes correlate weakly with the solid Earth tides: effects of periodic stress on the rate and probability of earthquake occurrence. J. Geophys. Res. 108 (B2391). Available from: https://doi.org/10.1029/2001JB001518.

Ben-David, O., Rubinstein, S.M., Fineberg, J., 2010. Slip-stick and the evolution of frictional strength. Nature (Letters) 463, 76—79.

Blekhman, I.I., 1988. Synchronization in Science and Technology. A. S. M. E. Press, New York.

Bowden, F., Tabor, D., 1950. Friction and Lubrication of Solids. Clarendon, Oxford.

Brace, W.E., Byerlee, I.D., 1966. Stick slip as a mechanism for earthquakes. Science 153, 990—992.

Brodsky, E., Prejean, S., 2005. New constraints on mechanisms of remotely triggered seismicity at Long Valley Caldera. J. Geophys. Res. 110B. Available from: https://doi.org/10.1029/2004JB003211.

Broomhead, D., King, G., 1986. On the qualitative analysis of experimental dynamical systems. In: Sarkar, S. (Ed.), Nonlinear Phenomena and Chaos. Adam Hilger, Bristol, pp. 113—144.

Bureau, L., Baumberger, T., Caroli, C., 2000. Shear response of a frictional influence to a normal load modulation. Phys. Rev. E 62, 6810—6820.

Burridge, R., Knopoff, L., 1967. Model and theoretical seismicity. Bull. Seismo. Soc. Am 57, 341—371.

Capozza, R., M., Rubinstein, S.M., Barel, I., Urbakh, M., Fineberg, J., 2011. Stabilizing stick-slip friction. Phys. Rev. Lett. 107. Available from: https://doi.org/10.1103/PhysRevLett.107.024301.

Capozza, R., Vanossi, A., Vezzani, A., Zapperi, S., 2012. Triggering frictional slip by mechanical vibrations. Tribol. Lett. 48, 95−102.

Chao, K., Peng, Z., Wu, C., Tang, C.-C., Lin, C.-H., 2012. Remote triggering of non-volcanic tremor around Taiwan. Geophys. J. Int. 188, 301−324.

Chelidze, T., 1982. Percolation and fracture. Phys. Earth Planet. Inter. 28, 93−101.

Chelidze, T., Lursmanashvili, O., 2003. Electromagnetic and mechanical control of slip: laboratory experiments with slider system. Nonlinear Processes Geophys. 20, 1−8.

Chelidze, T., Matcharashvili, T., 2003. Electromagnetic control of earthquake dynamics? Comp. Geosci. 29, 587−593.

Chelidze, T., Matcharashvili, T., 2007. Complexity of seismic process, measuring and applications—A review. Tectonophysics 431, 49−61.

Chelidze, T., Matcharashvili, T., 2013. Triggering and synchronization of seismicity: laboratory and field data - a review. In: Konstantinou, K. (Ed.), Earthquakes − Triggers, Environmental Impact and Potential Hazards. Nova Science Pub, Hauppauge, New York, pp. 165−231.

Chelidze, T., Matcharashvili, M., 2015. Dynamical patterns in seismology. In: Webber, C., Marwan, N. (Eds.), Recurrence Quantification Analysis: Theory and Best Practices. Springer, Cham Heidelberg, pp. 291−335.

Chelidze, T., Varamashvili, N., 2010. Models of stick-slip motion: impact of periodic forcing. In: De Rubeis, V., Czechowski, Z., Teisseyre, R. (Eds). Geoplanet: Earth and Planetary Sciences, vol. 1. Synchronization and Triggering: from Fracture to Earthquake Processes. Springer, Berlin, Heidelberg, pp. 23−33.

Chelidze, T., Varamashvili, N., Devidze, M., Chelidze, Z., Chikhladze, V., Matcharashvili, T., 2002. Laboratory study of electromagnetic initiation of slip. Ann. Geophys. 45, 587−599.

Chelidze, T., Matcharashvili, T., Gogiashvili, J., Lursmanashvili, O., Devidze, M., 2005. Phase synchronization of slip in laboratory slider system. Nonlinear Processes Geophys. 12, 1−8.

Chelidze, T., Lursmanashvili, O., Matcharashvili, T., Devidze, M., 2006. Triggering and synchronization of stick slip: waiting times and frequency-energy distribution. Tectonophysics 424, 139−155.

Chelidze, T., Lursmanashvili, O., Matcharashvili, T., Varamashvili, N., Zhukova, N., Mepharidze, E., 2009. High order synchronization of stick-slip process: experiments on spring-slider system. Nonlinear Dyn. Available from: https://doi.org/10.1007/s11071-009-9536-6.

Chelidze, T., Lursmanashvili, O., Matcharashvili, T., Varamashvili, N., Zhukova, N., Mepharidze, E., 2010. Triggering and synchronization of stick-slip: experiments on spring-slider system. In: De Rubeis, V., Czechowski, Z., Teisseyre, R. (Eds.). Synchronization and Triggering: from Fracture to Earthquake Processes. Geoplanet: Earth and Planetary Sciences, vol. 1. Springer, Berlin, Heidelberg. pp. 123−164.

Chen, L.Y., Talwani, P., 1998. Reservoir-induced Seismicity in China. Pure Appl. Geophys 153, 133−149.

Corral, A., 2004. Long-term clustering, scaling, and universality in the temporal occurrence of earthquakes. Phys. Rev. Lett. 92, 108501. Available from: http://dx.doi.org/10.1103/PhysRevLett.92.108501.

De Arcangelis, L., Godano, C., Grasso, J.R., Lippiello, E., 2016. Statistical physics approach to earthquake occurrence and forecasting. Phys. Rep. 628, 1−91.

Dieterich, J.H., 1972. Time-dependent friction in rocks. J. Geophys. Res. 77 (872), 3690−3697.

Dieterich, J.H., 1979. Modeling of rock friction 1. Experimental results and constitutive equations. J. Geophys. Res. 84B, 2161−2168.

Eckmann, J.-P., Kamphorst, S., Ruelle, D., 1987. Recurrence plots of dynamical systems. Europhys. Lett. 4, 973−977.

García-Álvarez, D., Stefanovska, A., McClintock, P., 2008. High-order synchronization, transitions, and competition among Arnold tongues in a rotator under harmonic forcing. Phys. Rev. E. 77, 056203.

Ghil, M., Vautard, R., 1991. Interdecadal oscillations and the warming trend in global temperature time series. Nature 350, 324−327.

Ghil, M., Allen, M.R., Dettinger, M.D., Ide, K., Kondrashov, D., Mann, M.E., et al., 2002. Advanced spectral methods for climatic time series. Rev. Geophys. 40, 1003.

Grassberger, P., Procaccia, I., 1983. Characterization of strange attractors. Phys. Rev. Letters 50, 346−349.

Griffith, A.A., 1921. The phenomena of rupture and flow in solids. Phil. Trans. Royal Soc. London A 221, 163−198.

Gupta, H.K., 2002. A review of recent studies of triggered earthquakes by artificial water reservoirs with special emphasis on earthquakes in Koyna, India. Earth Sci. Rev. 58, 279−310.

Hill, D., Prejean, S., 2009. Dynamic triggering. In: Kanamori, H. (Ed.), Earthquake Seismology. Elsevier, pp. 257−293.

Hubbert, M., Rubbey, W., 1959. Role of fluid pressure in mechanics of overthrust faulting. Bull. Geol. Soc. Am. 70, 115−166.

Irwin, G., 1957. Analysis of stresses and strains near the end of a crack traversing a plate. J. Appl. Mech. 24, 361−364.

Iwata, T., 2002. Tidal stress/strain and acoustic emission activity at the Underground Research Laboratory, Canada. Geophys. Res. Lett. 29, 301−304.

Jeffrey, J., 1992. Chaos game visualization of sequences. Comput. Graphics 16 (1), 25−33.

Johnson, P.A., Kaproth, B.M., Scuderi, M., Ferdowsi, B., Griffa, M., Carmeliet, J., et al., 2013. Acoustic emission and microslip precursors to stick-slip failure in sheared granular material. Geophys. Res. Lett. 40, 5627−5631.

Kantz, H., Schreiber, T., 1997. Nonlinear Time Series Analysis. Cambridge University Press, Cambridge.

Karner, S.L. Marone, C., 2000. Effects of loading rate and normal stress on stress drop and stick-slip recurrence interval. In: Rundle, J., Turcotte, D., Klein, W. (Eds.), Amer. Geophys. Un., Geophys. Mono. 120, Geocomplexity and the Physics of Earthquakes, pp. 187−198.

Kasahara, K., 1981. Earthquake Mechanics. Cambridge Univ. Press.

Kononov, E., 2006. Visual Recurrence Analysis. <www2.netcom.com/~eugenek/page6.html> (assessed 21.12.15).

Lempel, A., Ziv, J., 1976. On the complexity of finite sequences. I. E. E. E. Trans. Infor. Theory IT-22, 75−81.

Lursmanashvili, O., Paatashvili, T., Gheonjian, L. 2010. Detecting quasi-harmonic factors synchronizing relaxation processes: application to seismology. In: De Rubeis, V., Czechowski, Z. and Teisseyre, R. (Eds.), Geoplanet: Earth and Planetary Sciences, vol. 1. Synchronization and Triggering: From Fracture to Earthquake Processes. Springer, Berlin, Heidelberg, pp. 305−322.

Marone, C., 1998. Laboratory-derived friction laws and their application to seismic faulting. Ann. Revs. Earth Plan. Sci. 26, 643−696.

Marwan, M., 2003. Encounters with Neighbourhood. PhD Thesis. Potsdam.

Marwan, N., Romano, M.C., Thiel, M., Kurths, J., 2007. Recurrence plots for the analysis of complex systems. Phys. Rep. 438, 237−329.

Matcharashvili, T., Chelidze, T., 2010. Nonlinear dynamics as a tool for revealing synchronization and ordering in geophysical time series: application to Caucasus seismicity. In: De Rubeis, V., Czechowski, Z. and Teisseyre, R. (Eds.), Geoplanet: Earth and Planetary Sciences, vol. 1. Synchronization and Triggering: from Fracture to Earthquake Processes. Springer, Berlin, Heidelberg, pp. 3−21.

Matcharashvili, T., Chelidze, T., Abashidze, V., Zhukova, N., Meparidze, E., 2010. Changes in dynamics of seismic processes around enguri high dam reservoir induced by periodic variation of water level. In: De Rubeis, V., Czechowski, Z. and Teisseyre, R. (Eds.), Geoplanet: Earth and Planetary Sciences, vol. 1. Synchronization and Triggering: from Fracture to Earthquake Processes. Springer, Berlin, Heidelberg, pp. 273−286.

Matsukawa, H., Saito, T., 2007. Friction, stick-slip motion and Earthquake. Lect. Notes Phys. 705, 169−189. Available from: https://doi.org/10.1007/3-540-35375-5_7.

Métivier, L., de Viron, O., Conrad, C., Renault, S., Diament, M., Patau, G., 2009. Evidence of earthquake triggering by the solid earth tides. Earth Planet. Sci. Lett. 278, 370–375.

Meyers, R. (Ed.), 2009. Encyclopaedia of Complexity and Systems Science. Springer.

Mitarai, N., Alon, U., Jensen, M.H., 2013. Entrainment of noise-induced and limit cycle oscillators under weak noise. Chaos . Available from: https://doi.org/10.1063/1.4808253.

Nasuno, S., Kudrolli, A., Bak, A., Gollub, J.P., 1998. Time-resolved studies of stick-slip friction in sheared granular layers. Phys. Rev. E 58, 2161–2171.

Nikolaev, V.A., 2003. Research of lithospheric stress state on the base of correlation of tidal forces and seismicity. Anakharsys, Moscow (in Russian).

Packard, N.H., Crutchfield, J.P., Farmer, J.D., Shaw, R.S., 1980. Geometry from a time series. Phys. Rev. Lett. 45, 712–716.

Peng, Z., Hill, D., Shelly, D., Aiken, C., 2010. Remotely triggered microearthquakes and tremor in central California following the 2010 Mw 8.8 Chile earthquake. Geophys. Res. Lett. 37, L24312. Available from: https://doi.org/10.1029/2010GL045462.

Pikovsky, A., Rosenblum, M.G., Kurths, J., 2003. Synchronization: Universal Concept in Nonlinear Science. Cambridge University Press, Cambridge.

Prejean, S., Hill, D., 2009. Dynamic triggering of earthquakes. In: Meyers, A. (Ed.), Encyclopaedia of Complexity and Systems Science. Springer, pp. 2600–2621.

Putelat, T., Dawes, J.H.P., Willis, J.R., 2007. Sliding interactions of two frictional interfaces. J. Mech. Phys. Solids 55 (10), 2073–2105.

Reid, H.F., 1910. The mechanics of the earthquake, The California Earthquake of April 18, 1906, Report of the State Investigation Commission, vol. 2. Carnegie Institution of Washington, Washington, D.C.

Rikunov, L., Khavroshkin, O., Tsiplakov, V., 1980. Lunar-solar periodicities in spectral lines of temporal variations of high-frequency microseisms. Reports of Ac. Sci. USSR. 252, 577–579. In Russian.

Rosenblum, M.G., Pikovsky, A., Kurths, J., 1996. Phase synchronization of chaotic oscillators. Phys. Rev. Lett. 76, 1804–1808.

Rosenblum, M.G., Pikovsky, A., Kurths, J., 1997. Effect of phase synchronization in driven chaotic oscillators. I. E. E. E. Trans. C. A. S.I 44, 874–888.

Rosenstein, M.T., DeLuca, C.J., 1993. A practical method for calculating largest Lyapunov exponents from small data sets. Physica D 65, 117–134.

Rubinstein, J.L., Shelly, D.R., Ellsworth, W.L., 2010. Non-volcanic tremors. In: Cloetingh, S., Negendank, J. (Eds.), New Frontiers in Integrated Solid Earth Sciences. Springer, Berlin, pp. 287–314. Available from: https://doi.org/10.1007/978-90-481-2737-5.

Rubinstein, S.M., Cohen, G., Fineberg, J., 2004. Detachment fronts and the onset of dynamic friction. Nature 430, 1005–1009.

Rubinstein, S.M., Cohen, G., Fineberg, J., 2009. Visualizing stick-slip: experimental observations of processes, governing the nucleation of frictional sliding. J. Phys. D: Appl. Phys. 42, 214016–214032.

Ruina, A., 1983. Slip instability and state variable friction laws. J. Geophys. Res 88B, 10359–10370.

Rundle, J., Turcotte, D., Klein, W. (Eds.), 2009. Geocomplexity and Physics of Earthquakes. AGU, Washington, DC.

Savage, H., 2007. The Effects of Friction on the Earthquake Triggering and Fault Zone Evolution. PhD Thesis. The Pennsylvania State University.

Savage, H., Marone, C., 2007. Effects of shear velocity oscillations on stick-slip behavior in laboratory experiments. J. Geophys. Res, Solid Earth 112B. Available from: https://doi.org/10.1029/2005JB004238.

Scholz, C.H., 1998. Earthquakes and friction laws. Nature 391, 37–42.

Scholz, C.H., 2002. The Mechanics of Earthquakes and Faulting. Cambridge Univ. Press, Cambridge.

Scholz, C.H., 2003. Earthquakes: good tidings. Nature 425, 670–671. Available from: https://doi.org/10.1038/425670a.

Schuster, A., 1897. On lunar and solar periodicities of earthquakes. Proc. R. Soc. London 61, 455–465.

Sornette, D., 2000. Critical Phenomena in Natural Sciences. Springer:, Berlin.

Sprott, J., 2003. Chaos and Time-Series Analysis. Oxford University Press.

Strogatz, S., 2000. Nonliear Dynamics and Chaos. Westview: Perseus Books Group.

Takens, F., 1981. Detecting strange attractors in turbulence. In: Rand, D.A., Young, L.S. (Eds.), Dynamical Systems and Turbulence, vol. 898. Springer, Berliner, pp. 366–381. , Springer Lecture Notes in Mathematics.

Telesca, L., 2010. Analysis of the cross-correlation between seismicity and water level in the Koyna Area of India. Bull. Seismol. Soc. Am. 100, 2317–2321. Available from: https://doi.org/10.1785/0120090392.

Telesca, L., Matcharasvili, T., Chelidze, T., Zhukova, N., 2012. Relationship between seismicity and water level in the Enguri high dam area (Georgia) using the singular spectrum analysis. Nat. Hazards Earth Syst. Sci. 12, 2479–2485.

Theiler, J., Farmer, J., 1992. Testing for nonlinearity in time series: the method of surrogate data. Physica D 58, 77–94.

Tsallis, C., 1988. Possible generalization of Boltzmann-Gibbs statistics. J. Stat. Phys 52, 479–487.

Vallianatos, F., Papadakis, G., Michas, G., 2016. Generalized statistical mechanics approaches to earthquakes and tectonics. Proceedings of the Royal Society A. DOI: 10.1098/rspa.2016.0497

Varamashvili, N., Chelidze, T., 2004. Sick-slip and electromagnetic field. J. Georgian Geophys. Soc. 9A, 3–11.

Varamashvili, N., Chelidze, T., Lursmanashvili, O., 2008. Phase synchronization of slips by periodical (tangential and normal) mechanical forcing in the spring-slider model. Acta Geophysica. 56, 357–371.

Vidale, J., Agnew, D., Johnston, M., Oppenheimer, D., 1998. Absence of earthquake correlation with Earth tides: an indicatiob of high preseismic fault stress rate. J. Geophys. Res. 103B, 24 567–24 572.

Webber, C., Marwan, N. (Eds.), 2015. Recurrence Quantification Analysis. Springer:, Heidelberg.

Zbilut, J.P., Webber Jr., C.L., 1992. Embeddings and delays as derived from quantification of recurrence plots. Phys. Lett. A 171, 199–203.

Further Reading

Chelidze, T., Gvelesiani, A., Varamashvili, N., Devidze, M., Chikchladze, V., Chelidze, Z., et al., 2004. Electromagnetic initiation of slip: laboratory model. Acta Geophys. Pol. 52, 49–62.

Chelidze, T., Lursmanashvili, O., Varamashvili, N., 2008a. High order synchronization of stick-slip process in the spring-slider model with periodic (tangential) mechanical forcing. J. Georgian Geophys. Soc. 12A, 16–24.

Chelidze, T., Matcharashvili, Lursmanashvili, O., Varamashvili, N., 2008b. Acoustics of stick-slip deformation under external forcing: the model of seismic process synchronization. In: Triantis, D., Jelenska, M., Vallianatos, F. (Eds.), Advanced Topics in Geology and Seismology. University of Cambridge, W. S. E. A. S. Press, pp. 36–43.

Huang, N.E., Shen, Z., Long, S.R., Wu, M.L., Shih, H.H., Zheng, Q., et al., 1998. The empirical mode decomposition and Hilbert spectrum for nonlinear and nonstationary time series analysis. Proc. Roy. Soc. A 454, 903–995.

Complexity in Earthquake Generation and Seismic Hazard Assessment

10

Complexity and Time-Dependent Seismic Hazard Assessment: Should We Use Fuzzy, Approximate and Prone-to-Errors Prediction Models to Overcome the Limitations of Time-Independent Models?

Costas B. Papazachos[1], Domenikos A. Vamvakaris[1], George F. Karakaisis[1], Christos A. Papaioannou[2], Emmanuel M. Scordilis[1], Basil C. Papazachos[1]

[1]ARISTOTLE UNIVERSITY OF THESSALONIKI, THESSALONIKI, GREECE
[2]INSTITUTE OF ENGINEERING SEISMOLOGY AND EARTHQUAKE ENGINEERING (ITSAK), THESSALONIKI, GREECE

CHAPTER OUTLINE

Complexity of Seismic Time Series. DOI: https://doi.org/10.1016/B978-0-12-813138-1.00010-9

10.1 Intermediate-Term Earthquake Prediction Based on Seismicity Patterns

The main problem in earthquake prediction has been an issue of constant controversy and dispute within the earthquake science community. It was about 40 years ago when Frank Press, in an article on earthquake prediction, noted that "Recent technical advances have brought this long-sought goal within reach. With adequate funding several countries, including the United States, could achieve reliable long-term and short-term forecasts in a decade" (Press, 1975). This optimism was strongly challenged more than 20 years later by Geller et al. (1997), who stated that "...each failed attempt at prediction lowers the a priori probability for the next attempt. The current probability of successful prediction is extremely low, as the obvious ideas have been tried and rejected for over 100 years." Geller and his colleagues based that statement on observations of the distribution of the magnitudes of earthquakes; all but the largest events show a scale-invariant distribution, which is a characteristic of self-organized critical systems (Bak, 1996). Since the Earth's crust can be considered as such a system and any small earthquake has some probability of cascading into a large event, the predictability of the latter is precluded because of a lack of knowledge of the initial conditions, i.e., the details of the physical conditions on the fault surface to be ruptured and the stresses in the surrounding area.

However, although the Earth's crust is a complex and may be considered as a chaotic system, with strong earthquakes being the culmination of complex physical processes, this does not necessarily mean that predictions are impossible. In their well-known paper entitled *Rethinking Earthquake Prediction*, Sykes et al. (1999) suggest that long-term (and perhaps intermediate-term) predictions for large earthquakes appear to be possible. Within this context, an intermediate-term earthquake prediction, e.g., of 5–10 years, would permit a number of efficient mitigation measures to be taken, e.g., strengthening of critical structures, etc.

In the course of analysing seismicity, numerous spatiotemporal and other patterns have been retrospectively "observed" to precede strong and large earthquakes. Detection and identification of seismicity patterns is of particular interest, not only because of their societal value regarding seismic hazard assessment (SHA) and risk mitigation, but as reliable evidence of our understanding of the earthquake generation physical processes. In the last several years a significant amount of work in intermediate-term prediction has been made (see Mignan, 2011; Tiampo and Shcherbakov, 2012 for detailed reviews).

In the following, we present some of the methods and algorithms which have had the most significant impact, along with typical application examples, in an attempt to describe the current knowledge status. Moreover, we present preliminary results on the effect of a generic prediction algorithm on time-dependent probabilistic seismic hazard assessment (PSHA). Finally, some ideas on the feasibility of earthquake prediction are briefly discussed.

10.2 The Algorithms M8, Mendocino Scenario and California-Nevada

In a series of articles during the 1980s, Keilis-Borok and Kossobokov presented an intermediate-term earthquake prediction algorithm (Keilis-Borok and Kossobokov, 1984, 1986, 1988). This algorithm, M8, was used to explore the possibility of identifying the times of increased probability (TIPs) of a large earthquake (of magnitude about 8.0 +, hence its name) by premonitory intermediate-term seismic activation in the lower magnitude range (Keilis-Borok and Kossobokov, 1990a).

The general outline of the algorithm M8, as originally described by Keilis-Borok and Kossobokov (1990b), concerns the earthquake "flow," i.e., the level of seismic activity, in a certain area with size dependent on the magnitude, M_0, of the strong earthquakes for which a diagnosis of TIP is sought. For example, searching for an $M_0 = 7$ earthquake, the radius, R (in degrees), of the circular area to be examined is initially set to $0.5 \bullet [\exp(M_0 - 5.6) + 1] = 2.5° \approx 280$ km.

Several quantities, which characterize the seismic activity, are estimated as functions of a sliding time window (e.g., 6 months). In the event of large values of these parameters within a certain narrow time interval a TIP is diagnosed for a few years (e.g., 5 years). The functions examined for each circular area, to diagnose a TIP for an M_0 earthquake (after the declustering of the earthquakes located within this area and the identification of mainshocks) are:

1. The cumulative number, N, of mainshocks above a certain cutoff magnitude, which describes an increase in seismic activity;
2. The rate differential, L, which describes the deviation of N from a longer-term average;
3. The spatial clustering, Z, which is a dimensionless indicator of concentration of mainshocks, i.e., the average source size divided by the average distance between sources; and
4. A characteristic measure of aftershock clustering, B, which shows the maximum number of aftershocks during the first several days (e.g., 2 days after a mainshock) and describes the bursts of aftershocks (Tikhonov and Rodkin, 2012).

Since only mainshocks are considered by M8, the original catalogue has to be properly declustered, i.e., all associated shocks (mostly aftershocks) are removed using appropriate spatial and temporal windows (e.g., Gardner and Knopoff, 1974; Gabrielov et al., 1986). Fig. 10−1 illustrates the basic characteristics of M8 (Keilis-Borok, 1996).

Functions N, Z, and B can be defined 6 years after the beginning of the earthquake catalogue. It should be noted that normalization of the earthquake flow requires adjustment of the mainshock magnitude threshold, M_c, so that the average yearly mainshock occurrence rate in an area is constant. For one set of functions (e.g., N_1, L_1, Z_1) the constant is 10 events per year and for the other set (N_2, L_2, Z_2) the constant is 20 events per year. The upper magnitude threshold, M_{up}, is set to M_0; for functions Z, $M_{up} = M_0 - 0.5$ and functions B, $M_{up} = M_0 - 0.2$, $M_c = M_0 - 2$ (Keilis-Borok and Kossobokov, 1990b). These constants may be modified according to the background seismicity level of the investigated area.

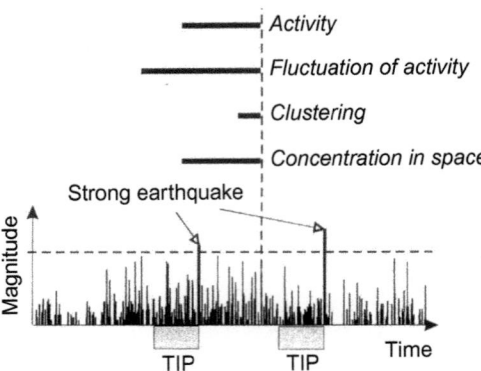

FIGURE 10–1 Main features of M8: Vertical lines show the earthquake sequence (mainshocks only, after declustering) in an area. Several functions of this sequence are defined in sliding time windows shown by horizontal bars. TIPs (alarms) are recognized by one or several such functions. The horizontal dashed line denotes the cutoff magnitude, M_0 (Keilis-Borok, 1996).

The examined area is scanned using overlapping circles with radii proportional to the magnitude of the target earthquake, M_0, and all seven functions (N_1, L_1, Z_1, N_2, L_2, Z_2, B) are calculated. To define a TIP for the area starting at time t, the following conditions should be satisfied over the preceding 3 years (including t and excluding $t-3$ years) and for two consecutive time windows: (1) each group (N_1, N_2), (L_1, L_2), (Z_1, Z_2) and B contains functions with extremely large values, and (2) at least six of the functions, including B, have extremely large values. The "extremely large" values are found within the upper 10% for the six functions (N_1, L_1, Z_1, N_2, L_2, Z_2) and the upper 25% for function B (Keilis-Borok and Kossobokov, 1990b).

Typical graphs with the results of the application of the M8 algorithm are shown in Fig. 10–2. The earthquakes that occurred during 1962–92 in a circular area ($R = 427$ km) centred at the Southern Kuril Islands (44°N, 149°E) were used. The functions were calculated every 6 months. Solid lines show the values calculated for large statistics (20 events or more per year) and dashed lines correspond to values calculated for small statistics (10 events or less per year). Stars denote the anomalies (extremely large values) which resulted in a TIP diagnosis for the following 5 years, 1993–97. A large earthquake occurred in the examined area after about 2 years of the issued diagnosis (4 October 1994, $M_w = 8.3$, Shikotan earthquake, epicentre: 43.5°N−147.4°E) (Tikhonov and Rodkin, 2012).

By 1990, the original algorithm M8 and its subsequent versions were applied retrospectively to segments of the Circum-Pacific (Central and South America, Western USA, Kamchatka-Kuril island, Japan, Taiwan) and Alpine−Himalayan Seismic Belts for mainshock magnitudes between 6.5 and 8.0 (Keilis-Borok and Kossobokov, 1990b). Forward tests of M8 have been also performed on these areas (Healy et al., 1992).

To reduce the spatial uncertainty of a retrospectively diagnosed TIP by the M8 algorithm over a square area with size $D = 560$ km, where the Eureka earthquake occurred on

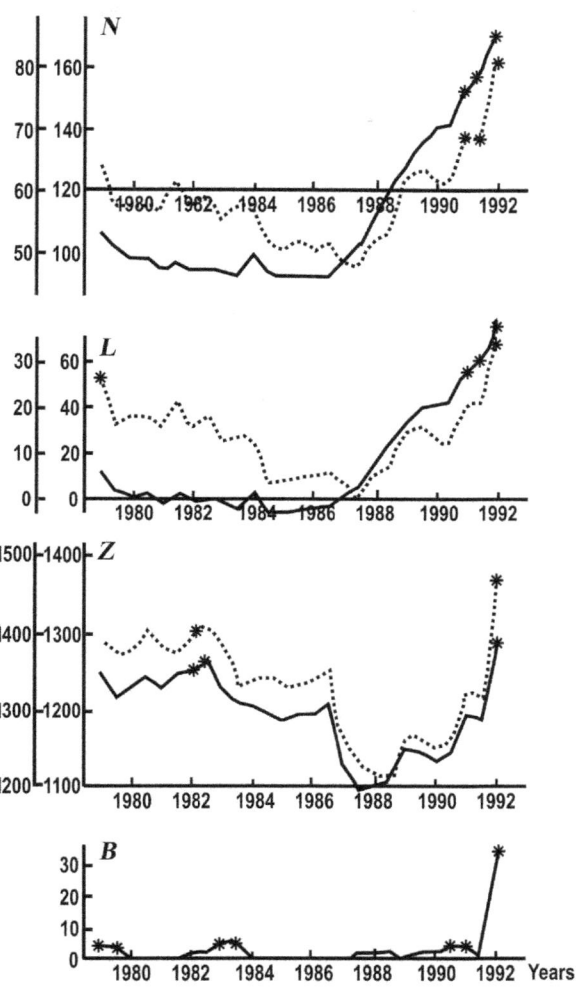

FIGURE 10–2 The seven functions calculated (in time windows of 6 months) by M8 on earthquakes that occurred in a circular area (*R* = 427 km) during the diagnosis period 1979–92. *Solid lines* show the values calculated for large statistics (20 events or more per year) and *dotted lines* show the values calculated for small statistics (10 events or less per year). *Stars* denote the anomalously large values. *Modified after Tikhonov, I.N., Rodkin, M.V., 2012. Current state of art in earthquake prediction, typical precursors and experience in earthquake forecasting at Sakhalin Island and Surrounding Areas. In: Earthquake Research and Analysis – Statistical Studies, Observations and Planning. (Ed.) Sebastiano D'Amico. InTech, pp. 43–78.*

8 November 1980 (M_w = 7.3) near Cape Mendocino in NW California, Kossobokov et al. (1990a) examined the smaller magnitude seismicity within the area for which the TIP alarm had been issued by using an algorithm called Mendocino Scenario (MSc). The purpose of the MSc algorithm was that, given a TIP diagnosed for a certain area *U* at time *T*, a smaller area *V* is to be found within *U*, where the predicted strong earthquake can be expected. For this reason, a complete catalogue is required with earthquake magnitudes smaller than those

employed in the M8 algorithm (e.g., $M_{min} \geq M_0-4$). The area U is divided into small squares with side, $s \sim 100$ km (typically $1° \times 1°$). For each small square the number of events (with magnitudes $M \geq M_{min}$, including aftershocks), is calculated for consecutive overlapping small time windows, u (e.g., 2 months), with a step of 1 month, starting at time $T-6$ years, in order to include the earthquakes that contributed to the diagnosis of the TIP. In this way, the space–time considered is divided into small boxes of size $(s \times s \times u)$. "Quiet" boxes are termed those where the number of earthquakes is below the 10% of the numbers of earthquakes of all boxes. When four or more such quiet boxes form a cluster, the area they occupy is the V area and it is in this area that the strong earthquake is expected to occur. This means that within the large area, for which a TIP alarm has been issued and where the seismicity is continuously high, the MSc algorithm outlines a smaller area where seismicity decreases (Kossobokov and Shebalin, 2003). Fig. 10–3 shows the square area, U, for which a TIP had been diagnosed by the M8 algorithm and the shaded area (V area) identified by the application of the MSc algorithm. The epicentre of the Eureka, 1980 earthquake is denoted by the black circle (Kossobokov et al., 1990b). An example of joint application of the M8 and MSc algorithms is the test performed on several large earthquakes that occurred in the Circum-Pacific during 1992–97 (Kossobokov et al., 1999).

A similar algorithm, proposed by Keilis-Borok and Rotwain (1990) for examining precursory seismicity associated with earthquakes of $M \geq 6.4$ in California and adjacent regions (e.g., Nevada), employed the quantities mentioned above, i.e., level of seismic activity and its time variation, clustering of earthquakes in space and time, as well as their long-range interactions. In this algorithm, called California-Nevada (CN), nine functions are calculated to describe the seismicity preceding strong earthquakes. Only large, medium and small values for each function are distinguished and an alarm for a TIP is identified when earthquake clustering is high, seismicity is irregular, high and growing (alternating time intervals of low and high seismic activity) and quiescence precedes the increase in seismicity (Keilis-Borok, 2002). Different cutoff values were determined by applying the CN algorithm worldwide, e.g., Italy (5.6), Northern and Southern Appalachians (5.0), Brabant-Ardennes (4.5), Gulf of California (6.6), etc. (Keilis-Borok and Rotwain, 1990). Fig. 10–4 shows the application of CN in California (Rotwain and Novikova, 1999).

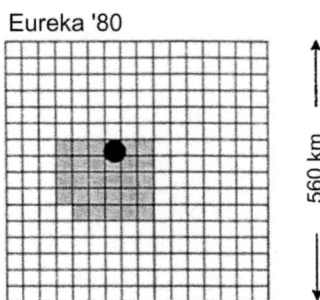

Eureka '80

560 km

FIGURE 10–3 The square area for which a TIP had been diagnosed for the Eureka, 1980 earthquake (*black circle*). The shaded area, where the earthquake might be expected to occur, was identified through the application of the MSc algorithm (Kossobokov et al., 1990b).

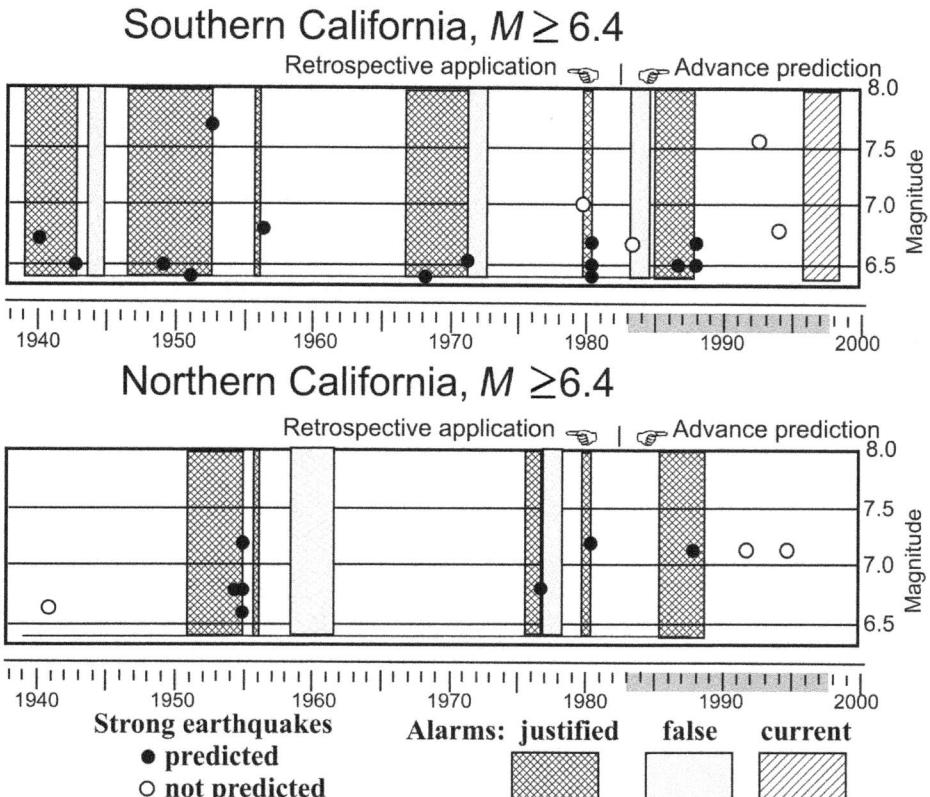

FIGURE 10–4 Times of increased probabilities (TIPs) of strong earthquakes in California determined using the CN algorithm. *Modified after Rotwain, I., Novikova, O., 1999. Performance of the earthquake prediction algorithm CN in 22 regions of the world. Phys. Earth Planet. Int. 111, 207–213.*

10.3 Accelerating (Accelerating Moment Release) and Decelerating Seismicity

Increasing seismic activity over a broad region prior to the occurrence of a strong main-shock has long attracted the interest of the scientific community (e.g., Imamura, 1937; Gutenberg and Richter, 1954; Tocher, 1959; Papadopoulos, 1986; Jaumé and Sykes, 1999; Papazachos et al., 2005a; Mignan, 2011 and references therein). The main interest is that this seismicity pattern may be considered as a rather easily observable manifestation of the seismogenic process which culminates in the mainshock (Jaumé and Sykes, 1999). The numerous attempts which have been made during the last three decades or so to describe quantitatively this pattern, known as "accelerating moment release" (AMR), fall generally into two categories (Fig. 10–5). In most relevant studies, AMR is considered as a critical process and AMR observations are modelled by analogy with the statistical mechanics of

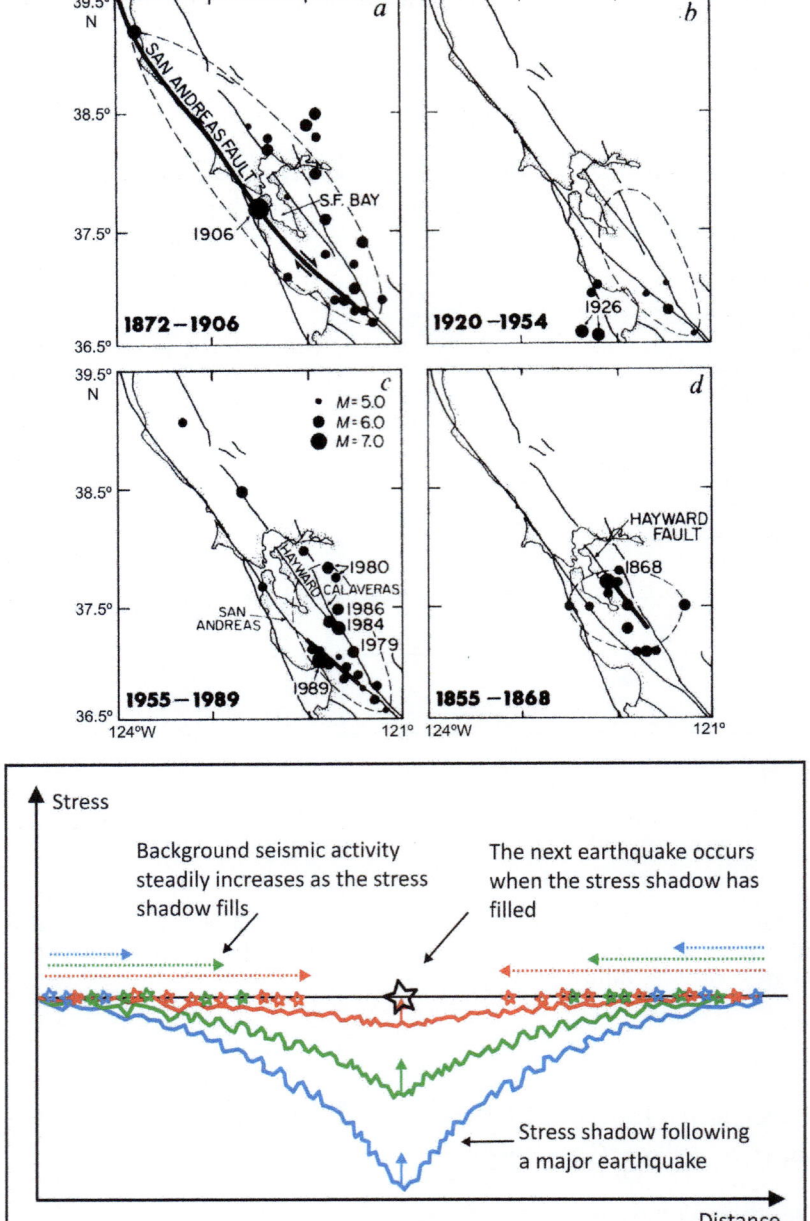

FIGURE 10–5 Two physical models for AMR: (Upper figure) Moderate magnitude earthquakes (M ≥ 5.0) in the San Francisco Bay region during four equal time periods. The areas in which earthquakes participate in the accelerating moment release prior to the mainshocks of 1868 (M ~ 7), 1906 (M = 7.8) and 1989 (M = 7.0) are outlined with a dashed line and scale with the magnitude of the oncoming mainshock. Moderate magnitude seismicity, shown in map b, was low in the area of the 1989 mainshock during 1920–54 (Sykes and Jaumé, 1990). (Lower figure) Following a large earthquake, a stress hole or shadow is created around the associated fault (lower line) and progressively fills until the next large earthquake (middle and upper lines). As the stress shadow fills, activity increases over a wide region. *Modified after King, G.C.P., Bowman, D.D., 2003. The evolution of regional seismicity between large earthquakes. J. Geophys. Res., 108, doi:10.1029/2001JB000783 and King, G.C.P., 2007. Fault interaction, earthquake stress changes, and the evolution of seismicity. In: Dziewonski, A., Romanowicz, B. (Eds.), Treatise on Geophysics, vol. 1Elsevier.*

phase transitions culminating in a critical point, i.e., a mainshock (e.g., Allègre et al., 1982; Sornette and Sornette, 1990; Sornette and Sammis, 1995; Rundle et al., 1999). The mainshock is considered as the end-result of a process in which the stress field in a broad area (critical area) becomes correlated over increasingly large scales. Before the mainshock the growing correlation length is manifested as an increase in the frequency and mainly in the magnitude of intermediate magnitude events (preshocks). After the mainshock the criticality is destroyed and a period of relative quiescence follows. Then, the process starts over by rebuilding growing correlation lengths towards criticality and the next mainshock (e.g., Smalley et al., 1985; Saleur et al., 1996; Sammis et al., 1996; Bowman et al., 1998; Jaumé and Sykes, 1999, among others). On the other hand, in the "stress accumulation model" (SAM), which has been proposed by King and Bowman (2003), AMR is attributed to the decrease in the size of a stress shadow left from one or more previous events (i.e., the accelerating seismicity is a secondary result of the loading of a large fault primarily by aseismic creep on this fault in the lower crust). This process is similar to an increase in the size of the region of background seismicity. Mignan (2011) presents a thorough review on the accelerating seismicity hypothesis.

Accelerating moment (or seismic strain) release has been reported to precede a considerable number of strong mainshocks in a variety of regions such as, e.g., in California (e.g., Sykes and Jaumé, 1990; Bufe and Varnes, 1993; Knopoff et al., 1996; Bowman et al., 1998; Jaumé and Sykes, 1999; Bowman and King, 2001), New Madrid (Brehm and Braile, 1998), Alaska (Bufe et al., 1994), Mexico (Sammis et al., 2004), Mediterranean (Papazachos and Papazachos, 2000, 2001; Papazachos et al., 2006), New Zealand (Robinson, 2000), Japan and central Asia (Papazachos et al., 2006), Italy (Scordilis et al., 2004; De Santis et al., 2015) and Sumatra (Jiang and Wu, 2005; Mignan et al., 2006). Accelerating seismic energy release is typically modelled by the power-law time-to-failure equation (Varnes, 1989; Bufe and Varnes, 1993):

$$S(t) = A + B(t_c - t)^m \tag{10.1}$$

where $S(t)$ is the cumulative seismic energy release (most often in terms of Benioff strain, i.e., square root of seismic energy), t_c is the mainshock origin time and A, B, m are model parameters calculated by the available data (with $m < 1$, $B < 0$). The quantity $S(t)$, which is considered as a measure of the preshock seismic deformation at time t, is defined as:

$$S(t) = \sum_{i=1}^{n(t)} E_i^{1/2} \tag{10.2}$$

where E_i is the seismic energy of the ith preshocks and $n(t)$ is the number of preshocks that occurred up to time t. It has to be noted that the term "preshocks" refers not only to the shocks which occur up to several days before a mainshock, close to its focus, but also to intermediate magnitude earthquakes (2−3 magnitude units smaller than the mainshock magnitude) which occur in a broad area (critical area) several years before its generation. Bowman et al. (1998) proposed the curvature parameter, C, to quantify the accelerating seismicity. C is the ratio of the root mean square error of the power-law fit (relation 1) to the corresponding linear fit error. This parameter, which is also used to quantify decelerating seismicity, takes positive values smaller than 1, becomes equal to 1 for a linear fit and decreases when the accelerating (or the decelerating) Benioff strain release becomes more intense.

Recent studies have also shown that in the narrow (focal) region of an oncoming mainshock, a seismic excitation occurs and is followed by a drop of seismicity, i.e., a quiescence period (Evison and Rhoades, 1997; Evison, 2001). Papazachos et al. (2005b) used global data to show that intermediate magnitude preshocks in the focal region (seismogenic region) form a decelerating pattern and that the time variation of the cumulative Benioff strain up to the mainshock also follows a power-law (relation 1) but with a power value larger than one ($m > 1$). That is, this pattern of decelerating strain in the focal region is formed of a seismic excitation followed by a decrease of seismicity of intermediate-magnitude preshocks. Papazachos et al. (2006) used global data to develop the decelerating–accelerating seismicity (D-AS) model for intermediate-term earthquake prediction. In this model, the magnitude of the oncoming mainshock is proportional to the size of the critical and seismogenic regions and its occurrence time can be calculated by the duration of the accelerating and decelerating sequences. In all calculations the Benioff strain rate, i.e., the seismicity level in the areas examined, is also taken into account.

In most cases where retrospective identification of accelerating seismicity has been identified, critical regions were either circles, centred at the epicentres of the investigated mainshocks (e.g., Bowman et al., 1998; Jiang and Wu, 2006, 2013; De Santis et al., 2010, among many others) or stress contours defined by the SAM, that optimized the precursory accelerating activity (e.g., Sammis et al., 2004). In other cases the approach employed circles of radii corresponding to the known mainshock magnitude and its uncertainty (e.g., ± 0.2 magnitude units) to search a grid over the broader epicentral area (e.g., ± 3.0 degrees with step 0.5 degrees) for accelerating seismicity (e.g., Papazachos and Papazachos, 2001; Papazachos et al. 2006).

Fig. 10−6 shows a map of California, with several circles (critical regions) centred on the epicentre of the 1952 Kern County earthquake (star, $M = 7.5$). The epicentres of the earthquakes $M \geq 5.5$ (black dots) which have occurred since 1910 are also shown. Next to the map the cumulative Benioff strain, released by the latter within the three circles, is plotted. It is evident that at a radius of 350 km, the seismicity produces a well-defined power-law, whereas in the larger area ($R = 600$ km) the energy release is essentially a linear function of time (Bowman et al., 1998). Fig. 10−7 presents the epicentres of the accelerating (small open circles) and decelerating preshocks (dots) which preceded the 1992 California (25 April, $M = 7.1$) and the 1994 Japan (4 October, $M = 8.3$) earthquakes (white stars). The critical (large thin line circles) and the seismogenic (smaller thick line circles) regions are also shown. The corresponding time variations of the cumulative Benioff strain for the accelerating and decelerating preshocks are plotted next to the maps (Papazachos et al., 2006).

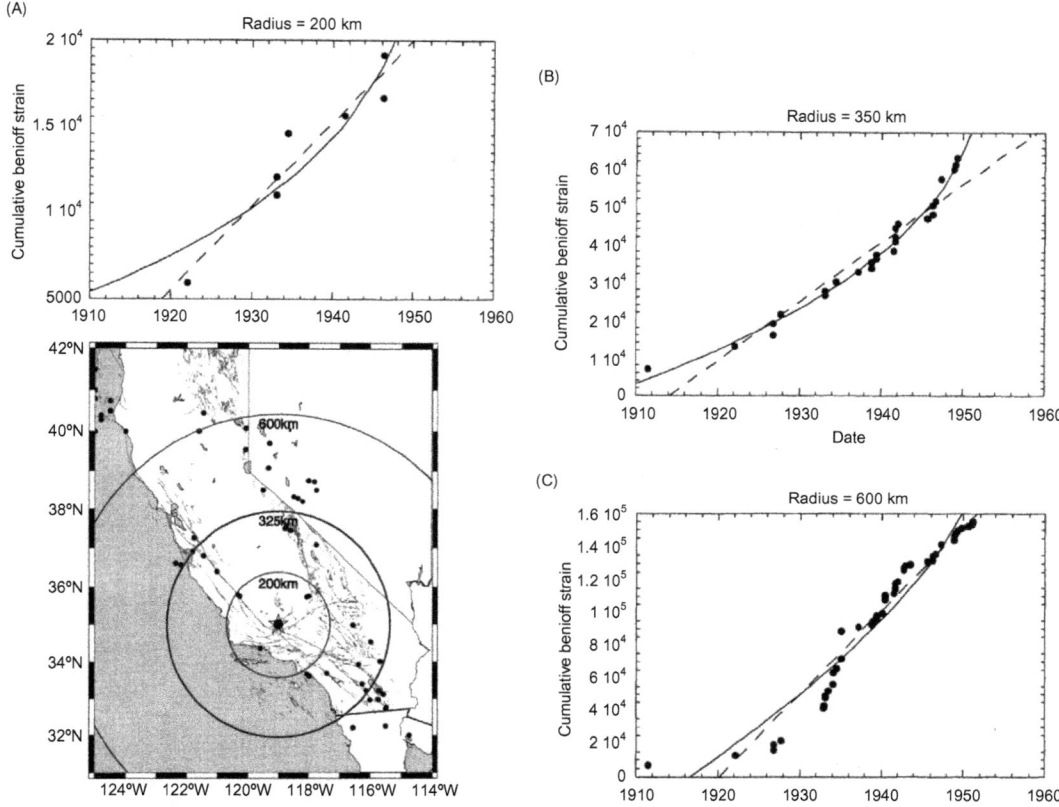

FIGURE 10–6 Earthquakes with $M \geq 5.5$ (dots) in California since 1910 before the 1952 Kern County earthquake ($M = 7.5$). Three circular regions were tested for precursory accelerating seismicity. The cumulative Benioff strain is plotted together with the linear and the power-law fits (*dashed* and *solid lines*, respectively) (Bowman et al., 1998).

10.4 Load/Unload Response Ratio

The LURR approach (load/unload response ratio) was proposed as a tool that might help short- and particularly intermediate-term earthquake prediction attempts (Yin, 1987, 1993; Yin et al., 1995). It is based on the idea that an earthquake is practically the failure or instability of a focal region on a fault, accompanied by a rapid energy release and that, for this reason, the earthquake preparation process is the deformation and damage process of the material in the focal region (Yin et al., 2000).

The concept of LURR is explained in Fig. 10–8, which describes the stress/strain behaviour of a brittle material (e.g., the Earth's crust) as a loading/response curve. For small and moderate loading (stress) there is elastic response of the material (strain) which is reversible, i.e., its loading response is equal to the unloading response. Higher loading will result in irreversible phenomena, namely initiation and growth of cracks (instabilities), and eventually in coalescence of these cracks and failure.

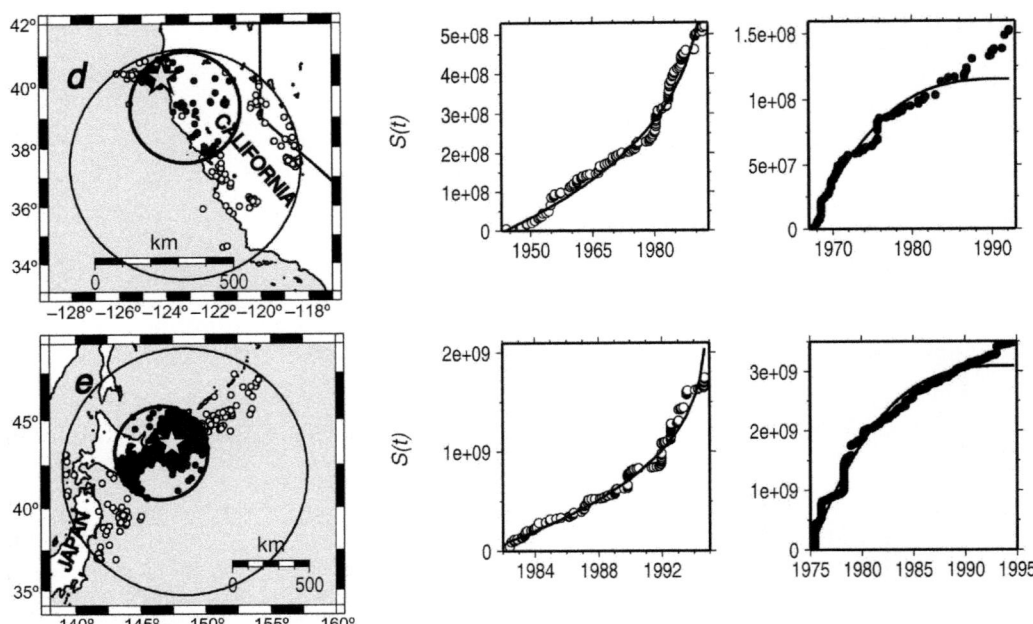

FIGURE 10–7 (Left) Spatial distribution of the epicentres of the accelerating (*small open circles*) and decelerating (*dots*) preshocks for the 1992 California (*M* = 7.1) and the 1994 Japan (*M* = 8.3) earthquakes (*white stars*), with the respective critical (larger, *thin line circles*) and seismogenic (smaller, *thick line circles*) regions. (Right) Corresponding time variations of the cumulative Benioff strain and the best-fit power-law curves for accelerating and decelerating preshocks (Papazachos et al., 2006).

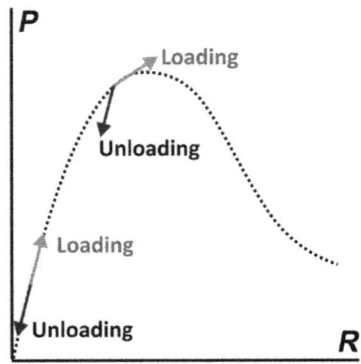

FIGURE 10–8 Loading/unloading response curve (*dotted line*) of a brittle material/system. The material response, *R*, is linear for load, *P*, (and unload) smaller than that of its strength. For higher load/unload values (close to the yielding point) the material response becomes nonlinear. *Modified after Yin, X.C., Zhang, L.P., Zhang, H.H., Yin, C., Wang, Y.C., Zhang, Y.X., et al., 2006. LURR's twenty years and its perspective. Pure Appl. Geophys., 163, 2317–2341.*

Two parameters are used to quantify this process: the response rate, X, and the LURR, Y. The response rate, X, is defined by Eq. (10.3):

$$X = \lim_{\Delta P \to 0} \left(\frac{\Delta R}{\Delta P} \right) \qquad (10.3)$$

where, ΔP, is a small change of loading that causes a small change, ΔR, of the response of the material (or the system on which loading is applied).

The LURR parameter, Y, is given by Eq. (10.4):

$$Y = \frac{X^+}{X^-} \qquad (10.4)$$

where X^+ and X^- correspond to the response rates during loading and unloading, respectively.

According to Yin et al. (1995), during the elastic phase (linear segment of the curve in Fig. 10−8) the response rate of the material, X^+, during loading ($\Delta P > 0$) is equal to the response rate, X^-, during unloading ($\Delta P < 0$) and consequently $Y = 1$. For higher loading, $X^+ > X^-$ and $Y > 1$ whereas close to failure of the material, $X^+ >> X^-$ and $Y \to \infty$. Therefore, growing LURR values can indicate that the material (or a system) approaches towards instability and that more energy is released in the loading period than in the unloading period.

The application of the LURR approach for earthquake prediction purposes requires: (1) understanding of the way a part of the Earth's crust can be loaded and unloaded; (2) knowledge of the seismicity properties that will help distinguishing loading from unloading; and (3) the choice of an appropriate measure quantifying LURR variations (Yin et al., 1995; Yuan et al., 2010).

Loading and unloading of a part of the Earth's crust may be caused by tidal forces, although there is a continuing controversy over earthquake occurrence and Earth tides in many relevant published articles. For instance, Vergos et al. (2015), analysing shallow and intermediate-depth earthquakes with magnitudes, $M_L = 2.5-6.2$, that occurred along the Hellenic Arc during 1964−2012, reported that the monthly variation of the frequencies of earthquake occurrence is in accordance with the period of the tidal lunar monthly variations, and the daily variation with the diurnal lunar−solar and semidiurnal solar variations. Lunar tidal forces may be an important contributing factor of strong ($M \geq 7.0$) seismicity worldwide (Su et al., 2012). On the other hand, Smith and Sammis (2004) after examining, within the context of the LURR approach, peak tidal stresses acting on faults of five strong earthquakes in California (e.g., Loma Prieta 1989 $M = 7.0$, Landers 1992 $M = 7.3$, Northridge 1994 $M = 6.7$, etc.), concluded that theses stresses did not activate precursory seismicity in the years prior to these earthquakes.

The criterion to distinguish loading or unloading is based on the Coulomb failure hypothesis, i.e., loading and unloading periods are determined by incrementally calculating Coulomb failure stress (*CFS*, e.g., Harris, 1998); when ΔCFS (a small change in *CFS*) is positive, it is defined as a loading period, and earthquakes occurring in this period are defined as loading earthquakes. Otherwise, when ΔCFS is negative, it is defined as an unloading period, and earthquakes occurring in this period are defined as unloading earthquakes (Zhang et al., 2015).

LURR is expressed in terms of seismic energy, E, calculated by the magnitudes of the earthquakes (e.g., Kanamori and Anderson, 1975) of the loading and unloading periods:

$$Y = \frac{\left(\sum\limits_{i=1}^{N^+} E_i^m\right)_+}{\left(\sum\limits_{i=1}^{N^-} E_i^m\right)_-} \tag{10.5}$$

The signs " $+$ " and " $-$ " correspond to the loading and the unloading periods, respectively. For $m = 1/2$, E^m denotes the Benioff strain and for $m = 0$, $Y = N^+/N^-$, i.e., the ratio of the numbers, N, of earthquakes during the two periods. The examined circular area, where loading and unloading events occur, scales with the magnitude of the future mainshock, as does the time period between the maximum LURR value and the time of occurrence of this mainshock (Yin et al., 2002; Zhang et al., 2005). Typically, the optimal radii for identification of the LURR anomaly are 75, 100, 200, 300 and 600 km, for earthquake magnitude values 5.0, 5.7, 6.5, 7.0 and 7.9, respectively (Yin et al., 2002), while the time period is about 6 months for an $M6$ earthquake, 1 year for an $M7$ earthquake and 3 years for an $M8$ event (Yin et al., 2006).

A typical example of LURR precursory variations prior to strong earthquakes in California is presented by Zhang et al. (2015). Using a 0.5 degrees grid over the examined area, circular areas with radius $R = 100$ km are defined and all complete earthquakes within each circular area during a 1-year time window are considered and a LURR value is calculated. Calculations are repeated with a time step of 1 month and circles of different radius are defined (e.g., up to 300 km with a step of 25 km). The whole set of calculations is repeated until the maximum LURR value is attained. Alternatively, the spatial window examined is the area delineated by the contour of a positive ΔCFS level (e.g., 0.01 bar, Yu et al., 2015) or the area where AMR is observed (Yin et al., 2002). Fig. 10−9 shows typical LURR precursory anomalies prior to the Loma Prieta, 1989, earthquake in California (left) and the Kobe, 1995 and Tottori, 2000, earthquakes in Japan (right).

The LURR approach has been used for about 30 years for retrospective predictions, as well as for predictions of future mainshocks in several areas worldwide (e.g., China, Yu et al., 2011; Iran, Zhang et al., 2007; Pakistan, Yu, 2004; Japan, Yin et al., 1996, among others). Recent detailed reviews on LURR can be found in Tiampo and Shcherbakov (2012) and Zhang et al. (2015).

10.5 The Region−Time−Length Algorithm

The region−time−length (RTL) algorithm, proposed and developed by Sobolev and Tyupkin (1996, 1997, 1999) and Sobolev et al. (1997), is a statistical method aiming at identifying

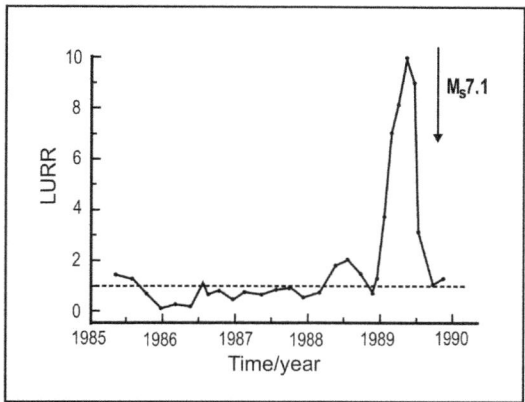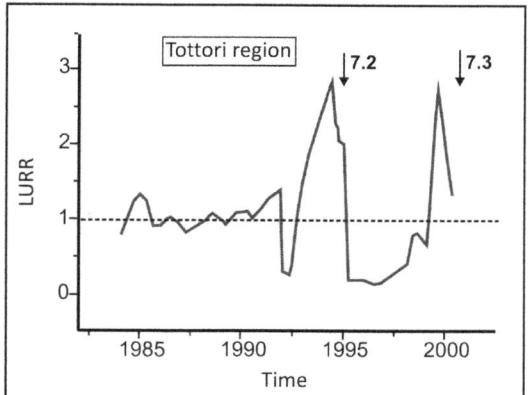

FIGURE 10–9 Precursory LURR variations (left) prior to the Loma Prieta, 1989, California earthquake (Zhang et al., 2010), and (right) the Kobe, 1995 and Tottori, 2000 earthquakes in Japan. *Modified after Yin, X.C., Mora, P., Peng, K.Y., Wang, Y.C., Weatherley, D., 2002. Load-Unload Response Ratio and accelerating moment/energy release critical region scaling and earthquake prediction. Pure Appl. Geophys. 159, 2511–2523.*

deviations from the background seismicity, i.e., decreasing or increasing values of RTL functions denote seismic quiescence or seismic excitation, respectively. Sobolev et al. (2002) noted that the reported cases of seismic quiescence, observed in a significant amount of seismicity studies, deserve particular attention and that these cases support the idea that during the earthquake preparation process several phases can be distinguished (Scholz, 1990). When a mainshock occurs an aftershock sequence follows, with the duration dependent on the mainshock magnitude. Then, seismicity in the vicinity of the ruptured fault and in the broader area drops to its background level due to stress redistribution. This phase is the longest in duration and may be about 50%–70% of the duration of a seismic cycle. Stress build-up follows and may be manifested by an increase in seismicity in the broader area, whereas a decrease in seismicity, i.e., seismic quiescence, may be observed in the fault zone of the oncoming strong earthquake (Mogi, 1969).

The three functions employed in the RTL algorithm, for a point with geographical coordinates x, y, depth z, and time t, are defined as follows:

$$R(x,y,z,t) = \left[\sum_{i=1}^{n} \exp\left(-\frac{r_i}{r_0} \right) \right] - R_{ltr}(x,y,z,t)$$

$$T(x,y,z,t) = \left[\sum_{i=1}^{n} \exp\left(-\frac{t-t_i}{t_0} \right) \right] - T_{ltr}(x,y,z,t) \qquad (10.6)$$

$$L(x,y,z,t) = \left[\sum_{i=1}^{n} \left(-\frac{l_i}{r_0} \right)^p \right] - L_{ltr}(x,y,z,t)$$

where: l_i and t_i are the rupture dimension (dependent on the earthquake magnitude by a relationship, e.g., $\log l_i = 0.5 M_i - 1.8$; Kasahara, 1981) and the origin time of the ith earthquake, satisfying certain criteria (e.g., completeness of the catalogue: $M_i \geq M_{min}$, with M_{min} being the cutoff magnitude), n is the number of events examined, $r_i \leq R_{max} (= 2r_0)$ and $(t - t_i) \leq T_{max} (= 2t_0)$ (with r_0 being a characteristic distance, i.e., the search distance, and the time window $[t - 2t_0, t]$ in which the summation is performed). r_0 characterizes the diminishing influence of more distant earthquakes, t_0 characterizes the rate at which preceding earthquakes are "forgotten" as the time of analysis moves on and the coefficient P characterizes the contribution of size of each preceding earthquake. For P values equal to 1, 2 and 3, this quantity is proportional to rupture length, square of rupture length or energy, respectively. r_i is the epicentral distance of an earthquake examined from the point selected for analysis, t_i is the event's origin time. The focal depths, d_i, of the earthquakes examined have to be smaller than a certain value, d_0 (e.g., $d_0 \leq 60$ km). Function $R(x,y,z,t)$ assigns a decreasing weight to each earthquake as a function of distance from the point of interest, $T(x,y,z,t)$ decreases the weight of each event as a function of the difference from the time of interest, and $L(x,y,z,t)$ weights the contribution of each event magnitude (expressed in terms of rupture length; Sobolev, 2011).

$R_{ltr}(x,y,z,t)$, $T_{ltr}(x,y,z,t)$ and $L_{ltr}(x,y,z,t)$ are long-term (background) values of the three quantities of Eq. (10.6), and their removal eliminates linear trends of these functions. Typically, the earthquake catalogue should be about 10 times longer than the expected anomaly duration (quiescence and preseismic activation; Sobolev et al., 2002). Functions $R(x,y,z,t)$, $T(x,y,z,t)$ and $L(x,y,z,t)$ are dimensionless and normalized to have unit variance, facilitating their use in various combinations. The product of the above three functions is calculated as the RTL parameter, describing the deviation from the background seismicity in standard deviation units (Gentili, 2010; Sobolev, 2011):

$$\text{RTL}(x, y, z, t) = \frac{R(x, y, z, t)}{\sigma_R} \cdot \frac{T(x, y, z, t)}{\sigma_T} \cdot \frac{L(x, y, z, t)}{\sigma_L} \tag{10.7}$$

Modified RTL versions have been proposed and decreasing or increasing RTL values (implying seismic quiescence and activation, respectively) are considered to be equally reliable precursors (e.g., Jiang et al., 2004; Chen and Wu, 2006), though some researchers only consider the decrease in the RTL function (Huang and Sobolev, 2001). The original RTL algorithm and its modified versions have been used to predict mainshocks with magnitudes 5.0–9.0 in several areas, e.g., China (Jiang et al., 2004; Liu and Su, 2006; Huang, 2008), Japan (Huang et al., 2001; Huang and Sobolev, 2001; Huang and Nagao, 2002; Huang and Ding, 2012), Sumatra (Sukrungsri and Pailoplee, 2015), Kamchatka and Kuriles (Sobolev and Tyupkin, 1997, 1999; Sobolev, 2011), Italy (Gentili, 2010) and elsewhere.

A typical example of RTL application was presented by Huang and Sobolev (2001), who identified, retrospectively, the seismic quiescence that preceded the Nemuro Peninsula mainshock (Japan, $M = 6.8$, 28 January 2000). For this mainshock rupture length (~ 40 km, Kasahara, 1981), the characteristic distance r_0 was set to 50 km and the threshold distance,

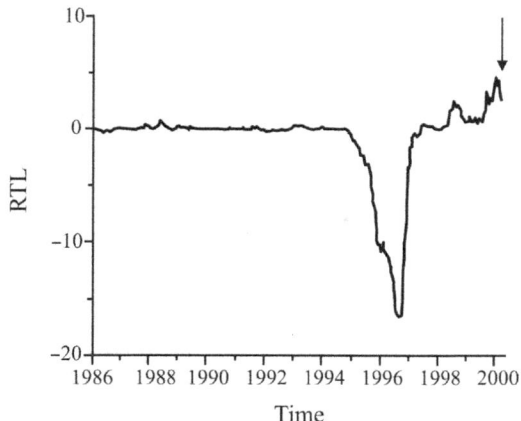

FIGURE 10–10 Time variation of the RTL parameter at the epicentral area of the 2000, M6.8 Nemuro Peninsula earthquake in Japan. Seismic quiescence is identified between 1995 and 1996, followed by seismic activation. The arrow denotes the mainshock origin time. *Modified after Huang, Q., Sobolev, G., 2001. Seismic quiescence prior to the 2000 M = 6.8 Nemuro Peninsula earthquake. Proc. Japan Acad., 77, Ser. B, pp. 6.*

$R_{max} = 2r_0 = 100$ km. The characteristic time interval, t_0, was set to 1 year and the threshold time window, $T_{max} = 2t_0 = 2$ years. Background RTL values R_{ltr}, T_{ltr} and L_{ltr} were estimated by a linear regression model.

Fig. 10–10 shows the obtained temporal variation of the RTL parameter before the $M = 6.8$ mainshock (time step of 10 days). Quiescence started in 1995 and attained its lowest value at the end of 1996, about 3 years before the mainshock, with a maximum deviation exceeding 16σ. The duration of the seismic quiescence was about 1.5–2 years and the subsequent activation, which started at the end of 1996, lasted about 0.7 year. To check the reliability of the procedure, Huang and Sobolev (2002) used different r_0 and t_0 values and obtained similar results.

10.6 Time-Dependent Probabilistic Seismic Hazard Assessment

PSHA essentially summarizes the rates of seismic ground-motion hazards at a site (Anderson and Biasi, 2016). The output of any SHA study can be presented by the seismic hazard recurrence curves, which reflect the annual rate of earthquakes that produce a ground motion amplitude, Y, higher than a given value, y. Such information is more useful than ill-defined single numbers such as the "probable maximum" or the "maximum credible" intensity, as even well-defined single numbers, such as the "expected lifetime maximum", are insufficient to provide engineers with an adequate understanding of how quickly the hazard (annual probability of exceedance) decreases as the ground motion intensity increases (Cornell, 1968). According to Anderson and Biasi (2016), the hazard curve can be improved by improving inputs and by identifying and resolving inconsistencies between observations and assessed hazard. Such discrepancies do not invalidate the hazard curve or the probability theory used to estimate it.

PSHA depends on a complete and accurate description of seismicity, combined with a model for ground motions, using standard probabilistic methods to estimate the hazard curve. The basic methodology used in most PSHA is the one proposed by Cornell (1968) and realized in several computer codes (e.g., McGuire, 1976, 1978; FRisk88M, 1995). This procedure is described by four basic steps, shown in Fig. 10–11 (TERA Corporation, 1978; Reiter, 1990), which correspond to: (1) identification of the seismic source model using historical

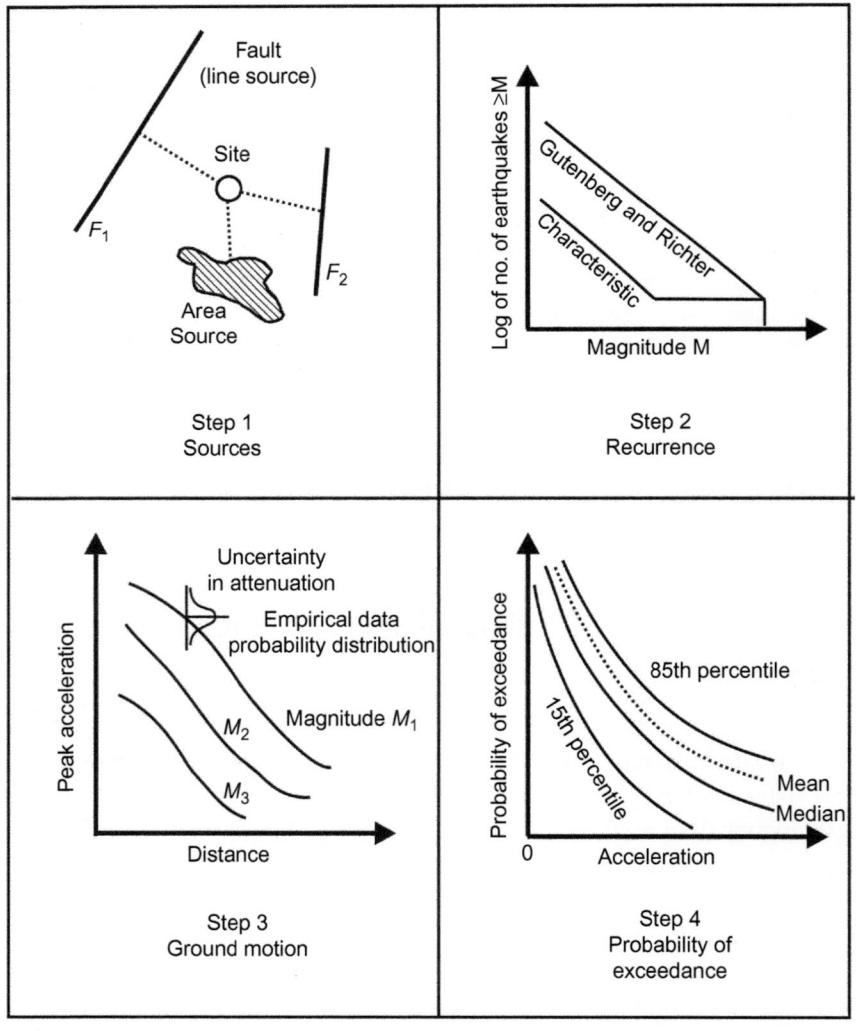

FIGURE 10–11 Basic steps of PSHA analysis (TERA Corporation, 1978; Reiter, 1990).

and instrumental period data combined with information on geology, tectonics, etc.; (2) adoption of a recurrence law for the calculation of the occurrence rate at which a given size of an earthquake will be exceeded; (3) implementation of a model for the prediction of the ground motion parameters; and finally, (4) determination of the probability of exceedance of the ground motion parameter of interest during a given time period by combining the uncertainties corresponding to the source geometry, location, distance, magnitude and ground motion parameter.

One of the most important tasks involved in PSHA is the examination of the temporal and spatial patterns of earthquake occurrence. In their pioneering work, Anagnos and Kiremidjian (1988) compiled a summary of the basic assumptions of the various earthquake occurrence models involved in hazard analysis. However, the widespread application of probabilistic hazard analysis has been typically based on the memoryless Poisson model of earthquake occurrence (Gardner and Knopoff, 1974; McGuire and Barnhard, 1981).

Modelling of faults in laboratory experiments (Byerlee and Brace, 1968; Brune, 1973) has shown that as two sides of a fault move in opposite directions they remain locked until sufficient shear stress builds up, then slip occurs and the fault subsequently locks again. Thus, a sequence of earthquakes can be represented by a process of strain accumulation interrupted by sudden releases. This is a laboratory representation of the elastic rebound theory (Reid, 1910), suggesting that the times of occurrence and magnitudes of a sequence of earthquakes in each source may not be stochastically independent. In other words, we may consider that the probability of a strong earthquake decreases when a seismogenic fault ruptures, and then increases with time as the local tectonic stresses rebuild.

Usually, there is not enough information for time-dependent modelling for all faults, therefore for such cases the use of Poisson or other time-independent PSH models is adopted. In a Poisson model, the probability remains constant for any time period. However, for the case of a small number of faults, for which adequate information supporting their time-dependent behaviour is available, a time-dependent PSH model may be better for identifying the short-term risks for economic loss assessment.

Though several papers have been published for *"Time Dependent Hazard"*, the actual number of publications that actually obtain hazard estimates is much smaller compared to those which determine earthquake occurrence probabilities (essentially time-dependent seismicity). A conditional probability requires knowledge of the interevent-time distribution. Typically, distribution functions as Gaussian, lognormal or Weibull are employed to fit recurrence time estimates, on the basis of a combination of historical and instrumental seismicity data.

Fitzenz and Nyst (2015) considered that the two major advancements, which contributed in the approach of PSHA since the first probabilistic seismic hazard studies were introduced (Cornell, 1968), are the hazard model improvement by the incorporation of uncertainties and the hazard model improvement by the incorporation of a time-dependent recurrence model for crustal faults. This consideration was supported by numerous research works, dealing with time-dependent seismicity and its application for SHA. A limitation of these models is that, in addition to the mean frequency (or recurrence time) of earthquakes, additional information is required regarding the variability of the frequency of events (its variance or standard deviation) and the occurrence time of the last event, which is not always known. Reviews for the research work performed during the last 30 years can be found in Harris (1998), Stein (1999), Cramer et al. (2000), King and Cocco (2001), among others.

During the last three decades, USGS and CGS have developed time-dependent source and ground-motion models for California using the elapsed time since the last mainshock (WGCEP, 1988, 1990, 1995 (led by the Southern California Earthquake Center), 1999, 2003; Cramer et al., 2000; Petersen et al., 2002). The probabilities of occurrence for the next event were assessed using Poisson, Gaussian, log-normal and Brownian passage time (BPT) statistical distributions. Previously, other working groups used a value of about 0.5 (± 0.2) for the ratio of the total sigma to the mean of the recurrence distribution. This ratio, known as the coefficient of variation, accounts for the periodicity in the recurrence times of an earthquake. According to Petersen et al. (2007) a value of 1.0 represents irregular behaviour (nearly Poissonian), while a value of 0 indicates periodic behaviour. They also calculated the time-dependent earthquake probabilities by employing simple models with parameters such as the mean-recurrence interval (T-bar), parametric uncertainty (Sigma-P), intrinsic variability (Sigma-P and Sigma-T), and the year of occurrence of the last earthquake. The parametric sigma was calculated from the uncertainties in mean displacement and mean slip rate of each fault (Cramer et al., 2000). The intrinsic sigma describes the randomness in the periodicity of the recurrence intervals. The total sigma of the log-normal distribution is the square root of the sum of the squares of the intrinsic and parametric sigmas. For the analysis, they assumed characteristic earthquake recurrence models with segment boundaries defined by previous working groups.

Petersen et al. (2007) used data for type A faults (faults with geologic evidence for long-term rupture histories and an estimate of the elapsed time since the last mainshock) and considered only the influence of the elapsed time since the last mainshock. They presented their results by both time-independent and time-dependent models. Their maps depict the value of peak ground acceleration with a 10% probability of exceedance for a period of 30 years starting in 2006. The time-dependent maps differ by 10%−15% from the time-independent maps near A-fault sources (Fig. 10−12).

However, for those regions of California that are located well away from the sources showing time-dependent behaviour, the ground motions are similar to the time-independent model. The southern San Andreas Fault, the Cascadia subduction zone, and the eastern San Francisco Bay area faults generally have elevated hazard relative to the time-independent maps. According to Petersen et al. (2007), this is due to the fact that it has been quite a long time since the last strong mainshock − about 150 years since the 1857 M7.9 Fort Tejon earthquake, more than 300 years since the 1700 M9 Cascadia earthquake, and nearly 140 years since the 1868 M6.8 on the southern Hayward fault. These faults are, most likely, in the latter half of their seismic cycles. On the other hand, the northern San Andreas fault, the southern San Jacinto fault, and the Imperial fault exhibit lower levels of time-dependent than time-independent hazard probably due to relatively recent activations (i.e., 1906 M7.8 San Francisco, 1968 M6.4 Borrego Mountain and 1971 M6.4 Imperial Valley earthquakes), a fact that places these faults in the first half of their seismic cycles.

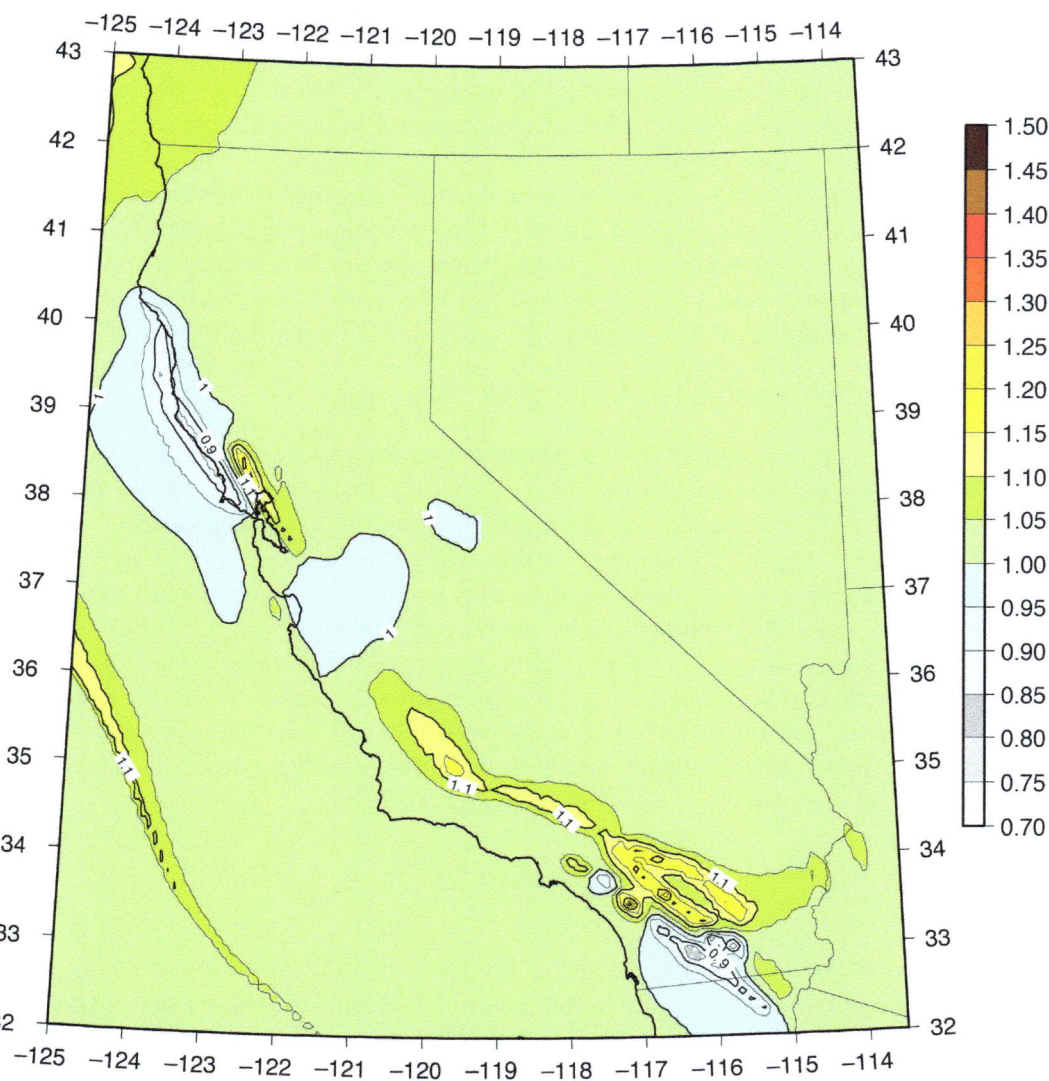

FIGURE 10–12 Ratio of PGA values (for a 10% probability of exceedance in 30 years, starting at 2006) between the time-dependent seismic hazard map and the time-independent map for rock site conditions (Petersen et al., 2007).

The implementation of time-dependent renewal models strongly depends on information about the date of the last event. However, in the case of unknown date of the last rupture, a time-independent Poisson model is applied to obtain earthquake probabilities (e.g., Field et al., 2009). According to Field and Jordan (2015), this substitution is not formally correct, because the Poisson distribution (for which the recurrence aperiodicity, α, equals 1) is not necessarily what one gets by integrating the time-dependent renewal model over all possible dates of the last event. The same authors found differences greater than 10%, relative to the

Poisson distribution, when the forecast duration exceeds about 20% of the mean recurrence interval, regardless of aperiodicity value. Thus, for the 50-year return period, which is relevant to standard building codes, the difference between BPT (which gives results similar to the lognormal model) and Poisson probabilities becomes important when the recurrence interval drops below ~ 250 years.

Since the work of Papazachos (1989) on time-dependent seismicity, several attempts have been made to test this model to single faults but also to regions that include several groups of faults (Papazachos and Papaioannou, 1993; Papazachos et al., 1997a,b, among others). Papazachos et al. (1997b) used data of strong ($M \geq 5.5$) mainshocks, which occurred in the fracture zone of the Alpine−Himalayan belt, and proposed the following relation:

$$logT_t = 0.19 \cdot M_{min} + 0.33 \cdot M_P - 0.39 \cdot logm_0 + 7.81 \tag{10.8}$$

where T_t is the interevent time of the mainshocks of each source, M_{min} is the magnitude of the smallest mainshock considered (e.g., $M_{min} \geq 6.0$), M_p is the magnitude of the preceding mainshock, and m_o is the seismic moment rate (dyn · cm/yr). Assuming a normal distribution of the variable $log_{10}(T/T_t)$, where T is the observed interevent time of strong mainshocks, they found a standard deviation, σ, equal to 0.29. Papaioannou and Papazachos (2000) used this time-dependent seismicity model and applied a hybrid procedure for the SHA of major towns and cities in Greece. As a measure of the seismic hazard they considered the probability of occurrence of macroseismic intensity value $I_{MM} \geq VII$ during the period 1996−2010. For the calculation of the probabilities of earthquakes with $M \geq 5.5$, which typically influence the seismic hazard results, the time-dependent seismicity model was used and the probability was calculated using the formula:

$$P_{(Dt)} = \frac{F\left(\frac{L_2}{\sigma}\right) - F\left(\frac{L_1}{\sigma}\right)}{1 - F\left(\frac{L_1}{\sigma}\right)} \tag{10.9}$$

where $L_1 = \log (t/T_t)$, $L_2 = \log (t + \Delta t/T_t)$, t is the years elapsed since the last strong earthquake, F is the complementary cumulative value of the standardized normal distribution with mean equal to zero and standard deviation equal to σ, of the quantity $\log(t/T_t)$, Δt is the prediction interval, and T_t is determined by relation (10.8). For smaller magnitude events the Poisson distribution was adopted and probabilities were calculated using the formula:

$$P = 1 - exp\left(-\frac{\Delta t}{T_m}\right) \tag{10.10}$$

Following this, a joint probability due to j sources was determined by the formula:

$$P = 1 - \prod_{i=1}^{i=j}(1 - p_i) \tag{10.11}$$

where P_i is the probability of occurrence of the predefined intensity by earthquakes of the ith source. This procedure, for the calculation of the probabilities, was similar to that proposed by Papaioannou et al. (1992). The same authors found that the application of time-independent and time-dependent seismicity models resulted in significant differences for short time windows (~ 50 years), while these differences seem to vanish for larger time windows (about 200 years). Their conclusions are depicted graphically in Fig. 10−13. To obtain robust results, they considered seismic sources with at least three main events. Their results were obtained for $M_{min} = 6.0$, as they considered that this magnitude corresponds to the observation of associated epicentral macroseismic intensity (damage level) of $I_{MM} \geq VII$.

Vamvakaris (2010) used synthetic catalogues to perform SHA for the area of Greece based on the results (expected earthquakes) of a time-dependent seismicity model. The results, presented in the form of maps, revealed areas with considerably high values of expected PGA (e.g., the North Aegean trough and the SE Hellenic Arc), while lower expected PGA values were found at SSW coasts of mainland Greece. Also, Gerstenberger et al. (2016) established a hybrid time-dependent seismic hazard model for the area of Canterbury, New Zealand. Their model is based on the combination of earthquake-clustering models of three timescales (short term, medium term and long term), in an attempt to develop a model that accounts for the significant epistemic uncertainty in the earthquake rates.

Recently, Akinci et al. (2017) attempted to improve PSHA estimates towards a time-dependent SHA by using new databases for the Calabria region in Italy. They employed a BPT model characterized by mean recurrence, aperiodicity or uncertainty in the recurrence distribution and elapsed time since the last characteristic earthquake, to express the time dependence of the seismic processes for the assessment of future ground motions in this region. They found that hazard might increase by more than 20% or decrease by as much as 50%, depending on the different recurrence model. PGA values decrease about 20% in the Messina Strait, where a recent major earthquake took place, with respect to traditional

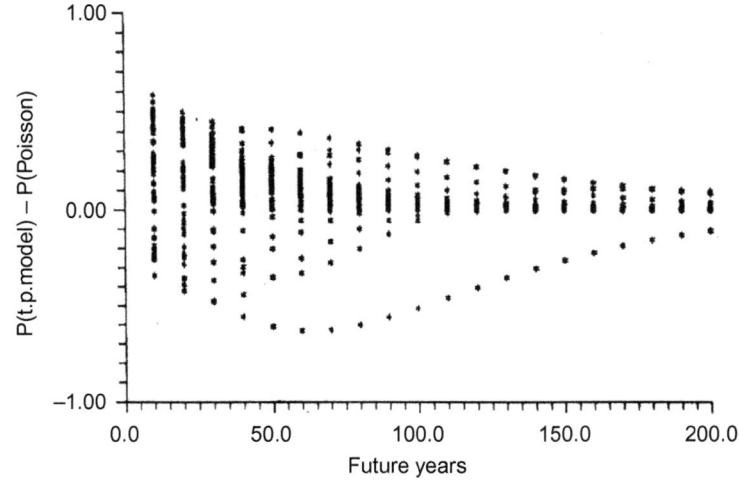

FIGURE 10–13 Variation of the difference between the calculated probabilities for the time-dependent and time-independent models for various future hazard time windows (Papaioannou et al., 1992).

time-independent estimates, whereas near the city of Cosenza they reach up to 0.36 g for the time-independent model and 0.40 g for the time-dependent one (i.e., a 15% increase). Furthermore, the same authors (Akinci et al., 2017) calculated for the time-dependent map PGA values differing by 50% from the time-independent ones close to fault sources. A second observation is that the time-dependent models lead to higher probabilities and seismic hazard than the traditional time-independent ones when the elapsed time since the last earthquake is greater than about half of the mean recurrence interval. Both the time-dependent and time-independent models for the period of 2015−65 demonstrate that the city of Cosenza and surrounding areas bears the highest seismic hazard in Calabria.

10.7 Time-Dependent Probabilistic Seismic Hazard Assessment for a Generic Prediction Model

To assess in a quantitative manner the impact of time-dependent seismicity models on PSHA, we examine here a hypothetical prediction scenario for a generic time-dependent, intermediate-term prediction model. As a test case, we use the broader Aegean area and consider a typical decade, namely 1995−2004, as the target time prediction window. During this period, six large earthquakes with magnitudes $M > 6.4$ have occurred in this area, presented in Table 10−1, which constitute the target mainshocks for our generic prediction algorithm. For this prediction period (10 years) and study area (a high seismicity region) we have considered the following generic prediction scenario, following Vamvakaris et al. (2008) and Vamvakaris (2010):

1. All mainshocks above a certain threshold (set to $M_{min} = 6.4$ in our case) can be predicted using the generic mid-term prediction model. This concept is based on the fact that almost all proposed prediction methods, as described previously, focus on predicting strong mainshocks, as they have the most significant impact on seismic hazard, but are also associated with the most prominent spatiotemporal seismicity phenomena.
2. The employed generic mainshock prediction method has specific uncertainties regarding the mainshock magnitude (dM), origin time (dT) and epicentre (dH), which should be

Table 10–1 Information on Six Strong Mainshocks ($M \geq 6.4$) of the Broader Aegean Area Which Occurred During the Period 1995−2004

No	Year	Date	Time	Area	Lat (°)	Lon (°)	Depth (km)	M
1	1995	13 May	08:47:13	Kozani	40.167	21.686	14	6.6
2	1995	15 June	00:15:47	Aigio	38.404	22.272	3	6.4
3	1997	13 Oct	13:39:36	Kalamata	36.348	22.105	13	6.4
4	1997	18 Nov	13:07:38	Strofades	37.482	20.692	10	6.6
5	1999	17 Aug	00:01:39	Izmit	40.756	29.955	17	7.5
6	2001	26 Jul	00:21:39	Skyros	39.097	24.268	19	6.4

known/determined a priori. In the present study, we consider that these errors exhibit a standard Gaussian distribution, though this may be adapted/changed according to each method's properties.

3. All other smaller-magnitude mainshocks ($M < 6.4$ in our case), as well as any "associated" shocks (foreshocks, aftershocks, etc.) and background seismicity cannot be predicted. The corresponding events are considered to be temporally random (e.g., Poissonian time distribution).

4. The model has a specific prediction failure probability (PF), essentially corresponding to the probability that a mainshock will occur completely outside the $dM/dT/dH$ windows. In other words, PF corresponds to the probability that the mid-term prediction model will completely fail to predict a mainshock that has occurred (model failure). We also assume that the model has a specific false alarm probability (FA), corresponding to the probability that a predicted mainshock will simply not occur at all. In other words, if a prediction model proposes that, e.g., five mainshocks will occur within the prediction time window (e.g., 10 years) and only three mainshocks actually occur in the predicted spatial windows, we consider the model to have a false alarm rate of 2/5 (40%).

The earthquake epicentres for the considered decade (1995–2004) are presented in Fig. 10–14, together with the seismicity zonation model of Vamvakaris et al. (2016). As can be seen from Table 10–1, five strong mainshocks have rather similar magnitudes ($M = 6.4$–6.6), typical for the study area (e.g., Papazachos, 1990). The only exception is the large Izmit earthquake ($M = 7.5$), which had a very heavy impact in the broader NW Turkey area.

To test the impact of the previously described generic time-dependent prediction on PSHA, we have performed seismic hazard estimations for different prediction scenarios. For each scenario, we generate a large number of synthetic catalogues (typically 500), which follow the properties of each seismicity scenario, similar to Vamvakaris et al. (2008). Each synthetic catalogue is generated using random numbers that follow the space/time/magnitude probability distributions of each considered scenario. For example, if a time-independent scenario is considered, all events have a Poissonian (random) temporal probability distribution.

We employed a dense grid of sites (spacing $0.25° \times 0.25°$) covering the area of Fig. 10–14, and perform for each site (grid-point) a direct computation of PGA and PGV values for each earthquake of each synthetic catalogue, using the GMPE proposed by Boore et al. (2014) only for bedrock-soil conditions. The distribution of maximum PGA and PGV values for each site and synthetic catalogue is employed to assess the corresponding seismic hazard values, for various probability levels. For example, for the typical PSHA probabilities of exceedance of 10% for 50 years (usually adopted in modern seismic codes), computations are performed using 50-year catalogues and the 10% percentile of the maximum PGA/PGV values from all catalogues is used to determine the probability level of exceedance for each site (grid-point).

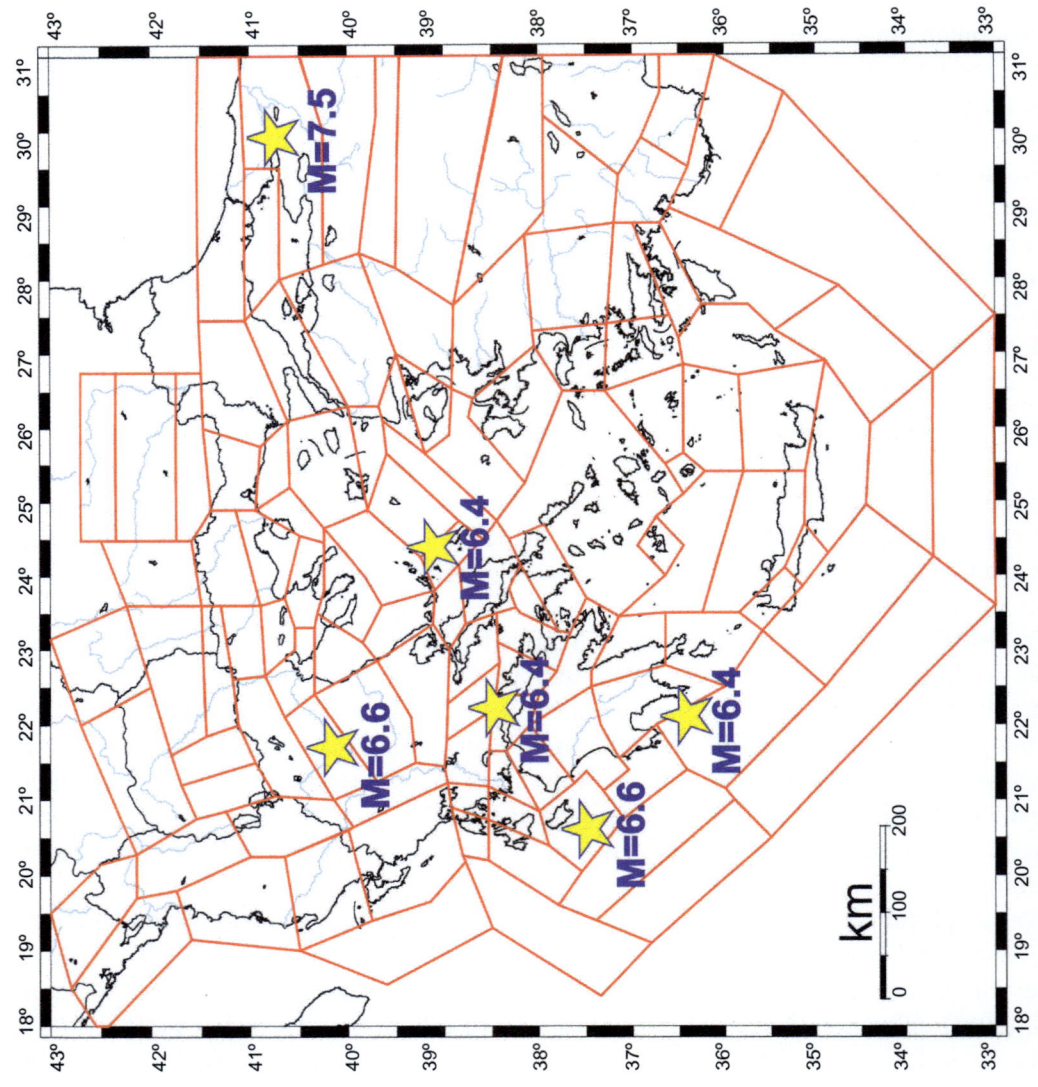

FIGURE 10–14 Mainshocks ($M > 6.4$) in the Aegean area for the period 1995–2004, which have been considered as target events for the generic prediction model. The seismic zones from Vamvakaris et al. (2016) are also shown.

For the tests, we have considered three typical scenarios:

1. *A time-independent scenario*: This case corresponds to synthetic catalogues following a Poissonian (random) time generation distribution, corresponding to the background PSHA assessment, used as a reference level. The magnitude distribution is constrained to follow the G-R magnitude distribution proposed by Vamvakaris et al. (2016) for each zone of Fig. 10−14, while all earthquakes are uniformly distributed within the spatial limits of each seismic zone.

2. *A "perfect" prediction scenario*: For this scenario, we consider the case of a perfect prediction, where all mainshocks are predicted (PF = 0), with no false alarms (FA = 0), and with very small uncertainties, comparable to the typical standard deviations associated with location errors (20 km in our case) and magnitude assessment ($dM = 0.2$) in modern catalogues. In other words, we assume that all mainshocks that occurred in the decade 1995−2004 have been predicted successfully by our generic but "perfect" prediction method, with a minimal level of uncertainty, similar to the real earthquake catalogue uncertainty.

3. *A "realistic" prediction scenario*: In this case, mainshock predictions can have nonzero FA and PF rates, hence we examine different "failure" levels. Moreover, the mainshock prediction can also have errors (even large) regarding its predicted time, location and magnitude, depending on the specific method/test considered. In other words, we consider realistic "fuzzy" mid-term predictions, prone to large errors, in an attempt to assess their impact on time-dependent PSHA.

The generation of synthetic catalogues that follow scenarios 1 (time-independent model) and 2 ("perfect" prediction) is straightforward: Catalogues are separated into two groups, namely mainshocks (that can be predicted by our generic algorithm) and all other (smaller) earthquakes (which cannot be predicted), using the adopted magnitude limit (6.4 in our test case). All earthquakes with smaller magnitudes ($M \leq 6.3$) are considered to have a random (Poissonian) occurrence time, while their spatial and magnitude distributions follow the model adopted in scenario (1) (time-independent scenario). Using these assumptions, a large number (e.g., 500) of synthetic catalogues for nonmainshocks is generated. On the other hand, all mainshocks ($M \geq 6.4$) are considered to exhibit Gaussian errors with respect to the predicted parameters of Table 10−1, using different $dM/dT/dH$ errors, assigned for each prediction model. An equal number of "mainshock" catalogues is also generated using random Gaussian deviates. If a mainshock scenario includes a certain FA rate (e.g., 20% false alarm rate), we randomly omit predicted mainshocks from the synthetic "mainshock" catalogues, using this target rate. If a mainshock scenario includes a specific PF rate (e.g., a 30% probability that a mainshock occurs completely outside the predicted time/space/magnitude windows), we generate 30% of the "mainshock" catalogues using a random time/space/magnitude pattern, similar to smaller (nonmainshock) events. Both "nonmainshock" and "mainshock" catalogues are merged, to generate the final synthetic catalogues, that have the desired space/time/magnitude seismicity pattern properties.

An issue to consider is the exceedance probability level to be adopted for the 10-year mid-term prediction interval. In the present work, we have adopted a 2% probability level for 10 years, which is equivalent to the 10% probability level for a 50-year return period usually adopted in seismic codes (e.g., EN1998) for standard (e.g., RC) buildings, if a Poissonian distribution is considered. While this probability level is very small, it renders values that can be easily compared with seismic code predictions, as well as PSHA analysis results from worldwide and national projects.

Fig. 10—15 depicts the distribution of PGA values using the adopted synthetic catalogue approach, for a 2% probability level in 10 years, corresponding to the time-independent seismicity model (scenario 1). The obtained PSHA results are very similar to other local (e.g., Papaioannou and Papazachos, 2000) and regional (e.g., Giardini et al., 1999) assessments, with the highest values observed in the Ionian islands, the Hellenic Arc, the gulf of Corinth and the North Aegean Trough. Fig. 10—16 depicts the same map for the perfect prediction scenario (2), i.e., a perfect generic mid-term prediction algorithm, with minimal location and magnitude uncertainties. It is clear that the PSHA results for the perfect prediction (Fig. 10—16) and time-independent model (Fig. 10—15) are quite similar, with differences

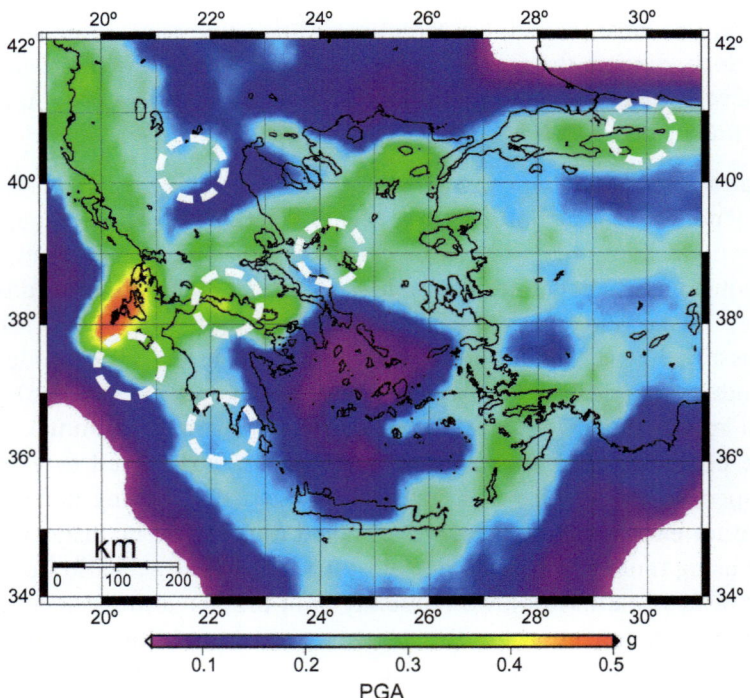

FIGURE 10–15 Distribution of PGA values for the time-independent seismicity zonation model of Vamvakaris et al. (2016), using the adopted synthetic catalogue approach, for a 2% probability level of exceedance in a time period of 10 years. We also present the positions of the six mainshocks of Table 10—1, for which a generic time-dependent prediction is tested.

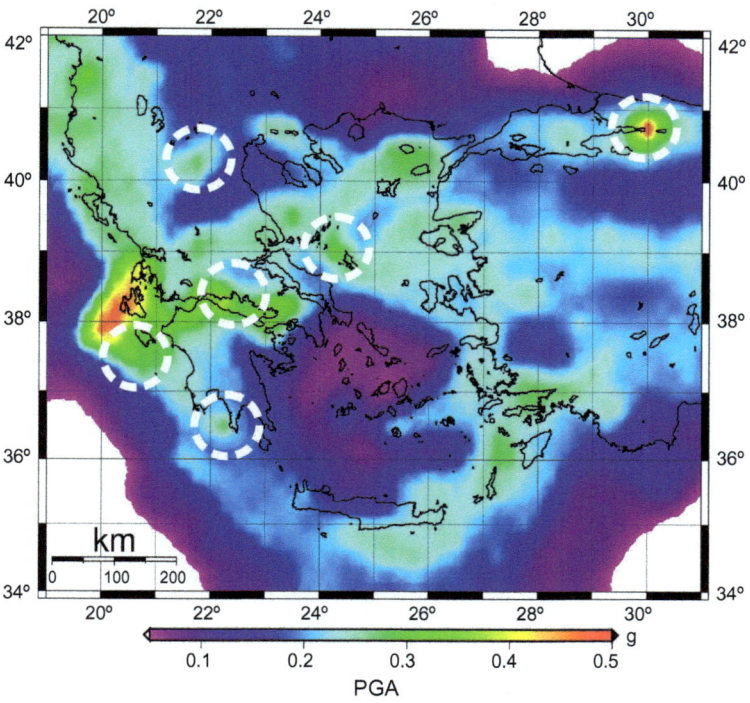

FIGURE 10–16 As in Fig. 10–15 for the perfect-prediction scenario (scenario 2).

mainly in the vicinity of the mainshocks region, as these are listed in Table 10–1 (dashed circles in Figs. 10–15 and 10–16). The most prominent differences can be observed for the broader area of the Izmit **M** = 7.5 event, where PGA levels increase from ~0.3 g to ~0.5 g in the epicentral area. For the other five strong but smaller mainshocks, PGA level changes are smaller but locally recognizable.

To study the effect of the main quantities involved in the generic prediction assessment, we have varied the prediction accuracy (*dM*, *dT*, *dH*), as well as the PF and FA levels. For each examined quantity, we used a large number of values, namely 0.2, 0.3, 0.4 and 0.5 for the predicted mainshock magnitude error, *dM*, 1–5 years for the predicted occurrence time error, *dT*, and mainshock epicentral errors ranging from 20 to 200 km, in an attempt to capture the main a posteriori error levels reported by most prediction algorithms previously described. Similarly, we varied PF and FA percentages from 0% to 95%. For each examined case, we have constructed diagrams similar to Fig. 10–17, where we compare the difference of the *ln(PGA)* values of the "real" prediction (scenario 3) and "perfect" prediction (scenario 2) from the background, time-independent (scenario 1) ln(PGA) levels. The plot in Fig. 10–17 corresponds to PF = 70% (the generic prediction model will fail to predict mainshocks in 70% of the examined cases), FA = 0% (no false alarms) and relatively large

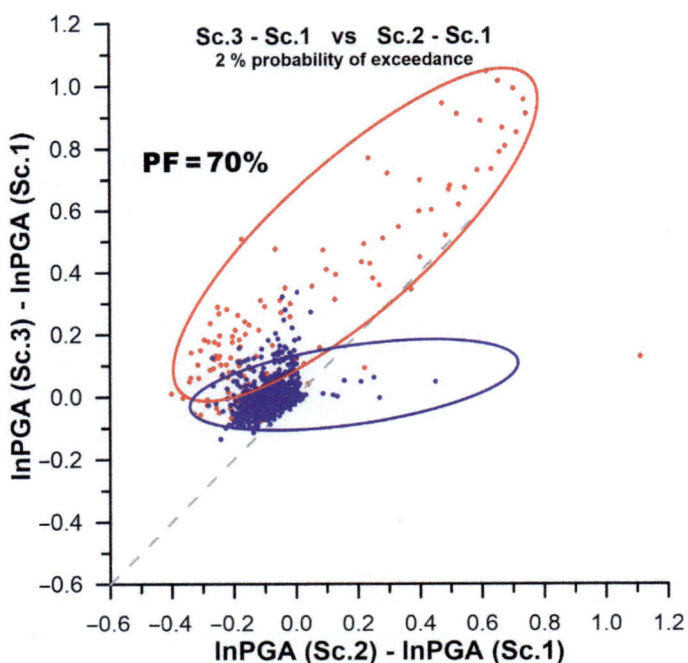

FIGURE 10–17 Variation of the ln(PGA) difference between a "realistic" prediction (scenario 3) and the time-independent model (scenario 1), against the corresponding difference between a "perfect" prediction (scenario 2) and the time-independent model results (scenario 1, light-colored dots: Izmit area, dark-colored dots: all other sites).

prediction errors ($dM = 0.4$, $dT = 3$ year and $dH = 100$ km), mimicking the large uncertainties typically associated with mid-term prediction models.

From Fig. 10–17, we can recognize several interesting features:

1. The "realistic" prediction scenario exhibits (in general) larger seismic hazard values (expressed in PGA here) compared to the perfect prediction values, always with respect to the time-independent SHA. This is an expected pattern, as larger uncertainties (involved in the realistic prediction case) typically lead to more conservative hazard estimates.

2. In Fig. 10–17 we have plotted results for all points in the broader Izmit ($M = 7.5$) earthquake area in red and all other sites (grid points) in blue. It is evident that for the Izmit area, even a large model failure (PF = 70%) has a minimal impact, with $ln(PGA)$ values for the "real" prediction scenario exhibiting a similar trend with the "perfect" prediction scenario (always with respect to the background, time-independent PSHA). In other words, even very poor mainshock predictions (with a prediction failure of PF = 70%) have a significant impact on the time-dependent PSHA results, when this concerns a very large ($M \sim 7.5$) event, such as the Izmit earthquake. On the other hand, for all other cases ("normal" strong $M \sim 6.5$ mainshocks), the slope of the "real" versus

"perfect" prediction comparison is small, showing that we recover only a small fraction of the actual PSHA impact of a "perfect" prediction.

The previous results suggested that the broader Izmit area should be examined separately from the remaining sites, as the impact of the $M = 7.5$ earthquake on PSHA is so strong that its contribution to SHA at the defined probability level (2% for a 10-year period) is practically always significant. For this reason, in Fig. 10−18 we present plots similar to Fig. 10−17 for various PF and FA levels, using the same large magnitude ($dM = 0.4$), time ($dT = 3$ year) and location ($dH = 100$ km) uncertainties in all cases, after omitting the broader Izmit area sites (grid-points). Notice that for the considered nonzero PF levels, FA was always set to 0, while for nonzero FA levels PF was also set to 0, in an attempt to separately study the effect of each factor. We have also obtained similar curves for dM, dT and dH prediction errors, assuming a PF = FA = 0 level in all cases.

The results of our parametric investigation can be summarized as follows:

1. The effect of the prediction errors, dM, dT and dH is minimal in most cases. In other words, due to the small probability level considered in this study (2% probability of exceedance for a 10-year prediction period), even large prediction errors do not have a significant impact on the information gain provided by such "fuzzy," prone to large errors, predictions.
2. The effects of PF and FA are completely different. This is depicted in Fig. 10−19, where we plot the slope of the least-squares fit on the plots that we have constructed (similar to Fig. 10−18), after excluding the broader Izmit area earthquakes. It is evident that large best-fit slope values suggest that the "real" prediction scenario considered is performing similar to the "perfect" prediction scenario, while values close to ~ 0, suggest that the "real" prediction scenario has values close to the time-independent one, therefore it is practically noninformative.

The results of Fig. 10−19 clearly suggest that the most important factor to consider is the PF level. As the PF probability increases, the "gain" of the "real" predictions in comparison to a "perfect" prediction (always with respect to the time-independent PSHA level) drastically drops. On the contrary, FA have a negligible effect on the performance of time-dependent SHA, at least at the probability level considered in this study (2% for a 10-year prediction period). Therefore, if a predicted mainshock does not occur, this has a negligible impact on the time-dependent PSHA. On the contrary, if the model fails to predict a mainshock that occurs well outside the prediction windows, this has a critical impact on the time-dependent PSHA.

It should be noted that while the above results correspond to a specific prediction period (10 years) and a specific mainshock ($M > 6.4$) definition, as well as for a certain high-seismicity study area (broader Aegean Sea region), the approach adopted here can be adapted to any model, depending on its prediction windows and target earthquake definitions.

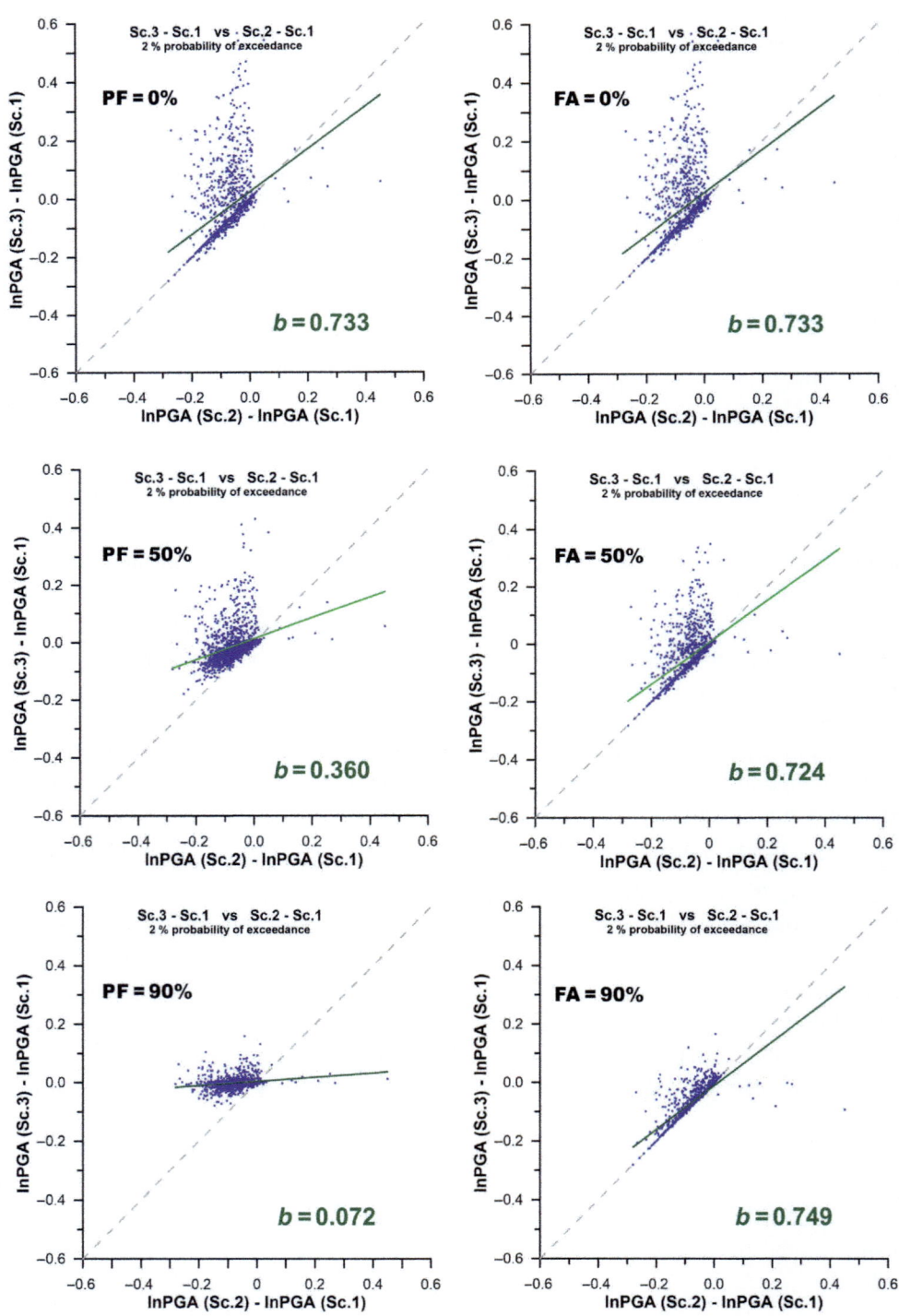

FIGURE 10–18 As in Fig. 10–17, for three prediction failure (PF) and false-alarm (FA) levels.

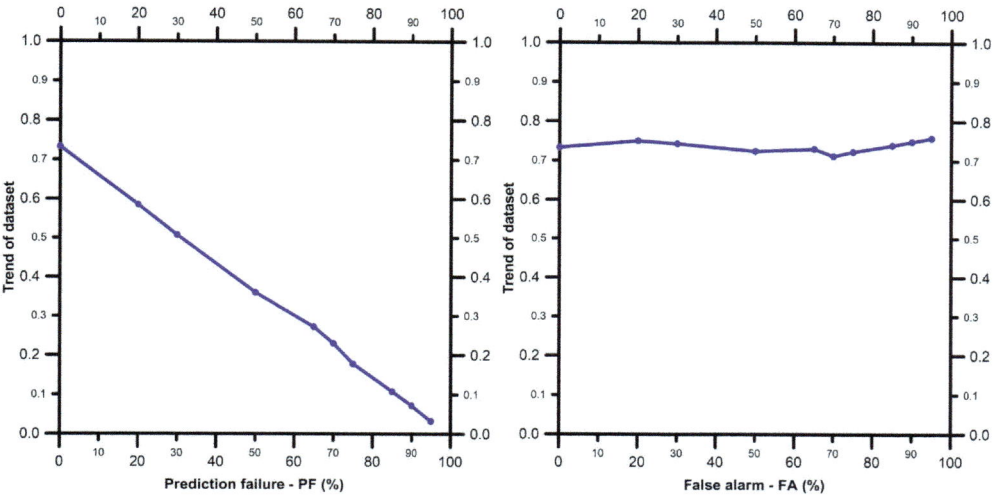

FIGURE 10–19 Plot of the slope of the best-fit line for the variation of ln(PGA) difference between a "realistic" prediction scenario and the time-independent model, against the corresponding difference between a "perfect" prediction scenario and the time-independent model results (see Fig. 10–17), for various PF and FA levels (see text for explanation). Note the significant drop of the prediction PSHA performance for large PF (prediction failure) values.

10.8 Conclusions

In this review, we have presented several intermediate-term earthquake prediction schemes which, in addition to their importance for seismological research, may potentially contribute to time-dependent SHA. For all methods, the observations concern seismicity variations in both time and space domains. Specifically, both decreases (quiescence, deceleration) and increases (activation, acceleration) in seismicity were found to precede strong and large earthquakes a few years prior to their occurrence. The methods have been tested by various approaches. The most common test involves the retrospective (a posteriori) prediction of strong earthquakes that have occurred in areas which had not been studied during the formulation of the relevant algorithms (e.g., M8/MSc/CN: Italy (Peresan et al., 2005); AMR/ Decelerating Seismicity: Sumatra (Jiang and Wu, 2005; Italy: De Santis et al., 2015; Papazachos et al. 2006); LURR: California (Yin, 1993); RTL: Japan (Huang, 2004), among many other publications. Other tests involve the application of the algorithms on random earthquake catalogues (e.g., M8 (Eneva and Ben-Zion, 1997); AMR/Decelerating seismicity (Papazachos et al., 2006); LURR (Zhuang and Yin, 1999); RTL (Huang, 2004), among others). It should be noted that several of the forward tests for the prediction of probably oncoming strong earthquakes were successful (e.g., M8 (Healy et al., 1992); AMR/Decelerating seismicity (Papazachos et al., 2002); LURR (Zhang et al., 2015); RTL (Sobolev and Tyupkin, 1999)).

We have also presented indicative test results regarding the application of a generic time-prediction model on PSHA. For these tests, we adopted a 2% probability of exceedance for a

10-year prediction period, to provide values that can be directly compared to standard time-independent PSHA levels used in seismic codes (i.e., 10% probability of exceedance for a return period of 50 years). The results suggest that, as expected, the most critical factor is the PF level of the algorithm, i.e., the probability that a mainshock will occur outside the joint time/space/magnitude prediction windows. Time-dependent PSHA is much less prone to the actual errors of the prediction (regarding the predicted mainshock magnitude, time and location), or to the probability that a predicted mainshock will actually never occur (false alarm). Moreover, predictions of large events seem to be important for time-dependent PSHA, even at large model failure levels, due to their significant impact for the low probability levels of exceedance considered here (2% for a 10-year prediction window).

The previous preliminary results indicate that the role of time-dependent prediction models should be further reviewed, as even approximate and "fuzzy" predictions can have a significant impact on time-dependent PSHA. This suggests that further work on the physical processes, assumptions and implications of time-dependent seismicity models should be carried out. Tiampo and Shcherbakov (2012) reviewed several seismicity-based forecasting techniques, i.e., techniques that may identify particular physical processes or filter (smooth) the seismicity to enhance precursory patterns, and suggest that intermediate-term forecasting of the order of 5−10 years is within reach. More specifically they propose that "...forecasting techniques is a response to the recognition that the evolving stress field in a regional fault system is the driving force behind a dynamic, if slow-moving, system and that small-to-medium magnitude seismicity is providing important information on the temporal and spatial evolution of the local and regional stress field."

On the other hand, the debate over the predictability of earthquakes has been intensified during the last two decades, and particularly since the failure of the Parkfield earthquake prediction experiment (e.g., Jackson and Kagan, 2006). Moreover, strong skepticism has been expressed on the applicability of some of the earthquake prediction algorithms under discussion. The criticism concerns either fundamental theories or data fitting. For instance, Geller et al. (2016) propose that there is no satisfactory physical model for earthquakes, none is in sight and a new paradigm of earthquake occurrence is required since the adoption of stick−slip and elastic rebound as the predominant mechanisms for generating large earthquakes is at odds with reality. Hardebeck et al. (2008) tested the hypothesis that accelerating seismic moment release (AMR) is a precursor to large earthquakes and suggest that spurious cases of AMR can arise from a combination of data fitting and normal foreshock and aftershock activity because the time period, area and magnitude range analysed before each examined mainshock are often optimized to produce the strongest AMR signal. On the same issue (i.e., data fitting), Mulargia (2001) suggests that since long timescales and the lack of any natural laboratory restrict research to retrospective analysis of data, such approaches lead to optimal selection of data, resulting in the introduction of significant biases which are capable of falsely representing simple statistical fluctuations as significant anomalies requiring physical explanation.

While we may still be a long way from achieving the goal of earthquake prediction, the precursory seismic activity preceding a major earthquake remains a subtle but exciting new possibility in earthquake physics (Chen et al., 2005). The results presented in this chapter suggest that this goal can have important implications on time-dependent PSHA. As a final remark, regarding the need for a new paradigm of earthquake occurrence (Geller et al., 2016), we would like to quote the words of Kuhn (1962) who wrote: "no theory ever solves all the puzzles with which it is confronted at a given time; nor are the solutions already achieved often perfect. If any and every failure to fit were ground for theory rejection, all theories ought to be rejected at all times. On the contrary, it is just the incompleteness and imperfection of the existing data-theory fit that, at any time, define many of the puzzles that characterize normal science."

Acknowledgement

This work has been partly supported by the project 'HELPOS — Hellenic System for Lithosphere Monitoring' (MIS 5002697) of the Operational Programme NSRF 2014—20, co-financed by Greece and the European Union (European Regional Development Fund).

References

Akinci, A., Vannoli, P., Falcone, G., Taroni, M., Tiberti, M.M., Murru, M., et al., 2017. When time and faults matter: towards a time-dependent probabilistic SHA in Calabria, Italy. Bull. Earthquake Eng. 15, 2497—2524. Available from: https://doi.org/10.1007/s10518-016-0065-7.

Allègre, C.J., Le Mouël, J.L., Provost, A., 1982. Scaling rules in rock fracture and possible implications for earthquake predictions. Nature 297, 47—49.

Anagnos, T., Kiremidjian, A.S., 1988. A review of earthquake occurrence models for seismic hazard analysis. Prob. Eng. Mech. 3, 3—11.

Anderson, J.G., Biasi, G.P., 2016. What is the basic assumption for probabilistic seismic hazard assessment? Seism. Res. Lett. 87, 323—326.

Bak, P., 1996. How Nature Works: The Science of Self Organized Criticality. Springer, New York, p. 212, 10.1007/978-14757-5426-1.

Boore, D.M., Stewart, J.P., Seyhan, E., Atkinson, G.M., 2014. NGA-West2 equations for predicting PGA, PGV, and 5% damped PSA for shallow crustal earthquakes. Earthquake Spectra 30 (3), 1057—1085.

Bowman, D.D., King, G.C.P., 2001. Accelerating seismicity and stress accumulation before large earthquakes. Geophys. Res. Lett. 28, 4039—4042.

Bowman, D.D., Quillon, G., Sammis, C.G., Sornette, A., Sornette, D., 1998. An observational test of the critical earthquake concept. J. Geophys. Res. 103, 24359—24372.

Brehm, D.J., Braile, L.W., 1998. Intermediate-termearthquake prediction using precursory events in the New Madrid seismic zone. Bull. Seismol. Soc. Am. 88, 564—580.

Brune, J.N., 1973. Earthquake modelling by stick-slip along precut surfaces in stressed foam rubber. Bull. Seism. Soc. Am. 63, 2105—2119.

Bufe, C.G., Varnes, D.J., 1993. Predictive modeling of seismic cycle of the Great San Francisco Bay Region. J. Geophys. Res. 98, 9871—9883.

Bufe, C.G., Nishenko, S.P., Varnes, D.J., 1994. Seismicity trends and potential for large earthquakes in the Alaska-Aleutian region. Pure Appl. Geophys. 142, 83—99.

Byerlee, J.D., Brace, W.F., 1968. Stick slip stable sliding and earthquakes-effects of rock type, pressure, strain rate and stiffness. J. Geophys. Res. 73, 6031−6037.

Chen, C.C., Wu, Y.X., 2006. An improved region-time-length algorithm applied to the 1999 Chi-Chi, Taiwan earthquake. Geophys. J. Int. 166, 1144−1147.

Chen, C.C., Rundle, J.B., Holliday, J.R., Nanjo, K.Z., Turcotte, D.L., Li, S.-C., et al., 2005. The 1999 Chi-Chi, Taiwan earthquake as a typical example of seismic activation and quiescence. Geophys. Res. Lett. 32. Available from: https://doi.org/10.1029/2005GL023991.

Cornell, C.A., 1968. Engineering seismic risk analysis. Bull Seism. Soc. Am. 58, 1583−1606.

Cramer, C.H., Petersen, M.D., Cao, T., Toppozada, T.R., Reichle, M., 2000. A time-dependent probabilistic seismic-hazard model for California. Bull. Seismol. Soc. Am. 90, 1−21.

De Santis, A., Cianchini, G., Qamili, E., Frepoli, A., 2010. The 2009 L'Aquila (Central Italy) seismic sequence as a chaotic process. Tectonophysics 496, 44−52.

De Santis, A., Cianchini, G., Di Giovambattista, R., 2015. Accelerating moment release revisited: examples of application to Italian seismic sequences. Tectonophysics 639, 82−98.

Eneva, M., Ben-Zion, Y., 1997. Techniques and parameters to analyze patterns associated with large earthquakes. J. Geophys. Res. 102, 17785−17795.

Evison, F.F., 2001. Long-range synoptic earthquake forecasting: an aim for the millennium. Tectonophysics 333, 207−215. Available from: https://doi.org/10.1016/S0040-1951(01)00076-2.

Evison, F.F., Rhoades, D.A., 1997. The precursory earthquake swarm in New Zealand. N. Z. J. Geol. Geophys. 40, 537−547.

FRisk88M, 1995. User's Manual, ver. 1.70. Risk Engineering Inc., Boulder CO., 69 pp., 2 Appendixes.

Field, E.H., Jordan, T.H., 2015. Time-dependent renewal-model probabilities when date of last earthquake is unknown. Bull. Seismol. Soc. Am. 105, 459−463.

Field, E.H., Dawson, T.E., Felzer, K.R., Frankel, A.D., Gupta, V., Jordan, T.H., et al., 2009. Bull. Seismol. Soc. Am. 99, 2053−2107. Available from: https://doi.org/10.1785/0120080049.

Fitzenz, D., Nyst, M., 2015. Building time-dependent earthquake recurrence models for probabilistic risk computations. Bul. Seism. Soc. Am. 105, 120−133. Available from: https://doi.org/10.1785/0120140055.

Gabrielov, A.M., Dmitrieva, O.E., Keilis-Borok, V.1, Kossobokov, V.G., Kouznetsov, I.V., Levshina, T.A., et al., 1986. Algorithrns of Long-Term Earthquakes' Prediction. CERESIS, Lima, Peru, p. 61.

Gardner, J.K., Knopoff, L., 1974. Is the sequence of earthquakes in Southern California, with aftershocks removed, Poissonian? Bull. Seismol. Soc. Am. 64, 1363−1367.

Geller, R.J., Jackson, D.D., Kagan, Y.Y., Mulargia, F., 1997. Earthquakes cannot be predicted. Nature 275, 1616−1617.

Geller, R.J., Mulargia, F., Stark, P.B., 2016. Why we need a new paradigm of earthquake occurrence. In: Mora, G., Yuen, D.A., King, S.D., Lee, S.M., Stein, S. (Eds.), Subduction Dynamics: From Mantle Flow to Mega Disasters, Geophysical Monograph 211, first ed. American Geophysical Union.

Gentili, S., 2010. Distribution of seismicity before the largest earthquakes in Italy in the time interval 1994-2004. Pure Appl. Geophys. 167, 933958.

Gerstenberger, M.C., Rhoades, R.A., McVerry, C.H., 2016. A hybrid time-dependent probabilistic seismic-hazard model for Canterbury, New Zealand. Seism. Res. Let. 87, 1311−1318.

Giardini, D., Grünthal, G., Shedlock, K.M., Zhang, P., 1999. The GSHAP global seismic hazard map. Ann. Geophys. 42 (6).

Gutenberg, B., Richter, C.F., 1954. Seismicity of the Earth and Associated Phenopmena. Princeton University Press, Princeton, NJ.

Hardebeck, J.L., Felzer, K.R., Michael, A.J., 2008. Improved tests reveal that the accelerating moment release hypothesis is statistically insignificant. J. Geophys. Res. 113, B08310. Available from: https://doi.org/10.1029/2007JB005410.

Harris, R.A., 1998. Introduction to special section: stress triggers, stress shadows and implications for seismic hazard. J. Geophys. Res. 103, 24347−24358.

Healy, J.H., Kossobokov, V.G., Dewey, J.W., 1992. A test to evaluate the earthquake prediction algorithm M8. USGS Open-File Report 92−401.

Huang, Q., 2004. Seismicity pattern changes prior to large earthquakes-An approach of the RTL algorithm. Terrest. Atmosph. Oceanic Sci. 15, 469−491.

Huang, Q., 2008. Seismicity changes prior to the Ms8.0 Wenchuan earthquake in Sichan, China. Geophys. Res. Lett. 35, L23308.

Huang, Q., Ding, X., 2012. Spatiotemporal variations of seismic quiescence prior to the 2011 M 9.0 Tohoku earthquake revealed by an improved region-time-length algorithm. Bull. Seism. Soc. Am. 102, 1878−1883.

Huang, Q., Nagao, T., 2002. Seismic quiescence before the 2000 M = 7.3 Tottori earthquake. Geophys. Res. Lett. 29, 1578.

Huang, Q., Sobolev, G.A., 2002. Seismic quiescence prior to the 2000 M = 6.8 Nemuro Peninsula earthquake. Proc. Jpn. Acad. 77 (Ser. B), 6.

Huang, Q., Sobolev, G.A., 2001. Precursory seismicity changes associated with the Nemuro Peninsula earthquake, January 28, 2000. J. Asian Earth Sci. 21, 135−146.

Huang, Q., Sobolev, G.A., Nagao, T., 2001. Characteristics of the seismic quiescence and activation patterns before the M = 7.2 Kobe earthquake, January 17, 1995. Tectonophysics 337, 99−116.

Imamura, A., 1937. Theoretical and Applied Seismology. Maruzen, Tokyo.

Jackson, D.D., Kagan, Y.Y., 2006. The 2004 Parkfield earthquake, the 1985 prediction, and characteristic earthquakes: Lessons for the future. Bull. Seismol. Soc. Am. 96, S397−S409.

Jaumé, S.C., Sykes, L.R., 1999. Evolving towards a critical point: a review of accelerating seismic moment /energy release rate prior to large and great earthquakes. Pure Appl. Geophys. 155, 279−306.

Jiang, C., Wu, Z., 2005. Test of the preshock accelerating moment release (AMR) in the case of the 26 December 2004Mw9.0 Indonesia earthquake. Bull. Seismol. Soc. Am. 95. Available from: https://doi.org/10.1785/0120050018.

Jiang, C., Wu, Z., 2006. Benioff strain release before earthquakes in China: accelerating or not? Pure Appl. Geophys. 163, 1965−1976.

Jiang, C., Wu, Z., 2013. Intermediate-term medium-range precursory accelerating seismicity prior to the 12 May 2008, Wenchuan earthquake. Pure Appl. Geophys. 170, 209−219.

Jiang, H., Hou, H., Zhou, H., Zhou, C., 2004. Region-time-length algorithm and its application to the study of intermediate-short term earthquake precursor in North China. Acta Seismol. Sinica 17, 164−176.

Kanamori, H., Anderson, D., 1975. Theoretical basis of some empirical relations in seismology. Bull. Seism. Soc. Am. 65, 1073−1095.

Kasahara, K., 1981. Earthquake Mechanics. Cambridge University Press, Cambridge, p. 248.

Keilis-Borok, V.I., 1996. Intermediate-term earthquake prediction. Proc. Natl. Acad. Sci. U.S.A. 93, 3748−3755.

Keilis-Borok, V.I., 2002. Earthquake prediction: state-of-the-art and emerging possibilities. Ann. Rev. Earth Planet. Sci. 30, 1−33.

Keilis-Borok, V.I., Kossobokov, V.G., 1984. A complex of long-term precursors for the strongest earthquakes of the world. In: Proceedings of the 27th Geological Congress, vol. 61, Earthquakes and Hazard Prevention, pp. 56−61, Nauka, Moscow.

Keilis-Borok, V.I., Kossobokov, V.G., 1986. Time of increased probability for the great earthquakes of the world. Cornput. Seismol. 19, 48–58.

Keilis-Borok, V.I., Kossobokov, V.G., 1988. Premonitory activation of seismic flow: Algorithm M8. In: Lecture Notes of the Workshop on Global Geophysical Informatics With Applications to Research in Earthquake Prediction and Seismic Risk, Rep. H4, $MR/303-10, 17 pp., Int. Cent. for Theor. Phys., Trieste, Italy.

Keilis-Borok, V.I., Kossobokov, V.G., 1990a. Times of increased probability of strong earthquakes (M ≥ 7.5) diagnosed by algorithm M8 in Japan and adjacent territories. J. Geophys. Res. 95, 12413–12422.

Keilis-Borok, V.I., Kossobokov, V.G., 1990b. Premonitory activation of earthquake flow: algorithm M8. Phys. Earth Planet. Inter. 61, 73–83.

Keilis-Borok, V.I., Rotwain, I.M., 1990. Diagnosis of time of increased probability of strong earthquakes in different regions of the world: algorithm CN. Phys. Earth Planet. Inter. 61, 57–72.

King, G.C.P., Bowman, D.D., 2003. The evolution of regional seismicity between large earthquakes. J. Geophys. Res. 108. Available from: https://doi.org/10.1029/2001JB000783.

King, G.C.P., Cocco, M., 2001. Fault interaction by elastic stress changes: new clues from earthquake sequences. Adv. Geophys. 44, 1–38.

Knopoff, L., Levshina, T., Kellis-Borok, V.J., Mattoni, C., 1996. Increased long-range intermediate-magnitude earthquake activity prior to strong earthquakes in California. J. Geophys. Res. 101, 5779–5796.

Kossobokov, V.G., Shebalin, P., 2003. Earthquake Prediction, In: Keilis-Borok, V.I., Soloviev, A.S. (Eds.), Nonlinear Dynamics of the Lithosphere and Earthquake Prediction. Springer, Berlin-Heidelberg, pp. 141–207.

Kossobokov, V.G., Keilis-Borok, V.I., Smith, S.W., 1990a. Reduction of territorial uncertainty of earthquake forecasting. Phys. Earth Planet Inter. 61, R1–R4.

Kossobokov, V.G., Keilis-Borok, V.I., Smith, S.W., 1990b. Localization of intermediate-term earthquake prediction. J. Geophys. Res. 95, 19,763–19,772.

Kossobokov, V.G., Romashkova, L.L., Keilis-Borok, V.I., Healy, J.H., 1999. Testing earthquake prediction algorithms: statistically significant advance prediction of the largest earthquakes in the CircumPacific 1992-1997. Phys. Earth Planet. Inter. 111, 187–196.

Kuhn, T.S., 1962. The Structure of Scientific Revolutions. University of Chicago Press, Chicago.

Liu, H., Su, Y.J., 2006. Application of region-time-length algorithm to Yunnan area. J. Seismol. Res. 29, 25–29.

McGuire, R.K., 1976. Fortran computer program for seismic risk analysis. Open-File Report 76–67 94.

McGuire, R.K., 1978. FRISK: computer program for seismic risk analysis using faults as earthquake sources. Open-File Report 78-1007 71.

McGuire, R.K., Barnhard, P., 1981. Effects of temporal variations in seismicity in seismic hazard. Bull. Seism. Soc. Am. 71, 321–334.

Mignan, A., 2011. Retrospective on Accelerating Seismic Release (ASR) hypothesis: controversy and new horizons. Tectonophysics 505, 1–16.

Mignan, A., King, G.C.P., Bowman, D., Lacassin, R., Dmowska, R., 2006. Seismic activity in the Sumatra-Java region prior to the December 26, 2004 (Mw = 9.0–9.3) and March 28, 2005 (Mw = 8.7) earthquakes. Earth Planet. Sci. Lett. 244, 639–654. Available from: https://doi.org/10.1016/j.epsl.2006.01.058.

Mogi, K., 1969. Some features of recent seismic activity in and near Japan (2). Activity before and after great earthquakes. Bull. Earthquake Res. Inst. Univ. Tokyo 47, 395–417.

Mulargia, F., 2001. Retrospective selection bias (or the benefit of hindsight). Geophys. J. Int. 146, 489–496.

Papadopoulos, G.A., 1986. Long term earthquake prediction in western Hellenic arc. Earthq. Predict. Res. 4, 131–137.

Papaioannou, Ch. A., Papazachos, B.C., 2000. Time independent and time dependent seismic hazard in Greece based on seismogenic sources. Bull. Seism. Soc. Am. 90, 22–33.

Papaioannou, Ch.A., Scordilis, E.M., Papazachos, B.C., 1992. Application of time dependent and non-time dependent seismicity models in seismic hazard assessment in Greece. Cahier du Centre Eur. Geod. Seism. 6, 53–69.

Papazachos, B.C., 1989. A time predictable model for earthquake generation in Greece. Bull. Seism. Soc. Am. 79, 77–84.

Papazachos, B.C., 1990. Seismicity of the Aegean and surrounding area. Tectonophysics 178 (2–4), 287–308.

Papazachos, B.C., Papaioannou, Ch.A., 1993. Long-term earthquake prediction in the Aegean area based on a time and magnitude predictable model. Pure Appl. Geophys. 140, 595–612.

Papazachos, B.C., Papazachos, C.B., 2000. Accelerated preshock deformation of broad regions in the Aegean area. Pure Appl. Geophys. 157, 1663–1681.

Papazachos, C., Papazachos, B., 2001. Precursory accelerated Benioff strain in the Aegean area. Annal. Di Geofisica 44, 461–474.

Papazachos, B.C., Papadimitriou, E.E., Karakaisis, G.F., Panagiotopoulos, D.G., 1997a. Long term earthquake prediction in the Circum-Pacific convergent belt. Pure Appl. Geophys. 149, 173–217.

Papazachos, B.C., Karakaisis, G.F., Papadimitriou, E.E., Papaioannou, C.A., 1997b. The regional time and magnitude predictable model and its application to the Alpine-Himalayan belt. Tectonophysics 271, 295–323.

Papazachos, B.C., Karakaisis, G.F., Savvaidis, A.S., Papazachos, C.B., 2002. Accelerating seismic crustal deformation in the Southern Aegean area. Bull. Seism. Soc. Am. 92, 570–580.

Papazachos, C.B., Karakaisis, G.F., Scordilis, E.M., Papazachos, B.C., 2005a. Global observational properties of the critical earthquake model. Bull. Seism. Soc. Am. 95, 1841–1855.

Papazachos, C.B., Scordilis, E.M., Karakaisis, G.F., Papazachos, B.C., 2005b. Decelerating preshock seismic deformation in fault regions during critical periods. Bull. Geol. Soc. Greece 36, 1490–1498.

Papazachos, C.B., Karakaisis, G.F., Scordilis, E.M., Papazachos, B.C., 2006. New observational information on the precursory accelerating and decelerating strain energy release. Tectonophysics 423, 83–96.

Peresan, A., Kossobokov, V., Romashkova, L., Panza, G.F., 2005. Intermediate-term middle range earthquake predictions in Italy: a review. Earth-Sci. Rev. 69, 97–132.

Petersen, M.D., Cramer, C.H., Frankel, A.D., 2002. Simulations of seismic hazard for the Pacific Northwest of the United States from earthquakes associated with the Cascadia subduction zone. Pure Appl. Geophys. 159, 2147–2168.

Petersen, M.D., Cao, T., Campbell, K.W., Frankel1, A.D., 2007. Time-independent and time-dependent seismic hazard assessment for the State of California: Uniform California Earthquake upture Forecast Model 1.0. Seism. Res. Lett. 78, 99–109.

Press, F., 1975. Earthquake prediction. Sci. Am. 232 (5), 14–23.

Reid, H.F. (1910). The mechanics of the earthquake, The California Earthquake of April 18, 1906, Report of the State Investigation Commission, 2, Carnegie Institution of Washington, Washington, DC. 1910.

Reiter, L., 1990. Earthquake Hazard Analysis. Columbia University Press, New York, p. 154.

Robinson, R., 2000. A test of the precursory accelerating moment release model on some recent New Zealand earthquakes. Geophys. J. Int. 140, 568–576.

Rotwain, I., Novikova, O., 1999. Performance of the earthquake prediction algorithm CN in 22 regions of the world. Phys. Earth Planet. Inter. 111, 207–213.

Rundle, J.B., Klein, W., Gross, S., 1999. Physical basis for statistical patterns in complex earthquake populations: models, predictions and tests. Pure Appl. Geophys. 155, 575–607.

Saleur, H., Sammis, C.G., Sornette, D., 1996. Discrete scale invariance, complex fractal dimensions, and log-periodic fluctuations in seismicity. J. Geophys. Res. 101, 17,661−17,677.

Sammis, C.G., Sornette, D., Saleur, H., 1996. Complexity and earthquake forecasting. In: Rundle, J.B., Klein, W., Turcotte, D.L. (Eds.), Reduction and Predictability of Natural Disasters, SFI Studies in the Sciences of Complexity, XXV. Addison-Wesley, Reading, MA, pp. 143−156.

Sammis, C.G., Bowman, D.D., King, G., 2004. Anomalous seismicity and accelerating moment release preceding the 2001 and 2002 earthquakes in Northern Baja California, Mexico. Pure Appl. Geophys. 161, 2369−2378.

Scholz, C.H., 1990. The Mechanics of Earthquakes and Faulting. Cambridge University Press, Cambridge, p. 439.

Scordilis, E.M., Papazachos, C.B., Karakaisis, G.F., Karakostas, B.G., 2004. Accelerating seismic crustal deformation before strong mainshocks in Adriatic and its importance for earthquake prediction. J. Seismol. 8, 57−70.

Smalley, R.F., Turcotte, D.L., Solla, S.A., 1985. A renormalization group approach to the stick-slip behavior of faults. J. Geophys. Res. 90, 1894−1900.

Smith, S.W., Sammis, C.G., 2004. Revisiting the tidal activation of seismicity with a damage mechanics and friction point of view. Pure Appl. Geophys. 161, 2393−2404.

Sobolev, G.A., 2011. Seismicity dynamics and earthquake predictability. Nat. Hazards Earth Syst. Sci. 11, 445−458.

Sobolev, G.A., Tyupkin, Y.S., 1996. New Method of Intermediate-Term Earthquake Prediction. *European Seismological Commission XXV General Assembly Seismology in Europe, Reykjavik, Iceland*, pp. 229−234, *9−13 September1996*.

Sobolev, G.A., Tyupkin, Y.S., 1997. Low seismicity precursors of large earthquakes in Kamchatka. Volc. Seis. 18, 433−446.

Sobolev, G.A., Tyupkin, Y.S., 1999. Precursory phases, seismicity precursors and earthquake prediction in Kamchatka. Volc. Seis. 20, 615−627.

Sobolev, G.A., Tyupkin, Y.S., Zavialov, A., 1997. Map of expectation earthquakes algorithm and RTL prognostic parameter: joint application. Russ. J. Earthquake Sci. 1, 301−309.

Sobolev, G.A., Huang, Q., Nagao, T., 2002. Phases of earthquake's preparation and by chance test of seismic quiescence anomaly. J. Geodynamics 33, 413−424.

Sornette, A., Sornette, D., 1990. Earthquake rupture as a critical point. Consequences for telluric precursors. Tectonophysics 179, 327−334.

Sornette, D., Sammis, C.G., 1995. Complex critical exponents from renormalization group theory of earthquakes: implications for earthquake predictions. J. Phys. I. 5, 607−619.

Stein, R.S., 1999. The role of stress transfer in earthquake occurrence. Nature 402, 605−609.

Su, Y., Fu, H., Hu, H., 2012. Correlation of tidal forces with global grate earthquakes. Int. J. Geosci. 3, 373−378.

Sukrungsri, S., Pailoplee, S., 2015. Precursory seismicity changes prior to major earthquakes along the Sumatra-Andaman subduction zone: a region-time = length algorithm approach. Earth, Planets Space . Available from: https://doi.org/10.1186/s40623-0269-0.

Sykes, L.R., Jaumé, S.C., 1990. Seismic activity on neighbouring faults as a long-term precursor to large earthquakes in the San Francisco Bay area. Nature 348, 595−599.

Sykes, L.R., Shaw, B.E., Scholz, C.H., 1999. Rethinking earthquake prediction. Pure Appl. Geophys. 155, 207−232.

TERA Corporation, 1978. Bayesian seismic hazard analysis, a methodology. In: TERA Report to Lawrence Livermore National Laboratory. Berkeley, CA, TERA Corporation.

Tiampo, K.F., Shcherbakov, R., 2012. Seismicity-based earthquake forecasting techniques: ten years of progress. Tectonophysics 522–523, 89–121.

Tikhonov, I.N., Rodkin, M.V., 2012. Current state of art in earthquake prediction, typical precursors and experience in earthquake forecasting at Sakhalin Island and surrounding areas. In: D'Amico, S. (Ed.), Earthquake Research and Analysis - Statistical Studies, Observations and Planning. InTech, pp. 43–78.

Tocher, D., 1959. Seismic history of the San Francisco bay region. Calif. Div. Mines Spec. Rep. 57, 39–48.

Vamvakaris, D., 2010. Contribution to the time-dependent seismicity and seismic hazard. PhD Thesis, Geophysical Lab., School of Geology, Aristotle University of Thessaloniki, 428 pp and 2 Appendices (in Greek).

Vamvakaris, D., Papazachos, C., Papaioannou, C., Scordilis, E., Karakaisis, G., 2008. Time-independent and time-dependent seismic hazard study using synthetic catalogues. In: Proc. 3rd National Conference of Earthquake Engineering and Engineering Seismology, Athens – Greece, 21 pp.

Vamvakaris, D.A., Papazachos, C.B., Papaioannou, C.A., Scordilis, E.M., Karakaisis, G.F., 2016. A detailed seismic zonation model for shallow earthquakes in the broader Aegean area. Nat. Hazards Earth Syst. Sci. 16 (1), 55–84.

Varnes, D.J., 1989. Predicting earthquakes by analyzing accelerating precursory seismic activity. Pure Appl. Geophys. 130, 661–686.

Vergos, G., Arabelos, D.N., Contadakis, M.E., 2015. Evidence for tidal triggering on the earthquakes of the Hellenic Arc, Greece. Phys. Chem. Earth 85–86, 210–215.

WGCEP, 1988. Probabilities of large earthquakes occurring in California on the San Andreas fault. USGS Open File Report 88-398.

WGCEP, 1990. Probabilities of large earthquakes in the San Francisco Bay Region, California. USGS Survey circular 1053.

WGCEP, 1995. Seismic hazards in southern California: probable earthquakes, 1994–2024. Bull. Seism. Soc. Am. 85, 379–439.

WGCEP, 1999. Earthquake Probabilities in the San Francisco Bay Region: 2000 to 2030—A Summary of Findings. USGS Open File Report 99-517.

WGCEP, 2003. Earthquake Probabilities in the San Francisco Bay Region: 2002-2031. USGS Open File Report 03-214.

Yin, X.C., 1987. A new approach to earthquake prediction. Earthquake Res. China 3, 1–7 (in Chinese with English abstract).

Yin, X.C., 1993. A new approach to earthquake prediction. PRERODA (Russia's Nature) 1, 21–27. in Russian.

Yin, X.C., Chen, X.Z., Song, Z.P., Yin, C., 1995. A new approach to earthquake prediction: the load/unload response ratio (LURR) theory. Pure Appl. Geophys. 145, 701–715.

Yin, X.C., Song, Z.P., Wang, Y.C., 1996. The temporal variation of LURR in Kanto and other regions of Japan and its application to earthquake prediction. Earthq. Res. China 10, 381–385.

Yin, X.C., Wang, Y.C., Peng, K.Y., BAI, Y.L., Wang, H.T., Yin, X.F., 2000. Development of a new approach to earthquake prediction: Load/Unload Response Ratio (LURR) theory. Pure Appl. Geophys. 157, 2365–2383.

Yin, X.C., Mora, P., Peng, K.Y., Wang, Y.C., Weatherley, D., 2002. Load-Unload Response Ratio and accelerating moment/energy release critical region scaling and earthquake prediction. Pure Appl. Geophys. 159, 2511–2523.

Yin, X.C., Zhang, L.P., Zhang, H.H., Yin, C., Wang, Y.C., Zhang, Y.X., et al., 2006. LURR's twenty years and its perspective. Pure Appl. Geophys. 163, 2317–2341.

Yu, H.Z., 2004. Experimental research on precursors of brittle heterogeneous media and earthquake prediction methods. Ph.D. Thesis, Institute of Mechanics, CAS (in Chinese).

Yu, H.Z., Cheng, J., Zhu, Q.Y., Wan, Y.G., 2011. Critical sensitivity of load/unload response ratio and stress accumulation before large earthquakes: example of the 2008 Mw7.9 Wenchuan earthquake. Nat. Hazards 58, 251–267.

Yu, H.Z., Zhou, F., Cheng, J., Wa, Y.G., Zhang, Y.X., 2015. Sensitivity of the Load/Unload Response Ratio and critical region selection before large earthquakes. Pure Appl. Geophys. 172, 2203–2214.

Yuan, S., Yin, X., Liang, N., 2010. Load-unload response ratio and its application to estimate future seismicity of Qiandao Lake region. Proc. Eng. 4, 333–339.

Zhang, H.H., Yin, X., Liang, N., 2005. The spatial variation of LURR and seismic tendency in western United States. Earthquake Pred. Res. China 19, 338–346.

Zhang, L.P., Yin, X.C., Liang, N.G., 2007. Application of load/unload response ratio in study of seismicity in the region of Iran. Earthquake Res. China 21, 147–155.

Zhang, L.P., Yin, X.C., Liang, N.G., 2010. Relationship between load/unload response ratio and damage variable and its application. Concurrency Computat: Pract. Exper. 22, 1534–1548.

Zhang, Y., Yikilmaz, M.B., Rundle, J.B., Yin, X., Liu, Y., Zhang, L., et al., 2015. Study of the potential earthquake risk in the Western United States by the LURR method, based on the seismic catalogue, fault geometry and focal Mechanisms. Pure Appl. Geophys. 172, 2265–2276.

Zhuang, J.C., Yin, X.C., 1999. Random distribution of the Load/Unload Response Ratio (LURR) under assumptions of Poisson model. Earthquake Res. China 15, 128–138.

Further Reading

King, G.C.P., 2007. Fault interaction, earthquake stress changes, and the evolution of seismicity. In: Dziewonski, A., Romanowicz, B. (Eds.), Treatise on Geophysics, 1. Elsevier, Amsterdam.

11

Are Seismogenetic Systems Random or Organized? A Treatise of Their Statistical Nature Based on the Seismicity of the North-Northeast Pacific Rim

Andreas Tzanis[1], Angeliki Efstathiou[1], Filippos Vallianatos[2]

[1]NATIONAL AND KAPODISTRIAN UNIVERSITY OF ATHENS, ATHENS, GREECE
[2]TECHNOLOGICAL EDUCATIONAL INSTITUTE OF CRETE AND UNESCO CHAIR ON SOLID EARTH PHYSICS AND GEOHAZARDS RISK REDUCTION, CHANIA, CRETE, GREECE

CHAPTER OUTLINE

Complexity of Seismic Time Series. DOI: https://doi.org/10.1016/B978-0-12-813138-1.00011-0

11.1 Introduction

Seismogenetic systems are generally thought to comprise a mixture of processes that express the continuum of tectonic deformation (*background process*) and a large population of after-shocks that express the short-term activity associated with the occurrence of significant earthquakes (*foreground process*). Although progress has been made in understanding the foreground process, the statistical physics of background seismicity, the nature of seismoge-netic system remains ambiguous.

There are two general theoretical frameworks to describe the statistics of (background) seismicity. The first (and currently most influential) postulates that the expression of the background process is Poissonian in time and space and obeys additive Boltzmann–Gibbs thermodynamics. In consequence, it expects background earthquakes to be statistically inde-pendent and while it is possible for one event to trigger another, it submits that this occurs in an unstructured way and does not to contribute to the long-term evolution of seismicity. Thus, according to the 'Poissonian viewpoint', seismogenesis should be a *memoryless* pro-cess. The most influential realization of the Poissonian paradigm is the ETAS model (Epidemic-Type Aftershock Sequence, e.g., Ogata, 1988, 1998; Zhuang et al, 2002; Helmstetter and Sornette, 2003; Touati et al, 2009; Segou et al, 2013), which essentially is a self-excited conditional Poisson process (Hawkes, 1972; Hawkes and Adamopoulos, 1973; Hawkes and Oakes, 1974). ETAS posits that randomly occurring background earthquakes trigger aftershocks, and aftershocks trigger their own aftershocks, thus spawning a short-term proliferation of clustered foreground events (aftershock sequences) whose number decays according to the Omori–Utsu power-law (e.g., Utsu et al., 1995). Proxy-ETAS models (Console and Murru, 2001), as well as point process models to address the problem of inter-mediate to long-term clustering, have also been developed, such as EEPAS (Each Earthquake is a Precursor According to Scale, e.g., Rhoades and Evison, 2006; Rhoades, 2007) and PPE (Proximity to Past Earthquakes, e.g., Marzocchi and Lombardi, 2008).

At this point, it is important to point out that Poissonian models are mainly concerned with the statistics of time and distance between events. The size (magnitude) distribution of both background and foreground processes is still thought to be governed by the time-honoured frequency–magnitude (F–M) relationship of Gutenberg and Richter. However, the scale-free grading between earthquake frequency and magnitude implied by the F–M relationship is a power-law that *cannot* be derived from the Boltzmann–Gibbs formalism. Likewise, the Omori–Utsu formula is a Zipf–Mandelbrot power-law, and is therefore incon-sistent with the Boltzmann–Gibbs formalism. The heavy reliance of Poissonian seismicity worldviews and models on irrefutable, yet evidently non-Poissonian empirical laws is an apparent contradiction (self-inconsistency) with no theoretical resolution; it shows that Poissonian seismicity models are effectively ad hoc conceptual constructs that try to recon-cile the (inherited) Poissonian worldview of statistical seismology with the obviously non-Poissonian dynamics of fault formation and clustering.

The second framework also comprises different classes of models and proposes that the seismogenetic system is complex. A well-studied class of models generically known as self-organized criticality (SOC) proposes that seismicity is the expression of a nonequilibrating, fractal active fault network that continuously evolves toward a stationary critical state with no characteristic spatiotemporal scale, in which events develop spontaneously and any small instability has a chance of cascading into global failure (e.g., Bak and Tang, 1989; Sornette and Sornette, 1989; Olami et al., 1992; Sornette and Sammis, 1995; Rundle et al., 2000; Bak et al., 2002; Bakar and Tirnakli, 2009, etc.). According to Hanken (1983), the macroscopic properties of a self-organized system may change with time due to perturbations in its possible microscopic configurations, but the system as a whole will remain in, or continuously try to reach, the critical state. The advantage and allure of SOC is that it is consistent both with itself and with several observed properties of earthquake occurrence: the Gutenberg–Richter law, the Omori–Utsu law and other power-law distributions of parameters pertaining to the temporal and spatial expression of a simulated fault network emerge *naturally* during the evolution of simulated fault networks. A variant of SOC is self-*organizing* criticality leading to critical point behaviour at the end of an earthquake cycle (e.g., Sornette and Sammis, 1995; Rundle et al., 2000; Sammis and Sornette, 2001; and many others). This has been influential during the late 1990s and early 2000s, but is no longer pursued as it made specific predictions (acceleration of seismic release rates) that could not be verified experimentally. In the context of criticality, the dependence between successive earthquakes (faults) is known as *correlation*; this involves a long-range interaction and endows the seismogenetic system with memory that should be manifest in power-law statistical distributions of energy release, temporal dynamics and spatial dependence.

A few authors have investigated models with alternative complexity mechanisms that do not involve criticality, yet maintain the fault system in a state of nonequilibrium: a list can be found in Sornette (2004) and a comprehensive discussion in Sornette and Werner (2009). In a more recent development, Celikoglu et al. (2010) applied the Coherent Noise Model (CNM) (Newman, 1996) based on the notion of external stress acting coherently on all agents of the system without having any direct interaction with them. The CNM was shown to generate power-law behaviour in the temporal expression of its agent interevent time distributions but has a rather weak point: it does not include some geometric configuration of the agents and it is not known how this would influence the behaviour of the system.

The Poissonian and complexity/criticality viewpoints agree that the foreground process comprises a set of dependent events but the former assigns only local significance to this dependence, while criticality considers them to be an integral part of the regional seismogenetic process. In practice, the fundamental difference between the two approaches is in their understanding of the background process. The former assumes that there is no correlation (interaction) between random background events and argues that their statistical manifestations should best be described with the exponential and Gaussian distributions (consistent with the Boltzmann–Gibbs thermodynamic formalism). Criticality requires short- and long-range interactions (correlation) between near or distal background/background, background/foreground and foreground/foreground events, leading to power-law distributions of its temporal and spatial dynamic parameters. Moreover, noncritical complexity models cannot develop power-law distributions unless they evolve in nonequilibrium states, meaning that even in this case correlation is unavoidable. It is, therefore, clear that if it is possible to identify and remove the foreground process (aftershocks), it would also be possible to clarify the nature and dynamics of the background process by examining its spatiotemporal characteristics for the existence of correlation. It should also be apparent that in order to successfully pursue this line of inquiry one must have a *natural* self-consistent general theoretical framework on which to base the search for the existence of correlation (and not model-based or ad hoc conceptual constructs). One also requires effective measures of correlation in the temporal and spatial expression of seismicity, as well as effective methods to separate the background and foreground processes. As it turns out, there are (nearly) satisfactory answers to all three requirements.

Nonextensive statistical physics (NESP) is a fundamental generalized conceptual framework to describe nonadditive (nonequilibrating) systems in which the total (systemic) entropy is not equal to the sum of the entropies of their components. The concept has been introduced by Constantino Tsallis (Tsallis, 1988, 2001, 2009; Tsallis and Tirnakli, 2010) as a generalization of the Boltzmann–Gibbs formalism of thermodynamics. As such, it comprises an appropriate tool for the analysis of complexity evolving in a fractal-like space-time and exhibiting scale invariance, long-range interactions and long-term memory (e.g., Gell-Mann and Tsallis, 2004). NESP predicts power-law cumulative probability distributions in nonadditive (nonextensive)[1] systems, which reduce to the exponential cumulative distribution in the limiting case of additive (extensive/random/point) processes. NESP has already been applied to the statistical description of seismicity with noteworthy results (see Section 11.2.2). It has also been shown to generate the Gutenberg–Richter frequency–magnitude distribution from first principles (Sotolongo-Costa and Posadas, 2004; Silva et al., 2006; Telesca, 2011, 2012). In conclusion, NESP provides a general, complete, consistent and model-independent context in which to investigate the nature and dynamics of seismogenetic processes.

A definite indicator of correlation (interaction) between faults is the lapse between consecutive earthquakes above a magnitude threshold and over a given area: this is referred to as *interevent time, waiting time, calm time, recurrence time,* etc. Understanding the statistics of earthquake frequency versus interevent time is apparently essential for understanding the dynamics of the active fault network. For that reason, the frequency–interevent time (F–T) distribution has been studied by several researchers. Almost every study hitherto has focused on foreground and mixed background/foreground processes. Empirical F–T distributions generally exhibit power-law characteristics and fat tails. For that reason, in the context of statistical seismology they have been analysed with tailed standard statistical models reducible to power laws in some way or another. Examples of this approach are the gamma distribution and the Weibull distribution (e.g., Bak et al., 2002; Corral, 2004; Davidsen and Goltz, 2004; Martinez et al, 2005; Talbi and Yamazaki, 2010). Some researchers, working from a statistical physics vantage point, proposed ad hoc mechanisms for the generation of power laws

[1] The term "extensive" (full/complete according to Merriam-Webster's definition), was used by Tsallis (1988) to designate systems that are equilibrating, as opposed to those that are not (incomplete, i.e., nonextensive). The terms "additive" and "nonadditive" are probably more appropriate but, for consistency, we adopt Tsallis's terminology.

by a combination of correlated aftershock and uncorrelated background processes (e.g., Saichev and Sornette, 2013; Hainzl et al., 2006; Touati et al., 2009). Nevertheless, Molchan (2005) has shown that for a stationary point process, if there is a universal distribution of interevent times, then it must be an exponential one! Investigations performed in the context of NESP are reviewed in Section 11.2.2. A second measure of fault interaction is the *hypocentral distance* between consecutive earthquakes above a magnitude threshold and over a given area (*interevent distance*). The statistics of the frequency−interevent distance (F−D) distribution should be related to the *range of interaction* over that area; unfortunately it is not fully understood as it has been studied by less than a handful of researchers (e.g., Eneva and Pavlis, 1991; Abe and Suzuki, 2003; Batak and Kantz, 2014; Schoenball et al., 2015). A third criterion of correlation, (albeit not commonly acknowledged as one), is the *b* value of the Gutenberg−Richter frequency−magnitude (F−M) distribution which expresses the scaling of the size-space of active faults over a given area (fault hierarchy) and conveys information about their distribution in space and the homogeneity of the domain they occupy. The F−M distribution is *static* and does not say much about the dynamics of the fault network, nor about correlations in the energy released by successive earthquakes. Nevertheless, this undisputable empirical relationship is a standard against which to compare and test any physical and statistical description of the scaling of earthquake sizes and as such will be used herein.

The discrimination between background and foreground processes is commonly referred to as *declustering* and can be carried out with deterministic or stochastic methods, the latter being generally more efficient. An excellent review of the subject can be found in van Stiphout et al. (2012). Herein we have chosen to implement the stochastic declustering method of Zhuang et al. (2002); full justification is given in Section 11.3.3 and is based on a significant (for our objective) property: the method is *paradigmatic* realization of the self-excited Poisson process, as it implements the ETAS model in order to optimize the probability of aftershock identification. Accordingly, if the background seismicity obeys Boltzmann−Gibbs statistics, then the Zhuang et al. (2002) method should be able to extract a nearly random background process against which to test alternative hypotheses. If it does not, the argument in favour of a complex background would be stronger.

This chapter is an attempt to examine the dynamics of seismogenesis by studying the local and regional statistical characteristics of earthquake occurrence in different seismogenetic areas and implementing the generalized NESP formalism for the search for signs of randomness or self-organization in the probability distributions of event size, interevent time and interevent distance. In order to ensure the rigour of our analysis, instead of considering only one-dimensional earthquake frequency distributions as almost all studies have done thus far, we will focus on *multivariate* distributions that express the joint probability of observing an earthquake larger than a given magnitude, after a given time lapse and beyond a given distance, thereby introducing additional mutual constraints on the permissible variation of the relevant parameters.

Our analysis will focus on the principal seismogenetic zones of the north and northeast Pacific Rim, specifically the Californian and Continental Alaskan transformational plate margins, and the Alaskan–Aleutian convergent plate margin. These areas were chosen not only for their longstanding, reliable earthquake monitoring services and seismological catalogues, but mainly because they comprise three different seismotectonic contexts in which there is: (1) lithospheric seismogenesis along transform faults; (2) lithospheric seismogenesis along a convergent margin; and (3) large-scale deep focus seismogenesis in and around a major subducting slab. The seismogenetic systems of California are all *crustal*: earthquakes occur mostly in the *schizosphere*, i.e., in the rigid, brittle part of the upper lithosphere. On the other hand, the Alaskan and Alaskan–Aleutian systems are both crustal and *subcrustal*; in consequence, the analysis of the matter will proceed by crudely separating *crustal* and *subcrustal* earthquakes according to the depth of the Mohorovičić discontinuity. This type of differential study will also provide the opportunity to begin an inquiry as to whether environmental conditions (e.g., temperature, pressure), or/and boundary conditions (free at the surface versus fixed at depth), have a role in the dynamic expression and evolution of the seismogenetic fault network. The comparison of results from such exercises may afford — for the first time — evidence as to the existence of differences between crustal and subcrustal seismogenesis and, in the case of an affirmative answer, as to the origin of the differences and the cause of complexity/criticality thereof.

11.2 Nonextensive Approach to the Statistical Physics of Earthquakes

11.2.1 Brief Exposé of NESP

In statistical mechanics, an N-component dynamic system may have $W = N!/\Pi_i N_i!$ microscopic states, where i ranges over all possible conditions (states). In classical statistical mechanics, the entropy of that system S is related to the totality of these microscopic states by the Gibbs formula $S = -k\sum_i p_i \ln(p_i)$, where k is the Boltzmann constant and p_i is the probability of each microstate. Furthermore, if the components of the system are all statistically independent and uncorrelated to each other (noninteracting), the entropy of the system factorizes into the product of N identical terms, one for each component; this is the Boltzmann entropy $S_B = -Nk\sum_i p_i \ln(p_i)$. It is easy to see that one basic property of the Boltzmann–Gibbs formalism is *additivity* (*extensivity*): the entropy of the system equals the sum of the entropy of their components. In the past few decades it has been widely appreciated that a broad spectrum of nonequilibrating natural and physical systems does not conform to this requirement. Such *nonadditive* systems, which are also commonly referred to as *nonextensive* after Tsallis (1988), include statistically dependent (*interacting*) components, in consequence of which they acquire memory and can no longer be described with Boltzmann–Gibbs (BG) statistical physics.

An appropriate thermodynamic description of nonextensive systems has been pioneered by Tsallis (1988, 2009), who introduced the concept of NESP as a direct generalization of Boltzmann–Gibbs statistical physics. Letting x be some dynamic parameter, the nonequilibrium states of nonextensive systems can be described by the Tsallis (1988) entropic functional:

$$S_q(p) = \frac{k}{q-1}\left[1 - \int_0^\infty p^q(x)dx\right],$$ (11.1)

where $p(x)dx$ is the probability of finding the value of x in $[x, x + dx]$ so that $\int_W p(x)dx = 1$, and q is the *entropic index*. In the limiting case $q \to 1$, Eq. (11.1) converges to the Boltzmann–Gibbs functional

$$S_{BG} = -k\int_W p(x)\ln(p(x))dX,$$ (11.2)

Like the Boltzmann–Gibbs, the Tsallis entropy is concave and fulfils the *H*-theorem but is not additive when $q \neq 1$. For a mixture of two statistically independent systems *A* and *B*, the Tsallis entropy satisfies

$$S_q(A, B) = S_q(A) + S_q(B) + (1 - q)S_q(A)S_q(B).$$

This property is known as *pseudoadditivity* and is further distinguished into *superadditivity* (*superextensivity*) if $q < 1$, *additivity* when $q \to 1$ (i.e., Boltzmann–Gibbs statistics) and *subadditivity* (*subextensivity*) if $q > 1$. Accordingly, the entropic index is a measure of *nonextensivity* in the system.

An additional feature of NESP is the generalization of the expectation value in accordance with the generalization of entropy. Thus, the *q-expectation* value of *x* is defined as

$$\langle x \rangle_q = \int_0^\infty x \cdot p_q(x)dx,$$ (11.3)

where

$$p_q(x) = \frac{[p(x)]^q}{\int_0^\infty [p(x')]^q dx'}.$$ (11.4)

is an *escort distribution*. The concept of escort distributions was introduced by Beck and Schloegl (1993) as a means of exploring the structures of (original) distributions describing fractal and multifractal nonlinear systems: the parameter q behaves as a microscope for exploring different regions of $p(x)$ by amplifying the more singular regions for $q > 1$ and the less singular for $q < 1$.

Maximization of the Tsallis entropy yields the probability density function:

$$\hat{p}(x) = \frac{1}{Z_q}\exp_q\left[-\frac{\lambda}{I_q}(x - \langle x \rangle_q)\right],$$ (11.5)

$$Z_q = \int_0^\infty \exp_q \left[-\frac{\lambda}{I_q} \cdot (x - \langle x \rangle_q) \right] dx,$$

$$I_q = \int_0^\infty [\hat{p}(x)]^q dx$$

where λ is an appropriate Lagrange multiplier associated with the constraint on the q-expectation value and $\exp_q(.)$ denotes the q-exponential function

$$\exp_q(z) = \begin{cases} (1 + (1-q)z)^{\frac{1}{1-q}} & 1 + (1-q)z > 0, \\ 0 & 1 + (1-q)z \leq 0 \end{cases} \tag{11.6}$$

that comprises a generalization of the exponential function: for $q \to 1$, $\exp_q(z) \to e^z$.

Eq. (11.5) is a *q-exponential distribution* and as is evident from the definition of Eq. (11.6), it is a power-law if $q > 1$ corresponding to *subextensivity* (*subadditivity*), an exponential if $q = 1$ corresponding to *extensivity* (*additivity*), and a power-law with cut-off if $0 < q < 1$ corresponding to *superextensivity* (*superadditivity*); in the last case the cutoff appears at

$$x_c = \frac{x_0}{1-q}, \qquad x_0 = I_q \lambda^{-1} + (1-q)\langle x \rangle_q. \tag{11.7}$$

Using the definitions of x_0 from Eq. (11.7) and the q-expectation value from Eq. (11.4), the probability $\hat{p}(x)$ can be expressed as

$$\hat{p}(x) = \frac{\exp_q(x/x_0)}{\int_0^\infty \exp_q(x'/x_0) dx'} \tag{11.8}$$

In the NESP formalism, the theoretical distribution to be fitted to the observed (empirical) distribution of x is not the original stationary distribution $\hat{p}(x)$ but the escort probability $\hat{p}_q(x)$. Accordingly, the *cumulative* probability function (CDF) becomes:

$$\hat{P}(>x) = \int_x^\infty \hat{p}_q(x') dx'. \tag{11.9}$$

By substituting Eq. (11.8) into Eq. (11.4) and evaluating the integral, Eq. (11.9) reduces to:

$$\hat{P}(>x) = \exp_q\left(-\frac{x}{x_0}\right) = \left[1 - (1-q)\left(\frac{x}{x_0}\right)\right]^{\frac{1}{1-q}} \tag{11.10}$$

which also is a q-exponential distribution that for $q > 1$, defines a CDF of the Zipf–Mandelbrot kind.

Fig. 11–1 illustrates the behaviour of a *q-exponential CDF* (Eq. 11.10) for different values of q. For $q > 1$ the CDF has a tail that becomes increasingly longer (fatter) with increasing q: this

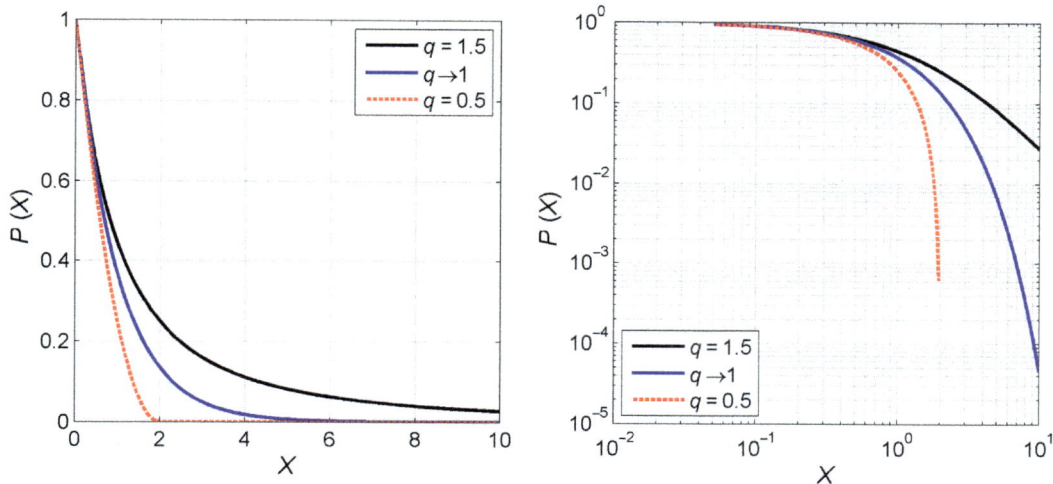

FIGURE 11–1 Three realizations of the *q*-exponential CDF for *q* < 1 (red line), *q* = 1 (*blue line*) and *q* > 1, plotted in linear (left) and double-logarithmic scale (right).

translates to increasing correlation (interaction) between its components and long-term memory. For $q \to 1$, the power law converges to the common exponential distribution so that the system comprises an uncorrelated and memoryless point (random) process. For $0 < q < 1$, the CDF is a power-law exhibiting a cutoff whenever the argument becomes negative, i.e., $\hat{P}(>x) = 0$, and is characterized by a bounded correlation radius.

11.2.2 Seismicity and NESP: An Overview

During the past several years, NESP has enjoyed increasing attention, with several researchers studying the properties of the F−T and F−M distributions (e.g., Vallianatos and Telesca, 2012). This includes studies of simulated *q*-exponential distributions emerging from critical seismicity models (e.g., Caruso et al., 2007; Bakar and Tirnakli, 2009), noncritical models, (e.g., Celikoglu et al., 2010), and rock fracture experiments (e.g., Vallianatos et al., 2012a). It also includes empirical studies of interevent time statistics based on the *q*-exponential distribution specified by Eq. 11.10 (e.g., Abe and Suzuki, 2005; Carbone et al., 2005; Vallianatos et al., 2012b, 2013; Michas et al., 2013, 2015; Papadakis et al., 2013, 2015; Vallianatos and Sammonds, 2013; Antonopoulos et al., 2014). A recent review of NESP applications over a broad spectrum of scales, from tectonic plates to country rock fractures and laboratory fragmentation experiments, is given by Vallianatos et al. (2016).

Nonextensive analysis of the F−M distribution has been undertaken by Sotolongo-Costa and Posadas (2004), Silva et al. (2006), and Telesca (2011, 2012). These authors proposed NESP generalizations of the Gutenberg−Richter law based on physical models that consider the interaction between two rough fault walls (asperities) and the fragments filling space between them (*fragment-asperity model*); this interaction is supposed to modulate earthquake triggering. In this model, the generalized Gutenberg–Richter law is approached by considering the size distribution of fragments and asperities and the scaling of size with energy. The transition from size to energy and magnitude distributions depends on how energy scales with size and with magnitude.

Sotolongo-Costa and Posadas (2004) assumed that the energy stored in the asperities and fragments scales with their linear characteristic dimension r ($E \propto r$) or, equivalently, with the square root of their areas σ ($E \propto \sigma^{1/2}$); they also assumed that the magnitude scales with energy as $M \propto \log(E)$. Darooneh and Mehri (2010) expand on the same model but assume that $E \propto \exp(\sigma^{1/a})$ and $M \propto \ln(E)$. We propose that the above assumptions are not compatible with the empirical laws of energy–moment and moment–magnitude scaling in particular (e.g., Lay and Wallace, 1995; Scholz, 2002). Silva et al. (2006) revisited the fragment-asperity model and expressed Eq. (11.10) as

$$\hat{p}(\sigma) = \left[1 - \frac{1-q}{2-q}\left(\sigma - \langle\sigma\rangle_q\right)\right]^{\frac{1}{1-q}}. \tag{11.11}$$

Assuming that the energy scales with the characteristic volume of the fragments ($E \propto r^3$), so that $E \propto \sigma^{3/2}$ because σ scales with r^2, it is easy to see that $(\sigma - \langle\sigma\rangle_q) = (E/\alpha)^{2/3}$ with α being a proportionality constant between E and r. This yields the energy density function

$$\hat{p}(E) = \left(\frac{2}{3} \cdot \frac{E^{-1/3}}{\alpha^{2/3}}\right) \cdot \left[1 - \frac{(1-q)}{(2-q)}\frac{E^{2/3}}{\alpha^{2/3}}\right]^{-\frac{1}{1-q}}$$

so that $\hat{P}(>E) = N(>E)N_0^{-1} = \int_E^\infty \hat{p}(E)dE$, where $N(>E)$ is the number of events with energy greater than E and N_0 is the total number of earthquakes. If the magnitude scales with energy as $M \propto 1/3\log(E)$, for $q > 1$,

$$\hat{P}(>M) = \frac{N(>M)}{N_0} = \left(1 - \frac{1-q_M}{2-q_M} \cdot \frac{10^{2M}}{\alpha^{2/3}}\right)^{\left(\frac{2-q_M}{1-q_M}\right)} \tag{11.12}$$

Eq. (11.12) has been used to investigate the seismicity of different tectonic regions (Telesca 2010a,b; Telesca and Chen, 2010; Esquivel and Angulo, 2015; Scherrer et al., 2015). Finally, assuming $E \propto r^3$ but that the magnitude scales with energy as $M \propto 2/3\log(E)$, Telesca (2011, 2012) has introduced a modified version of Eq. (11.12):

$$\hat{P}(>M) = \frac{N(>M)}{N_0} = \left(1 - \frac{1-q_M}{2-q_M} \cdot \frac{10^{M}}{\alpha^{2/3}}\right)^{\left(\frac{2-q_M}{1-q_M}\right)}. \tag{11.13}$$

We suggest that this model, by postulating that the energy released in the form of seismic waves scales with the effective area of the fault (fragments and asperities), is consistent with the empirical laws of energy–moment and moment–magnitude scaling and is also compatible with the well-studied rate-and-state friction laws of rock failure. In consequence, our analysis will be based on the F–M distribution specified by Eq. (11.13).

11.2.3 Multivariate Earthquake Frequency Distributions: Construction and NESP-Based Modelling

Our goal is to investigate whether seismicity is a Poissonian or complex/critical process by using the NESP formalism to search for the presence (or absence) of correlation in time, size and space. This can be done by determining the values and variation of the relevant entropic indices. To ensure rigour in our analysis, instead of considering only one-dimensional frequency distributions as almost all studies thus far have done, we focus on *multivariate* earthquake frequency distributions, thereby introducing additional mutual constraints on the permissible variation of the empirically determined entropic indices. The most general multivariate earthquake frequency distribution is one that expresses the joint probability of observing an earthquake larger than a given magnitude, after a given lapse time and beyond a given distance. This would require the construction and analysis of *trivariate* frequency−magnitude−interevent time−interevent distance (F−M−T−D) distributions, which live in a four-dimensional realm and would be more difficult to manage and interpret. Accordingly, we opted to use the easier to handle *bivariate* frequency−magnitude−interevent time (F−M−T) distributions in order to focus on correlations in earthquake size and time of occurrence. However, because this may not extract explicit information about the range of possible correlations, we shall also use the *interevent distance* as a spatial filter by which to separate and study the temporal correlation of *proximal* and *distal* earthquakes. The rationale behind this approach is that if distal earthquakes are correlated in time, then they have to be correlated in space via long-distance interaction and vice versa.

A bivariate F−M−T distribution can be constructed as follows: A *threshold* (cutoff) magnitude M_{th} is set and a bivariate frequency table (histogram) representing the empirical *incremental* distribution is first compiled. The empirical *cumulative* distribution is then obtained by backward bivariate summation, according to the scheme

$$N_{m\tau} = \sum_{j=D_T}^{\tau} \sum_{i=D_M}^{m} \left\{ \mathrm{H}_{ij} \Leftrightarrow \mathrm{H}_{ij} \neq 0 \right\}, \quad \tau = 1, \ldots D_T, \quad m = 1, \ldots D_M \tag{11.14}$$

where H is the incremental distribution, D_M is the dimension of H along the magnitude axis and D_T is the dimension of H along the Δt axis. In this construct, the cumulative frequency (earthquake count) can be written thus: $N(\{M \geq M_{th}, \Delta t : M \geq M_{th}\})$. Then, the empirical probability $P(>\{M \geq M_{th}, \Delta t : M \geq M_{th}\})$ is simply

$$\frac{N(>\{M \geq M_{th}, \Delta t : M \geq M_{th}\})}{N_0}, \qquad N_0 = N(M = M_{th}, 0) = \|N\|_\infty. \tag{11.15}$$

An empirical cumulative F−M−T distribution constructed according to Eq. (11.14) is presented in Fig. 11−2: it is based on a subset of 3653 events extracted from the NCSN earthquake catalogue published by the North California Earthquake Data Center, using a threshold magnitude $M_{th}=3.4$ over the period 1975−2012 and excluding the Mendocino Fracture Zone (MFZ) (for details see Section 11.3). The distribution is shown in linear (Fig. 11−2A) and logarithmic (Fig. 11−2B) frequency scales and comprises a well-defined

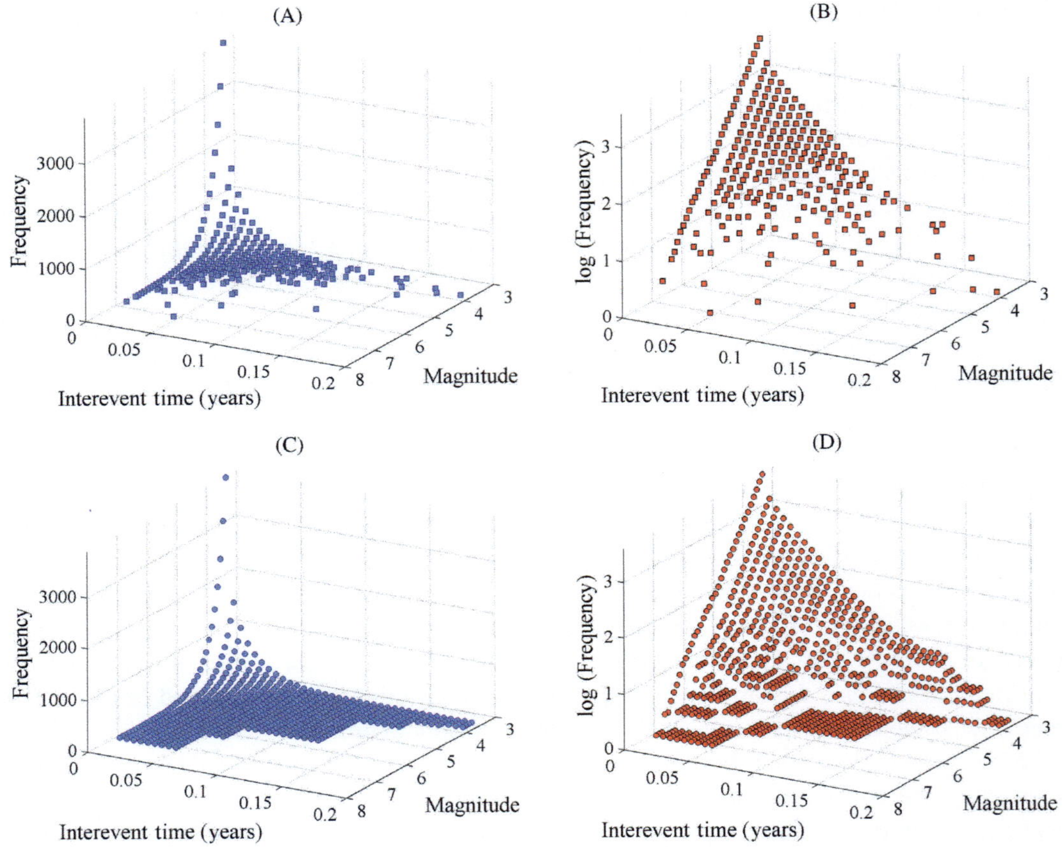

(A)

(B)

(C)

(D)

FIGURE 11–2 (A) Bivariate cumulative frequency–magnitude–interevent time (F–M–T) distribution constructed according to Eq. (11.14) on the basis of 3653 events with $M_L \geq 3.4$ extracted from the NCSN earthquake catalogue; see text for details. (B) As per (A) but in logarithmic frequency scale. (C) As per (A) but including unpopulated bins in the summation, i.e., using the scheme $N_{m\tau} = \sum_{j=D_T}^{\tau} \sum_{i=D_M}^{m} H_{ij}$ instead of Eq. (11.14). (D) As per (C) but in logarithmic frequency scale.

surface in which the end-member ($M \geq M_{th}$, $\Delta t = 0$) is the one-dimensional empirical Gutenberg–Richter law and the end-member ($M = M_{th}$, Δt) is the one-dimensional frequency–interevent time (F–T) distribution.

Assuming that magnitudes and interevent times are statistically independent, namely that the hierarchy of the active fault network does not influence the sequence of events, the joint probability $P(M \cup \Delta t)$ factorizes into the probabilities of M and Δt in the sense $P(M \cup \Delta t) = P(M) P(\Delta t)$. Then, by implicitly *identifying* the empirical and escort probabilities we obtain

$$\frac{N(> \{M \geq M_{th}, \quad \Delta t : M \geq M_{th}\})}{N_0} = \left(1 - \frac{1 - q_M}{2 - q_M} \cdot \frac{10^M}{\alpha^{2/3}}\right)^{\left(\frac{2 - q_M}{1 - q_M}\right)} \cdot \left(1 - (1 - q_T) \frac{\Delta t}{\Delta t_0}\right)^{\frac{1}{1 - q_T}}, \qquad (11.16)$$

where q_M and q_T are the entropic indices for the magnitude and interevent times, respectively, and Δt_0, is the *q-relaxation time*, analogous to the relaxation (characteristic) time often encountered in the analysis of physical systems. On taking the logarithm and setting $a = \log(N_0)$, Eq. (11.16) becomes

$$\log N(>\{M \geq M_{th}, \ \Delta t{:}M \geq M_{th}\}) =$$
$$= a + \left(\frac{2-q_M}{1-q_M}\right) \cdot \log\left(1 - \frac{1-q_M}{2-q_M} \cdot \frac{10^M}{\alpha^{2/3}}\right) + \frac{1}{1-q_T}\log(1 - \Delta t_0^{-1}(1-q_T)\Delta t) \qquad (11.17)$$

Eq. (11.17) is a generalized (bivariate) law of the Gutenberg−Richter kind in which

$$b_q = \frac{(2-q_M)}{(q_M-1)} \qquad (11.18)$$

is the NESP *generalization* of the *b* value (also see Telesca, 2012). Accordingly, Eq. (11.17) is the general model to be implemented in the ensuing analysis. It may also be worth noting that Eq. (11.17) has been applied to the analysis investigation of time dependence in the characteristics of complexity/criticality along the San Andreas Fault (SAF) (Efstathiou et al., 2015), as well as to a preliminary study of the spatiotemporal properties of seismicity in South California (Efstathiou et al., 2016).

The logarithmic form of the distribution shown in Fig. 11−2B can be approximated with Eq. (11.17) using nonlinear least-squares. Because the parameters are all positive and the entropic indices are bounded, we implemented the *trust-region reflective* algorithm (e.g., Moré and Sorensen, 1983; Steihaug, 1983), together with *least absolute residual* (LAR) minimization so as to suppress possible outliers. The result is shown in Fig. 11−3A. The quality of the approximation is excellent, with a correlation coefficient (R^2) of the order of 0.99. The magnitude entropic index $q_{M=}1.51$ so that $b_q \approx 1$, which compares well with *b* values computed with conventional one-dimensional techniques for the same data set. The temporal entropic index q_T is approximately 1.3 and indicates moderate subextensivity. Fig. 11−3B presents a succinct statistical appraisal of the result, performed by fitting a normal location-scale distribution (dashed line) and a Student's *t* test location-scale distribution (solid line) to the cumulative probability of the sorted residuals (*r*). Approximately 85% of the residual population, for which $|r| \leq 0.1$, is normally distributed. The short truncated tail forming at $r < -0.1$ consists of 39 residuals ($\sim 16\%$ of the population) and does not deviate significantly from normality. The long tail forming at $r > 0.2$ is fitted with neither the normal nor the *t*-location-scale distribution; however, it consists of only seven residuals (2.87%) and represents *outliers effectively suppressed* by the LAR procedure.

It is interesting to note that outliers are mainly observed at the intermediate and larger magnitude scales and longer interevent times. They frequently arise from minor flaws in the catalogue (e.g., omitted (sequences of) events, glitches in magnitude reporting, etc.), but in some cases they may comprise true exceptions to the continuum of the regional seismogenetic process: for instance, they may correspond to rare, externally triggered events. Herein,

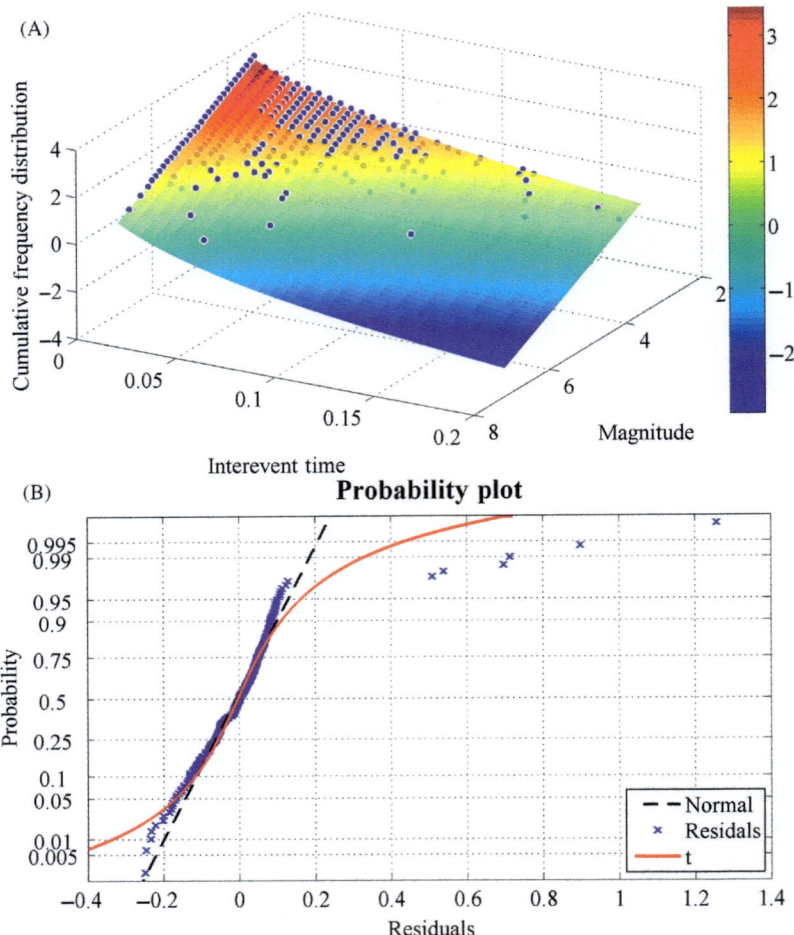

FIGURE 11–3 (A) The logarithmic scale F—M—T distribution of Fig. 11–2B together with the model fitted using Eq. (11.17); the colourbar represents the number of events in logarithmic (frequency) scale. (B) Probability analysis of the residuals (see Section 11.2.2 for details).

we shall not be concerned with such details but it is interesting to point them out. The existence of outliers has compelled us to introduce a significant constraint in the construction of the F—M—T distribution: according to Eq. (11.14), the cumulative distribution is formed by stacking only the populated (nonzero) bins of the incremental distribution. Regardless of the origin of the outliers, their inclusion in the summation would have generated a *stepwise* function in which the unpopulated bins (unknown probability densities) lying between the outliers and the populated bins would appear as patches of equal earthquake frequency (uniform probability), as illustrated in Fig. 11–2C,D. In this case, the high probability zones of the empirical bivariate distribution would comply with well-specified laws, but the lower probability zones would, for some 'unknown' reason, include uniform patches. In one-dimensional distributions this effect may not influence parameter estimation by a significant factor and is often neglected. In multivariate distributions however, in addition to the obvious absurdity, it would be numerically detrimental.

In a final note, in order to distinguish between proximal and distal earthquakes and assess their correlation, we apply the above modelling procedure to subsets of the catalogue in which earthquakes are grouped by interevent distance according to the rule

$$C \supset \{C_D : M > M_{th} \hat{\Delta} d_L \leq \Delta d \leq \Delta d_U\}, \tag{11.19}$$

where C is the catalogue, C_D is the subset catalogue, Δd is the interevent distance and Δd_L, Δd_U are the upper and lower group limits, respectively. This is equivalent to constructing and modelling the *conditional* bivariate cumulative distribution

$$P(> \{M \geq M_{th}, \Delta t : [M \geq M_{th} \hat{\Delta} d_L \leq \Delta d \leq \Delta d_U]\}) \tag{11.20}$$

as a proxy of the *trivariate* F−M−T−D distribution.

11.3 Earthquake Data and Analysis

11.3.1 Earthquake Source Areas and Catalogues

This study investigates the statistical nature of seismicity along the north and northeast Pacific Rim, focusing on the major earthquake source areas of California, Alaska and the Alaskan−Aleutian Arc and Trench System, as can be seen in the seismicity maps in Figs. 11−4 and 11−5. A brief description of the tectonic settings of these areas is given below, as we consider it to be necessary in understanding the rationale by which we categorize and treat our data.

11.3.2 California

The most prominent and well-studied seismogenetic feature of California is the SAF. This comprises a NW to NNW oriented, 1300 km long, right-lateral transformational boundary between the Pacific plate to the west and the North American plate to the east, and has generated several large ($M > 7$) earthquakes during the past two centuries (e.g., 1857, 1906, 1989, 1992 and 1999). The SAF system (main and 'sibling' faults) is generally thought to comprise three major segments: the Mojave segment in South California, between Salton Sea (approximately 33.36°N, 115.7°W at the SE corner of California) and Parkfield, Monterey County (approximately 35.9°N, 120.4°W); the central segment between Parkfield and Hollister (approximately 36.85°N, 121.4°W) and, finally, the northern segment between Hollister and through the San Francisco bay area up to the MFZ (offshore, approximately 40.36°N, 124.5°W).

The MFZ is a W−E right-lateral transformational plate boundary between the Pacific and Gorda plates, off the coast of Cape Mendocino in northern California (e.g., Dickinson

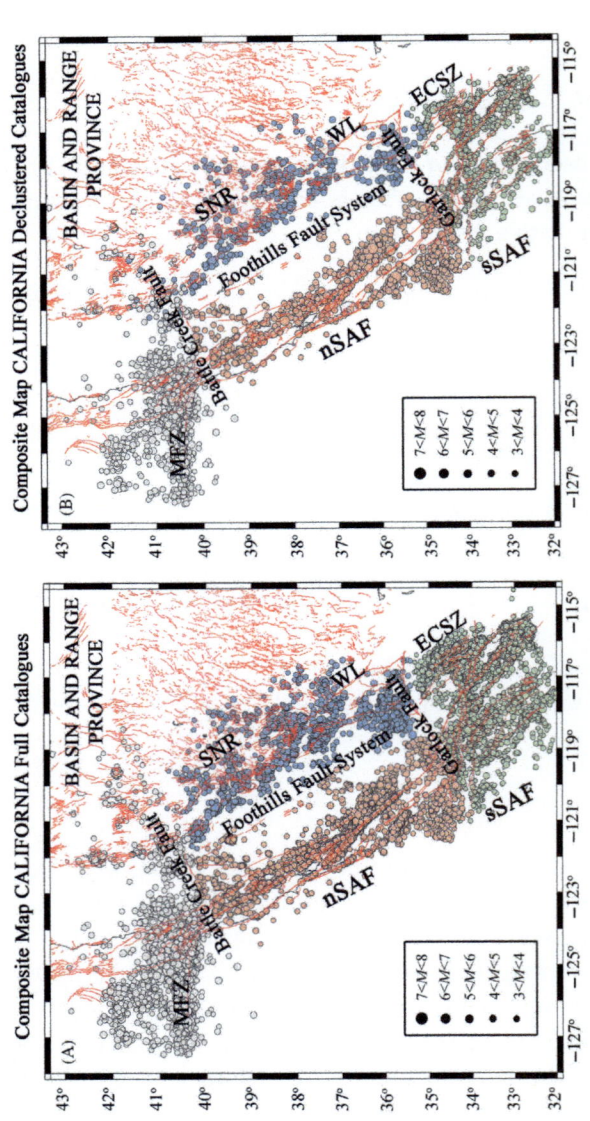

FIGURE 11–4 (A) The seismicity of California as illustrated by mapping the epicentres of earthquakes included in the full NCSN catalogue (1968–2015, $M \geq 3$) and the full SCSN catalogue (1980–2015, $M \geq 2.6$). The source areas treated herein are colour-coded as follows: gGrey, Mendocino Fracture Zone (MFZ); orange, north segment of the San Andreas Fault (nSAF); light blue, Central Valley–Sierra Nevada Range–Walker Lane (SNR); light green, South California Seismic Region. nSAF is the broader area of the San Andreas south segment; ECSZ is the East California Shear Zone. (B) as per (A) but for regional catalogues *declustered* at the $\phi \geq 70\%$ probability level.

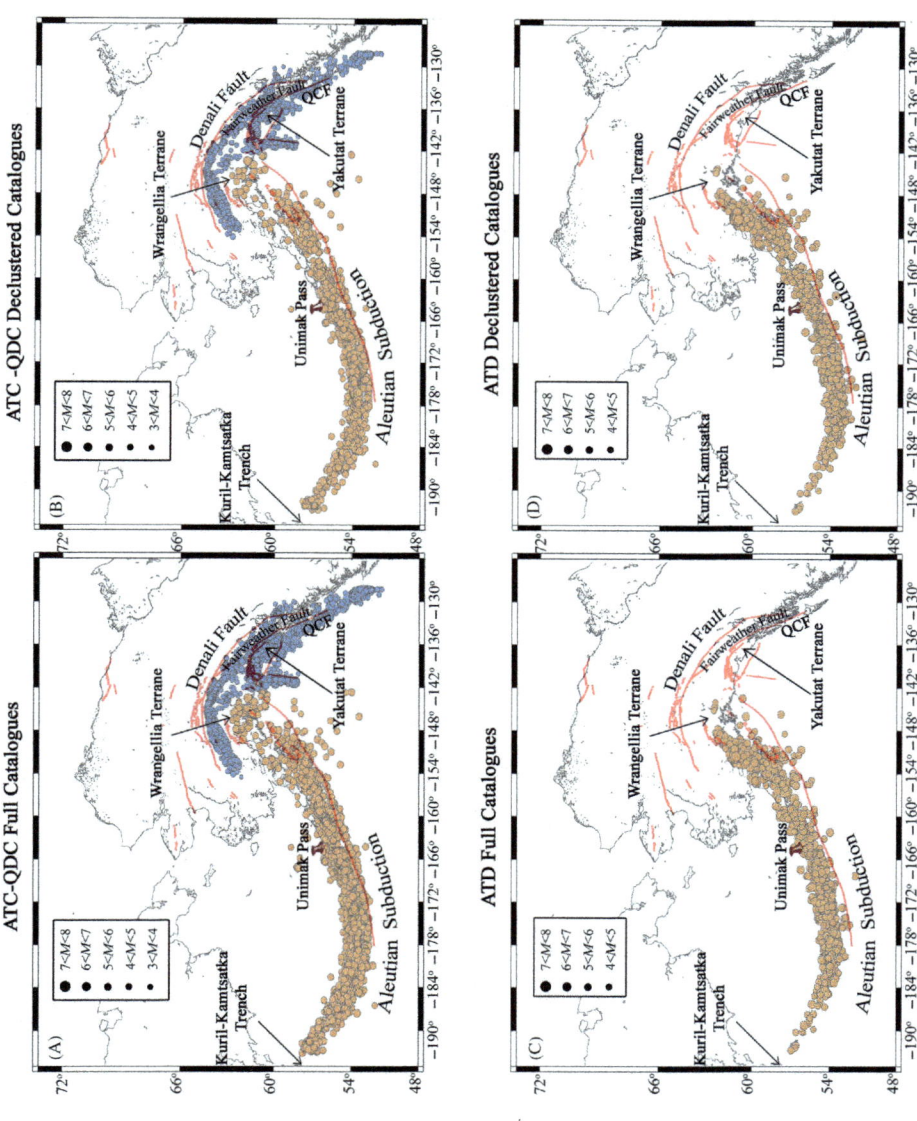

FIGURE 11-5 Seismicity recorded along the primary boundary of the Pacific and North American plates during 1968–2015 and used in the present analysis (AEC catalogue). The top row illustrates the epicentres of crustal (above Moho) earthquakes along the Queen Charlotte–Fairweather and Denali fault zones (light blue), and along the Alaskan–Aleutian Arc (orange). (A) shows the full catalogue and (B) the declustered catalogue ($\phi \geq 70\%$). The bottom row illustrates the epicentres of subcrustal (below Moho) earthquakes along the Aleutian Subduction zone. As before, (C) shows the full catalogue and (D) the declustered ($\phi \geq 70\%$).

and Snyder, 1979a; Furlong and Schwartz, 2004). It extends westward from its transform—transform—trench junction with the SAF and the Cascadia subduction zone (Mendocino Triple Junction), to the southern end of the Gorda Ridge at approximately 40.4°N, 128.7°W; it then continues on as an inactive segment for several hundred kilometres. The MFZ includes the most seismically active part of California (Yeats, 2013) and according to Dengler et al. (1995) the north coastal region accounted for about 25% of the seismic energy released in California in a 50-year period.

The SAF accommodates only about 75% of the total motion between the North American and Pacific plates. The rest is accommodated by NNW—SSE right-lateral deformation in an area east of the Sierra Nevada mountain range, called the Walker Lane or Eastern California Shear Zone (Wesnousky, 2005; Guest et al., 2007). The Walker Lane terminates between the Pyramid Lake in Nevada and Lassen Peak in California, approximately at 40.3°N, 120.6°W, where the Honey Lake Fault Zone meets the transverse tectonic zone forming the southern boundary of the Modoc Plateau and Columbia Plateau with the Great Basin. Pease (1965) observed that the alignment of that transverse zone and the MFZ suggests that the former might have once been the continental terminus of the MFZ.

To further complicate things, California is geologically divided into North and South by the SW—NE left-lateral Garlock fault which extends for approximately 250 km between its junction with the East California Shear Zone (ECSZ) at the north-eastern edge of the Mojave Desert (approximately 35.6°N, 116.4°W) and its junction with the SAF at Tejon Pass (approximately 34.8°N, 118.9°W). This major tectonic boundary is believed to have developed in order to accommodate the strain differential between the almost W—E extension of the Great Basin eastwards of the ECSZ (e.g., Wernicke et al., 1988), and the NW—SE right-lateral transformation of the ECSZ and SAF. Thus, the right-lateral motion on the SAF and ECSZ locks up in the area of the Garlock, where local variations in the mode of deformation and earthquake focal mechanisms are observed (e.g., Jones, 1988; Hardebeck and Hauksson, 2001; Becker et al., 2005; Fialko, 2006). Between 37.7°N and 35.1°N, the left-lateral motion of the Galrlock fault generates a restraining bend and a broad S-shaped westward displacement of the SAF, known as the 'Big Bend'.

The above-outlined tectonic setting results in four distinct earthquake source areas, as shown in Fig. 11—4:

1. The MFZ, bounded by the coordinates 40°N to 43°N and 123°W to 128°W.

2. The central and northern SAF segments (henceforth nSAF), north of the Garlock Fault between Parkfield and the MFZ. For the purpose of this study, the geographic borders of nSAF are defined to the north by the line joining the northern terminus of the SAF/Shelter Cove section (40.2°N, 124.3°W), the northern terminus of the Bartlett Springs Fault System (Lake Mountain fault) and the Battle Creek Fault (40.5°N, 121.9°W); to the east by the Battle Creek Fault, the Foothills Fault system (roughly 39.3°N, 118.8°W) and the Kern Gorge fault and White Wolf fault zone (35.3°N, 118.6°W) and to the West by an imaginary line running offshore parallel to the Pacific Coast.

3. The Central Valley and Sierra Nevada Range, up to and including the Walker Lane (henceforth SNR). This extends northward of the Garlock Fault and behaves as a semirigid

microplate (Sierran microplate) whose interior (Central Valley) is characterized by the absence of significant faults and large earthquakes (Goter et al., 1994; Dixon et al., 2000; McCaffrey, 2005; Saleeby et al., 2009; Hammond et al., 2012). In this study, the geographic boundaries of SNR are defined to the north by the line joining the Battle Creek Fault and the northern termini of the Butt Creek and Almanor fault zones (roughly 44.5°N, 121.2°W) and then up to 116°W; to the east by the 116°W meridian; to the south by the Garlock Fault and to the west by the White Wolf and Kern Gorge fault zones, the Foothills Fault system and the Battle Creek Fault.

4. In contrast to their distinct nature north of the Garlock Fault, the SAF and ECSZ converge and are not as easy to distinguish south of the fault. In consequence, we will consider that area (southern SAF segment and ECSZ) to comprise an integral seismogenetic entity and henceforth refer to it as the South California Seismic Region (SCSR). The north boundary of the SCSR begins at the western terminus of the Santa Ynez Fault Zone—Pacific Section, which is a virtual extension of the Garlock fault (34.5°N, 120.5°W); it then runs south of Tejon Pass and parallel to the Garlock Fault up to approximately 35.5°W, 116.3°W, past its eastern terminus. It then turns south and runs eastward of the South Bristol Mts. Fault (34.6°N, 115.6°W), to Yuma at the US—Mexico border (32.7°N, 114.6°W). It continues westwards to approximately 32°N, 117°W, which is south of Tijuana, Mexico, and then to 32°N, 119°W off the west coast of Mexico. Finally, it turns north and runs parallel to the coastline and west of the San Clemente and Santa Cruz islands to 34.5°N.

The earthquake data we utilized for the nSAF, SNR and MFZ source areas, were extracted from the regional earthquake catalogue of the North California Seismic Network (NCSN @ http://www.NCSN.org).The data utilized for the SCSR source area were extracted from the regional catalogue of the South California Earthquake Data Centre (SCSN @ http://www.data.scec.org). Details are given in Table 11−1. In both NCSN and SCSN catalogues, most earthquakes are reported in the M_L and M_w magnitude scales, while there is a considerable number of events in the duration (M_d) and amplitude (M_x) scales. The latter two have been exhaustively calibrated against the M_L scale: Eaton (1992) has shown that they are within 5% of the M_L scale for magnitudes in the range 0.5−5.5 and that they are virtually independent of the distance from the epicentre to at least 800 km. In consequence, M_d and M_x are practically equivalent to M_L. For the purpose of the present analysis, M_w magnitudes were also converted to M_L using the empirical formula of Uhrhammer et al. (1996): $M_w = M_L \cdot (0.997 \pm 0.020) - (0.050 \pm 0.131)$. Thus, both the NCSN and SCSN catalogues were reduced to the M_L scale and are homogeneous and complete for $M_L \geq 3.0$ and $M_L \geq 2.6$, respectively.

11.3.2.1 Alaska and the Alaskan—Aleutian Arc and Trench System

The Aleutian Arc and Continental (mainland) Alaska source areas are bounded by the coordinates 50°N to 70°N and 196°W to 126°W. The principal structural and geodynamic feature of this area − which also defines the geographical borderline of the north Pacific Rim − is the boundary between the North American and Pacific plates (Fig. 11−5). The eastern plate boundary is defined by the Queen Charlotte—Fairweather (QC−F) dextral transform fault

Table 11–1 Summary of the Earthquake Catalogues Used in the Present Analysis

Source Area		Source Area Code	Source Catalogue	Period	Full Catalogues		Declustered Catalogues ($\phi \geq 70\%$)	
					M_{comp}	No. Events	M_{comp}	No. Events
South California Seismic Region		SCSR	SCSN	1980–2015	2.6	20,088	2.6	3339
San Andreas Fault – North Segment		nSAF	NCSN	1968–2015	3.0	8596	3.2	943
Sierra Nevada Range – Walker Lane		SNR	NCSN	1968–2015	3.0	4982	3.2	591
Mendocino Fracture Zone		MFZ	NCSN	1968–2015	3.0	3706	3.0	1755
Continental Alaska: Queen Charlotte – Fairweather and Denali Fault Zones		QCD	AEC	1968–2015	3.0	4332	3.0	1639
Aleutian Arc	Crustal earthquakes	ATC	AEC	1968–2015	4.4	4775	4.4	1608
	Subcrustal earthquakes	**ATD**				1720	4.4	1381

system, parallel to which the Pacific plate moves N-NW relative to the North American plate at a rate of approximately 50 mm/year. The plate boundary transits from transformational to convergent along a zone extending between 57.5°N 137°W and 59°N 145.5°W, in which the Yakutat terrane accretes to the North American plate and complicates the interaction between the two plates; the boundary then continues westwards as the Aleutian Arc and Trench system. Landward of the QC–F system, and apparently related to the plate boundary, lies the right-lateral Denali transform fault. This is an arcuate feature running in a northwesterly direction for approximately 750 km, from about 59°N, 135.3°W to about 63.5°N 147°W; it then bends westwards and continues almost parallel to the plate boundary for an additional 500 km, to approximately 63°N, 155.2°W. The Aleutian Arc and Trench extends for approximately 3400 km, from the northern end of the Queen Charlotte–Fairweather fault system in the east (near 58.5°N, 137°W), to a triple junction with the Ulakhan Fault and the northern end of the Kuril–Kamchatka Trench in the west (near 56°N, 196°W). Westward of the Alaska Peninsula (Unimak Pass, 55.7°N, 164°W), it transits from continental in the east to intraoceanic in the west. Subduction along the Arc generates the Aleutian Volcanic Arc that extends as far as 182°W. The motion of the Pacific plate is always to the N-NW but due to the arcuate geometry of the trench, the relative velocity vector of the convergence changes from almost trench-normal in the east (Gulf of Alaska) to almost trench-parallel in the west. Along the continental part of the subduction the rate of convergence varies from 56 mm/year in the east (Gulf of Alaska), to 63 mm/year in the west (near Unimak Pass); along the oceanic part of the subduction the rate varies from 63 mm/year in the east to 74 cm/year in the west (e.g., DeMets and Dixon, 1999).

For the most part, seismicity in Alaska can be attributed to the plate boundary between the Pacific and North American plates. Most of the seismic energy is released by large events that rupture large segments of the boundary and accommodate most of the motion between the two plates. Within the North American plate (Continental Alaska), the highest seismicity rates are observed in southern Alaska, parallel to the plate boundary and decrease northwards, away from it. Fault-plane solutions of moderate earthquakes in south-central, central, and northern Alaska typically exhibit strike—slip kinematics with northwesterly to northerly compressional axes, whereas solutions in west-central Alaska generally exhibit normal faulting with northerly oriented tensional axes. Thus, with the exception of west-central Alaska, both the distribution of earthquake activity and the available focal mechanisms are qualitatively consistent with the hypothesis that the seismicity of Continental Alaska originates in the interaction of the Pacific Plate and North American plates (e.g., Page et al., 1991 and references therein). Moreover, it appears that the plate boundary is not composed of a single fault system but involves several secondary faults, both seaward and landward of the primary boundary, which accommodate a small fraction of the relative plate motion. The Aleutian Arc and Trench system generates large numbers of earthquakes in the crust, as well as in the subducting and overriding plates. Additionally, many earthquakes are associated with the activity of the Aleutian Volcanic Arc. Most large earthquakes in the region have thrust mechanisms indicating that they occur on the plate interface. However, some shallow (<30 km) events have either strike—slip or normal faulting mechanisms. Most of the normal faulting events occurring in the Aleutian outer rise region are caused by the bending of the Pacific plate as it enters the trench, while most of the shallow strike—slip events are concentrated along the island axis.

The earthquake data utilized for the source areas of Continental Alaska and the Aleutian Arc were extracted from the regional earthquake database of the Alaska Earthquake Center (http://www.aeic.alaska.edu/html_docs/db2catalog.html) and comprise a total of 48,995 events recorded in the area 50°N to 70°N and 196°W to 126°W over the period 1968—2015. In the AEC catalogue the overwhelming majority of events are reported in the M_L magnitude scale. However, a significant number are reported *only* in the surface (M_S) and body wave (m_b) magnitude scales. On the bright side, another significant number is reported in multiple magnitude scales and, of these, 1715 are jointly reported in the M_L, M_S and m_b scales. It is, therefore, straightforward to generate calibration tables by which to convert M_S and m_b to M_L. This exercise was carried out by robust reweighted linear regression with a redescending bisquare influence function. The M_L—M_S relationship is shown in Fig. 11—6A and the resulting regression (calibration) formula is

$$M_L = (1.074 \pm 0.018) \times m_b - (0.4099 \pm 0.0942), \quad 4 \le m_b \le 7.2.$$

The M_L—m_b relationship is shown in Fig. 11—6B and the corresponding regression formula is

$$M_L = (0.712 \pm 0.013) \times M_S + (1.651 \pm 0.066), \quad 3.5 \le M_S \le 7.5.$$

The relationships between M_L—m_b and M_L—M_S are obviously linear so that the regression coefficients are rather precisely determined. Thus, acknowledging the problems associated

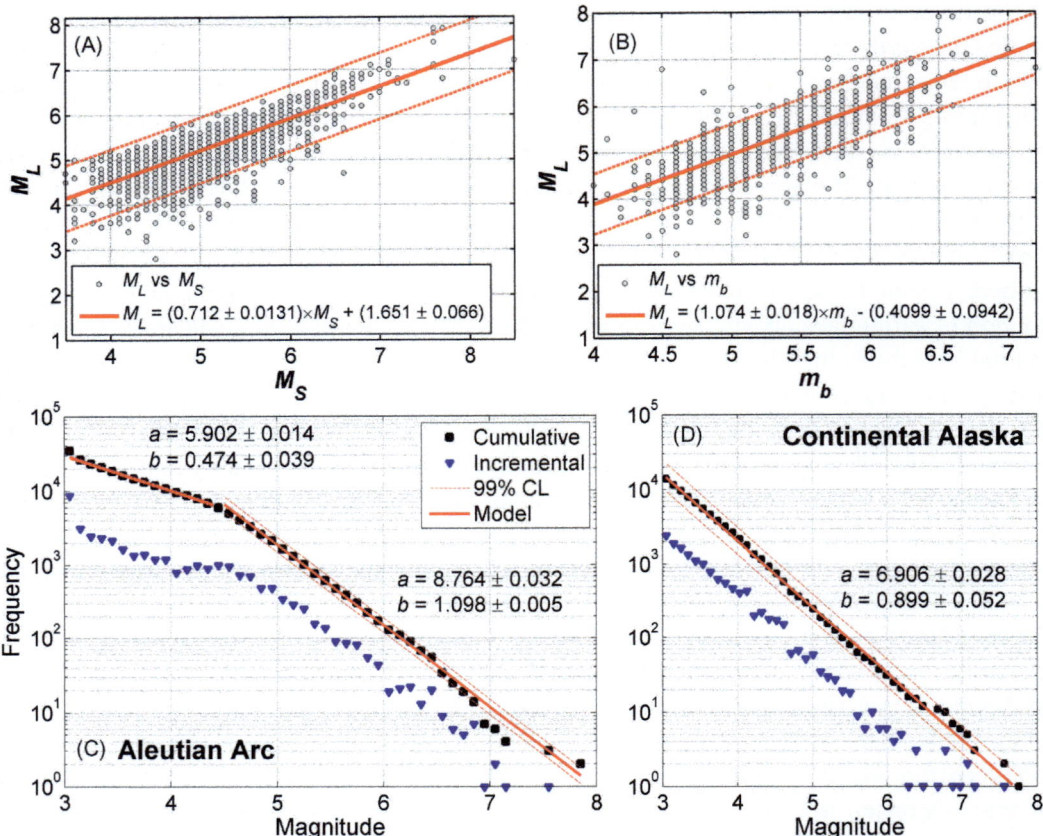

FIGURE 11–6 (A) Relationship between local and surface wave magnitude scales and (B) between the local and body wave magnitude scales, for the area of Alaska and the Aleutian Arc. Analysis based on 1715 events jointly reported in the M_L, M_S and m_b magnitude scales, in the catalogue of the Alaska Earthquake Center. The regression lines were fitted with robust linear least squares; broken lines mark the 95% confidence limits. (C) The frequency–magnitude distribution of seismicity along the Aleutian Arc and Trench. (D) As per (C) for continental Alaska. Down-pointing solid triangles represent the incremental distribution; solid squares represent the cumulative distribution; broken lines are 99% confidence limits.

with the saturation of the local and body wave scales at the large magnitude end of the spectrum, and assuming that both relationships can be linearly extrapolated to smaller magnitude scales, it is possible to construct a homogeneous version of the AEC catalogue with all events reported in the local magnitude scale.

The AEC catalogue presents a conundrum: Fig. 11–6C clearly shows that F–M distribution of seismicity recorded along the Aleutian Arc – as this is outlined in Fig. 11–5 – is bimodal, a feature not present in the seismicity of Continental Alaska (Fig. 11–6D). For magnitude scales between $M_{L}=3$ and 4.3 the b value is 0.47 and for $M_L \geq 4.4$ increases, almost abruptly, to 1.1. The origin of this bimodal distribution might be natural (different physical mechanisms operating at small and intermediate−large magnitude scales), although *b* values as low as 0.47 over so broad an area are not easy to explain. On the other hand, as can be seen in the incremental distribution (downward-pointing triangles in Fig. 11−6C) the escalation of frequency is faltering between $M_{L}=3.9$ and 4.3 (events missing) and there is a rather suspicious leap of about 5500 events between $M_{L}=3.0$ and 3.1 (event surplus), which is also difficult to explain naturally. Given, also, is the relative sparsity and almost one-dimensional geometry of the monitoring network along the Aleutians (see https://earthquake.alaska.edu/network) and the difficulties associated thereof, with the detection of small earthquakes. Finally, it is not difficult to verify that bimodality is definitely more pronounced in the western (oceanic) part of the convergence (west of Unimak Pass), where the network is most sparse. As a result, we cannot be certain that the differences between the small and intermediate−large magnitude scales are natural and we cannot investigate this rather nontrivial issue in the space available here. As a consequence, and as far as the Aleutian Arc and Trench is concerned, we shall only consider the intermediate and large earthquake population ($M_L \geq 4.4$), for which the F−M distribution, albeit imperfect, does raise concerns about its constitution. It is apparent that in that area, the homogenized version of the AEC catalogue is complete for $M_L \geq 4.4$ (Fig. 11−6C). Conversely, in Continental Alaska we shall consider all earthquakes with magnitudes $M_L \geq 3$, for which the catalogue appears to be complete (Fig. 11−6D).

As is evident in the foregoing, seismogenesis in Alaska and the Aleutian Arc develops in a rather complex tectonic background, extends over a very large area and range of depths and exhibits regional variation. For these reasons, it is not feasible to thoroughly examine the entire area of Continental Alaska and the Aleutian Arc. Rather, in keeping with the objective of studying the statistical nature of seismicity along the Pacific Rim, we will limit our inquiry to the area of the principal tectonic feature of the Rim: the broader boundary between the North American and Pacific plates. In this area it is possible to distinguish three classes of earthquake activity: (1) crustal earthquakes in Continental Alaska primarily associated with the eastern transformational plate margin, (2) crustal earthquakes along the Alaskan−Aleutian Arc primarily associated with the convergent plate margin, and (3) subcrustal earthquakes along the Alaskan−Aleutian Arc associated with the subducting slab. This provides an opportunity to study and compare the statistics of earthquakes generated in different seismotectonic settings, environmental (crust vs subducting slab) and boundary conditions (free in the crust vs fixed in the slab), and to inquire whether these differences affect the dynamic expression of the fault network.

Following the above reasoning, we will inquire into the statistical nature of crustal seismicity along the eastern transformational plate boundary defined by the Queen Charlotte−Fairweather and Denali faults, in which we include the transitional zone spanned by the Yakutat Terrane, as well as the Wrangelian Composite Terrane. This area will henceforth be referred to as the Queen Charlotte−Denali zone, or QCD. We will also inquire into the statistical nature of seismicity observed along the convergent plate boundary but in this case we will conduct a separate analysis of crustal and subcrustal earthquakes by crudely distinguishing them according to the depth of the Mohorovičićdiscontinuity; this is approximately

40 km beneath the Yakutat Terrane (Christeson et al., 2013) and approximately 38.5 km along the Aleutian Arc (Janiszewski et al., 2013). The crustal seismicity and earthquake catalogues will henceforth be referred to as ATC (Aleutian Trench Crustal), while their subcrustal counterpart will be referred to as ATD (Aleutian Trench Deep). The epicentral distributions of the QCD, ATC and ATD earthquakes are illustrated in Fig. 11−5; information about the respective catalogues is summarized in Table 11−1.

11.3.3 Declustering

The question of whether the background seismogenetic process is fundamentally random or correlated is open to debate and can be answered by analysing reduced versions of the earthquake catalogues, in which the aftershock sequences have been eliminated in as optimal a way as possible. The process of reducing an earthquake catalogue so as to separate background and foreground events is referred to as *declustering*. An excellent review of declustering methods and their evolution from deterministic (e.g., Gardner and Knopoff, 1974; Reasenberg, 1985) to stochastic (e.g., Zhuang et al., 2002; Marsan and Lengliné, 2008), is given in van Stiphout et al. (2012). The deterministic methods identify foreground events on the basis of temporal and spatial windows that scale with the magnitude of the mainshock while ignoring aftershocks triggered by aftershocks (higher-order events). The stochastic methods allow for multiple generations of aftershock triggering within a cluster and use Omori's law as a measure of the temporal dependence of aftershock activity. Both approaches ignore fault elongation and assume circular (isotropic) spatial windows. Stochastic declustering was introduced by Zhuang et al. (2002); their approach improves on previous methods because it optimizes the temporal and spatial window in which to search for aftershocks by fitting an ETAS model to the earthquake data. Furthermore, instead of assigning aftershocks to arbitrarily chosen mainshocks, it assigns each earthquake in the catalogue with a probability that it is an aftershock of its predecessor so that all earthquakes may be possible mainshocks to their short-term aftereffects. Marsan and Lengliné (2008) carried stochastic declustering one step further by introducing a generalized triggering process that does not require some underlying earthquake occurrence model; nevertheless, they still assume that background earthquakes occur at a constant and spatially uniform rate. Herein we have chosen to implement the method of Zhuang et al. (2002) because it has an additional and significant advantage for our objectives: it is a *paradigmatic* realization of the self-excited Poisson process. Thus, if the background seismicity obeys Boltzmann−Gibbs statistics, this method should be able to extract a nearly random background process against which to test the alternative hypotheses. If it is does not, the argument in favour of a non-Poissonian background would be stronger.

The Zhuang et al. (2002) method utilizes the following form of the normalized probability that one event will occur in the next instant, conditional on the history of the seismogenetic process:

$$\lambda(t,x,y,M|H_t) = \mu(x,y,M) + \sum_{i:t_i<t} \kappa(M_i) \cdot g(t-t_i) \cdot f(x-x_i, y-y_i|M_i) \cdot j(M|M_i)$$

where λ is the conditional intensity on the history of observation H_t until time t, $\mu(x, y, M)$ is the background intensity, $\kappa(M)$ is the expected number of foreground events triggered by a magnitude M mainshock and $g(t)$, $f(x,y|M_i)$ and $j(M|M_i)$ are, respectively, the probability distributions of the occurrence time, the location and the magnitude events triggered by a mainshock of magnitude M_i. If the catalogue is arranged in chronological order, then the probability of an event j having been triggered by an event $i < j$ can be estimated from the occurrence rate at its occurrence time and location as

$$p_{i,j} = \frac{\kappa(M_i) \cdot g(t_j - t_i) \cdot f(x_j - x_i, y_j - y_i|M_i)}{\lambda(t_j, x_j, y_j|H_t)}$$

and the probability that event j is an aftershock is given by

$$p_j = \sum_{i=1}^{j-1} p_{i,j}$$

Conversely, the probability that event j is background is given by

$$\phi_j = 1 - p_j = \frac{\mu(x_j, y_j|H_t)}{\lambda(t_j, x_j, y_j|H_t)}$$

The algorithm runs iteratively through the catalogue and by assigning probabilities $p_{i,j}$, p_j and ϕ_j to the jth event, generates the foreground subprocess associated with the ith event (i.e., its aftershock sequence). It thus separates the catalogue into a number of subprocesses whose initiating events comprise the background. As a general rule, events with $\phi_j \leq 50\%$ are considered to be foreground.

Since the output of stochastic declustering is not unique, it is useful to use the probabilities $p_{i,j}$ and ϕ_j to generate different realizations of the declustered catalogue at different probability levels and use them to test hypotheses associated with background seismicity and/or aftershock clustering. Our analysis herein will be based on the assumption that events with probability $\phi_j \geq 70\%$ are likely to be background. Results obtained from the NESP analysis of higher probability levels will not be shown here as they do not offer significant additional information with respect to the objectives of this chapter.

The results of our declustering exercise are summarized in Table 11–1 and illustrated in Fig. 11–7, where the cumulative earthquake counts of the full earthquake catalogues are shown with solid lines and the corresponding cumulative counts of their declustered versions with broken lines. It is apparent that all catalogues declustered at the $\phi \geq 70\%$ level are *almost* free of the time-local rate jumps that indicate the presence of aftershock sequences; therefore, they are fairly representative of the background process. It should be noted, however, that they are not always completely smooth and exhibit small fluctuations because a small portion of the remaining events are residuals of the foreground process.

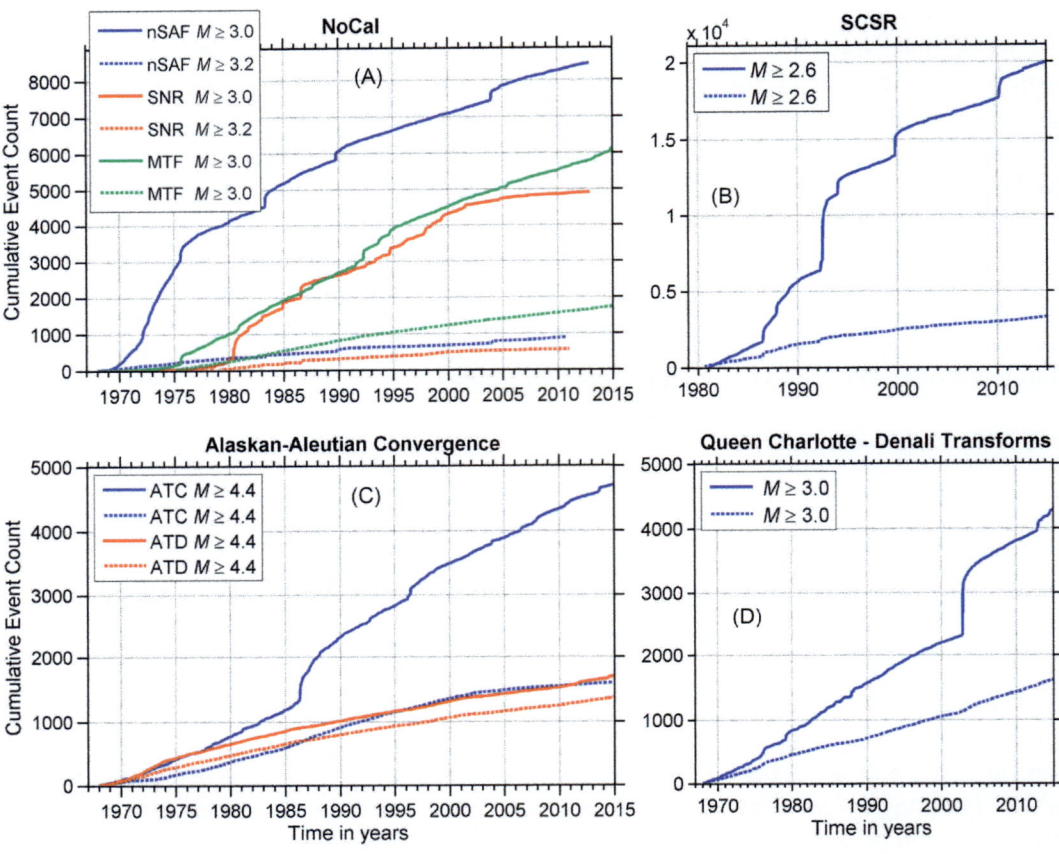

FIGURE 11–7 Cumulative event counts for the full (solid lines) and declustered (broken lines) earthquake catalogues we have used herein. (A) Full and declustered subcatalogues of North California (NoCal); see text for details. (B) As per (A) for the South California Seismic Region (SCSR). (C) As per (A) for the crustal (ATC) and subcrustal (ATD) catalogue subsets of the Aleutian Arc and Trench. (D) As per (A) for the Queen Charlotte–Fairweather and Denali zone of transform faults (Alaska).

11.4 Results

The analysis and appraisal of our results will be based on the fact that NESP predicts *correlation* (dependence) between successive earthquakes that involve long-range interaction, endows the seismogenetic system with memory and generates power-law statistical distributions of its dynamic parameters. The degree of correlation is measured by the entropic indices so that if $q \neq 1$, the system is nonextensive, whereas if $q \rightarrow 1$, the system is Poissonian (uncorrelated and memoryless). Because the appraisal of low-valued experimental realizations of q can be ambiguous for obvious reasons, we also require a measure (threshold) on the basis of which to confidently infer whether a seismogenetic system is nonextensive or Poissonian. Our answer to this problem is reported in Section 11.4.1.

11.4.1 Determination of Randomness Thresholds

In order to determine a threshold value above which it is safe to conclude that the temporal entropic index (q_T) indicates nonextensive seismogenetic processes, we apply Eq. (11.17) to the analysis of several background catalogues generated on the basis of the ETAS model: each of those catalogues should yield temporal entropic indices with an expectation value of unity. The synthetic catalogues were generated with the stochastic ETAS aftershock simulator program 'AFTsimulator' of Felzer (2007). The program uses the Gutenberg–Richter and Omori–Utsu laws to simulate the statistical behaviour background and foreground seismicity, and Monte Carlo methods to simulate background earthquakes, as well as multiple generations of aftershocks. Known mainshocks can be included as point or planar sources and background earthquakes are chosen randomly from observed or contrived spatial distributions (grids) of earthquake rates. This facilitates the generation of realistic synthetic background catalogues, consistent with the known long-term seismotectonic characteristics of a given area (for details see Felzer et al., 2002; Felzer and Brodsky, 2006). In our implementation of the AFTsimulator we have used the ETAS parameterizations for North and South California obtained (fitted) by declustering the NCSN and SCSN catalogues. We have also assumed a uniform background seismicity rate such that $b = 1$ and have set the maximum expected magnitude to be $M_{L=}7.2$, approximately the same as the maximum magnitudes observed in California during the 47-year period 1968–2015 (the Loma Prieta and Landers earthquakes of 1989 and 1992, respectively).

Fig. 11–8 illustrates results from NESP analysis of 40 synthetic background catalogues, 20 of which were compiled for the SCSR source area and 20 for the whole of North California (NoCal $\equiv$ nSAF + SNR + MFZ). Both sets of catalogues span a period of 47 consecutive years. Fig. 11–8A illustrates the variation of the mean values $\langle q_T \rangle$ and $\langle q_M \rangle$ computed from the analysis of the synthetic catalogues, together with their associated 3σ error margins, as a function of the threshold (cutoff) magnitude M_{th}. It is apparent that all $\langle q_T(M_{th}) \rangle$ are consistently lower than 1.1 without exception, so that $\max[\langle q_T(M_{th}) \rangle + 3\sigma] < 1.15$. Likewise, all $\langle q_M(M_{th}) \rangle$ exhibit an almost imperceptible variation around 1.5, so that $b_q \approx 1$, consistently with the assumption on which the synthetic ETAS catalogues were constructed. It is also apparent that the populations $\{q_T(M_{th})\}$ and $\{q_M(M_{th})\}$, from which $\langle q_T(M_{th}) \rangle$ and $\langle q_M(M_{th}) \rangle$ have been derived, are remarkably consistent: the 3σ error bars are generally very small and in many cases smaller than the size of the symbols representing the expectation values! Fig. 11–8B illustrates the variation of entropic indices computed by grouping the earthquakes of the synthetic catalogues according to interevent distance (Eq. 11.19) and modelling the conditional probability function expressed by Eq. (11.20). All results have been derived by considering earthquakes above a threshold magnitude $M_{th=}3.0$. As above, the figure shows mean values $\langle q_T(\Delta d) \rangle$ and $\langle q_M(\Delta d) \rangle$ with their associated 3σ error margins. All $\langle q_T(\Delta d) \rangle$ are consistently low for all interevent distance groups, so that $\max[\langle q_T(\Delta d) \rangle + 3\sigma] \leq 1.2$, while $\langle q_M(\Delta d) \rangle$ are also very stable and exhibit small fluctuations around 1.5, so that $b_q \to 1$ as expected.

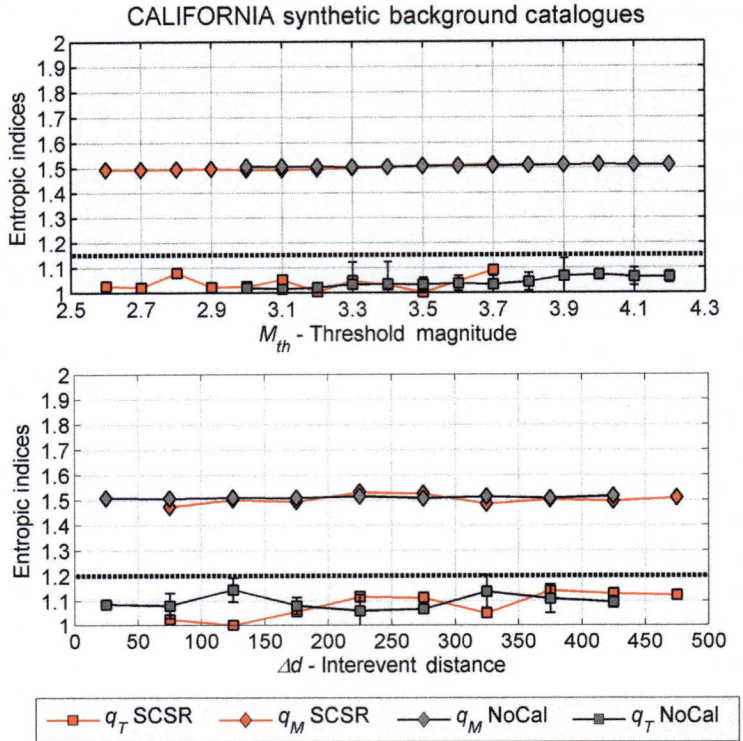

FIGURE 11–8 NESP analysis of 20 ETAS synthetic background catalogues constructed with the characteristics of South Californian (SCSR) and 20 constructed with the characteristics of North Californian seismicity (NoCal). Both catalogues span a period of 47 consecutive years. (A) Mean values $\langle q_T(M_{th})\rangle$ and $\langle q_M(M_{th})\rangle$ of the entropic indices and associated 3σ error margins, computed for different threshold magnitudes (M_{th}). The horizontal dashed line at $q_{T}=1.15$ marks the threshold above which $q_T(M_{th})$ can be *safely* assumed to indicate non-Poissonian processes. (B) Mean values $\langle q_T(\Delta d)\rangle$ and $\langle q_M(\Delta d)\rangle$ with associated 3σ error margins computed for different interevent distance groups Δd. The horizontal dashed line at $q_{T=}1.2$ marks the threshold above which $q_T(\Delta d)$ can be *safely* assumed to indicate non-Poissonian processes.

The above exercise was conducted with several random background catalogues generated on the basis of the ETAS model. In consequence, it can be concluded that the analytical procedure described in Sections 11.2.2 and 11.2.3 yields stable magnitude entropic indices and proxy *b-values* (b_q) absolutely consistent with the assumptions on which the synthetic ETAS catalogues were constructed. More importantly, however, the results establish that the systematic observation of experimental values $q_T(M_{th}) \geq 1.15$ and $q_T(\Delta d)\rangle > 1.2$ would be compelling evidence of nonextensive seismogenetic dynamics.

11.4.2 Entropic Indices

In order to conduct as comprehensive an analysis as possible, we analyse full and declustered catalogues of the source areas specified in Section 11.3.1. Basic information about the

Table 11–2 Range of Variation of the Entropic Indices Obtained From the Earthquake Source Areas of California

		$q_T(M_{th})$ Range	$q_T(\Delta d)$ Range		$q_M(M_{th})$ Range	$b_q(M_{th})$ RANGE	$q_M(\Delta d)$ Range	$b_q(\Delta d)$ Range
			$\Delta d < 100$ km	$\Delta d > 100$ km				
Full	SCSR	1.08–1.77	1.68–1.22	1.14–1.39	1.48–1.58	1.08–0.75	1.47–1.52	1.16–0.93
	nSAF	1.18–1.37	1.64–1.48	1.22–1.41	1.48–1.58	1.08–0.75	1.46–1.52	1.17–0.92
	SNR	1.32–1.52	1.46	1.54–1.68	1.52–1.56	0.93–0.80	1.46–1.58	1.16–0.78
	MFZ	1.06–1.32	1.23–1.40	1.23–1.33	1.56–1.60	0.78–0.67	1.53–1.57	0.89–0.75
Declustered $\phi \geq 70\%$	SCSR	1.51–1.42	1.64–1.42	1.42–1.52	1.51–1.53	0.96–0.89	1.49–1.51	1.08–0.96
	nSAF	1.48–1.23	–	–	1.51–1.52	0.96–0.92	–	–
	SNR	1.56–1.81	–	–	1.51–1.55	0.96–0.82	–	–
	MFZ	1.06–1.42	1.13–1.42	1.10–1.38	1.49–1.52	0.92–1.04	1.51–1.56	0.96–0.79

Table 11–3 Range of Variation of the Entropic Indices Obtained From the Earthquake Source Areas of Continental Alaska and Aleutian Arc and Trench

		$q_T(M_{th})$ Range	$q_T(\Delta d)$ Range		$q_M(M_{th})$ Range	$b_q(M_{th})$ Range	$q_M(\Delta d)$ Range	$b_q(\Delta d)$ Range
			$\Delta d < 150$ km	$\Delta d > 150$ km				
FULL	QCD ($M_{th} \geq 3$)	1.12–1.44	1.31–1.44	1.26–1.51	1.59–1.61	0.70–0.63	1.56–1.61	0.77–0.62
	ATC ($M_{th} \geq 4.4$)	1.10–1.31	1.31–1.34	1.00–1.33	1.52–1.47	0.92–1.14	1.54–1.51	0.84–0.96
	ATD ($M_{th} \geq 4.4$)	1.00–1.15	1.17	1.10–1.32	1.53–1.46	0.88–1.16	1.51–1.56	0.96–0.79
DECLUSTERED $\phi \geq 70\%$	QCD ($M_{th} \geq 3$)	1.1–1.38	NA	1.34–1.37	1.51–1.55	0.96–0.84	1.51–1.53	0.94–0.87
	ATC ($M_{th} \geq 4.4$)	1.29–1.52	NA	1.45–1.65	1.47–1.40	1.11–1.50	1.52–1.48	0.91–1.07
	ATD ($M_{th} \geq 4.4$)	1.00–1.10	NA	1.04–1.17	1.50–1.47	0.98–1.11	1.52–1.50	0.92–1.0

relevant earthquake catalogues is provided in Table 11–1. The analysis focuses on the variation of the entropic indices with respect to threshold magnitude, (M_{th}) and interevent distance (Δd). The results are summarized in Tables 11–2 and 11–3 and displayed in Figs. 11–9–11–15. In order to maintain experimental rigour, estimation of the entropic indices is *not* performed for catalogue subsets containing *less than* 300 events and results are *not* considered and displayed *unless* associated with a goodness of fit (R^2) *better* than 0.97.

11.4.2.1 California Full Catalogues

As can be seen in Fig. 11–9A, all magnitude entropic indices are quite stably and consistently determined. $q_M(M_{th})$ functions computed from the nSAF, SNR and SCSR catalogues are very comparable and vary between 1.48 at $M_{th} = 3.0$ ($b_q = 1.08$) and ~1.5 ($b_q = 1$) at $M_{th} = 3.5$, steadily increasing thereafter to 1.58 at $M_{th} = 4.3$ ($b_q = 0.72$). The entropic index q_M, like the *b*-value to which it is related, represents the scaling of the size distribution of earthquakes. Here it indicates a *subextensive* scalefree process, possibly associated with a change in the size distribution and spatial clustering of intermediate–large magnitude events that appears to become increasingly *tighter*. Notably, analogous changes are conspicuous in

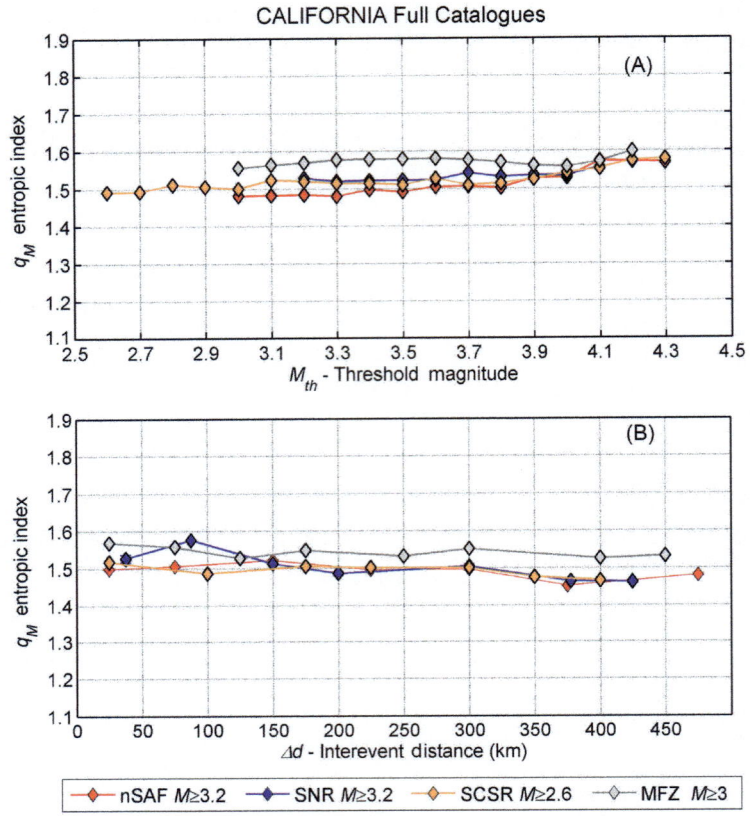

FIGURE 11–9 Analysis of the magnitude entropic index q_M for the full catalogues of the seismogenetic systems of California (see Section 11.3.1.1 and Fig. 11–4A for details). Panel (A) illustrates the variation of q_M as a function of threshold magnitude (M_{th}). Panel (B) illustrates the variation of q_M with interevent distance Δd; binning schemes vary so as to maximize statistical rigour. Confidence limits of 95% are also drawn but are not always visible as they can be smaller than the symbols.

conventional frequency–magnitude plots, where they appear to commence after about $M5$. Finally, for the MFZ area, q_M (M_{th}) is estimated at the markedly higher level of $1.56–1.60$ with a mean of $\langle q_M \rangle$-MFZ = 1.57 $\pm$ 0.01, so that $b_q(M_{th}) \in (0.78, 0.67)$. Such q_M and b_q values indicate rather high levels of clustering in the MFZ active fault network.

The variation of q_M with interevent distance Δd is shown in Fig. 11–9B. It is apparent that all $q_M(\Delta d)$ functions are rather stable over all interevent distances. As before, $q_M(\Delta d)$ functions computed for the nSAF, SNR and SCSR full catalogues are comparable and generally vary between 1.46 and 1.52 so that $b_q(\Delta d)$ varies between 1.17 and 0.92. Changes in scaling such as those observed in Fig. 11–9A are not evident because the threshold magnitude used in these calculations is considerably lower than the threshold of the changes. Finally, for the MFZ catalogue $q_M(\Delta d)$ is again higher than in all previous areas, as it varies between 1.57 and 1.53, so that $b_q(\Delta d) \in (0.75, 0.89)$. This shows that the high level of clustering

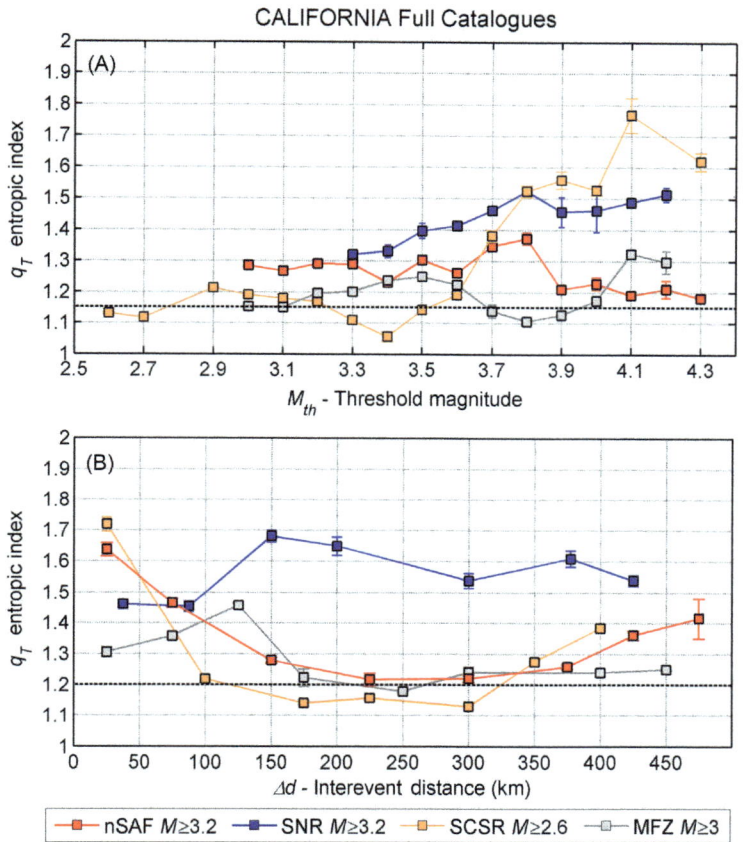

FIGURE 11–10 Analysis of the temporal entropic index q_T for the full catalogues of the seismogenetic systems of California (see Section 11.3.1 and Fig. 11–4A for details). Panel (A) illustrates the variation of q_T as a function of threshold magnitude (M_{th}). Panel (B) illustrates the variation of q_T with interevent distance Δd; binning schemes vary so as to maximize the statistical rigour. Error bars represent 95% confidence limits; they are not always visible as they are frequently smaller than the symbols.

inferred for the MFZ fault network from the analysis of Fig. 11–9A persists over distances of at least 400 km.

The variation of the temporal entropic index with threshold magnitude is shown in Fig. 11–10A. Let us begin with the results from the full SCSR catalogue. It is apparent that $q_T(M_{th})$ is lower than 1.2 at small magnitude scales, but for $M_{th} > 3.4$ increases steadily and steeply to higher than 1.6 at $M_{th} \geq 4.2$. Taken over the *entire* SCSR area, small earthquakes appear to be uncorrelated, possibly because very small events may be concurrently spawned by different parental earthquakes at different distant locations of an extended seismogenetic area; many of these events have no causal relationship and when mixed and chronologically ordered in a catalogue, they may randomize the statistics of interevent times. If this interpretation is correct, it is all the more significant to point out that the increase in correlation with

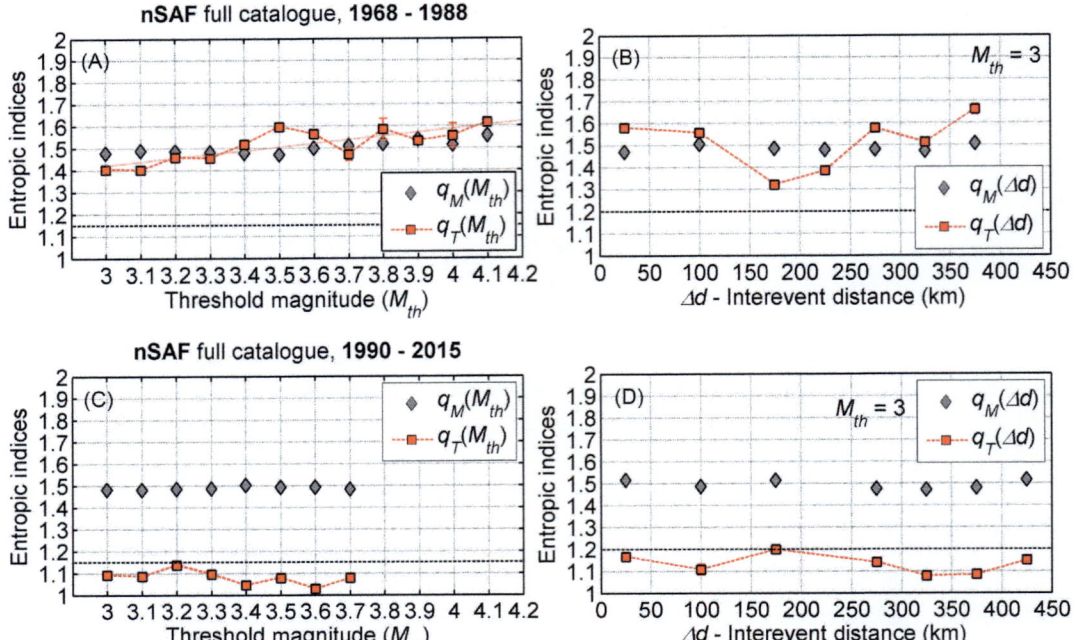

FIGURE 11–11 Analysis of the full nSAF catalogue for the periods 1968–88 (top row) and 1990–2015 (bottom row). Panels (A) and (C) illustrate the variation of the entropic indices with threshold magnitude (M_{th}). Panels (B) and (D) illustrate the variation of the entropic indices with interevent distance (Δd). In all cases error bars represent 95% confidence intervals.

magnitude — which involves faults distributed over the entire seismogenetic area — is compelling evidence of operational long-range interaction! The results obtained from the Sierra Nevada–Walker Lane (SNR) full catalogue are very similar, although here the increase in $q_T(M_{th})$ with magnitude is smoother and milder than in SCSR: the estimates of the temporal entropic index begin at the certainly higher level of 1.32 for $M_{th}=3.3$ and ends at the level of 1.51 for $M_{th} = 4.2$. Thus, the SNR system, which also bears evidence of long-range interaction, appears to exist in a state of correlation stronger than SCSR.

Results from the analysis of the nSAF and MFZ full catalogues are clearly different. In MFZ, $q_T(M_{th})$ fluctuates around 1.2 so that $\langle q_T(M_{th}) \rangle = 1.2 \pm 0.067$, but increases to 1.32 at larger threshold magnitudes ($M_{th} \geq 4.1$) exhibiting weak, albeit persistent, overall correlation. In nSAF, $q_T(M_{th})$ is stably determined around a mean value of 1.29 $\pm$ 0.04 for $M_{th} \leq$ 3.8, but decreases rapidly to 1.2 $\pm$ 0.02 for $M_{th} > 3.8$. It may come as a 'surprise' that the behaviour of the temporal entropic index of nSAF is quite unlike that of SCSR and SNR: it seems to imply that in adjacent 'sibling' tectonic settings, there can be fault systems simultaneously operating at very different levels of self-organization. This has prompted further scrutiny of the nSAF data set, whose results are presented below; as it turned out, there's more to this than meets the eye.

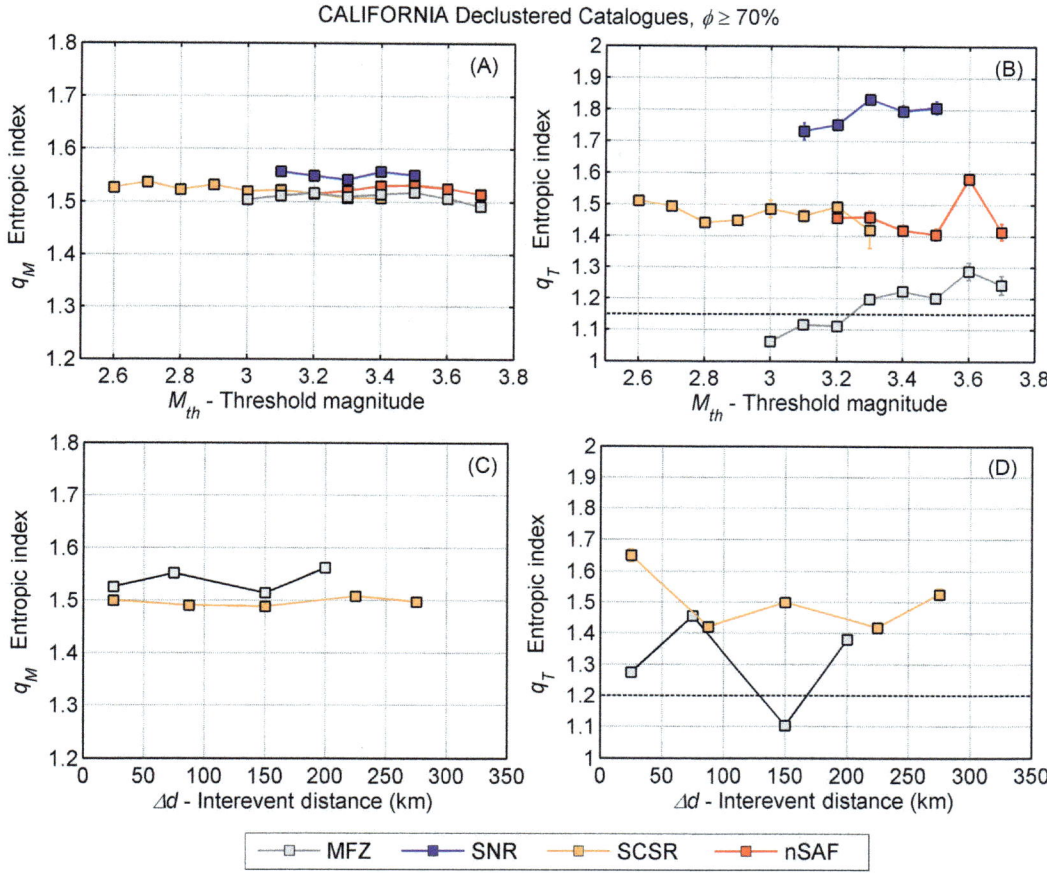

FIGURE 11–12 Analysis of the magnitude (q_M) and temporal (q_T) entropic indices for the *declustered* catalogues of the seismogenetic systems of California (see Section 11.3.1.1 and Fig. 11–4B for details). (A) Variation of q_M with threshold magnitude M_{th}. (B) Variation of q_T with threshold magnitude (M_{th}). (C) Variation of q_M with interevent distance Δd. (D) Variation of q_T with interevent distance Δd. In (C) and (D), binning schemes vary so as to maximize statistical rigour. Error bars represent 95% confidence limits; they are not always visible as they are frequently smaller than the symbols.

The variation of the entropic indices with earthquakes grouped according to *interevent distance* (Δd) is shown in Fig. 11–10B. When the analysis is carried out in this mode, it is expected that $q_T(\Delta d)$ will exhibit higher values at interevent distances shorter than 100 km due to the dominant effect of near-field interactions in aftershock sequences. Such behaviour is observed in SCSR and nSAF for which $q_T(\Delta d < 50$ km$)$ was determined to be 1.72 and 1.64, respectively. At longer interevent distances the results indicate moderate to weak correlation. In SCSR, $q_T(\Delta d > 100$ km$)$ varies from 1.14 to 1.39 with a mean value of 1.22 ± 0.1 and exhibits a clear tendency to increase for $\Delta d \geq 300$ km. In nSAF, $q_T(\Delta d > 100$ km$)$ varies from 1.22 to 1.41 with a mean value of 1.29 ± 0.08; it also tends to increase after 300 km. In

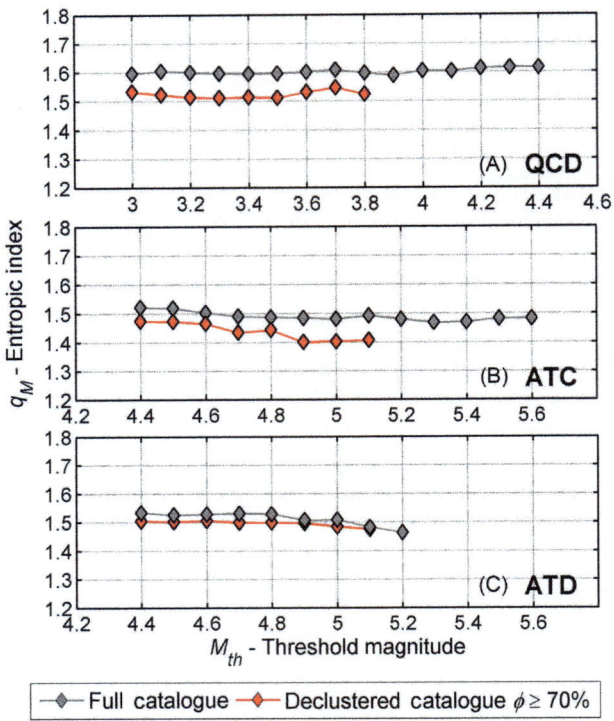

FIGURE 11–13 Analysis of the magnitude entropic index q_M versus threshold magnitude for full (grey) and declustered (red) earthquake catalogues along the Pacific–North American plate boundary in Alaska and the Alaskan–Aleutian Arc (see Section 11.3.1.2 and Fig. 11–5 for details). Panel (A) illustrates results from the Queen Charlotte–Fairweather–Denali transform zone (QCD). Panel (B) is the same for crustal seismicity in the Alaskan–Aleutian Trench (ATC). Panel (C) is the same for subcrustal seismicity of the Alaskan–Aleutian Wadati–Benioff zone (ATD). Confidence limits of 95% are also drawn but are not always visible as they are usually smaller than the symbols.

SNR, the correlation is merely significant ($q_T = 1.46$) at interevent distances shorter than 100 km, but increases to *strong* ($q_T > 1.54$) for all Δd longer than that. It is quite apparent that the SCSR and nSAF systems, which effectively are segments and branches of the SAF, are rather similar in their behaviour. The SNR system behaves in an opposite sense, which indicates that earthquake activity there, including aftershock sequences, is basically controlled by long-range interaction. Finally, in MFZ, $q_T(\Delta d)$ is only moderate (1.3–1.35) at short interevent distances, increasing to 1.46 in the interval 50–150 km, only to decrease again to the level of 1.25 for $\Delta d > 150$ km (moderate long-range correlation).

Let us now focus on nSAF in an attempt to explain the divergent (with respect to SCSR and SNR) behaviour observed in Fig. 11–10A. Fig. 11–11A,B illustrates the analysis of an nSAF subset catalogue spanning the period 1968–31 December 1988. Fig. 11–11C,D are the same for 1 January 1990–2015. The year not taken into consideration (1989) is the one leading to the M7 Loma Prieta earthquake of 17 October 1989 and including the bulk of its

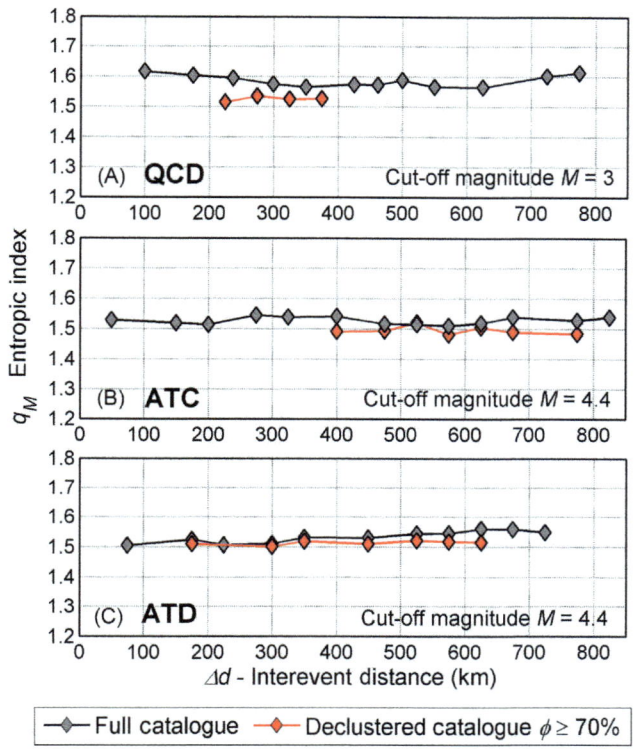

FIGURE 11–14 As per Fig. 11–13 but for q_M versus interevent distance Δd; binning schemes vary in order to maximize statistical rigour.

aftershock sequence. Prior to 1989, the full nSAF subcatalogue contains 5738 events while, after 1989, it has only 2862. It is interesting to observe (and certainly worthy of investigation), that during the first 20-year period, the full nSAF catalogue contains almost twice the number of events in comparison to the second 25-year period, meaning that there are significant differences in productivity rates. As is evident in Fig. 11–11A, for the period 1968–1988, $q_M(M_{th})$ exhibits a quasilinear trend from 1.48 at $M_{th} = 3$ to 1.56 at $M_{th} = 4.1$. In Fig. 11–11C, this trend has disappeared and $q_M(M_{th})$ seems to have stabilized just below the value of 1.5. However, because the number of earthquakes available for analysis at $M_{th} > 3.7$ is insufficient, it is not certain whether the 'trend' has altogether disappeared, or is simply unobservable. The estimation of entropic indices with respect to interevent distance is limited to $\Delta d < 450$ km due to the size of the SNR area. Still, one may observe that $q_M(\Delta d)$ is rather stably determined for both periods, slightly fluctuating about 1.5 (Fig. 11–11B,D). The temporal entropic index, however, is very different. For 1968–88, $q_T(M_{th})$ it behaves *exactly* like its SCSR and SNR 'siblings': it displays an upward linear trend, from 1.4 for $M_{th} = 3$ to higher than 1.6 for $M_{th} = 4.1$, at an average rate of 0.17 per magnitude unit obtained by fitting a straight line to the data (Fig. 11–11A). This indicates *strong*

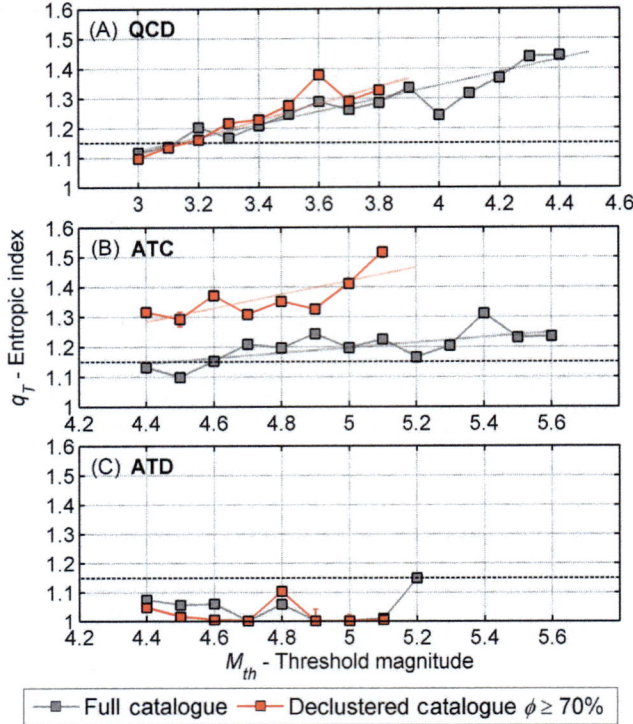

FIGURE 11-15 Analysis of the temporal entropic index q_T versus threshold magnitude for the full (grey) and declustered (red) earthquake catalogues along the Pacific–North American plate boundary in Alaska and the Alaskan–Aleutian Arc (see Section 11.3.1.2 and Fig. 11–5 for details). Panel (A) illustrates results from the Queen Charlotte–Fairweather–Denali transform zone (QCD). Panel (B) is the same for crustal seismicity of the Alaskan–Aleutian Trench (ATC). Panel (C) is the same for subcrustal seismicity in the Alaskan–Aleutian Wadati–Benioff zone. Confidence limits of 95% are also drawn but are not always visible as they are usually smaller than the symbols.

correlation and long-range interaction, particularly at the larger threshold magnitudes. Significant strong correlation over *all* ranges is also observed in Fig. 11–11B, where $q_T(\Delta d)$ varies from 1.58 to 1.55 for $\Delta d < 100$ km and consistently *increases* from 1.39 for 100 km $> \Delta d > 200$ km, to 1.66 for $\Delta d > 300$ km. Conversely, for the period 1990–2015, $q_T(M_{th})$ is consistently lower than 1.15 so that $\langle q_T(M_{th}) \rangle = 1.08 \pm 0.03$ (Fig. 11–11C). Likewise, $q_T(\Delta d)$ is consistently lower than 1.2 so that $\langle q_T(\Delta d) \rangle = 1.13 \pm 0.044$ (Fig. 11–11D). Thus, after 1990 nSAF turns out to be practically Poissonian.

The results above indicate that complexity/criticality may be dynamic/evolutionary and not stationary in the typical SOC sense. The same results indicate that if criticality had indeed been the cause of strong correlation in nSAF prior to 1989, it was probably not of the critical point 'variety' because there has never been unequivocal observation of CP point behaviour (accelerating seismic release rates) leading to the Loma Prieta event. It is also hard to imagine that the dramatic reduction in correlation and seismicity rates after 1990 is unrelated to aftereffects of the Loma Prieta event. Accordingly, such dynamic changes in the level of correlation may provide evidence as to how criticality develops, waxes and wanes, therefore as to the nature of a fault network in which criticality *can* develop. Additional, more detailed discussion on this subject is included in Section 11.5.

11.4.2.2 California Declustered Catalogues

One main objective of the present work is to investigate whether background seismicity is generated by non-Poissonian dynamic processes. Therefore, we examine declustered realizations of the earthquake catalogues, in which aftershock sequences have been removed by the stochastic declustering method of Zhuang et al. (2002) at the $\phi \geq 70\%$ probability level. It is important to note that the populations of earthquakes available for analysis after declustering is not always sufficient to maintain statistical rigour in the estimation of entropic indices. Accordingly, we reiterate that for the sake of experimental rigour, analysis is *not* performed for catalogue subsets containing *less* than 300 events and results are *not* considered, displayed and tabulated if associated with goodness of fits lower than, or equal to, 0.97. The results are illustrated in Fig. 11–12. Specifically, the variation of q_M and q_T with threshold magnitude M_{th} is shown in Figs. 11–12A and B, respectively, while the variation of q_M and q_T with interevent distance Δd is shown in Figs. 11–12C and D.

In Fig. 11–12A, it is apparent that $q_M(M_{th})$ does not have any trait worth commenting on. It should be noted, however, that results could not be obtained for $M_{th} > 3.7$, therefore it is not known whether q_M would behave as per the full catalogues. It is also worth noting that on removing dependent events, the $q_M(M_{th})$ determined for MFZ reduces to a mean value of 1.51 ± 0.008 indicating that the high level of active fault clustering observed in the full catalogue reduces to average levels.

Turning to the analysis of temporal entropic indices, in Fig. 11–12B we note that for SCSR, $q_T(M_{th})$ fluctuates smoothly and very stably around a mean value of 1.47 ± 0.03 which, for the interval $M_{th} \in [2.5, 3.7]$, is significantly higher than the mean value of 1.17 ± 0.083 observed in the full catalogue (Section 11.4.2.1). This is clear evidence of significantly correlated background seismicity. The same observation can be made in nSAF, where $q_T > 1.4$ for $M_{th} \in [3.2, 3.7]$ and $\langle q_T(M_{th}) \rangle = 1.45 \pm 0.07$, as opposed to the mean value of 1.25 ± 0.064 obtained for the full nSAF catalogue in the same magnitude interval. This also points to a significantly correlated background along the central and northern segments of the SAF, despite the relaxation observed after the Loma Prieta event. In SNR, q_T is characterized by an increasing trend from 1.73 to 1.81. This also represents a surprisingly large increase in the level of correlation in background seismicity, given also that the analysis of the full catalogue also indicated a strongly correlated seismogenetic process. Finally, and almost opposite results described so far, the declustered MFZ catalogue exhibits an upward quasilinear trend from no correlation ($q_T = 1.06$) at $M_{th} = 3.0$, to weak correlation $\langle q_T \rangle \approx 1.26$ at $M_{th} \geq 3.6$.

Due to population statistics in the declustered catalogues, determination of entropic indices with respect to interevent distance could be reliably performed only for SCSR and MFZ. In Fig. 11–12C, it is apparent that $q_M(\Delta d)$ determinations for both catalogues are rather unremarkable: for SCSR they are very consistent so that $\langle q_M(\Delta d)\rangle = 1.497 \pm 0.008$ ($b_q = 1.01$), while for MFZ they fluctuate rather significantly, so that $\langle q_M(\Delta d)\rangle = 1.54 \pm 0.022$ ($b_q = 0.85$). Interesting observations can be made with regard to the temporal entropic index of the declustered SCSR catalogue only. Here, $q_T(\Delta d)$ varies from 1.65 for $\Delta d \leq 50$ km to 1.52 for $\Delta d < 150$ km and between 1.42 and 1.52 for $\Delta d \geq 150$ km. The latter is also significantly and remarkably higher that the corresponding variation of q_T in the full catalogue, thereby confirming the existence of a significant to strong long-range correlation in South California. Finally, and presumably due to population statistics, $q_T(\Delta d)$ determinations from the declustered MFZ catalogue are limited to $\Delta d < 250$ km; they do not exhibit a pattern and fluctuate around a mean value of 1.3 ± 0.153, possibly indicating a system weakly correlated over short and intermediate ranges.

The analysis so far so far has shown that on removing aftershock sequences, significantly higher correlation is observed in comparison to the full catalogues. In fact, the SNR fault network exhibits such high correlation, that any background earthquake anywhere in the system would appear to be able of influencing the occurrence of future events anywhere else. Albeit to a lesser extent, the same appears to hold true for the SCSR and nSAF fault systems, although a note of caution applies to the latter. The MFZ catalogue has shown evidence of significantly lower correlation in comparison to the other fault networks of California, especially at small cutoff magnitudes and short ranges. Another interesting observation is the rather higher values of the magnitude entropic index which may indicate increased clustering of the fault network, and lower crustal heterogeneity in that area.

11.4.2.3 North Pacific Rim

As explained in Section 11.3.1.2, we shall conduct a comparative analysis of the seismicity observed along the two major components of the boundary between the North American and Pacific plates: Queen Charlotte–Fairweather and Denali zone of transform faults (QCD) and the Aleutian Arc and Trench system (AT) formed by the northerly subduction of the Pacific plate under the North American plate. The earthquakes caused by the former system occur primarily in the schizosphere. The earthquakes caused by the latter occur both in the crust and below the crust in association with the Aleutian Wadati–Benioff zone. In such a regional tectonic setting, we take our enquiry one step further by attempting to examine whether the environment in which seismogenesis occurs (pressure, material homogeneity, boundary conditions, etc.) has an effect on the dynamic expression of the seismogenetic system. Accordingly, we divide the Aleutian Arc and Trench seismicity into *crustal* and *subcrustal* based on published estimates of the Mohorovičić discontinuity, and conduct our analysis on two data subsets henceforth to be referred to as ATC (crustal seismicity) and ATD (subcrustal seismicity). We examine the full catalogues, as well as versions of all catalogues declustered at the 70% level. The results are summary presented in Table 11−3 and Figs. 11−13−11−16.

Fig. 11−13 illustrates the variation of the magnitude entropic index q_M with respect to threshold magnitude. Focusing first on the QCD catalogues it is straightforward to observe that $q_M(M_{th})$ is stable and exhibits minimal variation (Fig. 11−13A). However, while the full

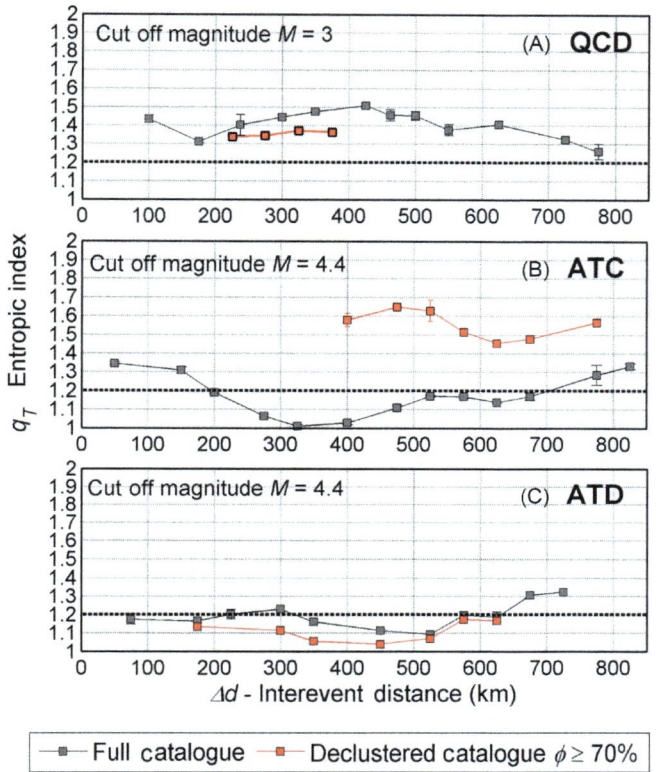

FIGURE 11–16 As per Fig. 11–15 but for q_T versus interevent distance Δd; earthquakes are binned according to Eq. (11.20) but binning schemes vary in order to maximize statistical rigour. Error bars (not always visible) represent 95% confidence limits.

catalogue yields a mean value of 1.60 $\pm$ 0.008 ($b_q = 0.67$), the declustered catalogue yields 1.52 $\pm$ 0.012 ($b_q = 0.92$). Analogous reduction of q_M with declustering has also been observed in SCSR and MFZ (California). As in those, activity in QCD is localized near the fault zones. Therefore, it is plausible that the reduction of q_M levels between the full and declustered catalogues implies a corresponding reduction in the level of activity localization from very high ($b_q = 0.67$) to nearly 'average' ($b_q = 0.92$), in direct consequence of clustered aftershock removal. This reduction appears to take place without the effects of scaling (hierarchical distribution) of the faults that does not change with magnitude.

We now turn our attention to the results of the Aleutian Arc and Trench catalogues which are shown in Figs. 11–13B and C. For the full ATC catalogue $q_M(M_{th})$ fluctuates slightly about the level of 1.5 so that $\langle q_M(M_{th}) \rangle = 1.49 \pm 0.016$ (Fig. 11–13B) and $b_q(M_{th})$ varies between 0.92 and 1.14. For the declustered ATC data, $q_M(M_{th})$ decreases smoothly from approximately 1.47 at $M_{th} = 4.4$, to approximately 1.40 at $M_{th=5}$, so that $\langle q_M(M_{th}) \rangle = 1.44 \pm 0.03$ (Fig. 11–13B); respectively, $b_q(M_{th})$ varies between 1.11 and 1.5. The small reduction of

q_M between the full and declustered catalogues might signify a corresponding reduction in the level of activity localization as per QCD but to a lesser degree. In Fig. 11–13C, stable and mutually consistent determination of $q_M(M_{th})$ is evident in both the full and declustered ATD catalogues: q_M fluctuates slightly about 1.5 so that the mean q_M is 1.51 $\pm$ 0.024 for the full catalogue and 1.49 $\pm$ 0.01 for the declustered (no statistical difference).

Fig. 11–14 illustrates the variation of q_M with respect to interevent distance $- q_m(\Delta d)$. The results obtained from the QCD catalogues are shown in Fig. 11–14A. It is straightforward to see that both the full and declustered catalogues yield results analogous to those shown for $q_M(M_{th})$ in Fig. 11–13A. For the full catalogue $q_M(\Delta d)$ is rather consistent over interevent distances of up to 800 km and varies between 1.56 and 1.61, so that $\langle q_T(\Delta d) \rangle =$ 1.58 $\pm$ 0.02; respectively, b_q varies in the range 0.62–0.77. In the declustered catalogue, earthquake populations sufficient for statistically significant results exist only for interevent distances between 150 and 450 km; these yield very consistent $q_M(\Delta d)$, which varies between 1.51 and 1.53, so that $\langle q_T(\Delta d) \rangle = 1.52$ $\pm$ 0.008 and $b_q(\Delta d)$ that varies between 0.94 and 0.87. It is again possible to observe a statistically significant reduction in the value of q_M, which can be interpreted in terms of a corresponding reduction in the level of activity localization upon aftershock removal. Focusing next on the Aleutian Arch and Trench, it is again straightforward to observe that the full and declustered catalogues have yielded very consistent, albeit *unremarkable*, results. In crustal seismicity (ATC), $q_M(\Delta d)$ is stable over interevent distances longer than 800 km and varies in the narrow range 1.48–1.54 for *both* full and declustered catalogues, so that b_q varies between 1.08 and 0.85 (Fig. 11–14B). The same is observed for subcrustal seismicity (ATD): $q_M(\Delta d)$ is stably determined over interevent distances longer than 700 km and varies between 1.50 and 1.56 for *both* the full and declustered catalogues, so that b_q varies in between 1.0 and 0.79 (Fig. 11–14C).

The analysis of the temporal entropic index with respect to threshold magnitude is shown in Fig. 11–15. Starting again with QCD, Fig. 11–15A illustrates the variation of $q_T(M_{th})$ for the full and declustered data sets. It is apparent that the temporal entropic index starts off low ($q_T \sim 1.1$), but demonstrates a steady linear increasing trend. For the full catalogue, it transcends the threshold of randomness at $M_{th} = 3.2$ and climbs to 1.44 at $M_{th} \geq 4.5$ (significant correlation); this variation can be fitted with a linear trend line giving an average rate of 0.22 per magnitude unit. For the declustered catalogue, q_T also transcends the randomness threshold at $M_{th} = 3.2$ and climbs to 1.33 at $M_{th} = 3.8$; the linear trend in this case has a rate of 0.29 per magnitude unit, noticeably higher than that of the full catalogue. It can also be seen that for all $M_{th} > 3.2$, the declustered catalogue q_T is consistently higher than the full catalogue q_T, indicating a more correlated background process. It is also worth remembering that a quasilinear increase of q_T with magnitude has been observed in the SCSR, SNR and pre-1989 nSAF catalogues and has been attributed to operational long-range correlation; therefore, the same interpretation should apply in the case of QCD.

Focusing now on the Aleutian Arc and Trench, Fig. 11–15B shows the variation of $q_T(M_{th})$ for the full and declustered datasets of crustal earthquakes (ATC). For the full catalogue, $q_T(M_{th})$ varies around 1.2 so that $\langle q_T(M_{th}) \rangle = 1.2$ $\pm$ 0.054. However, it can also be clearly seen that q_T increases steadily, from less than 1.15 at $M_{th=4.4}$ to over 1.2 for

$M_{th} \geq 5.1$, at an average rate 0.09 per unit magnitude (obtained by fitting a straight line to the data). It is also clearly seen that in the declustered catalogue, $q_T(M_{th})$ increases steadily from about 1.3 at $M_{th=}4.4$ to over 1.4 for $M_{th} \geq 5$ at a rate of 0.22 per magnitude unit, and the mean value also increases to 1.36 ± 0.07. As before, this implies a correlated background and points toward a long-range interaction. Exactly the opposite behaviour is observed in subcrustal seismicity (ATD). As evident in Fig. 11−15C, q_T is generally lower than 1.1 and on the basis of this evidence alone, the subcrustal fault network of the subduction zone would appear to be Poissonian.

In concluding the presentation of our results, Fig. 11−16 demonstrates the analysis of the temporal entropic index with respect to interevent distance. Fig. 11−16A illustrates results from the transformational plate boundary (QCD). A rather unexpected outcome is that inadequate earthquake populations prohibited the generation of dependable estimation of q_T at short interevent distances (less than 100−150 km) even for the full catalogue; it appears that even aftershocks are rather broadly spread out along the QCD fault zones. For interevent distances longer than 50 km, the full-catalogue $q_T(\Delta d)$ is rather significant: it always remains above the threshold of randomness, maximizing at ranges of the order 300 km to 600 km ($q_T > 1.45$), and thereafter slowly declining to moderate ranges of the order of 700 km ($q_T > 1.3$). Inadequate populations also did not allow determination of $q_T(\Delta d)$ from the declustered catalogue at distances shorter than 150 km and longer than 450 km. Yet, within this range $q_T(\Delta d)$ is consistently determined at the level 1.34−1.37, indicating moderate correlation. Given also the results obtained for $q_T(M_{th})$ in Fig. 11−15A, it can be concluded that within the period of observation, the QCD zone has existed in a persistent state of nonequilibrium.

Switching now to the analysis of the crustal seismicity along the Aleutian Arc and Trench, we note that for the full catalogue, weak to moderate correlation can be observed only at interevent distances shorter than 200 km and longer than 700 km; in all other cases q_T is lower than the threshold of 1.2 (Fig. 11−16B). However, declustering appears to unveil *strong* background correlation at long interevent ranges (300 km $< \Delta d <$ 800 km), where q_T varies between 1.45 and 1.65 and, notably, mirrors the variation of q_T in the full catalogue (Fig. 11−16B). Unfortunately, at short and intermediate ranges ($\Delta d <$ 300 km) q_T cannot be estimated due to dwindled earthquake populations (and consequent loss of statistical robustness). The analysis of subcrustal seismicity (ATD) shows nihil to marginal correlation over all interevent distances and up to 700 km (Fig. 11−16C). At the $M_{th=}4.4$ level, the full and declustered catalogues yield very comparable results. For the former, $q_T(\Delta d)$ determinations vary between 1.32 and 1.09 with a mean of 1.19 ± 0.07; and for the latter they vary between 1.04 and 1.17 with a mean of 1.11 ± 0.05. As per Fig. 11−15C, subcrustal seismicity appears to be Poissonian also with respect to interevent distance.

11.5 Discussion

The work reported herein begins with the question 'Are seismogenetic systems random or organized?' The question originates in a longstanding discourse between the two principal schools of thought (and epistemological paradigms) developed in the process of studying earthquake occurrence and quantifying the expectation of seismic activity. Accordingly, far from being purely academic, the problem of understanding the proper statistical nature of seismicity is also practical: the answer can have significant repercussions on forecasting intermediate-term earthquake hazards.

Seismicity comprises the superposition of a background process expressing the continuum of tectonic deformation, and a foreground process of prolific short-term activity associated with earthquake swarms or/and aftershock sequences. The first and historic school (doctrine) posits that background seismicity is produced by a self-excited conditional Poisson (point) process whose entropy is assumed to obey the Boltzmann–Gibbs formalism; background earthquakes are spontaneously and independently generated in the fault network and there is no interaction between faults, such that would influence their time and place of occurrence. The second and more recent doctrine posits that background seismicity is generated by a nonequilibrating fault network (system) in which background events are dependent due to correlations (interactions) developing and evolving between faults, which may extend over long spatiotemporal distances and influence their time and place of occurrence. Correlation effectively confers memory to the system and manifests itself in the form of power laws governing the temporal and spatial statistics of seismicity. Both Poisson and complex/critical doctrines, albeit from different vantage points, consider the earthquakes of an aftershock sequence to be dependent.

Herein, we attempt to explore the statistical nature of seismicity by using the generalized formalism of NESP (described in Section 44.2) as a universal context for the statistical description of earthquake occurrence, and trying to ascertain the existence and degree of correlation in active fault networks (or, equivalently, the level of nonequilibrium). The existence of correlation is assessed by evaluating the entropic index q appearing in the q-exponential distribution predicted by NESP for the dynamic parameters of nonequilibrating systems; q is bounded as $0 \leq q \leq 2$, with $q = 1$ corresponding to the pure exponential distribution expected for conservative Poissonian processes and $q > 1$ indicating complexity/criticality in nonconservative systems. Specifically, we evaluate an entropic index associated with the distribution of earthquake magnitudes, which conveys information about the size and space distribution of fault activity and is genetically related to the *b-value* of the Gutenberg–Richter law, and an entropic index associated with the distribution of the lapse between consecutive events (interevent time), which indicates the extent of interaction in a fault network. We refer to these as the magnitude (q_M) and temporal (q_T) indices, respectively, and we compute them by modelling bivariate empirical distributions of earthquake frequency versus magnitude *and* interevent time, or F–M–T for short; such distributions express the joint probability of observing earthquakes larger than a given magnitude after a given lapse time.

We examine seismogenetic systems along the NE and N boundary of the Pacific and North American plates. Specifically, we focus on the major transform fault systems of California (south and north segments of the SAF, SNR and MFZ) and Alaska (Queen Charlotte–Fairweather and Denali faults), as well as on the Alaskan–Aleutian convergence. With reference to the latter we examine *crustal* and *subcrustal* earthquakes by separating them according to the depth of the Mohorovičić discontinuity, so as to inquire whether environmental or/and boundary conditions affect the dynamics of a fault network. Finally, we apply our analysis to homogeneous and complete earthquake catalogues in which aftershock sequences are either included (full catalogues) or removed (declustered catalogues) with the efficient stochastic declustering method of Zhuang et al. (2002). If background seismicity is Poissonian, the removal of aftershocks should reduce the earthquake catalogue to an uncorrelated set of events; if it is does not, the argument against Poissonian seismicity would be compelling.

Turning now to the discussion of our results, it might be said that they comprise an 'expected' part and an 'interesting' part. The 'expected' part is the behaviour of the magnitude entropic index q_M, which after conversion to a proxy b-value through Eq. (11.18), turns out to be consistent with expectation from the Gutenberg–Richter law (see Tables 11–1 and 11–2). Naturally, q_M exhibits differences between seismogenetic systems. The analysis of full catalogues shows that some (e.g., MFZ, Fig. 11–9; QCD, Figs. 11–13A and 11–14A), exhibit rather high clustering of faulting activity and that in some cases (SCSR, nSAF, SNR, Fig. 11–9A) the degree of clustering (q_M) increases with threshold magnitude, i.e., it extends over long ranges. At any rate, in the general context of NESP, the Gutenberg–Richter law can be almost naturally derived from the q-exponential distribution (Section 11.2.2). Accordingly, a most significant outcome of q_M analysis is that it demonstrates that active fault networks may be classified as *subextensive* with a high degree of self-organization.

Since there can be little doubt that the time-honoured frequency–magnitude distribution of Gutenberg and Richter emerges from nonextensive fault networks, we shall concentrate the rest of our discussion on the temporal dynamics of seismicity, as indicated by the temporal entropic index q_T. Fig. 11–17A is a compact presentation and colour-coded classification of all $q_T(M_{th})$ functions shown in Figs. 11–10A, 11–11A, 11–11C, 11–12A, 11–12C and 11–15, and summarized in Tables 11–2 and 11–3. In the classification scheme, all values of $q_T(M_{th}) < 1.15$ as established in Section 11.4.1, are shown in red and are considered to indicate *nihil correlation*. Values higher than 1.15 generally indicate statistically significant correlation that is rated as *weak* (orange, $1.15 \leq q_T < 1.3$), *moderate* (light green, $1.3 \leq q_T < 1.4$), *significant* (green, $1.4 \leq q_T < 1.5$), *strong* (light blue, $1.5 \leq q_T < 1.6$) and *very strong* (blue, $1.6 \leq q_T$). The pie chart in Fig. 11–17B summarizes the proportions of q_T classes determined from the *full* crustal catalogues and provides a succinct picture of the existence, extent and relative strength of correlation in the crustal seismogenetic systems we have studied. Note that for nSAF, the results used in the compilation of the pie chart refer to the *entire* period 1968–2015; the results obtained from the analysis of the ante and post Loma Prieta subcatalogues have *not* been included separately. The pie chart in Fig. 11–17C is as per Fig. 11–17B but for the *declustered* crustal catalogues. Finally, Fig. 11–17D, as per Fig. 11–17A, is a compact presentation and colour-coded classification of all $q_T(\Delta d)$ functions from the full and declustered earthquake catalogues analysed herein and shown in Figs. 11–10B, 11–11B, 11–11D, 11–12B, 11–12D and 11–16. In this case however, the red class of Poissonian processes spans the interval $1 \leq q_T(\Delta d) < 1.2$ as has been established in Section 11.4.1.

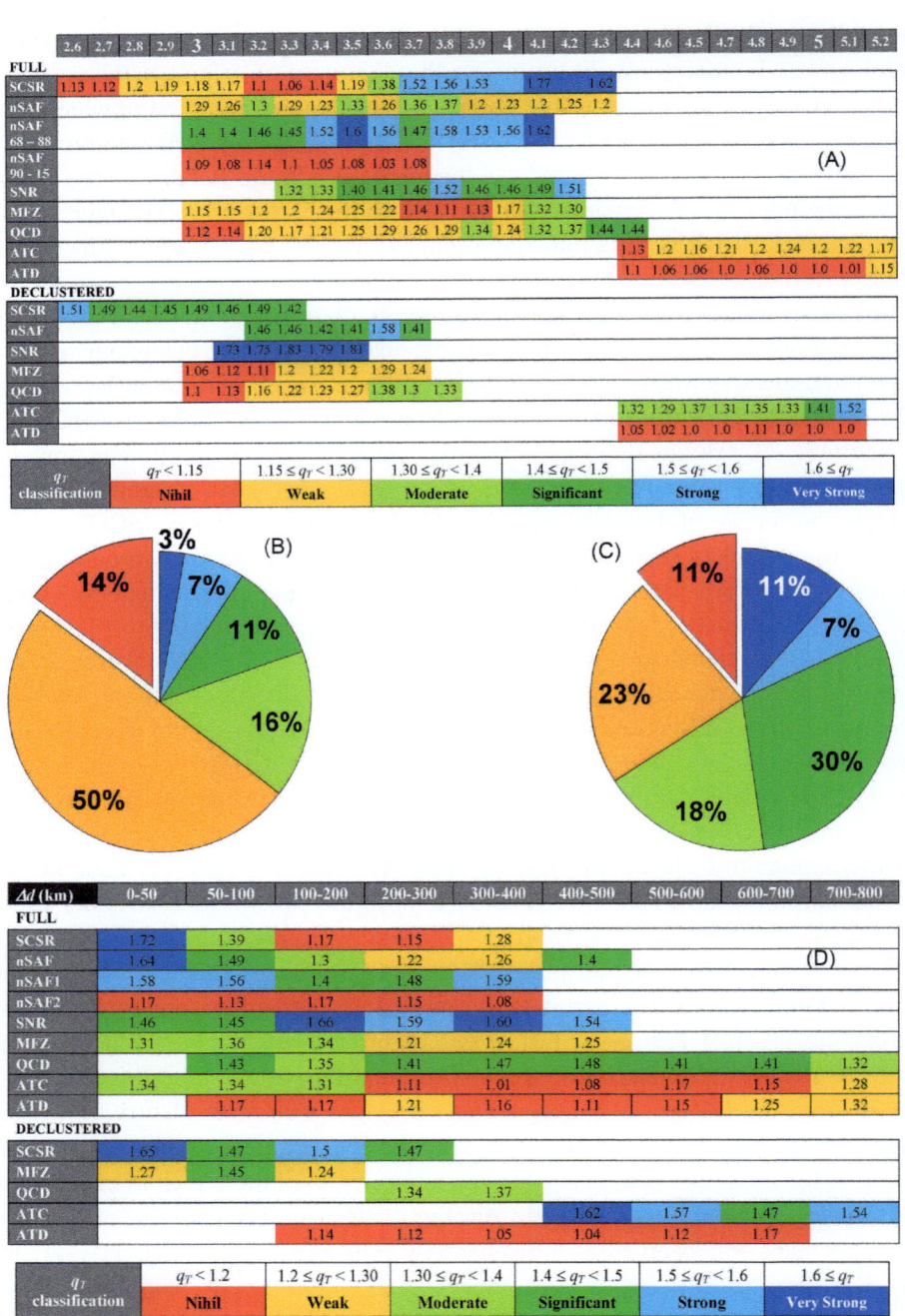

FIGURE 11–17 (A) Summarization and classification of all q_T versus M_{th} determinations shown in Figs. 11–10A, 11–11A, 11–11C, 11–12A, 11–12C and 11–15. (B) Proportions of $q_T(M_{th})$ classes determined from the analysis of *full crustal* catalogues; nSAF results used in the compilation of the pie chart refer to the entire period 1968–2015; ATD results are excluded. (C) Proportions of $q_T(M_{th})$ classes determined from the analysis of *declustered crustal* catalogues; ATD is again excluded. (D) Summarization and classification of all q_T versus Δd determinations shown in Figs. 11–10B, 11–11B, 11–11D, 11–12B, 11–12D and 11–16.

Mere inspection of Fig. 11−17 should suffice to satisfy one that the crustal seismogenetic systems we have studied are clearly correlated (Fig. 11−17B) and that correlation increases with declustering (Fig. 11−17C). This is a point of significance in that it demonstrates that removal of the clutter effected by the large numbers of time-local foreground events uncovers the existence of long-range interaction in the global background process. The extent of long-range correlation can be explicitly studied in Fig. 11−17D, where it becomes apparent that at intermediate and long interevent distances ($\Delta d > 150$ km), crustal seismicity is generally correlated and that declustering, either does not affect the degree of correlation, or causes it to *increase*. It should also be noted that even with full (clustered) earthquake catalogues, the correlation observed at interevent distances longer than 200 km can hardly be explained in terms of aftershock sequences: such ranges are several times larger than the characteristic dimensions of aftershock zones associated with M_w 6−6.7 earthquakes and significantly longer than zones associated with M_w 7−7.2 earthquakes (e.g., Kagan, 2002). Therefore, it is more reasonable to argue in favour of an operational long-range interaction. Significant to strong correlation is generally observed at short interevent distances ($\Delta d < 100$ km), a result easily explained by the overwhelming effect of (clustered and correlated) aftershock sequences.

Another point of significance is that ETAS-based stochastic declustering *fails* to reduce earthquake catalogues to sequences of independent events. van Stiphout et al. (2012) presented a study in which they compared declustering algorithms by applying the χ^2 goodness-of-fit test to determine whether the 'background' recovered by some declustering algorithm obeys a Poisson distribution in time. They found at the 5% significance level, that catalogues declustered by the methods of Zhuang et al. (2002) and Marsan and Lengliné (2008) follow a Poisson distribution in time; accordingly, they suggest that Poisson processes are in control of the background seismicity. We contend that this (and analogous) tests may be misleading because the distribution of occurrence times is *not* a measure of the interrelationship between distal successive events *whatsoever*, and does not relate the occurrence of an earthquake to its predecessor and successor events. On the other hand, the distribution of interevent times does, as adequately explained in the foregoing.

Based on our results, and as far as crustal seismogenetic systems are concerned, the answer to the question asked in the title appears to be that they are complex and that their complexity involves a long-range interaction, therefore, they are organized. However, there are different mechanisms by which complexity may arise. Inasmuch as power-law distributions and long-range effects are hallmarks of critical phenomena, SOC and self-organizing criticality (which naturally emerge from the inherent nonlinear dynamics of the fault system), are by far the principal candidates. However, complexity and criticality do not always go hand in hand and there are noncritical mechanisms that may generate power-laws (e.g., Sornette, 2004; Sornette and Werner, 2009). In one such example, Celikoglu et al. (2010) demonstrated that the CNM can generate q-exponential distributions of interevent times, although their simulation was incomplete in the sense that it did not include some spatial (geometric) configuration of interacting faults and could not assess the differences with an actual fault network.

Our results may be of some use in the course of understanding the origin and nature of complexity in the fault systems we have studied. First, let us make a list of some points we consider significant:

- To begin with, different fault systems may exhibit radically different attributes and degrees of complexity. A characteristic example is the adjacent/related MFZ and SNR systems, in which the first is partly accommodating the deformation effected by the second. However, the former exhibits marginal correlation while the second is strong or very strong. This may indicate that models calling for external driving forces that act upon all elements of a fault system, as for instance the CNM, may not be applicable to regional scales.

- Some systems, (SCSR, SNR, QCD and ATC) exhibit persistent and significant to strong long-range correlation over the entire period of 47 years since 1968; this reveals itself both implicitly, as an increase in the value of q_T with threshold magnitude (Fig. 11–17A), and explicitly (Fig. 11–17D); it is also observable in the full and declustered catalogues. Far from suggesting that the state of strong correlation may endure 'forever', we should, nevertheless, point out that this quasistationary state of high correlation has attributes of SOC.

- SOC is not a general rule. This is evident in nSAF undergoing enormous changes with respect to the large Loma Prieta earthquake and switching from a state of high correlation with strong attributes of criticality during the period leading up to the event (1968–89), to practically Poissonian in the period following the event (1990–present). This shows that criticality may be cyclic and possibly evolving in association with earthquake cycles. It also has attributes of the *self-organizing* variety, albeit without evident acceleration of seismic release rates as predicted by some models (e.g., Sammis and Sornette, 2001). It is very possible that the plain designation 'criticality' would suffice to characterize this case, or maybe all cases for that matter.

- A last important point we would like to make is that the only subcrustal system studied herein, the Alaskan–Aleutian subduction (ATD), is definitely Poissonian. Because it is the only one, it cannot serve as a basis for generalizations. The contrast with the crustal systems is rather impressive nonetheless, and may comprise a piece of information useful in the course of shaping up some preliminary understanding of the statistical (and physical) nature of seismogenesis. As seen in Fig. 11–6C, the Gutenberg–Richter a-value is 'normal' and the b-value is almost the same as the global average. Moreover, as seen in Fig. 11–13D and Table 11–3, q_M varies from 1.53 ($b_{q=0.88}$) at $M_{th}=4.4$, to 1.46 ($b_{q=1.16}$) at $M_{th}=5.2$ with a tendency to decrease (increase). On the other hand, there is (meagre at present) an indication that some crustal systems exhibiting strong long-range correlation, also exhibit long-range activity localization, therefore large-scale crustal homogeneity: for instance, in SCSR, nSAF and SNR and for $M_{th} > 3.5$, the full-catalogue q_M increases steadily with threshold magnitude from approximately 1.5 ($b_{q=1}$) to approximately 1.58 ($b_{q=0.72}$), while it remains persistently high in QCD ($q_M \sim 1.6$; $b_q \sim 0.67$). While the behaviour of q_M is clearly different between ATD and the crustal systems, possibly indicating different dynamics, gross earthquake productivity rates and large-scale domain heterogeneities are *not* dramatically dissimilar. Accordingly, the absence of temporal correlation in ATD may not have to do with the material properties of the subducting slab and should be sought elsewhere.

Based on the above observations, we believe that we can put together the basics of a plausible interpretation for our results, which will be based on fault networks with *small-world* topologies (e.g., Abe and Suzuki, 2004, 2007; Caruso et al., 2005, 2007). Given that active fault networks are *nonconservative* systems — friction is a nonconservative force — and therefore susceptible to nonlinearity, we are pointed to this direction by the documented existence of long-range interaction and (possible) criticality, fruitful studies based on nonconservative small-world Olami—Feder—Christensen models (Caruso et al., 2005, 2007), and suggestive evidence of small-worldness in the seismicity of California by Abe and Suzuki (2004, 2007).

In such networks each fault is a node that belongs to a local cluster where it occupies some hierarchical level according to its size and interacts with local or distal faults (nodes) according to the respective connectivity and range of its hierarchical level. Upon excitation by some (slow or fast) stress perturbation, a node responds by storing (accumulating) energy in the form of strain and subsequently transmitting it to *connected* nodes or/and releasing it at various rates; in other words, it operates as a delayed feedback loop inducing heterogeneity in the distribution of stress transfer and release rates. Finally, and more importantly, crustal fault networks are subject to free boundary conditions at the Earth—atmosphere interface; top-tier faults (which in transformational and convergent tectonic settings generally break at the surface), comprise primary boundary elements of the network.

It is documented that in Olami—Feder—Christensen networks, free boundary conditions compel the boundary elements to interact at a different (delayed) frequency with respect to the bulk of elements buried deeper in the network and that this inhomogeneity induces partial synchronization of the boundary elements, building up long-range spatial correlations and facilitating the development of a critical state (e.g., Lise and Paczuski, 2002; Caruso et al., 2005; Hergarten and Krenn, 2011). This effect should also be accentuated by heterogeneity and delayed feedback across the entire network, which also appear to be important for the development of criticality in small-world networks (Yang, 2001; Caruso et al., 2007). In the particularly interesting study by Hergarten and Krenn (2011), the dynamics of the network are governed by two competing mechanisms: synchronization, which pushes the system toward criticality, and desynchronization which prevents it from becoming overcritical and generates foreshocks and aftershocks. Once the system has reached the critical state, synchronized failure transfers more stress to connected nodes and this causes them to fail early, desynchronizing with the rest of the system. If, however, the time lag between desynchronized failures is short, the system can resynchronize and repeat the cycle. This mechanism generates sequences of foreshocks, mainshocks and aftershocks. Notably, the notion that aftershocks are generated by the desynchronization caused by large earthquakes is quite different — and more SOC — than that of spontaneous triggering advocated by the ETAS model.

In consequence of the above, it is plausible that the small-world character and subextensive critical state of crustal fault networks along the boundary of the Pacific and North

American plates, is induced by the high connectivity of synchronized top-tier faults, for instance the contiguous segments of the large transform faults. These may operate as 'hubs' that facilitate longitudinal interactions (transfer of stress) between distal clusters but inhibit interactions between distal or unconnected networks that operate quasiindependently and develop different levels of self-organization, as for instance between nSAF and SNR, or nSAF and MFZ. In addition, the intensity of the longitudinal interactions may vary in response to time-dependent changes in the external driving force and connectivity (stress transfer) between hubs, as for instance may have happened to nSAF before and after the Loma Prieta event. The interpretation posits that free boundary conditions are central to the development of complexity and criticality. By inference, it also implies that deep-seated fault networks, as for instance those of Wadati−Benioff zones, should be kept away from criticality as they are subject to fixed boundary conditions that inhibit synchronization. If this holds water, it might be the primary reason why subcrustal seismicity in the Alaskan−Aleutian subduction is Poissonian. Nevertheless, as we have already stated before, rigorous inferences and generalizations cannot be based on only one example. It follows that this detail, and the whole interpretation of our analysis for that matter, remains to be tested with future research.

In a final comment, we note that our analysis has been based on statistical physics for which the designation 'statistical' may not have the same meaning as that in 'statistical seismology'. As eloquently pointed out by Sornette and Werner (2009), statistical seismology is 'a field that has developed as a marriage between probability theory, statistics and the part of seismology concerned with empirical patterns of earthquake occurrences ... but not with physics'. On the other hand, statistical physics endeavours to generate the statistical models from first principles, respecting the laws of thermodynamics and taking into account physical laws such as those of friction, rupture, etc. In other words, it uses physics to support stochastic models, a quality often missing from traditional statistical seismology (Dieterich, 1994). In this respect, our NESP-based approach is constrained by physics and as such, it is analogously significant.

Acknowledgements

This work was partly supported by the THALES Program of the Ministry of Education of Greece and the European Union in the framework of the project 'Integrated understanding of Seismicity, using innovative methodologies of Fracture Mechanics along with Earthquake and Nonextensive Statistical Physics − Application to the geodynamic system of the Hellenic Arc − SEISMO FEAR HELLARC'.

References

Abe, S., Suzuki, N., 2003. Law for the distance between successive earthquakes. J. Geophys. Res. 108 (B2), 2113.

Abe, S., Suzuki, N., 2004. Complex network of earthquakes. In: Bubak, M., van Albada, G.D., Sloot, P.M.A., Dongarra, J. (Eds.), Computational Science - ICCS 2004. ICCS 2004. Lecture Notes in Computer Science, Vol, 3038. Springer, Berlin, Heidelberg. Available from: http://dx.doi.org/10.1007/978-3-540-24688-6_135.

Abe, S., Suzuki, N., 2005. Scale-free statistics of time interval between successive earthquakes. Physica A 350, 588–596.

Abe, S., Suzuki, N., 2007. Dynamical evolution of clustering in complex network of earthquakes. Eur. Phys. J. B 59, 93–97. Available from: https://doi.org/10.1140/epjb/e2007-00259-3.

Antonopoulos, C.G., Michas, G., Vallianatos, F., Bountis, T., 2014. Evidence of q-exponential statistics in Greek seismicity. Phys. A: Statist. Mech. Applicat. 409, 71–77. Available from: https://doi.org/10.1016/j.physa.2014.04.042.

Bak, P., Tang, C., 1989. Earthquakes as a self-organized critical phenomenon. J. Geophys. Res. 94, 15635–15637.

Bak, P., Christensen, K., Danon, L., Scanlon, T., 2002. Unified scaling law for earthquakes. Phys. Rev. Lett. 88, 178501. Available from: https://doi.org/10.1103/PhysRevLett.88.178501.

Bakar, B., Tirnakli, U., 2009. Analysis of self-organized criticality in Ehrenfest's dog-flea model. Phys. Rev. E 79, 040103. Available from: https://doi.org/10.1103/PhysRevE.79.040103.

Batak, R.C., Kantz, H., 2014. Observing spatio-temporal clustering and separation using interevent distributions of regional earthquakes, Nonlin. Processes Geophys. 21, 735–744. Available from: https://doi.org/10.5194/npg-21-735-2014.

Beck, C., Schloegl, F., 1993. Thermodynamics of Chaotic Systems: An Introduction. Cambridge University Press, pp. 88–93, Cambridge University Press.

Becker, T.W., Hardebeck, J.L., Anderson, G., 2005. Constraints on fault slip rates of the southern California plate boundary from GPS velocity and stress inversions. Geophys. J. Int 160 (2), 634–650.

Carbone, V., Sorriso-Valvo, L., Harabaglia, P., Guerra, I., 2005. Unified scaling law for waiting times between seismic events. Europhys. Lett. 71 (6), 1036. Available from: https://doi.org/10.1209/epl/i2005-10185-0.

Caruso, F., Latora, V., Rapisarda, A., Tadić, B., 2005. The Olami-Feder-Christensen model on a small-world topology, arXiv:cond-mat/0507643v1 [cond-mat.stat-mech] (last accessed April 2017).

Caruso, F., Pluchino, A., Latora, V., Vinciguerra, S., Rapisarda, A., 2007. Analysis of self-organized criticality in the Olami-Feder-Christensen model and in real earthquakes. Phys. Rev. E 75, 055101. Available from: https://doi.org/10.1103/PhysRevE.75.055101.

Celikoglu, A., Tirnakli, U., Duarte Queirós, S., 2010. Analysis of return distributions in the coherent noise model. Phys. Rev. E 82, 021124. Available from: https://doi.org/10.1103/PhysRevE.82.021124.

Console, R., Murru, M., 2001. A simple and testable model for earthquake clustering. J. Geoph. Res. 106 (B5), 8699–8711.

Corral, A., 2004. Long-term clustering, scaling, and universality in the temporal occurrence of earthquakes. Phys. Rev. Lett. 92, 108501.

Christeson, G.L., Van Avendonk, H.J.A., Gulick, S.P.S., Reece, R.S., Pavlis, G.L., Pavlis, T.L., 2013. Moho interface beneath Yakutat terrane, southern Alaska. J. Geophys. Res. Solid Earth 118, 5084–5097. Available from: https://doi.org/10.1002/jgrb.50361.

Darooneh, A.H., Mehri, A., 2010. A nonextensive modification of the Gutenberg-Richter law: q-stretched exponential form. Physica A 389 (3), 509–514. Available from: https://doi.org/10.1016/j.physa.2009.10.00.

Davidsen, J., Goltz, C., 2004. Are seismic waiting time distributions universal? Geophys. Res. Lett. 31, L21612. Available from: https://doi.org/10.1029/2004GL020892.

DeMets, C., Dixon, T., 1999. New kinematic models for Pacific-North America motion from 3 Ma to present, 1: Evidence for steady motion and biases in the NUVEL-1A model. Geophys. Res. Lett. 26, 1921–1924.

Dengler, L., Moley, K., McPherson, R., Pasyanos, M., Dewey, J.W., Murray, M.H., 1995. The September 1, 1994 Mendocino fault earthquake. California Geology 48, 43–53.

Dickinson, W.R., Snyder, W.S., 1979a. Geometry of triple junctions related to San Andreas transform. J. Geophys. Res. 84, 561–572.

Dieterich, J., 1994. A constitutive law for rate of earthquake production and its application to earthquake clustering. J. Geophys. Res. 99, 2601−2618.

Dixon, T.H., Miller, M., Farina, F., Wang, H., Johnson, D., 2000. Present-day motion of the Sierra Nevada block and some tectonic implications for the Basin and Range province, North American Cordillera. Tectonics 19, 1−24. Available from: https://doi.org/10.1029/1998TC001088.

Eaton, J.P., 1992. Determination of amplitude and duration magnitudes and site residuals from short-period seismographs in Northern California. Bull. Seism. Soc. Am. 82 (2), 533−579.

Efstathiou, A., Tzanis, A., Vallianatos, F., 2015. Evidence of Non-Extensivity in the evolution of seismicity along the San Andreas Fault, California, USA: An approach based on Tsallis Statistical Physics. Phys. Chem. Earth, Parts A/B/C 85−86, 56−68. Available from: https://doi.org/10.1016/j.pce.2015.02.013.

Efstathiou, A., Tzanis, A., Vallianatos, F., 2016. On the nature and dynamics of the seismogenetic system of South California, USA: an analysis based on Non-Extensive Statistical Physics. Bull. Geol. Soc. Greece 50 (3), 1329−1340. Available online in http://www.geosociety.gr/images/news_files/EGE_L/EGE2016_Proceedings_Volume_L_3.pdf (last accessed June 2017).

Eneva, M., Pavlis, L.G., 1991. Spatial Distribution of Aftershocks and Background Seismicity in Central California. Pure and Applied Geophysics 137 (1), 35−61.

Esquivel, F.J., Angulo, J.M., 2015. Non-extensive analysis of the seismic activity involving the 2011 volcanic eruption in El Hierro, 2015. Spatial Statistics 14 (B), 208−221. Available from: https://doi.org/10.1016/j.spasta.2015.08.001.

Felzer, K.R., 2007. Stochastic ETAS Aftershock Simulator Program (AFTsimulator), available at http://pasadena.wr.usgs.gov/office/kfelzer/AftSimulator.html; last access 20 October 2014.

Felzer, K.R., Brodsky, E.E., 2006. Evidence for dynamic aftershock triggering from earthquake densities. Nature 441, 735−738.

Felzer, K.R., Becker, T.W., Abercrombie, R.E., Ekstrom, G., Rice, J.R., 2002. Triggering of the 1999 Mw 7.1 Hector Mine earthquake by aftershocks of the 1992 Mw 7.3 Landers earthquake. J. Geophys. Res. 107, 2190. Available from: https://doi.org/10.1029/2001JB000911.

Fialko, Y., 2006. Interseismic strain accumulation and the earthquake potential on the South San Andreas fault system. Nature 441, 968−971. Available from: https://doi.org/10.1038/nature04797.

Furlong, K.P., Schwartz, S.Y., 2004. Influence of the Mendocino triple junction on the tectonics of coastal California. Annu. Rev. Earth Planet. Sci. 32, 403−433. Available from: https://doi.org/10.1146/annurev.earth.32.101802.120252.

Gardner, J.K., Knopoff, L., 1974. Is the sequence of earthquakes in Southern California, with aftershocks removed, Poissonian? Bull. Seism. Soc. Am. 64 (5), 1363−1367.

Gell-Mann, M., Tsallis, C. (Eds.), 2004. Nonextensive Entropy − Interdisciplinary Applications. Oxford University Press, New York.

Goter, S.K., Oppenheimer, D.H., Mori, J.J., Savage, M.K., Masse, R.P., 1994. Earthquakes in California and Nevada, U.S. Geological Survey Open-File Report 94-647, scale 1:1,000,000, 1 sheet.

Guest, B., Niemi, N., Wernicke, B., 2007. Stateline fault system: A new component of the Miocene-Quaternary Eastern California shear zone. Geol. Soc. Am. Bull. 119 (11−12), 1337−1347. Available from: https://doi.org/10.1130/0016-7606(2007)119[1337:SFSANC]2.0.CO;2.

Hainzl, S., Scherbaum, F., Beauval, C., 2006. Estimating background activity based on interevent-time distribution. Bull. Seismol. Soc Am. 96 (1), 313−320. Available from: https://doi.org/10.1785/0120050053.

Hammond, W.C., Blewitt, G., Li, Z., Plag, H.-P., Kreemer, C., 2012. Contemporary uplift of the Sierra Nevada, western United States, from GPS and InSAR measurements. Geology 40 (7), 667−770. Available from: ﮡs://doi.org/10.1130/G32968.1.

1983. Advanced Synergetics: Instability Hierarchies of Self-organizing Systems and Devices. ﮡrlin Heidelberg New York.

Hardebeck, J.L., Hauksson, E., 2001. Crustal stress field in southern California and its implications for fault mechanics. J. Geophys. Res. 106, 21,859–21,882.

Hawkes, A.G., 1972. Spectra of some mutually exciting point processes with associated variables. In: Lewis, P. A.W. (Ed.), Stochastic Point Processes. Wiley, pp. 261–271.

Hawkes, A.G., Adamopoulos, L., 1973. Cluster models for earthquakes - regional comparisons. Bull Internat. Stat. Inst. 45, 454–461.

Hawkes, A.G., Oakes, D., 1974. A cluster representation of a self-exciting process. J. Apl. Prob. 11, 493–503.

Helmstetter, A., Sornette, D., 2003. Predictability in the Epidemic-Type Aftershock Sequence model of interacting triggered seismicity. J. Geophys. Res. 108 (B10), 2482. Available from: https://doi.org/10.1029/2003JB002485.

Hergarten, S., Krenn, R., 2011. Synchronization and desynchronization in the Olami-Feder-Christensen earthquake model and potential implications for real seismicity, Nonlin. Processes Geophys. 18, 635–642. Available from: https://doi.org/10.5194/npg-18-635-2011.

Janiszewski, H.A., Abers, G.A., Shillington, D.J., Calkins, J.A., 2013. Crustal structure along the Aleutian island arc:new insights from receiver functions constrained by active-source data. Geochem. Geophys. Geosyst. 14. Available from: https://doi.org/10.1002/ggge.20211.

Jones, L.M., 1988. Focal Mechanisms and the state of San Andreas Fault in Southern California. J. Geophys. Res. 93 (B8), 8869–8891.

Kagan, Y.Y., 2002. Aftershock zone scaling. Bull. Seismol. Soc. Am. 92 (2), 641–655.

Lay, T., Wallace, T.C., 1995. Modern Global Seismology. Academic Press, New York, pp. 383–387.

Lise, S., Paczuski, M., 2002. A nonconservative earthquake model of self-organized criticality on a random graph. Phys. Rev. Lett. 88 (22), 228301. Available from: https://doi.org/10.1103/PhysRevLett.88.228301.

Marsan, D., Lengliné, O., 2008. Extending earthquakes's reach through cascading. Science 319, 1076. Available from: https://doi.org/10.1126/science.1148783.

Martinez, M.D., Lana, X., Posadas, A.M., Pujades, L., 2005. Statistical distribution of elapsed times and distances of seismic events: the case of the Southern Spain seismic catalogue. Nonlinear Proc. Geophys. 12, 235–244.

Marzocchi, W., Lombardi, A.M., 2008. A double branching model for earthquake occurrence. J. Geophys. Res. 113, B08317. Available from: https://doi.org/10.1029/2007JB005472.

McCaffrey, R., 2005. Block kinematics of the Pacific-North America plate boundary in the southwestern United States from inversion of GPS, seismological, and geologic data. J. Geophys. Res. 110, B07401. Available from: https://doi.org/10.1029/2004JB003307.

Michas, G., Vallianatos, F., Sammonds, P., 2013. Non-extensivity and long-range correlations in the earthquake activity at the West Corinth rift (Greece). Nonlin. Proc. Geoph. 20, 713–724.

Michas, G., Vallianatos, F., Sammonds, P., 2015. Statistical mechanics and scaling of fault populations with increasing strain in the Corinth Rift. Earth Planet. Sci. Lett. 431, 150–163. Available from: https://doi.org/10.1016/j.epsl.2015.09.014.

Molchan, G., 2005. Interevent time distribution in seismicity: A theoretical approach. Pure appl. geophys. 162, 1135–1150. Available from: https://doi.org/10.1007/s00024-004-2664-5.

Moré, J.J., Sorensen, D.C., 1983. Computing a trust region step. SIAM J. Sci. Statist. Comput. 3, 553–572.

Newman, M.E.J., 1996. Self-organized criticality, evolution and the fossil extinction record. Proc. Roy. Soc. Lond. B 263, 1605–1610.

Ogata, Y., 1988. Statistical models for earthquake occurrences and residual analysis for point processes. J. Am. Stat. Assoc. 83 (401), 9–27.

Ogata, Y., 1998. Space-time point-process models for earthquake occurrences. Ann. I. Stat. Math. 50 (2), 379–402.

Olami, Z., Feder, H.J.S., Christensen, K., 1992. Self-organized criticality in a continuous, nonconservative cellular automation modeling earthquakes. Phys. Rev. Lett. 68, 1244–1247.

Page, R.A., Biswas, N.N., Lahr, J.C., Pulpan, H., 1991. Seismicity of continental Alaska. In: Slemmons, D.B., Engdahl, E.R., Zoback, M.D., Blackwell, D.D. (Eds.), Neotectonics of North America: Boulder, Colorado, Volume l. Geological Society of America, Decade Map.

Papadakis, G., Vallianatos, F., Sammonds, P., 2013. Evidence of nonextensive statistical physics behaviour of the hellenic subduction zone seismicity. Tectonophysics 608, 1037–1048.

Papadakis, G., Vallianatos, F., Sammonds, P., 2015. A nonextensive statistical physics analysis of the 1995 Kobe, Japan Earthquake. Pure Appl. Geophys. 172 (7), 1923–1931.

Pease, R.W., 1965. Modoc County; University of California Publications in Geography, v. 17. University of California Press, Berkeley and Los Angeles, pp. 8–9.

Reasenberg, P., 1985. Second-order moment of central California seismicity, 1969-82. J. Geophys. Res. 90, 5479. 5495.

Rhoades, D.A., 2007. Application of the EEPAS model to forecasting earthquakes of moderate magnitude in Southern California. Seismol. Res. Lett. 78 (1), 110–115.

Rhoades, D.A., Evison, F.F., 2006. The EEPAS forecasting model and the probability of moderate-to-large earthquakes in central Japan. Tectonophysics 417 (1/2), 119–130.

Rundle, J.B., Klein, W., Turcotte, D.L., Malaud, B.D., 2000. Precursory seismic activation and critical point phenomena. Pure appl. Geophys. 157, 2165–2182.

Saichev, A., Sornette, D., 2013. Fertility heterogeneity as a mechanism for power law distributions of recurrence times. Phys. Rev. E 97, 022815. also available at arXiv:1211.6062 [physics.geo-ph] (last access 20 October 2014).

Saleeby, J., Saleeby, Z., Nadin, E., Maheo, G., 2009. Step-over in the structure controlling the regional west tilt of the Sierra Nevada microplate: eastern escarpment system to Kern Canyon system. Int. Geol. Rev. 51 (7–8), 634–669.

Sammis, C.G., Sornette, D., 2001. Positive feedback, memory and the predictability of earthquakes, e-print at http://arXiv.org/abs/cond-mat/0107143v1; last accessed December 2015.

Scherrer, T.M., França, G.S., Silva, R., de Freitas, D.B., Vilar, C.S., 2015. Nonextensivity at the circum-pacific subduction zones – preliminary studies. Phys. A: Statist. Mech. Applicat. 426, 63–71. Available from: https://doi.org/10.1016/j.physa.2014.12.038.

Schoenball, M., Davatzes, N.C., Glen, J.M.G., 2015. Differentiating induced and natural seismicity using space-time-magnitude statistics applied to the Coso Geothermal field. Geophys. Res. Lett. 42, 6221–6228. Available from: https://doi.org/10.1002/2015GL064772.

Scholz, C., 2002. The Mechanics of Earthquakes and Faulting, second ed. Cambrigde University Press, New York, pp. 198–211.

Segou, M., Parsons, T., Ellsworth, W., 2013. Comparative evaluation of physics-based and statistical forecasts in Northern California. J. Geophys. Res. Solid Earth 118. Available from: https://doi.org/10.1002/2013JB010313.

Silva, R., Franca, G.S., Vilar, C.S., Alcaniz, J.S., 2006. Nonextensive models for earthquakes. Phys. Rev. E 73, 026102. Available from: https://doi.org/10.1103/PhysRevE.73.026102.

Sornette, A., Sornette, D., 1989. Self-organized criticality and earthquakes. Europhys. Lett. 9, 197–202.

Sornette, D., 2004. Critical Phenomena in Natural Sciences: Chaos, Fractals, Self-organization and Disorder: Concepts and Tools, second ed. Springer, Berlin, 529 pp.

Sornette, D., Sammis, C.G., 1995. Complex critical exponents from renormalization group theory of earthquakes: Implications for earthquake predictions. J. Phys. 1 (5), 607–619.

Sornette, D., Werner, M.J., 2009. Statistical Physics Approaches to Seismicity, in Complexity in Earthquakes, Tsunamis, and Volcanoes, and Forecast, W.H.K. Lee (Ed), in the Encyclopedia of Complexity and Systems Science, R. Meyers (Editor-in-chief), 7872-7891, Springer, ISBN: 978-0-387-755888-6; available at arXiv:0803.3756v2 [physics.geo-ph] (last access 20 October 2014).

Sotolongo-Costa, O., Posadas, A., 2004. Tsalli's entropy: A non-extensive frequency-magnitude distribution of earthquakes. Phys. Rev. Letters 92 (4), 048501. Available from: https://doi.org/10.1103/PhysRevLett.92.048501.

Steihaug, T., 1983. The conjugate gradient method and trust regions in large scale optimization. SIAM J. Numer. Anal. 20, 626−637.

Talbi, A., Yamazaki, F., 2010. A mixed model for earthquake interevent times. J. Seismol 14, 289−307. Available from: https://doi.org/10.1007/s10950-009-9166-y.

Telesca, L., 2010a. Nonextensive analysis of seismic sequences. Phys. Stat. Mech. Appl. 389, 1911−1914.

Telesca, L., 2010b. A nonextensive approach in investigating the seismicity of L'Aquila area (central Italy), struck by the 6 April 2009 earthquake (M_L 5:8). Terra Nova 22, 87−93.

Telesca, L., 2011. Tsallis-based nonextensive analysis of the Southern California seismicity. Entropy 13, 1267−1280.

Telesca, L., 2012. Maximum likelihood estimation of the nonextensive parameters of the earthquake cumulative magnitude distribution. Bull. Seismol. Soc. Am. 102, 886−891.

Telesca, L., Chen, C.-C., 2010. Nonextensive analysis of crustal seismicity in Taiwan. Nat. Hazards Earth Syst. Sci. 10, 1293−1297.

Touati, S., Naylor, M., Main, I.G., 2009. Origin and nonuniversality of the earthquake interevent time distribution. Phys. Rev. Letters 102, 168501. Available from: https://doi.org/10.1103/PhysRevLett.102.168501.

Tsallis, C., 1988. Possible generalization of Boltzmann-Gibbs statistics. J. Stat. Phys. 52, 479−487. Available from: https://doi.org/10.1007/BF01016429.

Tsallis, C., 2001. Nonextensive statistical mechanics and thermodynamics: historical background and present status. In: Abe, S., Okamoto, Y. (Eds.), Nonextensive Statistical Mechanics and Its Applications. Springer, Berlin, Heidelberg, pp. 3−98. Available from: https://doi.org/10.1007/3-540-40919-X.

Tsallis, C., 2009. Introduction to Nonextensive Statistical Mechanics: Approaching a Complex World. Springer Verlag, Berlin, p. 378.

Tsallis, C., Tirnakli, U., 2010. Nonadditive entropy and nonextensive statistical mechanics − Some central concepts and recent applications. J. Phys. Conference Series 201 (2010), 012001. Available from: https://doi.org/10.1088/1742-6596/201/1/012001.

Uhrhammer, B.R.A., Loper, S.J., Romanowicz, B., 1996. Determination of local magnitude using BDSN Broadband Records. Bull. Seism. Soc. Am. 86 (5), 1314−1330.

Utsu, T., Ogata, Y., Matsu'ura, R.S., 1995. The centenary of the Omori formula for a decay law of aftershock activity. J. Phys. Earth 43, 1−33.

Statistical mechanics in earth physics and natural hazards. In: Vallianatos, F., Telesca, L. (Eds.), Acta Geophys., 60. pp. 499−501.

Vallianatos, F., Sammonds, P., 2013. Evidence of non-extensive statistical physics of the lithospheric instability approaching the 2004 Sumatran- Andaman and 2011 Honshu mega-earthquakes. Tectonophysics . Available from: https://doi.org/10.1016/j.tecto.2013.01.009.

Vallianatos, F., Benson, P., Meredith, P., Sammonds, P., 2012a. Experimental evidence of a non-extensive statistical physics behaviour of fracture in triaxially deformed Etna basalt using acoustic emissions. Europhy. Let. 97, 58002. Available from: https://doi.org/10.1209/0295-5075/97/58002.

Vallianatos, F., Michas, G., Papadakis, G., Sammonds, P., 2012b. A non-extensive statistical physics view to the spatiotemporal properties of the June 1995, Aigion earthquake (M6.2) aftershock sequence (West Corinth Rift, Greece). Acta Geophys. 60 (3), 758–768.

Vallianatos, F., Michas, G., Papadakis, G., Tzanis, A., 2013. Evidence of non-extensivity in the seismicity observed during the 2011–2012 unrest at the Santorini volcanic complex, Greece. Nat. Hazards Earth Syst. Sci. 13, 177–185. Available from: https://doi.org/10.5194/nhess-13-177-2013.

Vallianatos, F., Papadakis, G., Michas, G., 2016. Generalized statistical mechanics approaches to earthquakes and tectonics. Proc. R. Soc. A 472, 20160497. Available from: https://doi.org/10.1098/rspa.2016.0497.

van Stiphout, T., Zhuang, J., Marsan D., 2012. Seismicity declustering, Community Online Resource for Statistical Seismicity Analysis, https://doi.org/10.5078/corssa-52382934. Available at http://www.corssa.org.

Wernicke, B., Axen, G.J., Snow, J.K., 1988. Basin and range extensional tectonics at the latitude of Las Vegas, Nevada. Geol. Soc. Am. Bull. 100 (11), 1738–1757. Available from: https://doi.org/10.1130/0016-7606 (1988)100 < 1738:BARETA > 2.3.CO;2.

Wesnousky, S., 2005. Active faulting in the Walker Lane. Tectonics 24 (3), TC3009. Available from: https://doi.org/10.1029/2004TC001645.

Yang, X.S., 2001. Chaos in small-world networks. Phys. Rev. E 63, 046206. Available from: https://doi.org/10.1103/PhysRevE.63.046206.

Yeats, R., 2013. Active Faults of the World. Cambridge University Press, Cambridge.

Zhuang, J., Ogata, Y., Vere-Jones, D., 2002. Stochastic declustering of space-time earthquake occurrences. J. Amer. Stat. Assoc. 97, 369–380.

12

Phase Space Portraits of Earthquake Time Series of Caucasus: Signatures of Strong Earthquake Preparation

Tamaz Chelidze, Natalya Zhukova, Temur Matcharashvili

IVANE JAVAKHISHVILI TBILISI STATE UNIVERSITY, TBILISI, GEORGIA

CHAPTER OUTLINE

12.1 Introduction

Detailed statistical analysis of earthquake time series (ETS) showed that seismic catalogues contain both independent and correlated events (clusters), thus suggesting that they are complex time series. In some earlier works (Goltz, 1997; Matcharashvili et al., 2000) it was shown that the interevent or waiting time (WT) series, have low fractal dimensions, indicating that seismic catalogues could contain some hidden nonlinear structures.

In recent years some studies have been published on the identification of attractors in seismic time series, meaning that they can be modelled by deterministic chaos (Bhattacharya and Srivastava 1992; Pavlos et al., 1994; Srivastava et al., 1996; Thanassoulas et al., 2009; Sobolev, 2011); however, Beltrami and Mareschal (1993) claim that such an ordered structure does not exist.

Several decades ago, geophysical objects and events in Earth sciences were mainly considered as either random or deterministic. Complexity analysis reveals the enormous domain

Complexity of Seismic Time Series. DOI: https://doi.org/10.1016/B978-0-12-813138-1.00012-2

of structures and processes, located between completely random (white noise) and deterministic (Newton) extreme patterns and allows treating them in a quantitative manner. In this mesoscale domain, processes, though seeming to be random, possess certain nonlinear (hidden) temporospatial structures, which are invisible for routine statistical analysis, but which we can reveal by application of modern tools of nonlinear dynamics. The appropriate methods of complex time series analysis are developed in nonlinear dynamics (complexity) theory (Grassberger and Procaccia, 1983; Abarbanel and Tsimring,1993; Strogatz, 1994; Eckmann et al.,1987; Sornette, 2000; Pykovsky et al., 2003; Sprott, 2006). The new tools developed in complexity theory reveal a lot of important information, contained in seismic catalogues, considered as discrete earthquake (EQ) time series (Bak et al., 1989; Goltz, 1997; Newman and Turcotte, 2002; Chelidze et al., 2006; Rundle et al., 2009; Chelidze and Matcharashvili, 2015). Methods, developed in complexity theory allow visualization and quantification of seismic rate (SR) patterns and their variation in time, what enriches significantly the traditional statistical approach (Marsan and Nalbant, 2005; Marsan and Wyss, 2011). One of many methods, developed for the analysis of complex systems, is construction of the phase space plots of the dynamic process. The basis of the qualitative approach for reconstruction and testing of phase space objects, equivalent to unknown dynamics, is using Taken's fundamental time delay theorem (Takens, 1981). It is used for reconstruction of two- and three-dimensional phase portraits (strange attractors), Poincare sections, calculation of iterated function systems (IFSs) and recurrent plots (RPs) (Eckmann et al., 1987; Sprott, 2006; Webber and Marwan, 2015). These methods preserve the general topological peculiarities of investigated dynamics and allow carrying visual, preliminary analysis of an unknown dynamical process. One example, shown in Fig. 12−1, illustrates the potential of complexity analysis: it presents the attractor-like pattern in the phase space of SR, revealed by Sobolev (2011). The appearance of regular recurrences in ETS and in their laboratory models (stick−slip experiments) can be connected with the natural stick−slip mechanism of fault activity under tectonic stress and for the action of cyclic triggering factors, such as tides, seasonal changes, etc. (Scholz, 2003; Savage and Marone, 2007; Chelidze et al., 2010; Bartlow et al., 2012; Beeler and Lockner, 2013).

12.2 Methodology for ETS Analysis

The statistical analysis of SR is mainly aimed at statistically reliable assessment of rate change using long enough EQ time series, e.g., when comparing SR before and after strong earthquakes (Marsan and Wyss, 2011). Several methods like phase space portraits (PSPs), RPs, recurrence quantification analysis (RQA) (see, e.g., Webber and Marwan, 2015) are used to visualize and quantify hidden structures in ETS. PSP and RP are two different, but related, methods that reveal recurrence in time series: e.g., what is represented by diagonal lines in PR is parallel traces.

The application of these methods is even more complicated when applied to ETS, because this time series contains a strong noise component that makes it very difficult to retrieve the regular component. Therefore, the phase space is 'noisy': this situation is ubiquitous for

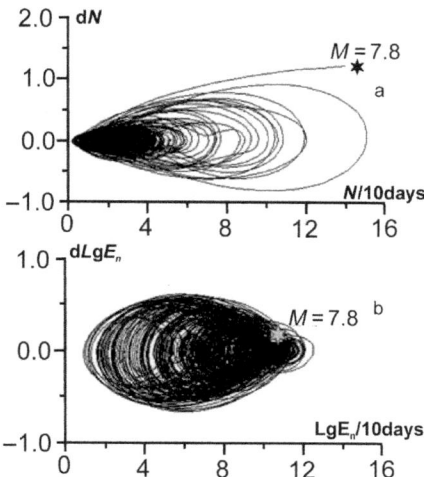

FIGURE 12–1 Phase portrait of seismicity within a radius of 100 km from the epicenter of the Kronotskoe earthquake for a 35-year period (1 January 1962–5 December 1997). The smoothed number N of earthquakes (A) or energy LgEn, Joules (B) for 10 consecutive days is marked on the X-axis. Rates at which these parameters change (dN or dLgEn), i.e., the difference between the following and the preceding values are shown on the Y-axis. Clockwise movement along the curve corresponds to an increase in time. The star marks the position of the main shock (Sobolev, 2011).

experimental data of various origin. Since the noise destroys the regularity of the trajectories in the phase space, especially when the signal-to-noise ratio is low, in order to retrieve the regular component, several methods have been suggested. For instance, Molaie et al. (2013) suggested projecting experimental trajectory points not on a one-degree curve in the space plot, but on an n-degree curve (e.g., with $n = 3$ or 4); Ralaivola and d'Alché-Buc (2005) suggested using the Kernel Kalman filter for effective time series filtering, and noise extraction (smoothing).

We have analysed SRs, which, according to general statistics terminology, correspond to conditional rate (conditional intensity) models in point process theory (Scholz, 1990; Ogata, 1999; Newman and Turcotte, 2002). We used the rates, obtained either by simple averaging (Sobolev, 2011) or by Savitzky–Golay (S–G) filtering of time series (Press et al., 1997). Savitzky–Golay filtering helps to resolve effectively the smoothing problem in the time domain. It approximates the data locally (corresponding to some user-chosen window) with an nth degree polynomial, preserving up to the nth moments of the data. Hence, it has an advantage over, for instance, a moving average filter, as the magnitude of the variations in the data, i.e., the value of the local extremes, is to a large extent preserved (Press et al., 1997). The optimal lag for PSP reconstruction from daily series of EQ occurrences by mutual information (MI) testing in our case is close to 50 days.

The PSP plots of EQ time series are compiled using the following methods:

1. The whole EQ datasets from the catalogue were declustered using Reasenberg's algorithm (Matthews and Reasenberg, 1988) and, then, smoothed in the following way: the X-axis represents the mean values of the number N of EQs per n days ($n = 10, 20, 50$) or N/n and the Y-axis represents the differential of X, i.e., $(N_{i+1} - N_i)/n = dN$ (see Sobolev, 2011).

2. The whole EQ datasets from the catalogue were declustered using Reasenberg's algorithm and smoothed by Savitzky—Golay (S—G) filter. Then the X-axis represents the S—G smoothed value of number N of EQs for a given day $N/$(day) and the Y-axis represents the S—G smoothed N value with a delay of some days ($N + Lag$). In our case, we used three different values of Lag: 10, 20 and 50 days.

3. Combined approach: PSP is compiled daily (like in the Sobolev approach) as dN versus $N/$day, declustered using Reasenberg's algorithm, and smoothed by S—G filter (contrary to Sobolev's approach, where just the mean value of N in the sliding window is used).

The trajectories of the PSP plots are obtained by connecting the consecutive phase states. The consecutive phase space points are plotted in a clockwise direction, which corresponds to increasing time. For plotting the PSPs we used either standard MATLAB scripts: seism_port and phase_portrait (hereafter indicated as 'standard') or Sobolev's (2011) approach (hereafter indicated as 'Sobolev'). Both approaches sometimes produce negative values of phase states, which means that the smoothed lagged values are smaller than previous ones. As an example of processing, in Fig. 12—2 phase space plots of daily EQ occurrence sequence of EQs are shown, compiled for original nonsmoothed data (A) and for the same data, but smoothed by Savitzky—Golay filter (B). It is evident that the former is less informative and the latter reveals some interesting attractor-like structure in the phase space.

The resulting 'noisy' phase space plot (Fig. 12—2B) manifests the presence of some source area ('noisy' basin), as well as evolving trajectories (orbits), which deviate from the source and finally return to it.

12.3 Study Area

For nonlinear analysis the seismic catalogue of Caucasus (1960—2011) has been used; the representative magnitude for the period is M2 (Fig. 12—3). The following parameters of ETS were varied: (1) the radius R of the area, where ETS were obtained; (2) the length of the time window for rate counting; (3) the years' span (periods of the catalogue); and (4) periods before and after strong events.

In the analysed period of 1960—2011 the two largest Caucasian EQs, Spitak and Racha (M6.9—7) struck the region in 1988 and 1991, respectively. Thus, three areas were selected: (1) Batumi (in order to show the pattern of ETS in a relatively quiet region); (2) Spitak; and (3) Racha (Fig. 12—3).

12.4 Results and Discussion

Nonlinear analysis of datasets obtained from the seismic catalogue of Caucasus for the period 1960—2011 was performed; the representative magnitude for the period was M2. We analysed both the original and declustered (by applying Matthews and Reasenberg (1988) approach) catalogues.

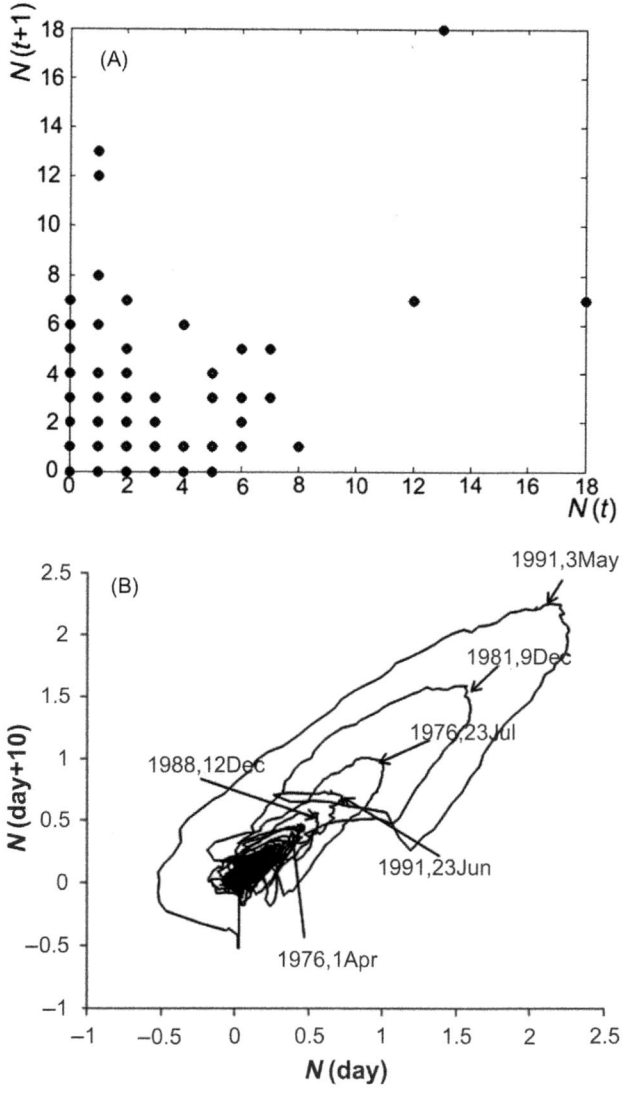

FIGURE 12–2 Phase space plots of daily EQ occurrence sequence in Caucasian earthquake catalogue (1961–91) for the area in a 200-km radius around Racha EQ compiled by the above-mentioned standard scripts for original nonsmoothed data (A) and for the same data, smoothed by Savitzky–Golay filter, i.e., for the seismic rate (B).

12.4.1 Batumi Area

Batumi area (Fig. 12–3) was chosen as a relatively seismically quiet area: though there were no strong EQs within a distance of 100 km from Batumi, two strong EQs occurred at a distance of about 200 km (Chkhalta M6.4, 16 July 1963 and Erzurum M6.9, 30 October 1983). Fig. 12–4 presents original and Savitzky–Golay filtered daily series of EQ occurrences in the

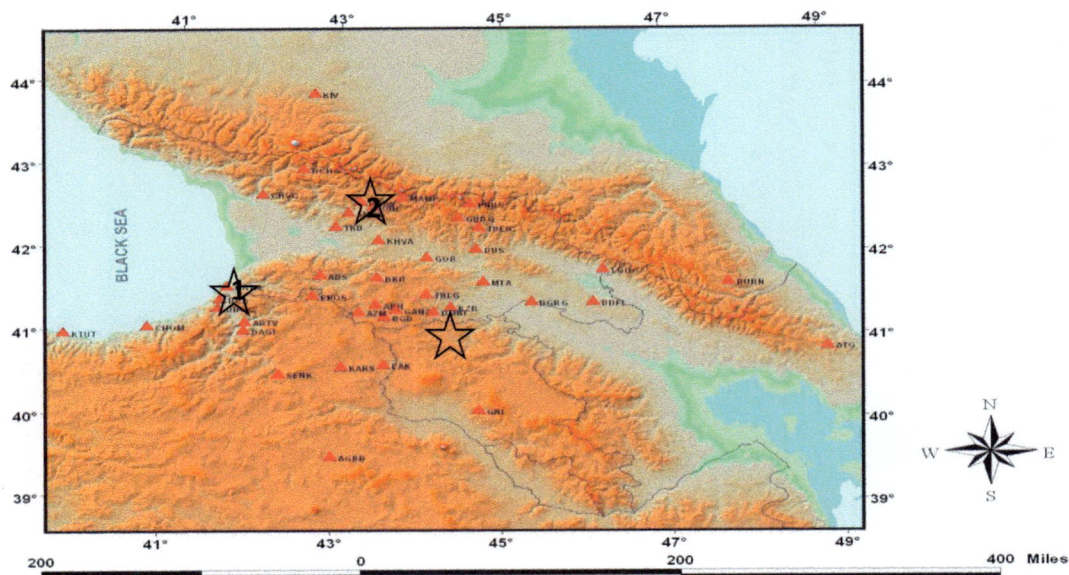

FIGURE 12–3 Areas in Caucasus, where the analysis of seismic data sets was carried out for revealing possible attractors. Stars are centres of test areas: 1 — Batumi; 2 — Racha; 3 — Spitak. Triangles show seismic station locations.

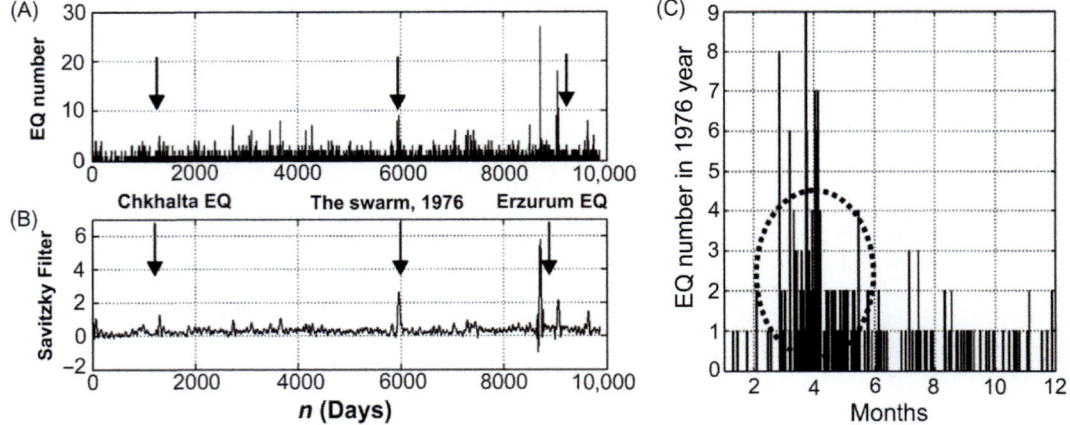

FIGURE 12–4 Daily series of EQ occurrences in the Batumi area ($R = 200$ km) in 1960—86, original (A), and declustered by Reasenberg's algorithm (B); (C) the swarm in 1976.

Batumi area ($R = 200$ km) in 1960−86, declustered by Reasenberg's algorithm. Fig. 12−5A−G show standard PSPs of declustered daily series of EQ occurrences in the Batumi area, smoothed by the S−G filter for various Lags. The X-axis represents the S−G filtered daily numbers of EQs and the Y-axis − the same values for 10-, 20- and 50-day lags (Fig. 12−5A, D and F, respectively). These PSP plots demonstrate two main details: a highly populated area between 0 and 1, which can be considered as a relatively stable domain (or a source area), due to a background seismic activity (these areas are shown by arrows for successive enlargement scales in Fig. 12−5B, C, E and G) and strongly deviating from the source trajectories (Fig. 12−5A, D, F). These latter orbit-like patterns should reflect deviations from the background activity due to some extremes − swarms, foreshocks and aftershocks. This looks a bit strange, as the declustering procedure should eliminate such effects. Still it seems that Reasenberg's procedure does not eliminate all correlated events, as was shown in Matcharashvili et al. (2015) and Telesca et al. (2016). Therefore, these orbits can be related to correlated events, still there despite the fact that the catalogue has been declustered. Matcharashvili et al. (2015) and Telesca et al. (2016) show that even in declustered catalogues of different regions the amount of correlated events is quite large.

Some of the anomalous orbits seem to be related to: (1) in 1968 and 1976 − possibly to swarms (see Fig. 12−4); (2) in 1983 − to the Erzerum EQ (30 October 1983, Ms 6.9); (3) in 1984 − probably to the extended aftershocks of Erzerum EQ. Note that, for $R = 100$ km around Batumi, for the Erzerum EQ (30 October 1983, Ms 6.9) the length of the orbit (Fig. 12−6) is less significant than that of the 1968 swarm, in contrast to the data for $R = 200$ km. This could be due to a lower density of Erzurum EQ aftershocks for radius $R = 100$ km. The length of the whole deviating from the basin trajectory (orbit) corresponds (approximately) to the period of foreshocks and aftershocks of the Erzerum EQ (Table 12−1).

Surprisingly, the strong 1963 Chkhalta EQ, M_l6.4 and 175 km from Batumi caused a relatively small deviation in the trajectory from a source area (see trajectories for Chkhalta (1963) in Fig. 12−5A, D). This could be due to relatively small deviations in the SR from the background value (Fig. 12−4A, B).

The PSPs of trajectories for $R = 100$ km from Batumi (Fig. 12−6) differ from the PSPs for $R = 200$ km. Anomalous orbits are possibly related: (1) in 1968 − to a swarm in 1968 (the length of this most outlying trajectory for the 1968 event is 62 days); (2) in 1983 − to the Erzerum EQ on 30 October 1983, Ms 6.9. Note that for $R = 100$ km, the orbit corresponding to Erzerum EQ on 30 October 1983 is less significant than that for a local swarm, in contrast to data for $R = 200$ km, due to a smaller density of aftershocks farther from the Erzerum epicentre area. According to Table 12−1, the duration of the half-trajectory till the Erzurum EQ is approximately 40 days; this can be considered as a precursory sign of an impending strong event.

Finally, Fig. 12−7 illustrates the impact of processing methodology of EQ time series on the structure of PSP. Note that trajectories of orbits in PSPs, plotted using differentials of the current and the previous values of N (differential $N_{i+1} − N_i)/n$) versus 10 days smoothed data ($N/10$ days), presented in Fig. 12−7 are much smoother and more ordered compared to the results obtained with larger steps (Figs. 12−5 and 12−6). The reason is that in the

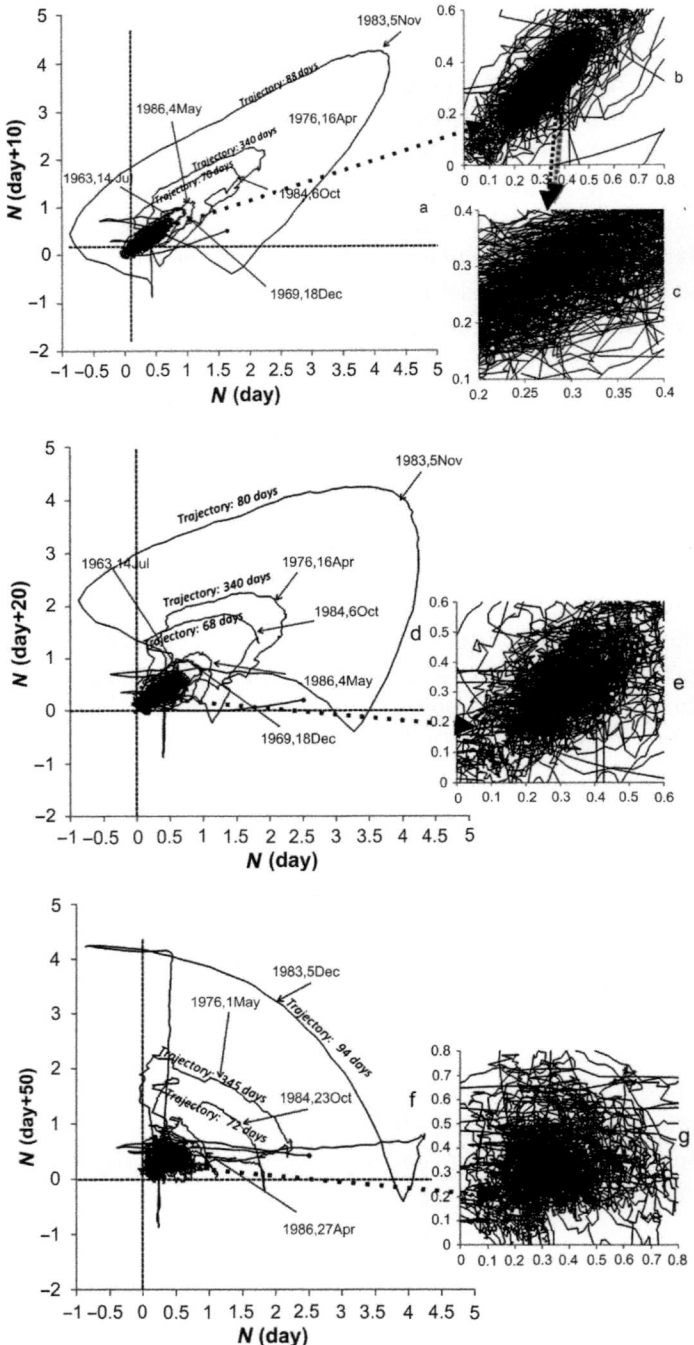

FIGURE 12–5 Phase space portraits (standard) of daily series of EQ occurrences in the Batumi area for $R = 200$ km, declustered and smoothed by the S–G filtered datasets from catalogues (data from Fig. 12–4). Phase space plots smoothed by S–G filter lagged daily value (N + Lag) where the Lag is 10, 20 and 50 days, versus daily value N(day) with: N in 10 days (A); N in 20 days (D) and N in 50 days (F). Panels (B, C, E and G) show the source area (marked by arrows) at successive enlargement scales.

latter case the successive samples do not differ significantly − by only 2 days (the first and the last days' data in the sample); while the rest of data in the samples are the same.

In contrast, the datasets in Figs. 12−5 and 12−6 do not contain identical data (the successive datasets are not overlapping), which results in a more jagged trajectory. It is evident that for strongly overlapping datasets the PSP structure is close to that of an attractor − we can see almost ordered orbits in the expanded source area (Fig. 12−7B).

In order to test whether the methodology used to obtain Figs. 12−5−12−7 is really informative, we apply this procedure to a random sequence of numbers (Fig. 12−8). It is evident

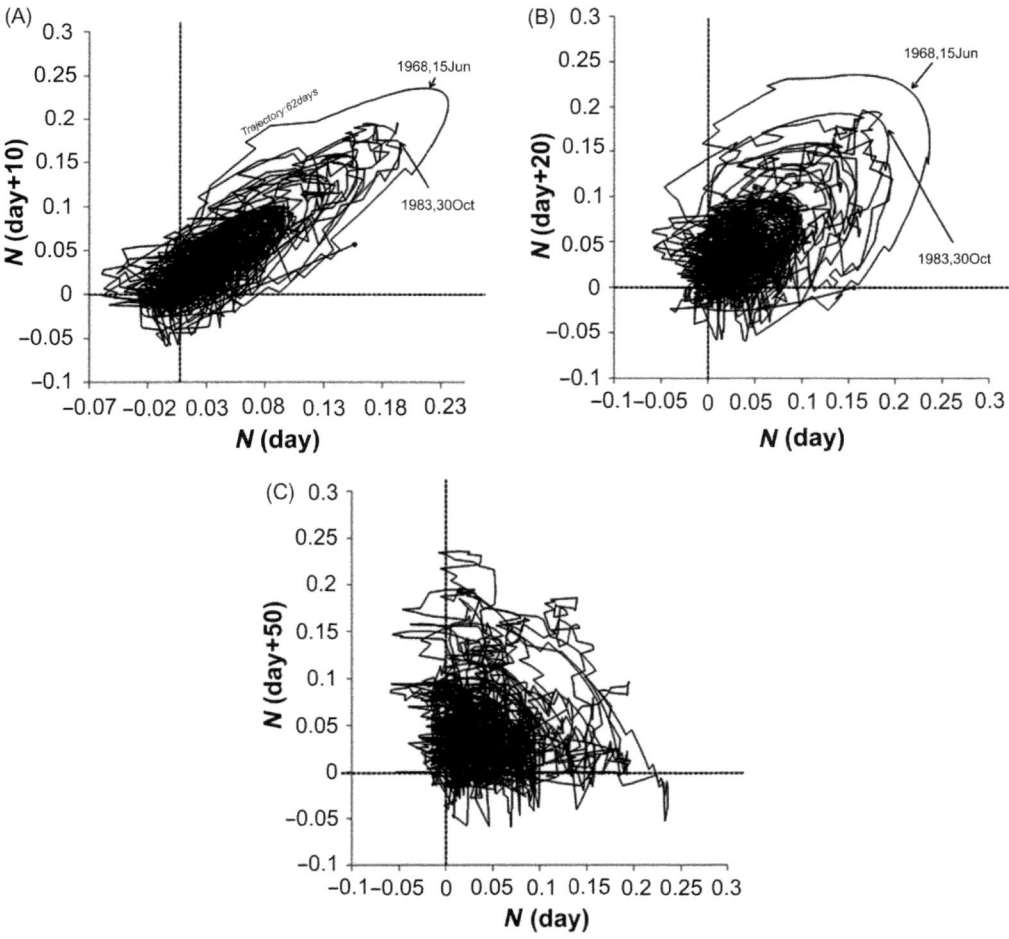

FIGURE 12–6 Phase space portraits (standard) of daily series of EQ occurrences ($M > 2$) in the Batumi area for $R = 100$ km, declustered and smoothed by the S–G filtered datasets (data from Fig. 12–4). Phase space plots smoothed by S–G filter lagged daily value (N + Lag), where the Lag is 10, 20 and 50 days, versus daily value N(day) with: (A) N in 10; (B) N in 20 and (C) N in 50 days.

Table 12–1 Duration of the Most Outlying Trajectories in Fig. 12–5A, D and F

Trajectories on Fig. 12–5A, D and F	Duration of a Full Trajectory (days)	Half Trajectory Duration
Most outlying (1983)	80–94	40
Second most distant (1976)	340–345	210
Third most distant (1984)	68–72	40

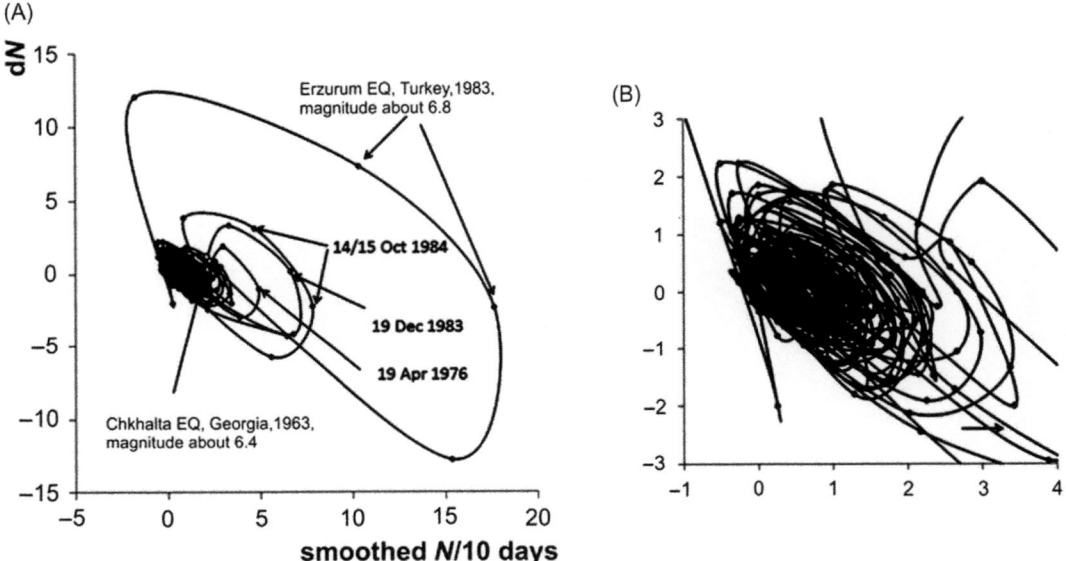

FIGURE 12–7 (A) PSP of dN versus declustered and S–G smoothed for N/10 days' data (catalogue 1960–86) in the Batumi area (here we apply S–G smoothing and 1 day lag in contrast to Figs. 12–5 and 12–6, i.e., here we combine standard and Sobolev approaches), R = 200 km; (B) the expanded view of a source area, limited by a small circle in (A).

that a combination of smoothing with small successive steps (*lags*) led to the appearance of smooth orbit-like trajectories even for random number sequences, similar to ordered trajectories in EQ rate time series (Fig. 12–7), as a result of a definite smoothing procedure. At the same time we can mark two main differences between PSRs of random sets and ETS. First, in the 'seismic' PSPs there are dense source areas (diffuse analogues of basin areas of deterministic chaos), which are absent in random PSP plots. Second, there are significant ordered reversible deviations in PSPs of ETS from the 'noisy' source areas (due to swarms and strong events), clearly revealed by both (standard and Sobolev) approaches, which are also absent in the random PSP plots (compare Figs. 12–5–12–8). In the last case, the pattern of orbits differs clearly from the PSP of the SR.

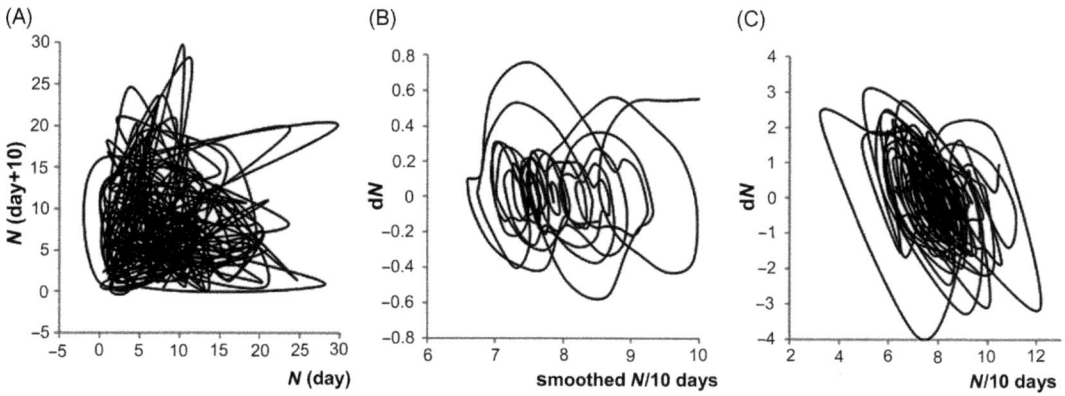

FIGURE 12–8 PSP compiled for a 'rate' of a random sequence of numbers in the range [0,1] considered as a proxy to a number of EQ random occurrence in 10 days; (A) standard PSP, no smoothing applied; (B) plot of dN versus $N/$10 'days' for original (nonsmoothed) data; (C) plot of dN versus $N/$10 'days' for S−G smoothed data; standard plot.

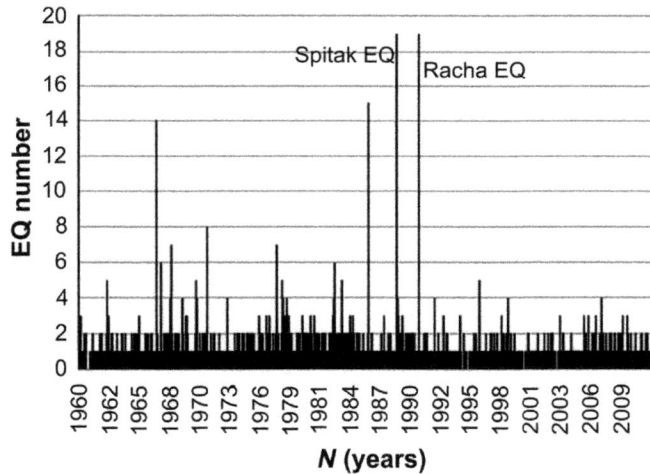

FIGURE 12–9 Daily occurrence of EQ in the Spitak area in 1960–2011 for $R = 100$ km; the two largest spikes are related to the 1988 Spitak and 1991 Racha EQs.

12.4.2 Spitak Earthquake Area

The Spitak earthquake of magnitude $M_s6.9$ (Fig. 12−2) occurred in Armenia, on 7 December 1988. We calculated PSPs of EQ time series in the Spitak EQ area, considering datasets for an area of radius 100 km around Spitak EQ epicentre (catalogues1960−88 and 1960−2011); both original and declustered ETSs were analysed. Fig. 12−9 shows the daily occurrence of EQs in the Spitak EQ epicentre area for $R = 100$ km. PSPs in Fig. 12−10A, D, F are plotted for

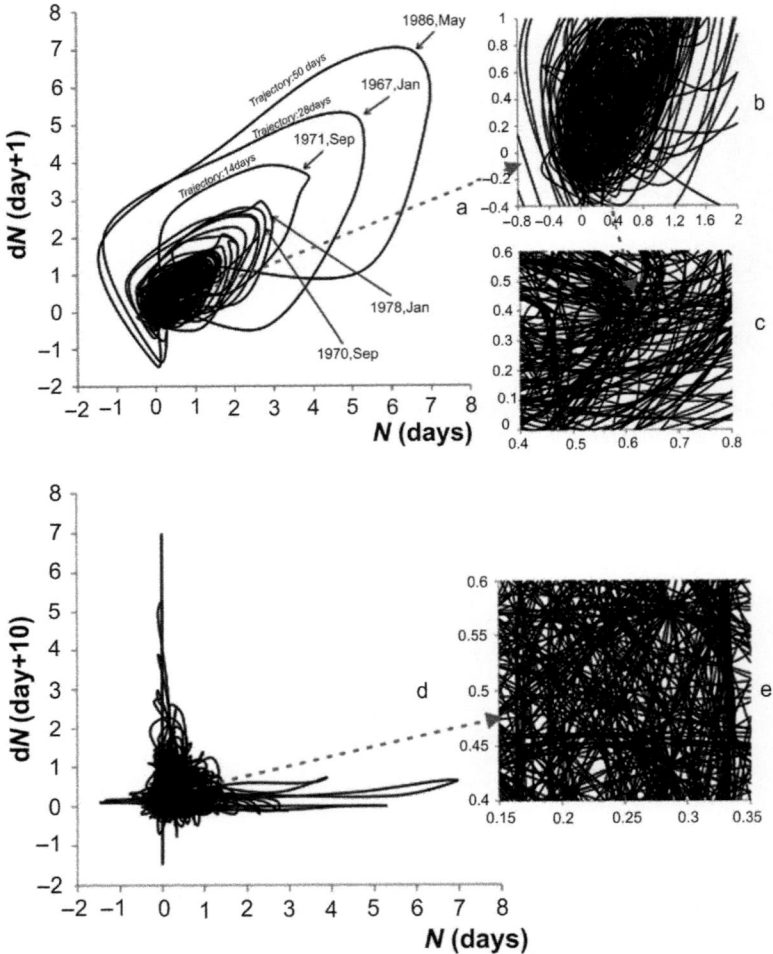

FIGURE 12–10 Phase space portraits of daily series of EQ occurrences in the Spitak area for $R = 100$ km, declustered and smoothed by the S–G filtered datasets from the following catalogues; (A, D, F) phase space plots (standard) for catalogue 1960–88, not including the Spitak EQ. Here the plot is smoothed by S–G filter lagged value ($N + $ Lag), where the Lag is 1, 10, 20 and 50 days versus the daily value N(day) with N 10, 20 and 50 days; (G, H, I) PSPs of dN versus declustered and smoothed for N/10 days' data in the same area for catalog 1960–2011, $R = 100$ km (Sobolev approach). Note a large difference in the structure between the PSP plots in A, D, F and G, H, I, due to different methods of catalogue processing and inclusion of foreshock/aftershock activity of the Spitak EQ in G, H, I. Panels (B, C and E) show the source area at successive enlargement scales, marked by arrows.

ETS, without the Spitak main event data and those in Fig. 12–10G, H, I are for the time series, which includes the Spitak EQ period. Note that there is a large difference in the structure of PSPs for the two analysed catalogues, which can be explained by the strong influence of seismicity, caused by foreshock/aftershock activity of the Spitak EQ included in the plots in Fig. 12–10G, H, I. The deviating orbits are visible in 1967, 1971, 1978 and 1986 in

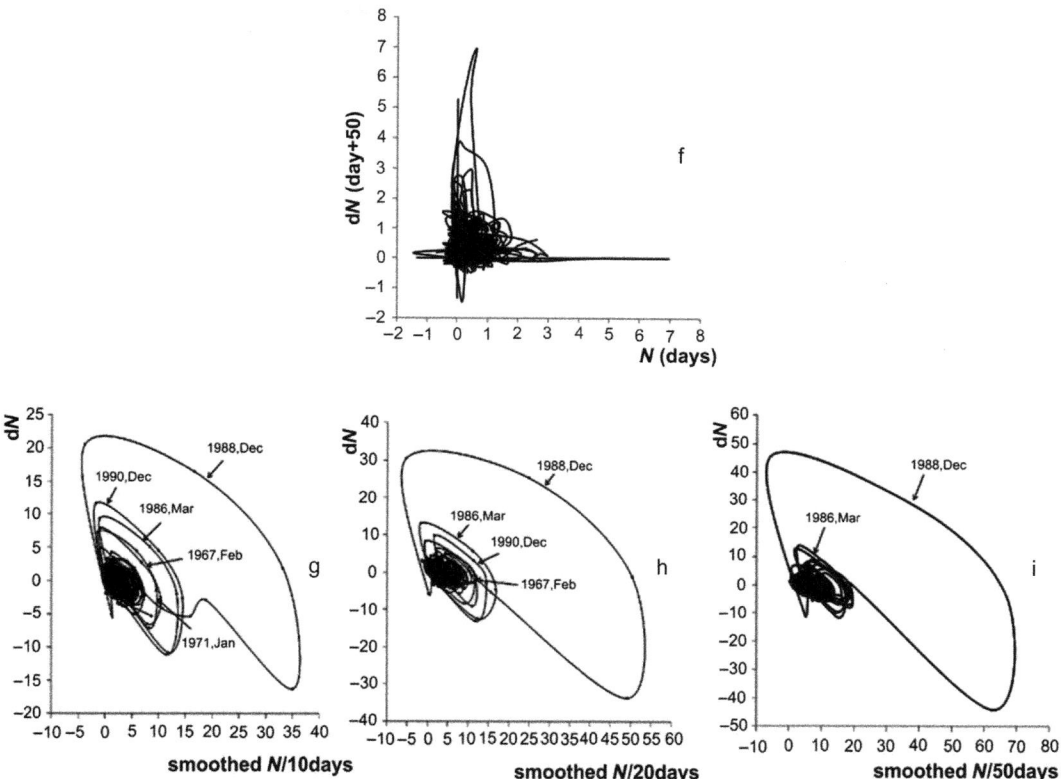

FIGURE 12–10 *Continued.*

Fig. 12−10A, D, F, and the most deviating orbit in 1988 in Fig. 12−10G, H, I. The last orbit is definitely related to the Spitak EQ foreshock/aftershock activity. It seems informative to divide the most outlying orbit in Fig. 12−10G, H, I into pre- and post-Spitak parts in order to assess the 'precursory' part of the trajectory. The full duration of the orbits in Fig. 12−10D, E, F is approximately 120−200 days and the duration of the 'precursory' part is 30−50 days for various lags. Thus, a strong deviation of the orbit from the source area can be considered as a precursor of the strong event, due probably to foreshock activity (not excluded fully by Reasenberg declustering − see Matcharashvili et al., 2015).

12.4.3 Racha EQ Area

The Racha earthquake occurred in the Racha province of Georgia at 9:12 UTC on 29 April 1991. It was centred on the districts of Oni and Ambrolauri at the southern foothills of the Greater Caucasus mountains and had a magnitude of 7.0. It was the most intense earthquake recorded in the Caucasus.

Table 12–2 Duration of the Most Outlying Trajectory on Fig. 12–11A, C and E

Trajectory on Fig. 12–11A, C and E	Duration of a Full Trajectory (days)	Trajectory Duration Till Racha 1991 EQ (days)
Most outlying (1991)	133–138	64

The most extended orbits reach the following maximal deviations at lags of 10 and 20 days: (1) the point 3 May 1991 May is close to the mainshock moment of Racha EQ, which occurred on 29 April; (2) the mark June 1991 corresponds to the Java strong aftershock of the Racha EQ, 15 June, 1991, M6.2; (3) the mark October 2009 is related to the Racha EQ, on 8 September 2009, M6.

The length of the most extended trajectory with the label 3 May 1991 is 133 days (starting at the source cluster). The time from a moment, when a significant deviation from the background seismicity ('noisy' basin) begins, to that with the label 3 May 1991 is approximately half of the full orbit duration. Possibly, this time, the need to form a half-orbit — approximately 60 days (Table 12–2) — can be considered as a precursor of the Racha 1991 mainshock.

The PSPs of Fig. 12–11B, D, F seem to be the most interesting ones: here in the radius $R = 100$ km in the detailed plots of the source cluster some clear recurrent orbits are visible with strange configurations — parabolas, right angles. We cannot see such recurrent configurations at PSPs for the larger test area, namely, for $R = 200$ km (not shown here).

We can point out that the PSPs of ETS contain attractor-like structures, which are observed for all three different processing methods (Section 12.2). The relatively smooth attractor-like trajectories in PSPs appear, when the step of the smoothing window is small (1 day): in this case the input data in successive windows, shifted by 1 day differ insignificantly as the content of windows is almost the same, except for the first and last days' data.

12.5 Discussion and Conclusions

The trajectories in the PSPs of ETS generally manifest two main features: a dense 'source/basin' area formed by background seismicity and (reversible) anomalous orbit-like deviations from the source area, related to swarms, and foreshock and aftershock activity. Unlike the case of theoretical attractors for deterministic chaos, the 'basin' area of the ETS is much more diffuse ('noisy'), than in the case of chaotic attractors, obtained by solution of model nonlinear equations. The same 'noisy' phase portraits are obtained during processing experimental data of various origin, e.g., in biology (Ralaivola and d'Alché-Buc, 2005; Molaie et al., 2013).

The patterns of PSPs of 10-day smoothed earthquake rate time series at volcanic area (Kronotskoe EQ), obtained earlier by Sobolev, seem to be very similar to attractors: this can be related either to the processing methodology (using 1-day step and overlapping 10-day windows), or to the high sensitivity of such areas to small cyclic perturbations.

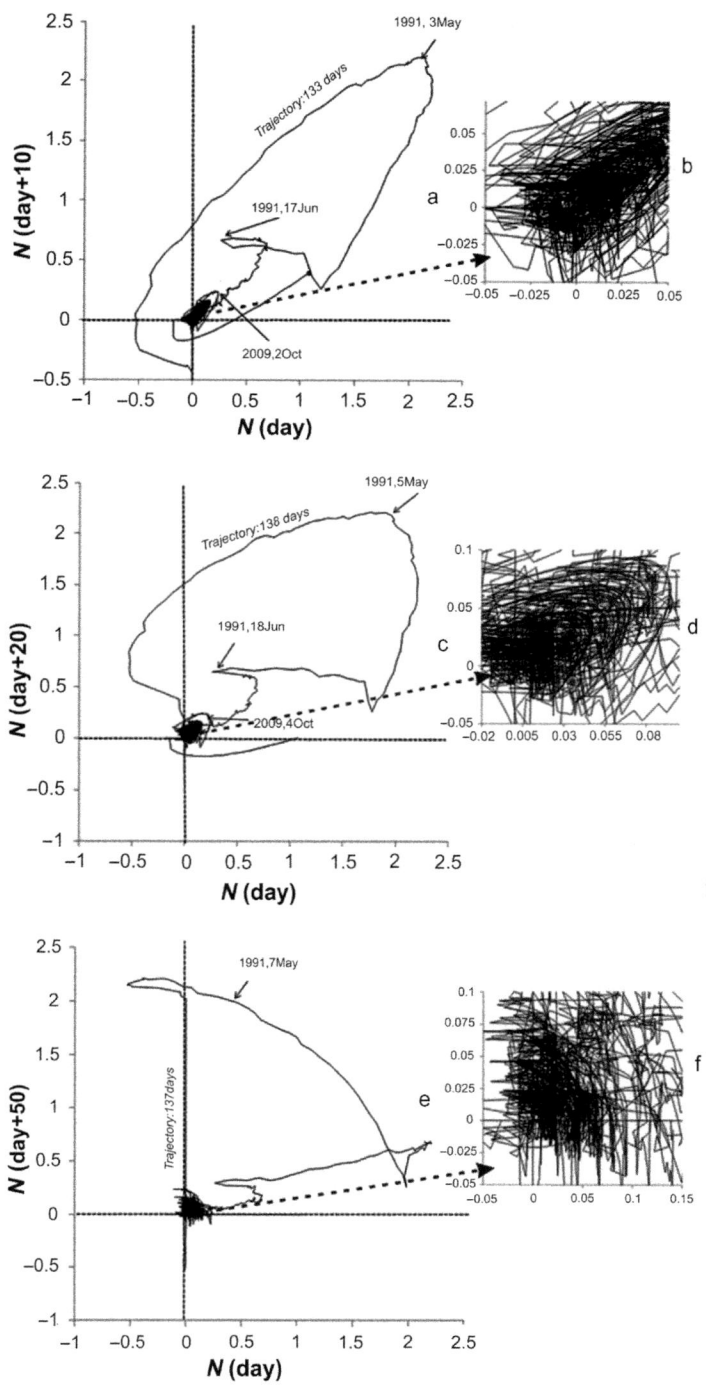

FIGURE 12–11 Phase space portraits of the dataset from the Racha test area for declustered S–G filtered catalogue 1960–2011 ($R = 100$ km, $M > 3$), for lags at 10, 20 and 50 days; note ordered structures in the enlarged images of the source area (B, D, F).

In this chapter the possibility of the existence of seismic attractors and, correspondingly, of a deterministic chaos regime in nonvolcanic areas (here Caucasus), is studied. Analysis shows that trajectories in the phase space are not very regular, except the smoothing (e.g., by Savitzky–Golay filter) procedure with a 10-day window and 1-day step applied to the data. The 1-day step for overlapping windows means that the content of the window is essentially the same, which results in the appearance of smooth attractor-like trajectories in the phase portraits of the ETS. When we use nonoverlapping windows, the attractor-like pattern is still evident in the phase space, though trajectories are not as smooth as in the former case. At the same time even these deformed PSPs of ETSs look quite different from the PSPs of random number 'time series': the ETS sequences have a 'basin' (though a 'noisy' one), to which the trajectories return after orbit-like deviations. Besides, the pattern of orbits in ETS phase portraits differs from those for random sets. The characteristic 'noisy' attractor-like structures emerge in seismic PSPs regardless of the methods used for SR calculations. The difference is only in the smoothness of the obtained structures.

PSP can be considered as an interesting visualization tools for analysis of seismicity dynamics. On phase space plots of smoothed (for 10, 20, 50 days) SR sequences in the Racha and Spitak areas some attractor-like structures emerge, which begin to deviate from the 'basin' area approximately 60 days before the 1991 Racha mainshock and 30–50 days before the Spitak event and return to the 'basin' after the mainshock.

It seems that before/after the strong Spitak and Racha earthquakes there were some anomalies in ETS (large deviations from the background pattern), which generated orbit-like structures in PSPs, even using declustered by Reasenberg approach catalogues. In principle, this effect can be used to search for strong earthquake precursors in the time domain, though the spatial resolution of the anomalous area is not high, as remote (separated by hundreds of km) strong earthquakes can result in significant changes in phase space plots of ETS even in seismically quiet areas. Further investigations are needed to reveal the EQ prediction potential of seismic phase space plotting.

Acknowledgements

The authors express their gratitude to the Rustaveli National Science Foundation of Georgia (Project FR/567/9–140/12) for financial support.

References

Abarbanel, H., Tsimring, L.S., 1993. The analysis of observed chaotic data in physical systems. Rev. Mod. Phys. 65, 1331–1392.

Bak, P., Tang, S., Winsenfeld, K., 1989. Earthquakes as self-organized critical phenomenon. J. Geophys. Res. 94, 15635–15637.

Bartlow, N.M., Lockner, D.A., Beeler, N.M., 2012. Laboratory triggering of stick-slip events by oscillatory loading in the presence of pore fluid with implications for physics of tectonic tremor. J. Geophys. Res. 117, B11411. Available from: https://doi.org/10.1029/2012JB009452.

Beeler, N.M., Lockner, D.A., 2013. Why earthquakes correlate weakly with the solid Earth tides: effects of periodic stress on the rate and probability of earthquake occurrence. J. Geophys. Res. 108, B2391. Available from: https://doi.org/10.1029/2001JB001518.

Beltrami, H., Mareschal, J.C., 1993. Strange seismic attractors? Pure Appl. Geophys. 141, 71−81.

Bhattacharya, S.N., Srivastava, H.N., 1992. Earthquake predictability in Hindukush region using chaos and seismicity pattern. Bull. Indian Soc. Earth Tech. 29, 23−25.

Chelidze, T., Matcharashvili, T., 2015. Dynamical patterns in seismology. In: Webber, C., Marwan, N. (Eds.), Recurrence Quantification Analysis: Theory and best practices. Springer, Cham Heidelberg, pp. 291−335.

Chelidze, T., Kolesnikov, Yu, Matcharashvili, T., 2006. Seismological criticality concept and percolation model of fracture. Geophys. J. Int. 164, 125−136. Available from: https://doi.org/10.1111/j.1365-246X.2005.02818.x.

Chelidze, T., Matcharashvili, T., Lursmanashvili, O., Varamashvili, N., Zhukova, N., Meparidze, E., 2010. Triggering and synchronization of stick-slip: experiments on Spring-Slider System. In: de Rubeis, V., Czechowski, Z., Teisseyre, R. (Eds.), Geoplanet: Earth and Planetary Sciences, Volume 1: Synchronization and Triggering: From Fracture to Earthquake Processes. Springer, Heidelberg, pp. 123−164.

Eckmann, J.P., Kamphorst, S., Ruelle, D., 1987. Recurrence plots of dynamical systems. Europhys. Lett. 4, 973−977.

Goltz, C., 1997. Fractal and Chaotic Properties of Earthquakes. Springer, Berlin.

Grassberger, P., Procaccia, I., 1983. Characteristic of strange attractors. Phys. Rev. Lett. 50 (5), 346−349.

Marsan, D., Nalbant, S.S., 2005. Methods for measuring seismicity rate changes: a review and a study of how the m-w 7.3 Landers earthquake affected the aftershock sequence of the Mw 6.1 Joshua tree earthquake. Pageoph. 162 (6−7), 1151−1185.

Marsan, D., Wyss, M., 2011. Seismicity rate changes. Community Online Resource for Statistical Seismicity Analysis . Available from: https://doi.org/10.5078/corssa-25837590. Available at http://www.corssa.org.

Matcharashvili, T., Chelidze, T., Javakhishvili, Z., 2000. Nonlinear analysis of magnitude and interevent time interval sequences for earthquakes of the Caucasian region. Nonlin. Processes Geophys. 7, 9−19.

Matcharashvili, T., Chelidze, T., Zhukova, N., 2015. Assessment of a ratio of the correlated and uncorrelated waiting times in the Southern California earthquake catalogue. Physica A 433, 291−303.

Matthews, M., Reasenberg, P., 1988. Statistical methods for investigating quiescence and other temporal seismicity patterns. Pure Appl. Geophys. 126, 357−372.

Molaie, M., Jafari, S., Moradi, M.H., Sprott, J.C., Golpayegan, S., 2013. A chaotic viewpoint on noise reduction from respiratory sounds. Biomed. Signal Process. Control. Available from: https://doi.org/10.1016/j.bspc.2013.10.009.

Newman, W., Turcotte, D., 2002. A simple model for the earthquake cycle combining self-organized complexity with critical point behavior. Nonlin. Processes Geophys. 9, 453−461. Available from: https://doi.org/10.5194/npg-9-453-2002.

Ogata, Y., 1999. Seismicity analysis through point-process modeling: a review. Pageoph. 155, 471−507.

Pavlos, G., Karakatsanis, L., Lattoussakis, L., Dialetis, N., Papaioannou, G., 1994. Chaotic analysis of time series composed of seismic events recorded in Japan. Int. Jr. Bifurcat. Chaos 4, 87−98.

Press, W., Teukolsky, S., Vetterling, W., Flannery, B., 1997. Numerical Recipes in C. The Art of Scientific Computing. Cambridge Univ. Press, Cambridge.

Pykovsky, A., Rosenblum, M., Kurths, J., 2003. Synchronization: A Universal Concept in Nonlinear Science. Cambridge Univ. Press, Cambridge.

Ralaivola, L., d'Alché-Buc, F., 2005. Time Series Filtering, Smoothing and Learning using the Kernel Kalman Filter. In: Proc. of IEEE Int. Joint Conference on Neural Networks, Vol. 3, pp. 1449−1454, Montreal, Canada.

Rundle, J., Turcotte, D., Klein, W. (Eds.), 2009. Geocomplexity and Physics of Earthquakes. American Geophysical Union, Washington, DC.

Savage, H., Marone, C., 2007. Effects of shear velocity oscillations on stick-slip behavior in laboratory experiments. J. Geophys. Res: Solid Earth. 112B. Available from: https://doi.org/10.1029/2005JB004238.

Scholz, Ch, 1990. The Mechanics of Earthquakes and Faulting. Cambridge Univ. Press, New York.

Scholz, Ch, 2003. Good Tidings. Nature 425, 670–671.

Sobolev, G., 2011. Seismicity dynamics and earthquake predictability. Nat. Hazards Earth Syst. Sci. 11, 445–458. Available from: https://doi.org/10.5194/nhess-11-445-2011.

Sornette, D., 2000. Critical Phenomena in Natural Sciences. Springer, Berlin.

Sprott, J., 2006. Chaos and Time-Series Analysis. Oxford University Press, Oxford, New York.

Srivastava, H., Bhattacharya, S., Sinha Ray, K., 1996. Strange attractor characteristics of earthquakes in Shillong Plateau and adjoining regions. Geophys. Res. Lett. 23, 3519–3522.

Strogatz, S., 1994. Nonlinear Dynamics and Chaos. Westview Press, Cambridge.

Takens, F., 1981. Detecting strange attractors in fluid turbulence. In: Rand, D., Young, L.-S. (Eds.), Dynamical Systems and Turbulence. Springer, Berlin, pp. 366–381.

Telesca, L., Lovallo, M., Golay, J., Kanevski, M., 2016. Comparing seismicity declustering techniques by means of the joint use of Allan Factor and Morisita index. Stoch. Environ. Res. Risk Analysis 30, 77–90.

Thanassoulas, C., Klentos, V., Verveniotis, G., Zymaris, N., 2009. Preseismic oscillating electric field "strange attractor" like precursor, of T = 14 days, triggered by M1 tidal wave. Application on large (Ms 6.0R) EQs in Greece (March 18th, 2006 – November 17th, 2008). arXiv:0901.0467v1 [physics.geo-ph].

Webber, C., Marwan, N., 2015. Recurrence Quantification Analysis. Theory and Best Practices. Springer, Cham Heidelberg.

Further Reading

Becker, T.W., 2000. Deterministic chaos in two state-variable friction sliders and the effect of elastic interactions, In: Rundle, J.B., Turcotte, D.L., Klein,W., (Eds.) Geo Complexity and the Physics of Earthquakes, Geoph. Monog. Series, 120: l5–26.

Huang, Y., Saleur, H., Sammis, C., Sornette, D., 1998. Precursors, aftershocks, criticality and self- organized criticality. EPL Europhys. Lett. 41, 43–48.

Lyubushin, A., Pisarenko, V., Ruzich, V., Buddo, V., 1998. A new method for identifying seismicity periodicities. Volcanol. Seismol. 20, 73–89.

Sobolev, G., Spetzler, H., Koltsov, A., Chelidze, T., 1993. An experimental study of triggered stick-slip. Pure Appl. Geophys. 140 (1), 79–94.

Rosenau, M., Corbi, F., Dominguez, S., 2017. Analogue earthquake and seismic cycles: experimental modelling across time scales. Solid Earth 8, 597–635.

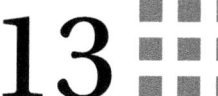

13

Four-Stage Model of Earthquake Generation in Terms of Fracture-Induced Electromagnetic Emissions: A Review

Konstantinos Eftaxias[1], Stelios M. Potirakis[2], Yiannis Contoyiannis[2]

[1]NATIONAL AND KAPODISTRIAN UNIVERSITY OF ATHENS, ATHENS, GREECE
[2]UNIVERSITY OF WEST ATTICA, ATHENS, GREECE

CHAPTER OUTLINE

Complexity of Seismic Time Series. DOI: https://doi.org/10.1016/B978-0-12-813138-1.00013-4

13.1 Introduction (The State of the Art at the Beginning of the Investigation)

A vital problem in material science is the identification of precursors of macroscopic defects or shocks. An opening crack of a stressed rock behaves like a *stress-electromagnetic (EM) transducer*. More specifically, EM emissions (EMEs) in a wide frequency spectrum ranging from the kHz to the MHz bands are produced by opening cracks, which can be considered as the so-called precursors of general fracture.

Improvements in the *MHz–kHz EME technique* have permitted real-time monitoring of the fracture process at the laboratory scale (Baddari et al., 1999, 2011; Fukui et al., 2005; Kumar and Misra, 2007; Chauhan and Misra, 2008; Baddari and Frolov, 2010; Lacidogna et al., 2010, 2011; Schiavi et al., 2011; Carpinteri et al., 2012, 2015). However, the MHz–kHz EM precursors are detectable not only at the laboratory but also at the geological scale. A stressed rock behaves like a *stress-EM transducer*. The idea that the fracture-induced MHz–kHz EM fields should also permit the monitoring of gradual damage to stressed materials in the Earth's crust, as happens in laboratory experiments, in real-time and step-by-step, cannot, in principle, be excluded.

An interesting experimental research direction would be the parallel monitoring of the corresponding observable manifestations of both laboratory and geophysical scale fracture phenomena. We consider earthquakes (EQs) as large-scale fracture phenomena. One cannot ignore the profound analogies between failure precursors at the laboratory and geophysical scales, and thus it has been early suggested that 'the mechanism of EQs is apparently some sort of laboratory fracture process' (Mogi, 1962a,b, 1968, 1985; Ohnaka and Mogi, 1982; Lockner and Madden, 1991; Lockner et al., 1991; Kuksenko et al., 1996, 2005, 2007, 2009; Ponomarev et al., 1997; Scholz, 2002; Sobolev and Ponomarev, 2003; Rumi and Ananthakrishna, 2004 and references therein; Lei and Satoh, 2007; Muto et al., 2007; Chauhan and Misra, 2008; Baddari et al., 2011). On the other hand, it has been emphasized that it is often difficult to study the kinetics of the fracture of brittle rocks in the laboratory due to rapid unstable fracture growth in the last and more interesting stages of this process (Lockner et al., 1991; Ponomarev et al., 1997). At the laboratory scale the fault growth process normally occurs violently in a fraction of a second (Lockner et al., 1991). *Therefore, crucial information probably is lost* (Main and Naylor, 2012). A major difference between the laboratory and natural processes is the order-of-magnitude differences in scale (in space and time), allowing the possibility of experimental observation at the geophysical scale for a range of physical processes which are not observable at the laboratory scale (Main and Naylor, 2012). Thus, the idea that field observations by means of EM anomalies will probably reveal features of the last crucial stages of EQ generation, which are not clearly observable at the laboratory scale, cannot, in principle, be excluded.

Based on this idea we have installed a field experimental station using the same instrumentation as in laboratory experiments for the recording of geophysical-scale EME. An exemplary telemetric station has been operating on Zakynthos (Zante) Island (Greece) since 1994, mainly aimed at the detection of kHz–MHz EM precursors. It has been installed in a carefully selected mountainous site in the southwest part of the island ($37.76°N–20.76°E$) providing low EM background noise. The measurement system (Fig. 13–1) is mainly comprised (1) six loop antennas detecting the three components (East-West (EW), North-South (NS) and vertical) of the variations of the magnetic field at 3 and 10 kHz, respectively and (2) three vertical $\lambda/2$ electric dipole antennas detecting the electric field variations at 41, 54 and 135 MHz, respectively. Moreover, two short thin wire antennas (STWAs), oriented at EW and NS directions at lengths of 100 m each, have been installed. The aim of the last installation is the detection of a different type precursor, namely, an ultra-low-frequency (ULF) (<1 Hz) EM precursor rooted in a preseismic lithosphere–atmosphere–ionosphere (LAI) coupling. All the time series are sampled once per second, i.e., with a sampling frequency of 1 Hz.

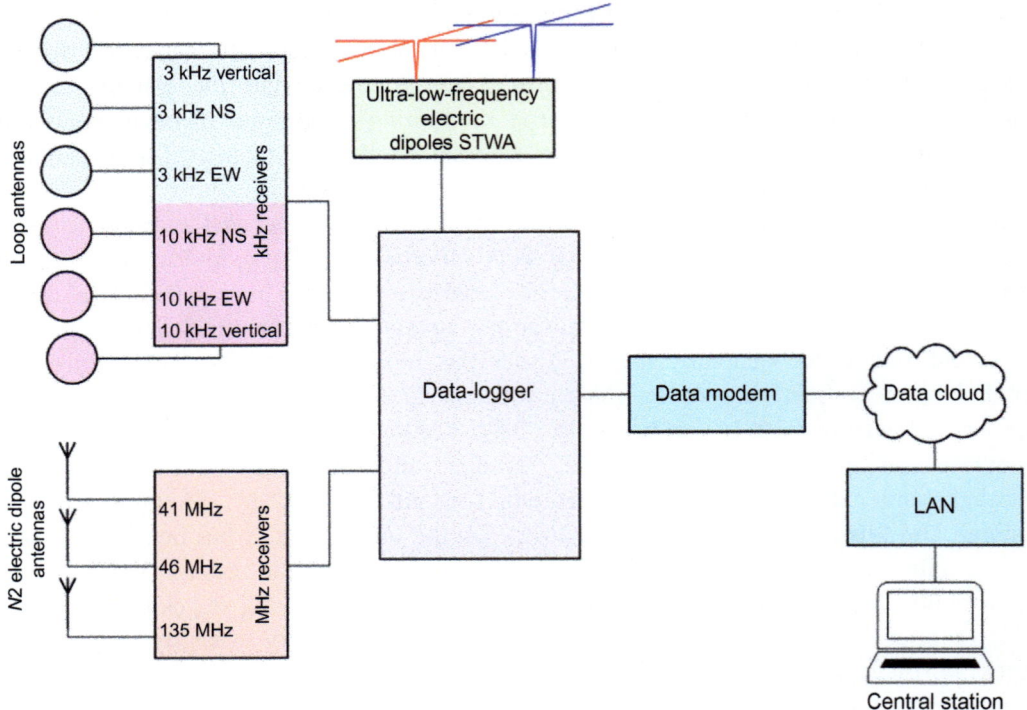

FIGURE 13–1 Block diagram of the measurement system installed at Zakynthos station.

Such an experimental setup helps to specify not only whether or not a single fracture-induced MHz or kHz EM anomaly or an LAI-coupling rooted activity is preseismic in itself, but mainly whether a sequence of three different EM disturbances at different frequencies emerging one after the other in a short time period with each one satisfying an austere set of criteria, could be characterized as a preseismic one.

13.2 A Proposed Strategy for the Study of MHz and kHz EM Precursors

The wind prevailing in the scientific community does not appear to be favourable for EQ prediction research, in particular for research related to short-term prediction (Uyeda et al., 2009). Sometimes the arguments have been extended to the extreme claim that any EM precursory activity is impossible. 'Are there credible EM EQ precursors?' This is a question debated in the scientific community (Eftaxias, 2012). Despite fairly abundant evidence, EM precursors have not been adequately accepted as real physical quantities.

We believe that a 'preseismic-EME' shift in thinking towards basic science can lead to their strict definitions; the thorough understanding of fracture-induced EM precursors in terms of physics is a path to achieve deeper knowledge of the last stages of the EQ prepara-tion process and thus a path to more credible short-term EQ prediction. *No scientific prediction is possible without the exact definition of the anticipated phenomenon and the rules, which define clearly in advance of it whether the prediction is confirmed or not* (Kossobokov, 2006).

Based on these ideas, we have focused, in an appropriately critical spirit, on asking the following crucial questions:

1. *How can we recognize a MHz or kHz EM observation as a preseismic one?* One wonders whether necessary and sufficient criteria have yet been established, which permit the characterization of an EM anomaly as a real EM precursor. One of the main purposes of this contribution is to suggest a procedure for the designation of observed kHz/MHz EM anomalies as seismogenic ones.

2. *How can we link an individual MHz and kHz EM precursor with a distinctive stage of the EQ preparation?* This is a crucial question. *Scientists ought to attempt to link the available various precursory EM observations, which appear one after the other, to the consecutive processes occurring in Earth's crust.* It is well established that as total instability approaches, the observed spectrum of acoustic emission (AE) and EME moves to lower frequencies. An important feature, observed both at laboratory and geophysical scales, is that the MHz radiation systematically precedes the kHz one (Ohnaka and Mogi, 1982; Qian et al., 1994; Eftaxias et al., 2002, and references therein; Eftaxias et al., 2004; Kapiris et al., 2004a,b; Contoyiannis et al., 2005; Kumar and Misra, 2007; Baddari and Frolov, 2010). The remarkable asynchronous appearance of these MHz and kHz precursors indicates that they refer to different stages of the EQ preparation process. Recently, based on synergetic principles of physical mesomechanics, we have shown that the observed transition from the 'mild' MHz to the 'strong' kHz fracto-EM activity marks the transition from small-scale (mesolevel) fragmentation to large-scale (macrolevel) fragmentation (Eftaxias et al., 2007). This result seems to justify the above-mentioned suspicion that the observed MHz and kHz precursors refer to different stages of the EQ preparation process. *But still, which are these last stages?*

3. *How can we identify precursory symptoms in EM observations which signify that the occurrence of the prepared EQ is unavoidable?* This is a crucial issue, as well. A question effortlessly arising is *whether the appearance of any preseismic signal is always accompanied by an EQ or not.*

4. *Are the MHz–kHz EM precursors consistent with others precursors?* EQ's preparatory process has various facets which may be observed before the final catastrophe. Therefore, the science of EQ prediction should, from the start, be multidisciplinary. Therefore, the precursors under study should be compatible with other precursors.

5. *Are the systematically observed preseismic EME characteristics which are commonly considered as 'puzzling features' really 'puzzling' ones or are they crucial precursory features of the EQ preparation process?* The negative views concerning the existence of real EM precursors are supported by the fact that specific 'puzzling features' are

systematically observed in candidate preseismic EME. Characteristically *(1) EM silence in all frequency bands appears before the main seismic shock occurrence. (2) Although strain changes are largest at the time of EQ there are not coseismic EME.* The systematic observation of potentially precursory EM signals before but not at the time of the main shock is considered a paradox on the grounds that any mechanism must explain why the emerged EM signals are not accompanied by large precursory strain changes, much larger than those taking place during the main shock. *(3) EM silence is also observed during the aftershock period. (4) Are the fracture-induced EME, if they really exist, detectable by ground-based observatories?*

13.2.1 Our Proposal: The Four-Stage Model of Earthquake Generation by Means of Fracture-Induced EM Activities

A 'preseismic-EME' shift in thinking towards basic science, which is based on a multidisciplinary analysis, has led to the proposal of a *four-stage model for EQ generation. We think that this model provides answers to all the crucial questions raised in the previous section.* The proposed model is summarized as follows:

1. *First stage:* The initially observed MHz EM anomaly is due to the fracture of the highly heterogeneous system that surrounds the formation of strong brittle and high-strength entities (asperities) distributed along the rough surfaces of the main fault sustaining the system. The MHz EME can be described by means of a second-order phase transition in equilibrium.

2. *Second stage:* The appearance of tricritical behaviour in the final stage of MHz EME, or in the initial stage of kHz EME, or in both, signals the next, distinct, state of the EQ preparation process.

3. *Third stage:* The final abruptly emerging strong sequence of kHz EM avalanches originates in the stage of stick–slip-like plastic flow, namely, the fracture of asperities themselves. The burst-like kHz EME does not present any footprint of a second-order transition in equilibrium.

4. *Fourth stage:* Finally, the systematically observed EM silence in all frequency bands before the time of the EQ occurrence is sourced in the process of preparation of the dynamic slip which results in the fast, even super-shear, mode that surpasses the shear wave speed and corresponds to the observed EQ tremor.

13.3 Focus on the First Stage Reflected in the Observed Preseismic MHz EM Field

A significant EQ is what happens when the two surfaces of a major fault slip one against the other under the stresses rooted in the motion of tectonic plates. However, large stresses siege the major fault after the activation of a population of smaller faults in the heterogeneous region that surrounds the major fault. EQ triggering is driven by the smallest EQs at all

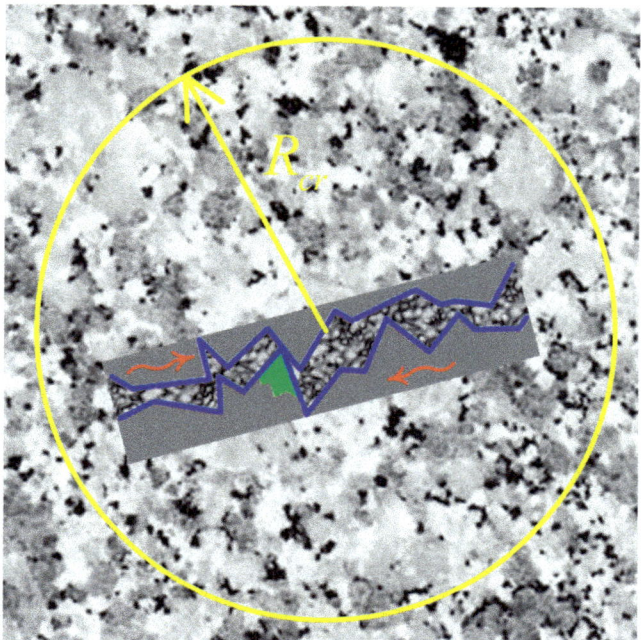

FIGURE 13–2 A fault (blue lines) is embedded in a heterogeneous environment. The MHz EME is emitted during the fracture of a disordered medium surrounding the major fault over a critical circle (yellow). The kHz EME is emitted during the fracture of the asperities (green highlighted area). *EME*, electromagnetic emission.

scales, even for the largest EQs (Helmstetter, 2003). *Small EQs are the agents by which longer stress correlations are established* (Bowman et al., 1998).

We suggest that the initially observed MHz EM anomaly is due to the fracture of the highly heterogeneous system that surrounds the formation of strong brittle and high-strength entities (asperities) distributed along the rough surfaces of the main fault sustaining the system (Fig. 13–2).

13.3.1 The MHz EM Phenomenon in Terms of Criticality

The compressive failure of a disordered medium appears as a complex cumulative process involving long-range correlations, interactions and coalescence of microcracks. In analogy to the study of critical-phase transitions in statistical physics, it has been proposed that the fracture of heterogeneous materials could be viewed as a critical phenomenon (Allegre et al., 1982; Chelidze, 1982, 1986; Herrmann and Roux, 1990; Sornette, 1991, 1999, 2000; Sornette and Sammis, 1995; Buchel and Sethna, 1997; Bowman et al., 1998; Sornette and Andersen, 1998; Kossobokov et al., 1999; Kun and Herrmann, 1999; Moreno et al., 2000; Gluzman and Sornette, 2001; Guarino et al., 2002; Rundle et al., 2003, and references therein). Nature seems to paint the following picture in the fracture of a heterogeneous system (Garcimartin et al., 1997; Bowman et al., 1998; Sornette, 2000; Halasz et al., 2012; Girard et al., 2010,

2012): In the early stages of deformation, when the disordered medium is subjected to external load, the weak components break immediately and serve as nucleation centres for the growth of broken clusters. The load transferred to the nearest neighbours of broken components gives rise to further breaking. As deformation proceeds, cooperative effects appear, cracking areas cluster in space according to scale-free patterns and dynamically interact with each other. As the external load increases, larger clusters are formed and long-range correlations buildup through local interactions until they extend throughout the entire system. *All these results advocate for a critical point interpretation of failure.* The challenge in the analysis of a recorded MHz EME time-series is to show that this anomaly includes the above-mentioned critical features and especially to detect the 'critical epoch' during which the 'short-range' correlations evolve into 'long-range' ones, as well as the epoch of localization of the damage. We argue that the aforementioned two crucial epochs can be identified in the recorded MHz EME time-series: the MHz EM activity behaves as a second-order phase transition in equilibrium.

13.3.2 The Fracture Control Mechanism is Characterized by a Negative Feedback

Based on a fractal spectral analysis of the MHz EME time-series, it has been shown that the associated Hurst-exponent, H, lies in the range $0 < H < 0.5$ indicating that the dynamics of the observed MHz EM field is characterized by *antipersistency* (Kapiris et al., 2004a; Contoyiannis et al., 2005), namely, if the EM fluctuations increase in one period, it is likely to reduce in the period immediately following, and vice versa. The associated physical information is that the control mechanism regulating the fracture is a negative feedback one that 'kicks' the cracking rate away from extremes, providing adaptability to the system that is the ability to respond to various external stresses. The existence of antipersistency supports the suggestion that the MHz EME could be described in analogy with a thermal second-order phase transition in equilibrium (Contoyiannis et al., 2005).

13.3.3 The Analysis of MHz EME by Means of the Method of Critical Fluctuations

Characteristic features at a critical point of a second-order transition are (1) the existence of strongly correlated fluctuations, right at the 'critical point' the subunits/cracking clusters are well-correlated even at arbitrarily large separation, this means that the correlation function $C(r)$ follows long-range power-law decay. *In the present case, small EQs are the agents by which longer stress relation is established: they effectively smooth the stress field at large-scale lengths* (Bowman et al., 1998). (2) The appearance of self-similar structures both in time and space. This fact is mathematically expressed through power law expressions for the distributions of spatial or temporal quantities associated with the aforementioned self-similar structures (Stanley, 1987, 1999). Below and above the critical point a dramatic breakdown of critical characteristics, in particular long-range correlations, appears; the correlation function turns into a rapid exponential decay (Stanley, 1987, 1999).

Recently, the method of critical fluctuations (MCF) has been introduced, which can reveal the critical state, the tricritical state, as well as the departure from critical state (Contoyiannis and Diakonos, 2000, 2007; Contoyiannis et al., 2002). It has been shown (Contoyiannis and Diakonos, 2000) that the dynamics of the order parameter fluctuations ϕ at the critical state for a second-order phase transition can be theoretically formulated by the nonlinear intermittent map:

$$\phi_{n+1} = \phi_n + u\phi_n{}^z, \tag{13.1}$$

where ϕ_n is the scaled order parameter value at the time interval n; u denotes an effective positive coupling parameter describing the nonlinear self-interaction of the order parameter; z stands for a characteristic exponent associated with the isothermal exponent δ for critical systems at thermal equilibrium ($z = \delta + 1$). The marginal fixed-point of the above map is the zero point, as expected from critical phenomena theory.

However, it has been shown that in order to quantitatively study a real (or numerical) dynamical system one has to add an unavoidable 'noise' term, ε_n, to Eq. (13.1), which is produced by all stochastic processes (Contoyiannis and Diakonos, 2007). Note that, from the intermittency mathematical framework point of view, the 'noise' term denotes ergodicity in the available phase space. In this respect, the map of Eq. (13.1), for positive values of the order parameter, becomes:

$$\phi_{n+1} = |\phi_n + u\phi_n{}^z + \varepsilon_n|. \tag{13.2}$$

Based on the map of Eq. (13.2), MCF has been introduced as a method capable of identifying whether a system is in a critical state of intermittent type by analysing time-series corresponding to an observable of the specific system. In a few words, MCF is based on the property of maps of intermittent type, like those in Eqs (13.1) and (13.2), that the distribution of properly defined laminar lengths (waiting times) l follow a power-law $P(l) \sim l^{-p_l}$ (Schuster, 1998), where the exponent p_l is $p_l = 1 + (1/\delta)$ (Contoyiannis et al., 2002). However, the distribution of waiting times for a real data time-series, which is not characterized by critical dynamics, follows an exponential decay, rather than a power-law one (Contoyiannis et al., 2004a,b), due to stochastic noise and finite size effects. Therefore, the dynamics of a real time-series can be estimated by fitting the distribution of waiting times (laminar lengths) to a function $\rho(l)$ combining both power-law and exponential decay (Contoyiannis and Diakonos, 2007):

$$\rho(l) \sim l^{-p_2} e^{-lp_3}. \tag{13.3}$$

The values of the two exponents p_2 and p_3, which result after fitting laminar length distribution in a log–log scale diagram, reveal the underlying dynamics. Exact critical state calls for $p_3 = 0$; in such a case $p_2 = p_l > 1$. As a result, in order for a real system to be considered to be at critical state, *both criticality conditions $p_2 > 1$ and $p_3 \approx 0$ have to be satisfied.*

Note that the choice of the function $\rho(l)$ of Eq. (13.3), which combines both power-law and exponential decay, to model the distribution of waiting times was deliberately made in order to include both these fundamentally different behaviours, i.e., the critical dynamics (Contoyiannis et al., 2002) and the complete absence of specific dynamics (stochastic processes) (Contoyiannis et al., 2004a), respectively. In addition, the specific function also models intermediate behaviours (Contoyiannis and Diakonos, 2007).

In applying the MCF the corresponding factors of $\rho(l)$ appear to be competitive: any increase of the p_2 exponent value corresponds to a p_3 exponent value reduction and vice versa. However, this is expected because, for example, any increase of the value of the p_3 exponent signifies a departure from critical dynamics and thus a reduction of the p_2 exponent value.

The analysis of MHz EME by means of the MCF reveals the above-mentioned critical features. Specifically, it reveals:

1. The time-window in the MHz EME time-series that corresponds to the 'critical window', i.e., the epoch during which the short-range correlations between the cracking areas have been evolved to long-range ones. More precisely, *the laminar lengths (waiting times) fit a power-law type distribution* (Contoyiannis et al., 2004b, 2005, 2010, 2013; Eftaxias et al., 2009; Potirakis et al., 2015, 2016a). Importantly, the 'critical window' in the MHz time-series is characterized by strong antipersistency, the system has the ability *to respond to various external stresses* (Kapiris et al., 2004a; Contoyiannis et al., 2005).

2. The 'noncritical window' in the MHz EM time-series which emerges after the appearance of the 'critical window' (Contoyiannis et al., 2004b, 2005, 2010, 2013). The timescale invariance that characterizes the critical window has been lost; *the laminar lengths (waiting times) fit an exponential type distribution*. This means that short-range correlations between the cracking areas have emerged. Moreover, this window shows lower antipersistency, which means that it becomes less anticorrelated, as the associated H-exponents are closer to 0.5; the system has lost a part of its adaptability which is the ability to respond to stresses.

The above-mentioned transition from the critical epoch to the noncritical one constitutes a crucial feature of second-order phase transition known as 'symmetry breaking' (Fig. 13–3) (Contoyiannis et al., 2005): its appearance reveals the transition from the phase of nondirectional, almost symmetrical, cracking distribution to a directional localized cracking zone (Fig. 13–2). *Therefore, the MCF also reveals the time window in the MHz EME time-series where the fracture process is strongly localized along the main fault.*

13.3.4 Is an Earthquake Unavoidable After the Appearance of an MHz EM Anomaly?

We emphasize that the completion of the 'symmetry breaking' implies that the rupture process has already been obstructed along the backbone of strong asperities sustaining the fault

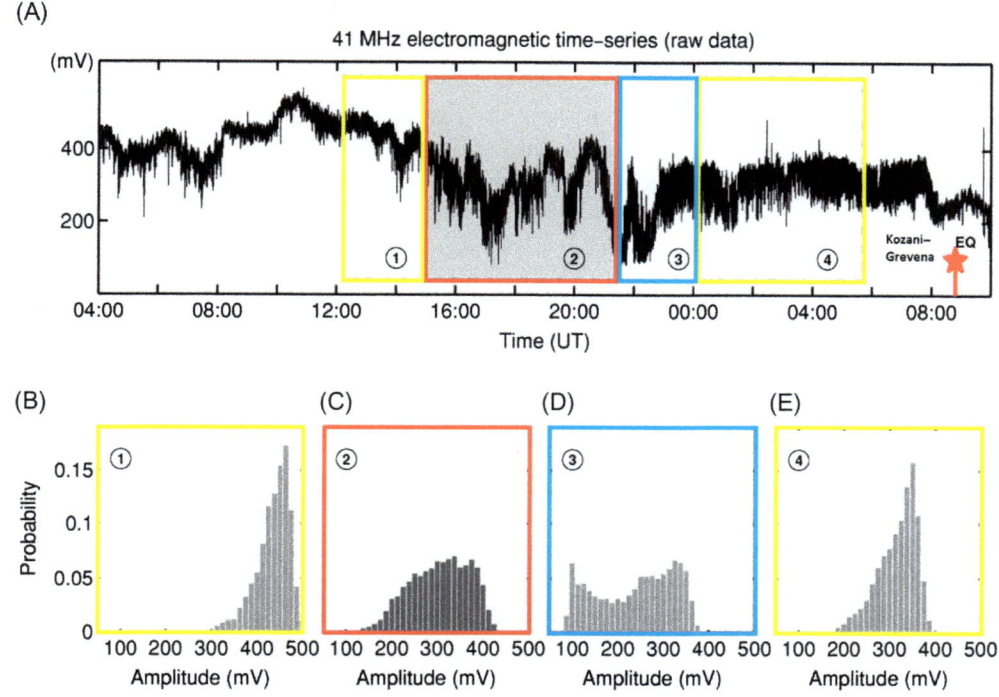

FIGURE 13–3 (A) The 41-MHz time-series associated with the Kozani–Grevena earthquake. The red star indicates the time of the earthquake occurrence. Bar graphs (B)–(E) show the distribution of the amplitude of EM pulses for four consecutive time intervals marked in (A). The second (red-framed and shaded) time interval determines, in terms of the MCF, the crucial time interval during which the short-range correlations evolve to long range (critical window); the corresponding distribution (C) might be considered to be a precursor of the impending symmetry breaking readily observable in the (blue-framed) distribution (D) of the subsequent time interval. The distribution in (E) is very similar to that of (B), while here there is an upward shift of the values to the range of the second lobe of the distribution in (D); the appearance of the distribution in (E) may indicate that the symmetry breaking in the underlying fracto-EM mechanism has been almost completed. The aforementioned evolution is expected in the framework of the hypothesis that the fracture in the highly disordered media develops as a kind of generalized continuous-phase transition. *EM*, electromagnetic; *MCF*, method of critical fluctuations.

surfaces (Fig. 13–2). The 'siege' of asperities has already been started. *However, this does not mean that the EQ is unavoidable.*

We note that Girard et al. (2012) conclude that *at peak load, the largest damage cluster does not yet span the heterogeneous system.* Then, during the postpeak phase cracking events are localized in the vicinity of one or a few large damage clusters that eventually evolve into a spanning cluster. *The last feature is consistent with that of 'symmetry breaking'.* Girard et al. (2010) using simulations conclude that *the spatial correlation length associated with damage events reaches the size of the system at peak load.* This means that the divergence of the correlation length precedes the final failure of the disorder medium. Strain-driven, compressive failure experiments on rocks have resulted in a similar observation: the failure plane is not fully formed at peak load (Lockner et al., 1991). In consistency, experiments in terms of synergetic principles of physical mesomechanics show that the 'mild' MHz fracto-EM activity marks the small-scale (mesolevel) fragmentation, while the 'strong' kHz fracto-EM activity marks the fragmentation to large-scale (macrolevel) (Eftaxias et al., 2007). Finally, the MHz EME reflects, in terms of percolation theory, the 'hydraulic threshold', where the transition from impermeable to permeable occurs, and not the 'mechanical' or 'damage threshold' (see Section 13.3.6). The above-mentioned results support the analysis by means of MCF and enhance the conclusion that the launch of MHz fracto-EME does not mean that the EQ under preparation is unavoidable. *The abrupt emergence of strong avalanche-like kHz EM activity reveals the fracture of asperities, if and when the local stresses exceed their fracture stress* (Contoyiannis et al., 2005) *indicating that the EQ is unavoidable* (see Section 13.5).

13.3.5 On the Physical Mechanism that Organizes the Heterogeneous System in its Critical State

A crucial question refers to what is the physical mechanism that organizes the heterogeneous system in its critical state. Lévy flights and Lévy walks are applied in modelling physical systems with spatiotemporal fractality (Bouchaud and Georges, 1990). The characteristic feature of Lévy flight is that it does not converge to the Gaussian stochastic process; instead it is 'attracted' towards the Lévy stable process with infinite variance. Lévy stable distributions, although they play an important role in mathematics are basically nonphysical, because in the real world there exist no processes ('Lévy flights') which would produce empirical data with infinite moments (Kwapieńa and Drożdża, 2012). For this reason, a family of more realistic distributions called 'truncated Lévy distributions' was introduced by Mantegna and Stanley (1994) where an upper cutoff to the values of random variables was introduced.

Combining the ideas of *truncated Lévy* statistics, nonextensive Tsallis statistical mechanics and criticality with features hidden in the precursory MHz EME time-series we have shown that *a truncated Lévy walk type mechanism can organize the heterogeneous system to criticality*. Intuitively, the proposed Lévy walk mechanism could be the result of a feedback 'dialogue' between the stresses and heterogeneity (Contoyiannis and Eftaxias, 2008).

13.3.6 The MHz EM Anomaly by Means of Percolation Theory

In this subsection, we attempt to explain the observed cessation of the MHz EME before the EQ occurrence based on the pioneering work of Chelidze, who first studied the EQ preparation process by means of the percolation theory (Stauffer, 1985).

It has been shown that the problem of the fracture of heterogeneous media mathematically corresponds to problems of the *percolation theory*, which quantitatively describe the connectivity of components in a nonhomogeneous system (Chelidze, 1979, 1980a,b, 1982, 1986, 1987, 1993; Chelidze et al., 1984, 1988, 1990, 2006; Arbadi and Sahimi, 1990). This theory predicts the existence of 'critical points' or *percolation transitions* in various properties of the material, when its characteristics change dramatically. Recently we have studied the emergence of a sequence of preseismic MHz and kHz EME anomalies in terms of percolation theory (Eftaxias et al., 2013a), discriminating the following thresholds (Chelidze, 1986):

1. The 'hydraulic threshold', x_c, where the transition from impermeable to permeable occurs.

2. The 'mechanical' or 'damage threshold', x_m, where the infinite cluster (IC) is formed, and the solid disintegrates. As a rule, the relation $x_m > x_c$ holds.

3. The EQ is considered as a shear displacement along the fault plane. However, a layer of thickness h of the material should contain the IC. Chelidze (1986) has proposed that the associated problem of shear displacement along the fault can be formulated in terms of mechanical percolation as follows: 'find the concentration x_{mf} at which the *flat IC (FIC) of voids (cracks) spanning the layer of thickness h* would be formed in a 3-D body'. The required critical concentration for the formation of FIC, x_{mf}, is higher in comparison with x_m (Chelidze, 1986).

It might be considered that the precursory MHz EME, which is emitted during the last week before the main shock occurrence, is associated with the mechanical threshold x_m, namely with the formation of a flat cluster of thickness h (Eftaxias et al., 2013). During this formation, the fracture or grinding of weak heterogeneous material distributed around the main fault emits the antipersistent MHz radiation. Its behaviour as a second-order phase transition as well as the observed symmetry breaking during the temporal evolution of this radiation strongly supports the aforementioned hypothesis, especially the association of the MHz precursor with the formation of the flat cluster.

The kHz EME, which is emitted from a few tens of hours up to a few minutes before the EQ occurrence, might emerge around the critical mechanical threshold x_{mf}, i.e., during the formation of the FIC, in other words, during the fracture of high-strength and brittle asperities sustaining the system (Eftaxias et al., 2013a).

The facts presented in this subsection suggest that *the MHz EM anomaly should cease before the EQ occurrence*; this anomaly in terms of percolation theory is associated with the 'hydraulic threshold', where the transition from impermeable to permeable occurs. This conclusion is supported by the analysis by means of critical phenomena (see Section 13.3.4). We recall that in terms of physical mesomechanics this emission is rooted in the small-scale fragmentation (Eftaxias et al., 2007), while an important feature observed at the laboratory is that the fracture-induced MHz radiation systematically precedes the kHz one. *Thus, the noncoseismic character of this signal, namely the absence of MHz EME during the main shock occurrence, is not a puzzling feature.*

13.3.7 On the Compatibility of the MHz EM Anomaly with the Corresponding Foreshock Activity

A strong negative view concerning the existence of EM precursors refers to the absence of simultaneous *seismological precursors* (Geller et al., 1997). This seems to be a legitimate negative view. But is it valid? Are the EM precursors unrelated to foreshock seismicity?

Based on the widely documented notion of the self-affine nature of faulting and fracture (Mandelbrot, 1982; Huang and Turcotte, 1988; Turcotte, 1997; Sornette, 2000) it is expected that the foreshock seismic activity and the precursory fracture-induced MHz EME should constitute two sides of the same coin, the same way that the AE and EME are considered as such for fractures on the laboratory scale. Therefore, a proof for the relation between EM precursors and seismological precursors calls for evidence that the expected similarity between them is indeed valid. As has been shown, the MHz EME phenomenon behaves as a critical phenomenon. Then, a question which logically arises is *whether the associated foreshock activity also behaves as a critical phenomenon*. A recent analysis leads to a positive answer.

Based on the recently introduced concept of the 'natural time' by Varotsos and his colleagues (Varotsos et al., 2001, 2005, 2006; Varotsos et al., 2011 and references therein) (see also Chapter 7: Natural Time Analysis of Seismic Time Series), we have proved that (Potirakis et al., 2013b, 2015, 2016a): *the foreshock seismic activity that occurs in the region around the epicentre of the upcoming significant shock a few days up to one week before the main shock occurrence, and the observed MHz EME precursor which emerges during the same period, both behave as critical phenomena.*

The aforementioned result strongly supports (1) the seismogenic origin of the observed MHz EME precursor; (2) the suggestion that the MHz EME precursor can be described in terms of criticality as it has been revealed by the application of the MCF analysis (see Section 13.3.3); and (3) that there are simultaneous EM precursors and seismological precursors, and thus the above-mentioned negative view concerning the existence of EM precursors is not valid.

13.3.8 On the Compatibility of the MHz EM Anomaly With the Corresponding Geodetic Precursors

Geller et al. (1997) also emphasize that the absence of simultaneous *geodetic precursors* means that the observed EM anomalies are not EQ-related ones. This also seems to be a legitimate negative view concerning the existence of precursory EM anomalies.

Nevertheless, new results show that the appearance of MHz EME precursors is also compatible with geodetic results in terms of synthetic aperture radar (SAR) interferometry, an imaging technique for measuring the topography of a surface, its changes over time, and other changes in the detailed characteristics of the surface (Potirakis et al., 2016a). This method has demonstrated potential to monitor and measure surface deformations associated with EQs. Characteristically, SAR measurements have shown deformations in the case of the two significant shallow EQs that occurred in January–February 2014 in Cephalonia (Kefalonia) island, Greece: (1) the $M_w = 6.0$ EQ which occurred on 26 January 2014 at (38.22°N, 20.53°E) and (2) the $M_w = 5.9$ EQ which occurred on 3 February 2014 at (38.25°N, 20.39°E). The source modelling results from multiple SAR techniques are consistent with the view that both of these were main events located on two different active faults that belong to the same seismic source zone (Merryman Boncori et al., 2015). Importantly, a pair of MHz EME signals was recorded at 41 MHz, prior to each of the above-mentioned significant shallow EQs, namely, on 24 January 2014, 2 days before the first ($M_w = 6.0$) EQ, and on 28 January 2014, 6 days prior to the second ($M_w = 5.9$) EQ. Both MHz EMEs present the critical features of a second-order phase transition in equilibrium, as resulted from the analysis in terms of the MCF method (see Section 13.3.3). We note that the corresponding foreshock seismic activity, as another manifestation of the same complex system, should be at a critical state as well, before the occurrence of a main event. We have shown in terms of the method of natural time (see Section 13.3.7 and Chapter 7) that this really happens for both the studied significant EQs (Potirakis et al., 2016a).

The above-mentioned results indicate that the recorded MHz EME were accompanied not only by simultaneous seismological but by geodetic precursors, as well.

13.3.9 Is the Observed MHz EM Anomaly in Accordance With Different Precursors?

The EQ preparatory process has various facets which reflect correspondingly different precursors. The MHz EM anomaly seems to be compatible with other precursors in the frame of the first stage of the proposed four-stage model of EQ generation.

Characteristic precursors are the short-lived seismo-ionospheric EM precursors and EM anomalies rooted in preseismic LAI-coupling (Pulinets et al., 2003; Pulinets and Boyarchuk, 2004; Uyeda et al., 2009). Pulinets et al. (2003) have provided strong evidence for the occurrence of ionospheric precursors well before the main shock: ionospheric precursors within 5 days before the seismic shock were registered in 73% of the cases for EQs with a magnitude ~5, and in 100% of the cases for EQs with a magnitude ~6. We focus on the fact that the 'critical window' included in the MHz EM anomaly emerges a few days before the EQ occurrence and disappears before the EQ occurrence, while precisely this happens in the case of the aforementioned anomalies too. The critical character of the observed MHz EM anomaly seems to provide a possible explanation for this consistency. Indeed, the generation of a preseismic ionospheric anomaly requires physical and chemical transformations which occur in a spatially extended preparation (activation) zone of an impending EQ. Such a requirement is satisfied during the appearance of the 'critical window', i.e., the epoch during which the short-range correlations between the cracking areas have evolved to long-range ones in an extended area; where the 'critical radius R' is given by the empirical relation $\log R \approx 0.5M$, where M is the EQ magnitude (Bowman et al., 1998).

Additionally, ULF magnetic field variations are regularly recorded by ground-based magnetic observatories before significant EQs. Just as a reference to some recent cases, the Sichuan EQ (12 May 2008; $M = 8.0$) (Hayakawa et al., 2015b), the Tohoku EQ (11 March 2011, $M = 9$) (Hayakawa et al., 2015a and references therein; Contoyiannis et al., 2016; Potirakis et al., 2017), and the Kobe EQ (12 April 2013, $M = 6.3$) (Potirakis et al., 2016b and references therein; Hayakawa et al., 2016) can be mentioned. Such ULF variations, either of lithospheric or ionospheric origin, are also a result of spatially extensive processes. The time frame during which ULF magnetic anomalies are observed is consistent with the 'critical window' occurrence of the fracture-induced MHz EME. Importantly, analyses performed for the aforementioned EQ cases in terms of the MCF method (see Section 13.3.3) or the method of natural time (see Section 13.3.7 and Chapter 7) reveal that these ULF magnetic fields include exactly the same critical characteristics as those included in the fracture-induced MHz EME (Hayakawa et al., 2015a,b; Contoyiannis et al., 2016; Potirakis et al., 2016b). Therefore, these anomalies are also in agreement with the first stage of the proposed four-stage model. Both EQ-related MHz and ULF EM anomalies precede the EQ by a few days, while as the EQ occurrence approaches such EM anomalies are no longer observed (e.g. Skeberis et al., 2015; Hayakawa et al., 2015a,b).

Within the framework of our research, a short-lived ULF (<1 Hz) EM anomaly is detected by the STWA antennas a few days before the EQ and disappears before the time of the main shock occurrence, as it happens in the case of the preseismic MHz EM anomaly (Kapiris et al., 2003; Eftaxias et al., 2004). During quiet periods, there is a standard diurnal variation of the EM data. The emergence of this LAI-coupling seismo-ionospheric anomaly is recognized as a strong perturbation of the characteristic bay-like morphology in the chain of daily data: the systematically observed minimum around midday during the quiet days significantly decreases, sometimes even becoming a local maximum, as the EQ approaches. A fractal spectral analysis reveals that the observed anomaly is characterized by (1) a progressive acceleration event rate; (2) gradual predominance of larger EM events; and (3) gradual appearance of higher frequencies in the spectrum with a simultaneous increase in the amplitudes at each emission rate, mainly characterizing lower emission rates.

We emphasize that the disappearance all of the above-mentioned anomalies before the EQ occurrence, as happens in the case of the MHz EM anomalies, is also in agreement with the first stage of the proposed four-stage model. As has been described in Section 13.3.3, the appearance of 'symmetry breaking' reveals the transition from the phase of nondirectional, almost symmetrical, cracking distribution in an extensive area to a directional localized cracking zone. The completion of the 'symmetry breaking' implies that the rupture process has already been obstructed along the backbone of strong asperities distributed across the surfaces of the main fault. The strong localization of the fracture process leads to the corresponding localization of the induced physical and chemical transformations, which justifies the disappearance of the corresponding precursors.

Precursory anomalies of hydrothermal parameters in the coversphere and atmosphere are also observed before the EQ occurrence and disappear prior to the EQ occurrence. For example, this has been reported for the 6 April 2009 M_w 6.3 L'Aquila EQ. Critical MHz EME anomalies were detected on 26 March 2009 and 2 April 2009, while the kHz EME anomalies emerged on 4 April 2009 (Eftaxias et al., 2009, 2010; Contoyiannis et al., 2010). Anomalies of hydrothermal parameters in the coversphere and atmosphere before the 2009 L'Aquila EQ appeared in significant quasisynchronous time windows, namely, on 29–31 March 2009 (Wu et al., 2016). We emphasize the fact that the critical MHz EM anomaly corresponds, by means of percolation theory, to the critical 'hydraulic threshold', where the transition from impermeable to permeable occurs (see Section 13.3.6), which strongly supports the hypothesis that the observation of precursory anomalies of hydrothermal parameters should be in time consistency with the launch of MHz EM anomaly.

It might be concluded that the observed fracture-induced MHz EME is related not only with seismological and geodetic precursory behaviours but also with other EM precursors and precursory hydrothermal phenomena.

13.4 Focus on the Second Stage Reflected in the Observed Preseismic EM Fracto-Emission With Tricritical Crossover Dynamics

As has been proposed in Section 13.3, the first-appearing MHz fracto-EME, rooted in the fracture of a highly heterogeneous system, behaves as a second-order phase transition in equilibrium. As the fracture process continues, the heterogeneity of the system decreases. It is known that the second-order region in the phase diagram may be separated from the first-order region by a tricritical point (Huang, 1987). By means of phase transition the heterogeneity is a control parameter (like the chemical potential in the Blume—Emery—Griffith Ising model) which controls the distance to a so-called tricritical transition as the disorder decreases from a critical 'second-order' phase transition to an abrupt 'first order' regime (Sornette and Andersen, 1998).

Therefore, two questions which effortlessly arise are (1) whether the characteristic property of 'tricritical crossover dynamics' is hidden in the precursory MHz EME time-series or/and the kHz EME time-series observed after the cease of the critical part of MHz EME which behaves as a second-order phase transition and (2) whether the finally abruptly emerging pulse-like strong kHz EME behaves as a first-order phase transition. We note that the second question has a positive answer (see Section 13.5.1). Herein, we focus on the first question.

Within the framework of the MCF analysis method (see Section 13.3.3), the specific dynamics of tricritical crossover is proved to be expressed by the map (Contoyiannis et al., 2015):

$$\phi_{n+1} = \left| \phi_n - u\phi_n^{-z} + \varepsilon_n \right|, \tag{13.4}$$

where ϕ is the order parameter. This map differs from the critical map of Eq. (13.2) in the sign of the parameter u and exponent z. Note that for reasons of unified formulation we use for these parameters the same notation as in the critical map of Eq. (13.2). At the level of MCF analysis this dynamics is expressed by the estimated values for the two characteristic exponent p_2, p_3 values, which satisfy *the tricriticality condition* $p_2 < 1, p_3 \approx 0$. These values have been characterized in Contoyiannis and Diakonos (2007) as a signature of tricritical behaviour.

We note that the tricritical behaviour can be approached either from the edge of the first-order phase transition (characterizing the strong avalanche-like kHz EME attributed to the third stage of the four-stage model) or from the edge of the second-order phase transition (characterizing the critical MHz EME attributed to the first stage of the four-stage model).

As a result, it can be found either in MHz time-series, following the emission of a critical MHz EME, or in kHz time-series, preceding the emission of avalanche-like kHz EME.

Recently, such a tricritical crossover has been identified: (1) in the kHz EM time-series recorded just before the launch of the strong avalanche-like kHz emissions of the 1999 Athens (see Fig. 13−4) and the 2009 L'Aquila EQs (Contoyiannis et al., 2015). We note that before the appearance of this crossover a MHz EME was detected (with a second-order phase transition in equilibrium characteristics) in both of the aforementioned cases, while

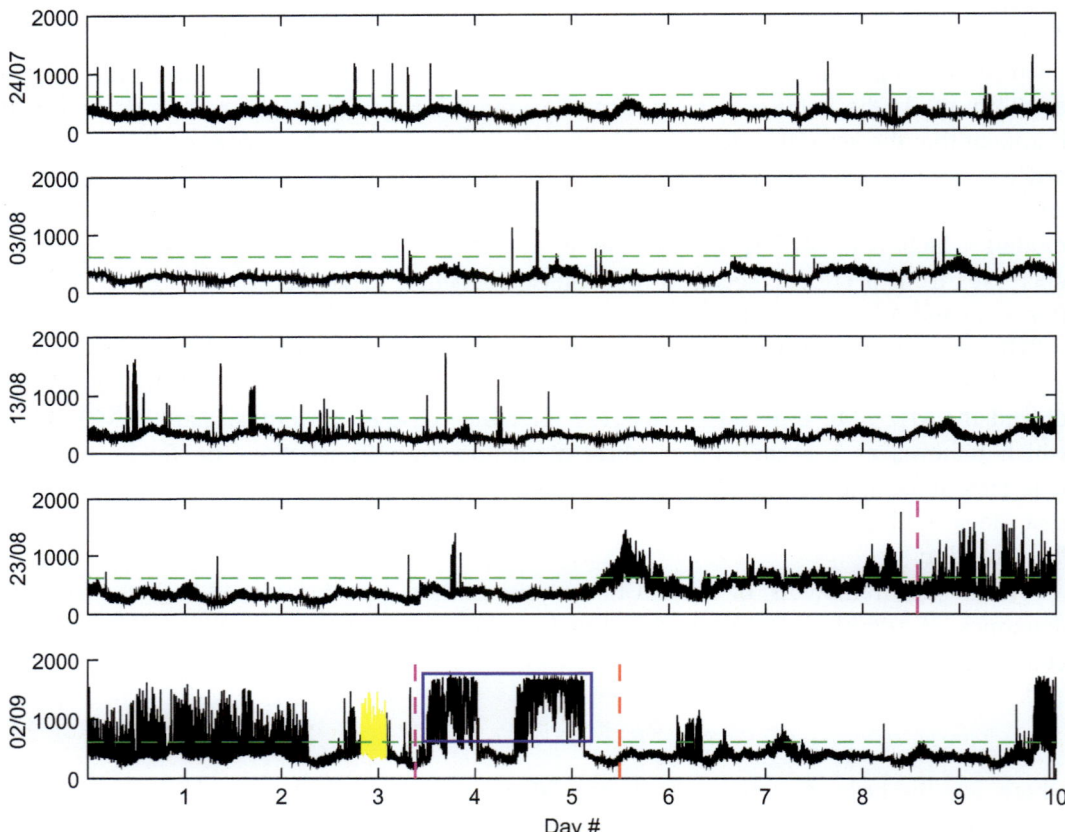

FIGURE 13–4 The 10-kHz vertical component time-series recorded prior to the Athens EQ. In this 50-day long recording, the kHz EME precursory activity starts during the day of 31 August 1999, as indicated by the vertical magenta dashed line in the second from bottom panel, with a signal presenting antipersistent characteristics which extends until 5 September 1999, indicated by the second vertical magenta dashed line in the left-hand side of the bottom panel. This antipersistent part is followed by the *strong avalanche-like kHz EME precursor* (blue rectangle), presenting persistent characteristics and corresponding to the *third stage* of the proposed four-stage model (see Section 13.5). The *tricritical EME precursor* (yellow part), corresponding to the *second stage* of the proposed four-stage model (see Section 13.4), emerges just before the strong avalanche-like kHz EME precursor. The vertical red dashed line indicates the time of occurrence of the EQ, while the horizontal green dashed line corresponds to the background noise threshold value. *EME*, electromagnetic emission; *EQ*, earthquake.

the precursory emissions were completed with the launch of first-order-type strong avalanche-like kHz EME and (2) in the MHz EM time-series recorded prior to the 2014 Cephalonia EQs (Potirakis et al., 2016a). In this case too, the observed tricritical MHz signal followed critical MHz EME.

It should be stressed here that tricritical behaviour is expected to be rarely observed in the recorded precursory EME time-series since it is expected to be strongly dependent on the degree of heterogeneity of the system that produces the MHz and kHz EME (Contoyiannis et al., 2015). We note that the tricritical windows included in the observed candidate precursory EME time-series are usually not detectable by visual inspection of the time-series. Sophisticated time-series analysis is necessary to detect them.

To summarize, in analogy to the study of critical phase transitions in statistical physics, it has been proposed that the fracture of heterogeneous materials could be studied as a critical phenomenon. The tricritical behaviour could be considered as a crucial natural working point of the EQ preparation process. We argue that the appearance of an EME phenomenon (either in the MHz or in the kHz band) presenting 'tricritical crossover dynamics', which emerges in a short time interval after the appearance of an MHz EME phenomenon presenting 'second-order-phase transition dynamics', supports the suggestion that both kinds, in terms of criticality, of observed EME are precursory ones with different critical characteristics.

Finally, we clarify that in the framework of our approach *even the launch of EM activity presenting tricritical crossover dynamics does not mean that the EQ under preparation is unavoidable.*

13.5 Focus on the Third Stage Reflected in the Observed Preseismic Strong Avalanche-Like kHz EME

The notably crucial character of the suggestion that the abrupt emergence of strong avalanche-like kHz EME activity (e.g. see the blue-rectangle-framed part of the time-series in Fig. 13−4) reveals the fracture of asperities distributed along the fault sustaining the system, namely the stick−slip-like plastic flow stage of EQ preparation, implying that the occurrence of the imminent EQ is unavoidable as soon as kHz EME have been observed, requires strong support by well-established fundamental arguments.

Laboratory experiments of rock fracture and frictional sliding have shown that the relative slip of two fault surfaces takes place in two phases. A stick−slip-like fracture-sliding precedes the dynamic fast global slip (Bouchon et al., 2001; Baumberger et al., 2002; Rubinstein et al., 2004, 2007; Chang et al., 2012; Kammer et al. 2012). Recent studies also reveal that physical systems under slowly increasing stress may respond through abrupt events. Such jumps in observable quantities are abundant, being found in systems ranging from complex social networks to EQs (Papanikolaou et al., 2012). We propose that the abruptly emerging sequence of kHz EM avalanches originates in the stage of stick–slip-like plastic flow in accordance with the above-mentioned studies.

Our understanding of different rupture modes is still very much in its infancy (Ben-David et al., 2010). We note that the pioneering laboratory friction experiments of Rabinowicz (1951) showed that the transition between static and dynamic friction occurs over a characteristic slip. We propose that the abruptly emerging sequence of kHz EM avalanches (Figs 13–4 and 13–5) originates in the stage of stick–slip-like plastic flow. This precursor shows persistent behaviour and does not include any signature of phase transition of second order in equilibrium.

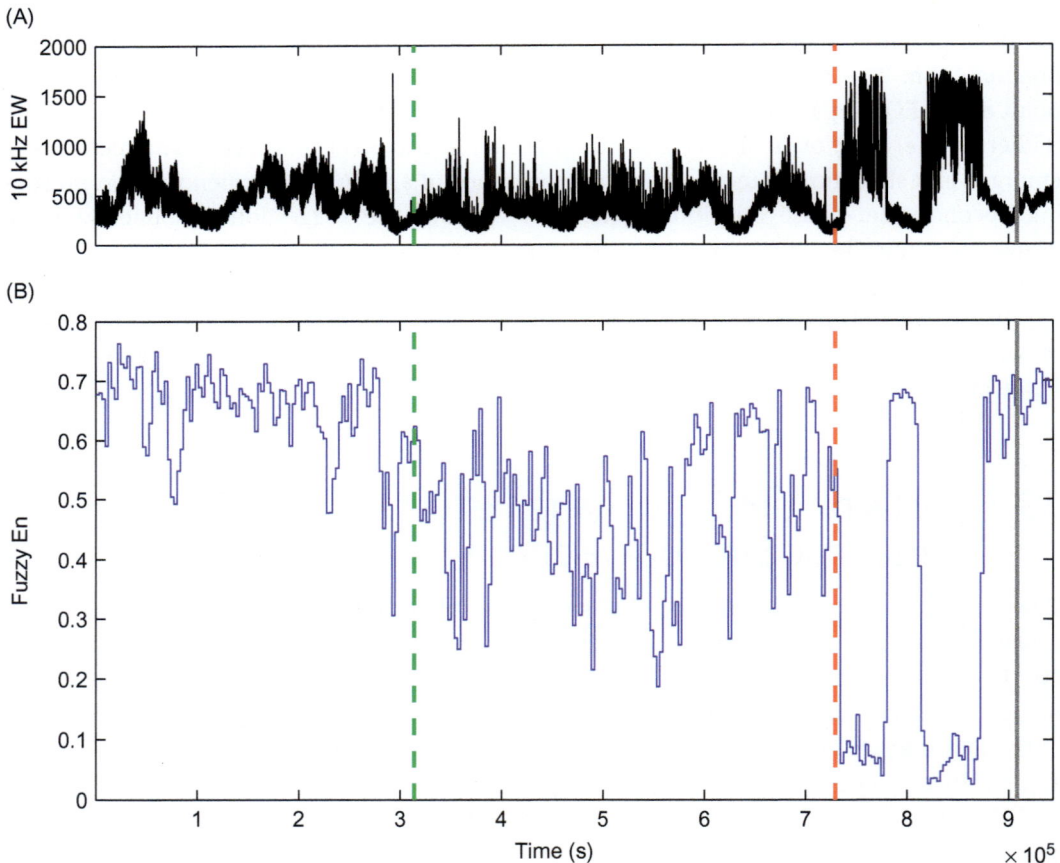

FIGURE 13–5 (A) Part of the recorded time-series of the 10-kHz (East–West) magnetic field strength (in arbitrary units) covering the 11-day period from 28 August 1999, 00:00:00 (UT), to 7 September 1999, 23:59:59 (UT), associated with the Athens EQ. (B) The corresponding FuzzyEn. The common horizontal axis is the time (in s), denoting the relative time position from the beginning of the analysed part of the kHz EME recording. The (left) vertical broken green line roughly indicates the start of the candidate precursor. The (right) vertical broken red line indicates where the high-organization, persistent, strong, avalanche-like kHz EME precursor starts. The (right) vertical solid grey line indicates the time of occurence of the Athens EQ. *EME*, electromagnetic emission; *EQ*, earthquake.

13.5.1 Arguments in Terms of Statistical Analysis

First of all, the pulse-like abruptly emerged kHz EME does not present a footprint of a second-order phase transition in equilibrium, as is happens in the case of the MHz EME. In terms of the MCF analysis method as presented in Section 13.3.3, it is also possible to detect first-order phase transitions. It has been shown (Contoyiannis et al., 2015) that the MCF analysis of kHz EM precursors results in the distribution of laminar lengths (waiting times) that diverge from power-law behaviour, due to increased p_3 exponent values, as well as $p_2 < 1$. This is explained by the fact that abrupt behaviours dominate inside a strong burst and so a 'cut' in long stay times of a power-law distribution occurs. This means that kHz EME seems to behave in analogy to a first-order phase transition where the abrupt behaviours do indeed dominate. Based on a multidisciplinary statistical analysis, we have verified that the strong avalanche-like kHz EME time-series are characterized by the following crucial symptoms of an extreme, out of equilibrium, phenomenon:

1. *High organization, high information content, low complexity.* A challenge in this field of research is to distinguish characteristic epochs in the evolution of candidate precursory kHz EME time-series including signatures that imply the transition from a normal state to a main catastrophic event, i.e., an EQ. It is well accepted that an important organization of a physical system precedes a catastrophic event. A way to examine transient phenomena is to divide the measured time-series into time windows of short duration and analyse them. If this analysis yields different results for specific epochs, then a transient behaviour can be extracted. In this context, we have shown that various organization, information and complexity measures provide evidence of a clear change of state leading to the point of global instability: they detect the pattern of alterations in the preseismic kHz EME signals and they are able to discriminate the 'injury levels' of the focal area. Analysing the raw data or the sequences of symbols which are generated by the initial measurements using the concept of symbolic dynamics, it has been shown that the launch of kHz EME is combined by significantly lower: dynamical (Shannon-like) block entropy, Shannon n-block entropy per letter, conditional entropy, and Kolmogorov−Sinai entropy (Karamanos et al., 2005; Eftaxias et al., 2009), approximate entropy (Karamanos et al., 2006; Eftaxias et al., 2009; Potirakis et al., 2012b), fuzzy entropy (e.g. Fig. 13−5), eigenvalue entropy (Donner et al., 2015), and q-Tsallis entropy organization (e.g. Fig. 13−6) (Kalimeri et al., 2008; Papadimitriou et al., 2008; Potirakis et al., 2011, 2012a,c). These findings show that the kHz EME precursor is characterized by high organization.

   Complementary analyses by means of T-complexity and correlation dimension suggest, as the main result, that the appearance of the kHz EME precursor is accompanied by a clear transition from higher to lower complexity, as well (Karamanos et al., 2006; Eftaxias et al., 2009).

   Finally, it has been shown that the precursor under study is characterized by high information content by means of Fisher information (Potirakis et al., 2012a,b).

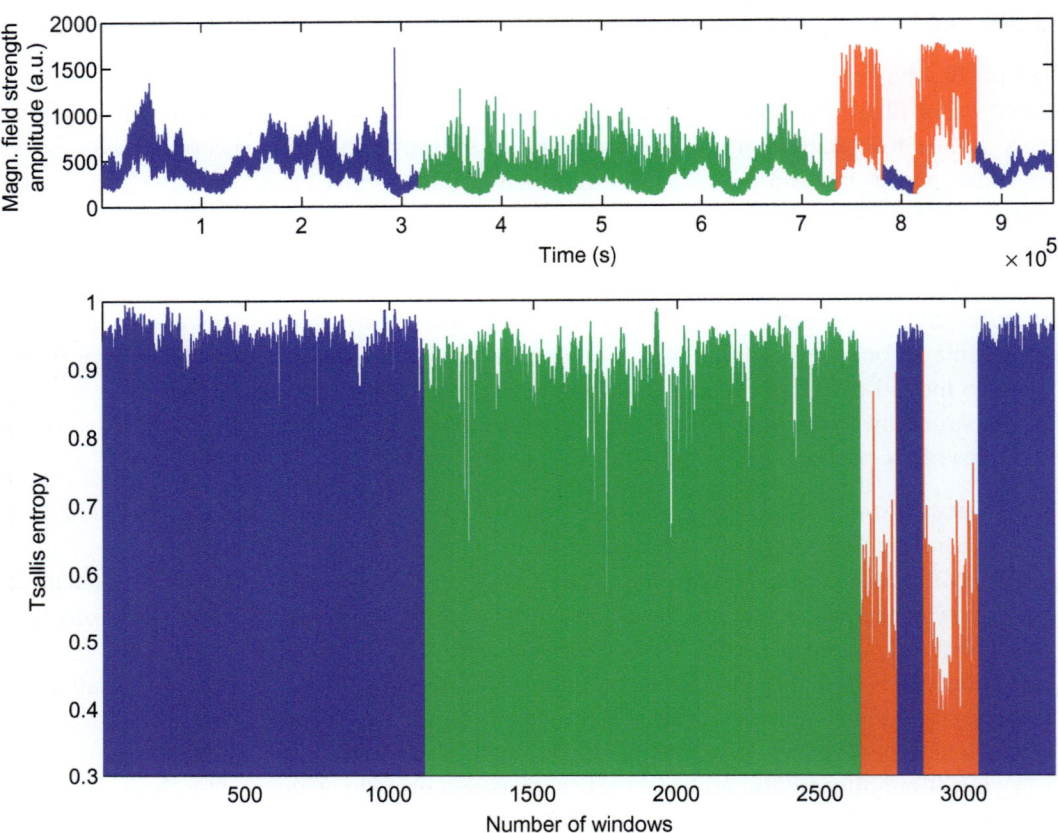

FIGURE 13–6 The last part of the 10 kHz EME time-series with the Athens EQ (upper graph). Tsallis entropy (bottom graph), for the population of the nonoverlapping data windows of 1024 samples each, which proved to be stationary. The nonextensive parameter value $q = 1.8$ (see Section 13.5.2.1) was used.

2. *Strong persistency.* This means that increases in the value of a time-series are likely to be followed by a further increase, namely, the underlying fracto-EM mechanism has been organized by a positive feedback process that leads the systems out of equilibrium. The positive feedback influence of the fracture events on each other leads to instability in the system. The specific symptom has been verified by means of fractal spectral analysis, R/S analysis and DFA analysis (Contoyiannis et al., 2005; Eftaxias et al., 2009) (e.g. Fig. 13−7).

It is worth mentioning that laboratory experiments by means of AE and EME also show that the main rupture occurs after the appearance of persistent behaviour (Karamanos et al., 2006 and reference therein). We consider that the temporal sequence of the antipersistent−persistent behaviour of the observed MHz−kHz EME time-series is very important. More precisely, we observe the existence of strong antipersistent behaviour in the first emerged MHz EME time-series (first stage), especially into the 'critical window', then a decrease in the antipersistent behaviour with time, especially

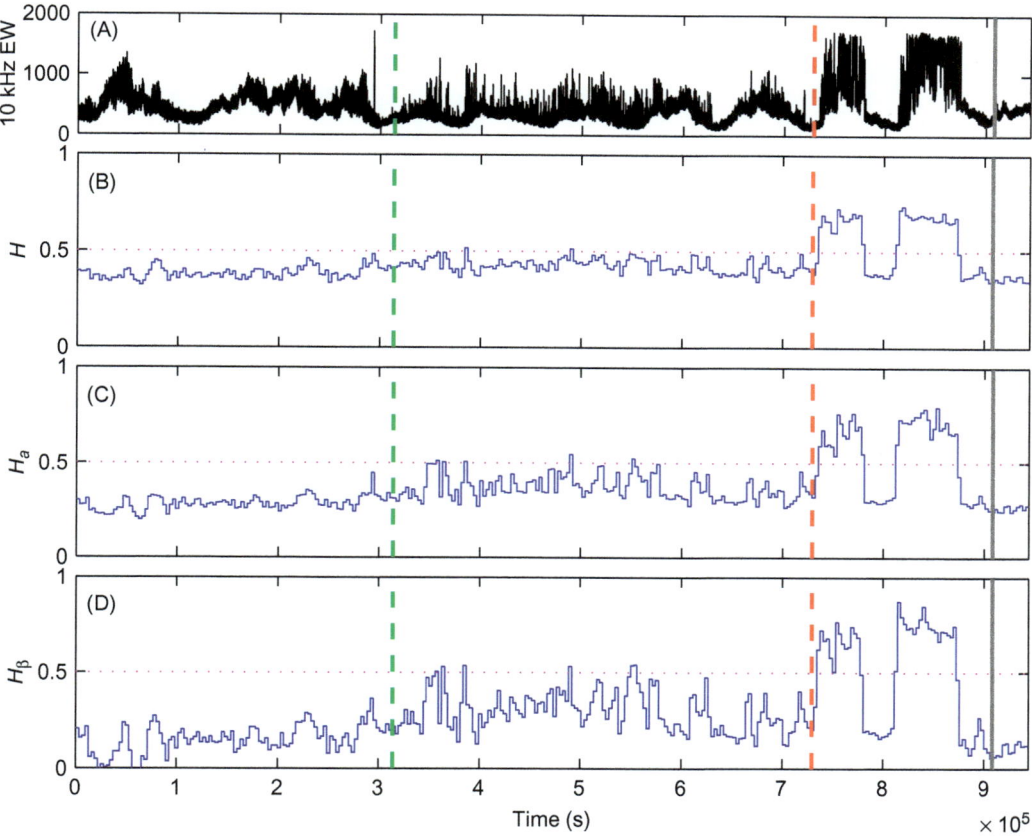

FIGURE 13–7 (A) Part of the recorded time-series of the 10-kHz (East–West) magnetic field strength (in arbitrary units) covering the 11-day period from 28 August 1999, 00:00:00 (UT), to 7 September 1999, 23:59:59 (UT), associated with the Athens EQ. The corresponding variation vs time of Hurst exponent, (B) H, resulting from R/S method, (C) H_a, estimated via DFA, and (D) H_β calculated from spectral power law (Potirakis et al., 2014). The common horizontal axis is the time (in s), denoting the relative time position from the beginning of the analysed part of the kHz EME recording. The vertical lines have the same position and meaning as in Fig. 13–5. *EME*, electromagnetic emission; *EQ*, earthquake.

after the appearance of 'symmetry breaking', conservation of the antipersistency during the tricritical window (second stage), and finally emergence of highly persistent properties in the strong avalanche-like kHz EME (third stage). In terms of scalogram analysis (e.g. Fig. 13–8), which depicts the temporal evolution of the wavelet spectrum, we observe features of a catastrophic phenomenon in the recorded strong pulse-like kHz EME signal (Kapiris et al., 2005a). First of all, we identify a predominance of large fracto-emission events in the kHz EME time-series with the launch of the strong pulse-like kHz EME precursor, signified by the fact that the higher part of the emitted energy is localized to the lower frequencies (higher scales). In parallel, we observe a significant acceleration of the energy release as the main event approaches, i.e., an increase in the susceptibility

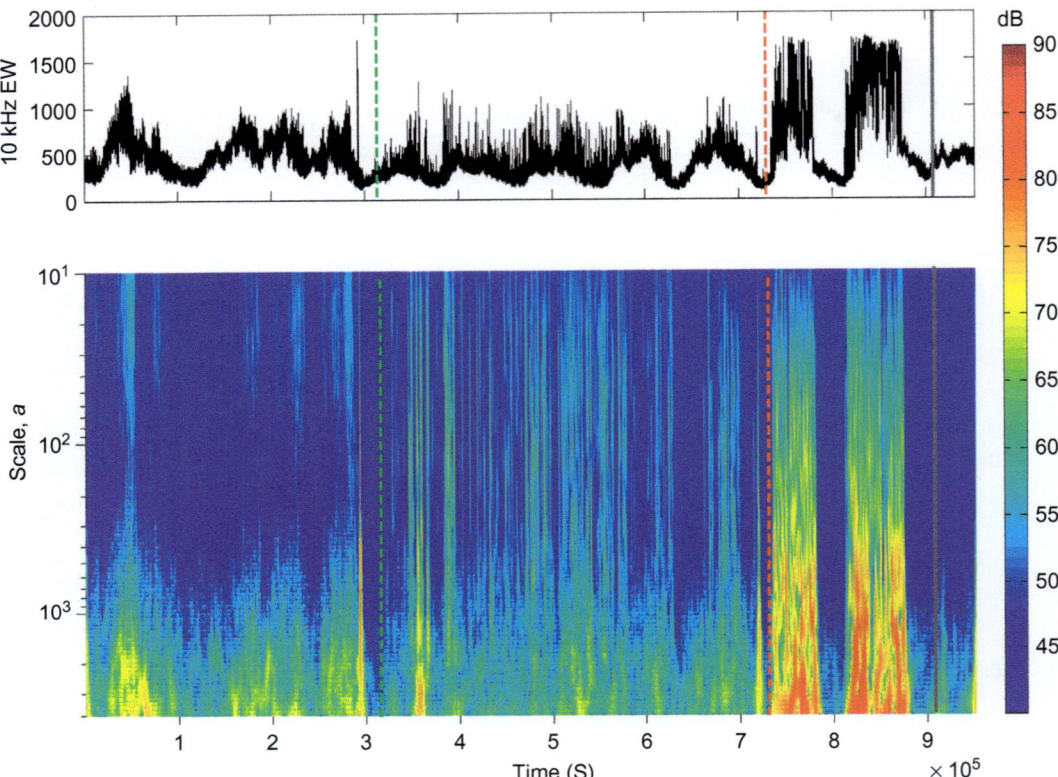

FIGURE 13–8 (A) Part of the recorded time-series of the 10-kHz (East–West) magnetic field strength (in arbitrary units) covering the 11-day period from 28 August 1999, 00:00:00 (UT), to 7 September 1999, 23:59:59 (UT), associated with the Athens EQ. (B) The corresponding morlet wavelet scalogram, with the vertical axis corresponding to the scale, *a*, of the wavelet (time scale, reciprocal to the wavelet 'frequency') and the colour representing the power spectral level in dB (side colour bar). The common horizontal axis is the time (in s), denoting the relative time position from the beginning of the analysed part of the kHz EME recording. The vertical lines have the same position and meaning as in Fig. 13–5. *EME*, electromagnetic emission; *EQ*, earthquake.

of the system. There is also a gradual appearance of higher frequencies (lower scales) in the spectrum resulting in an extension of kHz EME to all scales as the EQ approaches, although, as already mentioned, the emitted energy is localized to the lower frequencies corresponding to spatially larger fracture events. Importantly, a similar sequence of the catastrophic phenomenon evolution with time by means of antipersistency–persistency is also observed as an epileptic seizure (ES) (Kapiris et al., 2005b; Li et al., 2005; Eftaxias et al, 2006, 2013b) or a magnetic storm (Balasis et al., 2006) approaches/occurs. Note that a similar physical picture, but with more details, is achieved through the detection of statistically significant components of the kHz EME by using multispectral analysis methods (Kalimeris et al., 2016).

3. *Existence of clear preferred direction of fracture activities.* The transition from a dispersed fracture mode to a clustered one along a distinct inclined zone is a characteristic of

faulting nucleation (Reches and Lockner, 1994). Laboratory, theoretical and numerical studies show that the damage localizes in a narrow band at the end of the failure process (e.g. Mogi, 1968; Dodze et al., 1996; Ponomarev et al., 1997; Reasenberg, 1999; Li et al., 2002; Malakhovsky and Michels, 2007 and references therein). Notice that Malakhovsky and Michels (2007), based on simulation results, conclude that *macroscopic localization and strong damage anisotropy set in around the maximum stress point, leading to the final crack formation.* Both the temporal and spatial activity can be described as different cuts in the same underlying fractal (Maslov et al. 1994; Ponomarev et al., 1997). The kHz EME precursor satisfies the key feature of localization characterizing extreme phenomena: a sudden drop in the fractal dimensions of the profile of kHz EME time-series emerged prior to the Athens EQ (see Section 13.5.4), to the value $D \approx 1.3$ is observed when the strong pulse-like anomaly is emerged (e.g. Fig. 13−9) (Potirakis et al. 2012b, 2014). Seismological measurements, as well as theoretical studies, suggest that a surface trace of a single fault might be characterized by $D \approx 1.2$ (Sahimi et al., 1993). A few days prior to the Athens EQ, the seismicity was centred at a distance of about one source dimension (<30 km) from the epicentre, while the seismic energy release exhibited a power-law-type increase. A similar power-law-type has exhibited the preseismic kHz EM energy release recorded prior to the Athens EQ (see Section 13.5.4) (Kapiris et al., 2005a and references therein; Potirakis et al., 2011). We note that the aforementioned *sudden-drop of fractal dimension* is also accompanied by a *sudden-drop of Tsallis entropy* (e.g. Fig. 13−6) (Potirakis et al., 2012c). The detection of a sudden drop of fractal dimension and entropy has been proposed as a precursor for an impending catastrophic event in disordered media (Lu et al., 2005).

The aforementioned footprints included in the observed, abruptly emerging, strong avalanche-like kHz EME precursor, namely, the absence of any feature of a second-order phase transition in equilibrium, the existence of high organization, high information content, low complexity, strong persistency, and finally, the existence of a clearly preferred direction of fracture activities, not only clearly discriminate the candidate kHz EM anomaly from the background EM noise, but are consistent with the endorsement of the kHz EM anomaly as an image of an underlying extreme event; the possible seismogenic origin of such an emerged kHz EME cannot be excluded.

13.5.2 Arguments in Terms of Universal Structural Patterns of the Fracture Process

In contrast to previously prevailing views, from our point of view, even the aforementioned multidisciplinary statistical analysis per se does not link the kHz EME phenomenon with the fracture of asperities. *The aforementioned statistical results are likely to offer necessary but not sufficient criteria in order to recognize a kHz EM anomaly as an indicator of the fracture of asperities.* The crucial question is whether different approaches could provide additional information that would allow one to accept that the kHz EME anomalies signal the fracture of asperities.

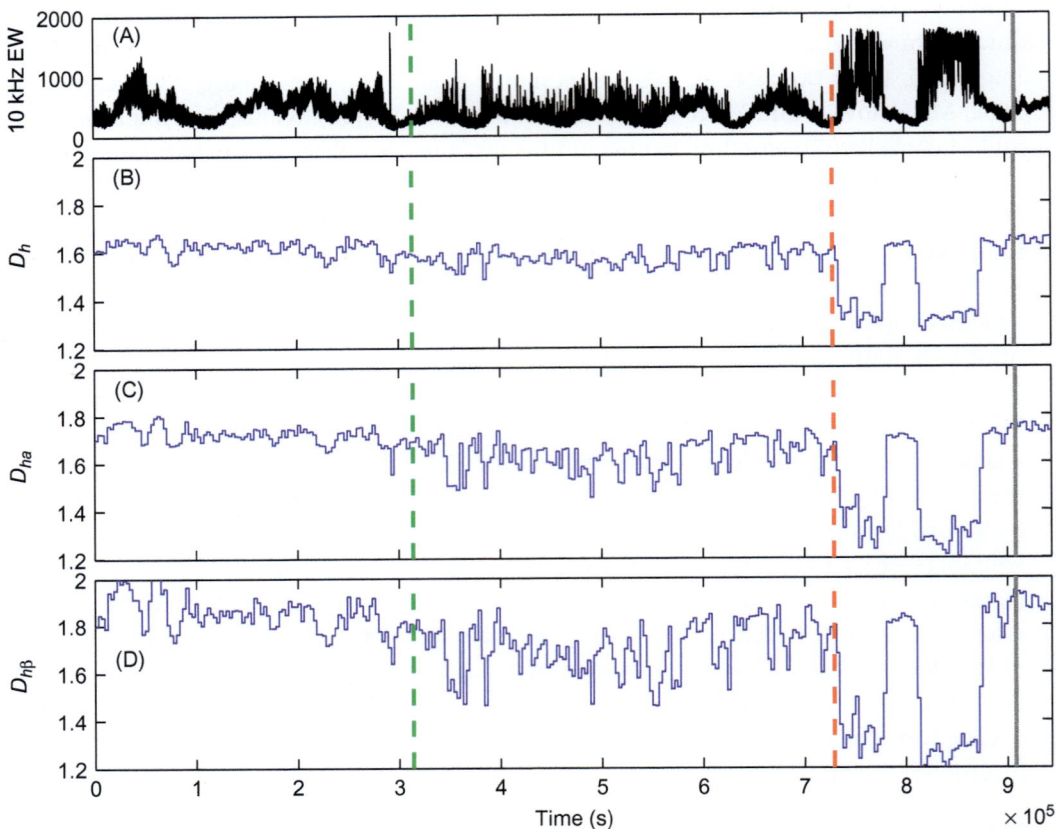

FIGURE 13–9 (A) Part of the recorded time-series of the 10-kHz (East–West) magnetic field strength (in arbitrary units) covering the 11-day period from Aug 28, 1999, 00:00:00 (UT), to Sep 07, 1999, 23:59:59 (UT), associated with the Athens EQ. The corresponding temporal variation of the Hausdorff–Besicovitch FD (B) D_h, as resulting from H (C) D_{ha} as resulting from H_a, and (D) $D_{h\beta}$ calculated from H_β, all calculated using the relation $D_h = 2 - H = (5 - \beta)/2$ (Potirakis et al., 2014). The common horizontal axis is the time (in s), denoting the relative time position from the beginning of the analysed part of the EM recording. The vertical lines have the same position and meaning as in Fig. 13–5. *EQ*, earthquake; *EM*, electromagnetic; *FD*, Fractal Dimension.

The basic information that guided our thinking in this direction was the following well-established evidence: *Despite the complexity of the fracture process, there are universally holding scaling relations.* The aspect of the self-affine nature of faulting and fracture is widely documented (Mandelbrot, 1982; Huang and Turcotte, 1988; Turcotte, 1997; Sornette, 2000). Thus, from our point of view, *universal structural patterns of the fracture process should be included in a precursor associated with the fracture of asperities.* We have shown that the above-mentioned requirement is satisfied. Indeed, the kHz EME precursor includes the following universal features.

13.5.2.1 The Activation of a Single Fault is a Reduced Self-Affine Image of Regional Seismicity and a Magnified Image of Laboratory Seismicity in Terms of Precursory kHz EME

Huang and Turcotte (1988) have stated that, in the framework of the self-affine nature of the faulting and fracture process, *the statistics of regional seismicity could be merely a macroscopic reflection of the physical processes in EQ sources, namely, the activation of a single fault is a reduced self-affine image of regional seismicity.*

A model of EQ dynamics, which is based on first principles of nonextensive statistical mechanics (Sotolongo-Costa and Posadas, 2004; Silva et al., 2006), leads to a Gutenberg–Richter-type (G–R-type) law for the magnitude distribution of EQs:

$$\log[N(>M)] = \log N + \left(\frac{2-q}{1-q}\right)\log\left[1 - \left(\frac{1-q}{2-q}\right)\left(\frac{10^{2M}}{a^{2/3}}\right)\right], \tag{13.5}$$

where N is the total number of EQs, $N(>M)$ the number of EQs with magnitude higher than M, $M \propto \log\varepsilon$, a is the constant of proportionality between the EQ energy, ε, and the size of fragment, r, $(\varepsilon \propto r^3)$. It should be remembered that the entropic index q characterizes the degree of nonextensivity (Tsallis, 1988, 1998, 2009). The proposed nonextensive G–R-type law provides an excellent fit to seismicities generated in various large geographic areas usually identified as 'seismic regions', each of them covering many geological faults. It is very important to observe the similarity in the value of the nonextensivity parameter q for various catalogues used: the q-values are restricted in the narrow region from 1.60 to 1.71.

According to our proposal, an EM burst (referred to as 'fracto-EM EQ' or simply 'EM-EQ') with energy ε included in the observed burst-like kHz EME precursor is emitted during the fracture of an asperity. A question effortlessly arising is whether the magnitude distribution of EM-EQs ($M \propto \log\varepsilon$) included in a kHz EME precursor also follows the distribution of Eq. (13.5) with similar q-value.

It has been shown that this really happens, characteristically: a very clear kHz EME precursor emerged prior to the Athens (Greece) EQ ($M = 5.9$) that occurred on 7 September 1999 (see Section 13.5.4) (Eftaxias et al., 2001; Kapiris et al, 2004a,b, 2005a; Karamanos et al., 2006). The included 'EM-EQs' in this precursor follow Eq. (13.5) with $q \approx 1.80$ (Papadimitriou et al., 2008). The kHz EME precursor recorded prior to the L'Aquila EQ ($M = 6.3$) that occurred on 6 April 2009 in Italy also follows Eq. (13.5) with $q \approx 1.82$ (Eftaxias 2009; Eftaxias et al., 2010). We note that the estimated nonextensive q parameter is in full agreement with the upper limit $q < 2$ obtained from several independent studies involving the Tsallis nonextensive framework (Carvalho et al., 2008; Zunino et al., 2008). Moreover, it is in harmony with an underlying sub-extensive system, i.e., $q > 1$, verifying the emergence of strong interactions in the Earth's crust during the EQ preparation process.

The above finding indicates, in terms of the precursory kHz EME, *that the physical processes that govern the regional seismicity really could be merely a macroscopic reflection of the physical process in a specific EQ source.* This finding actually answers to a question open for a number of years: Hallgass et al. (1997) have emphasized that what was missing was the description of what happened locally, i.e., as a consequence of a single event.

Another question arising is whether *the physical process that governs the physical process in a specific EQ source is a magnified image of the laboratory seismicity*, as well. This really happens.

Vallianatos et al. (2012) have applied the concepts of nonextensive statistical physics to analyse AE data. They found that the AE scalar moment distribution reflects a subadditive system with thermodynamic q-values of $q = 1.82$. We recall that the magnitude distribution of 'EM-EQs' included in the recorded kHz EMEs rooted in a specific activated fault is characterized by $q \approx 1.80 - 1.83$ (e.g. Fig. 13−10).

We focus on the following result. The cumulative number $N(>A)$ of 'kHz EME events' (the number of 'kHz EME events' having amplitudes larger than A in an emergent precursor) follows the power-law: $N(>A) \sim A - c$, with $c = 0.62$ (Kapiris et al., 2004b). Rabinovitch et al. (2001) have studied the fractal nature of EM radiation induced by uniaxial and triaxial rock fracture. The analysis of the prefracture EME time-series reveals that the cumulative distribution function of the amplitudes A also follows the power-law $N(>A) \sim A - c$, with $c = 0.62$, as well.

The above-mentioned results suggest that the activation of a single fault by means of the observed kHz EME behaves as a reduced self-affine image of the regional seismicity and as a magnified self-affine image of laboratory seismicity. We consider that these findings strongly enhance the seismogenic origin of the observed kHz EME precursors.

13.5.2.2 *Arguments in Terms of Universal Characteristics of Rock Surfaces: The kHz EME Includes the Signature that Rock Surfaces Can Be Represented by a Persistent Fractional Brownian Motion (fBm) Model*

Many authors have pointed out that natural rock surfaces can be represented by fractional Brownian surfaces over a wide scale range (Huang and Turcotte, 1988; De Rubeis et al., 1996; Turcotte, 1997; Hallgass et al., 1997 and references therein; Chakrabarti and Benguigui, 1997). Maslov et al. (1994) have formally established that both the temporal and spatial activity can be described as different cuts in the same underlying fractal, while laboratory experiments support this view (Ponomarev et al., 1997). Based on an analysis in terms of fractal spectral analysis, R/S analysis and DFA analysis it has been proved that the profile of the observed kHz EME follows the persistent fractional Brownian motion (fBm) model, while the associated local H exponents are distributed in the region from 0.7 to 0.8 (e.g. Fig. 13−7) (Eftaxias et al., 2008).

13.5.2.3 *The kHz EME is Consistent with the Interpretation that Surface Roughness is a Universal Indicator of Surface Fracture*

Surface roughness has also been interpreted as a universal indicator of surface fracture, weakly dependent on the nature of the material and on the failure mode (Lopez and Schmittbuhl, 1998; Hansen and Schmittbuhl, 2003; Zapperi et al., 2005; Mourot et al., 2006; Ponson et al., 2006) The roughness of fracture surfaces is quantitatively characterized by the Hurst

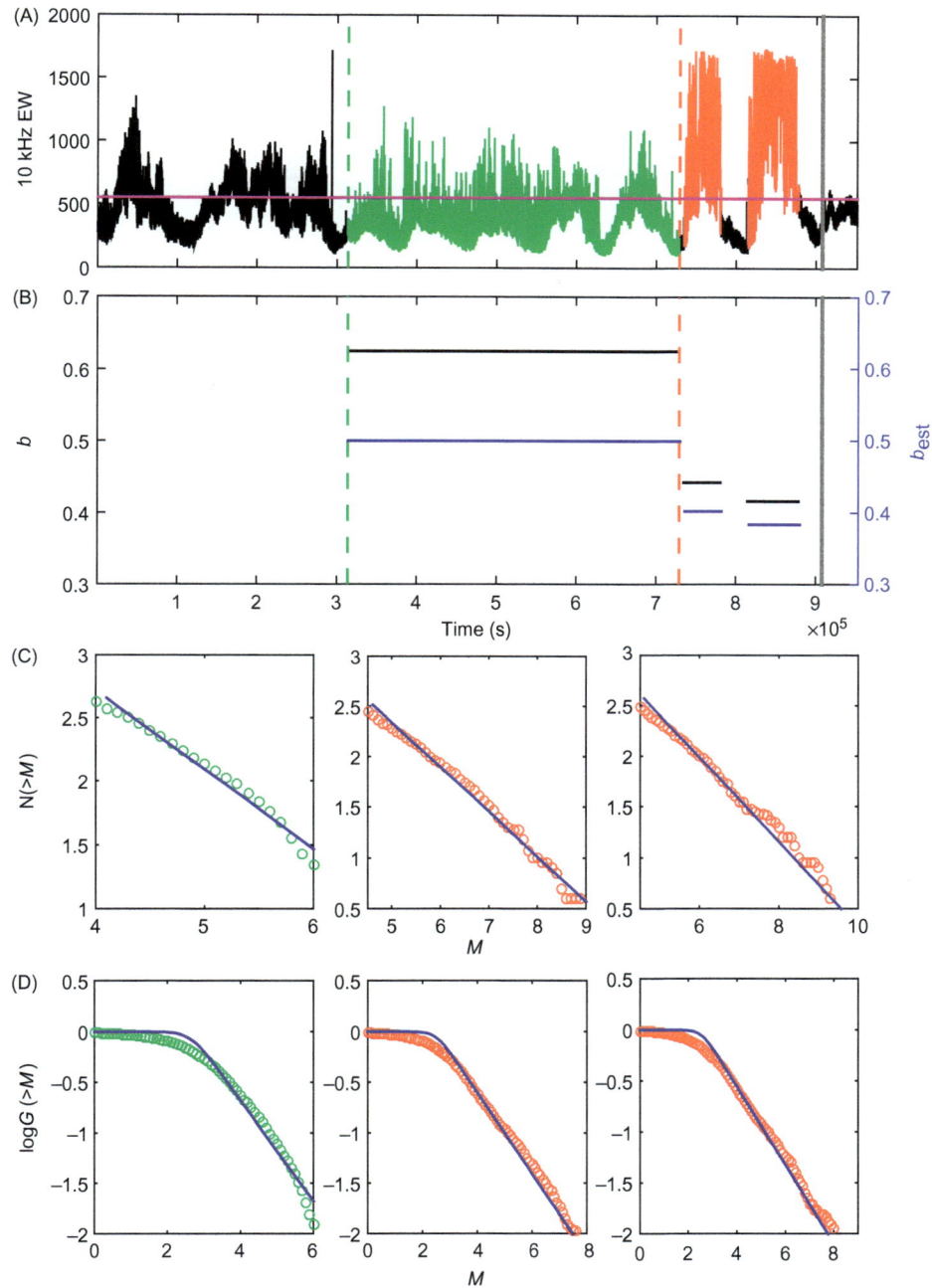

FIGURE 13–10 (A) Part of the recorded time-series of the 10 kHz (East–West) magnetic field strength (in arbitrary units) covering the 11-day period from 28 August 1999, 00:00:00 (UT), to 7 September 1999, 23:59:59 (UT), associated with the Athens EQ. (B) Temporal variation of the b-value, and the b_{est}, estimated from the calculated nonextensive q parameter ($q \approx 1.83 \Rightarrow b \approx 0.4$) using the relation $b_{est} = 2 \cdot \left((2 - q)/(q - 1)\right)$ (Sarlis et al., 2010). (C) Fitting of the Gutenberg–Richter law and (D) the nonextensive Gutenberg–Richter law, on the three parts of the analysed signal (colour and position correspondence from left to right). The common horizontal axis is the time (in s), denoting the relative time position from the beginning of the analysed part of the EM recording. The magenta horizontal line on Fig. 13–4A indicates the noise level threshold $A_{noise} = 620$ a.u. (see also Fig. 13–4). The vertical lines have the same position and meaning as in Fig. 13–5. *EQ*, earthquake; *EM*, electromagnetic.

exponent, H, since the average height difference $\langle y(x) - y(x + L) \rangle$ between two points on a profile increases as a function of their separation, L, like L^H with $H \sim 0.75$. The roughness of a recently exhumed strike-slip fault plane has been measured. Statistical scaling analyses showed that the striated fault surface exhibits a self-affine scaling invariance that can be described by a scaling roughness exponent $H = 0.7$ in the direction of slip (Renard et al., 2006).

The aforementioned universal indicator of surface roughness should characterize the roughness of the profile of a seismogenic kHz EME time-series. This really happens. We have shown by means of fractal spectral analysis, R/S analysis and DFA analysis that the roughness of the profile of the observed kHz EM anomaly is consistent with the aforementioned universal H-value (e.g. Fig. 13–7) (Kapiris et al., 2004a; Contoyiannis et al., 2005; Eftaxias et al., 2009; Minadakis et al., 2012a).

It might be concluded that the above-mentioned crucial universal features of the fracture and faulting process are included in the observed kHz EME supporting the seismogenic origin of an observed candidate kHz EME precursor.

13.5.3 On the Morphology of the Observed KHz EME Precursors

1. *On the avalanche-like morphology of the kHz EME precursor.*

A systematically observed characteristic of the recorded kHz EME precursor is its burst-like morphology. The large fluctuation in the amplitude of the included avalanches constitutes by itself an important feature mimicking the EQ generation. EQs have long been recognized as resulting from a stick–slip frictional instability; it is known that when the static friction resistance is overcome, then the frictional surfaces suddenly slip, lock and then slip again in a repetitive manner. The unstable transition from the static friction to a kinetic one is known as a 'stick–slip' state. The asperities on a fault are treated as individual elements with various strengths. If one element fails, the stress is transferred to the adjacent element on which an induced failure could occur. McGarr and Fletcher (2003) have shown that the maximum slips inferred for major EQs are consistent with those measured in the laboratory during large-scale biaxial stick–slip friction experiments (Lockner and Okubo, 1983) if differences in the state of stress and loading stiffness are taken into account, while the recent works of McGarr and Fletcher (2003) and McGarr et al. (2010) suggest that stick–slip friction events observed in the laboratory and EQs in continental settings, even with large magnitudes, have similar rupture mechanisms. As has already been mentioned, laboratory experiments have shown that stick–slip-like fracture-sliding precedes dynamic fast global slip (Bouchon et al., 2001; Baumberger et al., 2002; Rubinstein et al., 2004, 2007; Chang et al., 2012; Kammer et al., 2012), while recent studies also reveal that physical systems under slowly increasing stress may respond through abrupt events. Such jumps in observable quantities are abundant, being found in systems ranging from complex social networks to EQs (Papanikolaou et al., 2012). Based on the above-mentioned studies, we suggest that the avalanche-like morphology of the kHz EME precursor is justified in the framework of the proposed four-stage model of EQ preparation.

The completion of the fracture of the family of strong brittle asperities, which is signalled, within the framework of our proposal, by the observed avalanche like kHz EME, should be accompanied by corresponding sharp drops in stress. Thus a vital question emerges as to whether laboratory studies support such a situation. *Indeed, recent laboratory studies on specimens made of different heterogeneous materials suggest that EME is observed only when a sharp stress drop occurs* (see Figs 13−4−13−7 in Lacidogna et al., 2010), (Fukui et al., 2005; Tsutsumi and Shirai, 2008; Carpinteri et al., 2011, Carpinteri et al., 2012; Lacidogna et al., 2011).

2. *On the sharp onset of the detected kHz EME precursory activities.*

Another characteristic of the detected strong multipeaked kHz signal is its sharp onset. This characteristic is consistent with the abrupt transition to the nucleation stage. Numerical (e.g. Lockner and Madden, 1991) and laboratory (e.g. Reches and Lockner, 1994) studies suggest an abrupt transition to the nucleation stage.

3. *On the time of launch of the detected kHz EME precursory activities.*

The kHz EME precursor emerges at the tail of the precursory MHz−kHz sequence, from a few tens of hours up to a few minutes before the EQ occurrence. Accumulated evidence justifies the aforementioned short time interval. As has already been mentioned (see Sections 13.2 and 13.3.6), laboratory experiments reveal that the kHz EME also emerges at the tail of the fracture-induced EME sequence of MHz-kHz emission. Laboratory experiments by Ohnaka and Mogi (1982) on rocks at 1 MHz and 30, 250 and 400 kHz have shown changes in the frequency content of AE activity during progressive deformation of rocks: the kHz activity appears in the stage from 97% to 100% of the failure strength, after the MHz one. Laboratory experiments performed by Reches and Lockner (1994) reveal that the abrupt transition to the nucleation stage occurs at a differential stress above about 0.995 of maximum load. The kHz EME might emerge around the critical mechanical threshold, x_{mf}, i.e., during the formation of the FIC, by means of percolation theory (see Section 13.3.6). Based on synergetic principles of physical mesomechanics, we have shown that the observed transition from the 'mild' MHz to the 'strong' kHz fracto-EM activity marks the transition from small-scale (mesolevel) fragmentation to large-scale (macrolevel) fragmentation (Eftaxias et al., 2007). Its time of appearance by itself connects the kHz EME activity with the last stage of the fracture process.

4. *On the abrupt ceasing of the kHz EM anomaly.*

Another morphological characteristic of the detected strong multipeaked kHz EME signal is its sharp ceasing. Park and Song (2013) have presented a new numerical method for the determination of contact areas of a rock joint under normal and shear loads. They report that, at the peak stage, the normal dilation was initiated, which led to a *sharp drop in the contact area.* Approximately 53% of the surface area remained in contact, supporting the normal and shear loads. The active zone was partially detached, and the inactive zone was partially in contact. After the peak stage, the contact area ratio *decreased rapidly* with increasing shear displacement, and few inactive elements came into contact until the residual stage. At the residual stage, only small fractions, 0.3%, were involved in contact. The rapid decrease in the contact area ratio is in full agreement with the observed abrupt ceasing of the kHz EME. Thus, the last appearing kHz EM avalanche (EM EQ) of the EME precursor may reveal the damage of the last strong asperities which sustain the system in the sticking regime.

13.5.4 On the Consistency of the kHz EME Precursor With Other Precursors

An EQ's preparatory process has various facets which may be observed before the final catastrophe. Therefore, it is a reasonable requirement that a candidate precursory kHz EME activity should be consistent with other precursors that are imposed by data from other disciplines. Geller et al. (1997) emphasize that: 'The absence of simultaneous geodetic or seismological precursors means that the observed EM anomalies are not earthquake-related ones'. As has already been discussed (see Sections 13.3.7 and 13.3.8) the observed MHz EME precursor is in full agreement with simultaneous seismological and geodetic precursors. A question also naturally arises as to whether the kHz EME precursor is combined by simultaneous seismological or geodetic precursors.

13.5.4.1 Focus on the Existence of Simultaneous kHz EME and Seismological Precursors

An observed kHz EME, which is truly related to the fracture of asperities, should be consistent with the fault modelling of the occurred EQ which has been the result of studies from different disciplines. It is a crucial question whether such a vital relation exists.

Two strong avalanche-like kHz EM anomalies were detected before the shallow Athens EQ ($M_w = 5.9$, 7 September 1999) with the following characteristics (Eftaxias et al., 2001; Kapiris et al, 2004b): The first and second anomalies lasted for 12 and 17 hours, respectively, with an intermediate cessation of 12 hours; the second anomaly ceased at about 9 hours before the EQ. The larger anomaly, the second that was recorded, contained approximately 80% of the total EM energy released. The observation of two precursory events in a short time gap, with different energy content, offered a rare opportunity to answer the aforementioned question.

A seismic data analysis in the case of the Athens EQ, performed using the now standard methodology (Kikuchi and Kanamori, 1991), indicates that a two-event solution, for the Athens EQ, with magnitudes $M_w = 5.5$ and $M_w = 5.8$ is more likely than a single event solution with magnitude $M_w = 5.8$. According to Kikuchi, there was probably a subevent ($M_w = 5.5$) which happened about 3.5 seconds before the main event ($M_w = 5.8$) (Eftaxias et al., 2001), while the moment ratio of the two seismic events 1:3 is not in conflict with the ratio of the EM energy of the two recorded strong avalanche-like kHz EME signals.

The above-mentioned experimental findings show that the two kHz EME signals are consistent with the fault modelling of the EQ that occurred which was the result of seismic data analysis. Importantly, these two kHz EME signals are also consistent with the fault modelling of the Athens EQ, as was shown by geodetic data analysis (see Section 13.5.4.2).

We emphasize that the aforementioned compatibility in terms of energy is further extended by means of information content (Fisher information) and organization (approximate entropy) as well (Potirakis et al., 2012b).

We recall (see Section 13.5.2.1) that in terms of a model of EQ dynamics, which is based on first principles of nonextensive statistical mechanics, the population of (1) the kHz fracto-EM-EQs that emerge during the fracture of the asperities of a single fault and (2) the population of EQs generated in various 'seismic areas', each of them covering many geological faults, both follow the same statistics, namely, the same relative cumulative number of EQs/EM-EQs vs magnitude (Papadimitriou et al., 2008; Eftaxias, 2009; Minadakis et al., 2012a,b). This finding further supports the above-reported consistency between the twokHz EME precursors observed before the Athens EQ and the corresponding fault modelling of the actual EQ as shown by seismic data analysis. Notice, as has been reported in Section 13.3.7, that the MHz EME precursor and the associated foreshock seismic activity constitute two sides of the same coin.

13.5.4.2 Focus on the Existence of Simultaneous kHz EM and Geodetic Precursors

In the framework of the proposed four-stage model of EQ preparation the observed precursory kHz EME activity is associated with the stick−slip mode (see Section 13.5.3). Slip fluctuates spatially because of pinning on local asperities (Perfettini et al., 2001). Sensitive and fast measurements have been performed by Nasuno et al. (1998) on sheared layers undergoing stick−slip motion with simultaneous optical imaging; measurements of vertical displacements revealed dilation of material associated with each slip event. The hypothesis that these vertical displacements are analogous to deformations on the Earth's surface during EQ preparation is reasonable. Consequently, the hypothesis that precursory kHz EME activity should be consistent with other precursors rooted in the deformation of the Earth's surface is very sensible. This requirement has been fulfilled. Indeed, *SAR interferometry* is an imaging technique for measuring the topography of a surface, its changes over time, and other changes in the detailed characteristics of the surface. This method has demonstrated potential to monitor and measure surface deformations associated with EQs. Such deformations have also been reported in the case of the Athens EQ (see Section 13.5.4.1). Interferometric analysis of satellite ERS2 SAR images leads to the fault model of the Athens EQ (Kontoes et al., 2000). This model predicts the activation of two faults; the main fault segment is responsible for 80% of the total seismic energy released, while the secondary fault segment is responsible for the remaining 20%. Importantly, as was described in Section 13.5.4.1, the observed kHz EME precursor is in agreement with the above-mentioned findings resulting from SAR or seismological techniques: *the larger kHz EM anomaly, the second one, contains approximately 80% of the total EM energy released, with the first one containing the remaining 20%.*

The above-mentioned exceptional observational evidence, which is beyond any analysis, strongly enhances the hypothesis that the kHz EME precursor is associated with the fracture of asperities via a stick−slip mechanism.

We consider that based on the findings of these last two subsections, the proposal that the observed avalanche-like kHz EM anomaly before the Athens EQ is a seismogenic one, rooted in the fracture of family of asperities distributed along the main fault, is not groundless.

13.6 Focus on the Fourth Stage Reflected in the Observed Quiescence in all EM Frequency Bands Following the Strong Avalanche-Like kHz EME

As was described in Section 13.2, EM silence in all frequency bands is systematically observed before EQ occurrence. This feature is puzzling and it is used as a strong argument in support of the nonexistence of preseismic EM precursors. As has already been mentioned, laboratory experiments of rock fracture and frictional sliding have shown that the relative slip of two fault surfaces takes place in two phases. Stick–slip-like fracture-sliding precedes dynamic fast global slipping (Bouchon et al., 2001; Baumberger et al., 2002; Rubinstein et al., 2004, 2007; Chang et al., 2012; Kammer et al., 2012). Recent accumulated evidence has enhanced the view that the emergence of preseismic EM silence is not a puzzling feature but, on the contrary, it is the final precursory signal indicating the stage of transition to the last stage of the EQ preparation process, namely, the dynamic slip which results in the fast, even super-shear, mode (Eftaxias and Potirakis, 2013). Recent laboratory experiments and numerical studies justify the observed preseismic EM silence by means of the sharp drop in the contact area at the peak stage, the behaviour of the elastic moduli as damage increases, and the recently reached clarification that the recorded AE and EME are not always two sides of the same coin. We focus on these topics in the following subsections.

13.6.1 Focus on the EM Silence in Terms of Numerical Experiments

As has already been described in Section 13.5.3, Park and Song (2013) have presented a new numerical method for the determination of contact areas of a rock joint under normal and shear loads. We focus on their conclusion that at the peak stage, normal dilation was initiated, which led to a sharp drop in the contact area. Approximately 53% of the surface area remained in contact, supporting the normal and shear loads. The active zone was partially detached, and the inactive zone was partially in contact. After the peak stage, the contact area ratio *decreased rapidly* with increasing shear displacement, and few inactive elements came into contact until the residual stage. 'At the residual stage, only small fractions, 0.3%, were involved in contact.' The dramatic rapid decrease in the contact area ratio is in full agreement with both the observed abrupt ceasing of the kHz EME as well as the following emergence of fracture-induced EM silence. Thus, the last emerged kHz EM avalanche (EM EQ) of the EME precursor may reveal damage by the last strong asperity which sustains the system in the sticking regime. The above-mentioned scenario, which bridges the observed kHz EM anomaly with recent experimental results and leads to the crucial hypothesis that the observed EME precursor reflects the damage of a critical number of asperities that leads to the transition to the preparation of the final fast slip, calls for further documentation. We note that McGarr and Fletcher (2003) and McGarr et al. (2010) suggest that stick–slip friction events observed in the laboratory and EQs in continental settings, even with large magnitudes, have similar rupture mechanisms.

13.6.2 Focus on the EM Silence by Means of AE and EME Laboratory Experiments

Laboratory experiments in terms of AE have shown that this emission continues to increase up to the time of the final collapse. On the other hand, it was well accepted that the laboratory-recorded AE and EME are, in general, two sides of the same coin. Thus, the until recently existing impression that the appearance of EM silence just before the EQ occurrence is really a puzzling feature was strongly supported. However, recent accumulated experimental evidence has enhanced some older ones indicating that the above-mentioned well-accepted view was false: *the recorded AE and EME are not always two sides of the same coin.* Indeed, simultaneous laboratory measurements of AE and EME reveal the existence of *two categories of AE signals* (Yamada et al., 1989; Mori et al., 1994, 2004a,b, 2006, 2009; Morgounov, 2001; Mori and Obata, 2008; Baddari and Frolov, 2010; Lacidogna et al., 2010; Carpinteri et al., 2012):

1. *AE signals which are associated with EME signals.* Both emissions are simultaneously generated during the creation of new fresh surfaces which is accompanied by the rupture of interatomic bonds and charge separation.

2. *AE signals which are not associated with EME signals.* It has been proposed that this category of AE is rooted in frictional noises that appear during the rearrangements of the previously created fragments which are not accompanied by significant production of new surfaces.

Laboratory studies reveal that strong AE and EME are simultaneously observed during stress drops that occurred close to the peak stage in the load vs time diagram (Fukui et al., 2005; Lacidogna et al., 2011; Carpinteri et al., 2011; Carpinteri et al., 2012). At this stage new fresh surfaces, which are accompanied by the rupture of interatomic bonds and charge separation, are produced. Thus the simultaneous emission of strong kHz AE and EME is reasonable. However, during the last phase of the postpeak stage, that is the softening branch in the load vs time diagram, which is not accompanied by significant production of new surfaces, no EME is detected, while, on the contrary, the most intense AE emerges (e.g. Yamada et al., 1989; Morgounov, 2001; Baddari and Frolov, 2010; Carpinteri et al., 2012).

13.6.3 Focus on EM Silence in Terms of Elastic Moduli

Stick–slip events rooted in the damage of strong contacts are characterized by sudden shear stress drops that range from 10% to 30% of the maximum frictional strength (Johnson et al., 2008). This sharp drop in stress means that the strain resistance dramatically decreases. Elastic moduli, characteristics of the solid being considered, are the key parameters for defining the relationships between stress and strain and evaluating strain resistance. We note that the higher the index of brittleness, compressive strength, elastic moduli and volume of the damaged asperity, the higher is the EME energy generated (Wang and Zhao, 2013). Laboratory and theoretical studies show that the breaking of an element is associated with a decrease in the elastic modulus of the damaged material (Amitrano and Helmstetter, 2006); the elastic modulus significantly decreases as damage increases, approaching zero as the global fracture approaches (Lin et al., 2004; Shen and Li, 2004; Chen, 2012). On the other hand, elastic moduli also constitute crucial parameters for the detection of AE or EME from material experiencing 'damage'. An increase in Young modulus and strength enhances the EME amplitude (Nitsan, 1977; Khatiashvili, 1984; Rabinovitch et al., 2002; Fukui et al., 2005). It might be concluded that the observed kHz EME gap just before the EQ is further supported by the above-mentioned well-established behaviour of the elastic moduli.

In summary, recent laboratory and numerical evidence exactly correspond to the observations obtained at the geophysical scale where EME is detected just before the EQ, while there is no EME at the time of the EQ.

13.6.4 The Heat-Flow Paradox and the EM Silence Paradox: Two Sides of the Same Coin

In terms of EME, the observed last stage of EM silence just before the EQ requires the appearance of a corresponding stage with low dynamic friction coefficient. It is well known that such a stage that prepares the final fast, even super-shear, slip exists. A well-known phenomenon also calls for similar conditions. Indeed, one of the unresolved controversies in the field of EQ generation is a phenomenon which is referred to by geophysicists as 'the heat-flow paradox' (Sornette, 1999; Alonso-Marroquin et al., 2006 and references therein). To allow for large EQs, a fault should have a large friction coefficient so that it can restore a large amount of elastic energy and overpass large barriers. It is common sense that when two blocks grind against one another, there will be friction, and this will produce heat. Thus, large EQs should generate a large quantity of heat due to the rubbing of the two fault surfaces. However, measurements of heat flow during EQs were not able to detect the amount of heat predicted by simple frictional models. Calculations using the value of rock friction measured in the laboratory, i.e., a typical friction coefficient of between 0.6 and 0.9, led to overestimation of the heat flux. As an example, one refers in this context to the heat flow observations made around the San Andreas fault, which show that the effective friction coefficient must be around 0.2 or even less (Alonso-Marroquin et al., 2006 and references therein).

It might be concluded that the paradox of EM silence and the heat-flow paradox are two sides of the same coin. Both paradoxes originate from the appearance of low dynamic friction coefficient during the last stage of EQ generation, namely, the stage that prepares the final fast, even super-shear, slip. We cannot deny the existence of possibly EQ-related EME because of the EME silence paradox, in the same sense that we cannot deny the existence of EQs because of the heat-flow paradox.

13.6.5 EM Silence in Light of the Notion of Granular Packings

Herein, we focus on the mechanism which justifies the existence of a wide range of sliding velocities (or shear rates), even super-shear rupture modes that surpass the shear wave speed in the final stage of a fast dynamic slip (Scholz, 2002; Xia et al., 2004, 2005; Coker et al., 2005; Ben-David et al., 2010). *The appearance of fast sliding implies the existence of a kind of 'lubrication' mechanism between fault plates.*

Recent studies have verified that gouge formation, which behaves like bearings (lubrication mechanism), is found to be ubiquitous in brittle faults at all scales, and most slipping along mature faults is observed to have been localized within gouge zones, while gouges included in various faults display similar characteristics. Such a bearing-like mechanism has been proposed to explain the 'lubrication' of the fault surfaces (e.g. Chester and Chester, 1998; Sornette 1999; Åström et al., 2000, 2001; Baram et al., 2004; Wilson et al, 2005; Reches and Dewers, 2005; Alonso-Marroquin et al, 2006; Baker and Warner, 2012;Åström and Timonen, 2012). Wilson et al. (2005) proposed that the observed fine-grain gouge is formed by dynamic rock pulverization during the propagation of a single EQ; a gouge zone quickly develops with progressive slips reaching thicknesses larger than the height of the asperities, and further grain-size reduction occurs by systematic grain crushing due to amplified grain-contact stresses enhanced by the formation of stress-chains (Reches and Dewers, 2005 and references therein). In this case, the system can be regarded as granular matter that is sheared by the two surfaces (Kawamura et al., 2012 and references therein): tectonic faults are a characteristic example of shear failure in narrow zones (Åström et al., 2001; Alonso-Marroquin et al., 2006).

Numerical studies have also shown that the so-called shear bands appear, for example, in granular packings (Åström et al., 2000; 2001) while local 'rotating bearings' are formed spontaneously (Åström et al., 2000). Many authors report the discovery of a self-similar space-filling bearing in which an arbitrary chosen sphere can rotate around any axis and all the other spheres rotate accordingly with negligible torsion friction (Baram et al., 2004; Verrato and Foffi, 2011; Åström and Timonen, 2012; Reis et al., 2012).

We note that not only granular dynamics simulations but also laboratory Couette experiments (Veje et al., 1999) demonstrate the spontaneous formation of bearings processes.

The above-mentioned 'lubrication' mechanism by means of granular packings justifies the appearance of pre-EQ kHz EM silence and reduces the gap between what is 'paradox' and 'practice' as concerns EQ prediction.

13.6.6 Focus on the Duration of the Observed EM Silence

Two regimes for granular friction have been proposed: *the quasistatic and dynamic regimes* (Midi, 2004; Da Cruz et al., 2005; Mizoguchi et al., 2006, 2009; Forterre and Pouliquen, 2008; Hayashi and Tsutsumi, 2010; Kawamura et al., 2012 and references therein). Laboratory and numerical studies show that a time interval is needed for the formation of a shear band in the granular medium and thus for the transition from quasistatic to dynamic surface flow of a granular system. Numerical studies have revealed that this transition is characterized by intermittent local dynamic rearrangements and can be described by an order parameter defined by the density of critical contacts, namely, contacts where the friction is fully mobilized. Analysis of the spatial correlation of critical contacts shows the occurrence of 'fluidized' clusters which exhibit a power-law divergence in size at the approach of stability limit, as predicted by recent models that describe the granular systems during static/dynamic transition as a multiphase system (Sharon et al., 2002 and references therein). Laboratory studies also show local rearrangements. For example, quantitative X-ray diffraction analyses indicate that strain localization and grain size reduction are also accompanied by changes in the nature and abundance of phases at rock localities (Boulton et al., 2012). Laboratory studies by means of acoustic measurements (Khidas and Jia, 2012) reveal that when a granular medium is sheared, the shear strain is essentially localized in a narrow zone location at the mid-height of the box where a shear band is formed. Such a shear localization zone exhibits distinct features compared to the rest of the medium, including extremely large voids and the presence of a highly anisotropic network of force gains (Khidas and Jia, 2012 and references therein). Welker and McNamara (2011) have studied a numerical simulation of granular assemblies subjected to a slowly increasing deviator stress. They found that during the first half of the simulation, sliding contacts are uniformly distributed throughput the packing, but in the second half, they become concentrated in certain regions. This suggests that the loss of homogeneity occurs well before the appearance of shear bands.

In summary, laboratory, theoretical and numerical studies indicate that the stage of preparation of the fast dynamical slip is associated with the appearance of a rolling-type 'lubrication' mechanism of the included gouge between the fault surfaces. This phase is not accompanied by significant damage (breaking bonds) of brittle and strong material. A time interval is needed for the formation of a shear band in the granular medium and thus for the transition from quasistatic to dynamic surface flow of a granular system. The absence of fracture-induced kHz EME just before and at the time of the EQ occurrence is therefore fully justified.

13.7 On the Paradox of the Association of EME Signals with Small Precursory Strain Changes but not with much Larger Coseismic Strains: Shedding Light From Nanoscale Plastic Flow on the Geophysical Scale

The general observation that all the preseismic EM signals are stopped before the EQ is actually an important issue because strain changes are largest at the time of the EQ. Any mechanism of EME precursor generation must explain why EME signals are associated with small preseismic strain changes but not with much larger coseismic ones. Concerning the first emerged MHz EME, its nature justifies by itself the observed preseismic silence: this is rooted in the fracture of the heterogeneous system surrounding the main fault (see Section 13.3.6). The problem is focusing in the observed preseismic silence of the finally emerged kHz EM anomaly. This is obviously much more important since in the frame of the proposed four-stage model of EQ generation the kHz EME is rooted in the fracture of asperities.

We argue that a size-scale effect explains the aforementioned crucial, considered as paradox, feature (Eftaxias and Potirakis, 2013). Recent progress in experimental techniques, allowing one to test and probe materials at sufficiently small lengths, or time scales, or in three dimensions, has led to a quantitative understanding of the physical processes involved from the micro- to the geophysical scales (Bouchaud and Soukiassian, 2009; Papanikolaou et al., 2012). A size-scale effect is defined as a change in material properties which is rooted in a change in either the dimensions of an internal feature or structure or in the overall physical dimensions of a sample.

It is now well established that plastic flow is size-dependent; characteristically, flow stress or hardness increase with decreasing volume of material under load (Miguel et al., 2001; Dimiduk et al., 2006; Ward et al., 2009 and references therein). Plastic deformation in macroscopic samples is described as a *smooth* process occurring in an elastic continuum. However, recent experiments on micron-sized crystals reveal *step-like stress–strain* curves. A dislocation dynamic model suggests that the onset of plastic flow corresponds to a nonequilibrium phase transition, controlled by the external stress that separates a jammed phase, in which dislocations are immobile, from a flowing phase (Miguel et al., 2002). Plastic flow proceeds through a sequence of intermittent slip avalanches (Uchic et al., 2004; Richeton et al., 2005, 2006; Dimiduk et al., 2006; Miguel and Zapperi, 2006; Csikor et al., 2007; Dahmen et al., 2009; Zapperi, 2012). The resulting irreversible deformations intermittently change the microscopic material shape, while the isolated slip events lead to jumps in the stress–strain curves (strain bursts).

The statistics of the aforementioned discrete changes can reveal the underlying processes. Importantly, the emerging population of discrete slip events of microplasticity follows a scale-free (power-law) size distribution. In contrast, in macroscopic samples plasticity appears as a smooth process. Therefore, an intriguing question that has been refers to the nature of the cutoff which truncates scale-free behaviour in large avalanches. More precisely one wonders (Csikor et al., 2007) if there is no intrinsic limit to the magnitude of dislocation avalanches, why we do not see them in deformation curves of macroscopic samples. Are the properties of dislocation avalanches truly universal?

Through ultraprecise nanoscale measurements on pure metal crystals loaded above the elastic–plastic transition, Dimiduk et al. (2006) directly determined the size of the emerged discrete slip events; the displacement events, Δl, follow a scale-free distribution with probability density function $p(\Delta l) \sim \Delta l^{-a}$ with $a \sim 1.5$. The scaling relationship found is independent of the sample size over the range examined, as well as the gradually increasing stress over the range of the test, namely, there is no work-hardening effect for single slip-plane flow. Based on an alternative approach suggested by Newman (2005) the authors estimated a power-law slope of 1.60 ± 0.02 by a bootstrap method. On the other hand, a statistical characterization of intermittent plastic strain bursts has also been performed by means of

AE. Dynamic processes associated with nucleation, motion and emergence of dislocation groups and regular dislocation pileups (such as slip bands and cracks) on the crystal surfaces, produce AE. Experimental studies through AE have revealed that the plastic flow in crystalline solids is characterized by temporal intermittency. The emerging AE consists of a sequence of intermittent avalanches. The energy E of the acoustic bursts follows a scale-free distribution, having a probability density function $p(E) \sim E^{-\kappa}$ with $\kappa \sim 1.5 - 1.6$ (Weiss and Grasso, 1997; Miguel et al., 2001; Weiss and Marsan, 2003; Zaiser and Moretti, 2005; Richeton et al., 2005; 2006). The scale-free behaviour is extended up to over eight orders of magnitude (eight decades). The absence of any cutoff is characteristic. We pay attention to the finding that the exponents associated with the probability density function of both discrete slip events and AE events are practically identical. This implies that a fixed fraction of the work done by the external stresses during an elongation jump is released in the form of acoustic energy (Schwerdtfeger et al., 2007).

Dimiduk et al. (2006) conclude that the aforementioned results support an emerging view that a statistical framework that creates a coarse-grained description of dislocation response is needed to bridge the gap between the behaviour of individual dislocations and the ensemble of dislocations that govern macroscopic metal plasticity. Importantly, Sethna et al. (2001) propose that the existence of a scale-free set of variables that describe deformation suggests that such a coarse-graining variable set really exists. This assessment puts dislocation motion in the same class as EQs, sand pile avalanches, magnetic domain dynamics, and a wide variety of other dynamic systems. Dislocated nanocrystals are a model system for studying EQ generation; in analogy to plate tectonics, smooth macroscopic-scale crystalline glide arises from the spatial and time averages of disruptive EQ-like events at the nanometer scale (Dimiduk et al., 2006).

Csikor et al. (2007) determine the distribution of strain changes during dislocation avalanches by combining three-dimensional simulations of the dynamics of interacting dislocations with statistical analysis of the corresponding behaviour, and establish the dependence of this distribution on microcrystal size. More precisely, according to their study, the avalanche strain distributions obey the general form $P(s) = Cs^{-\tau}\exp\left[-(s/s_0)^2\right]$, where C is a normalization constant, τ is a scaling exponent, and s_0 is the characteristic strain of the largest avalanches. The authors tested the robustness of the former equation in various physical situations and concluded that the distributions can be described with a universal exponent $\tau = 1.5$.

To elucidate the physical origin of the observed cutoff, Csikor et al. (2007) consider the suggestion that during the progress of an avalanche, two processes reduce the effective stress upon the dislocations (Zaiser and Moretti, 2005; Zaiser, 2006): (1) Because of intrinsic hardening coefficient Θ, a higher driving stress is needed to sustain the avalanche. The stress required to sustain plastic flow increases with deformation, as if an additional back-stress $\sigma_b = -\Theta\gamma$ was building up inside the crystal. The back-stress opposes the propagation of large plastic avalanches, including a finite characteristic size (Miguel and Zapperi, 2006). (2) In the case of displacement-controlled deformation, the driving stress decreases due to relaxation of the elastic strain. Based on these considerations they conclude that $s_0 \propto bE/L(\Theta + \Gamma)$, where Γ is the effective stiffness of the specimen-machine system (for a cubic compression specimen with rigid boundaries Γ equals the elastic modulus E), L is the characteristic specimen size, and b the dislocation Bungers vector modulus. Rescaling the experimental data points by setting $s \rightarrow S = sL\Theta/bE$ and using a hardening coefficient $\Theta = E/1000$, Csikor et al. (2007) found that the scaled experimental data and simulated results are described by a single, universal scaling function $P(S) \sim S^{-3/2}\exp\left[-\left(S/0.6\right)^2\right]$. Therefore, their results demonstrate the *universality* of avalanche behaviour in plastic flow and elucidate the crossover between episodic and smooth plasticity.

The fact that the avalanche strains decease in inverse proportion to the sample size explains why it is difficult to observe strain bursts in macroscopic samples. In contrast, in AE measurements, the acoustic energy is recorded. The energy release associated with a dislocation avalanche may be assumed to be proportional to the dissipated energy e, which is related to the strain by $e \approx \sigma s V$, where σ is the stress and V is the volume. Hence, the cutoff of the AE energy distribution is expected to increase with sample size as $e_0 \propto L^2$. On the other hand, a strong correlation between fracture-induced AE and EME events has been well documented, demonstrating that during the plastic flow (damage) both AE and EME are radiated as two sides of the same coin (Hadjicontis et al., 2007 and references therein).

The above-described considerations explain why it is easy to observe the precursory strong intermittent avalanche-like EME at the geophysical scale while it is not easy to observe the associated intermittent strain bursts.

13.8 On the Paradox of the Systematically Observed EM Silence During the Aftershock Period

An EM silence in all frequency bands is systematically observed during the aftershock period. Herein we focus on this silence, which is considered to be a paradox too.

Irreversible deformation of rocks is accompanied by the experimentally and theoretically well-established Kaiser effect: 'If the heterogeneous material is loaded, then unloaded before fracture, and loaded again, only a small number of micro-fractures are detected before attaining the previous load. Micro-fracturing activity increases dramatically as soon as the largest previously experienced stress level is exceeded'. AE studies of cracking in rocks have demonstrated that, in general, new AE is generated only once the previous maximum stress has been exceeded during cyclic loading. This phenomenon was first reported in metals (Kaiser, 1953) and is now known as the Kaiser 'stress-memory' effect (Lavrov, 2005 and references therein).

EME, as a phenomenon rooted in the damage process, should also be an indicator of memory effects. Indeed, laboratory studies verify that the Kaiser effect is also observed for the EME: during cyclic loading, the level of EME increases significantly when the stress exceeds the maximum previously reached stress level; since the cracking does not practically happen before the previously maximum stress level is reached again, fracto-emission (electron, ion, neuron and photon emissions) accompanying the formation of fresh fracture surfaces (fresh charge separation) are therefore not observed (Khatiashvili, 1984; Yamshchikov et al., 1994; Shkuratnik and Lavrov, 1997; Lavrov, 2005 and references therein; Hadjicontis et al., 2005; Mori and Obata, 2008; Mavromatou et al., 2008).

We emphasize that the Kaiser effect by means of AE often appears at a stress level not exactly equal to, but lower than the largest stress of the previous loading (Li and Nordlund, 1993; Lavrov, 2005 and references therein). This characteristic phenomenon is known as the 'felicity effect' (Li and Nordlund, 1993). Importantly, *the EME does not follow the felicity effect: EME does not appear at a stress level lower than the largest stress of the previous loading.* This evidence, beyond supporting the proposal that EME and AE represent different phases of the destruction process (see Section 13.6.2), implies a superiority of EME as an indicator of damage development in rocks. Recently, simultaneous measurements of AE and EME during repeated loading tests of rock sample have been conducted by Mori and Obata (2008). The results verified that AE appeared at stress levels below the previous maximum stress, and thus the Kaiser effect could not be clearly recognized in the rock tested. In contrast, the EME appeared only when the loading stress was approaching and exceeding the previous maximum stress.

The existence of a clear Kaiser effect by means of EME in the geological scale can justify the systematically observed absence of EME during the aftershocks that occurred within the wide fractured region surrounding that main fault. The stress during the aftershock period does not exceed the previously reached maximum stress level associated with the main shock occurrence. On the contrary, the generation of AE during the aftershock period cannot be excluded either due to the felicity effect or due to frictional-noise-type local rearrangements/rolling between the previously formed fracture surfaces/fragments (Mori et al., 1994, 2004a,b; Mori and Obata, 2008). The rocks in the focal area seem to have the capacity to retain EME 'imprints' and not AE 'footprints' from previous treatments and, under certain conditions, to reproduce information about those treatments.

A crucial question: was the significant EQ the main shock? A crucial question arises immediately after any significant EQ referring to whether the specific EQ was the main shock or a foreshock of an ensuing larger EQ. If we accept that the EME Kaiser effect is extended to the geological scale, then, as long as the silence is continued after the significant EQ occurrence we can consider that the occurred EQ was the main shock. In contrast, the observation of a new stronger kHz EME signifies that the EQ was not the main event.

We refer characteristically to the case of the Athens EQ ($M_w = 5.9$, 7 September 1999) (see Section 13.5.4): Before this catastrophic EQ two strong avalanche-like kHz EM anomalies were detected, the larger anomaly was the second one, which contained approximately 80% of the total EM energy released. Two EQs with magnitudes $M_w = 5.5$ and $M_w = 5.8$ occurred one after the other (Eftaxias et al., 2001), while the two activated faults were very close to each other. The first EQ was not the main shock.

13.9 On the Traceability of the EM Precursors

A critical view often raised concerns the traceability of the fracture-induced EME at the geophysical scale: 'even if one accepts the generation of the EME before, and not at, the EQ occurrence, an EME produced in the Earth's crust should be strongly attenuated by the Earth or, much more, by the sea before reaching the surface and being launched to the atmosphere.' (Johnston, 1997).

First of all, we clarify that the observed EME precursors are associated with shallow EQs that occurred on land or near the coastline with a magnitude ~ 6 or larger (e.g. Kapiris et al., 2002, 2003; Eftaxias et al., 2004; Karamanos et al., 2006). Moreover, it is known that for an EQ with magnitude ~ 6 the fracture process extends to a critical radius of $\sim 120\,\mathrm{km}$ (Bowman et al., 1998). Therefore, in such cases the fracture process is extended up to the Earth's surface; the hypothesis that the fracture-induced EME are launched to the atmosphere is reasonable. Accumulated evidence supports the possibility of detecting preseismic fracture-induced EM anomalies.

Most of the released energy is consumed in creating the fault zone: (1) McGarr et al. (1979) conclude that 'most of the released energy is consumed in creating the fault zone, with less than 1% being radiated seismically.' (2) Boler (1990) found that the proportion of radiated elastic wave energy to the energy associated with *new areas* was low (0.001). (3) Chester et al. (2005) conclude that the energy required to create the *fracture surface area* in the fault is about 300 times greater than seismological estimates would predict for a single large EQ. New surface areas generated during an EQ are $S = 10^3 - 10^6\,\mathrm{m}^2$ for each m^2 of fault area. We recall that during the formation of new surfaces EM radiations are emitted. Therefore, the hypothesis that a high amount of EME is radiated during the creation of the fault zone cannot be excluded.

We note that the principal feature of a fracture is its fractal organization in both space and time. The EQs occur on a fractal structure of faults: fault displacements, fault and fracture trace lengths, and fracture apertures follow a power-law distribution. Fractals are highly convoluted, irregular shapes. The number of breaking bonds is dramatically higher in the fractal fracture process in comparison to those of the Euclidean fracture process. This situation justifies why a high amount of energy is consumed when the fault zone is created.

The self-affine nature of the fracture process may also imply an answer as to why nature plays meaningful '$1/f$ music' during the EQ preparation process: a huge amount of energy is consumed in risk-free (hazard-free) ruptures.

We have focused on the fact that the rupture fault lengths follow a power-law distribution. An opening crack, due to emitting, diffusing and recombination charge, can act as an EM emitter. In this view, an active crack or rupture can be simulated by a 'radiating element'. Of particular interest is the research area known as 'fractal electrodynamics'. The term fractal electrodynamics was first suggested by Jaggard (1990) to identify the newly emerging branch of research, which combines fractal geometry with Maxwell's theory of electrodynamics. The creation of the aforementioned network of traces/new surfaces forms a fractal network of EM emitters which radiate in a cooperative way at the last stages of EQ preparation. The idea is that 'A fractal geo-antenna (FGA) can be formed as an array of line elements having a fractal distribution on the ground surface as a strong and shallow EQ is approached'. (Eftaxias et al., 2004). The fractal tortuous structure can significantly increase the radiated power density, as compared to a single dipole antenna. The tortuous path increases the effective dipole moment, since the path length along the emission is significantly longer than the Euclidean distance.

Optimal paths play a fundamental role in fractures. Andrade et al. (2009) explored the path that is activated once an optimal path fails and what happens when this new path also fails and so on, until the system is completely disconnected. The authors conclude that for all disorders the path along which all minimum energy paths fracture is a fractal of dimension $D = 1.22$. Interestingly, the fractal dimension of the observed kHz EME precursors is $D \approx 1.3$ (see Section 13.5.1), while a surface trace of a single major fault might be characterized by fractal dimension $D = 1.2$ (Sornette, 1991; Sahimi, 1993; Sahimi et al., 1993).

It might be concluded that there is no reason why a high amount of fracture-induced EME should not be directly launched through a fractal geo-antenna to the atmosphere in the case of large shallow EQs that occur on land or near the coastline.

13.10 The Earth as a Living Planet by Means of Precursory EM Activities

The Earth is a living planet where many complex systems run perfectly without stopping at all. The EQ generation is a fundamental sign that the Earth is a living planet. In this section we show human-type diseases appearing in the Earth's crust during the EQ generation process, enhancing the consideration that the Earth really is a living planet.

More precisely, we have shown that during the EQ preparation process the Earth's crust, sequentially: (1) first undergoes a transition similar to that which the human heart undergoes during its transition from the healthy state to the pathological one (this transition occurs during the first stage of the proposed model), and then (2) experiences an epileptic-seizure-type crisis (this crisis occurs during the third stage of the proposed model). In the following we briefly refer to the above-mentioned two human-type diseases.

13.10.1 The Earth's Crust and Human Heart Undergo Similar Second-Order Phase Transitions from the Healthy State to a Pathological One

In this subsection we refer to the fact that an analysis by means of MCF (see Section 13.3.3) shows that the Earth's crust undergoes a transition similar to that which the human heart undergoes during its transition from the healthy state to the pathological one (Contoyiannis et al., 2013). This transition happens during the first stage of the proposed model of EQ generation.

As was mentioned in Section 13.3.3, an analysis by means of criticality, namely, the MCF method, reveals that the fracture of the heterogeneous system surrounding a main fault undergoes a second-order phase transition in equilibrium reflected in the associated fracture-induced MHz EME. During the 'critical epoch' the fracture events/EM emitters show self-similar structures both in time and space and are well correlated even at arbitrarily large separation. This means that the 'short-range' correlations have evolved into 'long-range' ones, more precisely, the laminar lengths (waiting times) fit a power-law type distribution. Beyond the appearance of a broad range of integrated self-affine outputs and long-range correlations, the fracture-induced MHz EME time-series of the 'critical epoch' are also characterized by strong antipersistent behaviour, namely, an underlying negative feedback mechanism that kicks the system far from extreme operation. The aforementioned two crucial critical features characterize a 'healthy system', since such a mechanism provides the system with *adaptability*, namely, the ability to respond to various stresses and stimuli (Ivanov et al., 1999; Goldberger et al., 2002) keeping the system away from extreme states. Analysing human electrocardiogram (ECG) time-series of healthy individuals in terms of the MCF method we have provided evidence that the healthy heart dynamics include the aforementioned two crucial critical characteristics as well, which characterizes the 'critical epoch' of a second-order phase transition (Contoyiannis et al., 2013).

Following the 'critical epoch' of a continuous thermal phase transition a 'noncritical' epoch appears, which is characterized by a dramatic breakdown of the above-mentioned two healthy crucial critical features: (1) the correlation function turns into a rapid exponential decay, the laminar lengths (waiting times) fit an exponential type distribution; and (2) the system shows lower antipersistent behaviour. This means that the system has lost a part of its adaptability, namely, the ability to respond to all stresses and stimuli; it has been fallen to an 'injury state'. Analysing MHz EME time-series we have shown that such a 'noncritical'/ 'injury' epoch follows the appearance of the 'critical epoch' (Contoyiannis et al., 2005). On the other hand, analysing human ECG time-series of cardiac infarction patients we have shown that the underlying heart dynamics is also characterized by the aforementioned two crucial characteristics of the 'non-critical'/'injury state': the laminar lengths (waiting times) fit an exponential type distribution, while the injured heart shows lower antipersistent behaviour in comparison to that of the healthy heart (Contoyiannis et al., 2013).

13.10.2 Earthquakes Can Be Considered as Epileptic Seizures of the Earth's Crust in Terms of Complexity

In this subsection we show that the Earth's crust experiences an epileptic-seizure-type crisis. This crisis happens during the third stage of the proposed EQ generation model, namely, during the stick−slip-like plastic flow stage of EQ preparation, as it is reflected in the associated kHz EME.

Authors have suggested that the dynamics of EQs and ESs can be analysed within similar mathematical frameworks (Hopfield, 1994; Herz and Hopfield, 1995; Rundle et al., 2002). In the framework of this suggestion, Osorio et al. (2010) in a pioneering work have recently shown that a dynamic analogy supported by scale-free statistics exists between ESs and EQs, analysing populations of different ESs and EQs, respectively.

A question effortlessly arising is whether a dynamic analogy also exists between ESs and EQs at the level of a single fault/seizure activation, namely, whether a dynamic analogy may exist for the ways in which firing neurons/fracture events produce a single ES/EQ. We have shown that such a dynamic analogy really exists (Nikolopoulos et al., 2004, 2011; Kapiris et al, 2005b; Li et al.,2005; Eftaxias et al, 2006, 2013b), as follows.

A central property of EQ and ES generation is the occurrence of large-scale collective behaviour with a very rich structure, resulting from repeated nonlinear interactions among the constituents, namely, firing neurons/opening cracks, of the system. Consequently, nonextensive statistical mechanics (Tsallis, 2009) is the appropriate framework in order to investigate the existence of a common process for launching of the two considered shocks. We performed the analysis mainly based on a recently introduced nonextensive model for EQ dynamics (see Section 13.5.2.1) which leads to a G–R-type law for the relationship between the frequency and magnitude of EQs (Sotolongo-Costa and Posadas, 2004; Silva et al., 2006). We have shown that the populations of (1) electric pulses included in a single ES and (2) kHz fracto-EM pulses (EM-EQs), rooted in the activation of a single fault, follow the above-mentioned nonextensive statistical law expressed by Eq. (13.5) with similar nonextensive q-parameter ($q \approx 1.8$ for kHz EME vs $q \approx 1.7 - 1.8$ for ES electric pulses).

In terms of energy the G–R law states that the probability density function of having an EQ energy E is denoted by the power-law $P(E) \sim E^{-B}$, where $B \sim 1.4 - 1.6$. Importantly, the probability of an ES in a population of different events having energy E is proportional to E^{-B}, where $B \sim 1.5 - 1.7$ (Osorio et al., 2010). A reasonable question refers to whether the sequences of electrical pulses/kHz EM pulses included in a single human ES/kHz EME precursor follow the aforementioned power-law with a compatible exponent. The associated B-exponents are 1.72 (Eftaxias et al., 2013b) and 1.31 (Eftaxias et al., 2004), correspondingly.

Power-law correlations in both space and time are at least required in order to verify dynamic analogies between different catastrophic events. Hence, one may ask how the population of EM bursts included in a single kHz EME associated with the activation of a fault and the population of electric bursts included in a single ES present similarity in their time domain correlations. The existence of a rather common power-law distribution of burst life-time (duration) in the above mentioned two populations has been shown: the analysis of ESs and kHz EME time series leads to a negative exponent $\approx -(1.7 - 1.8)$ and ≈ -1.6, correspondingly (Eftaxias et al., 2013b).

Our analysis also revealed common 'pathological symptoms' of a transition to the emergence of the two seemingly different extreme events under study. The emergence of a single ES/kHz EME precursor is accompanied by the appearance of (1) *higher organization*, namely, significant lower nonextensive Tsallis entropy; (2) *persistence*, i.e., a mechanism which is characterized by a positive feedback (Nikolopoulos et al., 2004, 2011; Kapiris et al, 2005b; Eftaxias et al, 2006, 2013a). The appearance in both cases of a high organization dynamics, which is simultaneously characterized by a positive feedback mechanism indicating a strong influence of excitation of an event on succeeding events, is consistent with the emergence of a catastrophic phenomenon.

The above-mentioned findings are consistent with the basic notion of *the relatively new field of complexity* that a dynamic analogy among the extreme events of various complex systems exists, which may be considered as a footprint of universality among them. The field of study of complex systems considers that the dynamics of complex systems is founded on universal principles that may be used to describe disparate problems (Bar-Yam, 1997). This is a basic reason for our interest in complexity (Stanley, 1999, 2000; Sornette, 2000; Vicsek, 2001, 2002; Balasis et al., 2013). In this direction we note that the trade volume events of different shares/economic indices prior to a collapse, the price fluctuation (considered as the difference of maximum minus minimum price within a day) events of different shares/ economic indices prior to a collapse, X-ray flux events during the preparation process of powerful solar flares, and D_{st} events in the course of the preparation of intense magnetic storms also follow the same nonextensive frequency-size law of Eq. (13.5) with the nonextensive q-parameter similar to that of precursory kHz EME pulses and epileptic pulses, while present the aforementioned 'pathological symptoms' of a catastrophic phenomenon (Balasis et al., 2006, 2008, 2011a,b,c,d; Potirakis et al., 2013a). These results enhance the view that the dynamics of complex systems are founded on universal principles and that the dynamics of the kHz EME precursor satisfies such common universal principles.

We claim that the performed analysis captured common principal laws behind the seizures can be considered as 'quakes of the brain' (Osorio et al., 2010) *but also vice versa: the EQs can be considered as ESs of the Earth's crust.*

13.11 An Open Issue of the Materials Science Community: Do the Scaling Laws Associated with the Fracture and Faulting Processes Emerge From Geometrical and Material Built-In Heterogeneities or From the Critical Behaviour Inherent to the Nonlinear Equations Governing Earthquake Dynamics?

One of the largest open issues of the materials science community is the interpretation of scaling laws associated with the fracture and faulting processes (Carpinteri and Pugno, 2005). Especially, an important open question is whether the spatial and temporal complexity of EQs and fault structures and, above all, the interpretation of the observed scaling laws, emerge from geometrical and material built-in heterogeneities or from the critical behaviour inherent to the nonlinear equations governing the EQ dynamics.

As has already been mentioned (see Section 13.3.3), the analysis in terms of the MCF analysis method reveals that *the fracture of the highly heterogeneous system surrounding the main fault*, as it is reflected in the emitted MHz EME, can be described by means of a thermal second-order phase transition, while its analysis by means of the new method of natural time verifies the aforementioned critical behaviour. Importantly, the associated preseismic activity behaves as a critical phenomenon (see Section 13.3.7). At the critical state self-similar structures appear both in time and space. This fact is mathematically expressed through power-law expressions for the distributions of spatial or temporal quantities associated with the aforementioned self-similar structures (Stanley, 1987, 1999).

Based on the above-mentioned results, we suggest that the scaling laws associated with the fracture of a highly heterogeneous medium are rooted in the critical behaviour inherent to the nonlinear equations governing their critical dynamics.

In contrast, the analysis of kHz EME, which in the framework of our approach is rooted in the fracture of the 'backbone' of asperities distributed across the main fault, supports the hypothesis of the geometric nature of the scaling laws associated with the fracture of asperities, while accumulating theoretical and experimental evidence further enhances the proposal that these scaling laws result from preexisting fractal geometry.

In the following, we briefly summarize the main relevant arguments:

1. The topology of fracture surfaces has been found to be self-affine following the persistent fBm model over a wide range of scale lengths (see Section 13.5.2.2). We note that the profile of the kHz EME precursor follows this universal indicator, i.e., the fBm model (see Section 13.5.2.2).

2. The spatial roughness of fracture surfaces has been interpreted as a universal indicator of surface fracture, weakly dependent on the nature of the material and on the failure mode (see Section 13.5.2.3). We recall that the above-mentioned universal indicators are embedded in the kHz EME precursor (see Section 13.5.2.3).

3. A self-affine asperity model that mimics the friction by means of two fractional Brownian profiles that slide one over the other exhibits the G−R law with an exponent related to the roughness index of the profiles (De Rubeis et al., 1996; Hallgass et al., 1997).

   The self-affine asperity model predicts that the distribution of areas of the broken asperities A follows a power law $P(A) \sim A^{-\delta}$, with an exponent δ which could be related to the Hurst exponent $0 < H < 1$ that controls the roughness of the fault. The former relation is obtained by supposing that the area of the broken asperities scales with its linear extension l as $A_{asp} \sim l^{(1+H)}$. The Hurst exponent $H \sim 0.75$ characterizes the profile of the kHz EME precursor. Based on the above, if we accept that the kHz EME reflects the fracture of asperities, it is reasonable to expect that the area of the broken asperities scales with its linear extension l as $A_{asp} \sim l^{(1.75)}$. Numerical studies indicate that the number of bonds that break scales during the whole process of fracture as $l^{1.7}$ with the system size l (de Arcangelis and Herrmann, 1989). The observed similarity verifies that the scaling laws hidden in the kHz EME precursor result from the pure preexisting fractal geometry of the fault.

   The self-affine asperity model also reproduces the G−R law. This predicts that a seismic event releases energy in the interval $[E, E + dE]$ with a probability $P(E)dE$, $P(E) \sim E^{-B}$, where $B = a + 1$ and $a = 1 - H/2$ with $a \in [1/2, 1]$. We recall that the distribution of energies released at any EQ is well described by the power-law,

$P(E) \sim E^{-B}$, with $B \sim 1.4 - 1.6$ (Gutenberg and Richter, 1954). In our case, we have $H \sim 0.75$ for the observed kHz EME precursor, which leads to $a \sim 0.67$, and thus, the fracture of asperities release EM energies following the distribution $P(E) \sim E^{-B}$, with $B \sim 1.67$. The above findings further verify the hypothesis that the scaling hidden in the detected kHz EME precursor could be rooted in pure preexisting fractal geometry of the fault.

4. Authors have reported that the G−R relation for EQs can be derived if a power-law relation holds for the asperity size distribution (Ohnaka, 2000).

5. Authors (Sotolongo-Costa and Posadas, 2004; Silva et al., 2006) have studied the phenomenon of fault slipping from *a geometric viewpoint*, offering an idealized representation of the fragmented core of a fault (gouge). The proposed nonextensive formula for the magnitude distribution of EQs (see Section 13.5.2.1, Eq. (13.5)), which exhibits a power-law behaviour of the G−R-type as an asymptotic limit, successfully describes various seismicities with q-nonextensive parameter, which is the main parameter of this formula, distributed in a narrow range.

Based on the above presented arguments, we suggest that, concerning the process of EQ generation: (1) the scaling laws associated with the fracture of a highly heterogeneous system surrounding the main fault (stage 1 of our model) emerge from the critical behaviour of the corresponding fracture dynamics inherent to the nonlinear equations governing EQ dynamics of this stage, while in contrast, (2) the scaling laws associated with the family of asperities, which are distributed along the main fault and sustain the system, are rooted in the preexisting fractal geometry of the fault.

13.12 Conclusions

Understanding how EQs occur is one of the most challenging questions in fault and EQ mechanics (Shimamoto and Togo, 2012). We consider an EQ essentially as a large-scale fracture. Our research focuses on the last preparatory stage of EQ generation by means of fracture-induced MHz−kHz EME using only ideas and tools from physics. A field experimental network has been installed using the same instrumentation as in laboratory experiments for fracture-induced EME for the recording of fracture-induced EME at the geophysical scale. We consider that the conducted research has contributed in shedding light on the final stages of the EQ preparation process, through the proposal of the following 'Four Stage Model of EQ Dynamics'.

The first stage refers to the fracture of the highly heterogeneous system surrounding the main fault and this is reflected in the observed fracture-induced MHz EME. Such an EM precursor should satisfy the following criteria: (1) The underlying control mechanism regulating the fracto-emission should be a negative feedback one that 'kicks' the cracking rate away from extremes, providing adaptability to the system that is the ability to respond to various external stresses. (2) It should behave as a second-order phase transition in equilibrium including the associated crucial features of 'critical epoch' and 'symmetry breaking' that should appear one after the other. (3) The analysis of such a candidate precursor should lead to the conclusion that a truncated Lévy walk type mechanism can organize the heterogeneous system to criticality as a result of a feedback 'dialogue' between the stresses and heterogeneity. (4) The candidate MHz EME and the associated foreshock seismic activity should constitute two sides of the same coin, namely, the corresponding foreshock seismic activity should behave as a critical phenomenon, as well. (5) The candidate MHz EME should be consistent with geodetic measurements. (6) A large number of precursory anomalies are observed before an EQ, for example, short-lived seismo-ionospheric EM precursors, EM anomalies rooted in preseismic LAI-coupling, precursory anomalies of hydrothermal parameters in the coversphere and atmosphere, and TIR anomalies. The generation of such a large number of them requires the existence of physical and chemical transformations in a spatially extensive area. The MHz EME precursor includes the 'critical window', i.e., the epoch during which the short-range correlations between the cracking events have been evolved to long-range ones. Thus, the aforementioned precursors should be observed more or less during the same time period with the MHz EM anomaly well before the EQ occurrence, approximately during the last week prior to the main event, and it is found that this does happen. (7) The MHz EM anomaly, in the frame of critical phenomena, should cease by its nature before the EQ occurrence, namely, when the 'noncritical window' emerges after the 'critical window' indicating the appearance of 'symmetry breaking', i.e. (1) the transition from the phase of nondirectional, almost symmetrical, cracking distribution in a spatially extensive area to a directional localized cracking zone, which has been obstructed along the backbone of strong asperities distributed across the surfaces of the main fault; (2) the transition from the phase of long-range correlations to that of short-range correlations between the cracking event.

Base on the above, we consider that the noncoseismic character of the MHz EME precursor is not a puzzling feature. This conclusion is supported (1) by means of percolation theory due to the fact that this precursor is associated with the 'hydraulic threshold', where only the transition from impermeable to permeable occurs; (2) in terms of physical mesomechanics, which predicts that this emission is rooted in the small-scale fragmentation; and (3) in terms of laboratory experiments, which reveals that the fracture-induced MHz radiation systematically precedes the kHz one.

We emphasize that *the appearance of MHz EME precursors*, within the framework of our approach, *does not mean that the EQ under preparation is inevitable*. Note that cases for which valid MHz EME (first stage) anomalies have been observed while no significant EQ (main event) occurred have already been identified, and an article on this subject is currently being prepared.

We consider that the above multidisciplinary study: (1) establishes a strict set of criteria which safely recognize an emerging MHz EME candidate precursor as a seismogenic one, strongly associating this anomaly with the first stage of our model; (2) reveals that the MHz EME activity and the associated foreshock activity constitute two sides of the same coin; and finally (3) justifies MHz EME absence during the EQ.

The second stage refers to the final stage of fracture of a highly heterogeneous system surrounding the main fault (reflected in the tail of the observed MHz EM radiation) or to the initial stage of fracture of the fault (reflected in the observed initial mild kHz EM radiation). The appearance of such a MHz or kHz EM precursor should show that the underlying fracto-EM mechanism undergoes a tricritical phase transition. The appearance of an EM anomaly with tricritical behaviour after the appearance of a MHz EME precursor, which behaves as a second-order transition, in a short time interval, strongly supports the seismogenic origin of the observed two precursors by means of critical phenomena. The fact that the aforementioned mild MHz or kHz EM radiation is followed by the strong avalanche-like kHz EM anomaly, which does not have any symptoms of a second-order phase transition, but, on the contrary, behaves like a first-order phase transition, further enhances the existence of the second stage of our model and the seismogenic origin of the associated EM precursors.

The third stage refers to the fracture of a family of asperities distributed along the fault sustaining the system, i.e., the stick–slip-like plastic flow stage of EQ preparation (reflected in the finally observed strong avalanche-like kHz EME precursor), implying that the occurrence of the imminent EQ is inevitable as soon as kHz EME have been identified. This suggestion is supported by laboratory experiments, which suggest that the fracture-induced kHz EME emerges from the tail of the fracture process, studies by means of percolation theory, which connect this precursor with the formation of the flat infinite cluster disintegrating the system, and finally in terms of synergetic principles of physical mesomechanics, which connect the kHz fracto-EM activity with the large-scale (macrolevel) fragmentation.

Such a kHz EME precursor should satisfy the following criteria: (1) It should include the following crucial features of a catastrophic phenomenon: high organization, high information content, low complexity, strong persistency, existence of clearly preferred directions of activities, absence of any footprint of a second-order phase transition. (2) The kHz EME precursor should include universal patterns rooted in the well-documented aspect of the self-affine nature of faulting and fracture, more precisely, the included sequence of fracture-induced 'EM-EQs', which reflects the fracture of asperities, should be in consistency with the requirement that fracture of a single fault should be a reduced self-affine image of regional seismicity and a magnified image of laboratory seismicity, its profile should follow the persistent fBm model, and finally, its roughness should be consistent with the universal indicator of surface fracture. (3) The morphology of the observed kHz EME precursor should be consistent with characteristics of the stick–slip-like plastic flow stage of EQ preparation, namely, this should include a sequence of strong avalanche-like EME events, and this sequence should be characterized by a sharp onset and cease. (4) The characteristics of the kHz EME precursor, in terms of the associated energy, organization and information content, should be consistent with simultaneous seismological or geodetic data which refer to the fault modelling of the impending EQ.

We emphasize that within the framework of our approach, *the appearance of a kHz EME precursor, means that the EQ under preparation is unavoidable.*

We consider that the above multidisciplinary study: (1) establishes a strict set of criteria which safely recognize a strong pulse-like kHz EME candidate precursor as a seismogenic one; (2) strongly associates this anomaly with the third stage of our model; (3) reveals that the kHz EME data and the associated seismological and geodetic data which refer to the fault modelling constitute two sides of the same coin; and finally (4) justifies kHz EME absence during the EQ.

The fourth stage refers to the transition to the last stage of the EQ preparation process, namely, the dynamic slip which results in the fast, even super-shear, mode reflected in the observed quiescence in all EM frequency bands, which follows the abrupt ceasing of the emerged strong avalanche-like kHz EME. Accumulated evidence, in terms of laboratory experiments of EME, numerical studies, elastic moduli, heat-flow paradox, and granular packing's notions, enhance the view that the emergence of this silence is not a puzzling feature but, on the contrary, it is the final precursory signal indicating the transition to the stage of the dynamic fast slip.

It is difficult to prove associations between any two events separated in time (such as MHz−kHz EME precursors and EQ). However, we consider that in the case of an emerged sequence of MHz and kHz EM anomalies in a short time interval, namely, a few days, each of them satisfying all the above-mentioned strict criteria in terms of the proposed four-stage model of EQ dynamics, it is more difficult to prove their association groundless.

A basic goal of our study was to examine whether systematically observed preseismic EME characteristics, which are commonly considered as 'puzzling features', are really 'puzzling' or whether they are crucial precursory features of the EQ preparation process. Sometimes the arguments based on these 'puzzling features' were extended to the extreme claim that any EM precursory activity is impossible. We consider that recent accumulated evidence implies that the considered 'puzzling features' are crucial EM precursory features of an impending EQ. Characteristically (1) the observed silence of fracture-induced EME during the EQ occurrence reveals the transition to the final stage of dynamical slip which results in the fast, even super-shear, mode that surpasses the shear wave speed. (2) Shedding light from nanoscale plastic flow on the geophysical scale, we conclude that the association of EME signals with small precursory strain changes but not with much larger coseismic strains also does not constitute a 'puzzling feature'. (3) The arguments that the MHz−kHz EME precursors are not combined with seismological and geodetic precursors, seem to be groundless. In terms of the well-documented Kaiser effect the systematically observed EM silence during the aftershock period is also justified. Finally, it might be concluded, in terms of the introduced notion of the 'FGA', that there is no reason why a high amount of fracture-induced EME should not be directly launched through a FGA to the atmosphere in the case of large shallow EQs that occur on land or near coastlines, and thus are detectable.

In any case, readers should be aware that the results presented in this contribution represent a 'snapshot' of a rapidly evolving research field. The future will show whether the proposed model is correct or wrong. The complexity of the seismogenic EM systems is obvious, as is the amount of research necessary before we clearly understand them; the path appears to be long and challenging.

'Are EQs predictable?' This is a question debated in the EQ research community. Its answer leads to another question: 'Are there credible EQ precursors?' To the extent that our proposal is correct, it will contribute to the efforts towards answering the aforementioned crucial question, and thus to the efforts towards EQ prediction, since this depends on the degree of comprehension of the involved processes. The presented results suggest that there is no need for conflicts among seismologists, specialists in seismo-EM phenomena, or specialists in other EQ-related observables. The occurrence of an EQ is a result of the cooperation of cracking events. An EQ has many different facets. In our opinion, the best policy requires close cooperation of researchers who put efforts into different disciplines for EQ prediction. As has been emphasized, the Earth is a living planet, and its crust suffers from human-type diseases during the last stage of EQ generation, namely, it first undergoes a transition similar to that which the human heart undergoes during its transition from the healthy state to the pathological one and then experiences an epileptic-seizure-type crisis. The early and valid diagnosis of a human disease/'earthly disease' often requires the collaboration of medical doctors/'geophysical doctors' of various specialties!

References

Allegre, C., Le Mouell, J., Provost, A., 1982. Scaling rules in rock fracture and possible implications for earthquake prediction. Nature. 297, 47−49.

Alonso-Marroquin, F., Vardoulakis, I., Herrmann, H., Weatherley, D., Mora, P., 2006. Effect of rolling on dissipation in fault gauges. Phys. Rev. E 74, 031306.

Amitrano, D., Helmstetter, A., 2006. Brittle creep, damage, and time to failure in rocks. J. Geophys. Res. 111, B11201.

Andrade, J.S., Oliveira, E.A., Moreira, A.A., Herrmann, H.J., 2009. Fracturing the optimal Paths. Phys. Rev. Lett. 103 (22), 225503.

Arbadi, S., Sahimi, M., 1990. Test of universality for three-dimensional models of mechanical breakdown in disordered solids. Phys. Rev. B 41, 772−775.

de Arcangelis, L., Herrmann, H.J., 1989. Scaling and multiscaling laws in random fuse networks. Phys. Rev. B 39, 2678. Available from: https://doi.org/10.1103/PhysRevB.39.2678.

Åström, J., Timonen, J., 2012. Spontaneous formation of densely packed shear bands of rotating fragments. Eur. Phys. J. E. 35, 40.

Åström, J., Herrmann, H., Timonen, J., 2000. Granular packing and fault zone. Phys. Rev. Lett. 84, 638−641.

Åström, J., Herrmann, H., Timonen, J., 2001. Fragmentation dynamics within shear bands-a model for aging tectonic faults. Eur. Phys. J. E. 4, 273−279.

Baddari, K., Frolov, A., 2010. Regularities in discrete hierarchy seismo-acoustic mode in a geophysical field. Ann. Geophys. 53, 31−42.

Baddari, K., Sobolev, G.A., Frolov, A.D., Ponomarev, A.V., 1999. An integrated study of physical precursors of failure in relation to earthquake prediction, using large scale rock blocks. Ann. Geophys. 42 (5), 771−787.

Baddari, K., Frolov, A., Tourtchine, V., Rahmoune, F., 2011. An integrated study of the dynamics of electromagnetic and acoustic regimes during failure of complex macrosystems using rock blocks. Rock Mech. Rock Eng. 44, 269−280.

Baker, K., Warner, D., 2012. Simulating dynamic fragmentation processes with particles and elements. Eng. Fracture Mech. 84, 96−110.

Balasis, G., Daglis, I.A., Kapiris, P., Mandea, M., Vassiliadis, D., Eftaxias, K., 2006. From pre-storm activity to magnetic storms: a transition described in terms of fractal dynamics. Ann. Geophys. 24, 3557–3567.

Balasis, G., Daglis, I., Papadimitriou, C., Kalimeri, M., Anastasiadis, A., Eftaxias, K., 2008. Dynamical complexity in D_{st} time series using non-extensive Tsallis entropy. Geophys. Res. Lett. 35, L14102.

Balasis, G., Daglis, I.A., Anastasiadis, A., Papadimitriou, C., Mandea, M., Eftaxias, K., 2011a. Universality in solar flare, magnetic storm and earthquake dynamics using Tsallis statistical mechanics. Physica A 390, 341–346.

Balasis, G., Papadimitriou, C., Daglis, I.A., Anastasiadis, A., Athanasopoulou, L., Eftaxias, K., 2011b. Signatures of discrete scale invariance in D_{st} time series. Geophys. Res. Lett. 38, L13103.

Balasis, G., Papadimitriou, C., Daglis, I.A., Anastasiadis, A., Sandberg, I., Eftaxias, K., 2011c. Similarities between extreme events in the solar-terrestrial system by means of nonextensivity. Nonlinear Processes Geophys. 18, 563–572.

Balasis, G., Daglis, I.A., Papadimitriou, C., Anastasiadis, A., Sandberg, I., Eftaxias, K., 2011d. Quantifying dynamical complexity of magnetic storms and solar flares via nonextensive Tsallis entropy. Entropy 13, 1865–1881.

Balasis, G., Donner, R.V., Potirakis, S.M., Runge, J., Papadimitriou, C., Daglis, I.A., et al., 2013. Statistical mechanics and information-theoretic perspectives on complexity in the earth system. Entropy 15, 4844–4888.

Bar-Yam, Y., 1997. Dynamics of Complex Systems. Addison-Wesley, Reading.

Baram, R., Herrmann, H., Rivier, N., 2004. Space-filling bearings in three dimensions. Phys. Rev. Lett. 92, 044301.

Baumberger, T., Caroli, C., Ronsin, O., 2002. Self-healing slip pulses along a gel/glass interface. Phys. Rev. Lett. 88 (1–4), 075509.

Ben-David, O., Cohen, G., Fineberg, J., 2010. The dynamic of the onset of frictional slip. Science 330, 211–214.

Boler, F., 1990. Measurements of radiated elastic wave energy from dynamic tensile cracks. J. Geophys. Res. 95, 2593–2607.

Bouchaud, J.-P., Georges, A., 1990. Anomalous diffusion in disordered media: statistical mechanisms, model and physical applications. Phys. Rep. 195, 127–293.

Bouchaud, E., Soukiassian, P., 2009. Fracture: from the atomic to the geophysical scale. J. Phys. D: Appl. Phys. 42, 210301.

Bouchon, M., Bouin, M.-P., Karabulut, H., Toksoz, M., Dietrich, M., Rosakis, A., 2001. How fast is rupture during an earthquake? New insights from the 1999 Turkey earthquakes. Geophys. Res. Lett. 28, 2723–2726.

Boulton, C., Carpenter, B.M., Toy, V., Marone, C., 2012. Physical properties of surface outcrop cataclastic fault rocks, Alpine Fault, New Zealand. Geochem. Geophys. Geosyst. 13, 1–13.

Bowman, D., Quillon, G., Sammis, C., Sornette, A., Sornette, D., 1998. An observational test of the critical earthquake concept. J. Geophys. Res. 103, 24359–24372.

Buchel, A., Sethna, J.P., 1997. Statistical mechanics of cracks: Fluctuations, breakdown, and asymptotics of elastic theory. Phys. Rev. E 55 (6), 7669–7690.

Carpinteri, A., Pugno, N., 2005. Are scaling laws on strength of solids related to mechanics or to geometry? Nat. Mater. 4, 421–423.

Carpinteri, A., Lacidogna, G., Manuello, A. (Eds.), 2015. Acoustic, Electromagnetic, Neutron Emissions from Fracture and Earthquakes. Springer International Publishing, Heidelberg. Available from: https://doi.org/10.1007/978-3-319-16955-2.

Carpinteri, A., Cornetti, P., Sapora, A., 2011. Brittle failures at rounded V-notches: a finite fracture mechanics approach. Int. J. Fracture 172 (1), 1–8.

Carpinteri, A., Lacidogna, G., Manuello, A., Niccolini, G., Schiavi, A., Agosto, A., 2012. Mechanical and electromagnetic emissions related to stress-induced cracks. SEM Exp. Tech. 36, 53–64.

Carvalho, J., Silva, R., do Nascimento Jr., J.D., De Medeiros, J., 2008. Power law statistics and stellar rotational velocities in the Pleiades. Eur. Phys. Lett. 84, 59001.

Chakrabarti, B., Benguigui, L., 1997. Statistical Physics of Fracture and Breakdown in Disordered Systems. Oxford University Press, Oxford.

Chang, J., Lockner, D., Reches, Z., 2012. Rapid acceleration leads to rapid weakening in earthquake-like laboratory experiments. Science 338, 101–105.

Chauhan, V., Misra, A., 2008. Effects of strain rate and elevated temperature of electromagnetic radiation emission during plastic deformation and crack propagation in ASTM B 265 grade 2 Titanium sheets. J. Mat. Sci. 43, 5634–5643.

Chelidze, T., 1979. Percolation model of fracture of solids and earthquake prediction. Rep. Acad. Sci. USSR 246, 51–54.

Chelidze, T., 1980a. The model of fracture process of solids. Solid State Phys. (Moscow) 22, 2865–2866.

Chelidze, T., 1980b. Theory of percolation and fracture of rocks. In: Proc. of VI All Union Conf. on Mech. of Rocks, Frunze, 1978, 107–117, (in Russian).

Chelidze, T., 1982. Percolation and fracture. Phys. Earth Planet. Inter. 28, 93–101.

Chelidze, T., 1986. Percolation theory as a tool for imitation of fracture process in rocks. Pure Appl. Geophys. 124, 731–748.

Chelidze, T., 1993. Fractal damage mechanics of geomaterials. Terra Nova 5, 421–437.

Chelidze, T., Kolesnikov, Yu, 1984. On physical interpretation of transitional amplitude in percolation theory. J. Phys. A. 17, L791–L793.

Chelidze, T., Reusche, T., Darot, M., Gueguen, Y., 1988. On the elastic properties of depleted refilled solids near percolation. J. Phys. C: Solid State Phys. 21 (30), L1007–L1010.

Chelidze, T., Spetzler, H., Getting, L., Avaliani, Z., 1990. Experimental investigation of the elastic modulus of a fractal system – a model of fractured rock. Pure Appl. Geophys. 134, 31–43.

Chelidze, T., Kolesnikov, Yu, Matcharaahvili, T., 2006. Seismological criticality concept and percolation model of fracture. Geophys. J. Int. 164, 125–136.

Chelidze, T., 1987. Percolation Theory in Mechanics of Geomaterials. Nauka, Moscow, 273 p. (in Russian).

Chen, Y.Z., 2012. A novel solution for effective elastic moduli of 2D cracked medium. Eng. Fract. Mech. 84, 123–131.

Chester, F., Chester, J., 1998. Ultra cataclasite structure and friction processes of the Punchbowl fault, San Andreas system, California. Tectonophysics 295, 199–221.

Chester, J., Chester, F., Kronenberg, A., 2005. Fracture surface energy of the Punchbowl fault, San Andreas system. Nature 437, 133–136.

Coker, D., Lykotrafitis, G., Needleman, A., Rosakis, A., 2005. Frictional sliding modes along an interface between identical elastic plates subject to shear impact loading. J. Mech. Phys. Solids 53, 884–922.

Contoyiannis, Y., Diakonos, F., 2000. Criticality and intermittency in the order parameter space. Phys. Lett. A. 268, 286–292.

Contoyiannis, Y., Diakonos, F., 2007. Unimodal maps and order parameter fluctuations in the critical region. Phys. Rev. E 76, 031138.

Contoyiannis, Y., Eftaxias, K., 2008. Tsallis and Levy statistics in the preparation of an earthquake. Nonlinear Processes Geophys. 15, 379–388.

Contoyiannis, Y., Diakonos, F., Malakis, A., 2002. Intermittent dynamics of critical fluctuations. Phys. Rev. Lett. 89, 035701.

Contoyiannis, Y., Diakonos, F., Kapiris, P., Peratzakis, A., Eftaxias, K., 2004b. Intermittent dynamics of critical pre-seismic electromagnetic fluctuations. Phys. Chem. Earth 29, 397–408.

Contoyiannis, Y., Kapiris, P., Eftaxias, K., 2005. A monitoring of a pre-seismic phase from its electromagnetic precursors. Phys. Rev. E 71, 061123.

Contoyiannis, Y., Potirakis, S.M., Eftaxias, K., Contoyianni, L., 2015. Tricritical crossover in earthquake preparation by analysing preseismic electromagnetic emissions. J. Geodyn. 84, 40–54.

Contoyiannis, Y.F., Diakonos, F.K., Papaefthimiou, C., Theophilidis, G., 2004a. Criticality in the relaxation phase of a spontaneously contracting atria isolated from a Frog's Heart. Phys. Rev. Lett. 93, 098101.

Contoyiannis, Y.F., Nomicos, C., Kopanas, J., Antonopoulos, G., Contoyianni, L., Eftaxias, K., 2010. Critical features in electromagnetic anomalies detected prior to the L'Aquila earthquake. Physica A 389, 499–508.

Contoyiannis, Y.F., Potirakis, S.M., Eftaxias, K., 2013. The Earth as a living planet: human-type diseases in the earthquake preparation process. Nat. Hazard. Earth Syst. Sci. 13, 125–139.

Contoyiannis, Y, Potirakis, S.M., Eftaxias, K., Hayakawa, M., Schekotov, A., 2016. Intermittent criticality revealed in ULF magnetic fields prior to the 11March 2011 Tohoku earthquake ($M_W = 9$). Physica A 452, 19–28.

Csikor, F., Motz, C., Weygand, D., Zaiser, M., Zapperi, S., 2007. Dislocation avalanches, strain bursts, and the problem of plastic forming at the micrometer scale. Science 318, 251–254.

Da Cruz, F., Eman, S., Prochnow, M., Roux, H.-N., Chevoir, F., 2005. Rheophysics of dense granular materials: discrete simulation of plane shear flows. Phys. Rev. E. 72, 021309.

Dahmen, K., Ben-Zion, Y., Uhl, J., 2009. Micromechanical model for deformation in solids with universal predictions for stress–strain curves and slip avalanches. Phys. Rev. Lett. 102, 175501.

De Rubeis, V., Hallgas, R., Loreto, V., Paladin, G., Pietronero, L., Tosi, P., 1996. Self-affine asperity model for earthquakes. Phys. Rev. Lett. 76, 2599–2602.

Dimiduk, D., Woodward, C., LeSar, R., Uchic, M., 2006. Scale-free intermittent flow in crystal plasticity. Science 312, 1188–1190.

Dodze, D., Beroza, G., Ellsworth, W., 1996. Detailed observations of California foreshock sequences: implications for the earthquake initiation process. J. Geophys. Res. 101, 22371–22392.

Donner, R.V., Potirakis, S.M., Balasis, G., Eftaxias, K., Kurths, J., 2015. Temporal correlation patterns in pre-seismic electromagnetic emissions reveal distinct complexity profiles prior to major earthquakes. Phys. Chem. Earth 85–86, 44–55.

Eftaxias, K., 2009. Footprints of nonextensive Tsallis statistics, selfaffinity and universality in the preparation of the L'Aquila earthquake hidden in a pre-seismic EM emission. Physica A 389, 133–140.

Eftaxias, K., 2012. Are there pre-seismic electromagnetic precursors? A multidisciplinary approachEarthquake Research and Analysis – Statistical Studies, Observations and Planning 460 pages. InTechMarch . Available from: https://doi.org/10.5772/28069.

Eftaxias, K., Potirakis, S.M., 2013. Current challenges for pre-earthquake electromagnetic emissions: shedding light from micro-scale plastic flow, granular packings, phase transitions and self-affinity notion of fracture process. Nonlinear Processes Geophys. 20, 771–792.

Eftaxias, K., Potirakis, S.M., Chelidze, T., 2013a. On the puzzling feature of the silence of precursory electromagnetic emissions. Nat. Hazard. Earth Syst. Sci. 13, 2381–2397. Available from: https://doi.org/10.5194/nhess-13-2381-2013.

Eftaxias, K., Minadakis, G., Potirakis, S.M., Balasis, G., 2013b. Dynamical analogy between epileptic seizures and seismogenic electromagnetic emissions by means of nonextensive statistical mechanics. Physica A 392 (3), 497–509.

Eftaxias, K., Panin, V.E., Deryugin, Y.Y., 2007. Evolution-EM signals before earthquakes in terms of meso-mechanics and complexity. Tectonophysics 431, 273–300.

Eftaxias, K., Kapiris, P., Polygiannakis, J., Bogris, N., Kopanas, J., Antonopoulos, G., et al., 2001. Signature of pending earthquake from electromagnetic anomalies. Geophys. Res. Lett. 28, 3321–3324.

Eftaxias, K., Kapiris, P., Dologlou, E., Kopanas, J., Bogris, N., Antonopoulos, G., et al., 2002. EM anomalies before the Kozani earthquake: a study of their behavior through laboratory experiments. Geophys. Res. Lett. 29 (8), 1228.

Eftaxias, K., Frangos, P., Kapiris, P., Polygiannakis, J., Kopanas, J., Peratzakis, A., et al., 2004. Review-model of pre-seismic electromagnetic emissions in terms of fractal-electrodynamics. Fractals 12, 243–273.

Eftaxias, K., Kapiris, P., Balasis, G., Peratzakis, A., Karamanos, K., Kopanas, J., et al., 2006. A unified approach to catastrophic events: from the normal state to geological or biological shock in terms of spectral fractal and nonlinear analysis. Nat. Hazard. Earth Syst. Sci. 6, 205–228.

Eftaxias, K., Contoyiannis, Y., Balasis, G., Karamanos, K., Kopanas, J., Antonopoulos, G., et al., 2008. Evidence of fractional-Brownian-motion-type asperity model for earthquake generation in candidate pre-seismic electromagnetic emissions. Nat. Hazard. Earth Syst. Sci. 8, 657–669.

Eftaxias, K., Athanasopoulou, L., Balasis, G., Kalimeri, M., Nikolopoulos, S., Contoyiannis, Y., et al., 2009. Unfolding the procedure of characterizing recorded ultra-low frequency, kHz and MHz electromagetic anomalies prior to the L'Aquila earthquake as pre-seismic ones. – Part 1. Nat. Hazard. Earth Syst. Sci. 9, 1953–1971.

Eftaxias, K., Balasis, G., Contoyiannis, Y., Papadimitriou, C., Kalimeri, M., Athanasopoulou, L., et al., 2010. Unfolding the procedure of characterizing recorded ultra-low frequency, kHz and MHz electromagnetic anomalies prior to the L'Aquila earthquake as pre-seismic ones – Part 2. Nat. Hazard. Earth Syst. Sci. 10, 275–294.

Forterre, Y., Pouliquen, O., 2008. Flows of dense granular media. Ann. Rev. Fluid Mech. 40, 1–24.

Fukui, K., Ocubo, S., Terashima, T., 2005. Electromagnetic radiation from rock during uniaxial compression testing: the effects of rock characteristics and test conditions. Rock Mech. Rock Eng. 38 (5), 411–423.

Garcimartin, A., Guarino, A., Bellon, L., Ciliberto, S., 1997. Statistical properties of fracture precursors. Phys. Rev. Lett. 79, 3202–3205.

Geller, R.J., Jackson, D.D., Kagan, Y.Y., Mulargia, F., 1997. Earthquakes cannot be predicted. Science 275, 1616–1617.

Girard, L., Amitrano, D., Weiss, J., 2010. Failure as a critical phenomenon in a progressive damage model. J. Stat. Mech. P01013.

Girard, L., Weiss, J., Amitrano, D., 2012. Damage-cluster distributions and size effect on strength in compressive failure. Phys. Rev. Lett. 108, 225502.

Gluzman, S., Sornette, D., 2001. Self-consistent theory of rupture by progressive diffuse damage. Phys. Rev. E. 63, 066129.

Goldberger, A.L., Amaral, L., Hausdorff, A.N., Ivanov, J.M., Ch, P., Peng, C.K., et al., 2002. Fractal dynamics in physiology: alterations with disease and aging. Proc. Natl. Acad. Sci. U.S.A. Suppl 1, 2466–2472.

Guarino, A., Ciliberto, S., Garcimartin, A., Zei, M., Scorretti, R., 2002. Failure time and critical behaviour of fracture precursors in heterogeneous materials. Eur. Phys. J. B 26, 141–151.

Gutenberg, B., Richter, C.F., 1954. Seismicity of the Earth and Associated Phenomena. Princeton University Press, Princeton, NJ, USA.

Hadjicontis, V., Tombras, G.S., Ninos, D., Mavromatou, C., 2005. Memory effects in EM emission during uniaxial deformation of dielectric crystalline materials. IEEE Geosci. Remote Sens. Lett. 2 (2), 118–120.

Hadjicontis, V., Mavromatou, C., Antsygina, T.N., Chishko, K.A., 2007. Mechanism of electromagnetic emission in plastically deformed ionic crystal. Phys. Rev. B 76, 024106.

Halasz, Z., Danku, Z., Kun, F., 2012. Competition of strength and stress disorder in creep rupture, Phys. Rev. E. 85, 016116–1/8.

Hallgas, R., Loreto, V., Mazzela, O., Paladin, G., Pietronero, L., 1997. Earthquake statistics and fractal faults. Phys. Rev. E. 56, 1346−2602.

Hansen, A., Schmittbuhl, J., 2003. Origin of the universal roughness exponent of brittle fracture surfaces: stress-weighted percolation in the damage zone. Phys. Rev. Lett. 90, 45504−45507.

Hayakawa, M., Schekotov, A., Potirakis, S.M., Eftaxias, K., 2015a. Criticality features in ULF magnetic fields prior to the 2011 Tohoku earthquake. Proc. Jpn. Acad., Ser. B 91 (1), 25−30.

Hayakawa, M., Schekotov, A., Potirakis, S.M., Eftaxias, K., Li, Q., Asano, T., 2015b. An integrated study of ULF magnetic field variations in association with the 2008 Sichuan earthquake, on the basis of statistical and critical analyses. Open J. Earthquake Res. 4, 85−93.

Hayakawa, M., Yamauchi, H., Ohtani, N., Ohta, M., Tosa, S., Asano, T., et al., 2016. On the precursory abnormal animal behavior and electromagnetic effects for the Kobe earthquake (M∼6) on April 12, 2013. Open J. Earthquake Res. 5, 165−171.

Hayashi, N., Tsutsumi, A., 2010. Deformation textures and mechanical behavior of a hydrated amorphous silica formed along an experimentally produced fault in chert. Geophys. Res. Lett. 37 (1−5), L12305.

Helmstetter, A., 2003. Is Earthquake triggering driven by small earthquakes? Phys. Rev. Lett. 91, 058501.

Herrmann, H., Roux, S., 1990. Statistical Models for Fracture of Disordered Media. Elsevier, Amsterdam.

Herz, A., Hopfield, J., 1995. Earthquake cycles and neural reverberations: collective oscillations in systems with pulse-coupled threshold elements. Phys. Rev. Lett. 75, 1222−1225.

Hopfield, J., 1994. Neurons, dynamics and computation. Phys. Today 40, 40−46.

Huang, J., Turcotte, D., 1988. Fractal distributions of stress and strength and variations of b value. Earth Planet. Sci. Lett. 91, 223−230.

Huang, K., 1987. Statistical Mechanics. John Wiley and sons, New York.

Ivanov, P. Ch, Amaral, L.A.N., Goldberger, A.L., Havlin, S., Rosenblum, M.G., et al., 1999. Multifractality in human heartbeat dynamics. Nature 1999 (399), 461−465.

Jaggard, D., 1990. On fractal electrodynamics. In: Kritikos, H., Jaggard, D. (Eds.), Recent Advances in Electromagnetic Theory. Springer-Verlag, New York, pp. 183−224.

Johnson, P., Savage, H., Knuth, M., Gomberg, J., Marone, C., 2008. Effects of acoustic waves on stick-slip in granular media and implications for earthquakes. Nature 451, 57−60.

Johnston, M., 1997. Review of electric and magnetic fields accompanying seismic and volcanic activity. Surv. Geophys. 18, 441−475.

Kaiser, J., 1953. Erkenntnisse und Folgerungen aus der Messung von Geräuschen bei Zugbeanspruchung von metallischen Werkstoffen. Archiv für das Eisenhüttenwesen. Verlag Stahleisen 24 (1-2), 43−45. Available from: http://dx.doi.org/10.1002/srin.195301381.

Kalimeri, M., Papadimitriou, K., Balasis, G., Eftaxias, K., 2008. Dynamical complexity detection in pre-seismic emissions using nonadditive Tsallis entropy. Physica A 387, 1161−1172.

Kalimeris, A., Potirakis, S., Eftaxias, K., Antonopoulos, G., Kopanas, J., 2016. Multi-spectral detection of statistically significant components in pre-seismic electromagnetic emissions related with Athens 1999, $M = 5.9$ earthquake. J. Appl. Geophys. 128, 41−57.

Kammer, D., Yastebov, V., Spijker, P., Molinari, J.-F., 2012. On the propagation of slip at frictional interfaces. Tribol. Lett. 48, 27−32.

Kapiris, P., Polygiannakis, J., Peratzakis, A., Nomikos, K., Eftaxias, K., 2002. VHF-electromagnetic evidence of the underlying pre-seismic critical stage. Earth Planets Space 54, 1237−1246.

Kapiris, P., Eftaxias, K., Chelidze, T., 2004a. Electromagnetic signature of prefracture criticality in heterogeneous media. Phys. Rev. Lett. 92 (6), 065702.

Kapiris, P., Eftaxias, K., Nomicos, K., Polygiannakis, J., Dologlou, E., Balasis, G., et al., 2003. Evolving towards a critical point: a possible electromagnetic way in which the critical regime is reached as the rupture approaches. Nonlinear Processes Geophys. 10, 511–524.

Kapiris, P., Balasis, G., Kopanas, J., Antonopoulos, G., Peratzakis, A., Eftaxias, K., 2004b. Scaling similarities of multiple fracturing of solid materials. Nonlinear Processes Geophys. 11, 137–151.

Kapiris, P., Nomicos, K., Antonopoulos, G., Polygiannakis, J., Karamanos, K., Kopanas, J., et al., 2005a. Distinguished seismological and electromagnetic features of the impending global failure: did the 7/9/1999 M5.9 Athens earthquake come with a warning? Earth Planets Space 57, 215–230.

Kapiris, P., Polygiannakis, J., Li, X., Yao, X., Eftaxias, K., 2005b. Similarities in precursory features in seismic shocks and epileptic seizures. Europhys. Lett. 69, 657–663.

Karamanos, K., Peratzakis, A., Kapiris, P., Nikolopoulos, S., Kopanas, J., Eftaxias, K., 2005. Extracting pre-seismic electromagnetic signatures in terms of symbolic dynamics. Nonlinear Processes Geophys. 12, 835–848.

Karamanos, K., Dakopoulos, D., Aloupis, K., Peratzakis, A., Athanasopoulou, L., Nikolopoulos, S., et al., 2006. Pre-seismic electromagnetic signals in terms of complexity. Phys. Rev. E. 74, 016104.

Kawamura, H., Hatano, T., Kato, N., Biswas, A., Chakrabarti, B., 2012. Statistical physics of fracture, friction, and earthquakes. Rev. Mod. Phys. 84, 839–884.

Khatiashvili, N., 1984. The electromagnetic effect accompanying the fracturing of alcaline-halide crystals and rocks. Phys. Solid Earth 20, 656–661.

Khidas, Y., Jia, X., 2012. Probing the shear-band formation in granular media with sound waves. Phys. Rev. E 85, 051302.

Kikuchi, M., Kanamori, H., 1991. Inversion of complex body waves – III. BSSA 81, 2335–2350.

Kontoes, C., Elias, P., Sykioti, O., Briole, P., Remy, D., Sachpazi, M., et al., 2000. Displacement field and fault model for the September 7, 1999 Athens earthquake inferred from ERS2 satellite radar interferometry. Geophys. Res. Lett. 27 (24), 3989–3992.

Kossobokov, V., 2006. Testing earthquake prediction methods: the West Pacific short-term forecast of earthquakes with magnitude MwHRV5.8. Tectonophysics 413, 2531.

Kossobokov, V.G., Maeda, K., Uyeda, S., 1999. Precursory of seismicity in advance of the Kobe, 1995, $M = 7.2$ earthquake. Pure Appl. Geophys. 155, 409–423.

Kuksenko, V., Tomilin, N., Damaskinskaya, E., Lockner, D., 1996. A two stage model of fracture of rocks. Pure Appl. Geophys. 146, 253–263.

Kuksenko, V., Tomilin, N., Chmel, A., 2005. The role of driving rate in scaling characteristics of rock fracture. J. Stat. Mech.: Theory Exp. 2005 (06), P06012.

Kuksenko, V., Tomilin, N., Chmel, A., 2007. The rock fracture experiment with a drive control: a spatial aspect. Tectonophysics 431 (1-4), 123–129.

Kuksenko, V.S., Makhmudov, Kh. F., Mansurov, V.A., Sultanov, U., Rustamova, M.Z., 2009. Changes in structure of natural heterogeneous materials under deformation. J. Min. Sci. 45 (4), 355–358.

Kumar, R., Misra, A., 2007. Some basic aspects of electromagnetic radiation emission during plastic deformation and crack propagation in Cu–Zn alloys. Mater. Sci. Eng. A 454-455, 203–210.

Kun, F., Herrmann, H.J., 1999. Transition from damage to fragmentation in collision of solids. Phys. Rev. E 59, 2623–2632.

Kwapień, J., Drożdża, S., 2012. Physical approach to complex systems. Phys. Rep. 515, 115–226.

Lacidogna, G., Manuello, A., Carpinteri, A., Niccolini, G., Agosto, A., Durin, G., 2010. Acoustic and electromagnetic emissions in rocks under compression. In: Proceeding of the SEM Annual Conference, Indianapolis, IN, USA, 2010, Society for Experimental Mechanics Inc.

Lacidogna, G., Carpinteri, A., Manuello, A., Durin, G., Schiavi, A., Niccolini, G., et al., 2011. Acoustic and electromagnetic emissions as precursors phenomena in failure processes. Strain 47, 144–152.

Lavrov, A., 2005. Fracture-induced physical phenomena and memory effects in rocks: a review. Strain 41, 135–149.

Lei, X.L., Satoh, T., 2007. Indicators of critical point behavior prior to rock failure inferred from pre-failure damage. Tectonophysics 431, 97–111.

Li, C., Nordlund, E., 1993. Deformation of brittle rocks under compression with particular reference to microcracks. Mech. Mater. 15, 223–239.

Li, H., Jia, Z., Bai, Y., Xia, M., Ke, F., 2002. Damage localization, sensitivity of energy release and the catastrophe transition. Pure Appl. Geophys. 159, 19331950.

Li, X., Polygiannakis, J., Kapiris, P., Peratzakis, A., Eftaxias, K., Yao, X., 2005. Fractal spectral analysis of pre-epileptic seizures in terms of criticality. J. Neural Eng. 2, 1–6.

Lin, Q.X., Tham, L.G., Yeung, M.R., Lee, P.K., 2004. Failure of granite under constant loading. Int. J. Rock Mech. Min. Sci. 41, 362.

Lockner, D., Madden, T., 1991. A multiple-crack model of brittle fracture. Time-dependent simulations. J. Geophys. Res. 96 (B12), 19643–19654.

Lockner, D., Byerlee, J., Kuksenko, V., Ponomarev, A., Sidorin, A., 1991. Quasi-static fault growth and shear fracture energy in granite. Nature 350, 39–42.

Lockner, D.A., Okubo, P.G., 1983. Measurements of frictional heating in granite. J. Geophys. Res. 88 (B5), 4313–4320.

Lopez, J., Schmittbuhl, J., 1998. Anomalous scaling of fracture surfaces. Phys. Rev. E 57, 6405–6408.

Lu, C., Mai, Y.-W., Xie, H., 2005. A sudden drop of fractal dimension: a likely precursor of catastrophic failure in disordered media. Philos. Mag. Lett 85 (1), 33–40.

Main, I., Naylor, M., 2012. Extreme events and predictability of catastrophic failure in composite materials and in the earth. Eur. Phys. J. Spec. Top. 205, 183–197.

Malakhovsky, I., Michels, M.A.J., 2007. Effect of disorder strength on the fracture pattern in heterogeneous networks. Phys. Rev. B 76, 144201.

Mandelbrot, B.B., 1982. The Fractal Geometry on Nature. Freeman, NY, p. 1982.

Mantegna, R., Stanley, H.E., 1994. Analytic approach to the problem of convergence of truncated Levy flights towards the Gaussian stochastic process. Phys. Rev. Lett. 73, 2946–2949.

Maslov, S., Paczuski, M., Bak, P., 1994. Phys. Rev. Lett. 73, 2162–2165.

Mavromatou, C., Tombras, G.S., Ninos, D., Hadjicontis, V., 2008. Electromagnetic emission memory phenomena related to LiF ionic crystal deformation. J. Appl. Phys. 103, 083518.

McGarr, A., Fletcher, J., 2003. Maximum slip in earthquake fault zones, apparent stress, and stick-slip friction. Bull. Seismol. Soc. Am. 93, 2355–2362.

McGarr, A., Spottiswoode, S., Gay, N., Ortlepp, W., 1979. Observation relevant to seismic driving stress, stress drop, and efficiency. J. Geophys. Res. 84 (B5), 2251–2261.

McGarr, A., Fletcher, J., Boettcher, M., Beeler, N., Boatwright, J., 2010. Laboratory based maximum slip rate in earthquake rupture zones and radiated energy. Bull. Seismol. Soc. Am. 100, 3250–3260.

Merryman Boncori, J.P., Papoutsis, I., Pezzo, G., Tolomei, C., Atzori, S., Ganas, A., et al., 2015. The February 2014 Cephalonia earthquake (Greece): 3D deformation field and source modeling from multiple SAR techniques. Seismol. Res. Lett. 86, 1–14.

Midi, G., 2004. On dense granular flows. Eur. Phys. J. E 14, 341–365.

Miguel, M.-C., Zapperi, S., 2006. Fluctuations in plasticity at the microscale. Science 312, 1151–1152.

Miguel, M.-C., Vespignani, A., Zapperi, S., Weiss, J., Grasso, J.-R., 2001. Complexity in dislocation dynamics: model. Mater. Sci. Eng. A — Struct. Mater. Prop. Microstruct. Process. 309, 324–327.

Miguel, M.-C., Vespignani, A., Zaiser, M., Zapperi, S., 2002. Dislocation hamming and Andrade creep. Phys. Rev. Lett. 89, 165501.

Minadakis, G., Potirakis, S.M., Nomicos, C., Eftaxias, K., 2012a. Linking electromagnetic precursors with earthquake dynamics: an approach based on nonextensive fragment and self-affine asperity models. Physica A 391, 2232–2244.

Minadakis, G., Potirakis, S.M., Stonham, J., Nomicos, C., Eftaxias, K., 2012b. The role of propagating stress waves in geophysical scale: evidence in terms of nonextensivity. Physica A 391 (22), 5648–5657.

Mizoguchi, K., Hirose, T., Shimamoto, T., Fukuyama, E., 2006. Moisture-related weakening and strengthening of a fault activated at seismic slip rates. Geophys. Res. Lett. 33, L16319.

Mizoguchi, K., Hirose, T., Shimamoto, T., Fukuyama, E., 2009. Fault heals rapidly after dynamic weakening. Bull. Seismol. Soc. Am. 99, 3470–3474.

Mogi, K., 1962a. Study of the elastic shocks caused by the fracture of heterogeneous materials and its relations to earthquake phenomena. Bull. Earthquake Res. Inst. 40, 125–173.

Mogi, K., 1962b. Magnitude frequency relation elastic accompanying fractures of various materials and some related problems in earthquakes. Bull. Earthquake Res. Inst. 40, 831–853.

Mogi, K., 1968. Source locations of elastic shocks in the fracturing process in rocks. Bull. Earthquake Res. Inst. 46, 1103–1125.

Mogi, K., 1985. Earthquake Prediction. Academic Press, Tokyo.

Moreno, Y., Gómez, J.B., Pacheco, A.F., 2000. Fracture and second-order phase transitions. Phys. Rev. Lett. 85, 2865–2868.

Morgounov, V., 2001. Relaxation creep model of impending earthquake. Anali Di Geophysica 44, 369–381.

Mori, Y., Obata, Y., 2008. Electromagnetic emission and AE Kaiser Effect for estimating rock in-situ stress. Report of the Research Institute of Industrial Technology, Nihon University, No. 93.

Mori, Y., Saruhashi, K., Mogi, K., 1994. Acoustic emission from rock specimen under cyclic loading. Progress in Acoustic Emission VII: Proceedings of the 12th International Acoustic Emission Symposium, Sapporo, Japan, October 17–20, 1994. Japanese Society for Non-Destructive Inspection, JSNDI, Tokyo, pp. 173–178, 636 pp.

Mori, Y., Obata, Y., Pavelka, J., Sikula, J., Lolajicek, T., 2004a. AE Kaiser effect and electromagnetic emission in the deformation of rock sample. In: DGZ-Proceedings BB 90-CD, EWGAE, Lecture 14, 157–165.

Mori, Y., Obata, Y., Pavelka, J., Sikula, J., Lolajicek, T., 2004b. AE Kaiser effect and electromagnetic emission in the deformation of rock sample. J. Acoust. Emiss. 22, 91–101.

Mori, Y., Sedlak, P., Sikula, J., 2006. Estimation of rock in-situ stress by acoustic and electromagnetic emission. Adv. Mater. Res. 13–14, 357–362.

Mori, Y., Obata, Y., Sikula, J., 2009. Acoustic and electromagnetic emission from crack created in rock sample under deformation. J. Acoust. Emiss. 27, 157–166.

Mourot, G., Morel, S., Bouchaud, E., Valentin, G., 2006. Scaling properties of mortar fracture surfaces. Int. J. Fract. 140, 39–54.

Muto, J., Nagahama, H., Miura, T., Arakawa, I., 2007. Frictional discharge at fault asperities: origin of fractal seismo-electromagnetic radiation. Tectonophysics 431, 113–122.

Nasuno, S., Kudrolli, A., Bak, A., Gollub, J.-P., 1998. Time-resolved studies of stick-slip friction in sheared granular layers. Phys. Rev. E 58, 2161–2171.

Newman, M., 2005. Power laws, Pareto distributions and Zipf's law. Contemp. Phys. 46, 323–351.

Nikolopoulos, S., Kapiris, P., Karamanos, K., Eftaxias, K., 2004. A unified approach of catastrophic events. Nat. Hazard. Earth Syst. Sci. 4, 615−637.

Nikolopoulos, S., Kapiris, P., Karamanos, K., Eftaxias, K., 2011. In: Zeraoulia, Elhadj (Ed.), A Unified Approach of Catastrophic Events. Models and Applications of Chaos Theory in Modern Sciences. Science Publishers, 742 pages, Taylor & Francis Inc., Enfield, United States.

Nitsan, V., 1977. Electromagnetic emission accompanying fracture of quartz-bearing rocks. Geophys. Res. Lett. 4 (8), 333−336.

Ohnaka, M., 2000. A physical scaling relation between the size of an earthquake and its nucleation zone size. Pure Appl. Geophys. 157, 2259−2282.

Ohnaka, M., Mogi, K., 1982. Frequency characteristics of acoustic emission in rocks under uniaxial compression and its relation to the fracturing process to failure. J. Geophys. Res. 87 (B5), 3873−3884.

Osorio, I., Frei, M.G., Sornette, D., Milton, J., Lai, Y.-C., 2010. Epileptic seizures: quakes of the brain? Phys. Rev. E 82, 021919-1/13.

Papadimitriou, C., Kalimeri, M, Eftaxias, K., 2008. Nonextensivity and universality in the earthquake preparation process. Phys. Rev. E 77, 036101.

Papanikolaou, S., Dimiduk, D., Choi, W., Sethna, J., Uchic, M., Woodward, C., et al., 2012. Quasi-periodic events in crystal plasticity and the self-organized avalanche oscillator. Nature 490, 517−521.

Park, J.-W., Song, J.-J., 2013. Numerical method for determination of contact areas of a rock joint under normal and shear loads. Int. J. Rock Mech. Min. Sci. 58, 8−22.

Perfettini, H., Schmittbuhl, J., Vilotte, J., 2001. Slip correlations on a creeping fault. Geophys. Res. Lett. 28, 2133−2136.

Ponomarev, A., Zavyalov, A., Smirnov, V., Lockner, D., 1997. Physical modelling of the formation and evolution of seismically active fault zones. Tectonophysics 277, 57−81.

Ponson, L., Bonamy, D., Bouchaud, E., 2006. Two-dimensional scaling properties of experimental fracture surfaces. Phys. Rev. Lett. 96 (3), 035506.

Potirakis, S.M., Minadakis, G., Nomicos, C., Eftaxias, K., 2011. A multidisciplinary justification for traces of the last state of earthquake generation in preseismic electromagnetic emissions. Nat. Hazard. Earth Syst. Sci. 11, 2859−2879.

Potirakis, S.M., Minadakis, G., Eftaxias, K., 2012a. Analysis of electromagnetic pre-seismic emissions using Fisher information and Tsallis entropy. Physica A 391, 300−306.

Potirakis, S.M., Minadakis, G., Eftaxias, K., 2012b. Relation between seismicity and pre-earthquake electromagnetic emissions in terms of energy, information and entropy content. Nat. Hazard. Earth Syst. Sci. 12, 1179−1183.

Potirakis, S.M., Minadakis, G., Eftaxias, K., 2012c. Sudden drop of fractal dimension of electromagnetic emissions recorded prior to significant earthquake. Nat. Hazard. 64 (1), 641−650.

Potirakis, S.M., Zitis, P., Eftaxias, K., 2013a. Dynamical analogy between economical crisis and earthquake dynamics within the nonextensive statistical mechanics framework. Physica A 392, 2940−2954.

Potirakis, S.M., Karadimitrakis, A., Eftaxias, K., 2013b. Natural time analysis of critical phenomena: the case of pre-fracture electromagnetic emissions. Chaos 23 (2), 023117.

Potirakis, S.M., Eftaxias, K., Balasis, G., Kopanas, J., Antonopoulos, G., Kalimeris, A., 2014. Signatures of the self-affinity of fracture and faulting in pre-seismic electromagnetic emissions. Nat. Hazards Earth Syst. Sci. Discuss. 2, 2981−3013.

Potirakis, S.M., Contoyiannis, Y., Eftaxias, K., Koulouras, G., Nomicos, C., 2015. Recent field observations indicating an earth system in critical condition before the occurrence of a significant earthquake. IEEE Geosci. Remote Sens. Lett. 12 (3), 631−635.

Potirakis, S.M., Contoyiannis, Y., Melis, N.S., Kopanas, J., Antonopoulos, G., Balasis, G., et al., 2016a. Recent seismic activity at Cephalonia (Greece): a study through candidate electromagnetic precursors in terms of non-linear dynamics. Nonlinear Processes Geophys. 23, 223−240.

Potirakis, S.M., Eftaxias, K., Schekotov, A., Yamaguchi, H., Hayakawa, M., 2016b. Criticality features in ULF magnetic fields prior to the 2013 Kobe earthquake. Ann. Geophys. 59 (3), S0317.

Potirakis, S.M., Hayakawa, M., Schekotov, A., 2017. Fractal analysis of the ground-recorded ULF magnetic fields prior to the 11 March 2011 Tohoku earthquake ($M_W = 9$): discriminating possible earthquake precursors from Space-sourced disturbances. Nat. Hazard. 85, 59−86.

Pulinets, S., Boyarchuk, K., 2004. Ionospheric Precursors of Earthquakes. Springer, Berlin.

Pulinets, S., Legen'ka, A.D., Gaivoronskaya, T.V., Depuev, V. Kh, 2003. Main phenomenological features of ionospheric precursors of strong earthquakes. J. Atmos. Sol. Terr. Phys. 65, 1337−1347.

Qian, S., Yian, J., Cao, H., Shi, S., Lu, Z., Li, J., et al., 1994. Results of the observations on seismo-electromagnetic waves at two earthquake areas in China. In: Hayakawa, M., Fujinawa, Y. (Eds.), Electromagnetic Phenomena Related to Earthquake Prediction. Terrapub, Tokyo, pp. 205−211.

Rabinovitch, A., Frid, V., Bahat, D., 2001. Gutenberg−Richter-type relation for laboratory fracture-induced electromagnetic radiation. Phys. Rev. E 65, 11401.

Rabinovitch, A., Bahat, D., Frid, V., 2002. Similarity and dissimilarity of electromagnetic radiation from carbonate rocks under compression, drilling and blasting. Int. J. Rock Mech. Min. Sci. 39, 125−129.

Rabinowicz, E., 1951. The nature of static and kinetic coefficients of friction. J. Appl. Phys. 27, 1373−1379.

Reasenberg, P., 1999. Foreshock occurrence rates before large earthquakes worldwide. Pure Appl. Geophys. 155, 355−379.

Reches, Z., Dewers, T., 2005. Gouge formation by dynamic pulverization during earthquake rupture. Earth Planet. Sci. Lett. 235, 361−374.

Reches, Z., Lockner, D., 1994. Nucleation and growth of faults in brittle rocks. J. Geophys. Res. 99, 18159−18173.

Reis, S., Araújo, N., Andrade, J., Herrmann, H., 2012. How dense can one pack spheres of arbitrary size distribution. Europhys. Lett. 97, 1804.

Renard, F., Voisin, C., Marsan, D., Schmittbuhl, J., 2006. High resolution 3D laser scanner measurements of a strike-slip fault quantity its morphological anisotropy at all scales. Geophys. Res. Lett. 33, L04305.

Richeton, T., Weiss, J., Louchet, F., 2005. Breakdown of avalanche critical behaviour in polycrystalline plasticity. Nat. Mater. 4, 465−469.

Richeton, T., Dobron, P., Chmelik, F., Weiss, J., Louchet, F., 2006. On the critical behaviour of plasticity in metallic single crystals. Sci. Eng. A − Struct. Mater. Prop. Microstruct. Process. 424, 190−195.

Rubinstein, S.M., Cohen, G., Fineberg, J., 2004. Detachment fronts and the onset of dynamic friction. Nature 430, 1005−1009.

Rubinstein, S.M., Cohen, G., Fineberg, J., 2007. Dynamics of precursors to frictional sliding. Phys. Rev. Lett. 98, 226103.

Rumi, De, Ananthakrishna, G., 2004. Power laws, precursors and predictability during failure. Europhys. Lett. 66, 715−721.

Rundle, J., Tiampo, K., Klein, W., SaMartins, J., 2002. Selforganization in leaky threshold systems: the influence of near mean field dynamics and its implications for EQs, neurology, and forecasting. PNAS 99, 2514−2521.

Rundle, J.B., Turcotte, D.L., Shcherbakov, R., Klein, W., Sammis, C., 2003. Statistical physics approach to understanding the multiscale dynamics of earthquake fault systems. Rev. Geophys. 41, 1019−1049.

Sahimi, M., 1993. Flow phenomena in rocks: from continuum models to fractals, percolation, cellular automata, and simulated annealing. Rev. Mod. Phys. 65, 1393−1534.

Sahimi, M., Robertson, M., Sammis, C., 1993. Fractal distribution of earthquakes hypocenters and its relation to fault patterns and percolation. Phys. Rev. Lett. 70, 2186–2189.

Sarlis, N.V., Skordas, E.S., Varotsos, P.A., 2010. Nonextensivity and natural time: the case of seismicity. Phys. Rev. E 82, 021110.

Schiavi, A., Niccolini, G., Terrizzo, P., Carpinteri, A., Lacidogna, G., Manuello, A., 2011. Acoustic emissions at high and low frequencies during compression tests in brittle materials. Strain 47, 105–110.

Scholz, C.H., 2002. The Mechanics of Earthquakes and Faulting, second ed Cambridge University Press, Cambridge.

Schuster, H.G., 1998. Deterministic Chaos. VCH Publishers, Weinheim, New York.

Schwerdtfeger, J., Nadgorny, E., Madani-Grasset, F., Koutsos., V., Blackford, J., Zaiser, M., 2007. Scale-free statistics of plasticity-induced surface steps on KCl single crystals. J. Stat. Mech. 2007 (L04001), 1–6. Available from: https://doi.org/10.1088/1742-5468/2007/04/L04001.

Sethna, J., Dahmen, K., Myers, C., 2001. Crackling noise. Nature 410, 242–250.

Sharon, E., Cohen, G., Fineberg, J., 2002. Effects of crack front waves on dynamic fracture. Phys. Rev. Lett. 88, 085503.

Shen, L, Li, J., 2004. A numerical simulation for effective elastic moduli of plates with various distributions and sizes of cracks. Int. J. Solids Struct. 41, 7471–7492.

Shimamoto, T., Togo, T., 2012. Earthquakes in the lab. Science 338, 54–55.

Shkuratnik, V.L., Lavtov, A.V., 1997. Memory Effects in Rocks. Physical Features and Theoretical Models. Publishing House of the Academy of Mining Sciences, Moscow, 159 pp. (in Russian).

Silva, R., Franca, G., Vilar, C., Alcaniz, J., 2006. Nonextensive models for earthquakes. Phys. Rev. E 73, 026102.

Skeberis, C., Zaharis, Z.D., Xenos, T.D., Spatalas, S., Arabelos, D.N., Contadakis, M.E., 2015. Time–frequency analysis of VLF for seismic-ionospheric precursor detection: evaluation of Zhao–Atlas–Mark sand Hilbert–Huang transforms. Phys. Chem. Earth 85/86, 174–184.

Sobolev, G.A., Ponomarev, A.V., 2003. Physics of Earthquakes and Precursors. Nauka, Moscow, 270 pp.

Sornette, D., 1991. Self-organized criticality in plate tectonics. In: Riste, T., Sherrington, D. (Eds.), Proceedings of the NATO ASI Spontaneous Formation of Space-Time Structures and Criticality. Geilo, Norway 2–12 April 1991, vol. 349. Kluwer Academic Press, Dordrecht, Boston, pp. 57–106.

Sornette, D., 1999. Earthquakes: from chemical alteration to mechanical rupture. Phys. Rep. 313, 237–291.

Sornette, D., 2000. Critical Phenomena in Natural Sciences. Springer-Verlag, Berlin Heidelberg, Germany.

Sornette, D., Andersen, J.V., 1998. Scaling with respect to disorder in time-to- failure. Eur. Phys. J. B 1, 353–357.

Sornette, D., Sammis, C., 1995. Complex critical exponents from renormalization group theory of earthquakes: implications for earthquake predictions. J. Phys. I. 5, 607–619.

Sotolongo-Costa, O., Posadas, A., 2004. Fragment-asperity interaction model for earthquakes. Phys. Rev. Lett. 92, 048501.

Stanley, H.E., 1987. Introduction to Phase Transitions and Critical Phenomena. Oxford University Press, New York, USA.

Stanley, H.E., 1999. Scaling, universality, and renormalization: three pillars of modern critical phenomena. Rev. Mod. Phys. 71 (2), S358–S366.

Stanley, H.E., 2000. Exotic statistical physics: application to biology, medicine, and economics. Physica A 285, 1–17.

Stauffer, D., 1985. Introduction to Percolation Theory. Taylor, London.

Tsallis, C., 1988. Possible generalization of Boltzmann–Gibbs statistics. J. Stat. Phys. 52, 479–487.

Tsallis, C., 1998. Generalized entropy-based criterion for consistent testing. Phys. Rev. E 58, 1442–1445.

Tsallis, C., 2009. Introduction to Nonextensive Statistical Mechanics. Approaching a Complex Word. Springer, New York.

Tsutsumi, A., Shirai, N., 2008. Electromagnetic signals associated with stick-slip of quartz-free rocks. Tectonophysics 450, 79–84.

Turcotte, D., 1997. Fractals and Chaos in Geology and Geophysics, second ed Cambridge University Press, Cambridge, England.

Uchic, M., Dimiduk, D., Florando, J., Nix, W., 2004. Sample dimensions influence strength and crystal plasticity. Science 305, 986–989.

Uyeda, S., Nagao, T., Kamogawa, M., 2009. Short-term earthquake prediction: current status of seismo-electromagnetics. Tectonophysics 470 (3-4), 205–213.

Vallianatos, F., Benson, P., Meredith, P., Sammonds, P., 2012. Experimental evidence of a non-extensive statistical physics behaviour of fracture in triaxially deformed Etna basalt using acoustic emissions. EPL 97, 58002.

Varotsos, P., Sarlis, N., Skordas, E.S., 2011. Natural Time Analysis: The New View of Time. Springer, Berlin.

Varotsos, P.A., Sarlis, N.V., Skordas, E.S., 2001. Spatio-temporal complexity aspects on the interrelation between seismic electric signals and seismicity. Pract. Athens Acad. 76, 294–321.

Varotsos, P.A., Sarlis, N.V., Tanaka, H.K., Skordas, E.S., 2005. Similarity of fluctuations in correlated systems: the case of seismicity. Phys. Rev. E 72, 041103.

Varotsos, P.A., Sarlis, N.V., Skordas, E.S., Tanaka, H.K., Lazaridou, M.S., 2006. Entropy of seismic electric signals: analysis in the natural time under time reversal. Phys. Rev. E 73, 031114.

Veje, C., Howell, D.W., Behringer, R.P., 1999. Kinematics of a two-dimensional granular Couette experiment at the transition to shearing. Phys. Rev. E 59, 739–745.

Verrato, F., Foffi, G., 2011. Apollonian packing as physical fractals. Mol. Phys. 109, 2923–2928.

Vicsek, T., 2001. A question of scale. Nature 411, 421.

Vicsek, T., 2002. The bigger picture. Nature 418, 131.

Wang, E.Y., Zhao, E.L., 2013. Numerical simulation of electromagnetic radiation caused by rock deformation and failure. Int. J. Rock Mech. Min. Sci. 57, 57–63.

Ward, D., Farkas, D., Lian, H., Curtin, W., Wang, J., Kim, K.S., et al., 2009. Engineering size-effects of plastic deformation in nanoscale asperities. Nature 106, 9580–9585.

Weiss, J., Grasso, J.R., 1997. Acoustic emission in single crystals of ice. J. Phys. Chem. B101, 6113–6117.

Weiss, J., Marsan, D., 2003. Three-dimensional mapping of dislocation avalanches: clustering and space/time coupling. Science 299, 89–92.

Welker, P., McNamara, A., 2011. Precursors of failure and weakening in a biaxial test. Granular Matter 13, 93–105.

Wilson, B., Dewers, T., Reches, Z., Brune, J., 2005. Particle size and energetics of gouge from earthquake rupture zones. Nature 434, 749–752.

Wu, L., Zheng, S., De Santis, A., Qin, K., Di Mauro, R., Liu, S., et al., 2016. Geosphere coupling and hydrothermal anomalies before the 2009 M_W 6.3 L'Aquila earthquake in Italy. Nat. Hazard. Earth Syst. Sci. 16, 1859–1880.

Xia, K., Rosakis, A., Kanamori, H., 2004. Laboratory earthquakes: the sub-Rayleigh-to-supershear rupture transition. Science 303, 1859–1861.

Xia, K., Rosakis, A., Kanamori, H., Rice, J., 2005. Laboratory earthquakes along inhomogeneous faults; directionality and supershear. Science 308, 681–684.

Yamada, I., Masuda, K., Mizutani, H., 1989. Electromagnetic and acoustic emission associated with rock fracture. Phys. Earth Planet. Int. 57, 157–168.

Yamada, I., Masuda, K., Mizutani, H., 1989. Electromagnetic and acoustic emission associated with rock fracture. Phys. Earth Planet. Int. 57, 1570168.

Yamshchikov, V.S., Shkuratnik, V.L., Lavrov, A.V., 1994. Memory effects in rocks (review). J. Min. Sci. 30, 463–473.

Zaiser, M., 2006. Scale invariance in plastic flow of crystalline solids. Adv. Phys. 54, 185–245.

Zaiser, M., Moretti, P., 2005. Fluctuation phenomena in crystal plasticity-a continuum model. J. Stat. Mech. 2005 (P08004), 1–19. Available from: https://doi.org/10.1088/1742-5468/2005/08/P08004.

Zapperi, S., 2012. Current challenges for statistical physics in friction and plasticity. Eur. Phys. J. B. 85, 329.

Zapperi, S., Kumar, P., Nukala, V., Simunovic, S., 2005. Crack roughness and avalanche precursors in the random fuse model. Phys. Rev. E 71, 26106.

Zunino, L., Perez, D., Kowalski, A., Martin, M., Garavaglia, M., Plastino, A., et al., 2008. Fractional Brownian motion, fractional Gaussian noise, and Tsallis permutation entropy. Physica A 387, 6057–6088.

Further Reading

Cyranoski, D., 2004. A seismic shift in thinking. Nature 431, 1032–1034.

Gokhberg, M., Morgunov, V., Pokhotelov, O., 1995. Earthquake Prediction. Seismo-Electromagnetic Phenomena,. Gordon and Breach Publishers, Amsterdam, 193 pp.

Reches, Z., 1999. Mechanisms of slip nucleation during earthquakes. Earth Planet. Sci. Lett. 170, 475–486.

Varotsos, P., 2005. The Physics of Seismic Electric Signals. Terrapub, Tokyo.

Index

Note: Page numbers followed by "*f*" and "*t*" refer to figures and tables, respectively.

CPI Antony Rowe
Chippenham, UK
2018-06-05 18:21